AF182419

Hypoxic Pulmonary Vasoconstriction:

Cellular and Molecular Mechanisms

Developments in Cardiovascular Medicine

232. A. Bayés de Luna, F. Furlanello, B.J. Maron and D.P. Zipes (eds.): *Arrhythmias and Sudden Death in Athletes*. 2000 ISBN: 0-7923-6337-X
233. J-C. Tardif and M.G. Bourassa (eds): *Antioxidants and Cardiovascular Disease*. 2000. ISBN: 0-7923-7829-6
234. J. Candell-Riera, J. Castell-Conesa, S. Aguadé Bruiz (eds): *Myocardium at Risk and Viable Myocardium Evaluation by SPET*. 2000.ISBN: 0-7923-6724-3
235. M.H. Ellestad and E. Amsterdam (eds): Exercise Testing: New Concepts for the New Century. 2001. ISBN: 0-7923-7378-2
236. Douglas L. Mann (ed.): The Role of Inflammatory Mediators in the Failing Heart. 2001 ISBN: 0-7923-7381-2
237. Donald M. Bers (ed.): Excitation-Contraction Coupling and Cardiac Contractile Force, Second Edition. 2001 ISBN: 0-7923-7157-7
238. Brian D. Hoit, Richard A. Walsh (eds.): Cardiovascular Physiology in the Genetically Engineered Mouse, Second Edition. 2001 ISBN 0-7923-7536-X
239. Pieter A. Doevendans, A.A.M. Wilde (eds.): Cardiovascular Genetics for Clinicians 2001 ISBN 1-4020-0097-9
240. Stephen M. Factor, Maria A.Lamberti-Abadi, Jacobo Abadi (eds.): Handbook of Pathology and Pathophysiology of Cardiovascular Disease. 2001 ISBN 0-7923-7542-4
241. Liong Bing Liem, Eugene Downar (eds): Progress in Catheter Ablation. 2001 ISBN 1-4020-0147-9
242. Pieter A. Doevendans, Stefan Kääb (eds): Cardiovascular Genomics: New Pathophysiological Concepts. 2002 ISBN 1-4020-7022-5
243. Daan Kromhout, Alessandro Menotti, Henry Blackburn (eds.): Prevention of Coronary Heart Disease: Diet, Lifestyle and Risk Factors in the Seven Countries Study. 2002 ISBN 1-4020-7123-X
244. Antonio Pacifico (ed.), Philip D. Henry, Gust H. Bardy, Martin Borggrefe, Francis E. Marchlinski, Andrea Natale, Bruce L. Wilkoff (assoc. eds): Implantable Defibrillator Therapy: A Clinical Guide. 2002 ISBN 1-4020-7143-4
245. Hein J.J. Wellens, Anton P.M. Gorgels, Pieter A. Doevendans (eds.): The ECG in Acute Myocardial Infarction and Unstable Angina: Diagnosis and Risk Stratification. 2002 ISBN 1-4020-7214-7
246. Jack Rychik, Gil Wernovsky (eds.): Hypoplastic Left Heart Syndrome. 2003 ISBN 1-4020-7319-4
247. Thomas H. Marwick: Stress Echocardiography. Its Role in the Diagnosis and Evaluation of Coronary Artery Disease 2nd Edition. ISBN 1-4020-7369-0
248. Akira Matsumori: Cardiomyopathies and Heart Failure: Biomolecular, Infectious and Immune Mechanisms. 2003 ISBN 1-4020-7438-7
249. Ralph Shabetai: The Pericardium. 2003 ISBN 1-4020-7639-8
250. Irene D. Turpie; George A. Heckman (eds.): Aging Issues in Cardiology. 2004 ISBN 1-40207674-6
251. C.H. Peels; L.H.B. Baur (eds.): Valve Surgery at the Turn of the Millenium. 2004 ISBN 1-4020-7834-X
252. Jason X.-J. Yuan (ed.): Hypoxic Pulmonary Vasoconstriction: Cellular and Molecular Mechanisms. 2004 ISBN 1-4020-7857-9

Previous volumes are still available

Hypoxic Pulmonary Vasoconstriction:

Cellular and Molecular Mechanisms

edited by

Jason X.-J. Yuan, M.D., Ph.D.

Professor

Department of Medicine
University of California, San Diego
School of Medicine
San Diego, California

Springer Science+Business Media, LLC

Library of Congress Cataloging-in-Publication Data

Hypoxic Pulmonary Vasoconstriction: Cellular and Molecular Mechanisms edited by Jason X.-J. Yuan
p. ; cm.
Includes bibliographical references and index.
ISBN 978-1-4757-7992-9 ISBN 978-1-4020-7858-3 (eBook)
DOI 10.1007/978-1-4020-7858-3

1. Pulmonary hypertension—Pathophysiology. 2. Pulmonary veins.
I. Yuan, Jason X.-J., 1963- II. Series
[DNLM: WF 600 H998 2004]
RC776.P87 H98 2004
616.2/4—dc22 2004044223

Copyright © 2004 by Springer Science+Business Media New York
Originally published by Kluwer Academic Publishers in 2004
Softcover reprint of the hardcover 1st edition 2004

All rights reserved. No part of this work may be reproduced, stored in a retrieval system, or transmitted in any form or by any means, electronic, mechanical, photocopying, microfilming, recording, or otherwise, without the written permission from the Publisher, with the exception of any material supplied specifically for the purpose of being entered and executed on a computer system, for exclusive use by the purchaser of the work

Permission for books published in Europe: permissions@wkap.nl
Permissions for books published in the United States of America: permissions@wkap.com

Printed on acid-free paper.

The Publisher offers discounts on this book for course use and bulk purchases. For further information, send email to <melissa.ramondetta@wkap.com>.

TABLE OF CONTENTS

III. ROLE OF ION CHANNELS IN HYPOXIC PULMONARY VASOCONSTRICTION

IV. ROLE OF ENDOTHELIUM IN HYPOXIC PULMONARY VASOCONSTRICTION

V. MECHANISMS OF OXYGEN SENSING IN THE PULMONARY VASCULATURE

List of Contributors

Philip I. Aaronson, Ph.D., Reader, Department of Asthma, Allergy and Respiratory Science, King's College London, London, U.K. (*Chapter 12*)

Steven H. Abman, M.D., Professor, The Children's Hospital, Pediatric Heart-Lung Center, University of Colorado, Denver, Colorado (*Chapter 27*)

Helmut Acker, M.D., Professor, Facharzt für Physiologie, Max-Planck-Institut für Molekulare Physiologie, Dortmund, Germany (*Chapter 19*)

Serge Adnot, Ph.D., Professor, INSERM U492, Département de Physiologie, Hôpital Henri Mondor, Créteil, France (*Chapter 28*)

Stephen L. Archer, M.D., Professor and Director, Division of Cardiology, University of Alberta Hospital, Edmonton, Alberta, Canada (*Chapter 9*)

Pavel Babal, M.D., Associate Professor, Department of Pathology, Comenius University, Bratislava, Slovak Republic (*Chapter 29*)

Natalie R. Bauer, Ph.D., Fellow, Department of Medicine, University of Colorado Health Sciences Center, Denver, Colorado (*Chapter 24*)

Sébastien Bonnet, Ph.D., Laboratoire de Physiologie Cellulaire Respiratoire, Université Victor Ségalen Bordeaux2, Bordeaux, France (*Chapter 5*)

Navdeep S. Chandel, Ph.D., Assistant Professor, Department of Medicine, Northwestern University, Chicago, Illinois (*Chapter 15*)

Laura Conforti, Ph.D., Assistant Professor, Department of Internal Medicine, University of Cincinnati, Cincinnati, Ohio (*Chapter 22*)

Neil J. Davie, Ph.D., Fellow, Department of Pediatrics, University of Colorado Health Sciences Center, Denver, Colorado (*Chapter 26*)

Christopher A. Dawson, Ph.D., Professor, Department of Physiology, Medical College of Wisconsin, Milwaukee, Wisconsin (*Chapter 2*)

Michelle Dipp, Ph.D., Student, Worcester College, University of Oxford, Oxford, U.K. (*Chapter 6*)

Saadia Eddahibi, Ph.D., INSERM U492, Faculté de Médecine de Créteil,

Créteil, France (*Chapter 28*)

A. Mark Evans, Ph.D., Lecturer, School of Biology, University of St. Andrews, Fife, U.K. (*Chapters 6 and 18*)

Karen A. Fagan, M.D., Assistant Professor, Department of Medicine, University of Colorado Health Sciences Center, Denver, Colorado (*Chapter 24*)

Maria G. Frid, Ph.D., Instructor, Department of Pediatrics, University of Colorado Health Sciences Center, Denver, Colorado (*Chapter 26*)

Yuan-Cheng Fung, Ph.D., Professor Emeritus, Department of Bioengineering, University of California, San Diego, La Jolla, California (*Chapter 3*)

Maria García-Fernández, B.Sc., Student, Laboratorio de Investigaciones Biomédicas Edificio de Laboratorios, Hospital Universitario Virgen del Rocio, Sevilla, Spain (*Chapter 20*)

Sarah A. Gebb, Ph.D., Instructor, Department of Medicine, University of Colorado Health Sciences Center, Denver, Colorado (*Chapter 24*)

Mark N. Gillespie, Ph.D., Professor and Chairman, Department of Pharmacology and Center for Lung Biology, University of South Alabama, Mobile, Alabama (*Chapter 29*)

John B. Gordon, M.D., Associate Professor, Department of Pediatrics, Medical College of Wisconsin, Milwaukee, Wisconsin (*Chapter 31*)

Agnes Görlach, M.D., Department Leader, Abt. Experimentelle Kinderkardiologie, Deutsches Herzzentrum Muenchen, Muenchen, Germany (*Chapter 19*)

Sachin A. Gupte, M.D., Ph.D., Fellow, Department of Physiology, New York Medical College, Valhalla, New York (*Chapter 14*)

Alison M. Gurney, Ph.D., Professor, Department of Physiology and Pharmacology, University of Strathclyde, Glasgow, U.K. (*Chapter 11*)

Charles A. Hales, M.D., Professor and Chief, Pulmonary and Critical Care Medicine Unit, Massachusetts General Hospital, Harvard Medical School, Boston, Massachusetts (*Chapter 1*)

Nicholas S. Hill, M.D., Professor, Tufts University School of Medicine-New

England Medical Center, Boston, Massachusetts (*Chapter 30*)

Wei Huang, Ph.D., Associate Project Scientist, Department of Bioengineering, University of California, San Diego, La Jolla, California (*Chapter 3*)

Yadong Huang, M.D., Ph.D., Assistant Professor, Gladstone Institute of Cardiovascular Disease, University of California, San Francisco, California (*Chapter 32*)

Mallik R. Karamsetty, Ph.D., Assistant Professor, Department of Medicine, Brown University, Providence, Rhode Island (*Chapter 30*)

Thomas Kietzmann, M.D., Assistant Professor, Institut fuer Biochemie und Molekulare, Goettingen, Germany (*Chapter 19*)

James C. Leiter, M.D., Professor, Department of Physiology, Dartmouth Medical School, Hanover, New Hampshire (*Chapter 30*)

José López-Barneo, M.D., Ph.D., Professor, Laboratorio de Investigaciones Biomédicas Edificio de Laboratorios, Hospital Universitario Virgen del Rocio, Sevilla, Spain (*Chapter 20*)

Jane A. Madden, Ph.D., Professor, Department of Neurology, Medical College of Wisconsin, Milwaukee, Wisconsin (*Chapter 31*)

Roger Marthan, M.D., Ph.D., Professor, Laboratoire de Physiologie Cellulaire Respiratoire, INSERM 256, Université Victor Ségalen Bordeaux2, Bordeaux, France (*Chapter 5*)

Sean McMurtry, M.D., Fellow, Department of Medicine (Cardiology), University of Alberta, Edmonton, Alberta, Canada (*Chapter 17*)

Ivan F. McMurtry, Ph.D., Professor, Department of Medicine, University of Colorado Health Sciences Center, Denver, Colorado (*Chapters 7 and 24*)

Evangelos D. Michelakis, M.D., Associate Professor, Department of Medicine, University of Alberta, Edmonton, Alberta, Canada (*Chapters 9 and 17*)

David E. Millhorn, Ph.D., Professor and Director, Genome Research Institute and Department of Genome Science, University of Cincinnati College of Medicine, Cincinnati, Ohio (*Chapter 22*)

Rohit Moudgil, M.Sc., Student, Division of Cardiology, University of Alberta

Hospital, Edmonton, Alberta, Canada (*Chapter 9*)

Tetsutaro Nagaoka, M.D., Fellow, Department of Medicine, University of Colorado Health Sciences Center, Denver, Colorado (*Chapter 24*)

Lih-Chyuan Ng, Ph.D., Fellow, Department of Pharmacology, University of Nevada School of Medicine, Reno, Nevada (*Chapter 11*)

Colin A. Nurse, Ph.D., Professor, Department of Biology, McMaster University, Hamilton, Ontario, Canada (*Chapter 21*)

Masahiko Oka, M.D., Assistant Professor, Department of Medicine, University of Colorado Health Sciences Center, Denver, Colorado (*Chapter 24*)

Andrea Olschewski, M.D., Staff Anesthesiologist, Department of Anaesthesiology, Justus-Liebig-University Giessen, Giessen, Germany (*Chapter 16*)

Jack W. Olson, Ph.D., Professor, Department of Pharmacology and Center for Lung Biology, University of South Alabama, Mobile, Alabama (*Chapter 29*)

Patricia Ortega-Sáenz, Ph.D., Fellow, Laboratorio de Investigaciones Biomédicas Edificio de Laboratorios, Hospital Universitario Virgen del Rocio, Sevilla, Spain (*Chapter 20*)

Lo Chang Ou, Ph.D., Professor, Department of Physiology, Dartmouth Medical School, Hanover, New Hampshire (*Chapter 30*)

Lisa A. Palmer, Ph.D., Associate Professor, Department of Pediatrics, University of Virginia Health System, Charlottesville, Virginia (*Chapter 13*)

Ricardo Pardal, Ph.D., Fellow, Laboratorio de Investigaciones Biomédicas Edificio de Laboratorios, Hospital Universitario Virgen del Rocio, Sevilla, Spain (*Chapter 20*)

Hemal Patel, Ph.D., Fellow, Department of Pharmacology, University of California, San Diego, La Jolla, California (*Chapter 10*)

Ioana R. Preston, M.D., Assistant Professor, Tufts University School of Medicine-New England Medical Center, Boston, Massachusetts (*Chapter 30*)

Marlene Rabinovitch, M.D., Professor, The Wall Center for Pulmonary Vascular Diseases, Stanford University School of Medicine, Stanford, California (*Chapter 23*)

Bernadette Raffestin, M.D., Ph.D., INSERM U492, Département de Physiologie, Hôpital Ambroise Paré, Boulogne, France (*Chapter 28*)

Carmelle V. Remillard, Ph.D., Fellow, Department of Medicine, University of California, San Diego, California (*Chapters 4, 10 and 33*)

Thomas P. Robertson, Ph.D., Assistant Professor, Department of Physiology and Pharmacology, University of Georgia, Athens, Georgia (*Chapters 7 and 24*)

Mykhaylo Ruchko, Ph.D., Instructor, Department of Pharmacology, University of South Alabama, Mobile, Alabama (*Chapter 29*)

Jean-Pierre Savineau, Ph.D., Professor, Laboratoire de Physiologie Cellulaire Respiratoire, INSERM 356, Université Victor Ségalen Bordeaux2, Bordeaux, France (*Chapter 5*)

James S.K. Sham, Ph.D., Associate Professor, Division of Pulmonary and Critical Care Medicine, Johns Hopkins University School of Medicine, Baltimore, Maryland (*Chapter 4*)

Sergey V. Smirnov, Ph.D., Lecturer, Department of Pharmacy and Pharmacology, University of Bath, U.K. (*Chapter 8*)

Robin H. Steinhorn, M.D., Professor, Department of Pediatrics, Northwestern University, Chicago, Illinois (*Chapter 27*)

Kurt R. Stenmark, M.D., Professor, Department of Pediatrics, University of Colorado Health Sciences Center, Denver, Colorado (*Chapter 26*)

Roger J. Thompson, Ph.D., Fellow, Department of Cellular and Structural Biology, University of Colorado Health Sciences Center, Denver, Colorado (*Chapter 21*)

Rubin M. Tuder, M.D., Associate Professor, Departments of Pathology and Medicine, Johns Hopkins University School of Medicine, Baltimore, Maryland (*Chapter 25*)

Jeremy P.T. Ward, Ph.D., Professor, Department of Asthma, Allergy and Respiratory Science, King's College London, London, U.K. (*Chapter 12*)

E. Kenneth Weir, M.D., Professor, Department of Medicine, University of Minnesota, Minneapolis, Minnesota (*Chapter 16*)

Michael S. Wolin, Ph.D., Professor, Department of Physiology, New York Medical College, Valhalla, New York (*Chapter 14*)

Jason X.-J. Yuan, M.D., Ph.D., Professor, Department of Medicine, University of California, San Diego, California (*Chapters 10 and 33*)

Ari L. Zaiman, M.D., Ph.D., Fellow, Department of Medicine, Johns Hopkins University School of Medicine, Baltimore, Maryland (*Chapter 25*)

Wei-Min Zhang, M.D., Ph.D., Fellow, Division of Pulmonary and Critical Care Medicine, Johns Hopkins University School of Medicine, Baltimore, Maryland (*Chapter 4*)

Kathryn A. Ziel, Ph.D., Instructor, Department of Pharmacology and Center for Lung Biology, University of South Alabama, Mobile, Alabama (*Chapter 29*)

Preface

Hypoxic pulmonary vasoconstriction (HPV) serves a regulatory function by matching perfusion to ventilation and shunting blood flow away from the poorly oxygenated regions of the lung. HPV is a critical physiological mechanism of the lung to ensure maximal oxygenation of the venous blood in the pulmonary artery. Persistent alveolar hypoxia, however, causes pulmonary hypertension which is characterized by sustained pulmonary vasoconstriction and pulmonary vascular remodeling. The hypoxia-mediated pulmonary hypertension causes right heart failure in patients with a variety of cardiopulmonary diseases, including chronic obstructive pulmonary disease, congenital heart disease, and mountain sickness. Over the last decade, considerable progress has been made in understanding the cellular and molecular mechanisms involved in HPV and hypoxia-induced pulmonary vascular remodeling. These significant findings provide an essential basis to specify the precise sequences of events of HPV, to identify the etiology of hypoxia-mediated pulmonary hypertension, and to develop new therapeutic approaches for patients with pulmonary hypertension.

The major objective of this book is to provide a timely and long lasting guide for investigators in the fields of cardiovascular physiology and pathophysiology, pulmonary vascular disease, and high-altitude physiology and medicine. This will establish a solid scientific foundation for subsequent applications in clinical practice. The book is divided into eight sections: I. Physiology and pathophysiology of HPV; II. Role of intracellular Ca^{2+} and Ca^{2+} sensitivity in HPV; III. Role of ion channels in HPV; IV. Role of the endothelium in HPV; V. Mechanisms of oxygen sensing in the pulmonary vasculature; VI. Oxygen-sensing mechanisms in other organs and tissues; VII. Pathology and mechanisms of hypoxia-induced pulmonary hypertension; and VIII. Experimental models for the study of HPV.

Subsections in each of the main sections address critical aspects related to hypoxia-induced pulmonary vasoconstriction and pulmonary hypertension. Section I highlights the physiological function (Chapter 1) and heterogeneity (Chapter 2) of HPV, as well as the physical principles of pulmonary circulation, gas exchange, and HPV and their correlation with gene actions (Chapter 3). Intracellular Ca^{2+} is not only a major trigger for smooth muscle contraction, but also an important signal transduction element that mediates gene expression, protein synthesis, cell migration, and cell proliferation. Section II discusses how intracellular Ca^{2+} signals are regulated by hypoxia to induce HPV and pulmonary vascular smooth muscle cell proliferation. Four chapters are devoted to aspects of recent findings on the roles of Ca^{2+} sparks (Chapter 4), agonist-mediated Ca^{2+} transients (Chapter 5), Ca^{2+} mobilization from the sarcoplasmic reticulum (Chapter 6), and Ca^{2+} sensitization (Chapter 7) in the development of HPV. How acute hypoxia regulates cytoplasmic, nuclear, and intracellularly-stored Ca^{2+}

concentration in pulmonary artery smooth muscle cells is also discussed in this section. Section III is designed to explore the role of ion channels in HPV. Chapter 8 discusses the functionally expressed ion channels along with their regulation in the pulmonary vasculature. Two chapters focus on the regulation of K^+ channel activity by a mitochondrial redox sensor (Chapter 9) and by acute exposure to hypoxia (Chapter 10). The functional role of K^+ channels (especially voltage-gated K^+ channels) in regulating membrane potential and cytoplasmic Ca^{2+} concentration (via altering activity of voltage-gated Ca^{2+} channels) in pulmonary artery smooth muscle cells, the transcriptional regulation of K^+ channel genes by chronic hypoxia, and the role of dysfunctional K^+ channels in the development of hypoxia-induced pulmonary hypertension are also discussed extensively in Chapters 9 and 10. Furthermore, the contribution of transient receptor potential channels and capacitative Ca^{2+} entry to the development of HPV is reviewed in Chapter 11. Section IV includes an elegant discussion on the role of endothelium in HPV.

Section V is designed to describe the putative and potential mechanisms of oxygen sensing in the pulmonary vasculature. It is focused on oxygen radicals (Chapters 13 and 14), mitochondrial oxidative phosphorylation chain (Chapters 15 and 17), cellular redox status (Chapter 16), and cADPR accumulation (Chapter 18). In addition to the pulmonary vasculature, there are many tissues and cells whose function is regulated by oxygen. Section VI is focused on the cellular mechanisms involved in oxygen-sensitive gene expression (Chapter 19), as well as the oxygen sensing mechanisms and oxygen-sensitive ion channels in arterial chemoreceptor (Chapter 20), chromaffin cells (Chapter 21), and pheochromocytoma cells (Chapter 22). Section VII discusses current knowledge on the etiology and pathological characterization of hypoxia-induced pulmonary hypertension and right heart failure. It is focused on the hypoxia-sensitive agonists, mitogens, and transcription factors found in animal and human lung tissues (Chapters 25, 28, and 29); the role of the heterogeneity in hypoxia-induced pulmonary vascular smooth muscle cells proliferation (Chapter 26); the potential mechanisms involved in pulmonary vascular remodeling and hypoxic pulmonary hypertension (Chapters 23, 24, and 28); the pathophysiology and treatment of persistent pulmonary hypertension in the newborn (Chapter 27); and the strain difference of hypoxia-induced pulmonary hypertension (Chapter 30). Section VIII includes three chapters on how to use animal and *in vivo* models (Chapter 31) and transgenic animal models (Chapter 32) for studying HPV and hypoxia-mediated pulmonary hypertension. A chapter on patch clamp and fluorescence microscopy techniques (Chapter 33) for measuring ion channel currents and intracellular Ca^{2+} is included.

In summary, this book not only covers the current state-of-the-art findings relevant to cellular and molecular processes of hypoxic pulmonary vasoconstriction but also provides the underlying conceptual basis and knowledge regarding etiological mechanisms and experimental therapeutics for hypoxia-mediated pulmonary hypertension. I hope this book will be something of use not only to those who are experienced basic science investigators in the research fields of hypoxic cardiopulmonary physiology and pathophysiology and pulmonary vascular diseases, but also to a large community of clinicians or physician scientists whose primary subspecialty is in pulmonary and critical care medicine, cardiology, cardiothoracic surgery, environmental medicine, and sports medicine.

Acknowledgment:

The book is dedicated to Dr. Ayako Makino for continuously supporting me in pursuing an academic career and for her selfless love during the editing of the book, to my parents and grandparents who taught me how to overcome hurdles and difficulties, and to my mentors who guided me into the research field and taught me what HPV was. I would like to take this opportunity to thank all contributors for the excellent chapters and Ms. M. Ramondetta for her instruction in preparing and editing the text. I am indebted to Dr. C.V. Remillard for her diligence in preparing and editing the figures, to Dr. I.F. McMurtry for his suggestions in compiling this book, and to my colleagues and students at the University of California, San Diego for their dedication to sharing their knowledge with others. Finally, I would like to thank Drs. M.P. Blaustein and L.J. Rubin for their guidance and support throughout my career.

Jason X.-J. Yuan
San Diego, California
October, 2003

I. PHYSIOLOGY AND PATHOPHYSIOLOGY OF HYPOXIC PULMONARY VASOCONSTRICTION

Chapter 1

Physiological Function of Hypoxic Pulmonary Vasoconstriction

Charles A. Hales
Massachusetts General Hospital, Harvard Medical School, Boston, Massachusetts, U.S.A.

1. Introduction

Hypoxic pulmonary vasoconstriction (HPV) clearly has a role in reducing perfusion through the lung *in utero* to enhance O_2 delivery to the systemic circulation. It may also have a role in adults to improve the balance of blood perfusion to ventilation in the lung to optimize gas exchange, although some have suggested that this is a vestigial response (17). This chapter will look at the physiological role of HPV in the fetus and the adult.

2. History of Hypoxic Pulmonary Vasoconstriction

Investigators from as early as the 1930s were aware that men at altitude in the Andes mountains had an enlarged right heart as seen by electrocardiogram, chest X-ray and autopsy compared to men at sea level (29, 34). This was thought to be due to hypoxia, "anoxica anoxia". How the right heart became hypertrophied was unclear until the recognition in 1946 that pulmonary vessels constricted to hypoxia (46). von Euler and Liljestrand were exploring the regulators of pulmonary blood flow in cats into which they had implanted a rigid tube and flange through the side of a pulmonary artery to measure pulmonary artery pressure. In 9 cats they measured a mean pulmonary artery pressure of 17 mmHg when the cats spontaneously inhaled air or were artificially ventilated. They noted that when the cats were ventilated with 10-11% O_2 that there was a distinct rise in pulmonary artery pressure. They did not measure cardiac output with hypoxia but did note only a small rise in pulmonary artery pressure with occlusion of one main pulmonary artery, causing double the blood flow to the remaining right lung, or only a modest rise in pulmonary artery pressure with muscular exercise. They felt the rise in pulmonary artery pressure when the cats received hypoxic ventilation was out of proportion to the rise seen with the flow-induced change in pulmonary artery pressure, suggesting a direct effect of hypoxia to constrict the pulmonary vessels. They also examined CO_2 and found a lesser response. CO_2 may enhance HPV more than it constricts by itself (7). Subsequent to these classic studies, many other investigators have confirmed that there is a rise in pulmonary artery pressure in most cases in animals or humans with little or no rise in cardiac output caused by inhalation of 10% O_2 (31).

3. Physiologic Characteristics of Hypoxic Pulmonary Vasoconstriction

In humans and adult animals the alveolar oxygen tension (PO_2) needs to reach 60 mmHg or lower to initiate pulmonary vasoconstriction (5, 26). In newborn sheep, however, there is evidence of active hypoxic tone even when being ventilated with 30% F_IO_2 (5). This enhanced sensitivity to alveolar hypoxia in the newborn probably accounts for the well-known flip/flop of the circulation in the newborn when weaning from the ventilator during which suddenly pulmonary vascular resistance is high and the fetal shunts have opened again. At an alveolar oxygen of 60 mmHg or greater, there is little pulmonary vasoconstriction to hypoxemia even when the mixed venous PO_2 is as low as 10 mmHg (26). Although as alveolar hypoxia gets more severe, mixed venous hypoxemia may become a more important stimulus to pulmonary vasoconstriction. Nevertheless, the greater responsiveness of lung vessels to alveolar than to vascular hypoxia has led to the assumption that lung vessels autoregulate flow in response to local alveolar ventilation so that poorly ventilated alveoli with low O_2 concentrations produce vasoconstriction to shift perfusion to better ventilated alveoli.

HPV characteristically has an onset of action in seconds and can be sustained for hours if the hypoxia is regional (43). Diffuse hypoxia tends to reach an early peak rise in pulmonary artery pressure which then tails off over time (44). HPV can be as focal as a lobule (10) or in a larger area, including one lung or both lungs. The ability of the lung vessels to constrict and shift blood flow from one region to another depends on the size of the area made hypoxic (25). If the whole lung is hypoxic, then the lung vessels constrict diffusely and pulmonary artery pressure rises as the heart pumps harder to overcome the rise in pulmonary vascular resistance, allowing cardiac output and oxygen delivery to the tissues to stay as normal as possible. On the other hand, if the hypoxia is regional then the local vasoconstriction can effectively shift blood flow with only a very small rise in pulmonary artery pressure to other well ventilated areas of the lung which are compliant and can receive more flow. The anesthetized dog in Figure 1A shifted 54% of the perfusion from its left lung to the right well-ventilated lung in response to 100% N_2 for 7 mins to the left lung (13). The mean pulmonary artery pressure only increased from 14 to 15 mmHg to achieve this diversion and the stimulus to the HPV was alveolar hypoxia (PAO_2, 25 mmHg) since the arterial PaO_2 (from the oxygenated lung) was 89 mmHg and the mixed venous PO_2 was near normal at 34 mmHg (13). The strength of the HPV in this dog exceeded that of most dogs and people where the reduction in blood flow in response to HPV of the lung is usually about 30% (13). The strength of HPV is nevertheless sufficient to divert blood flow from a dependent to a nondependent lung (Fig. 1B) in dogs and man (1, 8, 13).

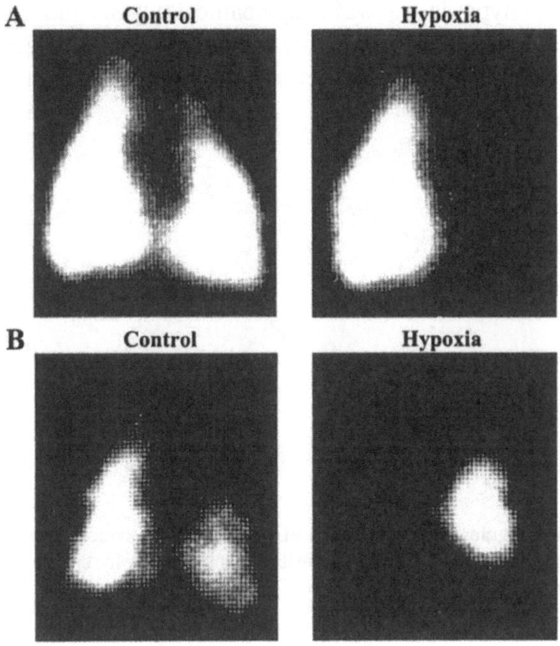

Figure 1. Examples of positron camera images of the distribution of perfusion in the lung using intravenous injections of ^{13}N at end-tidal expiration in the supine position (A) and in the left side dependent position (B) in the anesthetized dogs. (A) Left picture represents control perfusion on room air and the right picture represents the perfusion after 10 min of ventilation of the left lung with 100% N_2 while the right lung was on 100% O_2. There was a 54% reduction in perfusion to the hypoxic lung (PAO_2 25 mmHg) in this animal. (B) Left picture represents control perfusion with right side down on room air. Right picture is perfusion during ventilation of the dependent lung with 100% N_2 and demonstrates in this dog a 56% shift of perfusion away from the dependent lung and, therefore, directly against gravity (Reprinted from Ref. 13).

The site of HPV seems to be the precapillary 200-300 μm arterioles in pigs and dogs but perhaps as small as 25-50 μm vessels in cats based on microventilatory puncture techniques and sophisticated x-ray arteriography (Fig. 2) (33, 38). The veins may constrict but this seems to be modest and diffuse. This is hard to measure, especially radiologically, because of upstream resistance in the arteries causing less flow and distention of the veins during hypoxia. For years it was thought that the alveolar hypoxia sensor was in lung parenchyma. Now it is known from the work of Madden et al. that isolated pulmonary vessels (200 μm in diameter) themselves can constrict to hypoxia and further that pulmonary artery smooth muscle cells can contract although weakly in culture in response to hypoxia (23, 32). There is still a possible role for factors exogenous to the vessels to amplify or depress the regional hypoxic vasopressor response as recently reviewed (22). For example, the hypoxic vasoconstrictor response to alveolar hypoxia is less after sympathectomy in lambs though not so in sheep (5) and endothelin may play a role in pigs (21). Thus, exogenous factors may amplify the strength of vasoconstriction endogenous to the pulmonary artery smooth muscle cells.

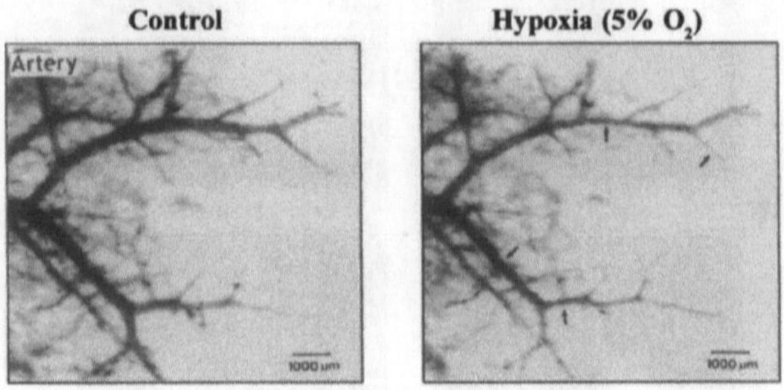

Figure 2. Typical arteriograms and venograms of left lower lobe under control and hypoxic conditions in the same cat are shown. Arteries constrict nonuniformly in series-arranged vessels, whereas veins constrict almost uniformly. Solid arrows indicate apparent vasoconstriction (Reprinted from Ref. 39).

4. Fetal and Neonatal Pulmonary Circulation

The placenta serves as the main source of oxygenation of fetal blood and it itself serves to autoregulate the distribution of blood flow through hypoxic fetoplacental vasoconstriction mediated via K^+ channels (15). Just prior to birth, blood flow to the airless lung is 8 to 10% of cardiac output and the Po_2 of the oxygenated placental blood (17-20 mmHg) is insufficient to decrease resistance to perfusion in the fetal lung. At birth there is a sudden and drastic reduction in pulmonary vascular resistance, a rise in pulmonary venous blood flow with an increase in left atrial pressure so that the patent foramen ovale closes and there is O_2-related closure of the patent ductus arteriosus such that there is an 8-10 fold increase in lung blood flow (35-37). The decrease in vascular resistance in the lung at birth is in part related to oxygenation of the lung but not entirely.

Hyperbaric oxygenation of the fetal lung *in utero* without expanding it, will increase blood flow (16), showing that HPV is active in decreasing lung perfusion. The increase in blood flow at most, though, rose only to 38% of cardiac output. However, the mixed venous Po_2 was only 44 mmHg which is probably too low to stop all vasoconstriction. Inflation of the fetal lung with hypoxic gas mixtures will also increase lung blood flow showing physical features of the collapsed lung contribute to pulmonary vascular resistance, and as expected, adding inhaled air to the hypoxic distended lung further reduces the pulmonary vascular resistance (4, 19, 41). Thus, HPV contributes substantially to reducing perfusion to the lung *in utero* although mechanical factors do so as well.

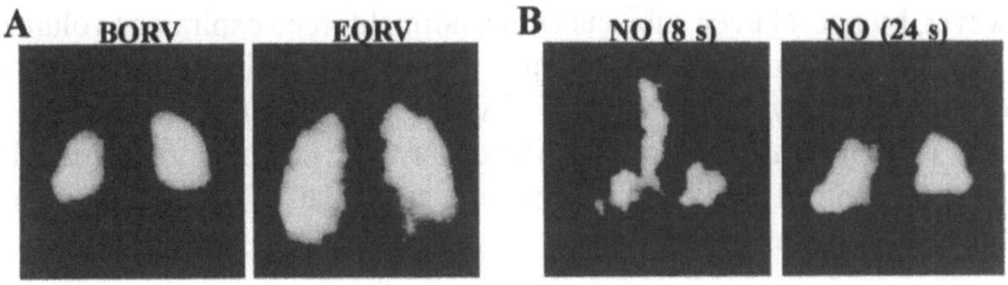

Figure 3. A: Scintigrams in an erect subject of the distribution of a 250 ml bolus of ^{13}N inhaled from residual volume (BORV) (left), and after rebreathing for 2 min and then breath holding at RV (EQRV) (right). B: Scintigrams were obtained during breath hold at RV as nitrous oxide was streaming into the lung through all patent airways to pulmonary capillary blood. Left scintigram made at 8 sec after injection of bolus of ^{13}N into the nitrous oxide stream shows the bolus in the trachea and major bronchi. Right scintigram 24 sec later shows that the bolus has largely cleared the major airways and is similar in distribution to BORV, both displaying basal decrease in activity compared to EQRV (Reprinted from Ref. 14).

5. Ventilation Distribution in Normal Subjects with Small Airways Dysfunction

The adult lung has over 300,000,000 alveoli and the distribution of ventilation is not uniform. Airway closure occurs in dependent lung at the diaphragm if the subject is erect or at the back if supine. In subjects with small airways disease such as from asthma or smoking, the airway closure is more prominent and occurs earlier in life. Figure 3 shows a scintigram taken by the Massachusetts General Hospital positron camera of the distribution of ^{13}N tracer gas during inhalation in a 29-year old man (14). In this case inhalation of the tracer occurred from near residual volume as is necessary at that age to collapse airways. The image in Figure 3A (left) is taken at residual volume after equilibration with air and the tracer ^{13}N which is almost insoluble. The image in Figure 3A (right) is of a bolus of 250 ml of ^{13}N labeled tracer inhaled from residual volume to total lung capacity. The tracer goes mainly to the apices of the lung. The image in Figure 3B (left) was made after the subject rebreathed the highly soluble gas nitrous oxide and then breath held at residual volume with an open glottis while the nitrous oxide streamed from the reservoir into the bloodstream. The image (Fig. 3B, left) was taken 8 seconds after injection of a bolus of ^{13}N in air into the mouth and showed the trachea and central airways. The image in Figure 3B (right) was 24 seconds later and showed a distribution of the gas similar to that seen with the inhaled bolus in the lower left. Waiting longer got no more tracer into the bases. Thus, the airways to the base were clearly closed, not allowing the inhaled gas to reach them even though a perfusion scan at residual volume showed persistent blood flow to the bases. This non-smoking young man had to exhale considerably to show airway closure. However, by age 44 even subjects with a normal forced expiratory volume in 1 second (FEV1) show airway collapse during tidal volume breathing while supine. Normals by age 65 show airway closure during tidal breathing even seated (18). These areas of low ventilation have the potential to decrease blood oxygenation considerably and it is in these areas that HPV could play a major role in reducing lung blood flow.

6. Extra-uterine Hypoxic Vasoconstriction

Nitroglycerin is a pulmonary artery vasodilator. We wondered if it would inhibit HPV, letting us demonstrate whether HPV was important in improving gas exchange in humans. In a dog model similar to that shown with the double lumen endotracheal tube in Figure 1A, we found a mean decrease in perfusion to the hypoxic lung of 28% in 8 animals (12). An infusion of 20 μg/min of nitroglycerin reduced the HPV so that only 9% of blood flow was diverted from the hypoxic lung resulting in a fall in $Pa{O_2}$ in the dogs from 89 to 59 mmHg. Knowing that HPV was inhibited by nitroglycerin, we then gave 0.6 mg sublingual nitroglycerin to a supine 53-year old man admitted to the hospital with suspected coronary artery disease but found to have gastritis with no coronary artery disease. He was a former smoker with an FEV_1 of 99% predicted but a flow at low lung volumes (V_{25}) of only 44% predicted, consistent with small airways dysfunction and susceptibility to increased airway collapse (Fig. 3). The arterial $P{O_2}$ fell from 80 to 65 mmHg six minutes after the nitroglycerin tablet was taken with no significant change in $Pa{CO_2}$, pH, shunt or cardiac output (Fig. 4). Thus, the nitroglycerin tablet disturbed ventilation-perfusion (V/Q) balance since neither shunt nor ventilation was changed, consistent with a loss of HPV allowing increased blood flow to hypoxic areas of lung (12).

We then looked at a series of individuals studied supine as it minimized any change in cardiac output by nitroglycerin, and maximized airway narrowing or closure since functional residual capacity falls by 15% in the supine position. Eight of the subjects had reduced airflow at low lung volumes (V_{25}) consistent with small airways dysfunction and six were completely normal. Normal subjects decreased their PaO_2 by 9 mmHg and those with small airways dysfunction by 14 mmHg after nitroglycerin (Fig. 5). Shunt fraction was assessed in 13 of these individuals and it changed by less than 1%, thus not being responsible for the fall in PaO_2. Subjects with advanced emphysema or pulmonary fibrosis were in generally much less responsive to nitroglycerin (Fig. 5). The DL_{CO} in these subjects was very low reflecting a likely significant loss of vascular bed. Perhaps in this state there is insufficient compliant vasculature anywhere in the lung for HPV to effectively divert blood. This display of the importance of HPV in humans with minimal or no lung disease has subsequently been shown by nitroprusside infusions in patients with congestive heart failure where PaO_2 fell in spite of an increase in cardiac output and a small rise in mixed venous PO_2 from 31 to 33 mmHg (30).

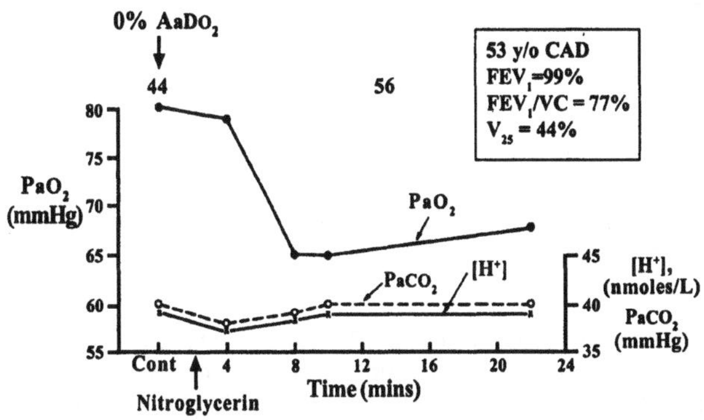

Figure 4. Sequential changes in arterial blood gases in one supine subject with small airways dysfunction given 0.6 mg of nitroglycerin sublingually at the arrow. Alveolar-arterial (A-a) gradient during O_2 breathing was obtained in the control period and at 16 mins after nitroglycerin administration (O_2 given from 9 to 16 mins). Twenty-two mins after the nitroglycerin was given, the PO_2 was still reduced at 67 mmHg (Reprinted from Ref. 12).

7. Hypoxic Vasoconstriction in Pneumonia

HPV appears to be ineffective in pneumonia. Light et al. measured blood flow with microspheres to the left lower lobe of the dog before and 3 days after inoculating the left lower lobe with *Streptococcus pneumoniae* (20). Using radiographs, they confirmed the presence of pneumonia confined to the left lower lobe. Perfusion to the left lower lobe showed a variable and nonsignificant decrease that was not affected by O_2 breathing. Utilizing blood samples from the lobar veins at terminal thoracotomy, they measured Q_S/Q_T through the left lower lobe with pneumonia as 0.69 compared to 0.08 through the control right lower lobe. Alveolar ventilation approached zero in some of the pneumonia lungs.

They concluded that HPV was ineffectual in pneumonia so that the magnitude of shunt and low V/Q perfusion was increased causing marked hypoxemia. Other authors with other models (27) have reached similar conclusions and Sostman et al. have shown that perfusion scans in consolidated pneumonia cases in humans may be normal (40).

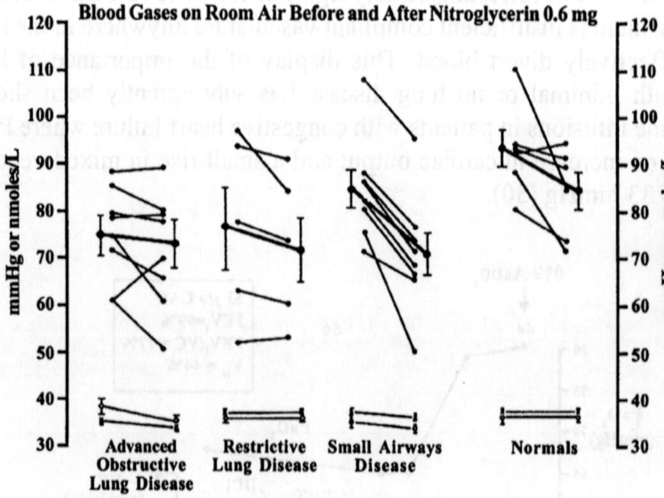

Figure 5. Control arterial blood gases and H^+ in 27 subjects before and the maximum change in arterial blood gases after the sublingual administration of 0.6 mg nitroglycerin. Individual values for PaO_2 and group mean values (means±SE) for PaO_2 (●), $PaCO_2$ (○), and H^+ (x) are shown. The decrease in PaO_2 in subjects with small airways disease was significant at $P<0.001$ and in the normal subjects at $P=0.05$ (Reprinted from Ref. 12).

8. Hypoxic Vasoconstriction in Lobar Collapse

As opposed to pneumonia where HPV does not seem to work well, there are abundant data in animals that HPV decreases blood flow through an atelectatic lobe. Benumof electromagnetically measured a 59% decrease in blood flow to the left lower lobes of dogs before and after atelectasis (3). Ventilation and re-expansion of that lobe with 95% N_2 and 5% CO_2 failed to increase blood flow to the left lower lobe whereas ventilation with O_2 fully restored perfusion. Others have also supported a major role for HPV in animals for diverting lung blood flow from atelectatic lobes, improving systemic oxygenation (28, 42).

The data in humans is sparse. Friedlander et al. inserted double lumen endotracheal tubes into anesthetized patients to separate ventilation between the two lungs (9). They then positioned the patients in the lateral dicubitus position and ventilated the non-dependent lung and not the dependent lung after letting the dependent lung collapse. The Pa_{O_2} fell from 531±42mmHg when both lungs were ventilated with O_2 to 285±42 mmHg when only the non-dependent lung was ventilated with O_2. An infusion of sodium nitroprusside increased cardiac output from 2.5 to 3.2 L/min ($P<0.05$) and increased Q_S/Q_T from 29±6.3 to 32.8±4.5 which was not significant in this small series leading the authors to conclude that there was no role for HPV in the dependent lung. It is apparent that the ability of HPV to effectively move blood flow from the collapsed dependent lung to the non-dependent lung was modest at best. Is this because HPV in the dependent collapsed lung in humans was inherently weak or was it weak because these studies were done under anesthesia?

I have anecdotal evidence of vasoconstriction in lobar collapse. A 58-year old woman was in the coronary care unit and on a ventilator for severe hypoxemia. She had collapse of the left lower lobe on chest x-ray and exam. On an F_IO_2 of 1.0 her PaO_2 was 220 mmHg and her mixed venous Po_2 was 36 mmHg. She was begun on intravenous nitroglycerin (TNG) for systemic blood pressure control. Her Po_2 fell to 40 mmHg. Mixed venous Po_2 was stable at 33mmHg. Upon cessation of the TNG her PaO_2 recovered to 216 mmHg. She was subsequently bronchoscoped with removal of a mucus plug in the left lower lobe bronchus. Her PaO_2 on an F_IO_2 of 1.0 rose to 580 mmHg and she was no longer sensitive to TNG. Thus, active vasoconstriction seemed to have been occurring in the collapsed lobe and animal data would suggest this was hypoxia-induced.

9. Diffuse Alveolar Hypoxia

Operation Everest II in 1985 involved 8 male athletes who were decompressed in a hypoxic chamber for 40 days to a barometric pressure of 240 Torr (11). The application of diffuse hypoxia (9.5% O_2) to the lungs of these atheletes caused a rise in pulmonary artery pressure acutely from 15±1 to 20±2 mmHg ($P<0.05$). Pulmonary capillary wedge pressure rose from 7±1 to 9±1 mmHg which was statistically insignificant. However, cardiac output rose from 6.7±0.4 to 8.1±0.7 L/min ($P<0.05$). Thus pulmonary vascular resistance rose only slightly from 1.19 to 1.35 mmHg/L/min. This was insignificant due to very low reactivity in response to hypoxia in some subjects. Interestingly, as we found in guinea pigs, the degree of pulmonary vasoreactivity at sea level did not correlate with the magnitude of the pulmonary hypertension over prolonged hypoxia (43). Thus the pulmonary hypertension of chronic hypoxia may be related not only to vasoconstriction but to an independent effect of hypoxia on the vascular wall (24).

Diffuse alveolar hypoxia induced pulmonary hypertension may also contribute to the pulmonary edema of altitude (2). Twenty one mountaineers were divided into two groups, one taking placebo and the other nifedipine before climbing to 4559 meters. Systolic pulmonary artery pressure at altitude averaged 53±16 mmHg in controls versus 41±8 mmHg in the nifedipine group and 7 of 11 controls developed X-ray evidence of pulmonary edema whereas only 1 of 10 nifedipine subjects did. The climb only lasted 22 hrs so acute vasoconstriction was involved, likely due to hypoxia as O_2 is known to reverse the edema if begun early. How HPV translates into non-cardiogenic pulmonary edema is speculative and, as with chronic hypoxia induced pulmonary hypertension, may require an independent effect of hypoxia on the vessel wall.

10. Summary

HPV clearly has a physiologic role in the fetus to divert blood flow from the lungs to the systemic circulation. Once out of the uterus, the human lung continues to display alveolar HPV. The usefulness of alveolar HPV to enhance gas exchange in normal people and patients with small airways dysfunction is based on limited studies but seems real. HPV, however, loses its effectiveness at preserving gas exchange in the presence of more diffuse hypoxia, as occurs in advanced chronic obstructive pulmonary disease or pulmonary fibrosis. Diffuse alveolar hypoxia may, however, contribute to the cor pulmonale seen in these diseases and to the pulmonary hypertension of altitude or hypoventilation. Since the strength of alveolar HPV acutely does not correlate with the magnitude of pulmonary hypertension in chronic diffuse hypoxia, hypoxia may induce pulmonary vascular remodeling by a direct effect on the vessel walls as well.

References

1. Arborelius M Jr, Lundin G, Svanberg L, and Defares JG. Influence of unilateral hypoxia on blood flow through the lungs in man in lateral position. *J. Appl. Physiol.* 1960; 15:595-597.
2. Bartsch P, Maggiorini M, Ritter M, Noti C, Vock P, and Oelz O. Prevention of high-altitude pulmonary edema by Nifedipine. *N. Engl. J. Med.* 1991; 325: 1284-1289.
3. Benumof JL. Mechanism of decreased blood flow to atelectatic lung. *J. Appl. Physiol.* 1979;46: 1047-1048.
4. Cassin S, Dawes GS, Mott JC, Ross BB, and Strang LB. The vascular resistance of the foetal and newly ventilated lung of the lamb. *J. Physiol.* 1963; 171: 61-79.
5. Custer J and Hales CA. Influence of alveolar oxygen on pulmonary vasoconstriction in newborn lambs versus sheep. *Am. Rev. Respir. Dis.* 1985; 132: 326-331.
6. Custer JR and Hales CA. Chemical sympathectomy decreases alveolar hypoxic vasoconstriction in lambs but not in sheep. *J. Appl. Physiol.* 1986; 60: 32-37.
7. Enson Y, Giuntini C, Lewis ML, Morris TQ, Ferrer MI, and Harvey RM. The influence of hydrogen ion concentration and hypoxia on the pulmonary circulation. *J. Clin. Invest.* 1964; 43: 1146-1161.
8. Fishman AP, Himmelstein A, Fritts, Jr. HW, and Cournand A. Blood flow through each lung in man during unilateral hypoxia. *J. Clin. Invest.* 1955; 34: 637-646.
9. Friedlander M, Sandler A, Kavanagh B, Winton T, and Benumof J. Is hypoxic pulmonary vasoconstriction important during single lung ventilation in the lateral decubitus position?

Can. J. Anesth. 1994; 41: 26-30.
10. Grant BJB, Davies EE, Jones HA, and Hughes JMB. Local regulation of pulmonary blood flow and ventilation-perfusion ratios in the coatimundi. *J. Appl. Physiol.* 1976; 40: 216-228.
11. Groves BM, Reeves JT, Sutton JR, Wagner PD, Cymerman A, Malconian MK, Rock PB, Young PM, and Houston CS. Operation Everest II: elevated high-altitude pulmonary resistance unresponsive to oxygen. *J. Appl. Physiol.* 1987; 63: 521-530.
12. Hales CA and Westphal D. Hypoxemia following the administration of sublingual nitroglycerin. *Am. J. Med.* 1978; 65: 911-918.
13. Hales CA, Ahluwalia B, and Kazemi H. Strength of pulmonary vascular response to regional alveolar hypoxia. *J. Appl. Physiol.* 1975; 38: 1083-1087.
14. Hales CA, Gibbons R, Burnham C, and Kazemi K. Determinants of regional distribution of a bolus inhaled from residual volume. *J. Appl. Physiol.* 1976; 41: 400-408.
15. Hampl V, Bibova J, Stranak Z, Wu X, Michelakis ED, Hashimoto K, and Archer SL. Hypoxic fetoplacental vasoconstriction in humans is mediated by potassium channel inhibition. *Am. J. Physiol. Heart Circ. Physiol.* 2002; 283: H2440-H2449.
16. Heymann MA, Rudolph AM, Nies AS, and Melmon KL. Bradykinin production associated with oxygenation of the fetal lamb. *Circ. Res.* 1969; 25: 521-534.
17. Hughs JMB and Morrell, NW, *Pulmonary Circulation: From basic mechanisms to clinical practice*. London: Imperial College Press, 2001.
18. LeBlanc P, Ruff F, and Milic-Emili J. Effects of age and body position on "airway closure" in man. *J. Appl. Physiol.* 1970; 28: 448-451.
19. Leffler CW, Hessler JR, and Green RS. The onset of breathing at birth stimulates pulmonary vascular prostacyclin synthesis. *Pediatr. Res.* 1984; 18: 938-942.
20. Light RB, Mink SN, and Wood LD. Pathophysiology of gas exchange and pulmonary perfusion in pneumococcal lobar pneumonia in dogs. *J. Appl. Physiol.* 1981; 50: 524-530.
21. Liu Q, Sham JSK, Shimoda LA, and Sylvester JT. Hypoxic constriction of porcine distal pulmonary arteries: endothelium and endothelin dependence. *Am. J. Physiol. Lung Cell. Mol. Physiol.* 2001; 280: L856-L865.
22. Madden JA. Focus on "Hypoxic constriction of porcine distal pulmonary arteries: endothelium and endothelin dependence". *Am. J. Physiol Lung Cell Mol Physiol.* 2001; 280: L853-L855.
23. Madden JA, Vadula MS, and Kurup VP. Effects of hypoxia and other vasoactive agents on pulmonary and cerebral artery smooth muscle cells. *Am. J. Physiol.* 1992; 263: L384-L393.
24. Mandegar M and Yuan JX-J. Role of K^+ channels in pulmonary hypertension. *Vasc. Pharmacol.* 2002; 38: 25-33.
25. Marshall BE and Marshall C. Continuity of response to hypoxic pulmonary vasoconstriction. *J. Appl. Physiol. Respir. Environ. Exercise Physiol.* 1980; 49: 189-196.
26. Marshall C and Marshall B. Site and sensitivity for stimulation of hypoxic pulmonary vasoconstriction. *J. Appl. Physiol. Respir. Environ. Exercise Physiol.* 1983; 55: 711-716.
27. McCormack DG, and Paterson NA. Loss of hypoxic pulmonary vasoconstriction in chronic pneumonia is not mediated by nitric oxide. *Am. J. Physiol.* 1993; 265: H1523-H1528.
28. Miller FL, Chen L, Malmkvist G, Marshall C, and Marshall BE. Mechanical factors do not influence blood flow distribution in atelectasis. *Anesthesiology.* 1989; 70: 481-488.
29. Miranda A, and Rotta A. Medidas del corazon en nativos de la altura. *Ann. Facultad. Medicina* 1944; 26: 49-58.
30. Mookherjee S, Keighley JFH, Warner RA, Bowser MA, and Obeid AI. Hemodynamic, ventilatory and blood gas changes during infusion of sodium nitroferricyanide (nitroprusside): Studies in patients with congestive heart failure. *Chest* 1977; 72: 273-278.
31. Motley HL, Cournand A, Werko L, Himmelstein A, and Dresdale D. The influence of short periods of induced acute anoxia upon pulmonary artery pressures. *Am. J. Physiol.* 1947; 150: 315-320.
32. Murray TR, Chen L, Marshall BE, and Macarak EJ. Hypoxic contraction of cultured pulmonary vascular smooth muscle cells. *Am. J. Respir. Cell. Mol. Biol.* 1990; 3: 457-465.
33. Nagasaka Y, Bhattacharya J, Nanjo S, Cropper MA, and Staub NC. Micropuncture

measurement of lung microvascular pressure profile during hypoxia in cats. *Circ. Res.* 1984; 54: 90-95.

34. Rotta A. Physiologic condition of the heart in the natives of high altitudes. *Am. Heart J.* 1947; 33: 669-676.
35. Rudolph AM. Fetal and neonatal pulmonary circulation. *Annu. Rev. Physiol.*1979; 41: 383-395.
36. Rudolph AM and Heymann MA. Circulatory changes during growth in the fetal lamb. *Circ. Res.* 1970; 26: 289-299.
37. Rudolph AM and Heymann MA. The circulation of the fetus in utero. Methods for studying distribution of blood flow, cardiac output and organ blood flow. *Circ. Res.* 1967; 21: 163-184.
38. Sada K, Shirai M, and Ninomiya I. X-ray TV system for measuring microcirculation in small pulmonary vessels. *J. Appl. Physiol.* 1985; 59: 1013-1018.
39. Shirai M, Sada K, and Ninomiya L. Effects of regional alveolar hypoxia and hypercapnia on small pulmonary vessels in cats. *J. Appl. Physiol.* 1986; 61: 440-448.
40. Sostman HD, Neumann RD, Gottschalk A, and Greenspan RH. Perfusion of nonventilated lung: Failure of hypoxic vasoconstriction? *AJR.* 1983; 141: 151-156.
41. Teitel DF, Iwamoto HS, and Rudolph AM. Effects of birth-related events on central blood flow patterns. *Pediatr. Res.* 1987; 22: 557-566.
42. Thomas HM III and Garrett RC. Strength of hypoxic vasoconstriction determines shunt fraction in dogs with atelectasis. *J. Appl. Physiol.* 1982; 53: 44-51.
43. Thompson BT, Hassoun PM, Kradin RL, and Hales CA. Acute and chronic hypoxic pulmonary hypertension in guinea pigs. *J. Appl. Physiol.* 1989; 66: 920-928.
44. Tucker A and Reeves JT. Nonsustained pulmonary vasoconstriction during acute hypoxia in anesthetized dogs. *Am. J. Physiol.* 1975; 228: 756-761.
45. Vejlstrup NG, O'Neill M, Nagyova B, and Dorrington KL. Time course of hypoxic pulmonary vasoconstriction: a rabbit model of regional hypoxia. *Am. J. Respir. Crit. Care Med.* 1997; 155: 216-221.
46. Von Euler US and Liljestrand G. Observations on the pulmonary arterial blood pressure in the cat. *Acta Physiol. Scand.* 1946; 12: 301-332.

Chapter 2

Hypoxic Pulmonary Vasoconstriction: Heterogeneity

Christopher A. Dawson
Medical College of Wisconsin, Marquette University, Milwaukee, Wisconsin, U.S.A.

1. Introduction

The concept of heterogeneity with regard to hypoxic pulmonary vasoconstriction (HPV) may evoke different notions depending on the level(s) of interest. These levels might include *i*) cellular and molecular mechanisms responsible for *a*) sensing low oxygen, *b*) linking the sensing mechanism to smooth muscle contraction, and *c*) integrating information about local conditions (e.g., local mechanical forces and chemical environment) or remote conditions (transmitted via neural or hormonal pathways) for effective modulation of the response; *ii*) cell types and anatomical locations wherein those mechanisms operate, and *iii*) both beneficial and unfavorable functional consequences. Heterogeneities observed at each of these levels include significant variations among species and experimental preparations. The latter may be partly the result of the highly modulated nature of the response, which makes it difficult to carry out the experimental manipulations necessary for addressing specific questions without inadvertently altering the balance of primary and modulating influences. These heterogeneities are discussed in most, if not all, chapters of this book, sometimes in the form of what may appear, at our present state of understanding (or confusion), to be contradictions. The emphasis of this chapter will be on some functional notions of heterogeneity arising mainly from studies directed at levels *ii*) and *iii*), including some of the same points raised in other chapters, but perhaps with some variation in context.

2. Heterogeneity in Lung Structure and the Physiological Role of Hypoxic Pulmonary Vasoconstriction

The remarkably close matching of the flow of air and blood to the thousands of functional gas exchange units in the normal lungs is mainly the result of the geometric and passive mechanical properties of the airways, vessels, and supporting tissues. The structures of the conducting airway and pulmonary vascular trees are quite heterogeneous with regard to the lengths and complexity of branching patterns of the various pathways traveled by the blood and air on route to the gas exchange surface. The microvascular and alveolar geometries are themselves quite heterogeneous even in normal lungs. Although there are similarities in the complex patterns of the airway and flow distributing arterial architecture down to the acinar level, the physics dominating blood and gas transport both above and particularly below that level are substantially different (16). As the resolution of methods for measuring regional ventilation and perfusion have increased, appreciation of the effectiveness of the system design in matching ventilation and perfusion has also increased (2), although knowledge as to exactly what the key design features are and how they develop using the available genomic information lags behind these functional observations.

Conjecture as to the path evolution might have taken to achieve this functional outcome may not provide much actual insight into evolution, but it can provide a framework for understanding some implications of the status quo, and this tactic has been used in other chapters. For this chapter, it is notable that the systemic circulation evolved such that local blood flow is determined primarily by modulation of local tone of vessels fed by a constant pressure head, with hypoxia being a dominant and locally self-serving vasodilatory stimulus. Conversely, in the pulmonary circulation, local flow is more or less simply proportional to right heart output (52), with hypoxia being a considerably less dominant and globally altruistic vasoconstricting stimulus without obvious direct benefit to the lungs themselves. Given the uniformity in the physiological role of the lung tissue, in comparison to the heterogeneity in the roles of systemic tissues, and the fact that the lungs must always accommodate the entire cardiac output, it is perhaps not surprising that local regulation of tone would have given way to a more structurally deterministic system during the evolutionary separation of the lungs from the systemic circulation.

HPV is critical for preventing wasted blood flow to the lungs in the mammalian fetus. Its existence in the postnatal mammalian lung is sometimes thought to be either a vestigial carryover from the prenatal genetic circuitry or a relatively minor adaptation of that circuitry for a role in air breathing. In fact, it is questionable whether HPV plays a significant role in matching perfusion to ventilation in the normal lung (27, 66, 74), and, as discussed further below, too tight a coupling between local gas exchange and pulmonary vasomotor tone might cause more problems than it could solve. Thus, the role of HPV in the postnatal lung seems to be to diminish the influence of more pathological heterogeneously distributed hypoventilation on the V/Q distribution, and its presence apparently requires extensive modulation, at least partly to offset potentially adverse consequences. Thus, HPV is not a high gain feedback system (29) consistent with evolutionary pressure to balance advantages of stabilizing the V/Q distribution in the face of disturbances in local lung mechanics (e.g., as the result of local airway obstruction, atelectesis, or inflammation) against the disadvantages of increasing pulmonary artery pressure in response to global alveolar hypoxia (e.g., ascent to high altitude, hypoventilation) or to the decrease in mixed venous Po_2 during exercise. The acute response to hypoxia has a rapid onset and is quickly reversed by an increase in Po_2 (43), whereas chronic hypoxia produces vascular remodeling and pulmonary hypertension that is sustained even after the alveolar Po_2 is returned to normal (75). A causal relationship between the acute and chronic response is generally assumed. What that relationship is has not been established, but the potential for contributing to chronic pulmonary vascular remodeling has been considered to be one of the potential risks of HPV as means of controlling the V/Q distribution.

3. Longitudinal Heterogeneity in Hypoxic Pulmonary Vasoconstriction

One of the first questions raised by the discovery of HPV by Von Euler and Liljestrand (90), and their immediate recognition of its potential for controlling regional V/Q, was that of the anatomical site of the constriction. This question evokes the concept of longitudinal (arterial to capillary to venous) heterogeneity in the ability of the vessels to respond to hypoxia. Clearly, the V/Q regulating potential of the response would have the finest resolution if the response were restricted to vessels no larger than those serving a functional ventilatory unit. Duke (19) carried out what has become an often-repeated (15) experimental approach to address the HPV site of action. In this experiment, the lungs are perfused using an extra-corporeal perfusion system and ventilating gas mixtures that allow pulmonary arterial Po_2 and alveolar (along with pulmonary venous) Po_2 to be controlled independently. Likewise, pulmonary venous Po_2 and alveolar (along with pulmonary arterial) Po_2 can be independently controlled while perfusing the lungs in the retrograde, pulmonary vein to pulmonary arterial, direction. Duke (19) found that decreasing alveolar Po_2 was sufficient and, under the conditions of her experiments, necessary to provoke HPV. Decreasing pulmonary arterial or venous Po_2 while maintaining high alveolar Po_2 produced no response. The conclusion was that HPV occurred in the pulmonary capillaries. Several studies have inferred that pulmonary capillary blood volume decreases in response to acute alveolar hypoxia (14, 15), and there is anatomical evidence that the capillaries are capable of responding to hypoxia (48). However, there is little direct evidence for hypoxia induced constriction of individual capillaries (92), and the concept presently has a low profile.

Subsequently, direct observations have revealed that hypoxia can cause pulmonary arteries and veins to constrict (3, 83), with the response being most intense in the small arteries (3, 25, 83, 84). It also seems clear, from studies on isolated arteries and arterial smooth muscle cells (59, 88), that the smooth muscle cells include both the sensing and contractile machinery. Thus, the search for a humoral mediator that carries the message from O_2 sensor cells to contracting cells has been essentially abandoned. The observation that alveolar hypoxia is so much more an effective stimulus than pulmonary arterial hypoxia has been a major reason that the humoral mediator concept has been difficult to give up. The fact that pulmonary arterial (mixed venous) blood is always hypoxic compared to alveolar gas is the reason that HPV has the potential for matching perfusion to ventilation in the first place. Thus, there would be no obvious benefit to having pulmonary arteries responsive to pulmonary arterial PO_2. However, from an anatomical and gas transport point of view, it is still not entirely clear how it is that the relevant arterial vascular smooth muscle is more affected by alveolar than pulmonary arterial Po_2. There is evidence that gas exchange can occur between alveolar air and blood in small arteries (15), as it can between tissue and arteriolar blood in systemic vascular beds (80), i.e., that the vessel wall Po_2 can be affected by alveolar Po_2. There has not been a clear demonstration that it is sufficient to explain the dominance of alveolar Po_2 as the stimulus.

Longitudinal heterogeneity in the pulmonary arterial hypoxic response is demonstrated by observations such as those in Figure 1 wherein the changes in vessel diameters in response to hypoxia can be compared with those in response to the infused vasoconstricting agent serotonin. The comparison reveals that while only the smaller vessel diameters decreased during hypoxia, the larger vessels were fully capable of constricting as much as smaller arteries in response to a different stimulus. Thus, the vessel size dependence in the hypoxia response cannot be attributed simply to a nonspecific size dependent contractility gradient. In the Figure 1B experiment wherein the flow was held constant, the larger vessels in the range studied were actually distended by the pressure increase resulting from the downstream constriction. That is not to say that hypoxia did not activate the smooth muscle in the larger vessels. In fact, with passive distension alone, the large vessels would have been even larger as indicated in Figure 1B. Thus, although more effective in the small vessels, hypoxia can activate arteries as large as 1.5 mm, and probably larger, in the dog lung.

The observation that rather large pulmonary arteries, and smooth muscle cells isolated from their walls, have the capacity to constrict within the normal range of mixed venous Po_2 led Marshall et al. (63) to consider the influence of the vasa vasorum, which supplies the larger vessel walls with systemic arterial blood. From observations using extracorporeal manipulation of the Po_2 of the bronchial arterial blood supply to the pulmonary arterial vasa vasorum in sheep, they concluded that the effect of the higher Po_2 of the vasa vasorum explains the relative insensitivity of pulmonary arteries to their own luminal Po_2 in vivo. Data such as those in Figure 1B in isolated lungs and observations on reactivity in different sized isolated pulmonary arteries (5, 59) suggest that the systemic arterial supply to the vessel wall is not the only reason for the vessel size dependence. Heterogeneity in the population of smooth muscle cell phenotypes probably contributes as well, as suggested by the observations of Archer et al. (5) that the vessel size dependence in hypoxia reactivity of isolated pulmonary arterial rings was correlated with the relative distributions of three electrophysiologically distinct smooth muscle cell phenotypes.

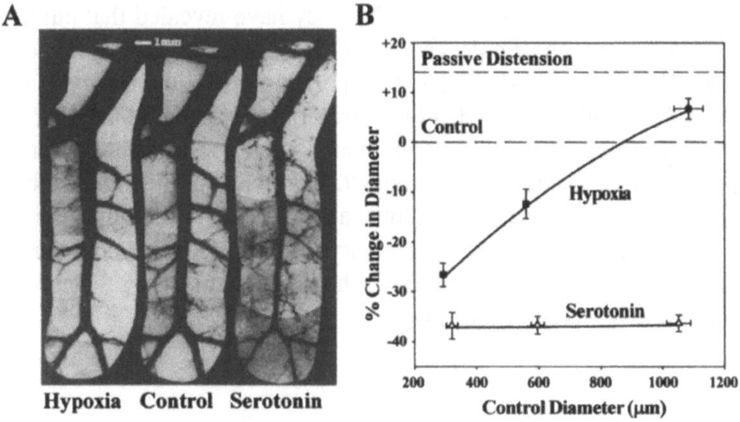

Figure 1. A: X-ray angiographic images of pulmonary arteries in a pump perfused dog lung obtained during the control condition (alveolar $Po_2 \approx 106$ Torr) and during hypoxia (alveolar $Po_2 \approx 36$ Torr) or during infusion of serotonin at a rate producing about the same increase in perfusion pressure as during hypoxia. During serotonin infusion, vessels throughout the observable size range are narrower than their respective control diameters. During hypoxia the smallest vessels are narrower than in the control image. Whereas, the largest vessels have been distended. B: The percent changes in arterial diameter produced by hypoxia or serotonin infusion obtained from experiments of the type producing the images in A. Again the smaller vessels narrowed in response to hypoxia. Whereas, when serotonin was infused, the vessels were narrowed throughout the size range studied. The average increase in perfusion pressure in these constant flow experiments was from a control pressure of about 7 to ~16 mmHg with both hypoxia and serotonin infusion. The dashed line labeled "Passive Distension" indicates the increase in diameter occurring with the same increase in pressure produced by increasing venous pressure instead of by vasoconstriction. It demonstrates that all vessels in the size range studied were activated by hypoxia even though some did not narrow and were actually distended by the increase in pressure (Modified from Ref. 3)

As indicated in other chapters, pulmonary arterial smooth muscle cells have become the model system for studying the mechanisms of HPV. However, normal pulmonary arteries smaller than about 100 μm in diameter, which have few if any smooth muscle cells, constrict in response to hypoxia as well as the larger vessels from which smooth muscle cells are normally harvested (37, 92). In fact, it may be that the smaller "non-muscular" but hypoxia sensitive vessels actually have more influence on local blood flow control than the larger vessels from which muscle cells are usually isolated (3).

As indicated above, the location of the flow diverting vessels upstream from the most relevant stimulus has been difficult to reconcile with a complete theory that includes the anatomical, gas transport, and cell signaling factors involved. In addition, the upstream site of action may have been important in guiding the evolution of modulating mechanisms. Several adaptations of the experiment for dissociating alveolar and vascular hypoxia (15, 19) have confirmed that alveolar hypoxia is the most effective stimulus, but they have revealed that pulmonary arterial hypoxia can contribute as well (15, 40). Using a version of that experiment, Hyman et al. (40) isolated the perfusion to a lobar artery of the cat lung so that they could independently control lobar arterial and alveolar Po_2 *in vivo*. When they maintained a constant lobar venous Po_2 (an approximation of, or a lower bound on alveolar Po_2) of about 92 Torr while perfusing the lobe with a constant flow of blood with varying lobar arterial Po_2, they obtained the results presented in Figure 2, wherein the pulmonary artery pressure began to rise when the lobar arterial Po_2 was decreased through the range including what might be considered normal mixed venous Po_2.

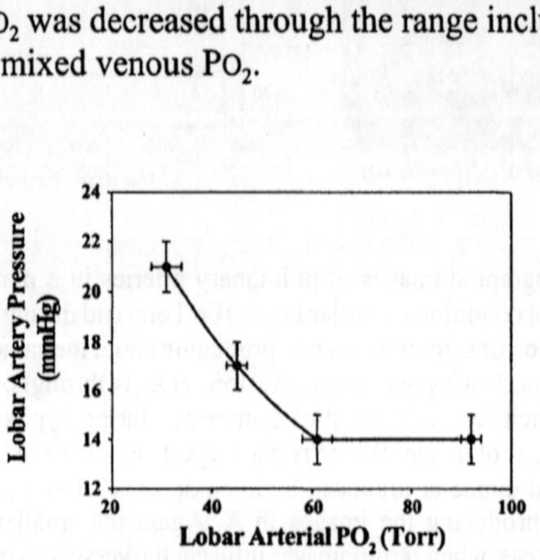

Figure 2. The lobar arterial pressure response to decreasing lobar arterial Po_2 in the left lower lobe of a cat lung perfused with constant flow in situ. The lobar venous Po_2 (as an approximation of lobar alveolar Po_2) was held constant at 92 Torr over the entire range of lobar arterial Po_2. (Modified from Ref. 40).

4. Heterogeneity in the Modulation of HPV

HPV in response to a decrease in pulmonary arterial Po_2 raises one of the potential risks associated with local V/Q control based on constriction of pulmonary arteries in response to low Po_2, namely, that of maintaining a low pulmonary vascular resistance during exercise. Part of the increase in oxygen consumption associated with exercise is accommodated by increased oxygen extraction by the working muscles, the result being a decrease in mixed venous Po_2. For example, in human subjects exercising at an intensity sufficient to increase cardiac output to 3.3 times the resting level, the mixed venous Po_2

decreased from about 38 to about 25 Torr (87). Thus, the mixed venous Po_2 can easily fall to levels for which data such as in Figure 2 predict significant HPV. Studies such as those represented in Figure 2, and (19) have been carried out at what might be considered resting to substantially below resting pulmonary arterial blood flow rates. Thus, given the expectation that, as the blood flow increases, the mixed venous Po_2 would penetrate further into the small pulmonary arteries and capillaries, the results of such studies probably represent lower bounds on the potential influence of the decrease in mixed venous Po_2 during exercise. Thus, it would appear that the complex modulation of the hypoxic response might be directed, at least in part, at preventing its potentially pulmonary hypertensive effects during exercise. In fact, exercise has been found to result in pulmonary vasoconstriction after NO synthase and β-adrenergic receptor blockade in sheep (47).

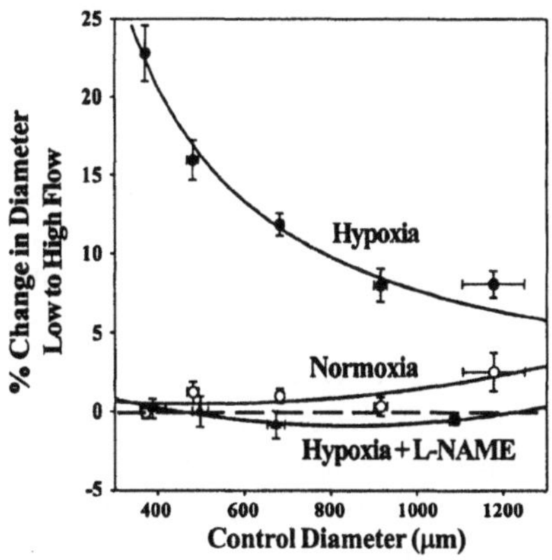

Figure 3. The change in arterial diameter for vessels of a given normoxic "Control Diameter" that resulted from increasing blood flow into the left lower lobar artery of the ferret lung while maintaining constant vascular pressure. The 3 study conditions were "Normoxia" (alveolar Po_2 ≈ 120 Torr), "Hypoxia" (56 Torr), and "Hypoxia+L-NAME." The high flow was 4 times the low flow. Under normoxic conditions, increasing the flow had little effect on the vessel diameters as would be expected given the vessels low tone under normoxic conditions. During hypoxia the constricted vessels dilated when the flow was increased. The smaller vessels dilated the most as would be expected given that they were the most constricted as in the Figure 1A example. During hypoxia with L-NAME, increasing flow had no effect on the diameters even though the vessels were constricted. Thus, the L-NAME interfered with the flow-induced dilation of the hypoxia constricted vessels. (Modified from Ref. 12).

Increased pulmonary flow has resulted in release of NO (31, 71) or prostacyclin (15, 89), and both of these vasodilators are apparently involved in modulating the hypoxic response even under what might be considered resting conditions as demonstrated by the enhancement of HPV after inhibition of their production (8, 55, 68). Increased lung NO output is observed during exercise (13, 73), although it is not clear that the pulmonary arterial endothelium makes a significant contribution to exhaled NO. In addition, Figure 3 shows that the hypoxic response can be attenuated by increased flow in ferret lungs and that the effect is eliminated by treatment with an NO synthase inhibitor. Thus, the argument can be made that the attenuation of HPV by vasodilator mechanisms called into play by mechanical stresses on the vessel walls (12) and/or blood (85) allows cardiac output to increase during exercise without overloading the right ventricle.

An interesting twist on this concept is provided by a study from Henderson et al. (36), wherein they compared the pulmonary arterial pressure response to exercise and hypoxia between two strains of rats one designated as having low exercise capacity, LCR, and the other high exercise capacity, HCR. They observed that the elevation in pulmonary arterial pressure with hypoxia during exercise in the LCR, was absent in the HCR, even though the HCR had higher cardiac output, and they noted the potential advantage that the attenuated HPV might afford the HCR. There is also evidence that exercise training can enhance endothelium-dependent (and presumably shear stress mediated) pulmonary vasodilation (44), which might suggest an adaptation that would enhance the ability of the pulmonary arteries to avoid HPV during exercise.

While shear stress, τ, modulation may serve to limit HPV during exercise or global hypoxia, it may serve to enhance the HPV response to localized hypoxia. This may be appreciated as follows. One common experimental approach for studying HPV is to perfuse lungs with constant flow and measure the changes in perfusion pressure in response to decreased inspired Po_2. Then, the effects of hypoxia on the diameters, *D*, of hypoxia-sensitive vessels are deduced. This experimental design is consistent with the fact that pulmonary blood flow is normally determined by the cardiac output independently of pulmonary vascular resistance. Thus, in this experimental design, as with low inspired Po_2 *in vivo*, HPV results in a rise in the shear stress on the walls of the constricted vessels. Using relationships for Poiseuille flow to make the point, the resistance, *R*, of a vessel is proportional to $D^{-1/4}$, and wall shear stress is proportional to $D^{-1/3}$. Thus, the fractional increase in wall shear stress, $(\tau_h - \tau_n)/\tau_n$, for a given fractional increase in vascular resistance, $(R_h - R_n)/R_n$, would be:

$$(\tau_h - \tau_n) / \tau_n = (1 + ((R_h - R_n) / R_n))^{3/4} - 1$$

where the subscripted *h* and *n* refer to the hypoxic and normoxic conditions, respectively. For example, a 100% increase, or doubling, of the resistance in a vessel would result in a 68% increase in wall shear stress. This experimental design does not reproduce what happens in the case of hypoventilation of a small region in the lungs, wherein the flow is dependent primarily on local resistance at any given cardiac output. In this case, the effect of the local alveolar hypoxia and HPV is to decrease the flow with virtually no change in the pressure at the regional arterial inlet. In that case, the shear stress on the walls of the constricted vessels decreases. For constant pressure driving the flow through the region, the fractional decrease in wall shear stress for a given fractional increase in vascular resistance would be:

$$(\tau_n - \tau_h) / \tau_n = (1 + ((R_h - R_n) / R_n))^{-1/4} - 1$$

e.g., a 100% increase in resistance would result in a 16% decrease in shear stress. Therefore, in the case of regional hypoxia, if the region is small enough, the stimulus for shear stress stimulated vasodilation decreases, releasing the HPV from shear stress dependent inhibitory modulation. As the hypoxic region becomes a larger fraction of the lung, the decrease in vessel wall shear stress within the region becomes smaller and then increases. Thus, in a sense the vessels within a hypoxic region of the lungs receive an integrated message regarding the number and/or extent of constriction of vessels within the hypoxic region(s), which is transmitted via the shear stress on the vessel walls (82). As the size of the hypoxic region increases, flow-diverting effectiveness of HPV decreases simply because the pulmonary arterial pressure begins to rise (63), but, in addition, shear stress activated modulation begins to attenuate the HPV.

Modulation of the hypoxic response has been revealed by the fact that inhibition of several vasodilator pathways (e.g., mediated by arachidonate metabolites, NO, and purine nucleotides) (8, 15, 51, 68, 85, 93) can potentiate hypoxic vasoconstriction under various experimental conditions. The hypoxic response has also been attenuated by inhibiting production of, or receptors for, various vasoconstrictors (5HT, histamine, endothelin, arachidonate metabolites, etc.), none of which is presently thought to be the mediator of the hypoxic response *per se* (15, 35, 45, 57). The role of endothelin receptor stimulation is probably the most timely in this context and is discussed elsewhere in this book. In isolated lung experiments, including or excluding various blood components has had dramatic although not always clearly explainable effects (9, 32, 64). Variations even within a species (49, 61) may be due to variability in the modulating (78) as well as in the mediating mechanisms. Thus, some of the heterogeneity and apparently contradictory responses observed in different preparations probably reflect the fact that the relative contributions of all these influences, known and unknown, are difficult to control or to control for.

Additional complexity is implied by the fact that, for the most extensively studied mechanical stress activated vasodilatory mechanisms involving NO or prostacyclin, both are potentially affected not only by the mechanical consequences of the constriction but also by hypoxia itself. In normoxia, NOS inhibition tends to produce small increases in pulmonary vascular resistance (68). However, NOS inhibition has typically produced substantial increases in HPV

(8, 68). The latter might be interpreted as suggesting that basal NO production attenuates HPV or that hypoxia increases either the production or effectiveness of NO. However, the endothelial NO synthase K_m for oxygen has been reported to be close enough to physiologically relevant Po_2 levels that hypoxia itself would be expected to directly depress NO production (42, 77). Suppression of NO production by hypoxia in whole lungs has been observed (54, 69), although the extent that this effect can be extrapolated to the small fraction of the tissue comprised of hypoxia responsive vessels and their endothelium is not clear. Hypoxia has also inhibited the uptake of the NO precursor L-arginine by pulmonary arterial endothelial cells (10).

There are parallels between the effects of NOS inhibition and cyclooxygenase inhibition on HPV. Cyclooxygenase inhibition also potentiates the hypoxic response, presumably by inhibiting prostacyclin production (15). In addition, the K_m of cyclooxygenases for oxygen are in the same range as that of NOS (46), suggesting that the direct effect of hypoxia would be to decrease prostacyclin production. However, acute hypoxia has both decreased (60) and increased (70) pulmonary arterial endothelial cell prostacyclin production.

These modulating influences are more or less autocoidal in nature. More remote modulation is not as clear. Investigations into reflex modulation of hypoxic vasoconstriction have come to mixed conclusions suggesting that it can occur (84), but is perhaps not a normally dominant influence (56, 58). The extent to which that might change under various conditions, e.g., exercise and/or high altitude exposure, may warrant further examination (21, 47).

5. Hypoxic Pulmonary Vasoconstriction Heterogeneity and High Altitude Pulmonary Edema

HPV is thought to play an important role in high altitude pulmonary edema (HAPE) (39) and the heterogeneity in susceptibility among individuals has been attributed to variability in the intensity of the hypoxic response. Several studies of individuals classified as HAPE susceptible (HAPE-S) on the basis of a previous incident of HAPE, and of individuals with similar high altitude exposure histories but without HAPE incident, have identified a correlation between the elevation in pulmonary artery pressure during an acute episode of hypoxia and HAPE susceptibility (49, 61). Thus, it is generally assumed that a particularly vigorous hypoxic vasoconstrictor response is a risk factor for HAPE, and that HPV is an important component of the mechanism of HAPE. The latter is also supported by the efficacy of pulmonary vasodilators in HAPE (4, 30, 81).

Another kind of heterogeneity considered a possible contributor to HAPE is heterogeneity in the hypoxic response among the parallel vessels throughout the lungs (39). Although low inspired Po_2 is often thought to decrease perfusion heterogeneity by reducing the gravity dependent perfusion gradient (15), the vertical gradient is only one contributor to the overall perfusion heterogeneity in the lung, and perhaps less dominant than once thought (26). In fact, hypoxia increased the heterogeneity of the flow distribution measured with high-resolution topographical mapping in the dog lung (62), and hypoxia has generally increased the coefficient of variation of pulmonary vascular transit times (14). One could imagine that it would take a rather complicated control system to produce a uniformly distributed increase in vascular resistance among parallel vessels given even a small degree of heterogeneity in the distribution of normoxic diameters, and wall thicknesses and composition. In studies such as those represented by Figures 1 and 2, the fractional changes in vessel diameters even with a uniform decrease in Po_2 have been quite heterogeneous (3, 83). With regard to HAPE, the concept is that these heterogeneities, combined with the moderate to intense muscular exertion that usually accompanies hypoxia in HAPE, result in regions of excessive perfusion that are subject to mechanically-induced vascular injury and edema (39).

The heterogeneous perfusion hypothesis is not universally accepted (53), and excessive pulmonary venous constriction during hypoxia has been considered as an alternative (61). The measurements of pulmonary arterial occlusion (wedge) pressure obtained during HAPE have been normal. Thus, explanations for HAPE have tended to disregard elevated pulmonary capillary pressure as a primary event. However, Naeije et al. (61) found evidence for hypoxia induced pulmonary venous constriction in HAPE-S subjects during a sojourn at 4559 m. They confirmed, using pulmonary arterial catheterization, that the HAPE-S individuals had a greater pulmonary arterial pressure response to hypoxia without elevated wedge pressures, but when they applied an analysis of the transient phase of the pulmonary arterial occlusion pressure data that provides an estimate of pulmonary capillary pressure, they found that the pulmonary capillary pressure was elevated in the HAPE-S. In other words, pulmonary venous resistance was increased during hypoxia in the HAPE-S. Thus, while pulmonary venous constriction is not normally considered to be a major component of HPV, heterogeneity in its contribution among individuals might help explain some of the heterogeneity in HAPE susceptibility.

There is no reason at present to think of these heterogeneities as mutually exclusive, and heterogeneity in modulation of HPV might be superimposed on other explanations for HAPE susceptibility. Eldridge et al. (22) compared HAPE-S and control subjects at sea level and 3810-m altitude. Their HAPE-S group did not have a greater pulmonary arterial pressure response to hypoxia, but rather a greater increase in pulmonary artery pressure in response to exercise at both altitudes. They suggested flow dependent vasoconstriction in the HAPE-S group as a possible explanation. However, their results would also appear to be consistent with the possibility that mechanical stress evoked attenuation of HPV may be less active in HAPE-S individuals. Lower exhaled NO excretion during hypoxia in HAPE-S individuals (11, 20), and an association between two particular NOS polymorphisms and HAPE-S (17) have been observed, although, the functional implications of the latter have not yet been determined. As an alternative to, or in concert with, a lower production of attenuators of the hypoxic response in HAPE-S, augmentation of HPV by modulators such as endothelin-1 may (79) or may not (18) be involved. Thus, there is at least a suspicion that heterogeneity in the modulation of HPV contributes to the heterogeneity in HAPE susceptibility among individuals.

6. Additional Heterogeneities

The flow diverting effectiveness of HPV may depend not only on some of the factors indicated above, but also on the cause of the poor ventilation. For example, in pneumonia, both the host and pathogen mediated inflammatory response would be expected to affect local HPV and/or its modulation (6, 41, 91). In addition, the extent to which the arteries serving the region are mechanically influenced by the alteration in local lung mechanics responsible for the local hypoventilation may have an effect as a consequence of way the vessels are imbedded within the lung parenchyma. Pulmonary vessels have been categorized as alveolar and extra-alveolar vessels. The designation is based on how they are affected by lung inflation. Due to interdependence between the vessel walls and surrounding parenchyma, lung inflation distends extra-alveolar vessels, which include the larger intrapulmonary arteries. Thus, an increase in lung volume would oppose extra-alveolar vessel HPV (15). The alveolar vessels include the capillaries within the alveolar septa that stretch and narrow as the alveolar wall circumference increases with inflation. Thus, an increase in volume would augment alveolar vessel HPV. The small hypoxia responsive arteries are located close to or within the transition region between the larger extra-alveolar arteries and the smaller alveolar capillaries. The extent to which these vessels behave more like extra-alveolar or alveolar vessels may determine the extent to which local distortions of the lung parenchyma will augment or oppose the HPV (15), and there are species variations, with hypoxia responsive vessels being mainly extra-alveolar in the dog (15), rat (7), and ferret (15, 86), but more alveolar in the sheep and pig (15, 86). In the rat at least, the growth of vascular smooth muscle into smaller vessels, that occurs during chronic hypoxia (50), is accompanied by a transition in the behavior of the hypoxic response from that indicative of an extra-alveolar site of action to that of an alveolar site of action (7). In addition, in any region in which the alveoli are disconnected from the flow of tracheal air, such as in a region of atelectasis or behind an obstructed airway, the vessels within the region are subject to the forces transmitted through the surrounding parenchyma in a manner similar to the interdependence affecting the extra-alveolar vessels. Thus, the extent to which the hypoxia responsive vessels are subjected to these forces will impact on the extent to which HPV is effective in reducing regional flow (15).

Between-species heterogeneities (72) in the intensity of HPV are difficult to evaluate at least in part because the modulation of the hypoxic response may vary among species depending on the preparation used in the study. For example in isolated lung experiments the dog came out as a poorly responding species (72), whereas in situ dog lungs responded well to hypoxia (62, 63), much like isolated dog lungs treated with cyclooxygenase inhibitors (3, 15). One between species heterogeneity that has a teleological rationale is the correlation between the effectiveness of the response and the extent of collateral ventilation, which holds for several species studied (33). The rationale is that the presence of collateral ventilation, which would tend to diminish the impact of small airway obstruction on local ventilation, would also decrease the importance of HPV as a means of controlling V/Q mismatch.

It may be surprising that, given the importance of HPV in the fetus, the intensity of HPV typically increases with post-natal age (24, 76). This may be at least partly because the transition from fetal to neonatal circulation, in which increased pulmonary blood flow is a key element, involves mechanisms that attenuate HPV (1, 28, 38, 51). Postnatal development also involves extension of smooth muscle into smaller arteries (67), in keeping with a transition of the hypoxic response from that providing the tone required to keep the flow to the lungs low during the globally and normally low lung oxygen levels before birth to that providing local control of blood flow in the air breathing lung. HPV has been found to be both more (22) and less intense (65) than normal in lungs from mature animals chronically exposed to hypoxia, and the functional implications are not entirely obvious either way.

These and other heterogeneities certainly add complexity to the problem of understanding the physiological and pathophysiological implications of HPV. In so doing they help to make the pursuit of that understanding challenging and exciting, as reflected in the many ingenious approaches that have been applied to the problem, only a very few of which could be represented in the available space.

Acknowledgments

Work from our laboratory presented in this chapter was supported by the NIH (HL-19298), the Department of Veterans' Affairs and the W. M. Keck Foundation.

References

1. Abman SH, Chatfield BA, Hall SL, and McMurtry IF. Role of endothelium-derived relaxing factor during transition of pulmonary circulation at birth. *Am. J. Physiol.* 1990; 259: H1921-

H1927.
2. Altemeier WA, Robertson HT, and Glenny RW. Pulmonary gas-exchange analysis by using simultaneous deposition of aerosolized and injected microspheres. *J. Appl. Physiol.* 1998; 85: 2344-2351.
3. Al-Tinawi A, Krenz GS, Rickaby DA, Linehan JH, and Dawson CA. Influence of hypoxia and serotonin on small pulmonary vessels. *J. Appl. Physiol.* 1994; 76: 56-64.
4. Anand IS, Prasad BAK, Chugh SS, Rao KRM, Cornfield DN, Milla CE, Singh N, Singh S, and Selvamurthy W. Effects of inhaled nitric oxide and oxygen in high-altitude pulmonary edema. *Circulation* 1998; 98: 2441-2445.
5. Archer AL, Huang JMC, Reeve HL, Hampl V, Tolarová S, Michelakis E, and Weir EK. Differential distribution of electrophysiologically distinct myocytes in conduit and resistance arteries determines their response to nitric oxide and hypoxia. *Circ. Res.* 1996; 78: 431-442.
6. Baboolal HA, Ichinose F, Ullrich R, Kawai N, Bloch KD, and Zapol WM. Reactive oxygen species scavengers attenuate endotoxin-induced impairment of hypoxic pulmonary vasconstriction in mice. *Anesthesiology* 2002; 97: 1227-1233.
7. Barer GR, Emery CJ, Bee D, and Wach RA. "Mechanisms of pulmonary hypertension: an overview." In *The Pulmonary Circulation in Health and Disease*, Will JA, Dawson CA, Weir EK, and Buckner CK, eds. Orlando, FL: Academic Press, Inc., 1987, pp. 409-422.
8. Barer G, Emery C, Stewart A, Bee D, and Howard P. Endothelial control of the pulmonary circulation in normal and chronically hypoxic rats. *J. Physiol.* 1993; 463: 1-16.
9. Berkov S. Hypoxic pulmonary vasoconstriction in the rat, the necessary role of angiotensin II. *Circ. Res.* 1974; 35:256-261.
10. Block ER, Herrera H, and Couch M. Hypoxia inhibits L-arginine uptake by pulmonary artery endothelial cells. *Am. J. Physiol.* 1995; 269: L574-L580.
11. Busch T, Bartsch P, Pappert D, Grunig E, Hildebrandt W, Elser H, Falke KJ, and Swenson ER. Hypoxia decreases exhaled nitric oxide in mountaineers susceptible to high-altitude pulmonary edema. *Am. J. Respir. Crit. Care Med.* 2001; 163: 368-373.
12. Chammas JH, Rickaby DA, Guarin M, Linehan JH, Hanger CC, and Dawson CA. Flow-induced vasodilation in the ferret lung. *J. Appl. Physiol.* 1997; 83:495-502.
13. Chirpaz-Oddou MF, Favre-Juvin A, Flore P, Eterradossi J, Delaire M, Grimbert F, and Therminarias A. Nitric oxide response in exhaled air during an incremental exhaustive exercise. *J. Appl. Physiol.* 1997; 82: 1311-1318.
14. Clough AV, Haworth ST, Ma W, and Dawson CA. Effects of hypoxia on pulmonary microvascular volume. *Am. J. Physiol. Heart Circ. Physiol.* 2000; 279: H1274-H1282.
15. Dawson CA. Role of pulmonary vasomotion in the physiology of the lung. *Physiol. Rev.* 1984, 64: 544-616.
16. Dawson CD, Krenz GS, and Linehan JH. "Complexity and Structure-Function relationships in the pulmonary arterial tree, Chapter 13." In *Lung Biology in Health and Disease. Complexity in Structure and Function of the Lung,* Hlastala MP and Robertson HT, eds. New York, NY: Marcel Dekker, Inc., 1998, pp. 401-427.
17. Droma Y, Hanaoka M, Ota M, Katsuyama Y, Koizumi T, Fujimoto K, Kobayashi T, and Kubo K. Positive association of the endothelial nitric oxide synthase gene polymorphisms with high-altitude pulmonary edema. *Circulation* 2002; 106: 826-830.
18. Droma Y, Ri-Li G, Tanaka M, Koizumi T, Hanaoka M, Miyahara T, Yamaguchi S, Okada K, Yoshikawa S, Fujimoto K, Matsuzawa Y, Kubo K, Kobayashi T, and Sekiguchi M. Acute hypoxic pulmonary vascular response does not accompany plasma endothelin-1 elevation in subjects susceptible to high altitude pulmonary edema. *Intern. Med.* 1996; 35: 257-260.
19. Duke HN. The site of action of anoxia on the pulmonary blood vessels of the cat. *J. Physiol.* 1954; 125: 373-382.
20. Duplain H, Sartori C, Leipri M, Egli M, Allemann Y, Nicod P, and Scherrer U. Exhaled nitric oxide in high-altitude pulmonary edema. Role in the regulation of pulmonary vascular tone and evidence for a role against inflammation. *Am. J. Respir. Crit. Care Med.* 2000; 162: 221-224.

21. Duplain H, Vollenweider L, Delabays A, Nicod P, Bartsch P, Sherrer U. Augmented sympathetic activation during short-term hypoxia and high altitude exposure in subjects susceptible to high -altitude pulmonary edema. *Circulation* 1999; 99: 1713-1718.
22. Eldridge MW, Podolsky A, Richardson RS, Johnson DH, Knight DR, Johnson EC, Hopkins SR, Michimata H, Grassi B, Feiner J, Kurdak SS, Bickler PE, Wagner PD, and Severinghaus JW. Pulmonary hemodynamic response to exercise in subjects with prior high-altitude pulmonary edema. *J. Appl. Physiol.* 1996; 81: 911-921.
23. Emery CJ, Bee, D, and Barer GR. Mechanical properties and reactivity of vessels in isolated perfused lungs of chronically hypoxic rats. *Clin. Sci.* 1981; 61: 569-583.
24. Fike CD and Hansen TN. Hypoxic vasoconstriction increases with postnatal age in lungs from newborn rabbits. *Circ. Res.* 1987; 60: 297-303.
25. Fishman AP. Hypoxia on the pulmonary circulation: how and where it acts. *Circ. Res.* 1976; 38: 221-231.
26. Glenny RW, Bernard S, Robertson HT, and Hlastala MP. Gravity is an important but secondary determinant of regional pulmonary blood flow in upright primates. *J. Appl. Physiol.* 1999; 86: 623-632.
27. Glenny RW, Robertson HT, and Hlastala MP. Vasomotor tone does not affect perfusion heterogeneity of gas exchange in normal primate lungs during normoxia. *J. Appl. Physiol.* 2000; 89: 2263-2267.
28. Gordon JB, Tod ML, Wetzel RC, McGeady ML, Adkinson NF Jr, and Sylvester JT. Age-dependent effects of indomethacin on hypoxic vasoconstriction in neonatal lamb lungs. *Pediatr. Res.* 1988; 23: 580-584.
29. Grant BJB. Effect of local pulmonary blood flow control on gas exchange theory. *J. Appl. Physiol.* 1982; 53: 1100-1109.
30. Hackett PH, Roach RC, Hartig GS, Green ER, and Levine BD. The effect of vasodilators on pulmonary hemodynamics in high altitude pulmonary edema: a comparison. *Int. J. Sports Med.* 1992; 13: S68-71.
31. Hakim TS. Flow-induced release of EDRF in the pulmonary vasculature: site of release and action. *Am. J. Physiol.* 1994; 267: H363-369.
32. Hakim TS and Malik AB. Hypoxic vasoconstriction in blood and plasma perfused lungs. *Respir. Physiol.* 1988; 72: 109-121.
33. Hanson WL, Boggs DF, Kay JM, Hofmeister SE, and Wagner WW Jr. Collateral ventilation and pulmonary arterial smooth muscle in the coati. *J. Appl. Physiol.* 1993; 74: 2219-2224.
34. He L, Chang SW, de Montellano PO, Burke TJ, and Voelkel NF. Lung injury in Fischer but not Sprague-Dawley rats after short-term hyperoxia. *Am. J. Physiol.* 1990; 259: L451-L458.
35. Helgesen KG and Bjertnaes L. The effect of ketanserin on hypoxia-induced vasoconstriction in isolated lungs. *Int J Microcirc Clin Exp* 1986; 5: 65-72.
36. Henderson KK, Wagner H, Favret F, Britton SL, Koch LG, Wagner PD, and Gonzalez NC. Determinants of maximal O_2 uptake in rats selectively bred for endurance running capacity. *J. Appl. Physiol.* 2002; 93: 1265-1274.
37. Hillier SC, Graham JA, Hanger CC, Godbey PS, Glenny RW, and Wagner WW. Hypoxic vasoconstriction in pulmonary arterioles and venules. *J. Appl. Physiol.* 1997; 82: 1084-1090.
38. Hislop AA, Springall DR, Buttery LDK, Pollock JS, and Haworth SG. Abundance of endothelial nitric oxide synthase in newborn intrapulmonary arteries. *Arch. Dis. Child Fetal Neonatal* 1995; 73: 17-21.
39. Hultgren HN. High altitude pulmonary edema: hemodynamic aspects. *Int. J. Sports Med.* 1997; 18: 20-25.
40. Hyman AL, Higashida RT, Spannhake EW, and Kadowitz PJ. Pulmonary vasoconstrictor responses to graded decreases in precapillary blood Po_2 in intact-chest cat. *J. Appl. Physiol.* 1981; 51: 1009-1016.
41. Ichinose F, Zapol WM, Sapirstein A, Ullrich R, Tager AM, Coggins K, Jones R, and Bloch KD. Attenuation of hypoxic pulmonary vasoconstriction by endotoxemia requires 5-lipoxygenase in mice. *Circ. Res.* 2001; 88: 832-838.

42. Ide H, Nakano H, Ogasa T, Osanal S, Kikuchi K, and Iwamoto J. Regulation of pulmonary circulation by alveolar oxygen tension via airway nitric oxide. *J. Appl. Physiol.* 1999; 87: 1629-1636.
43. Jensen KS, Micco AJ, Czartolomna J, Latham L, and Voelkel NF. Rapid onset of hypoxic vasoconstriction in isolated lungs. *J. Appl. Physiol.* 1992; 72: 2018-2023.
44. Johnson LR, Rush JWE, Turk,JR, Price EM, and Laughlin MH. Short-term exercise training increases ACh-induced relaxation and eNOS protein in porcine pulmonary arteries. *J. Appl. Physiol.* 2001; 90: 1102-1110.
45. Johnson W, Nohria A, Garrett L, Fang JC, and Igo J. Contribution of endothelin to pulmonary vascular tone under normoxic and hypoxic conditions. *Am. J. Physiol. Heart Circ. Physiol.* 2002; 283: H568-H575.
46. Juranek I, Suzuki H, Yamamoto S. Affinities of various mammalian arachidonate lipoxygenases and cyclooxygenases for molecular oxygen as substrate. *Biochim. Biophys. Acta* 1999; 1436: 509-518.
47. Kane DW, Tesauro T, Koizumi T, Gupta R, and Newman JH. Exercise-induced pulmonary vasoconstriction during combined blockade of nitric oxide synthase and beta adrenergic receptors. *J. Clin. Invest.* 1994; 93: 677-683.
48. Kapanci Y, Costabella PM, Cerutti P, and Assimacopoulos A. Distribution and function of cytoskeletal proteins in lung cells with particular reference to "Contractile interstitial cells. *Methods Achiev. Exp. Pathol.* 1979; 9:1 47-168.
49. Kawashima A, Kubo K, Kobayashi T, and Sekiguchi M. Hemodynamic responses to acute hypoxia, hypobaria and exercise in subjects susceptible to high-altitude pulmonary edema. *J. Appl. Physiol.* 1989; 67: 1982-1989.
50. Kay JM. "Pulmonary vasculature and experimental pulmonary hypertension in animals." In *The Pulmonary Circulation in Health and Disease*, Will JA, Dawson CA, Weir EK, and Buckner CK, eds. Orlando, FL: Academic Press, Inc, 1987, pp. 41-56.
51. Konduri GG and Mattei J. Role of oxidative phosphorylation and ATP release in mediating birth-related pulmonary vasodilation in fetal lambs. *Am. J. Physiol. Heart Circ. Physiol.* 2002; 283: H1600-H1608.
52. Krenz GS and Dawson CA. Flow and pressure distributions in vascular networks consisting of distensible vessels. *Am. J. Physiol. Heart Circ. Physiol.* 2003; 284: H2192-H2203.
53. Landolt CC, Matthay MA, Albertine KH, Roos PJ, Wiener-Kronish JP, and Staub NC. Overperfusion, hypoxia, and increased pressure cause only hydrostatic pulmonary edema in anesthetized sheep. *Circ. Res.* 1983; 52: 335-341.
54. LeCrass TD and McMurtry IF. Nitric oxide production in the hypoxic lung. *Am. J. Physiol. Lung Cell Mol. Physiol.* 2001; 280: L575-L582.
55. Leeman M, DeBeyl VZ, Delcroix M, and Naeije R. Effects of endogenous nitric oxide on pulmonary vascular tone in intact dogs. *Am. J. Physiol.* 1994; 266: H2343-H2347.
56. Lejeune P, Vachiery JL, Leeman M, Brimioulle S, Hallemans R, Melot C, and Naeije R. Absence of parasympathetic control of pulmonary vascular pressure-flow plots in hyperoxic and hypoxic dogs. *Respir. Physiol.* 1989; 78: 123-133.
57. Liu Q, Sham SK, Shimoda LA, and Sylvester JT. Hypoxic constriction of porcine distal pulmonary arteries: endothelium and endothelin dependence. *Am. J. Physiol. Lung Cell. Mol. Physiol.* 2001; 280: L856-L865.
58. Lodato RF, Micael JR, and Murray PA. Absence of neural modulation of hypoxic pulmonary vasoconstriction in conscious dogs. *J. Appl. Physiol.* 1988; 65: 1481-1487.
59. Madden J, Dawson CA, and Harder DA. Hypoxia-induced activation in small isolated pulmonary arteries from the cat. *J. Appl. Physiol.* 1985; 59: 113-118.
60. Madden MC, Vender RL, and Friedman M. Effect of hypoxia on prostaglandin production in cultured pulmonary artery endothelium. *Prostaglandins* 1986; 31: 1049-1062.
61. Maggiorini M, Melot C, Pierre S, Pfeiffer F, Greve I, Sartori C, Lepori M, Hauser M, Scherrer U, and Naeije R. High-altitude pulmonary edema is initially caused by an increase in capillary pressure. *Circulation* 2001; 103: 2078-2083.

62. Mann CM, Domino KB, Walther SM, Glenny RW, Polissar NL, and Hlastala MP. Redistribution of pulmonary blood flow during unilateral hypoxia in prone and supine dogs. *J. Appl. Physiol.* 1998; 84: 2010-2019.
63. Marshall BE, Marshall C, Benumof J, and Saidman LJ. Hypoxic pulmonary vasoconstriction in dogs: effects of lung segment size and oxygen tension. *J. Appl. Physiol.* 1981; 51: 1543-1551.
64. McMurtry IF, Hookway BW, and Roos SD. Red blood cells but not platelets prolong vascular reactivity of isolated rat lungs. *Am. J. Physiol.* 1978; 234: H186-H191.
65. McMurtry IF, Petrun MD, and Reeves JT. Lungs from chronically hypoxic rats have decreased pressor response to acute hypoxia. *Am. J. Physiol.* 1978; 235: H104-H109.
66. Melot C, Naeije R, Hallemans R, Lejeune P, and Mols P. Hypoxic pulmonary vasoconstriction and pulmonary gas exchange in normal man. *Resp. Physiol.* 1987; 68: 11-27.
67. Michel RP, Gordon JB, and Chu K. Development of the pulmonary vasculature in newborn lambs: structure-function relationships. *J. Appl. Physiol.* 1991; 70: 1255-1264.
68. Nelin LD and Dawson CA. The effect of N^{ω}-nitro-L-arginine methylester on hypoxic vasoconstriction in the neonatal pig lung. *Pediatr. Res.* 1993; 34: 349-353.
69. Nelin LD, Thomas CJ, and Dawson CA. Effect of hypoxia on nitric oxide production in neonatal pigs. *Am. J. Physiol.* 1996; 271: H8-H14.
70. North AJ, Brannon TS, Wells LB, Campbell WB, and Shaul PW. Hypoxia stimulates prostacyclin synthesis in newborn pulmonary artery endothelium by increasing cyclooxygenase-1 protein. *Circ. Res.* 1994; 75: 33-40.
71. Ogasa T, Nakano H, Ide H, Yamamoto Y, Sasaki N, Osanai S, Akiba Y, Kikuchi K, and Iwamoto J. Flow-mediated release of nitric oxide in isolated perfused rabbit lungs. *J. Appl. Physiol.* 2001; 91: 363-370.
72. Peake MD, Harabin AL, Brennan NJ, and Sylvester JT. Steady-state vascular responses to graded hypoxia in isolated lungs of five species. *J. Appl. Physiol.* 1981; 51: 1214-1219.
73. Phillips CR, Giraud GD, and Holden WE. Exhaled nitric oxide during exercise: site of release and modulation by ventilation and blood flow. *J. Appl. Physiol.* 1996; 80: 1865-1871.
74. Pison U, Lopez FA, Heidelmeyer CF, Rossaint R, and Falke KJ. Inhaled nitric oxide reverses hypoxic pulmonary vasoconstriction without impairing gas exchange. *J. App. Physiol.* 1993; 74: 1287-1292.
75. Reid LM. The pulmonary circulation: remodeling in growth and disease. *Am. Rev. Respir. Dis.* 1979; 119: 531-546.
76. Rendas A, Branthwaite M, Lennod S, Reid L. Response of the pulmonary circulation to acute hypoxia in the growing pig. *J. Appl. Physiol.* 1982; 52: 811-814.
77. Rengasamy A and Johns RA. Determination of Km for oxygen of nitric oxide synthase isoforms. *J. Pharmacol. Exp. Ther.* 1996; 276: 30-33.
78. Salameh G, Karamsetty MR, Warburton RR, Klinger JR, Ou LC, and Hill NS. Differences in acute hypoxic pulmonary vasoresponsiveness between rat strains: role of endothelium. *J. Appl. Physiol.* 1999; 87: 356-362.
79. Sartori C, Vollenweider L, Loffler BM, Delabays A, Nicod P, Bartsch P, and Scherrer U. Exaggerated endothelin release in high-altitude pulmonary edema. *Circulation* 1999; 99: 2665-2668.
80. Schacterle RS, Adams JM, and Ribando RJ. A theoretical model of gas transport between arterioles and tissue. *Microvasc. Res.* 1991; 41: 210-228.
81. Scherrer U, Vollenweider L, Delabays A, Savcic M, Eichenberger U, Kleger G-R, Fikrle A, Ballmer PE, Nicod P, and Bartsch P. Inhaled nitric oxide for high-altitude pulmonary edema. *N. Engl. J. Med.* 1996; 334: 624-630.
82. Secomb TW and Pries AR. Information transfer in microvascular networks. *Microcirculation* 2002; 9: 377-387.
83. Shirai M, Sada K, and Ninomiya I. Effects of regional alveolar hypoxia and hypercapnia in small pulmonary vessels in cats. *J. Appl. Physiol.* 1986; 61: 440-448.

84. Shirai M, Shindo T, and Ninomiya I. β-adrenergic mechanisms attenuate hypoxic pulmonary vasoconstriction during systemic hypoxia in cats. *Am. J. Physiol.* 1994; 266: H1777-H1785.
85. Sprague RS, Ellsworth ML, Stephenson AH, and Lonigro AJ. ATP: the red blood cell link to NO and local control of the pulmonary circulation. *Am. J. Physiol.* 1996; 271: H2717-H2722.
86. Sylvester, JT, Brower, RG, and Permutt, S. "Effects of hypoxic vasoconstriction on the mechanical interaction between pulmonary vessels and airways." In *The Pulmonary Circulation in Health and Disease*, Will JA, Dawson CA, Weir EK, and Buckner CK, eds. New York, NY: Academic Press, 1987, pp. 321-334.
87. Torre-Buenoi JR, Wagner PD, Saltzman HA, Gale GE, and Moon RE. Diffusion limitation in normal humans during exercise at sea level and simulated altitude. *J. Appl. Physiol.* 1985; 58: 989-995.
88. Vadula MS, Kleinman JG, and Madden JA. Effect of hypoxia and norepinephrine on cytoplasmic free Ca^{2+} in pulmonary and cerebral arterial myocytes. *Am. J. Physiol.* 1993; 265: L591-L597.
89. Van Grondelle A, Worthen GS, Ellis D, Mathias MM, Murphy RC, Strife RJ, Reeves JT, and Voelkel NF. Altering hydrodynamic variables influences PGI_2 production by isolated lungs and endothelial cells. *J. Appl. Physiol.* 1984; 57: 388-395.
90. Von Euler US and Liljestrand G. Observations on the pulmonary arterial blood pressure in the cat. *Acta Physiol. Scand.* 1946; 12: 301-320.
91. Yaghi A, Paterson NAM, and McCormack G. Nitric oxide does not mediate the attenuated pulmonary vascular reactivity of chronic pneumonia. *Am. J. Physiol.* 1993; 265: H943-H948.
92. Yamaguchi K, Suzuki K, Naoki K, Nishio K, Sato N, Takeshita K, Kudo H, Aoki T, Suzuki Y, Miyata A, and Tsumura H. Response of intra-acinar pulmonary microvessels to hypoxia, hypercapnic acidosis, and isocapnic acidosis. *Circ. Res.* 1998; 82: 722-728.
93. Zhu D, Birks EK, Dawson CA, Patel M, Falck JR, Presberg K, Roman, RJ, and Jacobs, ER. Hypoxic pulmonary vasoconstriction is modified by *P*-450 metabolites. *Am. J. Physiol. Heart Circ. Physiol.* 2000; 279: H1526-H1533.

The editor notes with sadness the passing of Dr. Christopher Dawson, an extraordinary mentor for young investigators and an eminent scientist in the field of pulmonary physiology, and was very grateful that this book contains one of his last works on the heterogeneity of hypoxic pulmonary vasoconstriction. His presence in the scientific community will be sorely missed.

Christopher A. Dawson, Ph.D.

Christopher A. Dawson, Ph.D., Professor and eminent research scientist died suddenly and unexpectedly on July 12, 2003 at his office. He is recognized as an inspirational mentor, model researcher and has had a profound influence as a teacher and faculty leader at the Medical College of Wisconsin. Dr. Dawson was born in 1942 in Long Beach, California and received his Ph.D. degree from the University of California, Santa Barbara in 1969. As a Professor of Physiology and Medicine at the Medical College of Wisconsin and at Marquette University, he was recognized as one of the world experts in the pulmonary circulation. As documented by more than 200 original research publications and 22 invited reviews and book chapters, he and his associates pioneered many novel technologies that revealed important functions of the lung that were previously unknown. He served as an Associate Editor of the major research journals in his field, *Journal of Applied Physiology* and *American Journal of Physiology*. His scholarly and multifaceted works were supported continuously since 1971 by the Department of Veterans Affairs and the National Institutes of Health. Among his many noteworthy discoveries was that the lining of the blood vessels in the lungs contribute critically to the regulation of hormones that modify pulmonary blood flow. He also was instrumental in the first studies to show that isolated small pulmonary arteries contracted and depolarized to hypoxia. His most recent studies using X-ray imaging of the pulmonary circulation have led to a new understanding of how multiple generations of the pulmonary vasculature function under normal and pathological conditions.

His quiet and humble manner, exacting scientific standards and selfless encouragement of other researchers made him a highly sought source of sound advice. Dr. Dawson collaborated closely with a number of bioengineers, physicians, and basic science investigators at Marquette University, the Medical College of Wisconsin, the Zablocki Veterans Hospital, the Froedtert Lutheran Memorial Hospital and the Children's Hospital of Southeast Wisconsin. For his internationally recognized contributions to lung research, Dr. Dawson was given the Medical Career Scientist Award by the Department of Veterans Affairs in 1999, and in the same year he received the Distinguished Service Award by the Medical College of Wisconsin.

Dr. Dawson is survived by his wife Michal Ann, his daughter Marcey Kay and her husband Keith Gulley and their son Dawson Gulley; his son Brian Christopher and wife Cecilia and their daughter Kana Rose; his mother Elvira and father Alfred and his brother Mark and wife Rebecca.

Chapter 3

The Physics of Hypoxic Pulmonary Hypertension and Its Connection with Gene Actions

Yuan-Cheng Fung and Wei Huang
University of California, San Diego, La Jolla, U.S.A.

1. Introduction

Blood pressure is a physical quantity. Blood pressure and flow obey the laws of physics. Hypoxic pulmonary hypertension (HPH) is a phenomenon caused by the reaction of the cells of the lung to hypoxia. The total reaction involves the pumping of the heart, the actions of the genes, and a unique set of changes in the cells, cell membranes, and ion channels in the pulmonary vascular smooth muscle cells. Other chapters of this book present important advances in the understanding of Ca^{2+} distribution, ion channels, endothelium, O_2 sensing, and pathology. In this chapter we consider the physics of HPH and the connection between the changes in vascular tissues with the actions of the genes. The following aspects will be discussed: the physical laws obeyed by the blood and blood vessels, the physical system of HPH, the role of microcirculation in HPH, the role of pulmonary veins, the meaning of the blood pressure–time history records according to physics, the handling of the blood pressure–time history records for medicine, the physiological changes of the vasculature including the changes of anatomical dimensions, the histological and mechanical properties of the lung in HPH, and the correlation of the activity of the genes with the physiological changes. Overall, we show the simplicity and rigor of physical views which are inseparable from the molecular and chemical views.

2. The Physical Laws Obeyed by the Blood and Blood Vessels

The physical laws of conservation of mass and energy, the first and second laws of thermodynamics, and the Newton's laws are universally accepted. Quantuum mechanics, relativity theory, and the concept of an array of genes in the chromosomes for heredity burst onto the scientific stage almost simultaneously about a century ago. In physics, there are two major activities.

One is to show that a minimal set of basic laws exist that are necessary and sufficient to explain everything observable. The other activity is to show that all observable events are solutions of boundary-value problems based on the basic laws. Use of such a physical approach often simplifies scientific investigations. The study of HPH should be no exception.

Since the lung contains a huge number of particles (atoms, molecules), it may be considered as a continuum. The concept of a continuum was introduced into mechanics a long time ago to simplify the application of basic laws (8, 15). By stating that a matter is a continuum means that the particles involved can be identified as *isomorphic* with the real number system in mathematics. Thus the molecules in our body are considered to be as densely distributed as the real numbers.

Classical continuum mechanics has been applied to the problems of pulmonary physiology in the last 30-40 years. In classical continuum mechanics, a set of universally accepted hypotheses is proposed. These hypotheses are called *axioms*. Axioms and boundary conditions define specific problems for investigation.

The axioms of the classical continuum mechanics are listed in Table 1. This includes the laws of conservation of mass and energy, the Newton's laws of motion, the immutable constitutive equations of the materials, the invariability of the zero-stress state of solids, the constancy of mechanical properties of solids, and the constancy of material composition of solids. Classically, these were considered as universal truth. Much of our modern civilization was developed on the basis of these axioms. Airplanes, ships, telescopes, microscopes, artificial heart valves, and kidney dialysis machines were designed on the basis of these classical axioms.

Table 1. Axioms in Classical Mechanics and Biomechanics

Classical Mechanics	Biomechanics
Conservation of mass	Conservation of atom, not molecules
Newton's law	Newton's law
Conservation of kinetic + potential energy	Sum of all energy categories conserved
Zero-stress state is invariable	Zero-stress state remodels
Constancy of mechanical properties	Mechanical properties remodels
Constancy of material composition	Chemical composition remodels

With the advancement of biomechanics, the truth of the axioms of classical continuum mechanics began to be doubted. When the problems of atherosclerosis and hypoxic pulmonary hypertension came into focus, a revolution in the system of axioms of mechanics became inevitable. Some of the reasons will be explained in detail in Section 8 *infra*. In fact, it appears that a new system of axioms has to be introduced to replace the old system in biomechanics. Our new system is listed in the right-hand column of Table 1 (10).

In this system, the atoms are conserved, but the molecules are not. The molecular composition may change with time. The zero-stress state of the solid part can change, and the mechanical properties of the solids may change with time. Thus, the study of HPH is indeed a major fundamental event that will change the course of biomechanics.

3. The Physical System of Hypoxic Pulmonary Hypertension

In a steady flow, a blood pressure drop equals the product of blood flow and resistance. To predict blood pressure, we must assess flow and resistance. For flow, we must know the heart and peripheral circulation. For resistance, we must know the geometry, elasticity, stress, strain, length and tension of the smooth muscle cells in every vessel in the circuit. This calls for a tremendous amount of macroscopic, microscopic, and nanoscale information. How well stocked are we with regard to the needed information? What pieces of information are missing? What steps can be taken to get every needed piece of information?

Consider a specific question: A person is flown to a high ski resort and left there, how would his/her pulmonary arterial blood pressure rise and how would his/her pulmonary blood vessels remodel? To answer, we need information on the structures, materials and dimensions of the pulmonary vascular tree, the mechanical properties of the smooth muscles and other tissues in the blood vessels, the tissue and cell remodeling process, as well as the behavior of the heart and systemic circulation under the circumstances. To solve the total problem, one must begin with morphometry. Weibel (26) published a fairly complete set of morphometric data on human pulmonary airways and arteries based on a bifurcation theory.

Cumming et al. (5) published their morphometric data on human pulmonary arteries on the basis of Strahler's theory which was originally used in geography. In later studies, Fung and his associates found it desirable to sharpen the definition of the generation changes in Strahler system by a simple formula based on the statistical distribution of the diameters of the blood vessels (9). Their system is called the *diameter-defined Strahler system*. Newer data on the morphometry of pulmonary arteries and veins of the cat, rat, dog, and human by Sobin, Fung, Yen, Jiang, Huang, and their associates are all organized according to the *diameter-defined Strahler system* (9, 22). Without morphometric data, no genuine physical analysis of pulmonary circulation can be done. The lack of progress in the morphometric department is alarming. At present, data on the effects of age, sex, race, body weight, health state, and disease states of the patients on their morphometrics of the pulmonary arteries and veins are unavailable. In the past, the required workload may have been daunting. But the fantastic progress in imaging and computing should have relieved the labor. We believe that a systematic supercomputer approach will initiate a renaissance of morphometry.

The next important job is to assess the mechanical properties of the blood vessels and the moduli of elasticity and viscoelasticity. Blood vessel materials are not isotropic, and not homogeneous. Hence it is necessary to assess the viscoelasticity of the different parts of the vasculature, in different directions, and in static and dynamic processes. Since the 1970's, a large amount of work has been done on this subject by many people (1). But as in the case of morphometry, there does not exist a set of mechanical properties data which list the effects of age, sex, race, body weight, health state, and disease states. Again, new imaging and computing technology should help.

With new data coming, we must make a complete computing program to analyze and synthesize the total system. With the change of axioms indicated in Table 1, new programs can be written for practical applications.

4. The Role of Microcirculation in Hypoxic Pulmonary Hypertension

The role of the capillary blood vessels of the lung in HPH is to reduce the resistance to flow in pulmonary microcirculation. The reason that these vessels can do it is because of their geometry. When the pulmonary capillary blood vessels of cat were photographed on face, they appeared as shown in Figure 1A. When these capillary vessels were photographed in cross section, they appear as shown in Figure 1B. From these photographs one sees that the capillaries form sheets of blood space that are about 5 to 10 μm thick, 200 to many 1000 μm wide. In the on-face view, the blood space is interrupted by more or less uniformly spaced roundish obstructions which are called *posts* (Fig. 1B). Fung and Sobin (12-14) called the geometry of pulmonary capillaries a *sheet model*, which consists of two thin membranes separated by posts which are about 3 μm in diameters. The *thickness* of the *blood sheet* space is a function of the capillary blood pressure. Experimental results of the sheet thickness versus the transmural pressure of the cat lung are shown in Figure 1C. The transmural pressure Δp is the difference of capillary blood pressure minus the alveolar gas pressure. When Δp is negative, the thickness is zero. When Δp is slightly positive the thickness jumps to 4.5 μm. The thickness increases linearly with increasing positive Δp until Δp reaches about 30 cm H_2O. For larger Δp, the rate of increase of sheet thickness decreases.

With the geometry and mechanical properties shown in Figure 1, the flow in the pulmonary capillaries as seen by a red blood cell might be said to be like a car moving in an underground parking garage with many posts. When the blood pressure increases, the ceiling height increases, and the resistance to flow decreases. The resistance of the pulmonary microcirculation is transferred to the pulmonary arteries and veins.

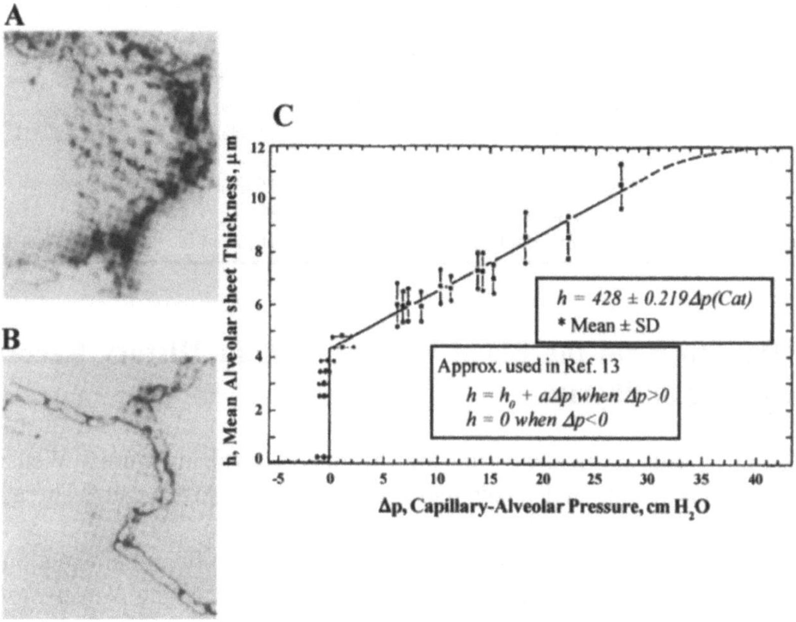

Figure 1. A plan perspective view (A) and a cross-sectional view (B) of the capillary blood vessels in the interalveolar septa of a cat lung (reprinted from Ref. 12). C: The variation of the thickness of a pulmonary alveolar sheet with the transmural pressure (Modified from Ref. 13).

5. The Role of Pulmonary Veins

The resistance to blood flow offered by pulmonary veins to the pulmonary circulation is far more important than the resistances to flow offered by peripheral veins to peripheral circulation. If the pressure drop from the largest artery to the smallest arterioles is compared with the drop in the capillaries and veins, we obtain the results shown in Table 2. It is seen that in the normal lung, we may expect that the pressure drop in the pulmonary veins be larger than that in the pulmonary arteries. If we assume that the vascular smooth muscles in pulmonary arteries and veins are similar, that their contractile mechanism, their chemistry, Ca^{2+} roles, and ion channels are similar, then the pulmonary veins are equally effective as the arteries in producing pulmonary arterial hypertension in hypoxia. The veins exert a powerful remote control on the arteries. The veins, by their contraction, can reduce the resistance of the pulmonary capillaries to blood flow, and increase the pressure in the pulmonary arteries. In the future, we should pay as much attention to the pulmonary venous smooth muscle (PVSM) as to the pulmonary arterial smooth muscle (PASM).

Table 2. Distribution of Pressure Drops in Pulmonary Blood Vessels of the Cat and Comparison with Data on the Dog in the Literature

		Arteries	Capillaries	Veins
	Pattern 1 (Fung, 1996), cat	35.9%	15.4%	48.7%
	Pattern 2 (Fung, 1996), cat	29.3%	21.9%	48.8%
Pulmonary	Brody et al. (1968), dog	46%	34%	20%
	Hakim et al. (1982), dog	40.2%	15.6%	44.2%
Mesentery	Fronek, Zweifach (1974), cat	73%	8%	19%
Skeletal Muscle	Fronek, Zweifach (1974), cat	65%	16%	19%

6. How to Read the Blood Pressure - Time History Records According to Physics

Some records of blood pressure-time history are given in Figure 2. We have two questions: how to understand these curves, and how to use them. These are discussed in the current and following sections.

According to physics, we can write down a partial differential equation to describe the motion of a fluid. This equation is known as the *Navier-Stokes equation* (one vector equation, 3 equations in components). Originally intended for applications to the flow of fluids like air and water (Newtonian fluids), the Navier-Stokes equation has been extended to the treatment of blood flow, with blood rheology taken into consideration (3, 7, 9, 23). The boundary conditions of blood flow are those imposed by blood vessels and the heart. There are solid bodies. The vectorial partial differential equation that describes the motion of elastic solids is called *Navier's equation*. Originally used to treat solids like steel, the Navier's equation has been extended to treat blood vessels, with full vascular rheology taken into consideration (1, 7, 9).

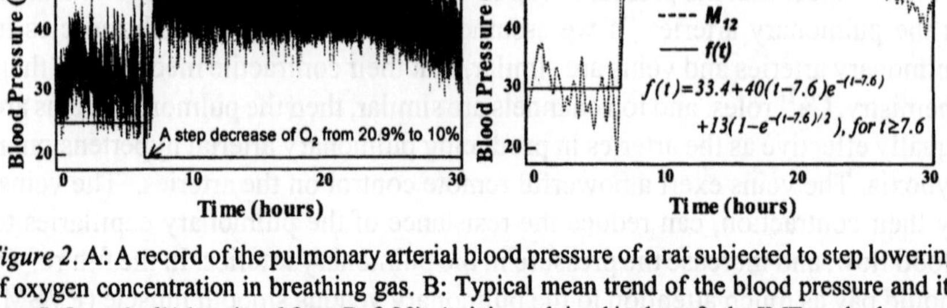

Figure 2. A: A record of the pulmonary arterial blood pressure of a rat subjected to step lowering of oxygen concentration in breathing gas. B: Typical mean trend of the blood pressure and its best-fit formula $f(t)$ (reprinted from Ref. 20). As it is explained in the ref., the IMF method, means of various orders can be defined rigorously, the one's with the higher order has fewer zero crossings. In this figure, the order of the mean is 12.

When nonlinear rheology and convective acceleration (acceleration originated by moving a particle in a nonuniform velocity field) are considered, both the Navier-Stokes and the Navier's equations are nonlinear. These nonlinear equations have few exact solutions. If we wish to read the blood pressure history illustrated in Figure 2 according to physics, we do not have the exact solutions to help us.

Without exact solutions of the full equations, are there exact solutions for simplified cases? The answer is, fortunately, yes. One important simplified case is the *Poiseuille flow* which assumes that the blood vessel is a circular cylindrical tube of uniform cross section with fixed, nonvarying inner wall (endothelium) of radius a, that the blood pressure, p, is fixed at the ends of a vessel of finite length, L, and that the velocity is zero on the wall. Then the pressure gradient is constant, the flow is steady, the velocity profile in the tube is parabolic, the shear strain gradient and the shear stress of the blood on the wall are constant, and the flow rate is governed by a formula frequently quoted in this book: *pressure gradient = flow × resistance.* Flow is the movement of volume per unit time across any cross section. Pressure gradient is the difference of the pressure at the ends divided by L. Under the simplifying assumptions named above, the resistance, R, is equal to:

$$R = (8/\pi) \times (\text{coefficient of viscosity}) \times (\text{vessel length}) \times (\text{radius of tube})^{-4}$$

The minus 4th power of the radius in this formula speaks for the significance of the change of the radius of the blood vessel on the arterial blood pressure. For example, a 10% reduction in radius would cause a 34.4% increase in resistance. It is this formula that laid the responsibility of hypoxic hypertension in pulmonary arteries on the contraction of the smooth muscle.

The second important simplified case is that of wave propagation of a nonviscous fluid in an elastic tube. The simplifying assumptions are that the tube is linearly elastic and cylindrical, the convective acceleration can be neglected, and that the viscosity effect is negligible. Under these assumptions the Navier-Stokes equations for the blood, and the Navier equations of the wall are linearized, and the equation of continuity (conservation of mass) becomes a simple equation relating the temporal rate of change of vessel cross sectional area to the spatial rate of change of blood flow, whereas the equation of motion is reduced to a linear equation between the local acceleration $\partial u/\partial t$ to the pressure gradient $\partial p/\partial x$. Combining these two equations yields a wave equation whose solution is an arbitrary wave moving forward or backward:

$$p = p_0 f(x - ct) + p_0' g(x + ct)$$
$$u = u_0 f(x - ct) - u_0' g(x + ct) \qquad [1]$$

Here p is blood pressure, p_0, p_0', u_0, u_0' are arbitrary constants, f and g are arbitrary functions of the variables $x\text{-}ct$ and $x+ct$, and c is the velocity of the waves

$$c=\sqrt{\frac{Eh}{2\rho(a_i+h/2)}} \qquad [2]$$

in which E is the Young's modulus of elasticity of the vessel wall, h is the thickness of the vessel wall, a_i is its inner radius, ρ is the density of the blood (see Ref. 9, p. 144). Notice the difference in the sign of the last terms in Eq. [1]. This leads to important relations:

$$p_0 = \rho c u_0, \qquad p_0' = -\rho c u_0' \qquad [3]$$

These formulas say that, for a wave propagating in the direction of flow, the pressure rise is equal to the velocity rise multiplied by the product of the density of the blood and the speed of sound in the blood. For a wave propagating in the opposite direction of the flow, the pressure rise has the same magnitude but opposite sign.

The Poiseuille solution takes care of the effect of blood viscosity in a steady flow. The wave solution takes care of the transient oscillation in a nonviscous fluid. The sum of these two solutions is not an exact solution of the total Navier-Stokes and Navier equations, because the two problems are linearized differently. Womersley and many other biomechanics researchers have worked very hard to find improved solutions (23, 24, 27, 28). Complexity can be expected, but the major features are given by these two simplified solutions. The principles are quite clear, immediate progress can be counted on with the use of supercomputers.

7. How to Handle the Blood Pressure-Time History Records for Physiology and Medicine

In order to relate tissue remodeling in HPH to blood pressure and flow, the simplified solutions discussed in the preceding section tell us to focus our attention on both the mean level of blood pressure and the oscillations about the mean. Looking at the actual recording as shown in Figure 2, we find that the task is not as easy as it appears. We know the signals vary with the location along the vascular tree. At any given point on the tree, the signals are never exactly periodic. Arrhythmia exists to some degree all the time at all places. Mathematicians call such signals nonstationary and stochastic, and a very large literature exists to deal with such signals. Well known methods are associated with names such as functional Fourier analysis, Fourier spectrum, Hilbert spectrum, wavelets, IMF, etc. We found the *intrinsic mode function (IMF)* approach of Huang et al. (17) most appropriate. The practical computation can be done on an existing software which analyses the digital signals into intrinsic modes of oscillations, each mode has the character that the local oscillations about the modal curve has an average of zero. Each given signal has a finite number of modes: successive modes have fewer and fewer zero crossings; the last mode has no zero-crossing at all, and it represents a trend. A pulmonary arterial pulse pressure record typically has 8-18 intrinsic modes, with the total number of models depending largely on the total length of time. Each complete record is resolved into a complete set of intrinsic modes. Thus we can define an average signal, or mean signal as the sum of a certain number of last intrinsic modes depending on an arbitrarily chosen number of zero-crossings you may wish to allow. The total signal is the sum of the mean signal and a signal of oscillations about the mean. Figure 2 illustrates the process. Other features and details are given in Huang et al (18, 19).

The advantage of the intrinsic mode approach is that we can get a much better understanding of the mean signals, the oscillations about the mean signals, and arrhythmia. Such understanding of signals will allow us to study tissue remodeling as a dynamic process. We would like to distinguish tissue remodeling in response to slowly varying mean stresses, from the remodeling of tissues in response to the oscillations about the mean.

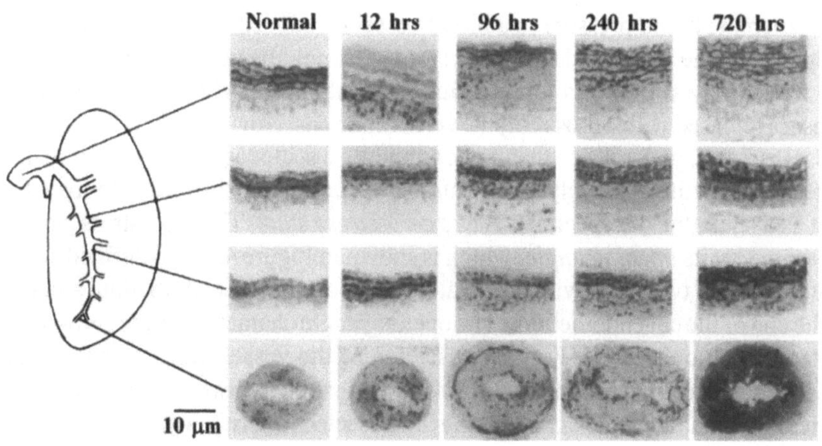

Figure 3. Photographs of histological slides from four regions of the main left pulmonary artery of a normal rat and hypertensive rats with different periods of 10% O_2 breathing (reprinted from Ref. 11).

8. The Physiological Changes in HPH: Changes in Anatomical Dimensions, Zero-Stress State, Chemical, Histological, and Mechanical Properties of the Lung in HPH

If we want to know how blood vessels change when the blood pressure is raised, it is worthwhile to learn the full story: by measuring the changes from time zero when an increase of blood pressure occurs. Let's use the rat for illustration. With a step decrease of the oxygen concentration from the normal sea level 20.9% to a level of 10% in the gas that a rat breathes, the pulmonary arterial blood pressure increases, the materials in the wall of the artery will change as shown in Figure 3 (11). Here the histological cross sections of the largest left pulmonary artery (of order 12) are shown in the first row, beginning with the normal state, then in a state after 2 hours of hypertension, then 12 hours, etc. In each panel, from top down, are the blood space, the endothelium, the basement membrane, the layers of vascular smooth muscle cells and elastin, and the adventitia. Looking at the *media layer* in the first row of Figure 3, we see a great thickening at 12 hours of hypoxia. Then the thickness decreased and finally stabilized. The lower rows of Figure 3 show the morphometric and histological changes of the smaller pulmonary arteries of the rat.

8.1. Zero-Stress State is not Permanent

When the blood pressure in a blood vessel is reduced to zero (i.e., when the pressures inside and outside the vessel are the same), the vessel is unloaded. But the vessel may still have a longitudinal residual stress. When the longitudinal stress is released with a cross cut, the vessel shortens. Now if you cut the isolated vessel radially, you will see that it opens up into a sector as shown in Figure 4A. Further cuts will cause only unmeasurable insignificant changes. We may say that as a shell, the cut specimen is at a *zero-stress state*. The zero-stress state is measured by its *opening angle*, which is shown in Figure 4A, and defined as the angle between two radii, with origin at the midpoint of the endothelium and tips at the end of the cut endothelium. The opening angle characterizes the zero-stress state of the blood vessel. Further dissection of the vessel wall into an intima-media layer and an adventitia layer shows that these two layers may have different opening angles. So, in the undissected arterial wall, some minor residual stress remains.

The interesting thing about the opening angle is that it changes with tissue remodeling. In the case of a pulmonary artery subjected to hypoxic hypertension, the opening angle changes with time as shown in Figure 4B. We find large opening angles in most curved cylindrical organs. In veins, airways, and intestines, opening angles are large. In round organs such as heart, the description of the zero-stress state is more complex.

We must conclude that living organs at the *in vivo* state have residual stress and residual strain, and the zero-stress state changes as the tissue remodels.

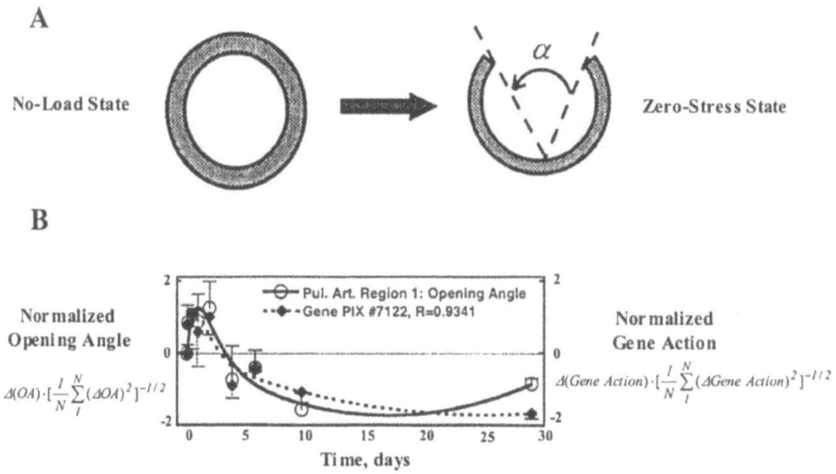

Figure 4. A: Definition of opening angle α, see text. B: Action of gene PIX #7122 (*pleckstrin 2 homolog gene*), vs. the change of opening angle (*ΔOA*) of the pulmonary arterial trunk vessel. Gene action equals to *(mean of gene expression in 4 rats at a specific day/mean of gene expression in 4 rats at day 0)-1*. The gene name and its identification number (PIX no.) are listed in the software GENEPIX PRO 3.0 (Axon Instruments). Symbols: small circle, ○, normalized change of opening angle, ɐ one SEM flag up; ♦, normalized gene action, ɐ one SEM flag down (Modified from Ref. 21).

8.2. Morphology as a Motion Picture

The history of the changes of the thickness of the media layer from that at normal condition is described by a solid curve in Figure 5A (21). Before time zero, the animal was in a normal stable equilibrium condition. The existence of such a condition *in vivo* is an axiom of biology. Thus, the change of media thickness is zero for $t \leq 0$. At $t > 0$, the media thickness increases with time. The corresponding change of the thickness of the adventitia layer of the pulmonary artery is shown in Figure 5B (21). The sum of the thickness of the intima, media, and adventitia layer is the thickness of the arterial wall. It is obvious that the thickness of the arterial wall changes with time in hypertension. The diameter of the artery also changes with time in hypertension.

8.3. Mechanical Properties Remodel

We measured the mechanical properties of the blood vessel material. We found that if we fit the experimental results with a stress-strain law, the material constants a_1, a_2, and b_1, b_2, b_4 will vary with time. An example is shown in Figure 5C (21), which shows the variation of the circumferential Young's modulus of elasticity with time after the onset of hypertension.

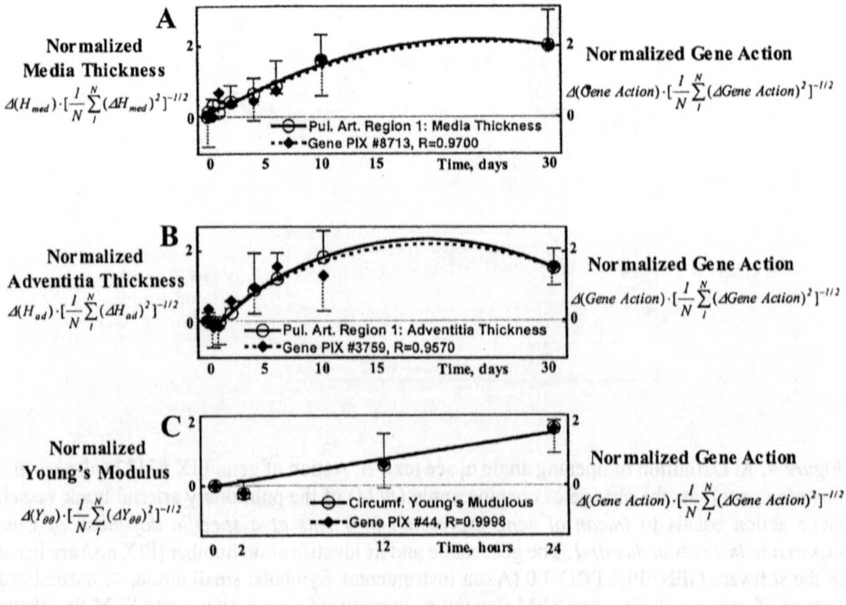

Figure 5. A: Action of gene PIX #8713 (*inorganic pyrophosphatase gene*), vs. the change of media thickness (H_{med}) of pulmonary arterial trunk. B: Action of gene PIX #3759 (*osteoblast-specific factor 2, fasciclin I-like gene*) vs. the change of adventitia thickness (H_{ad}) of pulmonary arterial trunk. C: Action of gene PIX #44 (*dynein, cytoplasmic light polypeptide gene*), vs. the change of Young's modulus of elasticity of the pulmonary artery, $Y_{\theta\theta}$, relating circumferential stress and strain. Symbols: small circle, o, normalized physiological changes, ɐ 1 SEM flag up; ♦, normalized gene action, ɐ 1 SEM flag down (reprinted from Ref. 21).

8.4. Physiological Changes

These examples show that tissue remodeling is a set of continuing processes. For medical reasons, we wish to know how reversible the processes are, how superposable the processes are, how linear the processes are, how nonlinearity can be described, and how important are these nonlinearities. For scientific reasons, we wish to know how these processes can be explained. Can they be explained quantitatively *in toto* (including histories of all aspects of physiology) by the distribution of ions and functions of ion channels? What other factors must be brought in?

There exists almost a limitless list of things to be learned, of ideas to be tried, and of theories to be checked. Yet experimental physiology is empiricism. Mechanics and physics provide a unifying theme for theoretical understanding. How can we go beyond the empiricism and current physics and mechanics, and advance to the next level of understanding and development? We think the next level lies in the study of genes. Someday in the future, we should be able to deduce all the features of tissue formation and remodeling theoretically from the molecular mechanics of the genes. To initiate this research, we shall first show that the physiological changes in tissue remodeling and the actions of the genes are related.

9. Correlation of Gene Actions with Physiological Changes

To study the relationship between the action of the genes and the physiological changes in tissue remodeling, we used the same hypoxic hypertension rat model described earlier, took the arterial specimens at scheduled times, extracted the total RNA in each specimen. Then the mRNA was isolated, polymerized cDNA, colored with biotin, and prepared for hybridization, all with known procedures and commercial kits described in detail in Refs. 4, 21, and 25. The genes in each of these processed specimen solutions are called *probes*. The solution of each specimen containing all the probes was used to measure gene expression in a microarray test equipment designed and constructed by Peck and his associates (4). In Peck's array, 9,600 selected genes were deposited as *targets* on a nylon membrane in a rectangular matrix pattern. They were PCR products of human cDNA clones rearrayed from the *IMAGE Consortium cDNA libraries* based on the Unigene clustering (4). When the specimen solution was spread over the array membrane, hybridization of the probes and targets occurred. After hybridization, genes on the nylon membrane were scanned by colorimetry. The changes of gene expression were expressed in terms of the changes of the intensity of color. Readings above a noise level are considered significant. The sum of all the significant readings is considered to be a measure of sample size. When several samples were taken for each set of experimental conditions, the sample sizes may vary. The colorimetric readings are normalized against sample size by dividing each reading with the sum of all significant readings of that sample. The normalized readings are considered as measures of gene expression. The ratio of the gene expression in active state divided by that of the *in vivo* state, minus one, is defined as the *action of the gene.*

9.1. Matching Gene Action with Physiological Changes

When we have the measured data on the action of the genes in the cells of the arteries whose tissue remodeling features were measured, we can compute the correlation coefficients of the gene actions and the physiological changes. The definition of the correlation coefficients is as follows: Let $y_j(t)$, $j=1, 2, 3...$, represent the history of the jth physiological feature, at a series of instants of time t_m, $m=1, 2, ...$ Let $z_k(t)$ be the action of the gene number k, $k=1, 2, 3...$ at time t_m, $m=1, 2, ...$ Then we define the correlation coefficient by the following formula:

$$R(y_j,z_k) = [\Sigma y_j(t_m)z_k(t_m)\Delta t_m]\,[\Sigma y_j^2(t_m)\Delta t_m \times \Sigma z_k^2(t_m)\Delta t_m]^{-1/2}, \qquad [4]$$

where t_m, with $m=1, 2, ...$, are the instants of time at which y_j, z_k are measured. Here Δt_m are the spacing of time intervals, $\Delta t_m = t_{m+1} - t_m$.

For each physiological feature, we computed its changes from the *in vivo* value, and the correlation coefficients with the action of every gene whose expression reading is significant. Then we lined up the genes in a row according to the size of the correlation coefficient. In a row, there is the first one, we call it the top gene. Plotting the action of the top gene and the physiological features together, we obtained the curves shown in Figures 4B and 5A, B and C.

9.2. Cause and Effect

When two important events in life have histories that have very good correlation, one wonders whether they are the cause and effect of each other, whether they are one and the same thing. We are at the crossroad!

10. Conclusion

It is plausible hypothesis that what we see in changes of physiology is gene action provoked by physiological activity. We can anticipate a new axiomatic biomechanics to be linked with genomic mechanics. Final clarification of the gene-physiology relationship will not only open up new avenues of practical applications, but also define the limitations to empiricism and imagination.

Acknowledgments

This work was supported by the National Heart, Lung, and Blood Institute of the National Institutes of Health (HL 43026). W. Huang would like to acknowledge the support over the years from the American Heart Association (California Affiliate, Postdoctoral Fellowship), the National Heart, Lung and Blood Institute of the National Institutes of Health, the National Institute of Standard and Technology, the NASA Goddard Space Flight Center, and the National Health Research Institutes (Taiwan).

References

1. Abè H, Hayashi K, and Sato M, eds. *Data Book on Mechanical Properties of Living Cells, Tissues, and Organs*. Tokyo: Spriner, 1996.
2. Brody JS, Stemmler EJ, and duBois AB. Longitudinal distribution of vascular resistance in the pulmonary arteries, capillaries, and veins. *J. Clin. Invest.* 1968; 47: 783-784.
3. Caro CG, Pedley TJ, Schroter RC, and Seed WA. *The Mechanics of the Circulation.* Oxford: Oxford University Press, 1978.
4. Chen JJW, Wu R, Yang PC, Huang JY, Sher YP, Han MH, Kao WC, Lee PJ, Chiu TF, Chang F, Chu YW, Wu CW, and Peck P. Profiling expression patterns and isolating differentially expressed genes by cDNA microarray system with colorimetry detection. *Genomics* 1998; 51: 313-324.
5. Cumming G, Henderson R, Horsfield K, and Singhal SS. "The functional morphology of the pulmonary circulation." In *The Pulmonary Ciculation and Interstitial Space*, Fisherman A, and Hecht H, eds. Chicago, IL: University of Chicago Press, 1968, pp. 327-338.
6. Fronek K and Zweifach BW. Pre- and postcapillary resistances in cat mesentery. *Microvasc Res.* 1974;7: 351-61.
7. Fung YC. *Biomechanics: Mechanical Properties of Living Tissues*. New York, NY: Springer-Verlag, 2nd ed., 1993.
8. Fung YC. *A First Course in Continuum Mechanics*. Englewood Cliffs, NJ: Prentice-Hall, 3rd ed., 1994.
9. Fung YC. *Biomechanics: Circulation*. New York, NY: Springer-Verlag, 2nd ed, 1996.
10. Fung YC. Biomechanics and gene activities - Celebrating the centennial of Prof. Chou Peiyuan. *Advances in Mechanics* 2002; 32: 484-494. (in Chinese, with English abstract)
11. Fung YC and Liu SQ. Changes of zero-stress state of rat pulmonary arteries in hypoxic hypertension, *J. App. Physiol.* 1991; 70: 2455-2470.
12. Fung YC and Sobin SS. Theory of sheet flow in lung alveoli. *J. Appl. Physiol.* 1969; 26: 472-488.
13. Fung YC and Sobin SS. Elasticity of the pulmonary alveolar sheet. *Circ. Res.* 1972; 30, 451-469.
14. Fung YC and Sobin SS. Pulmonary Alveolar Blood flow. *Circ. Res.* 1972; 30: 470-490.
15. Fung YC and Tong P. *Classical and Computational Solid Mechanics*. Singapore: World Scientific, 2001.
16. Hakim TS, Michael RP, and Chang HK. Partitioning of pulmonary vascular resistance in dogs by arterial and venous occlusion. *J. Appl. Physiol.* 1982; 52: 710-715.
17. Huang NE, Shen Z, Long SR, Wu ML, Shih HH, Zheng Q, Yen NC, Tung CC, and Liu HH. The empirical mode decomposition and the Hilbert spectrum for nonlinear and non-stationary time series analysis. *Proc. R. Soc. Lond. A.* 1998; 454: 903-995.
18. Huang W, Shen Z, Huang NE, and Fung YC. Engineering analysis of biological variables: an example of blood pressure over a day. *Proc. Natl. Acad. Sci. USA.* 1998; 95: 4816-4821.
19. Huang W, Shen Z, Huang NE, and Fung YC. Use of intrinsic modes in biology: Examples of indicial response of pulmonary blood pressure to ±step hypoxia. *Proc. Natl. Acad. Sci. USA* 1998; 95: 12766-12771.
20. Huang W, Sher YP, Peck K, and Fung YC. Correlations of gene expression with physiological functions: examples of pulmonary blood vessels rhealogy, hypoxic hypertension, and tissue remodeling, *Biorhealogy* 2001; 38: 75-87.
21. Huang W, Sher YP, Peck K, and Fung YC. Matching gene activity with physiological functions. *Proc. Natl. Acad. Sci, USA.* 2002; 99: 2603-2608.
22. Huang W, Yen RT, McLaurine M, and Bledsoe G. Morphometry of the human pulmonary vasculature. *J. Appl. Physiol.* 1996; 81: 2123-2133.
23. Pedley, TJ. *The Fluid Mechanics of Large Blood Vessels.* London: Cambridge University

Press, 1980.
24. Pythoud F, Stergiopulos N, Bertram CD, and Meister JJ. Effects of friction and nonlinearities on the separation of arterial waves into their forward and backward components. *J. Biomechnics* 1996; 29: 1419-1423.
25. Schena, M. *Microarray Analysis*. John Wiley & Sons, 2002.
26. Weibel, ER. *Morphometry of the Human Lung*. New York, NY: Academic Press, 1963.
27. Westerhof N, Sipkema P, Van den Bos GC, and Elzinga G. Forward and backward waves in the arterial system. *Cardiovasc. Res.* 1972; 6: 648-656.
28. Wormersley, JR. *An Elastic Tube Theory of Pulse Transmission and Oscillatory Flow in Mammalian Arteries*. Wright-Patterson Air Development Center: Wright-Patterson Air Force Base, Ohio. Technical Report 56-614, 1957.

II. ROLE OF INTRACELLULAR Ca^{2+} AND Ca^{2+} SENSITIVITY IN HYPOXIC PULMONARY VASOCONSTRICTION

Chapter 4

Ca^{2+} Sparks in Pulmonary Artery Smooth Muscle Cells: Implications For Hypoxic Pulmonary Vasoconstriction

Wei-Min Zhang, Carmelle V. Remillard, and James S.K. Sham
Johns Hopkins University School of Medicine, Baltimore, Maryland, U.S.A.

1. Introduction

Hypoxic pulmonary vasoconstriction (HPV) is an important regulatory mechanism for diverting pulmonary blood flow from poorly ventilated to well-ventilated lung regions to improve ventilation-perfusion matching and to optimize gaseous exchange. A huge body of literature indicates that this is a multifactorial process with an endpoint increase in cytosolic $[Ca^{2+}]$ ($[Ca^{2+}]_i$) and vasoconstriction in pulmonary arteries. Attention has been focused on the global elevation of $[Ca^{2+}]_i$ because it is responsible for the activation of Ca^{2+}/calmodulin dependent myosin light chain kinase to initiate actin-myosin interactions and smooth muscle contraction. However, the global increase of $[Ca^{2+}]_i$ in vascular tissues is mediated by multiple Ca^{2+} influx pathways, including L-type Ca^{2+} channels, receptor-operated and store-operated Ca^{2+} channels, Na^+/Ca^{2+} exchangers, and two types of Ca^{2+} release channels on the sarcoplasmic reticulum (SR) membrane, the IP_3 receptors (IP_3Rs) and ryanodine receptors (RyRs). Because of the spatial distributions of these Ca^{2+} transporters, diffusion kinetics of Ca^{2+} ions and subcellular micro-architecture, local heterogeneity in $[Ca^{2+}]_i$ is expected. Theoretical modeling has estimated that, in the vicinity of an open Ca^{2+} conducting channel (Ca^{2+} microdomain), local $[Ca^{2+}]$ can easily exceed 100 μM (47). Such large local gradients of $[Ca^{2+}]$ can provide fast and specific Ca^{2+} signals to neighboring effector molecules to trigger Ca^{2+} dependent processes that may not be responsive to global submicromolar increase in $[Ca^{2+}]_i$, yet which may modulate vascular reactivity.

Local Ca^{2+} events were first visualized as spontaneous local Ca^{2+} release transients, or "Ca^{2+} sparks", in cardiac myocytes by the use of laser scanning confocal microscopy (9). They originate from clusters of RyRs, which are functionally-coupled with L-type Ca^{2+} channels to form Ca^{2+} release units within the diadic junctions (43), and are considered to be the elementary Ca^{2+} release events underlying excitation-contraction coupling in cardiac and skeletal muscles. During an action potential, thousands of Ca^{2+} sparks are evoked via the "Ca^{2+}-induced-Ca^{2+} release" mechanism to generate global Ca^{2+} transients for muscle contraction. Ca^{2+} sparks have been identified in various types of vascular and non-vascular smooth muscle cells (SMCs). Recently, we have also identified and characterized Ca^{2+} sparks in rat intralobar pulmonary arterial SMCs (PASMCs) (39). Emerging evidence suggests that Ca^{2+} sparks of PASMCs may regulate pulmonary vascular reactivity in a unique tissue specific manner (24, 37, 39, 50). In this chapter, we will provide a brief review on Ca^{2+} sparks in systemic vascular smooth muscle (20), describe and contrast the properties of Ca^{2+} sparks of PASMCs, and discuss possible physiological roles of Ca^{2+} sparks in HPV. We hope this will arouse interest for future investigations on the role of local Ca^{2+} signaling in HPV.

2. Ca^{2+} Sparks of Systemic Vascular Smooth Muscle

SMC Ca^{2+} sparks were first identified in cerebral arterial myocytes (33), and subsequently in different types of vascular SMCs from coronary and mesenteric arteries and portal vein, and non-vascular SMCs from the trachea, ileum, and urinary bladder (20). In systemic arterial myocytes, they typically have a frequency of 0.24 Hz at rest, a peak amplitude of 100-200 nM, a rise-time of 20 ms, decay half-time of 48-56 ms, and a spread (full-width-half-maximum, FWHM) of 2.4 μm (Table 1). In contrast to cardiac or skeletal muscles, which exclusively express a single RyR subtype, all three RyR subtypes, RyR1 (skeletal), RyR2 (cardiac), and RyR3 (brain), are expressed in vascular smooth muscles (11, 27, 30, 34). Functional studies in portal vein myocytes using antisense oligonucleotides show that RyR1 and RyR2 are both required for Ca^{2+} spark generation presumably because these channels participate in RyR-cluster formation (11), whereas RyR3 channels only play a modulatory role on spark frequency or contribute to global $[Ca^{2+}]_i$ (27, 30).

Physiological roles of Ca^{2+} sparks have been studied originally and most thoroughly in cerebral arterial myocytes (33). In intact cerebral arteries, elevation of intraluminal pressure to 60-80 mmHg causes graded membrane depolarization above -40 mV to activate voltage-gated L-type Ca^{2+} channels and to sustain myogenic vasoconstriction. Pressurization also causes a concomitant increase in Ca^{2+} spark frequency in these arteries (19). There is substantial evidence suggesting that RyR clusters in these arterial myocytes are functionally coupled to Ca^{2+}-activated K^+ (K_{Ca}) channels in sarcolemmal invaginations known as caveolae (11, 23, 27). Ca^{2+} sparks originating within these complexes cause large local increases of $[Ca^{2+}]$ in the junctional cleft (~20 nm wide), activating a group of apposing K_{Ca} channels to generate spontaneous transient outward currents (STOCs), which cause membrane hyperpolarization, closure of L-type Ca^{2+} channels, and vasodilation (23, 32). Due to the large conductance of K_{Ca} channels (80 pS at physiological $[K^+]$) and the number of channels (at least 15 in number) being activated, a single Ca^{2+} spark can cause a substantial hyperpolarization of up to 20 mV (14). This robust Ca^{2+} spark/K_{Ca} channel interaction operates as a frequency-dependent negative feedback modulator of membrane potential (E_m) for counteracting and fine-tuning the myogenic tone. Moreover, because of the highly localized and transient nature of Ca^{2+} sparks, they are thought to have little contribution to the direct activation of myofilaments and smooth muscle contraction.

In addition to K_{Ca} channels, some vascular SMCs express Ca^{2+}-activated Cl^- (Cl_{Ca}) channels. Simultaneous activation of a group of Cl_{Ca} channels generates spontaneous transient inward currents (STICs) (17, 50), which cause membrane depolarization, increased Ca^{2+} influx and cell contraction. In tracheal myocytes, it has been demonstrated that Ca^{2+} sparks activate STICs in a manner similar to STOCs (58). Hence, the net physiological effect of Ca^{2+} sparks is dependent on the relative activity of the counteracting Ca^{2+} activated (K_{Ca} or Cl_{Ca}) channels and on the E_m (relative to the equilibrium potential of K^+ and Cl^-, E_K and E_{Cl} respectively) at which the Ca^{2+} sparks are generated. However, only a handful of vascular SMCs (e.g., from portal veins and pulmonary arteries) exhibit STICs (17, 50). Therefore, the default action of Ca^{2+} sparks in vascular SMCs is membrane hyperpolarization, and STICs may serve a special physiological function unique to the particular vascular bed.

Activity of Ca^{2+} sparks in vascular smooth muscle is regulated by many mechanisms, of which L-type Ca^{2+} channel activity appears to play a pivotal role. Activation of Ca^{2+} channels either by membrane depolarization or by Ca^{2+} channel agonists increases spark frequency in all SMC preparations. It has been postulated that L-type Ca^{2+} channels are also colocalized with RyRs in caveolae, and that the triumvirate of L-type Ca^{2+} channels, RyRs, and K_{Ca} channels operates as a functional unit to control $[Ca^{2+}]_i$ (23, 27), as exemplified by the enhancement of Ca^{2+} sparks in cerebral arteries during pressurization (19). However, the association between Ca^{2+} channels and RyRs in vascular smooth muscle is likely tissue-specific. In fact, "loose" coupling or uncoupling of Ca^{2+} channels and RyRs has been implied in a non-vascular smooth muscle (10).

Ca^{2+} spark activity in vascular SMC is also modulated by agonist-dependent and independent mechanisms, including SR Ca^{2+} loading, cyclic nucleotide- and protein kinase C (PKC)-dependent phosphorylation, and alkaline pH. Increasing SR Ca^{2+} content, either by elevating external $[Ca^{2+}]$ or SR Ca^{2+}-uptake through phospholamban phosphorylation, enhances Ca^{2+} spark frequency (52). It has been suggested that Ca^{2+} binding to the luminal sites of RyRs or some other intra-luminal molecules is responsible for the load-dependent activation of Ca^{2+} sparks to auto-regulate SR Ca^{2+} content (28). Vasodilating agents, such as nitric oxide (NO) and cyclic nucleotides, increase spark frequency and contribute to smooth muscle relaxation by increasing STOC frequency (7, 37). In contrast, vasoconstricting agonists, such as norepinephrine (NE), decrease Ca^{2+} sparks through PKC activation (6, 20, 29), enabling further contraction by reducing the negative feedback imposed by K_{Ca} channels. However, modulation of Ca^{2+} sparks by neuro-hormonal factors may vary between vascular beds and agonists. For example, angiotensin II, which is known to activate PKC, has been shown to trigger Ca^{2+} sparks in rat portal vein SMCs (2).

3. Biophysical Properties of Ca^{2+} Sparks in Pulmonary Artery Smooth Muscle Cells

Information on Ca^{2+} sparks in PASMCs is sparse. To date, there are four published papers reporting pulmonary Ca^{2+} sparks, and only two of them have quantified their physical properties (24, 37, 39, 50). The major properties of Ca^{2+} sparks of PASMCs and systemic myocytes are listed in Table 1 for comparison. In rat intralobar PASMCs, the resting spontaneous spark frequency is ~0.4 spark/s or 0.01 spark/μm/s, averaged amplitude ($\Delta F/F_0$) is ~0.5 (equivalent to $\Delta[Ca^{2+}]_i$ of 75 nM), the size (FWHM) is ~1.6 μm and duration (full-duration-half-maximum, FDHM) is ~35 ms (39). The spark frequency is comparable to those reported in canine PASMCs and other vascular myocytes, ranging between 0.1 to 0.4 sparks/s when measured using a scan line of similar length, and between 0.1 and 0.7 spark/cell/s when recorded using high speed frame scanning (23, 24, 36). The amplitude of Ca^{2+} spark in rat PASMCs is smaller than the 200 nM recorded in cerebral arteries (6, 22, 26, 32, 36), and canine PASMCs (24). The differences could be due to a lower SR Ca^{2+} content in rat PASMCs or to a bias in spark detection, since spark selection in rat PASMCs was guided by an automated algorithm (39) instead of by eye-detection as in some other studies. On the other hand, spark duration and size of rat PASMCs are similar to those of other vascular smooth muscle, ranging from 30-65 ms and 1.5 to 2.4 μm, respectively. Therefore, overall speaking, the spatio-temporal properties and Ca^{2+} sparks activities are similar in PASMCs and systemic vascular SMCs.

Table 1. Properties of Ca^{2+} Sparks in Systemic and Pulmonary Arterial SMC

	Systemic Myocytes	*Pulmonary Myocytes*
Unstimulated spark Fr.	0.24 sparks/cell/s	0.3 sparks/cell/s (0.01 sparks/μm/s)
Amplitude, $\Delta F/F_0$ (nM)	~1 (110-200 nM)	0.5-1 (75 nM)
Spread (FWHM)	1.5-2.4 μm	1.6 μm
Duration (half time)	30-65 ms	35-50 ms
RyR subtypes	RyR1-3	RyR1-3
Effect on E_m	Hyperpolarization	Depolarization (rat distal PASMC) Hyperpolarization (fetal PASMC)
Activation	Local Ca^{2+}, BAY K8644, Caffeine, NO	Local Ca^{2+}, BAY K8644, Caffeine, ET-1
Inhibition	Ryanodine, PKC, NE	Ryanodine, NE

4. Ryanodine Receptors, Ca^{2+} Channels and Ca^{2+} Sparks in Pulmonary Artery Smooth Muscle Cells

In pulmonary arteries, the expression of RyR subtypes has not been reported. Ongoing studies in our laboratory detected mRNA for all three RyR subtypes using reverse-transcription polymerase chain reaction, and RyR1 and RyR2 proteins by Western blot and immunostaining techniques (Yang and Sham, unpublished data), suggesting that multiple RyR subtypes are expressed in rat PASMCs as in systemic myocytes (11, 27, 30, 34). Even though it is unclear which RyR subtype is the major contributor for spark generation, there is no doubt that spontaneous Ca^{2+} sparks originate exclusively via RyRs in PASMCs. This is based on evidence that *i*) ryanodine concentration-dependently abolished Ca^{2+} sparks, *ii*) enhancement of RyR activity using a subthreshold concentration of caffeine increased spark frequency, an effect reversed by ryanodine, and *iii*) IP_3 receptor antagonists, 2-aminoethoxy-diphenylborate (2-APB) and xestospongin C, had no effects on resting spark frequency (24, 39). These observations are in agreement with most previous studies in other cell types, except in portal vein myocytes where inhibition of IP_3Rs or phosplipase C eliminated a population of larger Ca^{2+} release events, suggesting that RyRs are recruited by basal IP_3 dependent Ca^{2+} release to generate large Ca^{2+} sparks (16).

Similar to systemic myocytes, the L-type Ca^{2+} channel is an important modulator of Ca^{2+} spark in PASMCs. Enhancing Ca^{2+} influx via Ca^{2+} channels either by direct channel activation with BAY K8644, membrane depolarization with elevated extracellular K^+ concentration ($[K^+]_o$), or increasing extracellular Ca^{2+} concentration ($[Ca^{2+}]_o$) unequivocally increases Ca^{2+} spark frequency, and the effects can be reversed by nifedipine (39). However, it is undetermined if Ca^{2+} influx via Ca^{2+} channels activates PASMC RyRs directly because of the close coupling of the two sets of channels, as in cerebral arterial myocytes (23, 27), or indirectly via increasing SR Ca^{2+} load (52). Immunostaining shows that RyRs are located in sites close to the sarcolemma as well as within the cytoplasm (unpublished data). Thus, the uncoupled central SR is likely to be regulated by an indirect pathway, even if Ca^{2+} channels and RyRs are indeed coupled in peripheral junctions of PASMCs. In addition, resting spontaneous Ca^{2+} sparks in PASMCs are unaffected by nifedipine, suggesting the spontaneous Ca^{2+} sparks in resting PASMCs are not triggered by L-type Ca^{2+} channels (39), but initiated by the intrinsic random stochastic activity of RyRs instead. This concurs with observations in cardiac and other vascular SMCs (2, 5, 32).

5. Ca^{2+} Sparks Elicit Membrane Depolarization in Pulmonary Artery Smooth Muscle Cells

A major difference in the physiological function between Ca^{2+} sparks of pulmonary and systemic myocytes is their role in E_m regulation. E_m of adult rat intralobar PASMCs, when measured using perforated-patch to avoid disturbing subcellular Ca^{2+} dynamics, is usually less quiescent than those recorded under conventional whole-cell configuration. Small sporadic depolarizations are frequently observed. Activation of Ca^{2+} sparks with a subthreshold concentration (0.5 mM) (Fig. 1A) of caffeine elicits immediate membrane depolarizations from an averaged E_m of -45 to -37 mV that can be completely blocked by ryanodine (Fig. 1B) (39). This is in sharp contrast to the hyperpolarization elicited by Ca^{2+} sparks in systemic myocytes and raises the possibility that Ca^{2+} sparks may contribute to vasoconstriction, rather than vasorelaxation.

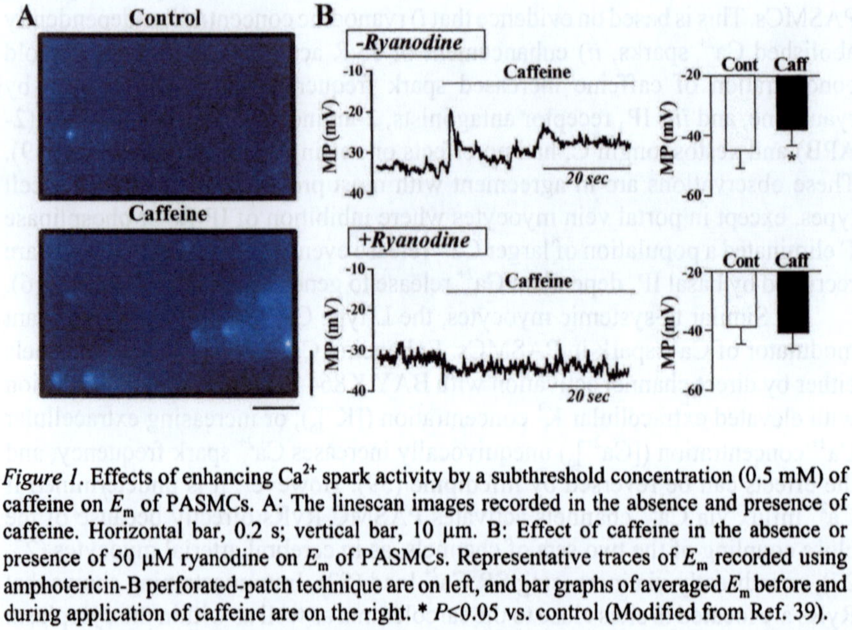

Figure 1. Effects of enhancing Ca^{2+} spark activity by a subthreshold concentration (0.5 mM) of caffeine on E_m of PASMCs. A: The linescan images recorded in the absence and presence of caffeine. Horizontal bar, 0.2 s; vertical bar, 10 μm. B: Effect of caffeine in the absence or presence of 50 μM ryanodine on E_m of PASMCs. Representative traces of E_m recorded using amphotericin-B perforated-patch technique are on left, and bar graphs of averaged E_m before and during application of caffeine are on the right. * $P<0.05$ vs. control (Modified from Ref. 39).

The disparity from systemic myocytes is likely due to *i)* diminished influence of K_{Ca} channels, *ii)* prominent expression of Cl_{Ca} channels, and *iii)* activation of other Ca^{2+} dependent membrane transporters in rat PASMCs. Developmental studies showed that Ca^{2+} sparks and STOCs are very active in fetal rabbit PASMCs (38). However, the occurrence of STOCs, the expression of K_{Ca} protein and mRNA and the responses of $[Ca^{2+}]_i$ to K_{Ca} channel blockers in distal PASMCs diminished with maturation (38, 40). The lower expression of K_{Ca} channels in adult PASMCs, and/or less effective coupling between RyRs and K_{Ca} channels may compromised the ability of Ca^{2+} spark to induce hyperpolarization. On the other hand, prominent Cl_{Ca} channels and STICs are found in PASMCs of several species. A recent study shows that metabolic inhibition with a low concentration of cyanide (1 mM) which has no effect on global $[Ca^{2+}]_i$ activates Ca^{2+} sparks and elicits STICs in rat PASMCs (50), suggesting a close association between Ca^{2+} sparks and Cl_{Ca} channels. Moreover, ion channels are differentially expressed in PASMCs, such that Cl_{Ca} and voltage-gated K^+ (K_V) channels are more abundant and K_{Ca} channels are much reduced in PASMCs of distal arteries (1, 46). The specific distributions of ion channels, hence, allow Ca^{2+} sparks to preferentially exert its depolarizing influence in distal resistant arteries. In addition, Ca^{2+} sparks may cause further depolarization by inhibiting K_V channels (15) and by activating the electrogenic Na^+/Ca^{2+} exchange.

In contrast to adult rat distal PASMCs, Ca^{2+} sparks primarily activate K_{Ca} channels to generate STOCs, causing membrane hyperpolarization and relaxation in fetal PASMCs (38, 40), similar to systemic arterial myocytes. Normoxia or high O_2 tension activates this Ca^{2+} spark/K_{Ca} channel mechanism to elicit O_2-dependent pulmonary vasodilation, which is critical for the transition to air breathing. Malfunction of this mechanism is found to be associated with an experimental model of pulmonary hypertension of the newborn. Moreover, STOCs have been recorded in small pulmonary arteries of adult rabbit (3). Therefore, the net effect of Ca^{2+} sparks on E_m may depend on developmental state, species, and location, all of which have significant bearings on the relative activities of the counteracting Ca^{2+}-activated channels in PASMCs.

6. ET-1 Enhances Ca^{2+} Sparks in Pulmonary Artery Smooth Muscle Cells

Under basal conditions, Ca^{2+} sparks are unlikely to elicit vasoconstriction because spark frequency is low. However, PASMCs are constantly under the influences of a wide variety of vasoactive factors, some of these factors may exert their effects in part by modulating Ca^{2+} sparks. Endothelin-1 (ET-1), a most potent vasoconstrictor, has been implicated as an important modulator/mediator of acute and chronic HPV (8, 35). When applied exogenously to rat intralobar PASMCs, it elicits dramatic global Ca^{2+} mobilization, involving Ca^{2+} release from RyR-gated Ca^{2+} stores (45). At the subcellular level, ET-1 at concentration between 10^{-10} to 10^{-8} M causes a concentration-dependent increase in spark frequency (Fig. 2B), with a 2-fold increase at 10^{-10} M and a 4-5-fold increase at the highest concentration (56, 57). The increase in spark frequency is associated with increases in spark size, duration, and amplitude. Pharmacological evidence shows that the enhancement of Ca^{2+} sparks by ET-1 is mediated through ET_A-receptor activation of phospholipase C, leading to IP_3 generation, Ca^{2+} release from IP_3Rs, and local cross activation of RyRs, but it is unrelated to L-type Ca^{2+} channel activation, increased SR Ca^{2+} content, or activation of protein kinase C (57). More importantly, ET-1 induced activation of Ca^{2+} sparks is agonist-specific and appears to be independent of a global increase in $[Ca^{2+}]_i$, because norepinephrine, at a concentration that elicits comparable global Ca^{2+} transients, actually reduces Ca^{2+} spark frequency (Fig. 2A) (39). The reduction of Ca^{2+} sparks by NE is similar to that in systemic vascular smooth muscle where vasoconstrictors typically reduce Ca^{2+} spark frequency via a PKC-dependent mechanism (6, 20, 29). The activation of the pro-constrictive Ca^{2+} sparks by ET-1 may provide a special mechanism for a unique physiological function in intralobar PASMCs, perhaps in modulating HPV.

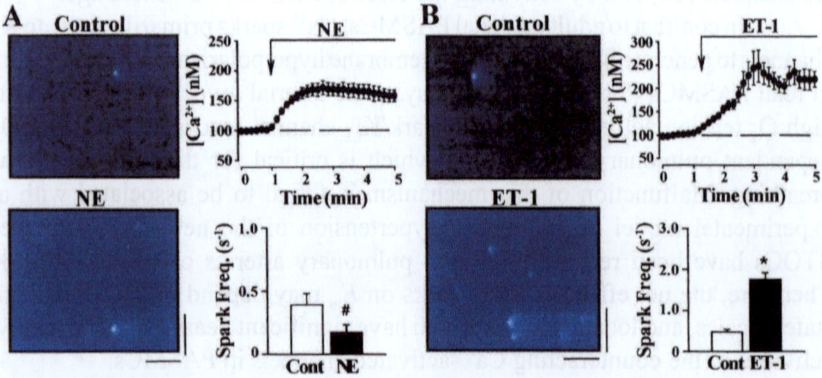

Figure 2. Effects of 10 nM norepinephrine (NE, A) and 3 nM endothelin-1 (ET-1; B) on global $[Ca^{2+}]_i$ and spark frequency in PASMCs. Representative linescan images in the absence (control) and presence NE or ET-1. Averaged time courses of changes in global $[Ca^{2+}]$ induced by NE and ET-1 and bar graphs of averaged spark frequency before and during application of the agonists. Horizontal bar, 0.2 s; vertical bar, 10 μm. * $P<0.05$ vs. control. # $P=0.53$ (Modified from Ref. 39).

7. Implications of Ca^{2+} Sparks for Hypoxic Pulmonary Vasoconstriction

Even though there are still uncertainties on whether Ca^{2+} release is an early event in HPV, it is rather clear that RyR-gated Ca^{2+} stores play a central role in the process. Inhibition of RyRs has been shown to completely abolish or to partially inhibit hypoxia-induced Ca^{2+} responses in PASMCs (15, 49) and hypoxia-induced vasoconstriction in isolated perfused lung (31), pulmonary arteries (18, 25). Recently, some have proposed that hypoxia causes a change in the cellular redox-state, in particular a reduction of the β-NAD$^+$:β-NADH ratio, which stimulates ADP-ribosyl cyclase and inhibits cyclic ADP-ribose hydrolase, leading to accumulation of cyclic ADP-ribose, an endogenous activator of RyRs to activate Ca^{2+} release from the SR (12, 54). Circumstantial evidence also suggest that increased reactive oxygen species (perhaps hydrogen peroxide) production in the proximal sites of electron transport chain in mitochondria during hypoxia may trigger Ca^{2+} release from the SR (51). Since RyR activity is modulated by multiple mechanisms, including Ca^{2+}-induced Ca^{2+} release, cyclic ADP-ribose, sulfhydryl oxidation, phosphorylation, calmodulin, FK506-binding proteins, and SR luminal Ca^{2+} content, detailed future studies are required to delineate the exact contributions of all these mechanisms in RyR activation during hypoxia.

Ca^{2+} release from RyR-gated stores may participate in the hypoxic response by *i*) providing Ca^{2+} to directly activate myofilaments, *ii*) acting as a trigger to initiate a chain of Ca^{2+} events, or *iii*) responding synergistically with other mechanisms to generate pulmonary vasoconstriction. Application of a high concentration of caffeine is known to generate large Ca^{2+} transients and to cause contraction of pulmonary arteries, indicating that the RyR-gated store of PASMCs is capable of providing sufficient global Ca^{2+} for direct myofilament activation. On the other hand, local Ca^{2+} release in the form of Ca^{2+} sparks can serve as the vehicle for the latter two possibilities.

In the case where hypoxia elicits a moderate activation of RyRs, the increase in Ca^{2+} spark frequency may cause a membrane depolarization which exceeds the activation threshold (-40 mV) of L-type Ca^{2+} channel to increase Ca^{2+} influx and contraction (33). This is analogous to the hypothesis that Ca^{2+} release is the initial event of HPV, and elevated $[Ca^{2+}]_i$ inhibits K_V channels to cause membrane depolarization and contraction (15). Activation of Ca^{2+} sparks may also lead to local SR Ca^{2+} depletion, resulting in activation of store-operated Ca^{2+} channels and capacitative Ca^{2+} entry (41, 42). This notion is consistent with the observation that the transient phase I of hypoxia-induced contraction in isolated pulmonary arteries is mainly dependent on capacitative Ca^{2+} entry, which can be blocked by a low concentration of La^{3+} (42). Moreover, because of the non-selective nature of store-operated channels, their activation may amplify vasoconstriction by causing further membrane depolarization, and Ca^{2+} influx via L-type voltage-dependent Ca^{2+} channel (53); providing a plausible explanation as to why the phase I hypoxic vasoconstriction can also be partially inhibited by nifedipine (41).

In the case of mild RyR activation, membrane depolarization induced by Ca^{2+} sparks may be insufficient to cross the threshold for L-type Ca^{2+} channel activation, and the reduction in SR Ca^{2+} may not be enough to elicit capacitative Ca^{2+} entry. However, it may set the E_m close to the activation threshold of L-type Ca^{2+} channels, to potentiate synergistically other hypoxia-induced voltage-dependent mechanisms, such as K_V channel inhibition (55) and Ca^{2+} channel activation (13), and/or help to sustain a vasoconstriction by reducing the rate of Ca^{2+} removal through Na^+/Ca^{2+} exchange. Indeed, it has been shown previously in some studies that a "priming" depolarization is required for hypoxia induced membrane depolarization and $[Ca^{2+}]_i$ elevation in PASMCs (4, 48). To this effect, we have found that a subthreshold concentration of ET-1 (10^{-10} M), which causes a 2-fold increase in spark frequency (56), but itself does not cause an increase in global $[Ca^{2+}]_i$ or contraction, potentiates hypoxia-induced contraction 6-fold in porcine PASMCs, and restores acute hypoxic constriction in endothelium-denuded porcine distal pulmonary arteries which are otherwise unresponsive to hypoxia (25, 44).

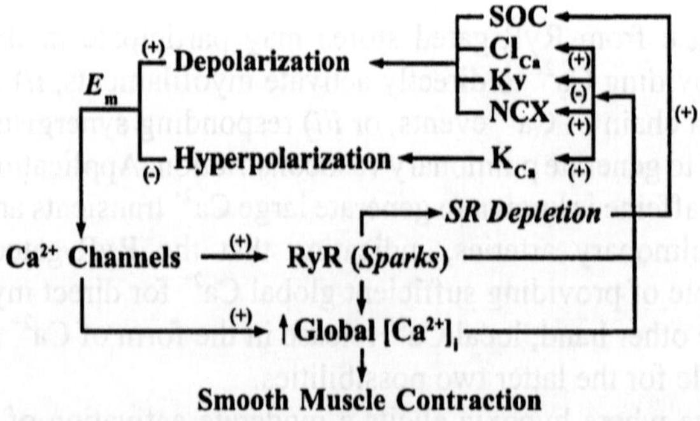

Figure 3. Schematic diagram depicting the possible Ca^{2+} signaling pathways in PASMCs. Abbreviations: SOC, store-operated Ca^{2+} channels; NCX, Na^+-Ca^{2+} exchanger; RyR, ryanodine receptors; Cl_{Ca}, Ca^{2+}-activated Cl^- channels, K_{Ca}, Ca^{2+}-activated K^+ channels, K_V, voltage-gated K^+ channels.

8. Concluding Remarks

Ca^{2+} signaling is a complex process, in which Ca^{2+} signals are generated by multiple specific Ca^{2+} transporters, delivered globally or locally, and decoded by different effectors according to the signal amplitude and frequency in their immediate vicinities. In this fashion, a single ionic species can serve as a ubiquitous messenger for numerous cellular functions (e.g., cell proliferation, ion channel regulation, and muscle contraction). In the pulmonary circulation, the research on the function(s) and regulation of local Ca^{2+} signals has just been started. In this chapter, we have described the biophysical properties and the unique pro-contractile function of Ca^{2+} spark-mediated membrane depolarization (Fig. 3) and their modulation by ET-1 in adult rat intralobar PASMCs, and discussed their possible relevance in HPV. We hope this information will serve as a starting point for the future pursuit in the understanding of local Ca^{2+} signaling in the regulation of pulmonary vascular reactivity.

Acknowledgments

This work is supported in part by grants from the NIH (HL 071835 and HL63813) and AHA to J.S.K. Sham. W.-M. Zhang was a visiting fellow from Liu Hua Qiao Hospital, P.R. China and supported by a Postdoctoral Fellowship Award from the AHA.

References

1. Archer SL, Huang JMC, Reeve HL, Hampl V, Tolarová S, Michelakis E, and Weir EK. Differential distribution of electrophysiologically distinct myocytes in conduit and resistance arteries determines their response to nitric oxide and hypoxia. *Circ. Res.* 1996; 78: 431-442.
2. Arnaudeau S, Macrez-Leprêtre N, and Mironneau J. Activation of calcium sparks by angiotensin in vascular myocytes. *Biochem. Biophys. Res. Comm.* 1996; 222: 809-815.
3. Bae YM, Park MK, Lee SH, Ho W-K, and Earm YE. Contribution of Ca^{2+}-activated K^+ channels and non-selective cation channels to membrane potential of pulmonary arterial smooth muscle cells of the rabbit. *J. Physiol.* 1999; 514: 747-758.
4. Bakhramov A, Evans AM, and Kozlowski RZ. Differential effects of hypoxia on the intracellular Ca^{2+} concentration of myocytes isolated from different regions of the rat pulmonary arterial tree. *Exp. Physiol.* 1998; 83: 337-347.
5. Bolton TB and Gordienko DV. Confocal imaging of calcium release events in single smooth muscle cells. *Acta Physiol. Scand.* 1998; 164: 567-575.
6. Bonev AD, Jaggar JH, Rubart M, and Nelson MT. Activators of protein kinase C decrease Ca^{2+} spark frequency in smooth muscle cells from cerebral arteries. *Am. J. Physiol. Cell Physiol.* 1997; 273: C2090-C2095.
7. Bychkov R, Gollasch M, Steinke T, Ried C, Luft FC, and Haller H. Calcium-activated potassium channels and nitrate-induced vasodilation in human coronary arteries. *J. Pharmacol. Exp. Ther.* 1998; 285: 293-298.
8. Chen SJ, Chen YF, Meng QC, Durand J, Dicarlo VS, and Oparil S. Endothelin-receptor antagonist bosentan prevents and reverses hypoxic pulmonary hypertension in rats. *J. Appl. Physiol.* 1995; 79: 2122-2131.
9. Cheng H, Lederer WJ, and Cannell MB. Calcium sparks: Elementary events underlying excitation-contraction coupling in heart muscle. *Science* 1993; 262: 740-744.
10. Collier ML, Ji G, Wang YX, and Kotlikoff MI. Calcium-induced calcium release in smooth muscle: Loose coupling between the action potential and calcium release. *J. Gen. Physiol.* 2000; 115: 653-662.
11. Coussin F, Macrez N, Morel J-L, and Mironneau J. Requirement of ryanodine receptor subtypes 1 and 2 for Ca^{2+}-induced Ca^{2+} release in vascular myocytes. *J. Biol. Chem.* 2000; 275: 9596-9603.
12. Dipp M and Evans AM. Cyclic ADP-ribose is the primary trigger for hypoxic pulmonary vasoconstriction in the rat lung in situ. *Circ. Res.* 2001; 89: 77-83.
13. Franco-Obregon A, and Lopez-Barneo J. Differential oxygen sensitivity of calcium channels in rabbit smooth muscle cells of conduit and resistance pulmonary arteries. *J. Physiol.* 1996; 491: 511-518.
14. Ganitkevich V and Isenberg G. Isolated guinea pig coronary smooth muscle cells. Acetylcholine induces hyperpolarization due to sarcoplasmic reticulum calcium release activating potassium channels. *Circ. Res.* 1990; 67: 525-528.
15. Gelband CH and Gelband H. Ca^{2+} release from intracellular stores is an initial step in hypoxic pulmonary vasoconstriction of rat pulmonary artery resistance vessels. *Circulation* 1997; 96:

3647-3654.
16. Gordienko DV and Bolton TB. Crosstalk between ryanodine receptors and IP_3 receptors as a factor shaping spontaneous Ca^{2+}-release events in rabbit portal vein myocytes. *J. Physiol.* 2002; 542: 743-762.
17. Hogg RC, Wang Q, and Large WA. Effects of Cl channel blockers on Ca-activated chloride and potassium currents in smooth muscle cells from rabbit portal vein. *Br. J. Pharmacol.* 1994; 111: 1333-1341.
18. Jabr RI, Toland H, Gelband CH, Wang XX, and Hume JR. Prominent role of intracellular Ca^{2+} release in hypoxic vasoconstriction of canine pulmonary artery. *Br. J. Pharmacol.* 1997; 122: 21-30.
19. Jaggar JH. Intravascular pressure regulates local and global Ca^{2+} signaling in cerebral artery smooth muscle cells. *Am. J. Physiol. Cell Physiol.* 2001; 281: C439-C448.
20. Jaggar JH and Nelson MT. Differential regulation of Ca^{2+} sparks and Ca^{2+} waves by UTP in rat cerebral artery smooth muscle cells. *Am. J. Physiol. Cell Physiol.* 2000; 279: C1528-C1539.
21. Jaggar JH, Porter VA, Lederer WJ, and Nelson MT. Calcium sparks in smooth muscle. *Am. J. Physiol. Cell Physiol.* 2000; 278: C235-C256.
22. Jaggar JH, Stevenson AS, and Nelson MT. Voltage dependence of Ca^{2+} sparks in intact cerebral arteries. *Am. J. Physiol. Cell Physiol.* 1998; 274: C1755-C1761.
23. Jaggar JH, Wellman GC, Heppner TJ, Porter VA, Pérez GJ, Gollasch M, Kleppisch T, Rubart M, Stevenson AS, Lederer WJ, Knot HJ, Bonev AD, and Nelson MT. Ca^{2+} channels, ryanodine receptors and Ca^{2+}-activated K^+ channels: a functional unit for regulating arterial tone. *Acta Physiol. Scand.* 1998; 64: 577-587.
24. Janiak R, Wilson SM, Montague S, and Hume JR. Heterogeneity of calcium stores and elementary release events in canine pulmonary arterial smooth muscle cells. *Am. J. Physiol. Cell Physiol.* 2001; 280: C22-C33.
25. Liu Q, Sham JSK, Shimoda LA, and Sylvester JT. Hypoxic constriction of porcine distal pulmonary arteries: endothelium and endothelin dependence. *Am. J. Physiol. Lung Cell. Mol. Physiol.* 2001; 280: L856-L865.
26. Löhn M, Fürstenau M, Sagach V, Elger M, Schulze W, Luft FC, Haller H, and Gollasch M. Ignition of calcium sparks in arterial and cardiac muscle through caveolae. *Circ. Res.* 2000; 87: 1034-1039.
27. Löhn M, Jessner W, Fürstenau M, Wellner M, Sorrentino V, Haller H, Luft FC, and Gollasch M. Regulation of calcium sparks and spontaneous transient outward currents by RyR3 in arterial vascular smooth muscle cells. *Circ. Res.* 2001; 89: 1051-1057.
28. Lukyanenko V, Viatchenko-Karpinski S, Smirnov A, Wiesner TF, and Györke S. Dynamic regulation of sarcoplasmic reticulum Ca^{2+} content and release by luminal Ca^{2+}-sensitive leak in rat ventricular myocytes. *Biophys. J.* 2001; 81: 785-798.
29. Mauban JRH, Lamont C, Balke CW, and Wier WG. Adrenergic stimulation of rat resistance arteries affects Ca^{2+} sparks, Ca^{2+} waves, and Ca^{2+} oscillations. *Am. J. Physiol. Heart Circ. Physiol.* 2001; 280: H2399-H2405.
30. Mironneau J, Coussin F, Jeyakumar LH, Fleischer S, Mironneau C, and Macrez N. Contribution of ryanodine receptor subtype 3 to Ca^{2+} responses in Ca^{2+}-overloaded cultured rat portal vein myocytes. *J. Biol. Chem.* 2001; 276: 11257-11264.
31. Morio Y and McMurtry IF. Ca^{2+} release from ryanodine-sensitive store contributes to mechanism of hypoxic vasoconstriction in rat lungs. *J. Appl. Physiol.* 2002; 92: 527-534.
32. Nelson MT, Cheng H, Rubart M, Santana LF, Bonev AD, Knot HJ, and Lederer WJ. Relaxation of arterial smooth muscle by calcium sparks. *Science* 1995; 270: 633-637.
33. Nelson MT, Patlak JB, Worley JF, and Standen NB. Calcium channels, potassium channels, and voltage dependence of arterial smooth muscle tone. *Am. J. Physiol.* 1990; 259: C3-18.
34. Neylon CB, Richards SM, Larsen MA, Agrotis A, and Bobik A. Multiple types of ryanodine receptor/Ca^{2+} release channels are expressed in vascular smooth muscle. *Biochem. Biophys. Res. Commun.* 1995; 215: 814-821.

35. Oparil S, Chen SJ, Meng QC, Elton TS, Yano M, and Chen YF. Endothelin-A receptor antagonist prevents acute hypoxia-induced pulmonary hypertension in the rat. *Am. J. Physiol.* 1995; 268: L95-100.
36. Pérez GJ, Bonev AD, Patlak JB, and Nelson MT. Functional coupling of ryanodine receptors to K_{Ca} channels in smooth muscle cells from rat cerebral arteries. *J. Gen. Physiol.* 1999; 113: 229-237.
37. Porter VA, Bonev AD, Knot HJ, Heppner TJ, Stevenson AS, Kleppisch T, Lederer WJ, and Nelson MT. Frequency modulation of Ca^{2+} sparks is involved in regulation of arterial diameters by cyclic nucleotides. *Am. J. Physiol. Cell Physiol.* 1998; 274: C1346-C1355.
38. Porter VA, Reeve HL, and Cornfield DN. Fetal rabbit pulmonary artery smooth muscle cell response to ryanodine is developmentally regulated. *Am. J. Physiol. Lung Cell. Mol. Physiol.* 2000; 279: L751-L757.
39. Remillard CV, Zhang W-M, Shimoda LA, and Sham JSK. Physiological properties and functions of Ca^{2+} sparks in rat intrapulmonary arterial smooth muscle cells. *Am. J. Physiol. Lung Cell. Mol. Physiol.* 2002; 283: L433-444.
40. Rhodes MT, Porter VA, Saqueton CB, Herron JM, Resnik ER, and Cornfield DN. Pulmonary vascular response to normoxia and K_{Ca} channel activity is developmentally regulated. *Am J Physiol Lung Cell Mol Physiol* 2001; 280: L1250-1257.
41. Robertson TP, Hague D, Aaronson PI, and Ward JPT. Voltage-independent calcium entry in hypoxic pulmonary vasoconstriction of intrapulmonary arteries of the rat. *J. Physiol.* 2000; 525: 669-680.
42. Salvaterra CG, and Goldman WF. Acute hypoxia increases cytosolic calcium in cultured pulmonary arterial myocytes. *Am. J. Physiol. Lung Cell. Mol. Physiol.* 1993; 264: L323-328.
43. Sham JSK, Cleemann L, and Morad M. Functional coupling of Ca^{2+} channels and ryanodine receptors in cardiac myocytes. *Proc. Natl. Acad. Sci. U.S.A.* 1995; 92: 121-125.
44. Sham JSK, Crenshaw BR Jr., Deng L-H, Shimoda LA, and Sylvester JT. Effects of hypoxia in porcine pulmonary arterial myocytes: roles of K_V channel and endothelin-1. *Am. J. Physiol. Lung Cell. Mol. Physiol.* 2000; 279: L262-272.
45. Shimoda LA, Sylvester JT, and Sham JSK. Mobilization of intracellular Ca^{2+} by endothelin-1 in rat intrapulmonary arterial smooth muscle cells. *Am. J. Physiol. Lung Cell. Mol. Physiol.* 2000; 278: L157-L164.
46. Smani T, Iwabuchi S, López-Barneo J, and Ureña J. Differential segmental activation of Ca^{2+}-dependent Cl^- and K^+ channels in pulmonary arterial myocytes. *Cell Calcium* 2001; 29: 369-377.
47. Stern MD. Theory of excitation-contraction coupling in cardiac muscle. *Biophys. J.* 1992; 63: 497-517.
48. Turner JL, and Kozlowski RZ. Relationship between membrane potential, delayed rectifier K^+ currents and hypoxia in rat pulmonary arterial myocytes. *Exp. Physiol.* 1997; 82: 629-645.
49. Vadula MS, Kleinman JG, and Madden JA. Effect of hypoxia and norepinephrine on cytoplasmic free Ca^{2+} in pulmonary and cerebral arterial myocytes. *Am. J. Physiol. Lung Cell. Mol. Physiol.* 1993; 265: L591-597.
50. Wang Y-X, Zheng Y-M, Abdullaev I, and Kotlikoff MI. Metabolic inhibition with cyanide induces calcium release in pulmonary artery myocytes and *Xenopus* oocytes. *Am. J. Physiol. Cell Physiol.* 2003; 284: C378-388.
51. Waypa GB, Marks JD, Mack MM, Boriboun C, Mungai PT, and Schumacker PT. Mitochondrial reactive oxygen species trigger calcium increases during hypoxia in pulmonary arterial myocytes. *Circ. Res.* 2002; 91: 719-726.
52. Wellman GC, Santana LF, Bonev AD, and Nelson MT. Role of phospholamban in the modulation of arterial Ca^{2+} sparks and Ca^{2+}-activated K^+ channels by cAMP. *Am. J. Physiol. Cell Physiol.* 2001; 281: C1029-1037.
53. Welsh DG, Morielli AD, Nelson MT, and Brayden JE. Transient receptor potential channels regulate myogenic tone of resistance arteries. *Circ. Res.* 2002; 90: 248-250.
54. Wilson HL, Dipp M, Thomas JM, Lad C, Galione A, and Evans AM. ADP-ribosyl cyclase

and cyclic ADP-ribose hydrolase act as a redox sensor. a primary role for cyclic ADP-ribose in hypoxic pulmonary vasoconstriction. *J. Biol. Chem.* 2001; 276: 11180-11188.

55. Yuan XJ, Goldman WF, Tod ML, Rubin LJ, and Blaustein MP. Hypoxia reduces potassium currents in cultured rat pulmonary but not mesenteric arterial myocytes. *Am. J. Physiol. Lung Cell. Mol. Physiol.* 1993; 264: L116-123.
56. Zhang W, Remillard CV, Shimoda LA, and Sham JSK. Endothelin-1 (ET-1) enhances Ca^{2+} spark activity in rat pulmonary artery smooth muscle cells. *Am. J. Respir. Crit. Care Med.* 2001; 163: A395.
57. Zhang WM, Yip KP, Lin MJ, Li WH, and Sham JSK. Local cross talk between IP_3- and ryanodine-receptors in endothelin-1 activation of Ca^{2+} spark in arterial myocytes. *Biophys. J.* 2003; 84: 384a.
58. ZhuGe R, Sims SM, Tuft RA, Fogarty KE, and Walsh JV Jr. Ca^{2+} sparks activate K^+ and Cl^- channels, resulting in spontaneous transient currents in guinea-pig tracheal myocytes. *J. Physiol.* 1998; 513: 711-718.

Chapter 5

Hypoxia-mediated Regulation of Ca^{2+} Transients in Pulmonary Artery Smooth Muscle Cells

Jean-Pierre Savineau, Sébastien Bonnet and Roger Marthan
Université V. Ségalen Bordeaux2, Bordeaux, France

1. Introduction

Hypoxia induces a selective pulmonary vasoconstriction due to the contraction of vascular smooth muscle cells contained in the pulmonary artery wall. When hypoxia is maintained (i.e., chronic hypoxia), the sustained pulmonary vasoconstriction leads to a selective pulmonary arterial hypertension (PAHT). An increase in the cytosolic Ca^{2+} concentration ($[Ca^{2+}]_i$) is a key step in initiating the contractile response (43, 46). In vascular smooth muscle, an increase in $[Ca^{2+}]_i$ can result from both an influx of extracellular calcium and a release of intracellular stored Ca^{2+} (28). Extracellular Ca^{2+} enters the smooth muscle cells mainly following the activation of voltage-gated Ca^{2+} channels (e.g., L-type Ca^{2+} channel), receptor-operated channels or store-operated channels, whereas intracellular Ca^{2+} mainly originates from the sarcoplasmic reticulum (SR) via the activation of two types of receptor/channels: the inositol 1,4,5 trisphosphate receptor/channel (IP_3R) and the ryanodine receptor/channel (RyR) (6, 7).

In normoxic conditions, pulmonary arterial tone is controlled by both membrane potential (35, 50) and a variety of circulating and locally released mediators (3). Under chronic hypoxic conditions, the concentration of some of these mediators (i.e., angiotensin II, endothelin-1, serotonin) is increased both in animals and in humans; thus their implication in the etiology of the PAHT has been questioned (14, 15, 24, 29, 32, 33). Although the exact mechanisms involved both in hypoxic pulmonary vasoconstriction (HPV) and in PAHT are not clearly defined, a direct effect of hypoxia on the calcium homeostasis of pulmonary artery smooth muscle cells (PASMCs) is generally suggested. The aim of this chapter is thus to review the effect of hypoxia (acute and chronic) on both the resting $[Ca^{2+}]_i$ value and the $[Ca^{2+}]_i$ response induced by agonists in PASMCs.

2. Agonist-induced Ca^{2+} Transients in Pulmonary Artery Smooth Muscle Cells Subjected to Normoxic Condition

2.1. Characteristics of Ca^{2+} Transients

Using microspectrofluorimetry to measure the $[Ca^{2+}]_i$ in single cells, studies from our laboratory and others have revealed that agonists controlling smooth muscle tone via activation of seven transmembrane domain, G-coupled surface membrane receptors (e.g., angiotensin II, endothelin-1, extracellular ATP and UTP, noradrenaline, and serotonin) induce a complex temporal $[Ca^{2+}]_i$ response in PASMCs. The pattern of the response depends on several factors, including the type of arteries (proximal vs distal), the species, the phenotype of the cells (i.e., fresh vs. cultured cells) and the agonist considered. For example, in short term cultured (24-48 h) vascular myocytes from intrapulmonary arteries (300-500 μm o.d.), endothelin-1 induces a transient increase in $[Ca^{2+}]_i$ followed by a small plateau. The amplitude of the peak is dependent on the endothelin-1 concentration from 0.1 to 10 nM (44). In freshly isolated and cultured myocytes from proximal extrapulmonary and intralobar arteries, the agonist-induced $[Ca^{2+}]_i$ response is composed of a series of cyclic increases in $[Ca^{2+}]_i$, so-called *Ca^{2+} oscillations*. The $[Ca^{2+}]_i$ rises, after a delay of 5 to 10 sec, from a resting value of 60-70 nM to a peak value of 400-800 nM (20, 25) before decreasing once again. This first increase is transient and followed by successive peaks of constant duration (Fig. 1). The oscillation frequency varies from 4-6 to 25-30/min, according to the cell type and the agonist concentration. On average, 50-80% of cells exhibit $[Ca^{2+}]_i$ oscillation under identical experimental conditions.

The concentration of agonist is the main factor that modulates the pattern of Ca^{2+} oscillations. However, this modulation still depends on both type of tissue and the nature of the agonist considered. In some cases (cultured canine PASMCs), both amplitude and the frequency of the Ca^{2+} oscillations increase with the mediator concentration (22). In some others cases (freshly isolated rat PASMCs), the overall pattern of oscillations appears to be independent of the mediator concentration, but the percentage of cells exhibiting Ca^{2+} oscillations in response to mediator stimulation does depend on the mediator concentration (20, 25). In this latter case, the combination of the number of responding cells with the amplitude of the first $[Ca^{2+}]_i$ peak also reveals a relationship between the concentration of the mediator and the $[Ca^{2+}]_i$ response (25).

In contrast to mediators acting at cell surface membrane receptors, caffeine and ryanodine, known to act directly on the SR and to potentiate Ca^{2+}-induced Ca^{2+} release, always induce a transient or monotonic increase of $[Ca^{2+}]_i$ (Fig. 3C) that is never followed by oscillations in PASMCs (18, 25, 27). The amplitude of this transient $[Ca^{2+}]_i$ -response is dependent on the concentration of caffeine used (0.1-10 mM).

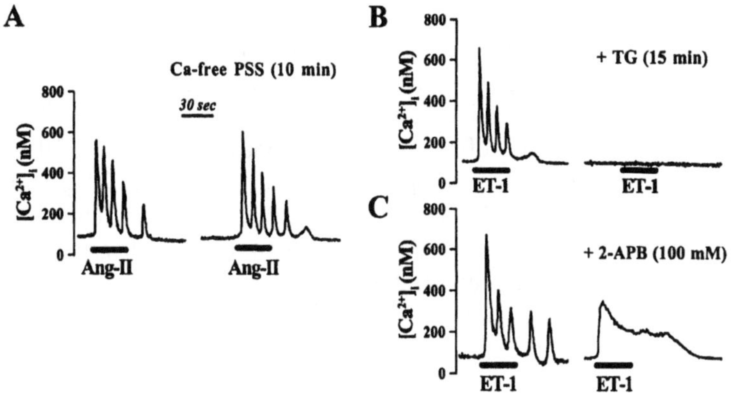

Figure 1. Agonist-induced Ca^{2+} oscillations in freshly isolated PASMCs from the rat main pulmonary artery. Short (30 s) ejection of angiotensin II (A, Ang II, 0.1μM) or endothelin-1 (B and C, ET-1, 0.01 μM) near the cell induced oscillations in $[Ca^{2+}]_i$. These Ca^{2+} oscillations were not altered when PASMCs were superfused with a Ca^{2+}-free physiological salt solution (PSS, A) but, disappeared when PASMCs are superfused for 15 min with 1 μM thapsigargin (TG, B). When PASMCs were superfused for 10 min with 1 μM of 2-APB, the ET-1-induced $[Ca^{2+}]_i$ oscillations disappeared (C). In each panel, the first record (*left trace*) is the control response. Each trace was recorded from a different cell and is typical of 8-10 cells.

2.2. Cellular Mechanisms of Agonist-induced $[Ca^{2+}]_i$ Responses in Pulmonary Artery Smooth Muscle Cells

In rat PASMCs, Ca^{2+} oscillations are modified neither by the presence of organic Ca^{2+}-channels blockers (verapamil or nifedipine), nor by La^{3+} or removal of extracellular Ca^{2+} (Fig. 1A) (18, 25) whereas, in canine PASMCs, Ca^{2+} oscillations progressively disappear when external Ca^{2+} is removed but are maintained in the presence of Ca^{2+} channel blockers (22). In freshly-dissociated or cultured PASMCs, agonists fail to induce Ca^{2+} oscillations when cells are pretreated with specific blockers of SR Ca^{2+}-ATPase pump, such as thapsigargin (TG) (Fig. 1B) or cyclopyazonic acid (CPA) (17). This result suggests that the main Ca^{2+} compartment involved in Ca^{2+} oscillations, or at least in the triggering of Ca^{2+} oscillations, is an intracellular store (mainly the SR). Hence, agonist-induced Ca^{2+} oscillations appear to be regulated by a cytosolic Ca^{2+} oscillator. Agonist-induced Ca^{2+} oscillations are inhibited by *i)* neomycin or U73122, potent inhibitors of the phosphoinositide phospholipase C (18, 22, 25), and *ii)* 2-aminoethyldiphenylborate (2-APB) and heparin, potent inhibitors of the IP_3R (Figure 1C). These results indicate that the IP_3 pathway is implicated in the mechanism of Ca^{2+} oscillations. The cyclic character of $[Ca^{2+}]_i$ variation could be due to either a discontinuous production of IP_3 (receptor-controlled oscillator) or an opening and closure of the IP_3R (second messenger controlled oscillator).

So far, no direct evidence for cyclical agonist-mediated production of IP_3 has been reported in smooth muscle cells. In contrast, the regulation of IP_3R is complex and involves both positive and negative feedback controls by cytosolic and/or luminal Ca^{2+} (26, 31). Ca^{2+} acts as a cofactor at the site of IP_3R. At low $[Ca^{2+}]_i$ (<300 nM), Ca^{2+} potentiates the IP_3 effect, whereas at high $[Ca^{2+}]_i$ (>300 nM), its antagonizes the IP_3 effect (26). This biphasic regulation of the IP_3R, together with the activity of the sarcoplasmic and plasmalemmal Ca^{2+} pumps, can explain the cyclical nature of the agonist-induced $[Ca^{2+}]_i$ increase. The amplitude of each $[Ca^{2+}]$ spike may represent the balance between Ca^{2+} release, the loss of Ca^{2+} from the cell, and the sequestration of Ca^{2+} into intracellular stores. The termination of the spike would occur when the release process is inactivated, allowing extrusion of Ca^{2+} from the cell and the sequestration of Ca^{2+} into intracellular store (Fig. 5).

Although caffeine-induced $[Ca^{2+}]_i$ response (CICR) disappears in PASMCs pretreated with thapsigargin, it is not altered by modulators of the IP_3 pathway but, conversely, is fully antagonized by tetracaine, a potent inhibitor of the CICR mechanism (20). It thus clearly appears that PASMCs exhibit two Ca^{2+}-release systems located at the site of the SR membrane. One is the IP_3R underlying Ca^{2+} oscillations through the IP_3-induced Ca^{2+} release (IICR) mechanism, the other is the ryanodine receptor/channel protein (RyR) responsible for the CICR mechanism (Fig. 5).

3. Effect of Hypoxia on Ca^{2+} Transients in Pulmonary Artery Smooth Muscle Cells

3.1. Effects of Acute Hypoxia

In pulmonary resistance arteries, it is well recognized that acute hypoxia *per se*, increases $[Ca^{2+}]_i$ in PASMCs in both freshly isolated and cultured cells as well as in intact arterial wall (13, 27, 39, 42). This $[Ca^{2+}]_i$ increase triggers the contractile response of PASMCs and thus underlies the subsequent HPV. Perfusion of PASMCs with hypoxic medium quickly increases $[Ca^{2+}]_i$ to its maximum within 1-2min, and is maintained during the duration of perfusion (49). Return to a normoxic environment quickly restores the initial $[Ca^{2+}]_i$ value. The $[Ca^{2+}]_i$ increase is graded from 0 to 500 nM as a function of the severity of hypoxia from Po_2 60-80 to 10-15 mmHg (34). A variety of studies have investigated the origin and the mechanism of hypoxia-induced $[Ca^{2+}]_i$ increase. Both intracellular and extracellular Ca^{2+} compartments are involved. It is generally admitted that hypoxia first induces a release of intracellular Ca^{2+} followed by an influx of extracellular Ca^{2+} (16, 37). Both voltage-dependent (23, 48) and voltage-independent (e.g., capacitative Ca^{2+} entry) mechanisms are involved in the Ca^{2+} influx (27, 34, 40). In addition the SR, mitochondria also seem to play a modulatory role in the hypoxia-induced Ca^{2+} release from intracellular stores (49). Detailed mechanisms involved in hypoxia-induced Ca^{2+} release are described in the next chapter (Chapter 6).

In proximal (conduit) arteries, acute hypoxia induces a decrease in the resting $[Ca^{2+}]_i$ value, and vasodilation. In one study (48), it has been shown that PASMCs can generate spontaneous Ca^{2+} oscillations that are mainly due to the activation of L-type voltage-dependent Ca^{2+} channels. Acute hypoxia attenuates and increases the frequency of Ca^{2+}-oscillations in conduit and resistance arteries, respectively. These results suggest that hypoxia modulates the activity of Ca^{2+} channels in opposite manner in different regions of the pulmonary arterial tree.

Finally, very few information is available about the interactions between acute hypoxia and agonist-induced Ca^{2+} response. In distal pulmonary artery myocytes from fetal lamb, angiotensin II induces small Ca^{2+} oscillations (30 nM in amplitude), which are attenuated by acute hypoxia (13).

3.2. Effect of Chronic Hypoxia on Agonist-induced $[Ca^{2+}]_i$ Responses

Some aspects of the modulation of cellular signaling by chronic hypoxia have been studied on cultured cells maintained in an hypoxic environment for 24 to 72 hrs. However, most of the studies investigating the effect of chronic hypoxia on $[Ca^{2+}]_i$ responses have been performed on pulmonary arteries obtained from the chronically hypoxic rats, where animals are exposed to a hypoxic environment either under normobaric (10% O_2) or hypobaric (0.5 atmosphere) conditions for 2-3 weeks. This procedure induces a selective PAHT with an increase of the mean pulmonary artery pressure from ~10 to 30 mmHg (Fig. 2A) (10).

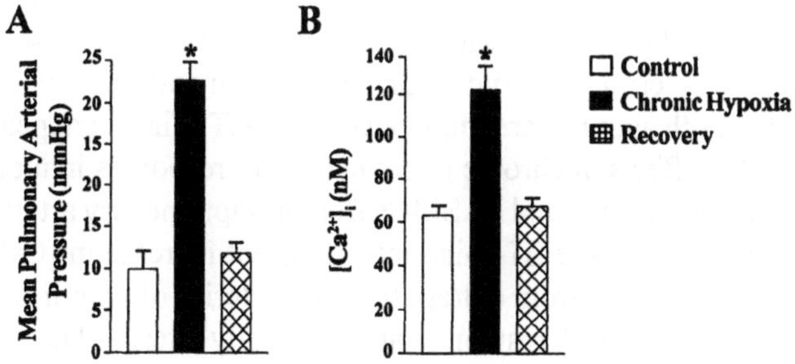

Figure 2. Effect of chronic hypoxia (CH) on pulmonary arterial pressure (PAP) and resting $[Ca^{2+}]_i$ in PASMCs in rats. CH (3 weeks) induced a significant increase in mean PAP (A) and in $[Ca^{2+}]_i$ (B) compared with control rats. After 3 weeks of normoxia recovery, mean PAP and PASMC $[Ca^{2+}]_i$ were not different from those of control rats. Results are mean±SE. * $P<0.05$ vs. control.

In general, chronic hypoxia increases the resting $[Ca^{2+}]_i$ value by ~60%, i.e., from 70-75 nM to 115-120 nM (Fig. 2B) (10, 11). This increase disappears in the absence of extracellular Ca^{2+} but is resistant to organic inhibitors of L-type voltage-dependent Ca^{2+} channels (e.g., nifedipine, verapamil) (11, 45), suggesting that a different Ca^{2+} entry pathway may be upregulated by chronic hypoxia. The exact nature of this Ca^{2+} influx has not yet been identified.

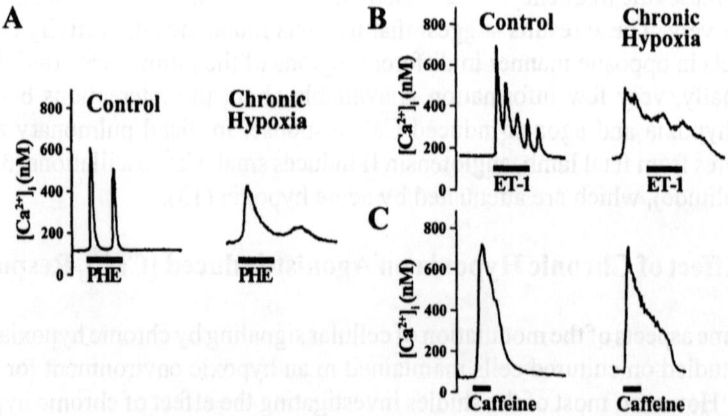

Figure 3. Effect of chronic hypoxia on agonist-induced $[Ca^{2+}]_i$ response in PASMCs. Short (30 s) application of 1 μM phenylephrine (PHE, A) or 0.1 μM endothelin (ET-1, B), induced oscillations in $[Ca^{2+}]_i$ in PASMCs obtained from control (normoxic) rats. The agonist-induced $[Ca^{2+}]_i$ oscillations disappeared in PASMCs from rats exposed to hypobaric hypoxia for 3 weeks. In contrast, the amplitude of the caffeine-induced transient $[Ca^{2+}]_i$ response was not altered by chronic hypoxia (C).

As for PAHT in humans, it has been demonstrated that chronic hypoxia in rat increases: *i*) ET-1 gene expression and plasma ET-1 levels (2, 14, 15, 30) and *ii*) angiotensin converting enzyme (ACE) expression and activity (32, 33, 51), suggesting that both agonists are implicated in PAHT. Thus, it appears relevant to investigate the effects of chronic hypoxia on Ca^{2+} responses induced by these agonists. In transiently cultured PASMCs from intrapulmonary arteries, chronic hypoxia largely inhibits the ET-1-induced transient increase in $[Ca^{2+}]_i$ (45). In freshly isolated PASMCs from extrapulmonary arteries of chronically hypoxic rats, agonist-induced $[Ca^{2+}]_i$ responses are drastically altered. The main change is a decrease in the percentage of responding cells (15-30%) and the disappearance of the oscillatory profile (Figs. 3A and B and Fig. 4). This change is not due to a chronic hypoxia-induced change in the Ca^{2+} source implicated in the $[Ca^{2+}]_i$ responses (8). As under control conditions, angiotensin II- or ET-1-induced $[Ca^{2+}]_i$ responses in PASMCs from chronically hypoxic rats are not altered in either Ca^{2+}-free solution or by the voltage-dependent Ca^{2+} channel blocker D600, but vanishes after pretreatment of the cells with thapsigargin.

These results indicate that agonist-mediated $[Ca^{2+}]_i$ responses in PASMCs also involve the mobilization of Ca^{2+} from an intracellular Ca^{2+} source, presumably the SR (8). In contrast, chronic hypoxia does not modify the percentage of cells responding to caffeine or the amplitude of the Ca^{2+} transient induced by caffeine (Fig. 3C) (8, 45). Collectively, these data demonstrate that chronic hypoxia alters Ca^{2+} oscillations via an action on the IP_3-signaling pathway, but not on the CICR mechanism.

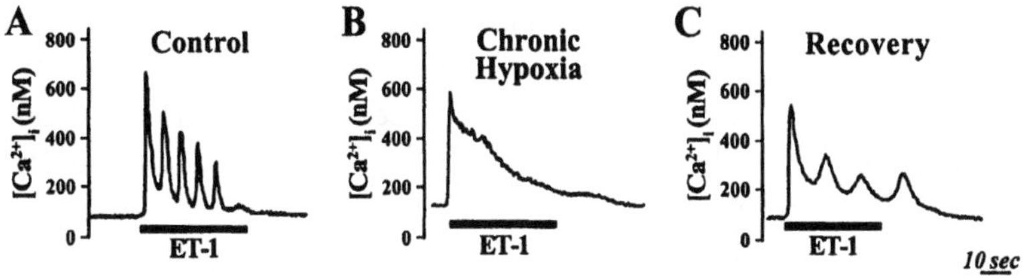

Figure 4. Reversal of chronic hypoxia-induced alteration in agonist-induced $[Ca^{2+}]_i$ response in PASMCs. Endothelin-1 (ET-1)-induced $[Ca^{2+}]_i$ oscillations in PASMCs from normoxic (control, A) and chronically-hypoxic (3 weeks, B) rats. The ET-1-induced $[Ca^{2+}]_i$ oscillations were partially recovered in PASMCs from CH rats after 3 weeks of recovery in normoxic conditions (C).

3.3. Potential Mechanism of Chronic Hypoxia-induced Disappearance of Ca^{2+} Oscillation

Chronic hypoxia-induced changes in agonist-induced $[Ca^{2+}]_i$ oscillations could result from: *i*) biochemical modulation of the agonist-receptor binding step resulting in a decrease in the number of receptors expressed at the surface membrane and/or a decrease in the agonist binding affinity, *ii*) functional alteration of IP_3R, and *iii*) different IP_3R subtypes involved in the response.

In smooth muscle, IP_3R is biphasically regulated by $[Ca^{2+}]_i$ (26) and this regulation accounts for the cyclical opening and closure of the associated Ca^{2+} channel (21) and, at least in part, for the so called Ca^{2+} oscillations. It is unlikely that the amplitude of chronic hypoxia-induced increase in resting $[Ca^{2+}]_i$ (~60%, corresponding to << 300 μM) could modify the negative feedback effect of $[Ca^{2+}]_i$ on IP_3R. Alternatively, the hypoxia-mediated change in agonist-induced $[Ca^{2+}]_i$ oscillation could be due to an alteration in the SR Ca^{2+} re-uptake mechanism which plays an important role in generating Ca^{2+} oscillation. Such a Ca^{2+} reuptake mechanism can be more easily examined by analyzing the falling part of the caffeine-induced $[Ca^{2+}]_i$ transients. Although chronic hypoxia does not modify the amplitude of the caffeine-induced $[Ca^{2+}]_i$ transients, it significantly decreases the rate of resting $[Ca^{2+}]_i$ restoration (8). These results suggest that chronic hypoxia affects Ca^{2+} reuptake into the SR by the SR Ca^{2+} pump

(SERCA) and/or Ca^{2+} extrusion by plasmalemmal Ca^{2+} pump (PMCA) and Na^+-Ca^{2+} exchanger. Indeed, the effect of chronic hypoxia on the kinetics of caffeine-induced $[Ca^{2+}]_i$ response is mimicked by CPA, suggesting that chronic hypoxia may inhibit SERCA to delay the recovery of resting $[Ca^{2+}]_i$. Interestingly, it has been shown in human cardiac muscle that SERCA expression changes during cardiac hypertrophy (1). We believe that the chronic hypoxia-induced decrease or disappearance of agonist-mediated Ca^{2+} oscillations in PASMCs is due to the downregulation of SERCA. Upon receptor activation, Ca^{2+} re-uptake into the SR is slowed after the first Ca^{2+} increase during hypoxia; therefore, cytosolic $[Ca^{2+}]$ remains elevated and subsequently modifies the function of IP_3R (Fig. 5).

In smooth muscle, as in non-muscle cells, three IP_3R isoforms (type 1-3) are encoded by three different genes (7, 36). Recent studies performed on IP_3R reconstituted in lipid bilayer have revealed important functional differences between the three isoforms. Interestingly, type 1-IP_3R is biphasically regulated by cytosolic Ca^{2+} (21, 38, 47). Preliminary results from our laboratory show that type 1-IP_3R is predominantly expressed in PASMCs from normoxic rats. It is thus tempting to speculate that chronic hypoxia could switch IP_3R from a biphasically to a non-biphasically regulated subtype. This hypothesis requires further molecular biological investigations.

4. Role of Ca^{2+} Oscillations and Consequences of Their Disappearance in Chronic Hypoxic Conditions

In smooth muscle, as in nonexcitable cells, the role of Ca^{2+} oscillations has not yet been clearly elucidated. As discuss in the first paragraph, in many cells types, the oscillation frequency is sensitive to the agonist concentration. It is thus suggested that oscillations may represent a digitalization of the Ca^{2+} signal, allowing a frequency-dependent control of the cellular response. Ca^{2+} oscillations are also implicated in the control of the membrane potential through regulation of Ca^{2+}-activated ion channels (19, 41).

Chronic hypoxia-induced disappearance of Ca^{2+} oscillations play a major role in chronic hypoxia-induced changes in main pulmonary artery (MPA) reactivity. In proximal PASMC, maximal ET-1 (Fig. 4) and angiotensin II-induced contraction were decreased as was the frequency of Ca^{2+} oscillations in response to the same mediator (5, 8). Interestingly, this effect is the opposite to that observed in another smooth muscle, the trachealis where chronic hypoxia increase both the cholinergic contractile response and the Ca^{2+} oscillation frequency in airway smooth muscle myocytes (4). The reason for the differential effect of chronic hypoxia on the reactivity of the two smooth muscle types remains to be established.

5. Reversal of Chronic Hypoxia-induced Alteration of Ca^{2+} Signaling

A distinguishing characteristic of chronic hypoxia-induced PAHT is its reversibility upon return to normoxia. In addition to pulmonary vasoconstriction (Fig. 2A), right ventricular hypertrophy, structural changes in elastic laminea, muscularization of distal pulmonary vessels, upregulation of angiotensin II receptors, and inhibition of K^+ channels are also fully reversed after 2-3 weeks of normoxia recovery (9, 10, 12). Furthermore, the sustained increase in resting $[Ca^{2+}]_i$ in PASMCs during chronic hypoxia is also reversed following normoxic recovery (Fig. 2B) (10). However, no detailed study has been performed on the reversibility of chronic hypoxia-induced alterations of agonist-induced $[Ca^{2+}]_i$ oscillation in PASMCs. Our preliminary studies show that, after 3 weeks of normoxic recovery, both the oscillating profile of ET-1-induced $[Ca^{2+}]_i$ response and percentage of cells exhibiting Ca^{2+} oscillations are restored (Fig. 5).

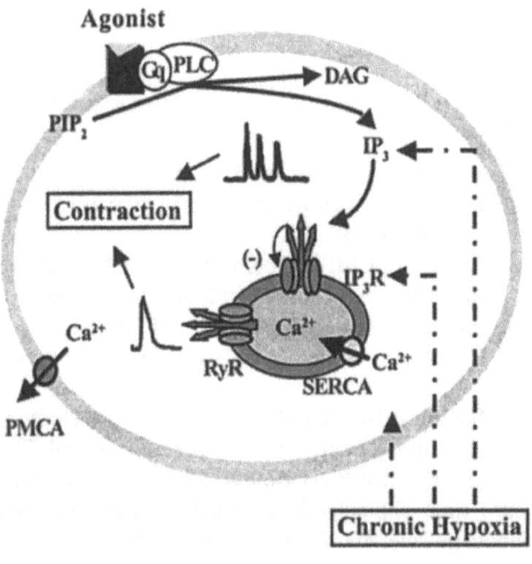

Figure 5. Proposed cellular mechanisms responsible for agonist-induced Ca^{2+} oscillations in PASMCs and their alteration by chronic hypoxia. Agonists bind to specific G-protein coupled receptors and activate phospholipase C (PLC) which hydrolyses phosphatidylinositol 4, 5 biphosphate (PIP_2) to diacyglycerol (DAG) and IP_3. IP_3 then binds to specific IP_3R on the SR membrane. The cyclical opening of the IP_3R-Ca^{2+} release channels due to its biphasic regulation by cytosolic Ca^{2+} induces $[Ca^{2+}]_i$ oscillations. Released Ca^{2+} is then sequestered into the SR by SERCA and/or extruded through the plasmalemmal Ca^{2+} pump (PMCA), thereby leading to a progressive diminution of$[Ca^{2+}]_i$ oscillations. Activation of RyR on the SR membrane by caffeine also induces a transient rise in $[Ca^{2+}]_i$ and contributes the overall Ca^{2+} signaling. Chronic hypoxia specifically alters $[Ca^{2+}]_i$ oscillations by regulating the IP_3 signaling pathway (e.g., IP_3production and/or IP_3R function and expression) and/or SERCA.

6. Conclusion

Both acute and chronic hypoxia alter resting $[Ca^{2+}]_i$ value and Ca^{2+} transients in PASMCs. These alterations are related to hypoxia-mediated effects on various pathways implicated in PASMC Ca^{2+} homeostasis: voltage-dependent and -independent Ca^{2+} influx, IP_3R, Ca^{2+} re-uptake, etc. These alterations also play a role in both structural and reactivity changes of the pulmonary circulation in the course of the development of PAHT, as well as in its potential reversal upon normal air breathing. Further studies are therefore required to provide a comprehensive description of the variety of hypoxia-induced effects on calcium signaling in PASMCs, and to identify the common mechanism(s) leading to these various effects in order to define new molecular therapeutic targets.

Acknowledgments

This work was supported in part by grants from the Conseil Régional d'Aquitaine (n° 20000301114).

References

1. Anger M, Lompre A-M, Vallot O, Marotte F, Rappaport L, and Samuel J-L. Cellular distribution of Ca^{2+} pumps and Ca^{2+} release channels in rat cardiac hypertrophy induced by aortic stenosis. *Circulation*. 1998; 98: 2477-2486.
2. Aguirre JI, Morrell NW, Long L, Clift P, Upton PD, Polak JM, and Wilkins MR. Vascular remodeling and ET-1 expression in rat strains with different responses to chronic hypoxia. *Am. J. Physiol. Lung Cell. Mol. Physiol.* 2000; 278: L981-L987.
3. Barnes PJ and Liu SF. Regulation of pulmonary vascular tone. *Pharmacol. Rev.* 1995; 47: 87-131.
4. Belouchi NE, Roux E, Savineau J-P, and Marthan R. Effect of chronic hypoxia on calcium signalling in airway smooth muscle cells. *Eur. Respir. J.* 1999; 14: 74-79.
5. Bialecki RA, Fisher CS, Murdoch WW, Barthlow HG, Stow RB, Mallamaci M, and Rumsey W. Hypoxic exposure time dependently modulates endothelin-induced contraction of pulmonary artery smooth muscle. *Am. J. Physiol Lung Cell. Mol. Physiol.* 1998; 274: L552-L559.
6. Berridge MJ. Elementary and global aspects of calcium signalling. *J. Physiol.* 1997; 499: 291-306.
7. Berridge MJ, Lipp P, and Bootman MD. The versatility and universality of calcium signaling. *Nat. Rev. Mol. Cell. Biol.* 2000; 1: 11-21
8. Bonnet S, Belus A, Hyvelin J-M, Roux E, Marthan R, and Savineau J-P. Effect of chronic hypoxia on agonist-induced tone and calcium signaling in rat pulmonary artery. *Am. J. Physiol. Lung Cell. Mol. Physiol.* 2001; 281: L193-L201.
9. Bonnet P, Bonnet S, Dubuis E, Vandier C, Savineau J-P. Reoxygenation enhanced iberiotoxin-sensitive current in chronic hypoxic rat pulmonary artery smooth muscle cells. *Circulation*. 2001; 104 (17, suppl.2): 403.
10. Bonnet S, Dubuis E, Vandier C, Marthan R, and Savineau J-P. Reversal of chronic hypoxia-induced alterations in pulmonary artery smooth muscle electromechanical coupling upon air

breathing. *Cardiovasc. Res.* 2002; 53: 1019-1028.
11. Bonnet S, Hyvelin J-M, Bonnet P, Marthan R, and Savineau J-P. Chronic hypoxia-induced spontaneous and rhythmical contractions in the rat main pulmonary artery. *Am. J. Physiol. Lung Cell. Mol. Physiol.* 2001; 281: L183-L192.
12. Chassagne C, Eddahibi S, Adamy C, Rideau D, Marotte F, Dubois-Randé JL, Adnot S, Samuel JL, and Teiger E. Modulation of angiotensin II receptor expression during development and regression of hypoxic pulmonary hypertension. *Am. J. Respir. Cell. Mol. Biol.* 2000; 22: 323-332.
13. Cornfield DM, Stevens T, McMurtry IF, Abman SH, and Rodman DM. Acute hypoxia increases cytosolic calcium in fetal pulmonary artery smooth muscle cells. *Am. J. Physiol.* 1993; 265: L53-L56.
14. Elton TS, Oparil S, Taylor GR, Hicks PH, Yang RH, Jin H, and Chen YF. Normobaric hypoxia stimulates endothelin-1 gene expression in the rat. *Am. J. Physiol.* 1992; 263: R1260-R1264.
15. Giaid A, Yanagisawa M, Langleben D, Michel RP, Levy R, Shennib H, Kimura S, Masaki T, Duguid WP, and Stewart DJ. Expression of endothelin-1 in the lungs of patients with pulmonary hypertension. *N. Engl. J. Med.* 1993; 328: 1732-1739.
16. Gelband GH and Gelband H. Ca^{2+} release from intracellular stores is an initial step in hypoxic pulmonary vasoconstriction of rat pulmonary artery resistance vessels. *Circulation.* 1997; 96: 3647-3654.
17. Gonzalez De La Fuente P, Savineau JP, and Marthan R. Control of pulmonary vascular smooth muscle tone by sarcoplasmic reticulum Ca^{2+} pump blockers: thapsigargin and cyclopiazonic acid. *Pflügers Arch.* 1995; 429: 617-624.
18. Guibert C, Marthan R, and Savineau JP. Angiotensin II-induced Ca^{2+}-oscillations in vascular myocytes from the rat pulmonary artery. *Am. J. Physiol.* 1996, 270: L637-L642.
19. Guibert C, Marthan R, and Savineau JP. Oscillatory Cl^- current induced by angiotensin II in rat pulmonary arterial myocytes: Ca^{2+} dependence and physiological implication. *Cell Calcium.* 1997; 21: 421-429.
20. Guibert C, Pacaud P, Loirand G, Marthan R, and Savineau JP. Effect of extracellular ATP on cytosolic Ca^{2+} concentration in rat pulmonary artery myocytes. *Am. J. Physiol.* 1996; 271: L450-L458, 1996.
21. Hagar RE, Burgstahler AD, Nathanson MH, and Ehrlich BE. Type III $InsP_3$ receptor channel stays open in the presence of increased calcium. *Nature.* 1998; 396: 81-84.
22. Hamada, H, Damron DS, Hong SJ, Van Wagoner DR, and Murray PA. Phenylephrine-induced Ca^{2+}oscillations in canine pulmonary artery smooth muscle cells. *Circ. Res.* 1997; 81: 812-823.
23. Harder DR, Madden JA, and Dawson C. Hypoxic induction of Ca^{2+}-dependent action potentials in small pulmonary arteries of the cat. *J. Appl. Physiol.* 1985; 59: 1389-1393.
24. Hervé P. Increased plasma serotonin in primary pulmonary hypertension. *Am. J. Med.* 1995; 99: 249-254.
25. Hyvelin J-M, Guibert C, Marthan R, and Savineau J-P. Cellular mechanisms and role of endothelin-1-induced calcium oscillations in pulmonary arterial myocytes. *Am. J. Physiol. Lung Cell. Mol. Physiol.* 1998; 275: L269-L282.
26. Iino M. Biphasic Ca^{2+} dependence of inositol 1,4,5-trisphosphate-induced Ca release in smooth muscle cells of the guinea pig taenia caeci. *J. Gen. Physiol.* 1990; 95: 1103-1122.
27. Kang TM, Park MK, and Uhm D-Y. Characterization of hypoxia-induced $[Ca^{2+}]_i$ rise in rabbit pulmonary arterial smooth muscle cells. *Life Sci.* 2002; 70: 2321-2333.
28. Karaki H, Ozaki H, Hori M, Mitsui-Saito M, Amano KI, Harada KI, Miyamoto S, Nakazawa H, Won K-J, and Sato K. Calcium movements, distribution, and functions in smooth muscle. *Pharmacol. Rev.* 1997; 49: 157-230.
29. Launay JM, Hervé P, Peoc'h C, Tournois C, Callebert J, Nebigil CG, Etienne N, Drouet L, Humbert M, Simonneau G, and Maroteaux L. Function of the serotonin 5-hydroxytryptamine 2B receptor in pulmonary hypertension. *Nature Med.* 2002; 8: 1129-1135.

30. Li H, Elton TS, Chen YF, and Oparil S. Increased endothelin receptor gene expression in hypoxic rat lung. *Am. J. Physiol.* 1994; 266: L553-L560.
31. Missiaen L, Parys JB, Weidema AF, Sipma H, Vanlingen S, De Smet P, Callewaert, G, and De Smedt H. The bell-shaped Ca^{2+} dependence of the inositol 1,4, 5-trisphosphate-induced Ca^{2+} release is modulated by Ca^{2+}/calmodulin. *J. Biol. Chem.* 1999; 274: 13748-13751.
32. Morrel NW, Atochina EN, Morris KG, Danilov SM, and Stenmark KR. Angiotensin converting enzyme expression is increased in small pulmonary arteries of rats with hypoxia-induced pulmonary hypertension. *J. Clin. Invest.* 1995; 96: 1823-1833.
33. Nong Z, Stassen JM, Moons L, Collen D, and Janssens S. Inhibition of tissue angiotensin-converting enzyme with quinapril reduces hypoxic pulmonary hypertension and pulmonary vascular remodeling. *Circulation.* 1996; 94: 1941-1947.
34. Olschewski A, Hong Z, Nelson DP, and Weir EK. Graded response of K^+ current, membrane potential, and $[Ca^{2+}]_i$ to hypoxia in pulmonary arterial smooth muscle. *Am. J. Physiol. Lung Cell. Mol. Physiol.* 2002; 283: L1143-L1150.
35. Osipenko ON, Alexander D, MacLean MR, and Gurney AM. Influence of chronic hypoxia on the contributions of non-inactivating and delayed rectifier K currents to the resting potential and tone of rat pulmonary artery smooth muscle. *Br. J. Pharmacol.* 1998; 124: 1335-1337.
36. Patel S, Joseph SK, Thomas AP. Molecular properties of inositol 1,4,5-trisphosphate receptors. *Cell Calcium* 1999; 25: 247-264.
37. Post JM, Gelband CH, and Hume JR. $[Ca^{2+}]_i$ inhibition of K^+ channels in canine pulmonary artery. Novel mechanism for hypoxia-induced membrane depolarization. *Circ. Res.* 1995; 77: 131-139.
38. Ramos-Franco J, Fill M, and Mignery GA. Isoform-specific function of single inositol 1,4,5-trisphosphate receptor channels. *Biophys. J.* 1998; 75: 834-839.
39. Robertson TP, Aaronson PI, and Ward JPT. Hypoxic vasoconstriction and intracellular Ca^{2+} in pulmonary arteries: evidence for PKC-independent Ca^{2+} sensitization. *Am. J. Physiol.* 1995; 268: H301-H307.
40. Robertson TP, Hague D, Aaronson PI, and Ward JPT. Voltage-independent calcium entry in hypoxic pulmonary vasoconstriction of intrapulmonary arteries of the rat. *J. Physiol.* 2000; 525: 669-680.
41. Salter KJ and Kozlowski RZ. Endothelin receptor coupling to potassium and chloride channels in isolated rat pulmonary arterial myocytes. *J. Pharmacol. Exp. Ther.* 1996; 279: 1053-1062.
42. Salveterra CG and Goldman WF. Acute hypoxia increases cytosolic calcium in cultured pulmonary arterial myocytes. *Am. J. Physiol.* 1993; 264: L323-L328.
43. Savineau J-P, and Marthan R. Modulation of the calcium sensitivity of the smooth muscle contractile apparatus: molecular mechanisms, pharmacological and pathophysiological implications. *Fundam. Clin. Pharmacol.* 1997; 11: 289-299.
44. Shimoda LA, Sylvester JT, and Sham JSK. Mobilization of intracellular Ca^{2+} by endothelin-1 in rat intrapulmonary arterial smooth muscle cells. *Am. J. Physiol. Lung Cell. Mol. Physiol.* 2000; 278: L157-L164.
45. Shimoda LA, Sham JSK, Shimoda TH, and Sylvester JT. L-type Ca^{2+} channels, resting $[Ca^{2+}]_i$, and ET-1-induced responses in chronically hypoxic pulmonary myocytes. *Am. J. Physiol. Lung Cell. Mol. Physiol.* 2000; 279: L884-L894.
46. Somlyo AP and Somlyo AV. Signal transduction and regulation in smooth muscle. *Nature.* 1994; 372: 231-236.
47. Trower EC, Hagar RE, and Ehrlich BE. Regulation of Ins(1,4,5)P_3 receptor isoforms by endogenous modulators. *Trends Pharmacol. Sci.* 2001; 11: 580-586.
48. Urena J, Franco-Obregon A, and Lopez-Barneo J. Contrasting effects of hypoxia on cytosolic Ca^{2+} spikes in conduit and resistance myocytes of the rabbit pulmonary artery. *J. Physiol.* 1996; 496: 103-109.
49. Waypa GB, Marks JD, Mack MM, Boriboun C, Mungai PT, and Schumacker PT.

Mitochondrial reactive oxygen species trigger calcium increases during hypoxia in pulmonary arterial myocytes. *Circ. Res.* 2002; 91: 719-726.

50. Yuan X-J. Voltage-gated K^+ currents regulate resting membrane potential and $[Ca^{2+}]i$ in pulmonary arterial myocytes. *Circ. Res.* 1995; 77: 370-378.
51. Zhao L, al-Tubuly R, Sebkhi A, Owji AA, Nunez DJ, and and Wilkins MR. Angiotensin II receptor expression and inhibition in the chronically hypoxic rat lung. *Br. J. Pharmacol.* 1996; 119: 1217-1222.

Chapter 6

Calcium Mobilization by Hypoxia in Pulmonary Artery Smooth Muscle

A. Mark Evans and Michelle Dipp
University of St Andrews, St Andrews, Fife, United Kingdom

1. Introduction

The anatomical drawings of the thirteenth century Islamic physician, Ibn Nafis, were the first to describe the pulmonary circulation (3). Four centuries later, William Harvey's experiments, recorded in De Motu Cordis (1628), confirmed that blood flowed through the lungs (64). The work of Nafis and Harvey provided the foundation on which further research flourished, and in 1661, Malpighi linked the right and left sides of the heart with his discovery of pulmonary capillaries (40).

Awareness of hypoxic pulmonary vasoconstriction (HPV) dawned in 1894, when Bradford and Dean recorded a rise in pulmonary arterial pressure upon asphyxia (8). However, it was another 50 years before von Euler and Liljestrand (19) showed that hypoxia without hypercapnia induced constriction within the pulmonary circulation. In fact, von Euler and Liljestrand were the first to hypothesize that HPV might aid ventilation perfusion matching at the alveoli, by diverting blood flow away from poorly ventilated areas of the lung. Thus, HPV was recognized as the critical and distinguishing characteristic of pulmonary arteries. In contrast, systemic arteries dilate in response to tissue hypoxemia, in order to match local perfusion to local metabolism (55).

When large areas of the lung are exposed to alveolar hypoxia, as is the case in diseases such as emphysema, obstructive airways disease, sleep apnea and cystic fibrosis, widespread HPV is triggered, which leads to hypoxic pulmonary hypertension (HPH) and eventually right heart failure (61). At present, the precise mechanisms that promote HPV remain obscure and although inhalation therapy (e.g., nitric oxide, O_2) and prostacyclin (PGI_2) injection have been considered, current therapies for HPH are poor (17). It is for this reason, in addition to scientific curiosity, that investigations on the mechanisms of HPV continue.

2. Nervous Regulation of Hypoxic Pulmonary Vasoconstriction

Investigations on the innervation of the pulmonary vasculature date back to the work of Bradford and Dean in 1894 (8), which demonstrated that gross excitation of the spinal cord caused vasorelaxation within the systemic circulation, without effect on the pulmonary circulation. Subsequently, Nissel (45) provided significant progress in our understanding of the basic mechanisms of HPV. In short, Nissel demonstrated that HPV was a local response largely, or entirely, independent of the autonomic nervous system. The limited influence on HPV of the innervation within the pulmonary vasculature was confirmed by denervation. First, chemical sympathectomy (using 6-hydroxy-dopamine) in the canine pulmonary vasculature was shown to be without effect on HPV (25). Naeije et al. (43) confirmed this finding, and also demonstrated that surgical denervation of the carotid and aortic chemoreceptors permitted HPV. More recently, the rise in pulmonary arterial pressure associated with HPV was shown to remain unchanged after bilateral cervical vagotomy (38).

Importantly, bilateral lung transplants allowed for the study of HPV after denervation in man. In these patients, HPV persists (54). We can conclude, therefore, that neither central nor local regulation of the autonomic nervous system plays a role in mediating HPV.

3. The Site of Hypoxic Pulmonary Vasoconstriction

In 1951, Duke demonstrated that HPV was not induced when the lung was perfused with hypoxic blood at a constant, normoxic alveolar O_2 tension (Po_2) (16). Later work confirmed that the Po_2 of the perfusate was not the determining factor, and instead that a fall in alveolar Po_2 consistently triggered a pronounced increase in pulmonary vascular perfusion pressure (4). Clearly, therefore, HPV is triggered by a reduction in O_2 to the airways and/or the alveoli. Thus, airway hypoxia, and not the Po_2 of the perfusate, should be the stimulus of choice for studies of HPV in isolated lungs.

Consistent with the finding that a fall in Po_2 at the terminal airways and alveoli triggers HPV, Kato and Staub (32) demonstrated, using unilobar hypoxia in the feline lung, that the small pre-capillary resistance arteries contributed most to the increase in pulmonary vascular perfusion pressure during alveolar hypoxia. This finding was confirmed by subsequent investigations (4). Furthermore, the magnitude of HPV in isolated pulmonary arteries, was also found to be inversely related to pulmonary artery diameter (32, 35). It seems likely, therefore, that the principal mechanism(s) that drives HPV would offer a degree of selectivity for small versus large pulmonary arteries, coupled with an even greater degree of selectivity for pulmonary versus systemic arteries.

4. Characteristics of the Hypoxic Response in Isolated Pulmonary Arteries

In isolated pulmonary arteries, HPV is biphasic when induced by switching from a normoxic to a hypoxic gas mixture (Fig. 1A). In agreement with Ward and co-workers (35, 51), we find that hypoxia induces an initial transient constriction (phase 1), followed by a slow tonic constriction (phase 2) (12, 13, 66). The initial transient constriction peaks within 5-10 min of the hypoxic challenge, whilst the secondary tonic constriction peaks after 30-40 min (12, 13, 35, 51, 66). Others describe a transient, monophasic constriction of isolated arteries by hypoxia, which has been referred to by some as biphasic (11, 23, 29, 69). Importantly, the time course of the records of HPV presented in the latter investigations, and/or the duration of exposure to hypoxia, indicate that they describe the initial phase 1 constriction only.

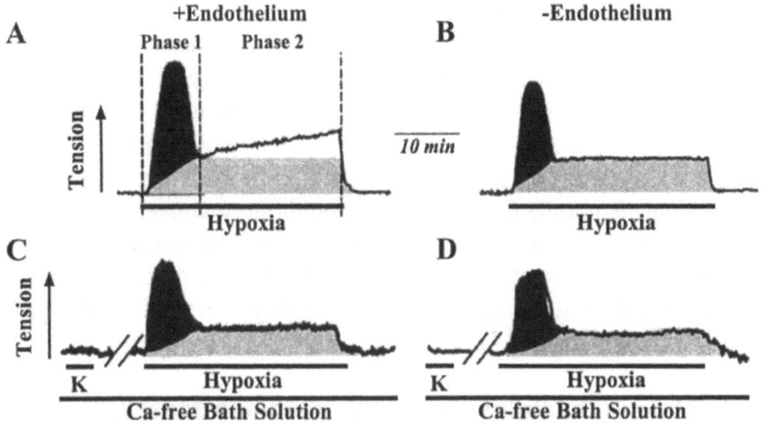

Figure 1. The effect on HPV of removing the endothelium and extracellular Ca^{2+}. A: A representative record indicating phase 1 and phase 2 of the response of an intact pulmonary artery ring to hypoxia. B: Constriction by hypoxia of a pulmonary artery ring without endothelium. C: Constriction by hypoxia of an intact pulmonary artery ring in the absence of extracellular Ca^{2+}. D: Constriction by hypoxia of a pulmonary artery ring without endothelium and in the absence of extracellular Ca^{2+}.

These variations in qualitative description of HPV demonstrate the significance of the basic methodology to our interpretation of events. We should be acutely aware, therefore, of the exact parameters within which each experiment was carried out before arriving at any definitive conclusion as to the mechanisms and/or mediators involved.

5. The Role of Pulmonary Artery Smooth Muscle and Endothelium

In 1976, Fishman questioned whether hypoxia induced constriction by modulating the release of local mediators, or by acting directly on vascular smooth muscle (20). This question has focused the minds of researchers in the field ever since. Consideration of the full facts before us now leaves us in no doubt that both local mediators and direct effects of hypoxia on the smooth muscle contribute to HPV.

5.1. The Pulmonary Artery Endothelium

The first clear evidence for endothelium-derived mediators in the hypoxic response was provided by Holden and McCall (27). Subsequent investigations provided support for this proposal, in that HPV was found to be attenuated following removal of the pulmonary artery endothelium (11-13, 28, 33, 35). These findings and those of others led to a general consensus in favor of a role for an endothelium-derived vasoconstrictor as a mediator of HPV (14, 18, 21, 39, 53; see Chapter 12). Significantly, and in marked contrast to our findings with respect to pulmonary artery smooth muscle (see below), our recent studies suggest that hypoxia releases an endothelium-derived vasoconstrictor by triggering transmembrane Ca^{2+} influx into pulmonary artery endothelial cells (Fig. 1D) (13). Furthermore, the endothelium-derived vasoconstrictor appears to promote HPV not by raising cytoplasmic Ca^{2+} concentration in the smooth muscle, but by inducing an increase in the sensitivity of the contractile apparatus to Ca^{2+} (51, 53; see Chapter 7).

5.2. The Role of the Pulmonary Artery Smooth Muscle

In 1976, McMurtry et al. showed that Ca^{2+} channel antagonists inhibited HPV in isolated rat lungs, indicative of a role for voltage-gated Ca^{2+} channels in HPV (42). Given that we now know that voltage-gated Ca^{2+} channels are not present in arterial endothelial cells, at least in resistance vessels, and that Ca^{2+} channel antagonists have little effect on the endothelium-dependent component of HPV (52), this finding may well have represented the first evidence in support of a direct effect of hypoxia on the smooth muscle. Further support for this conclusion may be derived from a variety of studies which have demonstrated that hypoxia causes constriction in isolated pulmonary arteries without endothelium (12, 13, 23, 28, 66). Moreover, hypoxia has been found to increase cytoplasmic Ca^{2+} concentration and cause contraction in isolated pulmonary artery smooth muscle cells (10, 13, 23, 58).

A physiological role for direct regulation of pulmonary artery smooth muscle by hypoxia has, however, been challenged. Firstly, the physiological significance of the transient phase 1 constriction of isolated pulmonary arteries by hypoxia was questioned, despite the fact that the transient constriction is consistently observed within the first 5-10 min of a hypoxic challenge in arteries with and without the endothelium (12, 13). When induced in the isolated lung, however, HPV may appear monophasic or biphasic, depending on the experimental conditions (e.g., nature of perfusate, basal perfusion pressure, and severity of hypoxic insult) (12, 61, 64). The relative contribution of the mechanisms that underpin the transient and maintained phase of pulmonary artery constriction by hypoxia, respectively, may therefore vary in a manner dependent on compounding physiological parameters.

Until recently it had also been argued that maintained (>15 min) HPV could not be induced in arteries without endothelium (1, 11, 27, 33, 35). It is important to note, however, that in all of these investigations, isolated pulmonary arteries had to be pre-constricted in order to record any constriction by hypoxia. This is a significant point, because HPV is quite rightly viewed as a vital homeostatic mechanism and, therefore, should be observed in the absence of pre-constriction. Consistent with this proposal, we have recently recorded maintained HPV in isolated pulmonary arteries in the absence of pre-constriction, both with and without endothelium (Fig. 1A and B) (12, 13, 66). These studies were the first to report that arteries without endothelium, when exposed to hypoxia, exhibit a transient constriction which declines to a plateau above baseline (Fig. 1C and D) (12, 13, 66). Furthermore, this plateau constriction was maintained whatever the duration of exposure to hypoxia, but declined rapidly to baseline, as one would expect, upon normoxia. There is compelling evidence, therefore, to support our view that hypoxia triggers maintained constriction of pulmonary arteries, in part, by acting directly on the smooth muscle.

6. Regulation by Hypoxia of Ca^{2+} Influx into Pulmonary Artery Smooth Muscle

In 1985, Harder et al. reported that hypoxia induced smooth muscle cell depolarization and concomitant constriction in isolated pulmonary arteries (26). The force generated was dependent on the extracellular Ca^{2+} concentration, and both depolarization and constriction were blocked by the Ca^{2+} channel antagonist, verapamil. It is notable that these investigations identified no change in smooth muscle K^+ permeability. It was concluded, therefore, that hypoxia activated a Ca^{2+} conductance, and that Ca^{2+} influx mediated both membrane depolarization and pulmonary vasoconstriction.

However, later electrophysiological investigations demonstrated that hypoxia inhibited a K^+ conductance in isolated pulmonary artery smooth muscle cells, and that hypoxia was without effect on the K^+ conductance in isolated systemic artery smooth muscle cells (47, 48, 50, 68). This led to the suggestion that HPV was initiated by smooth muscle cell depolarization due to inhibition of O_2-sensing K^+ channels and subsequent Ca^{2+} influx via voltage-gated Ca^{2+} channels. Initially, it was proposed that hypoxia inhibited Ca^{2+}-activated K^+ (BK_{Ca}) channels. In contrast, subsequent reports indicated that voltage-gated K^+ channels might underpin the principal O_2-sensitive K^+ conductance (47, 48, 56, 63, 68). Despite extensive research on this theme, however, recent investigations suggest that the inhibition of O_2-sensing K^+ channels by hypoxia does not trigger acute HPV, or contribute significantly to its maintenance (14, 18, 52). Moreover, the use of voltage-gated Ca^{2+} channel antagonists has proved disappointing in the treatment of HPH, and may even increase ventilation-perfusion mismatch (44).

7. Regulation by Hypoxia of Ca^{2+} Mobilization from Sarcoplasmic Reticulum Ca^{2+} Stores in Pulmonary Artery Smooth Muscle

Our most recent findings are contrary to the view that inhibition of O_2-sensing K^+ channels represents the primary trigger for HPV, and these findings question the pre-eminence of the endothelium in mediating maintained HPV. Rather, they point to a pivotal role for Ca^{2+} release from sarcoplasmic reticulum (SR) stores in the smooth muscle. Previous investigations on isolated pulmonary artery smooth muscle cells have reported an increase in intracellular Ca^{2+} by hypoxia after removal of extracellular Ca^{2+} (23, 30, 58). Furthermore, isolated pulmonary arteries have been shown to constrict in response to hypoxia in the absence of extracellular Ca^{2+} (23, 28), and evidence has been provided to support a role for SR Ca^{2+} release in mediating HPV (23, 29, 58). It is important to note, however, that the duration of exposure to hypoxia in all of the aforementioned studies was sufficient only to observe the initial transient constriction by hypoxia (i.e., phase 1). In short, these studies did not offer the opportunity to assess the mechanisms that underpin maintained HPV. In this respect, our investigations differed significantly.

7.1. The Effect of Removal of Extracellular Ca^{2+} on HPV in Isolated Arteries

In agreement with the investigations of others (23, 29, 58), and within a similar time scale, we find that hypoxia promotes phase 1 of HPV by stimulating Ca^{2+} release from ryanodine-sensitive SR stores in the smooth muscle (12, 13). However, we have now provided compelling evidence in support of a pivotal role for continued SR Ca^{2+} release in the maintenance of sustained HPV in isolated pulmonary arteries both with and without endothelium (12, 13, 66).

As mentioned previously, our experiments suggest that in arteries without endothelium the phase 1 constriction declines to a maintained plateau above baseline and that this plateau constriction remains throughout the period of exposure to hypoxia. Thus, the slow tonic constriction is endothelium-dependent (13), whilst phase 1 and a residual plateau constriction are likely mediated by mechanisms intrinsic to the smooth muscle (12, 13, 66). Even more surprising (considering the proposal that voltage-gated Ca^{2+} influx triggers HPV) (47, 48, 68) was our finding that HPV, in pulmonary arteries without endothelium, remained unaffected after removal of extracellular Ca^{2+} (Fig. 1C and D) (12). Importantly, and in marked contrast, constriction by K^+ (i.e., due to membrane depolarization and voltage-gated Ca^{2+} influx) was abolished (Fig. 1C and D) (13). Consistent with the idea that Ca^{2+} release from SR stores may mediate HPV, we found that depletion of SR Ca^{2+} stores by preincubation with ryanodine and caffeine abolished constriction by hypoxia (12, 13). Furthermore, we found constriction by K^+ to remain unaffected in the presence of ryanodine, caffeine, and hypoxia (12, 13). Clearly, these findings are in conflict with the proposal that HPV is triggered by inhibition of O_2-sensitive K^+ channels and voltage-gated Ca^{2+} influx (10, 47, 48, 50, 63, 68). On the contrary, they strongly suggest that hypoxia initiates (phase 1) and maintains (phase 2) acute HPV by triggering Ca^{2+} release from ryanodine-sensitive SR stores in the smooth muscle. Irrespective of this, our findings are consistent with the observation that Ca^{2+} channel antagonists may act as weak inhibitors of HPV (11, 29, 42, 58), because these compounds also inhibit SR Ca^{2+} release in arterial smooth muscle (see below) (57).

Our finding that a significant component of maintained HPV persists in the absence of extracellular Ca^{2+} may at first appear contrary to previous investigations, which showed that removal of extracellular Ca^{2+} markedly inhibits HPV. However, a number of investigations report a proportion of HPV to be insensitive to the removal of extracellular Ca^{2+} (23, 26, 28). Thus, our findings may be in closer agreement than may be apparent at first sight. This is due to the fact that we also found that removal of extracellular Ca^{2+} blocks the release of the endothelium-derived vasoconstrictor(s) by hypoxia (Fig. 1A, C and D) (13). Variations in the proportion of HPV which is sensitive to removal of extracellular Ca^{2+} may therefore be due to differences in experimental protocol, and hence variations in the relative contribution to HPV of the smooth muscle and of the endothelium, respectively (18).

It has been noted that an excess of pre-constriction, hypoxia or initial tension can lead to a reduction in tension to below baseline during phase 2 of HPV in isolated arteries (1). Consistent with these proposals, we find that applied pre-constriction reduces the component of maintained HPV that is dependent on SR Ca^{2+} release (13) and may therefore down-regulate SR Ca^{2+} release by hypoxia. Secondly, we find the relationship between the degree of hypoxia and constriction due to maintained SR Ca^{2+} release in smooth muscle to be bell-shaped (see below) (15). Thus, the contribution to HPV of smooth muscle SR Ca^{2+} release may vary depending on the degree of hypoxia applied. Finally, anaesthetics (e.g., pentobarbitone) inhibit and can even reverse HPV (46), and after pentobarbitone anaesthesia we find it impossible to induce HPV in isolated pulmonary arteries in the absence of pre-constriction (13,18). This may be due, in part, to disruption of Ca^{2+} signaling in the smooth muscle (13, 52). Taking the above into consideration, it is quite possible that when HPV is supported by pre-constriction in the absence or presence of pentobarbitone anaesthesia and/or extreme hypoxia, the balance of HPV may swing towards endothelium-dependent mechanisms of constriction. Clearly, under such conditions, maintained HPV could appear entirely dependent on extracellular Ca^{2+}, because the release of the endothelium-derived vasoconstrictor is driven by Ca^{2+} influx into endothelial cells (13).

At this juncture it is important to consider two further points. Firstly, when agents are used to pre-constrict pulmonary arteries, they may bias investigations towards the study of the modulation by hypoxia of those mechanisms activated by the applied vasoconstrictor, and away from the study of HPV *per se*. This is especially significant when using agents whose mechanism of constriction is unclear. Secondly, since HPV is a vital homeostatic mechanism, when hypoxia fails to constrict pulmonary arteries in the absence of pre-constriction in a given experimental model, it may be useful to focus on the physiological parameters of the model rather than artificial adjustment (e.g., pre-constriction) to procure a response. The significance of this point cannot be underestimated, given that the effects of hypoxia are likely mediated by changes in the metabolic status within O_2-sensing cells (see Chapter 18).

8. Selective Block of Phase 1 of Hypoxic Pulmonary Vasoconstriction in Isolated Arteries

Many studies have shown that phase 1 of HPV may be inhibited by Ca^{2+} channel blockers (11, 32, 42, 58), or by depletion of ryanodine-sensitive SR Ca^{2+} stores (12,13, 23, 29, 58). More recently, it has been proposed that inhibition of SR Ca^{2+} ATPase activity (12, 52) and subsequent activation of Ca^{2+} entry via a store-refilling current may underpin phase 1 of HPV (52).

Our investigations proved to be consistent with all of the above except for a role for Ca^{2+} entry via a store-refilling current. As mentioned previously, we found that depletion of ryanodine-sensitive SR Ca^{2+} stores abolished both phase 1 and the maintained plateau constriction in pulmonary arteries without endothelium (12, 13). In contrast, however, other pharmacological manoeuvres had discrete effects on phase 1 of HPV.

We found phase 1, but not phase 2 of HPV to be abolished by pre-incubation of isolated pulmonary arteries with nifedipine, a Ca^{2+} channel antagonist, in the absence of extracellular Ca^{2+} (Fig. 2C*a*) (Dipp and Evans, unpublished data). Thus, nifedipine is most likely blocking SR Ca^{2+} release by hypoxia, as predicted by previous investigations on systemic artery smooth muscle (57). Whatever the mechanism, however, nifedipine is clearly without effect on maintained smooth muscle constriction by hypoxia, and is therefore unable to block maintained smooth muscle SR Ca^{2+} release by hypoxia.

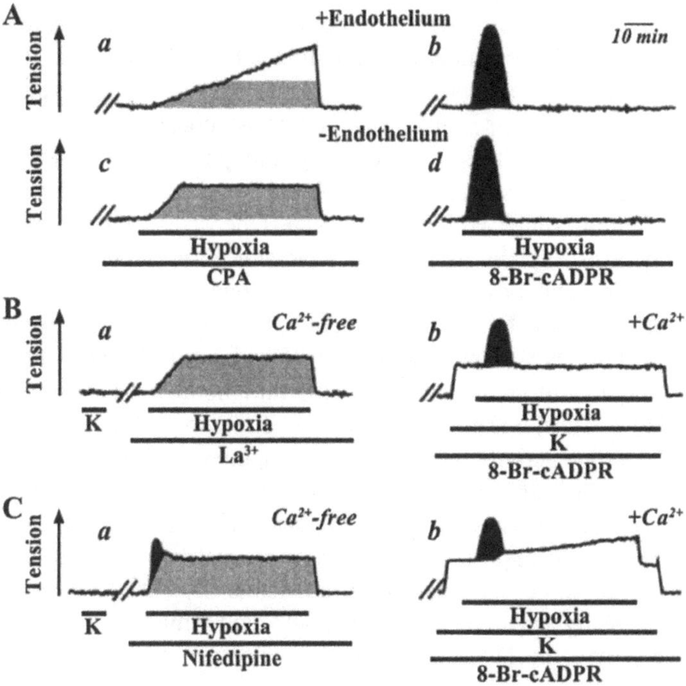

Figure. 2. Pharmacological separation of 3 components of HPV in isolated pulmonary arteries. Black indicates component 1, grey indicates component 2, and white indicates component 3. A: Constriction by hypoxia of an artery ring with (*a* and *b*) or without © and *d*) endothelium in the presence of cyclopiazonic acid (*a* and *c*) or 8-bromo-cADPR (*b* and *d*). B: Constriction by hypoxia of an artery ring without endothelium in the absence of extracellular Ca^{2+} and in the presence of La^{2+} (*a*) or in the presence of 8-bromo-cADPR and K^{+}-induced pre-constriction (*b*). C: Constriction by hypoxia of an artery ring with endothelium in the absence of extracellular Ca^{2+} and in the presence of nifedipine (*a*) or in the presence of 8-bromo-cADPR and K^{+}-induced pre-constriction (*b*).

Consistent with the findings of Robertson et al. (52), we also found that pre-incubation of pulmonary arteries with an SR Ca^{2+} ATPase antagonist, cyclopiazonic acid, abolished phase 1 of HPV, and without effect on maintained smooth muscle constriction by hypoxia (Fig. 2A*a* and A*c*) (12). In contrast to Robertson et al. (52), however, we find no evidence for the involvement of capacitative Ca^{2+} entry during phase 1, because removal of extracellular Ca^{2+} under the conditions of our experiments had no effect on HPV in arteries without endothelium (Fig. 1C) (13). Moreover, we found that lanthanum, which has been used as a selective blocker of the store-refilling current (52), also blocked phase 1 of HPV after removal of extracellular Ca^{2+}, raising the possibility that lanthanum is also capable of inhibiting a component of SR Ca^{2+} release at relatively low concentrations (Fig. 2B*a*) (Dipp and Evans, unpublished data). Despite these facts it is important to note that store refilling by some route likely occurs during prolonged exposure to hypoxia (60), will certainly be required to maintain SR Ca^{2+} release, and may be modulated by hypoxia (31).

Contrary to our findings, Robertson et al. (52) found no evidence of a ryanodine- and caffeine-sensitive component of constriction. We have no concrete explanation for these differences, but they may result from the use of pre-constriction, pentobarbitone anaesthesia, and/or the "physiological" parameters within which HPV was induced. One or a combination of these factors may also explain the varied dependence of phase 1 of HPV on extracellular Ca^{2+} under different experimental conditions.

Strikingly, we also found cyclopiazonic acid to be without effect on phase 2 of HPV in arteries without endothelium (Fig. 2A*c*) (12), even though Ca^{2+} release from ryanodine-sensitive SR stores in the smooth muscle underpins both phases of HPV. Thus, it is possible that phase 1 may be mediated by inhibition of SR Ca^{2+} ATPase activity by a fall in ATP supply (36, 59), and a consequent increase in net Ca^{2+} efflux from the SR via ryanodine receptors (RyRs). To achieve this whilst allowing for a second phase of maintained SR Ca^{2+} release, however, one would require the presence of at least two spatially segregated intracellular Ca^{2+} stores each served by a discrete SR Ca^{2+} ATPase (SERCA) subtype, one being sensitive to cyclopiazonic acid whilst the other is not (7). Alternatively, other mechanisms sensitive to cyclopiazonic acid may be involved (see below).

9. Cyclic Adenosine Diphosphate-Ribose (cADPR), SR Ca^{2+} Release and Maintained Hypoxic Pulmonary Vasoconstriction in Isolated Pulmonary Arteries

Given our finding that maintained HPV in arteries without endothelium was abolished by depletion of ryanodine-sensitive intracellular stores, we considered the possibility that an endogenous regulator of RyRs may be involved. This led us to investigate the role of a novel Ca^{2+} mobilizing pyridine nucleotide, namely cyclic adenosine diphosphate ribose (cADPR) (37). The enzymes that synthesize (ADP-ribosyl cyclase/CD38) and metabolize (cADPR hydrolase) cADPR are present in many cell types including arterial smooth muscle (37, 66), and, significantly, an elevation in cADPR levels has previously been shown to sensitize Ca^{2+}-induced Ca^{2+} release via RyRs in the SR and to induce SR Ca^{2+} release by RyR activation (22). Furthermore, we were struck by the fact that cADPR was a β-NAD^+ metabolite, since hypoxia had been shown to increase β-NADH levels in all O_2-sensing cells studied to date (5, 59, 67). This offered the possibility that cADPR synthesis itself may, in some way, be sensitive to changes in the metabolic state of pulmonary artery smooth muscle.

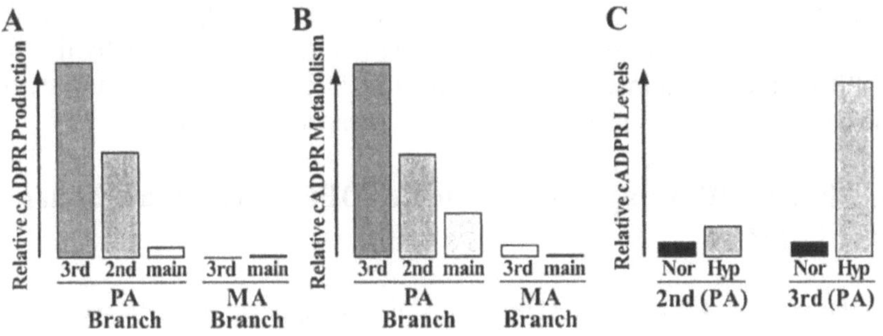

Figure 3. Synthesis and metabolism of cADPR in pulmonary arteries and the effect of hypoxia. A and B: Synthesis of cADPR from β-NAD^+ (A) and metabolism of cADPR (B) in smooth muscle homogenates from a series of pulmonary (PA) and mesenteric (MA) arteries. C: Relative cADPR levels in 2nd or 3rd order branches of the pulmonary artery in the presence of normoxia and hypoxia, respectively.

9.1. ADP-ribosyl Cyclase and cADPR Hydrolase Activities are Differentially Distributed in Pulmonary Versus Systemic Artery Smooth Muscle

Given the fact that HPV is the critical and distinguishing characteristic of pulmonary artery smooth muscle, our initial investigations into the role of cADPR in HPV focused on measurements of the enzyme activities for the synthesis and metabolism of cADPR in pulmonary versus systemic artery smooth muscle. Our findings were striking, in that the enzyme activities for the synthesis and metabolism of cADPR were at least an order of magnitude higher in homogenates of pulmonary artery smooth muscle than in those of aortic or mesenteric artery smooth muscle (Fig. 3A and B) (66). Thus, the differential distribution of these enzyme activities offered the pulmonary selectivity required of a mediator of HPV. Of perhaps even greater significance was the finding that the level of these enzyme activities was inversely related to pulmonary artery diameter (Fig. 3A and B) (66), given that the magnitude of constriction by hypoxia has also been shown to be inversely related to pulmonary artery diameter (32, 35).

9.2. Hypoxia Increases cADPR Content in Pulmonary Arterial Smooth Muscle

The possibility that hypoxia may promote HPV, in part, by increasing cADPR accumulation in pulmonary artery smooth muscle, gained strong support from direct measurements of cADPR content using a [^{32}P]cADPR binding assay. Hypoxia (16-21 Torr) increased cADPR levels 2-fold in 2nd order branches of the pulmonary artery, and 10-fold in 3rd order branches (Fig. 3C) (66). Thus, like constriction by hypoxia and the distribution of the enzyme activities for cADPR synthesis and metabolism, the increase in cADPR content induced by hypoxia was inversely related to pulmonary artery diameter (66).

9.3. The cADPR Antagonist 8-Bromo-cADPR Blocks Phase 2 but not Phase 1 of HPV

The effects of 8-bromo-cADPR, a cADPR antagonist, on HPV were quite different from the effects of ryanodine and caffeine, cyclopiazonic acid, nifedipine and lanthanum (La^{3+}), respectively (Fig. 2). In isolated pulmonary arteries with and without the endothelium, 8-bromo-cADPR had no effect on phase 1 of HPV. However, it abolished phase 2 in the presence of the pulmonary artery endothelium, and blocked the maintained plateau constriction observed in arteries without endothelium (Fig. 2A*b* and A*d*, respectively) (12, 66). Thus, while cADPR-dependent SR Ca^{2+} release may maintain acute HPV in isolated pulmonary artery rings, it does not mediate the phase 1 constriction (12, 66). This is the exact reverse of the effect of cyclopiazonic acid, and is consistent with our proposal that there may be two O_2-sensitive mechanisms of SR Ca^{2+} release in pulmonary artery smooth muscle.

Because 8-bromo-cADPR abolished phase 2 of HPV in pulmonary arteries with the endothelium intact, this raised the possibility that 8-bromo-cADPR also blocked the release or action of the endothelium-derived vasoconstrictor(s). However, when we pre-constricted arteries by depolarization using a submaximal concentration of K^+ (i.e., by Ca^{2+} from a source independent of the SR), the slow tonic constriction associated with phase 2 of HPV was recovered (Fig. 2B*b*) (12,66). In marked contrast, the plateau constriction induced by hypoxia in arteries without endothelium was not recovered with K^+-induced pre-constriction in the continued presence of 8-bromo-cADPR (Fig. 2C*b*) (12). Collectively, these data suggest that an increase in cADPR levels mediates maintained smooth muscle SR Ca^{2+} release by hypoxia, but is not a pre-requisite for vasoconstrictor release from pulmonary artery endothelial cells (12). The block of the endothelium-dependent component of HPV by 8-bromo-cADPR is therefore indirect.

9.4. 8-Bromo-cADPR Inhibits Hypoxic Pulmonary Vasoconstriction in the Rat Lung *in-situ*

When added to the perfusate of the isolated, ventilated and perfused rat lung *in-situ*, the cADPR antagonist, 8-bromo-cADPR, abolished acute HPV induced by alveolar hypoxia (2% O_2) (12). Thus, an increase in smooth muscle cADPR levels, and cADPR-dependent SR Ca^{2+} release may be the primary trigger for HPV by alveolar hypoxia in the lung. This is evidenced by the pharmacological separation of the three primary components of acute HPV in isolated pulmonary arteries (Figs. 1 and 2): *a*) The phase 1 constriction of the smooth muscle driven by a cADPR-independent mechanism of SR Ca^{2+} release; *b*) The plateau constriction associated with phase 2 of HPV and mediated by cADPR-dependent Ca^{2+} release via RyRs in the smooth muscle SR stores, and *c*) The slow tonic constriction during phase 2 of HPV that is mediated by the release of an endothelium-derived vasoconstrictor. Because component 1 and component 3 are insensitive to 8-bromo-cADPR, 8-bromo-cADPR likely blocks HPV in the lung by selectively inhibiting cADPR-dependent SR Ca^{2+} release in the smooth muscle.

Because phase 1 of HPV in isolated pulmonary arteries remains unaffected in the presence of 8-bromo-cADPR, the mechanisms involved do not appear to contribute to HPV in the lung under the conditions of our experiments. It is surprising, therefore, that phase 1 in isolated pulmonary arteries is so pronounced. One explanation for this could be that the fall in Po_2 around an isolated pulmonary artery in the sealed experimental chamber was faster than it was when associated with alveolar hypoxia in the lung. This may result in a more pronounced fall in smooth muscle ATP levels in isolated pulmonary arteries (36, 59), and greater inhibition of SR Ca^{2+} ATPase (SERCA) activity. Consequently, this may trigger a greater rate of Ca^{2+} release from the SR in isolated pulmonary arteries than observed in the lung. It is possible, therefore, that either a monophasic cADPR-dependent pulmonary vascular smooth muscle constriction, or a biphasic vasoconstriction (12, 64, 71) may be triggered in the lung in a manner dependent on the experimental conditions (e.g., perfusion pressure, severity of oxidative stress).

In addition, our findings strongly suggest that although physiological concentrations of the endothelium-derived vasoconstrictor may sensitize the pulmonary vascular smooth muscle contractile apparatus to Ca^{2+} (14, 18, 51, 53), Ca^{2+}-sensitization by this mechanism alone may not be sufficient to promote acute HPV in the absence of cADPR-dependent SR Ca^{2+} release from the smooth muscle SR stores. Our findings also suggest that changes in levels of other vasoactive substances (e.g., nitric oxide) may act as secondary modulators of HPV.

10. Metabolic Regulation of cADPR Accumulation

Our findings and those of others offer two possible mechanism by which hypoxia may regulate cADPR synthesis and thereby promote HPV. Our findings suggest that an increase in cADPR levels by hypoxia may be mediated, in part, by an increase in β-NADH levels, which has been shown to occur in pulmonary artery smooth muscle (59). Alternatively, cADPR synthesis could be increased as a result of superoxide generation by mitochondria (34), or by a previously unidentified mechanism (see Chapter 18).

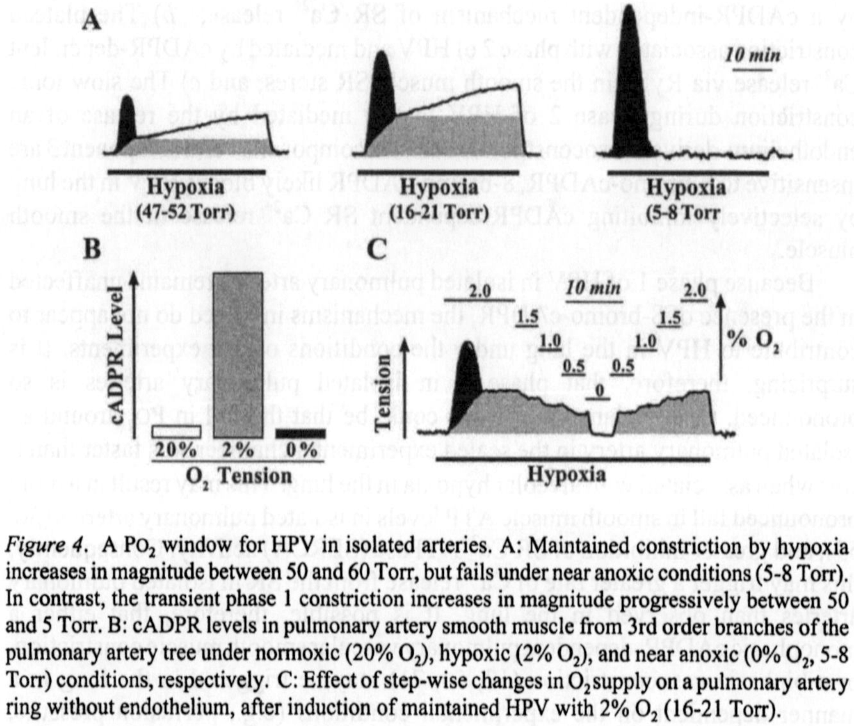

Figure 4. A PO_2 window for HPV in isolated arteries. A: Maintained constriction by hypoxia increases in magnitude between 50 and 60 Torr, but fails under near anoxic conditions (5-8 Torr). In contrast, the transient phase 1 constriction increases in magnitude progressively between 50 and 5 Torr. B: cADPR levels in pulmonary artery smooth muscle from 3rd order branches of the pulmonary artery tree under normoxic (20% O_2), hypoxic (2% O_2), and near anoxic (0% O_2, 5-8 Torr) conditions, respectively. C: Effect of step-wise changes in O_2 supply on a pulmonary artery ring without endothelium, after induction of maintained HPV with 2% O_2 (16-21 Torr).

11. PO_2 Dependence of Hypoxic Pulmonary Vasoconstriction

Our previous investigations established that hypoxia fails to induce constriction at PO_2>60 Torr, and that the magnitude of HPV increases in a manner dependent on the severity of hypoxia between 60 and 16 Torr, as one would expect (Fig. 4A) (12). However, further consideration of cADPR metabolism led us to propose that there may be a PO_2 window within which hypoxia can promote cADPR synthesis, cADPR-dependent SR Ca^{2+} release and, therefore, maintained HPV (18, 66). This is evident from the fact that cADPR is a β-NAD$^+$ metabolite, whilst β-NADH is a poor substrate for cADPR synthesis in pulmonary artery smooth muscle (66). Therefore, as β-NADH levels increase progressively with the severity of hypoxia there will be a consequent fall in β-NAD$^+$ levels, the substrate for cADPR. Thus, at a given PO_2, we would expect a fall in cADPR levels due to reduced substrate availability at a point determined by the kinetics of cADPR synthesis (see Chapter 18). Our most recent findings support this proposal. When we stepped from 20% O_2 to 0% O_2 (near anoxic conditions, 5-8 Torr) the transient phase 1 constriction increased in magnitude by approximately 10%. Surprisingly, however, phase 2 of HPV (i.e., maintained HPV) was abolished (Fig. 4A). This was associated with a concomitant reduction in cADPR accumulation in the smooth muscle under near anoxic conditions (5-8 Torr, Fig. 4B) (15), consistent with the view that the plateau constriction by hypoxia is mediated by cADPR-dependent SR Ca^{2+} release, and that this is required to support further endothelium-dependent constriction by hypoxia (15). Furthermore, when we induced maintained HPV in arteries without endothelium, by stepping from 20% O_2 to 2% O_2, further reductions in PO_2 resulted in a reversible decline in the maintained cADPR-dependent plateau constriction (Fig. 4C) (15). It would appear, therefore, that there is a PO_2 window within which cADPR accumulation in pulmonary artery smooth muscle and maintained HPV may be induced by smooth muscle SR Ca^{2+} release by hypoxia. We cannot be certain of the physiological relevance of the PO_2 window. Given that HPV is a vital homeostatic mechanism, however, it would seem sensible to have a failsafe. Thus a PO_2 window for HPV may protect against a life-threatening fall in O_2 supply to the airways. Under these conditions, failure of maintained HPV would maximize gaseous exchange and thereby help preserve life.

The presence of a PO_2 window for HPV is not, however, consistent with a role for O_2-sensing K^+ channels. This is clear from the fact that inhibition of K^+ currents by hypoxia is progressive, and peaks under near anoxic conditions (47), i.e., when maintained HPV fails.

12. Inhibition of cADPR-dependent Dilation by Hypoxia

Cyclopiazonic acid inhibits phase 1 of HPV without any effect on maintained constriction by hypoxia. We mentioned at the time that cyclopiazonic acid may act to deplete a Ca^{2+} store that is functionally discrete from that which serves cADPR-dependent SR Ca^{2+} release by hypoxia. Our most recent investigations provide strong support for this conclusion. We have demonstrated that cADPR mediates, in part, vasodilation by adenylyl cyclase-coupled receptors, such as β-adrenoceptors, by releasing Ca^{2+} from a cyclopiazonic acid-sensitive SR compartment proximal to the plasma membrane, and possibly via the activation of a discrete subtype of RyRs (7). This subplasmalemmal Ca^{2+} release recruits BK_{Ca} channels, leading to hyperpolarization and, ultimately, vasodilation (7). It is possible, therefore, that SR Ca^{2+} release by hypoxia serves two purposes. Hypoxia may primarily trigger constriction by cADPR-dependent Ca^{2+} release from a central SR compartment that is in close apposition to the contractile apparatus, and is served by a cyclopiazonic acid-insensitive SERCA. A secondary action of hypoxia may be, however, to deplete peripheral SR compartments that are in close apposition to the plasma membrane, served by a cyclopiazonic acid-sensitive SERCA, and which normally mediate vasodilation by releasing Ca^{2+} proximal to the plasma membrane. Consistent with this proposal, is our finding that cyclopiazonic acid induces a transient constriction in pulmonary arteries (7). Further support for this proposal may also be derived from the fact that hypoxia inhibits pulmonary vasodilation by β-adrenoceptors (41). Such inhibition may be facilitated if the metabolic status proximal to the plasma membrane confers a discrete Po_2 window for cADPR accumulation by limiting substrate (β-NAD^+) supply under hypoxic conditions (see Chapter 18). It is possible, therefore, that these processes may lead to a secondary reduction in BK_{Ca} conductance by hypoxia, as described by Post et al. (50).

Regulation by cADPR of Ca^{2+} release from an SR compartment proximal to the plasma membrane may, however, have wider implications. For example, a Po_2 window for cADPR accumulation by hypoxia may also underpin O_2-induced pulmonary artery dilation after birth, which has been shown to be mediated, at least in part, by opening of BK_{Ca} channels and smooth muscle hyperpolarization (49). In addition, the relatively low level of the enzyme activities for cADPR synthesis and metabolism in systemic artery smooth muscle, may point to a role for cADPR-dependent recruitment of BK_{Ca} channels in hypoxia-induced systemic artery dilation. If this is the case, however, the Po_2-dependence of cADPR accumulation, the spatial organization of the Ca^{2+} mobilizing pathways, and the ability of systemic artery smooth muscle to retain peripheral SR Ca^{2+} stores under hypoxia may also be determining factors.

13. Is There a Role for Other Ca^{2+} Mobilizing Messengers in Hypoxic Pulmonary Vasoconstriction?

We have recently demonstrated that nicotinic acid adenine dinucleotide phosphate (NAADP), a β-$NADP^+$ metabolite (37), is a potent Ca^{2+} mobilizing messenger in pulmonary artery smooth muscle (6). NAADP appears to trigger spatially restricted Ca^{2+} bursts from a thapsigargin-insensitive store, which may be a lysosomal-related organelle (9). These 'Ca^{2+} bursts' are then amplified into a global Ca^{2+} wave by Ca^{2+}-induced Ca^{2+} release (CICR) via RyRs in the SR, leading to smooth muscle contraction (6). Given that cADPR sensitizes RyRs to CICR, these findings raise the possibility that NAADP may act in concert with cADPR to promote HPV. Such tight coupling of NAADP-dependent CICR to cADPR levels could also explain the all-or-none block of HPV by the cADPR antagonist 8-bromo-cADPR (12). However, further investigations are required to determine whether or not NAADP plays a role in HPV.

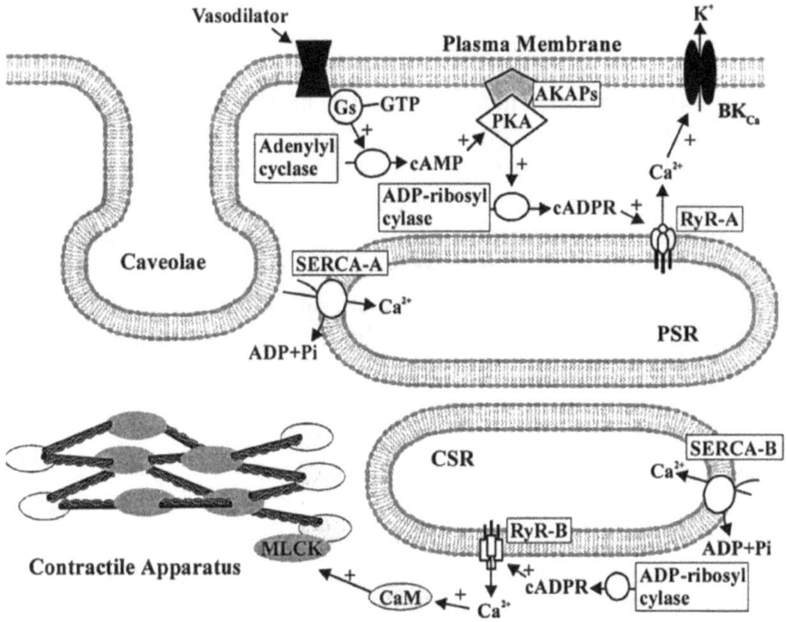

Figure 5. Proposed spatial organization of two functionally discrete pathways of cADPR-dependent SR Ca^{2+} release in pulmonary artery smooth muscle. MLCK, myosin light chain kinase; SERCA, SR Ca^{2+} ATPase subtype; PKA, protein kinase A; AKAPS, A kinase anchoring proteins; G_s, adenylyl cyclase-coupled heterotrimeric G-protein; BK_{Ca}, Ca^{2+}-activated K^+ channel; cADPR, cyclic adenosine diphosphate ribose; PSR, peripheral SR; CSR, central SR.

14. Summary

It is now clear that HPV is a multifactorial process. In isolated arteries an initial transient constriction is induced, in part, by depletion of an SR compartment proximal to the plasma membrane, and may lead to consequent inhibition of vasodilation by adenylyl cyclase coupled receptors. Depletion of this SR compartment may result from inhibition of a cyclopiazonic acid-sensitive SERCA pump and/or an as yet unidentified mechanism of SR Ca^{2+} release. Concomitant, cADPR-dependent Ca^{2+} release via RyRs in a central SR compartment may then promote maintained constriction by hypoxia (Fig. 5). Further progression of HPV will then be ensured by transmembrane Ca^{2+} influx into pulmonary artery endothelial cells, and subsequent vasoconstrictor release; whilst cADPR is not a pre-requisite for activation of the endothelium we cannot, however, rule out a modulatory role.

Despite the fact that acute hypoxia depresses membrane K^+ conductance in pulmonary artery smooth muscle cells, we find no evidence of a primary role for membrane depolarization in acute HPV. It seems likely, therefore, that this effect may serve to increase the susceptibility of smooth muscle to depolarization and/or to limit, by depolarization, Ca^{2+} sequestration by the Na^{+}/Ca^{2+} exchanger. However, the impact of reduced K^{+} channel expression and depolarization in pulmonary artery smooth muscle may be more significant during the development of chronic HPV (see Chapter 10).

Clearly, our studies and those of others have identified numerous and divergent pathways that may contribute in some way to HPV. Each of these divergent processes is likely regulated by a "primary metabolic sensor" and/or the subsequent activation of intermediates within a bifurcating enzymatic cascade. In the field of HPV, the principal challenge for the future must, therefore, be to identify this "primary metabolic sensor" and to characterize the signal transduction pathway (see Chapter 18).

Acknowledgments

We would like to thank Professor Antony Galione and Dr. Piers Nye for helpful discussion throughout our investigations described. In addition, we would like to thank the Wellcome Trust for providing financial support.

References

1. Aaronson PI, Robertson TP, and Ward JPT. Endothelium-derived mediators and hypoxic pulmonary vasoconstriction. *Resp. Physiol. Neurobiol.* 2002; 132:107-120.
2. Archer SL, Huang J, Henry T, Peterson D, and Weir EK. A redox-based O_2 sensor in rat pulmonary vasculature. *Circ. Res.* 1993; 73:1100-1112.
3. Batirel HF. Early Islamic physicians and thorax. *Ann. Thorac. Surg.* 1999; 67:578-580.
4. Bergofsky EH, Haas F, and Porcelli R. Determination of the sensitive vascular sites from which hypoxia and hypercapnia elicit rises in pulmonary arterial pressure. *Fed. Proc.* 1968; 27:1420-1425.
5. Biscoe TJ, and Duchen MR. Responses of type I cells dissociated from the rabbit carotid body to hypoxia. *J. Physiol. London*, 1990; 428:39-59.
6. Boittin F-X, Galione A, and Evans AM. Nicotinic acid adenine dinucleotide phosphate mediates Ca^{2+} signals and contraction in arterial smooth muscle via a two-pool mechanism. *Circ. Res.* 2002; 91:1168-1175.
7. Boittin F-X, Dipp M, Kinnear N P, Galione A, and Evans AM. Vasodilation by the calcium mobilising messenger cyclic ADP-ribose. *J. Biol. Chem.* 2003: 278:9602-9608.
8. Bradford JR, and Dean HP. The pulmonary circulation. *J. Physiol.* 1894; 16:34-96.
9. Churchill GC, Okada Y, Thomas JM, Genazzani AA, Patel S, and Galione A. NAADP mobilizes calcium from reserve granules, a lysosomal-related organelle, in sea urchin eggs. *Cell.* 2002; 111:703-708.
10. Cornfield D, Stevens T, McMurty I, Abman S, and Rodman D. Acute hypoxia causes membrane depolarization and calcium influx in fetal pulmonary artery smooth muscle cells. *Am. J. Physiol.* 1994; 266:L469-L475.
11. Demiryurek AT, Wadsworth RM, and Kane KA. Pharmacological evidence for the role of mediators in hypoxia-induced vasoconstriction in sheep isolated arteries. *Eur. J. Pharmacol.* 1991; 203:1-8.

12. Dipp M, and Evans AM. cADPR is the primary trigger for hypoxic pulmonary vasoconstriction in the rat lung *in-situ*. *Circ. Res.* 2001; 89:77-83.
13. Dipp M, Nye PCG, and Evans AM. Hypoxia induces sustained sarcoplasmic reticulum calcium release in rabbit pulmonary artery smooth muscle in the absence of calcium influx. *Am. J. Physiol.* 2001; 281:L318-L325.
14. Dipp M, Nye PCG, and Evans AM. The vasoconstrictor released by hypoxia into the rat pulmonary circulation acts through a rho-associated kinase pathway and is removed during alveolar normoxia. *J. Physiol.* 2001; 231P:96P.
15. Dipp M, Thomas JM, Galione A, and Evans AM. A Po_2 window for smooth muscle cADPR accumulation and constriction by hypoxia in rabbit pulmonary arteries. *J. Physiol.* 2003; 547 P:72C.
16. Duke HN. Pulmonary vasomotor responses of isolated perfused cat lungs to anoxia and hypercapnia. *Q. J. Eper. Physiol.* 1951; 36:75-88.
17. Dumas JP, Bardou M, Goirand F, and Dumas M. Hypoxic pulmonary vasoconstriction. *Gen. Pharmacol.* 1999; 33:289-297.
18. Evans AM, Dipp M. 2002. Hypoxic pulmonary vasoconstriction: cyclic adenosine diphosphate-ribose, smooth muscle calcium stores and the endothelium. *Resp. Physiol. Neurobiol.* 2002; 132: 3-15.
19. von Euler US, and Liljestrand G. Observations on the pulmonary arterial blood pressure in the cat. *Acta Physiol. Scand.* 1946; 12:301-320.
20. Fishman AP. Hypoxia on the pulmonary circulation. How and where it acts. *Circ. Res.* 1976; 38:221-231.
21. Gaine SP, Hales MA, and Flavahan NA. Hypoxic pulmonary endothelial cells release a diffusible contractile factor distinct from endothelin. *Am. J. Physiol.* 1998; 274: L657-L664.
22. Galione A, Lee HC, and Busa WB. Ca^{2+}-induced Ca^{2+} release in sea urchin egg homogenates: modulation by cyclic ADP-ribose. *Science.* 1991; 253:1143-1146.
23. Gelband CH, and Gelband H. Ca^{2+} release from intracellular stores is an initial step in hypoxic pulmonary vasoconstriction of rat pulmonary artery resistance vessels. *Circulation.* 1991; 96: 3647-3654.
24. Gonzalez C, Sanz-Alfayate G, Agapito MT, Gomez-Nino A, Rocher A, and Obeso A. Significance of ROS in oxygen sensing in cell systems with sensitivity to physiological hypoxia. *Resp. Physiol. Neurobiol.* 2002; 132:17-41.
25. Hales CA, and Westphal DM. Pulmonary hypoxic vasoconstriction: not affected by chemical sympathectomy. *J. Appl. Physiol.* 1979; 46:529-533.
26. Harder DR, Madden JA, and Dawson C. Hypoxic induction of calcium-dependent action potentials in small pulmonary arteries of the rat. *J. Appl. Physiol.* 1985; 59:1389-1393.
27. Holden WE, and McCall E. Hypoxia-induced contractions of porcine pulmonary artery strips depend on intact endothelium. *Exp. Lung Res.* 1984; 7:101-112.
28. Hoshino Y, Obara H, Kusunoki M, Fujii Y, and Iwai S. Hypoxic contractile response in isolated human pulmonary artery: role of calcium ion. *J. Appl. Physiol.* 1998; 65:2468-2474.
29. Jabr RI, Toland H, Gelband CH, Wang XX, and Hume JR. Prominent role of intracellular Ca^{2+} release in hypoxic vasoconstriction of canine pulmonary artery. *Br. J. Pharmacol.* 1997; 122: 21-30.
30. Kang TM, Park MK, and Uhm DY. Characterisation of hypoxia-induced calcium rise in rabbit pulmonary arterial smooth muscle. *Life Sci.* 2002; 70:2321-2333.
31. Kang TM, Park MK, and Uhm DY. Effects of mitochondrial inhibition on the capacitative calcium entry in rabbit pulmonary arterial smooth muscle cells. *Life Sci.* 2003; 72:1467-1479.
32. Kato M, and Staub NC. Response of small pulmonary arteries to unilobar hypoxia and hypercapnia. *Circ. Res.* 1966; 19:426-440.
33. Kovitz KL, Aleskowitch JT, Sylvester JT, and Flavahan NA. Endothelium-derived contracting and relaxing factors contribute to hypoxic responses of pulmonary arteries. *Am. J. Physiol.* 1993; 260:L516-L521.
34. Kumasaka S, Shoji H, and Okabe E. Novel mechanisms involved in superoxide anion radical-

triggered Ca^{2+} release from cardiac sarcoplasmic reticulum linked to cyclic ADP-ribose stimulation. *Antioxidants & Redox Signalling*. 1999; 1:55-69.
35. Leach RM, Robertson TP, Twort CHC, and Ward JPT. Hypoxic vasoconstriction in rat pulmonary and mesenteric arteries. *Am. J. Physiol.* 1994; 266:L223-L231.
36. Leach RM, Sheehan DW, Chacko VP, and Sylvester JT. Energy state, pH, and vasomotor tone during hypoxia in precontracted pulmonary and femoral arteries. *Am. J. Physiol.* 2000; 278: L294-L304.
37. Lee HC. Physiological functions of cyclic ADP-ribose and NAADP as calcium messengers. *Annu. Rev. Pharmacol. Toxicol.* 2001; 41:317-345.
38. Lejeune P, Brimioulle S, Leeman M, Hallemans R, Melot C, and Naeije R. Enhancement of hypoxic pulmonary vasoconstriction by metabolic acidosis in dogs. *Anaesthesiology*. 1990; 73:256-264.
39. Liu Q, Ham JSK, Shimoda A, and Sylvester JT. Hypoxic constriction of porcine distal pulmonary arteries: endothelium and endothelin dependence. *Am. J. Physiol.* 2001; 280: L856-L865.
40. Malphigi M. *Duae epistole de pulmonibus*. Florence, 1661.
41. McIntyre RC, Banerjee A, Bensard DD, Brew EC, Hahn AR, and Fullerton DA. Selective inhibition of cyclic adenosine monophosphate-mediated pulmonary vasodilation by acute hypoxia. *Am. J. Physiol.* 1994; 267:H2179-H2185.
42. McMurtry IF, Davidson AB, Reeves JT, and Grover RF. Inhibition of hypoxic pulmonary vasoconstriction by calcium channel antagonists in isolated rat lungs. *Circ. Res.* 1976; 38:99-104.
43. Naeije R, Lejeune P, Leeman M, Melot C, and Closset J. Pulmonary vascular responses to surgical chemodenervation and chemical sympathectomy in dogs. *J. Appl. Physiol.* 1989; 66: 42-50.
44. Neely CF, Stein R, Matot I, Batra V, and Cheung A. Calcium blockage in pulmonary hypertension and hypoxic vasoconstriction. *New Horiz.* 1996; 4:99-106.
45. Nissel O. Effects of oxygen and carbon dioxide on the circulation of isolated and perfused lungs of the cat. *Acta Physiol. Scand.* 1948; 16:121-127.
46. Nye PCG, and Robertson BE. Reversal of hypoxic pulmonary vasoconstriction in the isolated rat lung by pentobarbitone. *J. Physiol. London.* 1990; 424:59P.
47. Olschewski A, Hong Z, Nelson D, and Weir EK. Graded response of K^+ current, membrane, and $[Ca^{2+}]$ to hypoxia in pulmonary arterial smooth muscle. *Am. J. Physiol.* 2002; 283: L1143-L1150.
48. Osipenko ON, Evans AM, and Gurney AM. A novel O_2-sensing potassium channel may mediate hypoxic pulmonary vasoconstriction. *Br. J. Pharmacol.* 1997;120:1461-1470.
49. Porter VA, Rhodes, MT, Reeve HL, and Cornfield DN. Oxygen-induced fetal pulmonary vasodilation is mediated by intracellular calcium activation of K(Ca) channels. *Am. J. Physiol. Lung Cell Mol. Physiol.* 2001; 281: L1379-L1385.
50. Post J, Hume J, Archer S, and Weir E. Direct role for potassium channel inhibition in hypoxic pulmonary vasoconstriction. *Am. J. Physiol.* 1992; 262:C882-C890.
51. Robertson TP, Aaronson PI, and Ward JPT. Hypoxic vasoconstriction and intracellular Ca^{2+} in pulmonary arteries: evidence for PKC-independent Ca^{2+} sensitization. *Am. J. Physiol.* 1995; 268:H301-H307.
52. Robertson TP, Hague D, Aaronson PI, and Ward JPT. Voltage-independent calcium entry in hypoxic pulmonary vasoconstriction of intrapulmonary arteries of the rat. *J. Physiol. London.* 2000; 525:669-680.
53. Robertson TP, Ward JPT, and Aaronson PI. Hypoxia induces the release of a pulmonary-selective, Ca^{2+}-sensitising, vasoconstrictor from the perfused rat lung. *Cardiovasc. Res.* 2001; 50:145-150.
54. Robin ED, Theodore J, Burke CM, Oesterle SN, Fowler MB, Jamieson SW, Baldwin JC, Morris AJ, Hunt SA, and Vanskessel A. Hypoxic pulmonary vasoconstriction persists in the human transplanted lung. *Clin. Sci.* 1987; 72:283-287.

55. Roy CS, and Sherrington CS. The regulation of the blood supply of the brain. *J. Physiol.* 1890; 11:85.
56. Ruppersberg JP, Stocker M, Pongs O, Heinemann SH, Frank R, and Koenen M. Regulation of fast inactivation of cloned mammalian IK(A) channels by cysteine oxidation. *Nature.* 1991; 352:711-714.
57. Saida K, and van Breeman C. Mechanism of Ca^{++} antagonist-induced vasodilation. Intracellular actions. *Circ. Res.* 1983; 52:137-142.
58. Salvaterra CG, and Goldman WF. Acute hypoxia increases cytosolic calcium in cultured pulmonary arterial myocytes. *Am. J. Physiol.* 1993; 264:L323-L328.
59. Shigemori K, Ishizaki T, Matsukawa S, Sakai A, Nakai T, and Miyabo S. Adenine nucelotides via activation of ATP-sensitive K^+ channels modulate hypoxic response in rat pulmonary arteries. *Am. J. Physiol.* 1996; 270:L803-L809.
60. Vandier C, Delpech M, and Bonnet P. Hypoxia enhances agonist-induced pulmonary arterial contraction by increasing calcium sequestration. *Am. J. Physiol.* 1997; 273:H1075-H1081.
61. Vejlstrup NG, and Dorrington KL. Intense slow hypoxic pulmonary vasoconstriction in gas-filled and liquid-filled lungs: an in-vivo study in the rabbit. *Acta Physiologica Scand.* 1993; 148:305-313.
62. Voelkel NF. Mechanisms of hypoxic pulmonary vasoconstriction. *Am. Rev. Respir. Dis.* 1986; 133:1186-1195.
63. Weir E, and Archer S. The mechanism of acute hypoxic pulmonary vasoconstriction: the tale of two channels. *FASEB.* 1995; 9:183-189.
64. Welling KL, Sanchez R, Ravn JB, Larsen B, and Amtorp O. Effect of prolonged alveolar hypoxia on pulmonary arterial pressure and segmental vascular resistance. *J. Appl. Physiol.* 1993;75:1194-1200.
65. Whitteridge G. *The anatomical drawings of William Harvey.* Edinburgh: Livingstone, 1964.
66. Wilson HL, Dipp M, Thomas JM, Lad C, Galione A, and Evans AM. ADP-ribosyl cyclase and cyclic ADP-ribose hydrolase act as a redox sensor: a primary role for cADPR in hypoxic pulmonary vasoconstriction. *J. Biol. Chem.* 2001; 276:11180-11188.
67. Youngson C, Nurse C, Yeger H, and Katz E. Oxygen sensing in airway chemoreceptors. *Nature.* 1993; 356:153-155.
68. Yuan X-J, Goldman WF, Tod ML, Rubin LJ, and Blaustein MP. Hypoxia reduces potassium currents in cultured rat pulmonary but not mesenteric arterial myocytes. *Am. J. Physiol.* 1993; 264:L116-L123.
69. Yuan X-J, Tod ML, Rubin LJ, Blaustein MP. Contrasting effects of hypoxia on tension in rat pulmonary and mesenteric arteries. *Am. J. Physiol.* 1990; 259:H281-H289.

Chapter 7

Critical Role of Ca^{2+} Sensitization in Acute Hypoxic Pulmonary Vasoconstriction

Tom P. Robertson and Ivan F. McMurtry
University of Georgia, Athens, Georgia and University of Colorado, Denver, Colorado, U.S.A.

1. Introduction

Hypoxic pulmonary vasoconstriction (HPV) is a multi-factorial process pivotal to maintaining the optimum matching of perfusion to ventilation (54). The precise mechanisms that underlie HPV are the subject of much debate, as reflected by the number and diversity of putative mechanisms detailed within this book. However, a fundamental aspect of this and other vascular smooth muscle contractile responses is that HPV must involve an increase in one or both of the principal determinants of vascular smooth muscle tone, namely the level of cytosolic Ca^{2+} ($[Ca^{2+}]_i$) and the sensitivity of the contractile apparatus to $[Ca^{2+}]_i$.

Elevations in $[Ca^{2+}]_i$ elicit contraction primarily via activation of Ca^{2+}/calmodulin-dependent myosin light chain kinase (MLCK) and resultant phosphorylation of the 20 kDa myosin light chain (MLC_{20}). Phosphorylation of MLC_{20} increases the intrinsic ATPase activity of myosin, thereby enhancing the velocity and force of the actomyosin crossbridging cycle (49). However, it is well known that at any level of $[Ca^{2+}]_i$ the force generated by agonists is greater than that observed to depolarization (4, 25). Therefore, in the presence of pro-constrictor agonists, there is an apparent increase in the sensitivity of the contractile apparatus to $[Ca^{2+}]_i$, hence the term Ca^{2+} sensitization, also referred to as agonist-induced force enhancement (29).

2. Ca^{2+} Sensitization and Vascular Smooth Muscle Contraction

Vascular smooth muscle tone is determined primarily by the phosphorylation/ dephosphorylation ratio of MLC_{20}, which in turn is regulated by the relative activities of MLCK and myosin phosphatase (MLCP, also known as SMPP-1M) (49). Increases in the activity of MLCK or decreases in MLCP activity will therefore increase MLC_{20} phosphorylation and contraction. Ca^{2+}-

sensitization can therefore be due to Ca^{2+}-independent activation of MLCK, or Ca^{2+}-independent inhibition of MLCP. Although the former has been observed in isolated kinase studies (26), there is no evidence that Ca^{2+}-independent activation of MLCK occurs in intact arteries. However, the agonist-induced inhibition of MLCP leading to increased Ca^{2+}-sensitivity and contraction was first proposed over a decade ago (48), and it has become apparent that this is a key mechanism of Ca^{2+}-sensitization in vascular smooth muscle (50). MLCP is comprised of regulatory and catalytic (PP-1C) domains coupled to a third subunit of yet identified function, and can be regulated by several intracellular enzymes (11). The pathways to be discussed in this chapter that may lead to inhibition of MLCP during HPV are detailed in Figure 1.

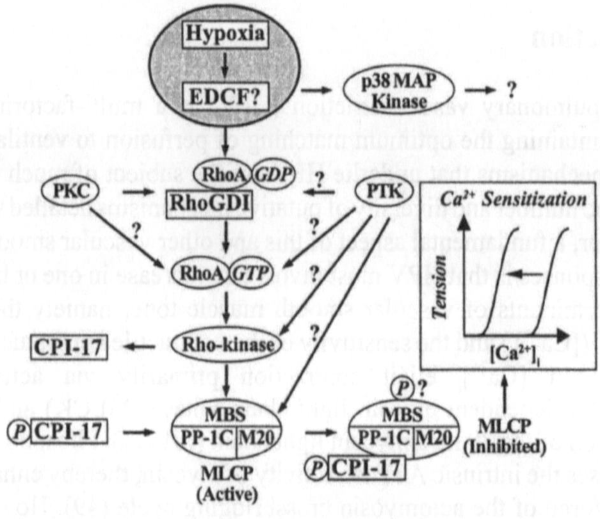

Figure 1. Putative pathways mediating Ca^{2+}-sensitization during HPV. Ca^{2+}sensitization in vascular smooth muscle is thought to be primarily mediated by inhibition of MLCP. Hypoxia, possibly via an endothelium-derived constricting factor (EDCF), elicits activation of Rho-kinase by RhoA. In its cytosolic (inactive) form, GDP-bound RhoA is complexed with RhoGDI (Rho-associated guanine nucleotide dissociation inhibitor). Upon agonist stimulation, RhoA is activated by Rho guanine nucleotide exchange factors that cause the replacement of GDP by GTP, in turn causing dissociation of RhoA from RhoGDI. Activated RhoA then translocates to the plasma membrane whereupon it activates Rho-kinase. Rho-kinase inhibits MLCP (myosin light chain phosphatase) by phosphorylating the regulatory subunit (MBS). However, it is unclear whether a direct interaction occurs in intact arteries; it is possible that Rho-kinase inhibits MLCP indirectly via phosphorylation of CPI-17. When phosphorylated, CPI-17 is a potent inhibitor of the catalytic subunit (PP-1C) of MLCP. PKC has also been shown to phosphorylate CPI-17. PKC may be an upstream regulator of Rho-kinase as PKC has been shown to phosphorylate RhoGDI, in turn, resulting in activation of RhoA. Inhibition of MLCP results in a net increase of MLC_{20} phosphorylation and contraction. An increase in Ca^{2+}-sensitivity results in a leftward shift in the tension/$[Ca^{2+}]_i$ curve (inset). The pathways that may mediate PTK and p38 MAP kinase associated Ca^{2+}-sensitization are yet to be determined. M20, small non-catalytic subunit.

3. Ca^{2+} Sensitization in Acute HPV

3.1. Acute HPV: A Transient or Sustained Constrictor Response?

HPV is a homeostatic mechanism within the lung. Since HPV occurs in isolated arteries, it is apparent that the sensor and effector mechanisms reside in the pulmonary artery smooth muscle and/or endothelium. *In vivo* and in isolated blood-perfused lungs, the pressor response to acute hypoxia is usually immediate in onset and sustained. In contrast, HPV in isolated arteries is often biphasic, consisting of an immediate transient constrictor response superimposed on a more slowly developing, sustained contraction (22). It has been proposed that the latter response is the more physiologically relevant process, since it is sustained. Indeed, when considering the role of HPV in the normal lung, it would appear important that HPV be a sustained response. It is apparent that HPV responds to localized alveolar hypoxia, thereby diverting blood flow to better oxygenated areas of the lung and maintaining ventilation/perfusion matching. If HPV were transient, then the re-direction of blood flow would also be transient, and ventilation/perfusion matching would not be maintained, presumably promoting hypoxemia, which does not seem to occur *in vivo*. Conversely if HPV is sustained, the re-direction of blood flow in response to localized hypoxia is maintained for the duration of the hypoxic episode and optimal ventilation/perfusion matching is maintained.

3.2. Ca^{2+} Sensitization During Acute HPV

The first studies designed to determine the relationship between $[Ca^{2+}]_i$ and tension development during acute HPV were performed in Dr. Jeremy Ward's laboratory in the early 1990's (34, 40). In these experiments, the Ca^{2+}-sensitive fluorophore Fura-2 was used to monitor $[Ca^{2+}]_i$ levels during HPV in rat isolated small intrapulmonary arteries mounted in small vessel myographs to simultaneously measure tension development. The experimental record from the first of these experiments in shown in Figure 2. The contractile response to acute hypoxia in rat intrapulmonary arteries is biphasic. During the initial transient constriction there is a simultaneous transient increase in $[Ca^{2+}]_i$. However, the pivotal result from these experiments was the demonstration of a dissociation between tension and $[Ca^{2+}]_i$ during the sustained constriction. Specifically, subsequent to the transient increases in $[Ca^{2+}]_i$ and tension, $[Ca^{2+}]_i$ remained constant (at a level higher than that prior to the induction of hypoxia) whereas tension increased for the remainder of the hypoxic challenge. Upon re-oxygenation, both vascular smooth muscle tension and $[Ca^{2+}]_i$ returned to their pre-hypoxic levels.

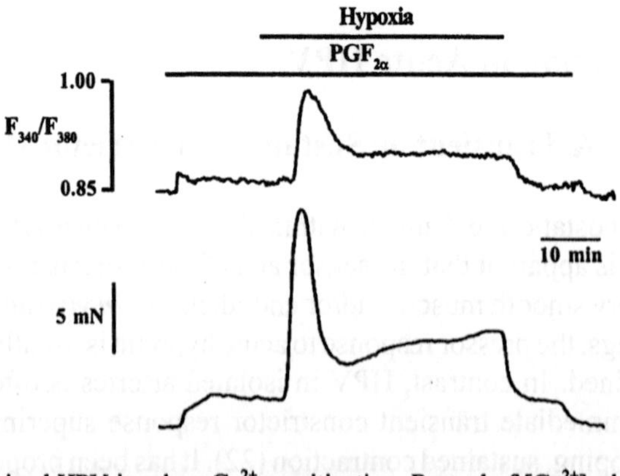

Figure 2. **Sustained HPV involves Ca^{2+} sensitization. Record of $[Ca^{2+}]_i$ changes (F_{340}/F_{380}) and tension in a Fura-2 loaded rat intrapulmonary artery during acute hypoxia. In the presence of a small degree of agonist-induced tone (3 μM prostaglandin $F_{2\alpha}$, $PGF_{2\alpha}$), hypoxia elicited a transient increase in both $[Ca^{2+}]_i$ and tension. However, whereas tension progressively increased for the remainder of the hypoxic challenge, $[Ca^{2+}]_i$ remained at a constant level.**

In subsequent work, it became apparent that the use of Fura-2 (and the related compound Fura-PE3) to measure $[Ca^{2+}]_i$ during hypoxia might overestimate the rise in $[Ca^{2+}]_i$ (21). Hypoxia increases NAD(P)H levels in pulmonary arteries, as determined by monitoring the ratio of autofluorescence in isolated pulmonary arteries when excited at 340 and 380 nm, respectively (21, 43). Coincidentally, these are the same excitation wavelengths employed when using Fura-2 or Fura-PE3 as Ca^{2+}-sensitive fluorophores. Therefore, the increase in fluorescence ratio observed in isolated arteries during HPV when using these probes will be due partly to the concomitant increase in NAD(P)H levels. Leach and co-workers (2001) calculated that this amounted to ~25% of the increase in fluorescence ratio during HPV, indicating that the rise in $[Ca^{2+}]_i$ during sustained HPV may be less and, in turn, that the dissociation between tension and $[Ca^{2+}]_i$ during sustained HPV greater than that reported previously using similar techniques (34). The dissociation between $[Ca^{2+}]_i$ and tension during sustained HPV can be readily explained by Ca^{2+} sensitization.

4. Mechanisms of Ca^{2+} Sensitization During HPV

4.1. Role of Protein Kinase C (PKC)

PKC exists as twelve isoforms that are subdivided into 4 groups according to their cofactor requirements, namely; classical, novel, atypical and recently described PKCs. Activation of classical and novel isoforms of PKC, via the application of phorbol esters, elicits Ca^{2+}-independent contractions in pulmonary arteries (42) and other vascular smooth muscle preparations. PKC elicits Ca^{2+}

sensitization by phosphorylating and activating CPI-17, which in turn inhibits the catalytic domain, PP-1C, of MLCP (7, 44). An example of PKC-induced Ca^{2+} sensitization and contraction is shown in Figure 3, where application of phorbol 12,13-dibutyrate to a rat small intrapulmonary artery elicited a slowly-developing contractile response, whilst $[Ca^{2+}]_i$ remained constant. It is noteworthy that the time-course of this contractile response is similar to that of sustained HPV. However, inhibition of all isoforms of PKC with the bisindoylmaleimide PKC inhibitor, Ro-31-8220 (at a concentration that abolished the contraction to phorbol ester), had no effect on the sustained HPV (34). Interestingly, this compound did reduce the transient constrictor response to hypoxia by ~42% (34) consistent with a role for PKC during the transient phase of HPV. However, the latter possibility is complicated by the fact that Ro-31-8220 has subsequently been shown to inhibit several intracellular kinases other than PKC (3, 12). The lack of specificity of this compound does not, however, undermine the important conclusion that the Ca^{2+} sensitization observed during sustained HPV is independent of activation of PKC in rat isolated intrapulmonary arteries (34).

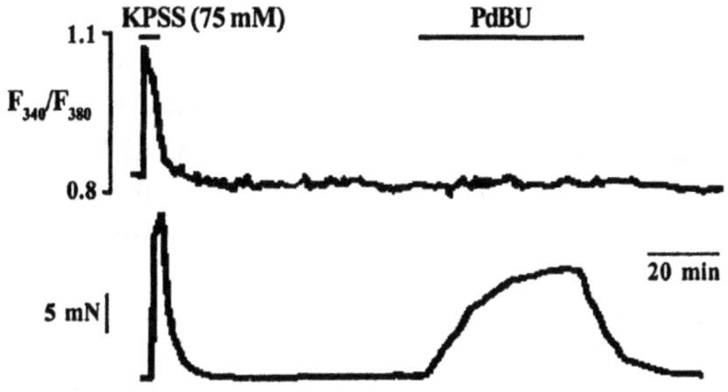

Figure 3. Activation of PKC elicits Ca^{2+} sensitization and contraction in pulmonary arteries. Experimental record of changes in $[Ca^{2+}]_i$ (upper panel) and tension (lower panel) in a rat pulmonary artery exposed to either 75 mM KPSS (physiological salt solution containing 75 mM KCl, isotonic replacement of NaCl) or phorbol 12,13-dibutyrate (PdBu). Depolarization of the artery with KPSS results in elevations of both tension and $[Ca^{2+}]_i$. In contrast, activation of PKC via exposure of the artery to PdBu results in a large sustained contraction that is not associated with any rise in $[Ca^{2+}]_i$.

The lack of specificity of conventional pharmacological agents, not only for PKC *vs* other intracellular kinases, but also among the various isoforms of PKC, makes interpretation of the effects of these compounds problematical. While inhibitors of PKC decrease HPV in a variety of preparations including isolated rat arteries (16) and rat (32), rabbit (57) and dog (2) isolated perfused lungs, the inhibitors used in these studies are not selective for PKC. For example, the inhibitors H-7 and staurosporine are compounds that inhibit virtually every kinase they have been tested against (30), and bisindolylmaleimide I and Go-6976 inhibit at least 7-10 intracellular kinases in addition to PKC (3). The extrapolation of the effects of these agents to a role for PKC activation during HPV may not, therefore, be justified. However, it is noteworthy that Barman (2) found that the inhibitor calphostin C, which appears to be less promiscuous in its non-PKC related actions, reduced HPV in isolated perfused dog lungs.

The lack of specificity of PKC antagonists has therefore made definitive studies of the possible role of PKC in HPV virtually impossible. However, the recent development of highly specific, cell-permeable peptide inhibitors of classical, novel and atypical PKC isoforms, namely myristoylated analogs of the respective pseudosubstrates for each of these PKC groups, offers promise that the role of PKC isoforms in the generation of HPV may yet be defined. Moreover, the availability of PKC isoform knockout mice will undoubtedly help in this regard. In a recent study, PKCε knockout mice had a blunted HPV response, whereas other vasoconstrictor responses were unaltered (24). Littler and co-workers reported that the PKCε knockout mice had an increased lung and pulmonary artery expression of the O_2-sensitive potassium channel Kv3.1, and they attributed the blunted HPV in isolated perfused lungs to this up-regulation. It is therefore unclear from this study whether an activation of PKCε is directly involved in HPV, or whether the deletion of this isoform leads to blunted HPV because of a functional antagonism associated with the upregulation of potassium channel expression. Nonetheless, the fact that HPV was preferentially blunted by deletion of PKCε is an exciting finding, and the application of such techniques to this field will undoubtedly prove enlightening.

4.2. Role of Protein Tyrosine Kinases (PTKs)

PTKs are usually associated with growth factor linked receptors and play pivotal roles in many cell-signaling processes (13). PTK activation has been associated with an increase in Ca^{2+}-sensitivity in smooth muscle (51) and evidence for involvement of PTKs in HPV has been found in sheep (53) and rat isolated pulmonary arteries (56). The evidence for the involvement of PTK activation in HPV is based upon the inhibitory effects of PTK antagonists, such as genistein and the tyrphostins, and the enhancement of HPV in the presence of the broad-spectrum tyrosine phosphatase inhibitor sodium orthovanadate. However, as with the majority of enzyme inhibitors, these compounds are similarly afflicted with non-specific effects (31, 47). The role of PTKs in HPV therefore remains unclear and experiments to determine whether there is an increase in tyrosine phosphorylation in intact arteries during HPV would help to address this issue.

A role of PTKs in the response of pulmonary arteries to hypoxia is an attractive possibility, especially when considering mechanisms that may be involved in the "transition" from the acute response (i.e., contraction) to the chronic response (i.e., proliferation and vascular remodeling). Since PTKs can be involved in both Ca^{2+} sensitization-associated contractions of vascular smooth muscle and in growth factor-induced proliferation, they would appear ideal candidates to be involved in both acute and chronic hypoxia. It is also interesting to note that tyrosine phosphorylation has been proposed as a convergent mechanism between PTKs and another key player, Rho-kinase, during Ca^{2+}-sensitization in vascular smooth muscle (27, 41).

4.3. Role of Rho-Kinase

Rho-associated coiled coil-forming serine/threonine kinase (Rho-kinase, also known as Rho-associated kinase) has recently been identified as a pivotal mediator of Ca^{2+} sensitization and vascular smooth muscle contraction (50). Since the first isolation of this kinase (23), it has become apparent that, upon activation by the binding of the small monomeric G-protein RhoA, Rho-kinase inhibits MLCP resulting in Ca^{2+}-independent contraction of vascular smooth muscle. Originally, the inhibition of MLCP was thought to be via a direct action of Rho-kinase upon the myosin binding subunit of MLCP (18). However, recent evidence suggests that, as with PKC, Rho-kinase can also inhibit MLCP via phosphorylation and activation of CPI-17 (19, 20, 28). Therefore, phosphorylation of CPI-17 and inhibition of MLCP may be a convergent mechanism through which both PKC and Rho-kinase elicit an increase in myofilament Ca^{2+}-sensitivity although the effect of PKC may be transient, at least with respect to agonist-induced Ca^{2+} sensitization (15). Rho-kinase is also known to directly phosphorylate MLC_{20} *in vitro*, but this mechanism does not appear to occur in intact arteries (14).

Subsequent to the initial identification of a role for Ca^{2+}-sensitization in HPV (34) and the exclusion of PKC from this process in rat isolated arteries, there was a hiatus in research directed at this mechanism until the "specific" Rho-kinase inhibitor, Y-27632 (52), became widely available to examine the possible role of Rho-kinase in this process. Indeed, the availability of Y-27632 has resulted in an explosion of reports in the literature implicating Rho-kinase in a diverse number of intracellular pathways in both smooth muscle and non-muscle cell types (50).

The first evidence that Rho-kinase may be involved in HPV, and specifically in the Ca^{2+} sensitization-mediated development of sustained HPV, came from studies of isolated arteries and *in situ* perfused lungs of the rat (36). It was found that Y-27632 preferentially inhibited sustained HPV, whilst having a minimal effect upon the transient contraction (Fig. 4). Indeed, the small attenuation of the transient phase of HPV may be due to the effect of Y-27632 upon the sustained portion of the response. As mentioned previously, the transient response is superimposed on the more-slowly developing sustained phase. Upon selective inhibition of the transient phase by sub-micromolar concentrations of the trivalent cation La^{3+}, it is apparent that the sustained phase of HPV in isolated arteries is immediate in onset (37), similar to HPV in perfused lungs and *in vivo*. Inhibition of this response by Y-27632 would therefore produce an apparent reduction in the transient phase without necessarily directly affecting the mechanisms responsible for the contraction. An important aspect of this study was that Y-27632 abolished the monophasic rise in pulmonary artery pressure in the perfused lung, whereas only the sustained phase was inhibited in small isolated pulmonary arteries. This observation would appear to question the physiological relevance of the large transient response to HPV *in vitro*, and raises the possibility that it may be, at least in part, an artifact of the experimental conditions employed to elicit HPV in isolated arteries. However, a similar transient phase of HPV, although markedly smaller, has been observed *in vivo* when the induction of hypoxia is rapid (38, 58).

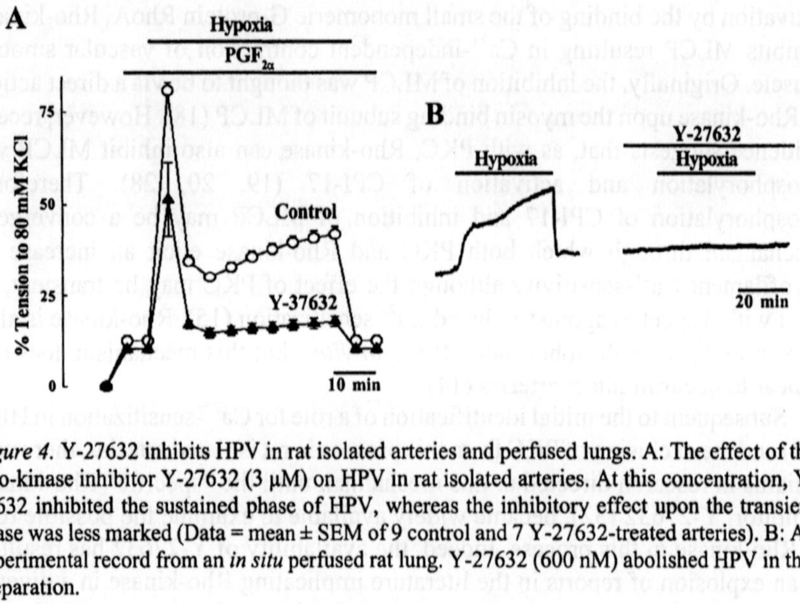

Figure 4. Y-27632 inhibits HPV in rat isolated arteries and perfused lungs. A: The effect of the Rho-kinase inhibitor Y-27632 (3 μM) on HPV in rat isolated arteries. At this concentration, Y-27632 inhibited the sustained phase of HPV, whereas the inhibitory effect upon the transient phase was less marked (Data = mean ± SEM of 9 control and 7 Y-27632-treated arteries). B: An experimental record from an *in situ* perfused rat lung. Y-27632 (600 nM) abolished HPV in this preparation.

The inhibition of HPV by Y-27632 in rat isolated intrapulmonary arteries, *in situ* perfused rat lungs (35) and perfused mouse lungs (Fagan K, personal communication) are consistent with a pivotal role for Rho-kinase-mediated Ca^{2+}-sensitization in the mechanism of sustained HPV. It is interesting that pretreatment of perfused rat lungs with Y-27632 completely prevents the hypoxic response. This raises the question of whether activation of Rho-kinase is involved in initiating HPV, or whether a basal level of Rho-kinase activity and suppression of MLCP activity, are necessary for the hypoxia-induced Ca^{2+} signal, and associated activation of MLCK, to increase MLC_{20} phosphorylation and elicit contraction. Moreover, as with all new pharmacological tools, the direct extrapolation of the effects of Y-27632 to Rho-kinase inhibition must be undertaken with caution and direct evidence for the activation of Rho-kinase during the response in question is important. This evidence was recently provided by Dr. R. Rhoades' laboratory (55). These investigators reported that hypoxia-induced activation of Rho-kinase was associated with a concomitant increase in MLC_{20} phosphorylation in cultured rat intrapulmonary artery smooth muscle cells maintained in 10% fetal bovine serum (54). In an elegant series of experiments, Wang and co-workers provided strong evidence that hypoxia can activate Rho-kinase, in a RhoA-dependent manner, and that inhibition of either RhoA (via application of exoenzyme C3) or Rho-kinase (via incubation of Y-27632) attenuated the increase in MLC_{20} phosphorylation in response to hypoxia. Furthermore, Y-27632 inhibited sustained HPV in rat isolated pulmonary arteries. These experiments, therefore, provided the first biochemical evidence for a role of Rho-kinase in HPV. However, this study also raised an important issue with respect to the role of the endothelium in sustained HPV (discussed in section 6).

4.4. Role of p38 Mitogen-activated Kinase (p38 MAP kinase)

Vascular smooth muscle and endothelial cells express a family of serine/threonine kinases collectively known as MAP kinases. These kinases include p38 MAP kinase, extracellular signal-regulated kinases (ERK1 and ERK2), and c-Jun-NH_2-terminal kinases/stress-activated protein kinase. It has become apparent that these kinases play vital regulatory roles in cell proliferation and differentiation and they may also be involved in the regulation of smooth muscle tone (1). It has recently been reported that activation of MAP kinases, and specifically p38 MAP kinase, may be involved in sustained HPV (17). Karamsetty and co-workers observed that hypoxia increased phosphorylation of p38 MAP kinase and that SB-202190, an inhibitor of p38 MAP kinase, abolished sustained HPV in rat isolated main pulmonary artery. However, at the concentration used, SB-202190 has also been found to directly inhibit Rho-kinase by ~40% (3). Whether the inhibition of sustained HPV by this compound is due to its effect upon p38 MAP kinase and/or Rho-kinase, and the possible role of p38 MAP kinase in acute HPV requires further investigation.

5. Glycolysis and Ca^{2+} Sensitization During HPV

It is probable the Ca^{2+} sensitization that occurs during acute HPV, as with Ca^{2+} sensitization in response to other agonists, involves the activation of one or more intracellular kinases, the functional endpoint(s) of which require ATP as the phosphate donor. One obvious consequence of hypoxia in any cell type is that the primary source of energy, oxidative phosphorylation, will be compromised to some degree. In this situation, the logical compensatory mechanism to maintain levels of ATP would be glycolysis. Indeed, glucose availability and glycolysis are absolute requirements for sustained HPV (21, 59, 61). The dependence of sustained HPV upon glucose availability and glycolysis could be explained if this process caused an increase in $[Ca^{2+}]_i$ during hypoxia. However, neither the abolition of sustained HPV by low extracellular glucose, nor the potentiation of HPV by high extracellular glucose, cause any change in $[Ca^{2+}]_i$ during HPV in isolated arteries, suggesting that glycolysis supports HPV by facilitating the associated Ca^{2+} sensitization (21).

Multiple lines of evidence suggest that, during hypoxia, glycolysis primarily supports membrane-associated processes (21, 59). It is interesting that certain isozymes of PKC, Rho-kinase and MLCP translocate to the plasma membrane upon agonist stimulation, and that activation of Rho-kinase by RhoA is a plasma membrane-associated event (5, 46, 50). Thus, it is tempting to speculate that such a co-ordination and localization of both effectors and energy production would be an elegant and efficient response of the pulmonary vasculature to hypoxia.

6. Role of the Endothelium in Ca^{2+} Sensitization During HPV

Perhaps the most contentious issue in the recent history of HPV has been the role, or lack thereof, of the endothelium in the generation of HPV. In general, the full expression of sustained HPV appears to require the presence of a functioning endothelium (see Chapter 12). It has been proposed that since the sustained phase of HPV is reduced (60) or abolished (22) upon endothelial denudation, that the associated Ca^{2+} sensitization may also be endothelium-dependent (34). In other words, whereas the elevation of $[Ca^{2+}]_i$ is a function of a direct effect of hypoxia upon the smooth muscle, the *coupling* of the elevation in $[Ca^{2+}]_i$ to tension development requires the presence of the endothelium. In support of this possibility, removal of the endothelium has no effect upon $[Ca^{2+}]_i$ during hypoxia in isolated intrapulmonary arteries, yet the sustained contractile response is abolished (33). These observations are consistent with the release of an endothelium-derived constrictor factor (EDCF) being pivotal to the generation of sustained HPV. One possible extension of this theory would be that the EDCF elicits an increase in Rho-kinase activity, possibly via interaction with a G protein-coupled receptor in pulmonary arterial smooth muscle cells, resulting in an increase in Ca^{2+}-sensitivity and contraction. In other words, sustained HPV may involve a classical agonist-induced force enhancement (discussed in section 7). However, Wang et al. (55) reported that Rho-kinase activity was increased in cultured pulmonary artery smooth muscle cells, consistent with a *direct* stimulation of Rho-kinase activity. Wang et al. (55) reported that the cell cultures utilized in their studies were >90% pure smooth muscle, but they did not report whether a portion of the non-smooth muscle cells included endothelial cells. Whether a cell population that includes <10% endothelial cells would be sufficient to generate enough EDCF to stimulate Rho-kinase activity in the vast number of neighboring smooth muscle cells is unclear. Wang et al. (55) also demonstrated that HPV in intrapulmonary arteries was inhibited by Y-27632, but they did not detail whether the sustained HPV in their arteries was endothelium-dependent. However, both the translocation of RhoA, a causal event in the activation of Rho-kinase (9), and the activation of Rho-kinase in rat small intrapulmonary arteries during HPV appear to be endothelium-dependent (Robertson T, unpublished observations).

7. Role of an EDCF-mediated Ca^{2+} Sensitization in HPV

The dependency of the full expression of sustained HPV upon the presence of a functional endothelium is consistent with the release of an EDCF. One obvious candidate for this EDCF is endothelin-1 (ET-1). Not surprisingly, the role of ET-1 in HPV is as contentious as that of the endothelium itself, with evidence for and against a role for ET-1 being as compelling as they are mutually exclusive (see Chapter 12). All the studies that have attempted to characterize an EDCF released during HPV have excluded ET-1 as the mediator of the constrictor activity observed (6, 8, 35, 39).

Since HPV is a phenomenon peculiar to the pulmonary circulation, it would seem reasonable to assume that the EDCF should be either selectively produced in the lung, or to have constrictor activity confined to pulmonary arteries. In addition, if the proposed mechanisms detailed within this chapter are correct, with respect to Rho-kinase-mediated Ca^{2+}-sensitization being an integral part of HPV, then one would predict that the isolated EDCF would elicit contraction via this pathway. Encouragingly, these facets have been reported to be present in several preliminary reports that attempted to isolate the elusive EDCF (6, 35, 39). Robertson et al. (35) published evidence that a constrictor factor is preferentially released during hypoxia from the perfused rat lung, and that the factor constricts pulmonary but not mesenteric arteries of the rat. This constrictor mechanism involved Ca^{2+} sensitization and the factor was not evident in the perfusate from the hypoxic perfused mesenteric vascular bed (39). Dipp et al. (6) reported that the hypoxic perfused rat lung produces a constrictor factor that elicits contraction of pulmonary arteries via activation of Rho-kinase. A particularly noteworthy aspect of the latter study was the finding that the constrictor activity was abolished upon the perfusate being passed through normoxia-ventilated lungs. This observation raises the possibility that the constrictor is synthesized by the pulmonary circulation in response to acute hypoxia and then degraded once normoxia is restored. Such properties are ideal when attempting to identify an EDCF with similar constrictor properties to those of HPV, i.e., a response that is rapid in onset and reversal.

8. Summary

Converging lines of evidence indicate that HPV involves an increase in Ca^{2+} sensitivity. Future work that addresses the mechanisms that underlie hypoxia-induced Ca^{2+} sensitization in the pulmonary circulation are therefore a priority in furthering our understanding of this vital physiological response. Such studies would include the determination of *i*) the precise role of the endothelium in supporting hypoxia-induced Ca^{2+} sensitization in pulmonary arteries, *ii*) the identity and mechanism of action of the putative EDCF involved in HPV and *iii*) the roles of PKC, PTK, p38 MAP kinase and Rho-kinase.

Whatever the relative contributions of the latter enzymatic pathways may be in HPV, it is probable that inhibition of MLCP is the pivotal step that results in an increase in MLC_{20} phosphorylation and contraction. The emergence of Rho-kinase as a key regulator of MLCP and vascular smooth muscle tone, coupled with the persuasive evidence that Rho-kinase activation is involved in HPV, makes this pathway a primary target for further investigation. It is tempting to speculate that Rho-kinase may be a link between Ca^{2+} sensitization during acute hypoxia and the pathogenesis of chronic hypoxia-associated pulmonary hypertension (see Chapter 24), especially since Rho-kinase antagonists appear to offer much promise for the treatment of a variety of systemic vascular diseases (45, 52).

References

1. Abdel-Latif AA. Cross talk between cyclic nucleotides and polyphosphoinositide hydrolysis, protein kinases, and contraction in smooth muscle. *Exp. Biol. Med.* 2001; 226: 153-163.
2. Barman SA. Effect of protein kinase C inhibition on hypoxic pulmonary vasoconstriction. *Am. J. Physiol. Lung Cell. Mol. Physiol.* 2001; 280: L888-L895.
3. Davies SP, Reddy H, Caivano M, and Cohen P. Specificity and mechanism of action of some commonly used protein kinase inhibitors. *Biochem. J.* 2000; 351: 95-105.
4. DeFeo TT and Morgan KG. Calcium-force relationships as detected with aequorin in two different vascular smooth muscles of the ferret. *J. Physiol.* 1985; 369: 269-282.
5. Dempsey EC, Newton AC, Mochly-Rosen D, Fields AP, Reyland ME, Insel PA, and Messing RO. Protein kinase C isozymes and the regulation of diverse cell responses. *Am. J. Physiol. Lung Cell. Mol. Physiol.* 2000; 279: L429-438.
6. Dipp M, Nye PCG, and Evans AM. The vasoconstrictor released by hypoxia into the rat pulmonary circulation acts through a Rho-associated kinase pathway and is removed during alveolar normoxia. *J. Physiol.* 2001; 531: 96.
7. Eto M, Ohmori T, Suzuki M, Furuya K, and Morita F. A novel protein phosphatase-1 inhibitory protein potentiated by protein kinase C. Isolation from porcine aorta media and characterization. *J. Biochem.* 1995; 118: 1104-1107.
8. Gaine SP, Hales MA, and Flavahan NA. Hypoxic pulmonary endothelial cells release a diffusible contractile factor distinct from endothelin. *Am. J. Physiol.* 1998; 274: L657-L664.
9. Gong MC, Fujihara H, Somlyo AV, and Somlyo AP. Translocation of RhoA associated with Ca^{2+} sensitization of smooth muscle. *J. Biol. Chem.* 1997; 272: 10704-10709.
10. Hardin CD, Wiseman RW, and Paul RJ. "Metabolism and energetics of vascular smooth

muscle." In *Physiology and Pathophysiology of the Heart*, 4th Edition, Sperelakis N, ed. Norwell, MA: Kluwer Academic Publishers, 2001, pp. 571-595.

11. Hartshorne DJ, Ito M, and Erdodi F. Myosin light chain phosphatase: subunit composition, interactions and regulation. *J. Muscle Res. Cell. Motil.* 1998; 19: 325-341.
12. Hers I, Tavare JM, and Denton RM. The protein kinase C inhibitors bisindolylmaleimide I (GF 109203x) and IX (Ro 31-8220) are potent inhibitors of glycogen synthase kinase-3 activity. *FEBS Lett.* 1999; 460: 433-436.
13. Hubbard SR and Till JH. Protein tyrosine kinase structure and function. *Annu. Rev. Biochem.* 2000; 69: 373-398.
14. Iizuka K, Yoshii A, Samizo K, Tsukagoshi H, Ishizuka T, Dobashi K, Nakazawa T, and Mori M. A μαφορ role for the Rho-associated coiled coil forming protein kinase in G-protein-mediated Ca^{2+} sensitization through inhibition of myosin phosphatase in rabbit trachea. *B.r J. Pharmacol.* 1999; 128: 925-933.
15. Jensen PE, Gong MC, Somlyo AV, and Somlyo AP. Separate upstream and convergent downstream pathways of G-protein- and phorbol ester-mediated Ca^{2+} sensitization of myosin light chain phosphorylation in smooth muscle. *Biochem. J.* 1996; 318:469-475.
16. Jin N, Packer CS, and Rhoades RA. Pulmonary arterial hypoxic contraction: signal transduction. *Am. J. Physiol.* 1992; 263: L73-L78.
17. Karamsetty MR, Klinger JR, and Hill NS. Evidence for the role of p38 MAP kinase in hypoxia-induced pulmonary vasoconstriction. *Am. J. Physiol. Lung Cell. Mol. Physiol.* 2002; 283: L859-L866.
18. Kimura K, Ito M, Amano M, Chihara K, Fukata Y, Nakafuku M, Yamamori B, Feng J, Nakano T, Okawa K, Iwamatsu A, and Kaibuchi K. Regulation of myosin phosphatase by Rho and Rho-associated kinase (Rho-kinase). *Science.* 1996; 273: 245-248.
19. Kitazawa T, Eto M, Woodsome TP, and Brautigan DL. Agonists trigger G protein-mediated activation of the CPI-17 inhibitor phosphoprotein of myosin light chain phosphatase to enhance vascular smooth muscle contractility. *J. Biol. Chem.* 2000; 275: 9897-9900.
20. Koyama M, Ito M, Feng J, Seko T, Shiraki K, Takase K, Hartshorne DJ, and Nakano T. Phosphorylation of CPI-17, an inhibitory phosphoprotein of smooth muscle myosin phosphatase, by Rho-kinase. *FEBS Lett.* 2000: 475: 197-200.
21. Leach RM, Hill HM, Snetkov VA, Robertson TP, and Ward JPT. Divergent roles of glycolysis and the mitochondrial electron transport chain in hypoxic pulmonary vasoconstriction of the rat: identity of the hypoxic sensor. *J Physiol.* 2001: 536: 211-224.
22. Leach RM, Robertson TP, Twort CH, and Ward JP. Hypoxic vasoconstriction in rat pulmonary and mesenteric arteries. *Am. J. Physiol.* 1994; 266: L223-L231.
23. Leung T, Manser E, Tan L, and Lim L. A novel serine/threonine kinase binding the Ras-related RhoA GTPase which translocates the kinase to peripheral membranes. *J. Biol. Chem.* 1995; 270: 29051-29054.
24. Littler CM, Morris KG Jr, Fagan KA, McMurtry IF, Messing RO, and Dempsey EC. Protein kinase C-ε-null mice have decreased hypoxic pulmonary vasoconstriction. *Am. J. Physiol. Heart Circ. Physiol.* 2003; 284: H1321-H1331.
25. Morgan JP and Morgan KG. Stimulus-specific patterns of intracellular calcium levels in smooth muscle of ferret portal vein. *J. Physiol.* 1984; 351: 155-167.
26. Morrison DL, Sanghera JS, Stewart J, Sutherland C, Walsh MP, and Pelech SL. Phosphorylation and activation of smooth muscle myosin light chain kinase by MAP kinase and cyclin-dependent kinase-1. *Biochem. Cell Biol.* 1996; 74: 549-557.
27. Nakao F, Kobayashi S, Mogami K, Mizukami Y, Shirao S, Miwa S, Todoroki-Ikeda N, Ito M, and Matsuzaki M. Involvement of Src family protein tyrosine kinases in Ca^{2+} sensitization of coronary artery contraction mediated by a sphingosylphosphorylcholine-Rho-kinase pathway. *Circ. Res.* 2002; 91 :953-960.
28. Niiro N, Koga Y, and Ikebe M. Agonist-induced changes in the phosphorylation of the myosin- binding subunit of myosin light chain phosphatase and CPI17, two regulatory factors of myosin light chain phosphatase, in smooth muscle. *Biochem. J.* 2003; 369: 117-128.

29. Nishimura J, Khalil RA, and van Breemen C. Agonist-induced vascular tone. *Hypertension.* 1989; 13: 835-844.
30. Nixon JS, Wilkinson SE, Davis PD, Sedgwick AD, Wadsworth J, and Westmacott D. Modulation of cellular processes by H7, a non-selective inhibitor of protein kinases. *Agents Actions.* 1991; 32: 188-193.
31. Ogura T, Imanishi S, and Shibamoto T. Activation of background membrane conductance by the tyrosine kinase inhibitor tyrphostin A23 and its inactive analog tyrphostin A1 in guinea pig ventricular myocytes. *Jpn. J. Pharmacol.* 2001; 87: 235-239.
32. Orton EC, Raffestin B, and McMurtry IF. Protein kinase C influences rat pulmonary vascular reactivity. *Am. Rev. Respir. Dis.* 1990; 141: 654-658.
33. Robertson TP, Aaronson PI, and Ward JPT. Ca^{2+} sensitization during sustained hypoxic vasoconstriction is endothelium-dependent. *Am. J. Physiol. Lung Cell. Mol. Physiol.* 2003; L1162-L1126.
34. Robertson TP, Aaronson PI, and Ward JPT. Hypoxic vasoconstriction and intracellular Ca^{2+} in pulmonary arteries: evidence for PKC-independent Ca^{2+} sensitization. *Am. J. Physiol.* 1995; 268: H301-H307.
35. Robertson TP, Clapham J, and Ward JPT. A vasoactive substance with a molecular wt less-than-3,000 released during hypoxic perfusion of isolated rat lungs, constricts isolated rat pulmonary but not mesenteric resistance arteries. *Br. J. Pharmacol.* 1994; 111: 277.
36. Robertson TP, Dipp M, Ward JPT, Aaronson PI, and Evans AM. Inhibition of sustained hypoxic vasoconstriction by Y-27632 in isolated intrapulmonary arteries and perfused lung of the rat. *Br. J. Pharmacol.* 2000; 131: 5-9.
37. Robertson TP, Hague D, Aaronson PI, and Ward JPT. Voltage-independent calcium entry in hypoxic pulmonary vasoconstriction of intrapulmonary arteries of the rat. *J. Physiol.* 2000; 525: 669-680.
38. Robertson TP and Lewis SJ. Hypoxic pulmonary vasoconstriction in conscious rats: lack of effects of endothelin-1 receptor blockade. *J. Physiol.* 2002; 547: PC85.
39. Robertson TP, Ward JPT, and Aaronson PI. Hypoxia induces the release of a pulmonary-selective, Ca^{2+}-sensitising, vasoconstrictor from the perfused rat lung. *Cardiovasc. Res.* 2001; 50: 145-150.
40. Robertson TP and Ward JPT. The effect of acute-hypoxia on intracellular calcium and tension in isolated small pulmonary-arteries of the rat. *J. Physiol.* 1993; 467: 55.
41. Sasaki M, Hattori Y, Tomita F, Moriishi K, Kanno M, Kohya T, Oguma K, and Kitabatake A. Tyrosine phosphorylation as a convergent pathway of heterotrimeric G protein- and Rho protein-mediated Ca^{2+} sensitization of smooth muscle of rabbit mesenteric artery. *Br. J. Pharmacol.* 1998; 125: 1651-1660.
42. Savineau JP, Robertson TP, Delafeunte PG, Twort CHC, Marthan R, and Ward JPT. Effect of protein kinase C stimulation on tension and calcium concentration in the rat pulmonary artery - modulation by forskolin. *Br. J. Pharmacol.* 1992; 107:390.
43. Scott DA, Grotyohann LW, Cheung JY, and Scaduto RC Jr. Ratiometric methodology for NAD(P)H measurement in the perfused rat heart using surface fluorescence. *Am. J. Physiol.* 1994; 267: H636-H644.
44. Senba S, Eto M, and Yazawa M. Identification of trimeric myosin phosphatase (PP1M) as a target for a novel PKC-potentiated protein phosphatase-1 inhibitory protein (CPI17) in porcine aorta smooth muscle. *J. Biochem.* 1999; 125: 354-362.
45. Shimokawa H. Rho-kinase as a novel therapeutic target in treatment of cardiovascular diseases. *J. Cardiovasc. Pharmacol.* 2002; 39: 319-327.
46. Shin H-M, Je H-D, Gallant C, Tao TC, Hartshorne DJ, Ito M, and Morgan KG. Differential association and localization of myosin phosphatase subunits during agonist-induced signal transduction in smooth muscle. *Circ. Res.* 2002; 90: 546-553.
47. Smirnov SV and Aaronson PI. Inhibition of vascular smooth muscle cell K^+ currents by tyrosine kinase inhibitors genistein and ST 638. *Circ. Res.* 1995; 76: 310-316.
48. Somlyo AP, Kitazawa T, Himpens B, Matthijs G, Horiuti K, Kobayashi S, Goldman YE, and

Somlyo AV. Modulation of Ca^{2+}-sensitivity and of the time course of contraction in smooth muscle: a major role for protein phosphatases? *Advances in Protein Phosphorylation.* 1989; 5: 181-195.

49. Somlyo AP and Somlyo AV. Signal transduction and regulation in smooth muscle. *Nature.* 1994; 372: 231-236.
50. Somlyo AP and Somlyo AV. Signal transduction by G-proteins, Rho-kinase and protein phosphatase to smooth muscle and non-muscle myosin II. *J. Physiol.* 2000; 522: 177-185.
51. Steusloff A, Paul E, Semenchuk LA, Di Salvo J, and Pfitzer G. Modulation of Ca^{2+} sensitivity in smooth muscle by genistein and protein tyrosine phosphorylation. *Arch. Biochem. Biophys.* 1995; 320: 236-242.
52. Uehata M, Ishizaki T, Satoh H, Ono T, Kawahara T, Morishita T, Tamakawa H, Yamagami K, Inui J, Maekawa M, and Narumiya S. Calcium sensitization of smooth muscle mediated by a Rho-associated protein kinase in hypertension. *Nature.* 1997; 389: 990-994.
53. Uzun O, Demiryurek AT, and Kanzik I. The role of tyrosine kinase in hypoxic constriction of sheep pulmonary artery rings. *Eur. J. Pharmacol.* 1998; 358: 41-47.
54. Von Euler US and Liljestrand G. Observations on the pulmonary arterial blood pressure in the cat. *Acta Physiol. Scand.*. 1946; 12: 301-320.
55. Wang Z, Jin N, Ganguli S, Swartz DR, Li L, and Rhoades RA. Rho-kinase activation is involved in hypoxia-induced pulmonary vasoconstriction. *Am. J. Respir. Cell Mol. Biol.* 2001; 25: 628-635.
56. Ward JPT, Hague D, and Aaronson PI. Effect of protein tyrosine kinase inhibition on hypoxic pulmonary vasoconstriction. *Am. J. Respir. Crit. Care Med.* 1999; 159: A570.
57. Weissmann N, Voswinckel R, Hardebusch T, Rosseau S, Ghofrani HA, Schermuly R, Seeger W, and Grimminger F. Evidence for a role of protein kinase C in hypoxic pulmonary vasoconstriction. *Am. J. Physiol.* 1999; 276: L90-L95.
58. Welling KL, Sander M, Ravn JB, Larsen B, Abildgaard U, and Amtorp O. Effect of alveolar hypoxia on segmental pulmonary vascular resistance and lung fluid balance in dogs. *Acta Physiol. Scand.* 1997: 161: 177-186.
59. Wiener CM and Sylvester JT. Effects of glucose on hypoxic vasoconstriction in isolated ferret lungs. *J Appl Physiol.* 1991; 70: 439-446.
60. Wilson HL, Dipp M, Thomas JM, Lad C, Galione A, and Evans AM. ADP-ribosyl cyclase and cyclic ADP-ribose hydrolase act as a redox sensor. A primary role for cyclic ADP-ribose in hypoxic pulmonary vasoconstriction. *J. Biol. Chem.* 2001; 276: 11180-11188.
61. Zhao Y, Packer CS, and Rhoades RA. The vein utilizes different sources of energy than the artery during pulmonary hypoxic vasoconstriction. *Exp. Lung Res.* 1996; 22: 51-63.

III. ROLE OF ION CHANNELS IN HYPOXIC PULMONARY VASOCONSTRICTION

Chapter 8

Regulation of Ion Channels in Pulmonary Artery Smooth Muscle Cells

Sergey V. Smirnov
University of Bath, Bath, U.K.

1. Introduction

In the pulmonary circulation, a moderate decrease in alveolar oxygen tension (Po_2) results in the development of hypoxic pulmonary vasoconstriction (HPV). This response is unique to the pulmonary circulation; systemic arteries generally respond to hypoxia with vasodilatation (10). Although the precise mechanisms of HPV are still not completely understood, it is commonly agreed upon that ion channels make an important contribution to the development of HPV (24, 32, 56, 57).

Based on their ion selectivity, ion channels can be divided into several major classes: K^+ channels, Ca^{2+} channels, Na^+ channels, Cl^- channels, and non-selective cation channels. The expression level of individual subtypes of ion channels varies in different vascular beds (7, 33, 54), probably reflecting diverse mechanisms which contribute to the control of vascular tone in various arteries. The unique response of the pulmonary circulation to hypoxia may suggest a unique expression and mechanisms of the regulation of ion channels in pulmonary artery (PA) smooth muscle cells (SMCs) and the current status of our knowledge about understanding these mechanisms will be the main focus of this chapter.

2. K^+ Channels

It is generally accepted that a balance between constitutively active K^+ channels and voltage-dependent Ca^{2+} channels (VDCC) is an important control mechanism of arterial tone (33). Under physiological conditions, the pulmonary circulation is a high flow, low resistance and low pressure system with measured resting membrane potential between -60 and -50 mV that does not reveal spontaneous electrical activity (10). K^+ channel blockers, e.g., 4-aminopyridine

(4-AP) or tetraethylammonium (TEA), depolarize the cell membrane, increase intracellular Ca^{2+} concentration ($[Ca^{2+}]_i$) and induce contraction of intact pulmonary arteries (24, 32, 56, 57), indicating a key role for K^+ channels in the maintenance of the negative resting potential in PASMCs. The first recording of K^+ currents in the pulmonary artery was performed by Buryi and Gurkovskaia in 1980 in a smooth muscle strip from the rabbit main pulmonary artery using the double sucrose gap method (12). Simultaneous development of the patch clamp technique, isolation of viable single vascular SMCs, and molecular biology techniques equipped investigators with potent experimental tools and enabled substantial progress in our understanding of the function and molecular nature of ion channels in vascular SMCs in general and in PASMCs in particular.

Functional K^+ channels are multimers of α-subunits which form the ionic pore. More than 70 α-subunits have now been identified which differ in structure, voltage-dependence, kinetics and pharmacology. Functional diversity of K^+ channels is further increased by the ability of different α-subunits to form heteromultimers and by the presence of α-subunits splice variants and regulatory (auxiliary) β-subunits. Not all types of K^+ channels are expressed in the vasculature. Currently, four main groups of K^+ currents have been described in vascular SMCs: voltage-gated (K_V) and large conductance Ca^{2+}-activated (BK_{Ca}) currents, which are encoded by α-subunits belonging to the six transmembrane segment and one pore domain class of K^+ channels, and ATP-sensitive (K_{ATP}) and inward rectifier (K_{ir}) currents which belong to the two transmembrane segment and one pore domain class of K^+ channels (see Ref. 7 for additional information about the molecular structure and properties of K^+ channels in vascular SMCs). K_V, and BK_{Ca}, and K_{ATP}, currents have been identified in single SMCs isolated from rat (4, 48, 59), mouse (5), rabbit (14, 15), dog (41), and human (19, 39) PASMCs. Currently, there is no evidence that K_{ir} channels are present in PASMCs. In addition, a novel non-inactivating K^+ current, termed I_{KN}, which is proposed to be formed by TASK-1 channels (25), has been described in PASMCs (21).

The complexity and heterogeneity of the K^+ channel expression pattern in PASMCs isolated from different species, and the lack of selective pharmacological tools, makes the investigation of the regulation of K^+ channels by oxygen and second messenger systems a challenging task. Our current understanding of these issues will now be discussed for each type of K^+ channel expressed in PASMCs.

2.1. K_V Channels

The main K_V current found in most types of PASMCs belongs to the delayed rectifier type (referred here and thereafter as K_V current) which is characterised by a relatively slow rate of activation and inactivation, in contrast to the rapidly-inactivating A-type voltage-gated current, also found in some types of PASMCs (32). Post et al. (41) first demonstrated that K_V currents were reversibly inhibited by hypoxia in canine PASMCs (41). Therefore, it was proposed that these channels could act as both potential oxygen sensors and mediators of HPV (57). Since then, research in pulmonary K^+ channel physiology has focused on determining the molecular identity of the K_V currents expressed in PASMCs, on whether molecular correlates of the K_V currents are modulated by oxygen, and on establishing the oxygen-sensing mechanism of K_V channels. Using RT-PCR and immunoblotting, the expression of mRNA and/or protein of $K_V1.1$, $K_V1.2$, $K_V1.3$, $K_V1.4$, $K_V1.5$ $K_V1.6$, $K_V2.1$, $K_V3.1b$ α-subunits and a "silent" $K_V9.3$ α-subunit has been demonstrated in PASMCs (8, 35, 38, 65). Electrophysiological and pharmacological characteristics of the heteromeric $K_V1.2/K_V1.5$, $K_V2.1/K_V9.3$ or homomeric $K_V3.1b$ α-subunits have proven to be the closest match to those of the native K_V currents in PASMCs (8, 17, 28, 35, 51). The presence of $K_V1.5$ and $K_V2.1$ isoforms in PASMCs was also confirmed by using specific antibodies to inhibit the native K_V currents (8, 26). Hence, a consensus is now emerging that $K_V1.2/K_V1.5$ and $K_V2.1/K_V9.3$ heteromultimers and $K_V3.1b$ homomultimers are the primary candidates underlying native K_V channels in PASMCs (17).

Moreover, when expressed in heterologous systems, both $K_V1.2/K_V1.5$, $K_V2.1$ (26), $K_V2.1/K_V9.3$ (28, 38) and $K_V3.1b$ (35) channel currents are inhibited by acute hypoxia, making these K_V isoforms suitable candidates for oxygen-sensitive K^+ channels in PASMCs. Since $K_V3.1b$ single channel currents are inhibited in inside-out patches, it has been proposed that $K_V3.1b$ α-subunits can sense oxygen directly (35). However, the dependence of oxygen sensitivity of the $K_V1.2/K_V1.5$ and $K_V2.1/K_V9.3$ heteromultimeric channels depends on the expression system, suggests that oxygen sensing by these K_V α-subunits depends on another as yet unidentified mechanism(s) (17). A potential role of auxiliary β-subunits, which interact primarily with α-subunits and are expressed in pulmonary arteries, has been suggested (17), but has not been directly confirmed. How oxygen inhibits homomeric $K_V2.1$ and heteromeric $K_V2.1/K_V9.3$ channels also remains unknown.

Hypoxic alteration in the cellular redox state defined by reactive oxygen species (ROS) or by the ratio GSH/GSSG or NAD(P)H/NAD(P)$^+$ may also affect K_V channel activity. Indeed, whole-cell K^+ currents in rat PASMCs are inhibited by the mitochondrial electron transport chain inhibitors rotenone and antimycin A (2), by the mitochondrial uncouplers FCCP (61), deoxyglucose and reduced glutathione (GSH) (62), and by cytochrome P-450 inhibitors (63) in PASMC. However, direct effects of diphenyleneiodonium, an NADPH-oxidase inhibitor, on K^+ and VDCC channels, as well as the presence of HPV in an NADPH oxidase deficient mice (lacking the gp 91 phox subunit) (6) argues against the involvement of NADPH oxidase in hypoxia-mediated inhibition of K_V channels and HPV. Hence, the role of the cellular redox state in the modulation of the pulmonary K_V channels remains to be established.

Despite the fact that pulmonary vasoconstriction is strongly modulated by a variety of humoral agents (10), our understanding of pulmonary K_V channels' regulation by vasoactive mediators and intracellular second messenger systems remains sparse. Activation of the K_V current by nitric oxide (NO), mediated by cGMP-dependent protein kinase, occurs in extrapulmonary arterial SMCs (3), while another report has proposed a direct effect of NO on K_V currents (64). Although no modulation via a protein kinase A (PKA)-dependent mechanism has yet been demonstrated, another serine/threonine protein kinase C (PKC) can alter K_V currents in PASMCs. Endothelin 1 (ET-1) causes PKC-dependent inhibition of the K_V current and accelerates its inactivation (45). We have previously demonstrated that activators of PKC, diacylglycerol and arachidonic acid (AA), produce a dual effect on K_V currents in rat intrapulmonary SMCs: an initial PKC-dependent increase in the current amplitude followed by a predominant PKC-independent inhibition of the K_V current amplitude. The latter prevailed overall since AA caused membrane depolarisation in these cells (50). It is noteworthy that PKC-dependent phosphorylation may be required for the association of $K_V1.5$ α-subunits and auxiliary Kvβ1.3 subunits which modulates the kinetics of the K_V current in a heterologous expression system (30). Whether a similar mechanism exists in PASMCs remains to be established. The information about the regulation of K_V currents by protein tyrosine kinases is practically absent except for one negative report (49).

Direct inhibition of the K_V currents by intracellular divalent cations, calcium and magnesium, has also been proposed in canine PASMCs (23). However, perfusion of rat single PASMCs with high (~0.5 μM) $[Ca^{2+}]_i$ does not significantly affect the K_V current amplitude (48, 53), suggesting that, at least in this preparation, the K_V current is not sensitive to intracellular Ca^{2+}.

It is worth mentioning that a number of factors could be responsible for the lack of information about the regulation of K_V channels not only in PASMCs, but in vascular SMCs in general (54). Broadly speaking, the native K_V channel can be regulated via changes in its permeability and/or voltage-dependent gating. To monitor these changes, the choice of the experimental protocol and elimination of other conductances, particularly BK_{Ca} and K_{ATP} currents, are crucial (discussed by Beech et al. in Chapter 23 in Ref. 7). Another important reason is that, despite the progress in molecular biology, it is still not clear whether multiple K_V channel isoforms *functionally* coexist in the same SMC. This issue is complicated by the absence of selective pharmacological tools allowing us to distinguish between hetero- and homo-multimeric K_V channels expressed in PASMCs; the most effective experimental pharmacological tools are still 4-AP and TEA. Nevertheless, the use of these two inhibitors has proven useful in functional discrimination between different K_V currents in the rat pulmonary arterial tree where at least three cell subtypes, distinguished electrophysiologically on the basis of the expressed K_V currents, have been identified (51). Two cells subtypes, termed I_{K1} and I_{K2} cells because their characteristics closely resemble those of the members of the K_V1 (*Shaker*) and K_V2 (*Shab*) subfamilies respectively, are found in the main PA. The third cell subtype, identified in resistance PASMCs, therefore termed I_{Kr}, also has properties of the K_V1 type current which are different from those described for I_{K1} cells in the main PA. The key pharmacological features of I_{K1} and I_{Kr} currents are their different sensitivity to 4-AP and TEA, whereas I_{K2} currents are blocked by TEA and are relatively insensitive to high concentrations of 4-AP (51). It is noteworthy that the properties of the K_V current in adult rat aortic SMCs are similar to those of I_{K2} cells, while the K_V currents in newborn animals strongly resemble those in I_{K1} cells (11). Moreover, chronic exposure of adult rats to carbon monoxide increases the sensitivity of the whole cell current to 4-AP and, reciprocally reduces its sensitivity to TEA in coronary arterial SMCs, although no thorough characterization of K_V currents has been given in this study (9). These results could suggest that K_V channel expression is dynamically regulated during vascular development and in pathological conditions. Understanding the mechanisms which control this process in PASMCs is particularly important in chronic hypoxia (CH) which causes the downregulation of K_V channel genes (34, 40, 42, 53) and may represent an intriguing area of the future research.

The recent evidence discussed above, suggests that the existence of "I_{K1}" and "I_{K2}" cell subtypes may be a more general phenomenon in the vasculature, and may be partly responsible for the variation in responses to hypoxia and modulation of K_V currents by vasoactive agents (10). Since mice are rapidly gaining popularity as an experimental model, we have compared some properties of K_V currents in mouse small intrapulmonary arterial SMCs (Fig. 1A) with those of rat main PASMC I_{K1} (an example of the TEA-sensitivity of which is shown in Fig.1B) and I_{K2} currents. The electrophysiological characteristics and sensitivity to TEA of the mouse K_V currents (Fig. 1C) revealed a remarkable similarity to those described in "I_{K2}" type cells in rat main PA (51) and adult rat aorta (11). Although it is premature to rule out the contribution of TEA-sensitive $K_V3.1b$ channels, mRNA expression of which has been demonstrated in mouse lungs (5), it is noteworthy that a component of current with the pharmacological properties of $K_V3.1b$ channels (i.e., blocked by TEA with an IC_{50} in the μM range (7), was not detected (Fig. 1C). The pharmacological characteristics of "I_{K2}" type current include a moderate sensitivity to TEA (IC_{50} ranged between 2 and 3 mM, Fig. 1C, open symbols) and sensitivity to 4-AP in the mM range (11, 51), currently match only the characteristics of K_V2 channels (see references in Ref. 51). Thus, although a small contribution of other K_V channels cannot be entirely excluded, it is likely that mouse PASMCs predominantly express $K_V2.1$ channels. This is also supported by the $K_V2.1$ mRNA expression in the mouse lung (5). It is noteworthy that the presence of a K_V2-type current in mouse PASMCs may explain, at least in part, the lack of the inhibition of acute hypoxic vasoconstriction in NADPH oxidase deficient mice (6), since these channels probably sense oxygen via a different mechanism. Therefore, rat and mouse small intrapulmonary arterial SMCs can be a useful experimental model for investigation of mechanisms of the regulation of the K_V1 and K_V2 channel currents, respectively.

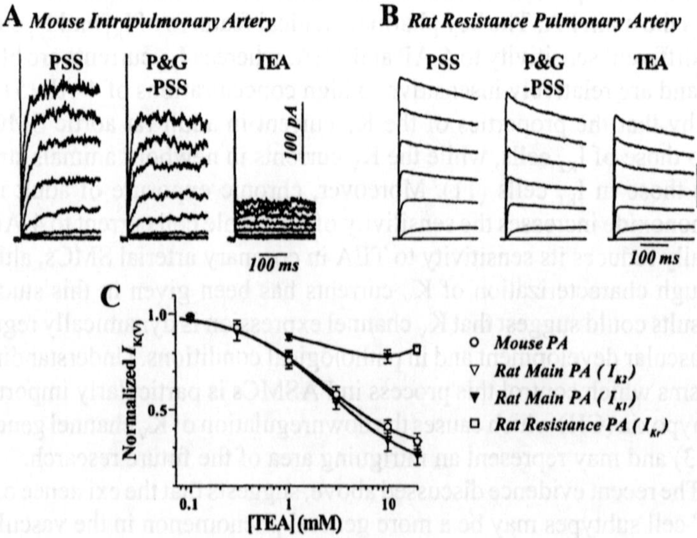

Figure 1. "I_{K1}" and "I_{K2}" cells types in rat and mouse PASMCs. Family of K_V currents shown in *A* and *B* were recorded between -40 and +80 mV in 20 mV increments from a holding potential of −80 mV using the same experimental conditions and protocols described in Refs. 11, 51. P&G-PSS is abbreviation for physiological saline solution (PSS) containing 1 μM paxilline and 10 μM glibenclamide. *C* Summary of TEA-sensitivity of K_V currents. TEA-sensitive ("I_{K2}" subtype) and TEA-insensitive ("I_{K1}" subtype) are shown by open and filled symbols, respectively. Smooth lines were drawn according to the equation described previously in Refs. 11 and 51, giving IC_{50} values of 2.3 (mouse PA) and 2.6 (for I_{K2} in main PASMCs) mM.

2.2 BK_{Ca} Channels

Another ubiquitous type of K^+ channel, BK_{Ca} channels, are activated by both voltage and intracellular Ca^{2+}. Although they are expressed in PASMCs (32) the relative contribution of the BK_{Ca} current to the whole cell current varies greatly in different species and different regions of the pulmonary arterial tree (1, 4, 14, 39, 41). The expression of a BK_{Ca} α-subunit has been demonstrated in rat PA (42). The single channel conductance of BK_{Ca} channels has been reported to lie in the range between 170 and 270 pS in symmetrical K^+ conditions. It is generally agreed that a functional BK_{Ca} channel in vascular SMCs exists as complex of α-subunits and regulatory β-subunits, which enhance Ca^{2+}-sensitivity of the channel (7). Additionally, BK_{Ca} channels, which originate from a single gene, undergo an extensive alternative splicing that may further add to the diversity of BK_{Ca} currents in vascular SMCs. Technically, therefore, regulation of the BK_{Ca} channels could occur via modulation of individual α- and/or β-subunits, or via functional uncoupling between α- and β-subunits in native BK_{Ca} channels. Thus, for example, the expression of different BK_{Ca} splice isoforms could explain the presence of high-conductance (245 and 185 pS) intracellular Mg^{2+}- and ATP-activated BK_{Ca} currents in rat PASMCs (1). The sensitivity of the BK_{Ca} current to $[Ca^{2+}]_i$ is apparently also low in rat intrapulmonary arterial SMCs (48, 53). It is not yet clear whether this is due to a low level of expression of the BK_{Ca} α-subunit or to decreased Ca^{2+} sensitivity of the expressed channels.

Owing to their Ca^{2+}-sensitivity, BK_{Ca} channels are activated due to the increased $[Ca^{2+}]_i$ caused by vasoconstricting mediators, resulting in vasodilatation. In PASMCs, BK_{Ca} channel activity is directly enhanced by membrane stretch and arachidonic acid (29, 52). Changes in the cell redox potential also affect BK_{Ca} currents. In rabbit small PASMCs, reducing agents such as dithiothreitol, GSH and NADH decreased, while an oxidizing agent (DTNB) increased the activity of BK_{Ca} channels (36). However, in SMC isolated rabbit conduit PA, BK_{Ca} channels were not affected by either reducing (GSH and NADH) or oxidizing (GSSG and NAD^+) agents except DTNB which enhanced the BK_{Ca} channel activity (55), similar to the effect of redox agents on BK_{Ca} currents in rabbit ear artery SMCs (36). The reason for such differences is not clear. However, it is possible that various BK_{Ca} isotypes, which respond differently to changes in the cellular redox state, are present in different regions of the pulmonary arterial tree. Since selective BK_{Ca} channel inhibitors do not affect acute HPV, it is unlikely that redox modulation of BK_{Ca} channels play a key role in HPV. However, their role can be enhanced in chronic hypoxia when K_V channel expression is reduced (46). Chronic hypoxia also decreases BK_{Ca} Ca^{2+} sensitivity and the ability of cGMP and NO to activate BK_{Ca} channels in cultured human PASMCs (39).

2.3. K_{ATP} Channels

The K_{ATP} channel is composed of four pore-forming inward rectifier K^+ channel subunits and four sulphonylurea receptors (SUR). Some voltage-independent K_{ATP} channels are inhibited by cytosolic ATP, while others are activated by nucleotide-diphosphates (NDP). In vascular SMCs, it is believed that native K_{ATP} and NDP-activated K_{ATP} channels (K_{NDP}) are formed by SUR2B/K_{ir}6.2 and SUR2BK_{ir}6.1, respectively. Both K_{ATP} and K_{NDP} are selectively blocked by sulphonylurea compounds such as glibenclamide, and are activated by levcromakalim (see Ref. 7 for details). Although the presence of K_{ATP} currents in PASMCs was demonstrated almost a decade ago (15), their molecular identity has only now become evident. Recently, the expression of K_{ir}6.1, and not K_{ir}6.2, and SUR2B, but not SUR1, has been detected in cultured human PASMCs (19), consisted with the presence of NDP-gated K_{ATP} currents in native cells. Interestingly, the reduction in cytosolic ATP enhanced the stimulatory effect of levcromakalim in PASMCs but not in HEK293 cells transfected with SUR2B/K_{ir}6.1, indicating the presence of specific regulatory mechanisms controlling native K_{ATP} currents in PASMCs (19). Although K_{ATP} channels can contribute to the control of the resting potential in isolated PASMCs (15, 53), it is generally believed that these channels are mostly inactive under normal conditions, but may be opened by changes in the intracellular ATP/ADP and/or GTP/GDP ratios, therefore providing a link between cellular metabolism and membrane excitability. In vascular SMCs, the activity of K_{ATP} channels is regulated via cAMP-dependent activation of protein kinase A (54). However, mechanisms of regulation, heterogeneity and molecular isoforms of K_{ATP} channels in various PASMCs, remains to be elucidated.

2.4. Two-pore Domain K^+ Channels

Two-pore domain K^+ channels are a rapidly growing family of K^+ channels. One member of this family, TASK-1, was recently identified in rabbit PASMCs (25). TASK-1 encodes a voltage-independent pH-sensitive background current and is also inhibited by hypoxia. In heterologous expression systems, TASK-1 currents are not sensitive to 4-AP and TEA, but are blocked by barium, quinine, quinidine and zinc (37). Similarities between some of the properties of TASK-1, including oxygen and pH sensitivity, and those of the non-inactivating I_{KN} have led to the suggestion that TASK-1 is a molecular correlate of I_{KN}, and that it may be a key oxygen-sensing K^+ channel in rabbit PASMCs (25).

3. Voltage-dependent Ca^{2+} Channels (VDCCs)

VDCCs are expressed in all vascular SMCs and are important players in the maintenance of vascular tone (33). Although two principal types of VDCCs, dihydropyridine-sensitive high-voltage-activated L-type and low-voltage-activated, rapidly-inactivating T-type, VDCCs have been characterised in vascular SMCs (33), electrophysiological evidence currently suggests that only L-type VDCCs are expressed in PASMCs (13, 22).

Despite the fact that L-type VDCC inhibitors are widely used to treat PPH, surprisingly little is known about the electrophysiological properties and regulation of L-type VDCCs in PASMCs. Clapp and Gurney (13) characterized nifedipine-sensitive Ca^{2+}- and Ba^{2+}-permeable VDCC currents in rabbit PASMCs (13). The amplitude of VDCC was reduced by ~45% by the vasosodilator sodium nitroprusside (SNP), suggesting that SNP can cause vasodilatation partly via inhibition of VDCC (13). Recently, an analysis of the current density of L-type VDCCs has revealed that it is greater in conduit than in resistance artery rabbit PASMCs (22). Intriguingly, acute hypoxia causes an increase in Ca^{2+}

currents (using Ba^{2+} as the permeating ion) and shifts the voltage-conductance dependence to the right along the voltage axes in resistance PASMCs; in conduit PASMCs, the current amplitude is decreased and the activation dependence is shifted in the opposite direction (22). The molecular mechanism of these effects has not been investigated. Similar upregulation of L-type VDCC currents also occurs in rat resistance PASMCs during CH (Fig. 2), suggesting that an increased expression of L-type VDCCs, together with simultaneous inhibition of K_V channels, contributes to membrane depolarisation and CH-induced pulmonary hypertension. Increased Ca^{2+} influx through L-type VDCC could also contribute to vascular remodelling in pulmonary arteries.

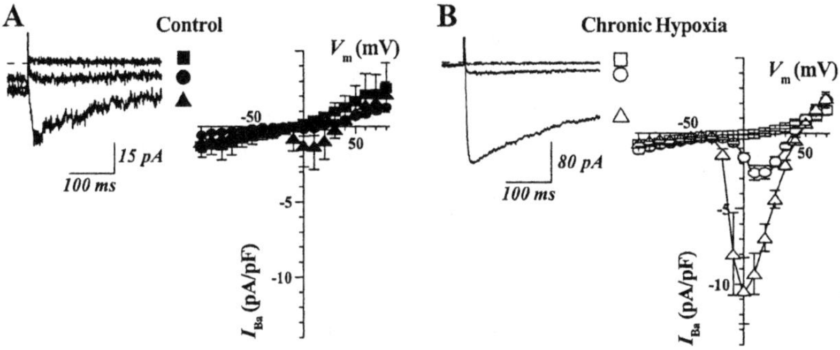

Figure 2. Up-regulation of the L-type VDCC currents by chronic hypoxia in rat resistance PASMCs. Mean current-voltage relationships (4-9 cells) for I_{Ba} recorded in the presence of 10 mM Ba^{2+} (circles), and 1 μM Bay K 8644 (triangles) or 10 μM nifedipine (squares), an activator and inhibitor of L-VDCCs, respectively, in control (A) and CH (3-4 weeks, B) animals. Inserts show I_{Ba} at 0 mV in representative cells in the absence and presence of Bay K 8644 and nifedipine as indicated by symbols. Currents were recorded using Cs^+-based, 10 mM EGTA-containing pipette solutions. Holding potential was -80 mV.

Another important class of calcium selective channels is transient receptors potential channels (TRPCs), which form the molecular basis of the store-operated or capacitative Ca^{2+} entry channels and agonist-activated nonselective cation channels in variety of tissues including Vascular SMCs. The role of TRPCs in the control of excitation and contraction of PASMCs has only recently been described and will be discussed in detail in a later chapter.

4. Voltage-gated Sodium (Na^+) Channels

Although small tetrodotoxin-sensitive Na^+ currents are found in some PASMCs (see Ref. 32 for references), it is generally agreed that voltage-gated Na^+ channels unlikely play any significant role in the regulation of pulmonary vascular tone.

5. Chloride (Cl⁻) Channels (ClCs)

The low intracellular Cl⁻ concentraction (30-50 mM) in vascular SMCs gives a calculated equilibrium potential of about ~-30 mV (31), such that activation of a Cl⁻ conductance at the resting potential leads to membrane depolarisation. Two types of ClCs, volume-sensitive ClC and Ca^{2+}-activated ClC (Ca^{2+}-ClC), have been identified so far in the pulmonary vasculature. Volume-sensitive ClC currents activated by external hypotonicity have been recently described in canine PASMCs (58). Since the ClC currents are suppressed by intracellular dialysis with anti-ClC-3 polyclonal antibody through the patch pipette, it has been proposed that the ClC-3 gene encodes volume-sensitive ClCs in canine PASMCs (20).

Ca^{2+}-ClCs are activated when $[Ca^{2+}]_i$ is increased via Ca^{2+} release from intracellular Ca^{2+} stores in response to vasoconstrictors such as ET-1 or ATP (43, 44), via Ca^{2+} entry through VDCC (60), or by flash photolysis from Nitr-5, which served as an artificial Ca^{2+} pool (16). In rat pulmonary and aortic SMCs activation of Ca^{2+}-ClC and BK_{Ca} currents by ET-1 was mediated via ET_B and ET_A receptors, respectively, (43), suggesting a different spatial organisation of the channels and/or receptors. Transient inward Cl⁻ currents, inhibited by 2-10 mM caffeine, were also recorded in rabbit PASMCs using the perforated patch technique (27). Ca^{2+}-ClC are also distributed differently along the rabbit pulmonary arterial tree being activated by $[Ca^{2+}]_i$ oscillations predominantly in resistance PASMCs in contrast to activation of BK_{Ca} currents in main PA myocytes (47). Because of the very small single channel conductance (1-3 pS, (31) and the unknown molecular identity of Ca^{2+}-ClC, the minimal $[Ca^{2+}]_i$ required for activation of these channels, and their general Ca^{2+}-sensitivity is not known. It has been proposed that Ca^{2+}-ClCs play an important role in agonist-induced membrane depolarisation (31, 32). Based on the above evidence, it is likely that physiological activation of Ca^{2+}-ClCs may require a relatively large local increase in $[Ca^{2+}]_i$ from intracellular stores in the close vicinity, as well as possible channel clustering.

6. Summary

Owning to the development of modern electrophysiological and molecular biology techniques, enormous progress in our understanding of the function, physiological significance and molecular identity of pulmonary ion channels has been made during last two decades since the first recording of the ion channel currents. This work has revealed a complexity and heterogeneity of ion channel expression in PASMCs, which depends on both the species and the site of the arterial vasculature under investigation, as well as multiple potential mechanisms for oxygen sensing by ion channels (24, 56). Therefore, the key questions in pulmonary ion physiology remain essentially the same: how does hypoxia causes HPV, which types of ion channels are involved in this process, and what is the mechanism(s) responsible for the oxygen sensitivity of ion channels involved? Since it is possible that hypoxia may not directly target the ion channels, but may alter their activity via unknown intracellular regulatory factors, the investigation of molecular mechanisms which specifically control the pulmonary ion channels, (particularly those regulating K^+ and Ca^{2+} transport) represents an important and challenging task for the future research in the field of pulmonary electrophysiology.

Acknowledgments

I would like to thank Dr Phil I. Aaronson for critical reading of this manuscript and the British Heart Foundation for supporting our research.

References

1. Albarwani S, Heinert G, Turner JL, and Kozlowski RZ. Differential K^+ channel distribution in smooth muscle cells isolated from the pulmonary arterial tree of the rat. *Biochem. Biophys. Res. Commun.* 1995; 208183-189.
2. Archer SL, Huang J, Henry T, Peterson D, and Weir EK. A redox-based O_2 sensor in rat pulmonary vasculature. *Circ. Res.* 1993; 73: 1100-1112.
3. Archer SL, Huang JMC, Hampl V, Nelson DP, Shultz PJ, and Weir EK. Nitric oxide and cGMP cause vasorelaxation by activation of a charybdotoxin-sensitive K channel by cGMP-dependent protein kinase. *Proc. Natl. Acad. Sci. USA.* 1994; 91: 7583-7587.
4. Archer SL, Huang JMC, Reeve HL, Hampl V, Tolarová S, Michelakis E, Weir EK, and Huang JM. Differential distribution of electrophysiologically distinct myocytes in conduit and resistance arteries determines their response to nitric oxide and hypoxia. *Circ. Res.* 1996; 78: 431-442.
5. Archer SL, London B, Hampl V, Wu X, Nsair A, Puttagunta L, Hashimoto K, Waite RE, and Michelakis ED. Impairment of hypoxic pulmonary vasoconstriction in mice lacking the voltage-gated potassium channel Kv1.5. *FASEB J.* 2001; 15: 1801-1803.
6. Archer SL, Reeve HL, Michelakis E, Puttagunta L, Waite R, Nelson DP, Dinauer MC, and Weir EK. O_2 sensing is preserved in mice lacking the gp91 phox subunit of NADPH oxidase. *Proc. Natl. Acad. Sci. USA.* 1999; 96: 7944-7949.
7. Archer SL, and Rusch NJ. *Potassium Channels in Cardiovascular Biology.* New York, NY: Kluwer Academic/Plenum Press, 2001.
8. Archer SL, Souil E, Dinh-Xuan AT, Schremmer B, Mercure JV, El Yaagoubi A, Nguyen-Huu L, Reeve HL, and Hampl V. Molecular identification of the role of voltage-gated K^+ channels, Kv1.5 and Kv2.1, in hypoxic pulmonary vasoconstriction and control of resting membrane potential in rat pulmonary artery myocytes. *J. Clin. Invest.* 1998; 101: 2319-2330.
9. Barbé C, Dubuis E, Rochetaing A, Kreher P, Bonnet P, and Vandier C. A 4-AP-sensitive current is enhanced by chronic carbon monoxide exposure in coronary artery myocytes. *Am. J. Physiol. Heart Circ. Physiol.* 2002; 282: H2031-H2038.
10. Barnes PJ and Liu SF. Regulation of pulmonary vascular tone. *Pharmacol. Rev.* 1995; 47: 87-131.
11. Belevych AE, Beck R, Tammaro P, Poston L, and Smirnov SV. Developmental changes in

the functional characteristics and expression of voltage-gated K^+ channel currents in rat aortic myocytes. *Cardiovasc. Res.* 2002; 54: 152-161.

12. Buryi VA and Gurkovskaia AV. Transmembrane ion currents in pulmonary artery smooth muscle. *Biull Eksper Biol. Med.* 1980; 90: 519-521.
13. Clapp LH and Gurney AM. Modulation of calcium movements by nitroprusside in isolated vascular smooth muscle cells. *Pflügers Arch.* 1991; 418: 462-470.
14. Clapp LH and Gurney AM. Outward currents in rabbit pulmonary artery cells dissociated with a new technique. *Exp. Physiol.* 1991; 76: 677-693.
15. Clapp LH and Gurney AM. ATP-sensitive K^+ channels regulate resting potential of pulmonary arterial smooth muscle cells. *Am. J. Physiol.* 1992; 262: H916-H920.
16. Clapp LH, Turner JL, and Kozlowski RZ. Ca^{2+}-activated Cl^- currents in pulmonary arterial myocytes. *Am. J. Physiol.* 1996; 270: H1577-H1584.
17. Coppock EA, Martens JR, and Tamkun MM. Molecular basis of hypoxia-induced pulmonary vasoconstriction: role of voltage-gated K^+ channels. *Am. J. Physiol. Lung Cell. Mol. Physiol.* 2001; 281: L1-L12.
18. Coppock EA and Tamkun MM. Differential expression of K_V channel α- and β-subunits in the bovine pulmonary arterial circulation. *Am. J. Physiol. Lung Cell. Mol. Physiol.* 2001; 281: L1350-L1360.
19. Cui Y, Tran S, Tinker A, and Clapp LH. The molecular composition of K_{ATP} channels in human pulmonary artery smooth muscle cells and their modulation by growth. *Am. J. Respir. Cell. Mol. Biol.* 2002; 26: 135-143.
20. Duan D, Zhong J, Hermoso M, Satterwhite CM, Rossow CF, Hatton WJ, Yamboliev I, Horowitz B, and Hume JR. Functional inhibition of native volume-sensitive outwardly rectifying anion channels in muscle cells and *Xenopus* oocytes by anti-ClC-3 antibody. *J. Physiol.* 2001; 531: 437-444.
21. Evans AM, Osipenko ON, and Gurney AM. Properties of a novel K^+ current that is active at resting potential in rabbit pulmonary artery smooth muscle cells. *J. Physiol.* 1996; 496: 407-420.
22. Franco-Obregón A and López-Barneo J. Differential oxygen sensitivity of calcium channels in rabbit smooth muscle cells of conduit and resistance pulmonary arteries. *J. Physiol.* 1996; 491: 511-518.
23. Gelband CH, Ishikawa T, Post JM, Keef KD, and Hume JR. Intracellular divalent cations block smooth muscle K^+ channels. *Circ. Res.* 1993; 73: 24-34.
24. Gurney AM. Multiple sites of oxygen sensing and their contributions to hypoxic pulmonary vasoconstriction. *Respir. Physiol. Neurobiol.* 2002; 132: 43-53.
25. Gurney AM, Osipenko ON, MacMillan D and Kempsill FE. Potassium channels underlying the resting potential of pulmonary artery smooth muscle cells. *Clin. Exp. Pharmacol. Physiol.* 2002; 29: 330-333.
26. Hogg DS, Davies AR, McMurray G, and Kozlowski RZ. $K_V2.1$ channels mediate hypoxic inhibition of I_{KV} in native pulmonary arterial smooth muscle cells of the rat. *Cardiovasc. Res.* 2002; 55: 349-360.
27. Hogg RC, Wang Q, Helliwell RM, and Large WA. Properties of spontaneous inward currents in rabbit pulmonary artery smooth muscle cells. *Pflügers Arch.* 1993; 425: 233-240.
28. Hulme JT, Coppock EA, Felipe A, Martens JR, and Tamkun MM. Oxygen sensitivity of cloned voltage-gated K^+ channels expressed in the pulmonary vasculature. *Circ. Res.* 1999; 85: 489-497.
29. Kirber MT, Ordway RW, Clapp LH, Walsh JV, Jr., and Singer JJ. Both membrane stretch and fatty acids directly activate large conductance Ca^{2+}-activated K^+ channels in vascular smooth muscle cells. *FEBS Lett.* 1992; 297: 24-28.
30. Kwak Y-G, Navarro-Polanco RA, Grobaski T, Gallagher DJ, and Tamkun MM. Phosphorylation is required for alteration of Kv1.5 K^+ channel function by the Kvβ1.3 subunit. *J. Biol. Chem.* 1999; 274: 25355-25361.
31. Large WA and Wang Q. Characteristics and physiological role of the Ca^{2+}-activated Cl^-

conductance in smooth muscle. *Am. J. Physiol.* 1996; 271: C435-C454.
32. Mandegar M, Remillard CV, and Yuan J-X. Ion channels in pulmonary arterial hypertension. *Prog. Cardiovasc. Dis.* 2002; 45: 81-114.
33. Nelson MT, Patlak JB, Worley JF, and Standen NB. Calcium channels, potassium channels, and voltage dependence of arterial smooth muscle tone. *Am. J. Physiol.* 1990; 259: C3-C18.
34. Osipenko ON, Alexander D, MacLean MR, and Gurney AM. Influence of chronic hypoxia on the contributions of non-inactivating and delayed rectifier K currents to the resting potential and tone of rat pulmonary artery smooth muscle. *Br. J. Pharmacol.* 1998; 124: 1335-1337.
35. Osipenko ON, Tate RJ, and Gurney AM. Potential role for Kv3.1b channels as oxygen sensors. *Circ. Res.* 2000; 86: 534-540.
36. Park MK, Lee SH, Lee SJ, Ho WK, and Earm YE. Different modulation of Ca-activated K channels by the intracellular redox potential in pulmonary and ear arterial smooth muscle cells of the rabbit. *Pflügers Arch.* 1995; 430: 308-314.
37. Patel AJ and Honoré E. Molecular physiology of oxygen-sensitive potassium channels. *Eur. Respir. J.* 2001; 18: 221-227.
38. Patel AJ, Lazdunski M, and Honoré E. Kv2.1/Kv9.3, a novel ATP-dependent delayed-rectifier K^+ channel in oxygen-sensitive pulmonary artery myocytes. *EMBO J.* 1997; 16: 6615-6625.
39. Peng W, Hoidal JR, Karwande SV, and Farrukh IS. Effect of chronic hypoxia on K^+ channels: regulation in human pulmonary vascular smooth muscle cells. *Am. J. Physiol.* 1997; 272: C1271-C1278.
40. Platoshyn O, Yu Y, Golovina VA, McDaniel SS, Krick S, Li L, Wang JY, Rubin LJ, and Yuan JX-J. Chronic hypoxia decreases K_V channel expression and function in pulmonary artery myocytes. *Am. J. Physiol. Lung Cell. Mol. Physiol.* 2001; 280: L801-L812.
41. Post JM, Hume JR, Archer SL, and Weir EK. Direct role for potassium channel inhibition in hypoxic pulmonary vasoconstriction. *Am. J. Physiol.* 1992; 262: C882-C890.
42. Reeve HL, Michelakis E, Nelson DP, Weir EK, and Archer SL. Alterations in a redox oxygen sensing mechanism in chronic hypoxia. *J. Appl. Physiol.* 2001; 90: 2249-2256.
43. Salter KJ and Kozlowski RZ. Endothelin receptor coupling to potassium and chloride channels in isolated rat pulmonary arterial myocytes. *J. Pharmacol. Exp. Ther.* 1996; 279: 1053-1062.
44. Salter KJ, Turner JL, Albarwani S, Clapp LH, and Kozlowski RZ. Ca^{2+}-activated Cl^- and K^+ channels and their modulation by endothelin-1 in rat pulmonary arterial smooth muscle cells. *Exp. Physiol.* 1995; 80: 815-824.
45. Shimoda LA, Sylvester JT, and Sham JSK. Inhibition of voltage-gated K^+ current in rat intrapulmonary arterial myocytes by endothelin-1. *Am. J. Physiol. Lung Cell. Mol. Physiol.* 1998; 274: L842-L853.
46. Shimoda LA, Sylvester JT, and Sham JSK. Chronic hypoxia alters effects of endothelin and angiotensin on K^+ currents in pulmonary arterial myocytes. *Am. J. Physiol. Lung Cell. Mol. Physiol.* 1999; 277: L431-L439.
47. Smani T, Iwabuchi S, López-Barneo J, and Ureña J. Differential segmental activation of Ca^{2+}-dependent Cl^- and K^+ channels in pulmonary arterial myocytes. *Cell Calcium.* 2001; 29: 369-377.
48. Smirnov SV and Aaronson PI. Alteration of the transmembrane K^+ gradient during development of delayed rectifier in isolated rat pulmonary arterial cells. *J. Gen. Physiol.* 1994; 104: 241-264.
49. Smirnov SV and Aaronson PI. Inhibition of vascular smooth muscle cell K^+ currents by tyrosine kinase inhibitors genistein and ST 638. *Circ. Res.* 1995; 76: 310-316.
50. Smirnov SV and Aaronson PI. Modulatory effects of arachidonic acid on the delayed rectifier K^+ current in rat pulmonary arterial myocytes. Structural aspects and involvement of protein kinase C. *Circ. Res.* 1996; 79: 20-31.
51. Smirnov SV, Beck R, Tammaro P, Ishii T, and Aaronson PI. Electrophysiologically distinct smooth muscle cell subtypes in rat conduit and resistance pulmonary arteries. *J. Physiol.*

2002; 538: 867-878.

52. Smirnov SV, Knock GA, and Aaronson PI. Effects of the 5-lipoxygenase activating protein inhibitor MK886 on voltage-gated and Ca^{2+}-activated K^+ currents in rat arterial myocytes. *Br. J. Pharmacol.* 1998; 124: 572-578.
53. Smirnov SV, Robertson TP, Ward JPT, and Aaronson PI. Chronic hypoxia is associated with reduced delayed rectifier K^+ current in rat pulmonary artery muscle cells. *Am. J. Physiol.* 1994; 266: H365-H370.
54. Standen NB and Quayle JM. K^+ channel modulation in arterial smooth muscle. *Acta Physiol. Scand.* 1998; 164: 549-557.
55. Thuringer D and Findlay I. Contrasting effects of intracellular redox couples on the regulation of maxi-K channels in isolated myocytes from rabbit pulmonary artery. *J. Physiol.* 1997; 500: 583-592.
56. Ward JP and Aaronson PI. Mechanisms of hypoxic pulmonary vasoconstriction: can anyone be right? *Respir. Physiol.* 1999; 115: 261-271.
57. Weir EK and Archer SL. The mechanism of acute hypoxic pulmonary vasoconstriction: the tale of two channels. *FASEB J.* 1995; 9: 183-189.
58. Yamazaki J, Duan D, Janiak R, Kuenzli K, Horowitz B, and Hume JR. Functional and molecular expression of volume-regulated chloride channels in canine vascular smooth muscle cells. *J. Physiol.* 1998; 507: 729-736.
59. Yuan X-J. Voltage-gated K^+ currents regulate resting membrane potential and $[Ca^{2+}]_i$ in pulmonary arterial myocytes. *Circ. Res.* 1995; 77: 370-378.
60. Yuan X-J. Role of calcium-activated chloride current in regulating pulmonary vasomotor tone. *Am. J. Physiol.* 1997; 272: L959-L968.
61. Yuan X-J, Sugiyama T, Goldman WF, Rubin LJ, and Blaustein MP. A mitochondrial uncoupler increases K_{Ca} currents but decreases K_V currents in pulmonary artery myocytes. *Am. J. Physiol.* 1996; 270: C321-C331.
62. Yuan X-J, Tod ML, Rubin LJ, and Blaustein MP. Deoxyglucose and reduced glutathione mimic effects of hypoxia on K^+ and Ca^{2+} conductances in pulmonary artery cells. *Am. J. Physiol.* 1994; 267: L52-L63.
63. Yuan X-J, Tod ML, Rubin LJ, and Blaustein MP. Inhibition of cytochrome P-450 reduces voltage-gated K^+ currents in pulmonary arterial myocytes. *Am. J. Physiol.* 1995; 268: C259-C270.
64. Yuan X-J, Tod ML, Rubin LJ, and Blaustein MP. NO hyperpolarizes pulmonary artery smooth muscle cells and decreases the intracellular Ca^{2+} concentration by activating voltage-gated K^+ channels. *Proc. Natl. Acad. Sci. USA.* 1996; 93: 10489-10494.
65. Yuan X-J, Wang J, Juhaszova M, Golovina VA, and Rubin LJ. Molecular basis and function of voltage-gated K^+ channels in pulmonary arterial smooth muscle cells. *Am. J. Physiol. Lung Cell. Mol. Physiol.* 1998; 274: L621-L635.

Chapter 9

Regulation of O_2-sensitive K^+ Channels by a Mitochondrial Redox Sensor: Implications for Hypoxic Pulmonary Vasoconstriction

Rohit Moudgil, Evangelos D. Michelakis, Stephen L. Archer
University of Alberta, Edmonton, Canada

1. Introduction

The survival of higher organisms is dependent on an adequate supply of O_2 and substrates. A variety of O_2 sensors underlie the various adaptive mechanisms (changes in regional blood flow, hormone release and ventilation) that ensure optimal O_2 supply. One such mechanism is hypoxic pulmonary vasoconstriction (HPV). This chapter will discuss evidence for a comprehensive vascular O_2 sensor mechanism which involves redox signaling between smooth muscle cell (SMC) mitochondria and plasmalemmal K^+ channels (Fig. 1). In adult mammals, under physiological conditions, the pulmonary vessels deliver deoxygenated blood to the lung's capillary bed where gas exchange takes place. In response to airway hypoxia, whether it is focal (e.g., pneumonia or atelectasis) or global (e.g., high altitude), the pulmonary artery (PA) supplying the hypoxic lung segment(s) constricts. If the airway hypoxia is localized, HPV redistributes perfusion to better oxygenated areas and optimizes systemic O_2 delivery (22) while minimally elevating net pulmonary vascular resistance (PVR). Regional blood flow is also determined by gravity (38). However, with global hypoxia less benefit is derived from HPV and the elevated PVR results in right ventricular hypertrophy. Inhibition of HPV, by sepsis (68) or drugs, results in ventilation perfusion mismatch and systemic hypoxemia. HPV can be exploited in single lung anesthesia to minimize blood flow to a lung that is intentionally made hypoxic to create a dry operative field (25, 127).

2. Comparative Physiology of O_2 Sensors

HPV is not the only homeostatic mechanism that optimizes systemic O_2 delivery. Mammals also have O_2 sensors in the systemic vasculature (carotid body) and airway (neuroepithelial body). In addition, there are powerful O_2 sensor systems that are primarily active in the fetus or during the transitional period at birth (e.g., the ductus arteriosus SMC and the adrenomedullary cell).

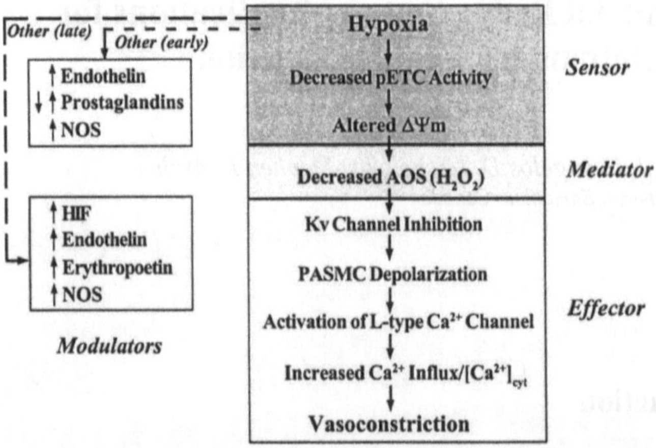

Figure 1. Proposed mechanism for HPV: An updated version of the redox hypothesis.

In adult mammals, PASMCs and the type 1 cells of the carotid body (CB) are the predominant O_2 sensing cells. Both these sensors respond rapidly (within seconds) to moderate hypoxia, in the airway (e.g., PASMC) or systemic (e.g., CB) circulation. The response of both cell types to hypoxia optimizes tissue O_2 delivery and ATP production. In the lung, hypoxia causes contraction of PASMC in small arteries, thereby redistributing blood to optimally-ventilated lobes; whilst in the carotid body, hypoxia activates type I cells leading to increased nerve discharge and hyperventilation. Although the response to hypoxia is tissue specific, most specialized O_2-sensitive systems share components of the mechanism underlying HPV. Specifically these rapidly responding cardiovascular and pulmonary sensors appear to involve a redox O_2 sensor (possibly the mitochondria or NADPH oxidase) and a membrane effector pathway (probably one or more O_2- and redox-sensitive K^+ channels (Fig. 2) (4). These acute compensatory pathways are distinct from but relevant to the slower onset mechanisms that are triggered when hypoxia persists, such as right ventricular hypertrophy, PA remodeling, and polycythemia, which result from activation of hypoxia inducible factor, HIF (131).

Additional O_2 sensors are also active in the fetus. The role of various O_2 sensors in fetus is somewhat different than in the adult, in large part because the normal fetal Po_2 is, by adult standards, very low (Po_2 ~20-30 mmHg). The ductus arteriosus (DA) connecting the pulmonary artery and aorta is one of the vessels that act as an O_2 sensor. In the normal fetus the placenta provides oxygenation. The DA is tonically open and shunts blood away from the nonexpanded fetal lungs. After the first breath, the increase in PO_2 triggers DA smooth muscle cell (DASMC) contraction, thereby directing blood to the pulmonary vasculature. In addition to the DA response, adrenomedullary cells also release catecholamines in response to hypoxia, which prepares the fetus for stress during labor (36, 48). It now appears that each of these O_2 sensing systems consists of a sensor that alters the production of a mediator in response to changes in PO_2. The mediator, in turn, alters the function of one or more effectors, which ultimately mediate the physiologic response of the system (73, 74). Teleologically it is optimal that the sensor responds to levels of hypoxia that are mild or brief enough so that ATP levels are preserved, averting tissue damage. An array of O_2 sensing systems has evolved to maintain the Po_2 within a tight physiological range, thus stabilizing ATP production and promoting survival during the periodic exposures to hypoxia that occurs in most aerobic lives. Often the sensor, mediator, and effector are linked in a functional unit. It appears that in many of these specialized tissues the sensor is the proximal portion of the mitochondrial electron transport chain (ETC), the mediators are ETC-derived activated oxygen species (AOS), and the effectors are redox-sensitive plasmalemmal K^+ channels. Although there may not be a single O_2 sensor, and despite the fact that response to hypoxia are modified by genetics and by neurohormonal factors, there is evidence suggesting that the proximal mitochondrial ETC is involved O_2 sensing in several tissues and species (7, 72).

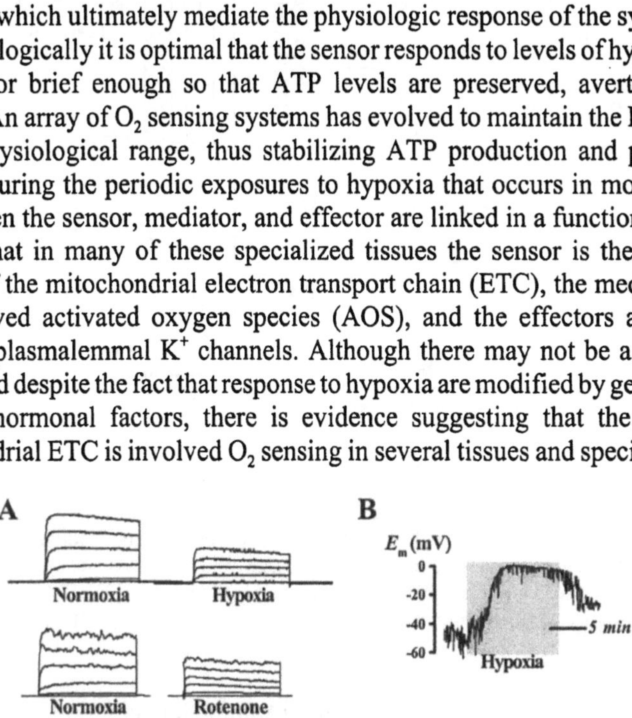

Figure 2. Hypoxia inhibits K^+ channels in rat resistance PASMCs. A: Whole cell K^+ current inhibition in response to hypoxia and rotenone (a complex I ETC blocker). B: PASMCs depolarize in response to hypoxia (Reproduced from Ref. 26).

3. Properties and Controversies of Hypoxic Pulmonary Vasoconstriction

Reductionist models are useful in defining subcellular mechanisms, however it is crucial that the defining characteristics of HPV are preserved in the experimental models. HPV is a response to the decrease in Po_2 of the terminal airways and alveoli. Thus, a nominal decrease in alveolar Po_2 giving rise to arterial Po_2 of over 55 mmHg can trigger HPV (117). Earlier studies have shown that mixed venous O_2 saturation failed to trigger pulmonary vasoconstriction (43), and HPV only occurs when a concomitant decrease in alveolar Po_2 is less than 50 mmHg (85). However a component of the hypoxic response may relate to pre-capillary PA Po_2 (67). HPV is strongest in resistance PA (43, 51). The hypoxic response is rapid and reversible within seconds (Fig. 3A) (46).

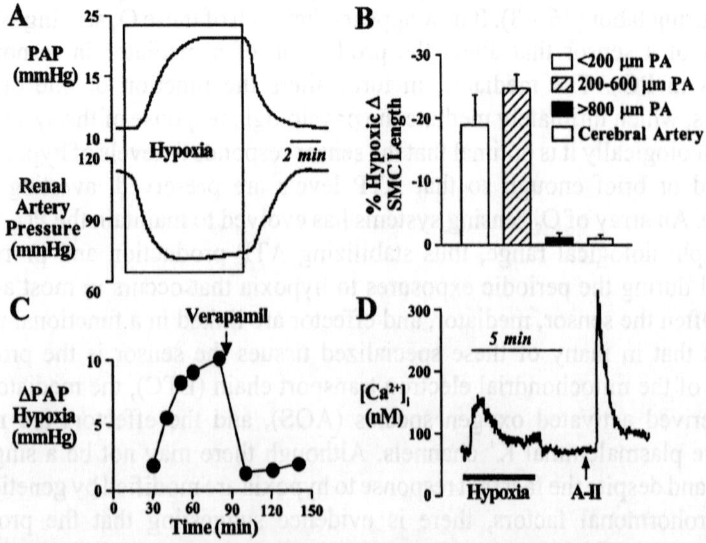

Figure 3. HPV is intrinsic to the resistance PA and depends on extracellular Ca^{2+}. A: When rat lungs and kidneys are perfused in series, hypoxia (induced by ventilating the lungs with hypoxic gas) causes pulmonary vasoconstriction and renal vasodilation in the presence of NO synthase and prostaglandin synthesis blockers. B: SMCs from small- (<200 μm) and medium- (200-600 μm) sized PAs constrict to hypoxia, whereas SMC from large (>800 μm) PAs and systemic (cerebral) arteries do not. C: The L-type Ca^{2+} channel blocker verapamil reverses HPV. D: $[Ca^{2+}]_i$ in PASMC increases in response to superfusion with a hypoxic solution or angiotensin II (A-II) (Reproduced from Refs. 17, 70).

Most studies of HPV in isolated lungs examine the initiation of HPV. Thus they focus on a short duration of hypoxic exposure and often differ in rationale and design from studies of sub-acute hypoxia in humans or in isolated PA rings. *In vivo*, it appears that HPV occurs in two continuous, constrictor phases. In humans, HPV increases in a progressive biphasic manner with a smaller initial rapid constrictor phase followed by a gradual increase in pulmonary vascular resistance over 2-3 hrs which then plateaus for at least 8 hours hypoxia (32). In PA rings, a biphasic response has also been noted, although there is more disagreement about the basis/importance of these two components of the response. Bennie et al. described a phase 1 constriction which was endothelium independent, and a later phase 2 constriction that was endothelium dependent (20). However, in another study of isolated arterial rings, Leach et al. found that hypoxia causes a phase 1 constriction reaching a peak within 2-3 min, followed by a slowly developing (>45 min) contraction (phase 2). They noted that the phase 1 transient contraction was similar in large and small PAs and mesenteric arteries whilst the sustained phase 2 response was seen only in the PAs (54). Removing the endothelium abolished phase 2 and had no significant effect on phase 1 in large PAs, but reduced phase 1 in small arteries by 40% (54). Although HPV *in vivo* is biphasic, the response is one of progressive constriction (i.e., there is not transient constriction which first abates before a more sustained phase begins). However, consistent with the proposed mechanism of HPV (126), it has been noted that verapamil substantially reduces phase 1 and abolished phase 2 constriction.

When resistance PAs were cannulated and pressurized to10-20 mmHg (similar to *in vivo* conditions), a monophasic contraction was achieved (59, 65). In the absence of an intact endothelium, HPV could be restored by pretreatment with endothelin (ET-1), suggesting that some priming effect by the endothelium may be necessary (59). However, we have found that a monophasic response can be elicited from non-pressurized resistance mouse PAs without pharmacological priming (12). Leach et al. also used priming to cause preconstriction (54). It is probable that intergroup differences in describing the endothelium-dependence of the response relate in part to issues such as the vascular segment studied and the use of priming (preconstriction). Moreover, removing endothelium from small PAs is challenging and may inadvertently damage the vessel. Nonetheless, numerous endothelium-derived vasoconstrictor substances enhance HPV (leukotrienes, endothelin) and numerous endothelium-derived vasodilators reduce HPV (nitric oxide, adrenomedullin, prostaglandin I_2) (6, 59). Consistent with this, hypoxic contraction has been elicited in isolated resistance PASMCs, and is notably absent in proximal PASMCs or SMCs from systemic arteries (122). Thus, while the endothelium unquestionably plays an important role in determining the amplitude and kinetics of HPV, there appears to be a core HPV response that can be elicited from the artery in the absence of endothelium. Likewise, the core of the carotid body's hypoxic response (63) and the O_2-response of the human DA (74) can be seen *ex vivo* in cellular preparations that lack endothelium. In the case of the ductus arteriosus, the O_2 sensor works normally in the presence of effective endothelin inhibition (73).

HPV and most other O_2 sensor systems are dependent on influx of extracellular Ca^{2+}, and the primary portal of entry is the L-type voltage-sensitive Ca^{2+} channel. L-type Ca^{2+} channel antagonists inhibit most (>80%) of the hypoxic constrictor response (Fig. 3C) whilst the Ca^{2+} channel agonist BAYK8644 enhances HPV (8, 36, 38, 69, 70). This dependence on extracellular Ca^{2+} for activation of the O_2 sensor pathway is also true in the ductus arteriosus (73, 120), the carotid body (78, 111) and the adrenomedullary cell (62). Although hypoxia also causes release of Ca^{2+} from intracellular pools (Fig. 3D), the relative role of intracellular release of Ca^{2+} remains controversial (123).

While the voltage-sensitive Ca^{2+} channels have some intrinsic O_2 sensitivity (33), they are largely responding to changes in membrane potential, as determined by K^+ channels (4). Lloyd demonstrated a Po_2-dependent contraction of PAs treated with the K^+ channel blocker procaine (60). Subsequently, it was shown that hypoxia depolarizes PASMCs. The implied ability of hypoxia to inhibit K^+ channels was directly demonstrated in 1992 (93) and was subsequently confirmed by other groups (113, 134). K^+ channel inhibition depolarizes the plasma membrane and activates the voltage-gated Ca^{2+} channel to increase intracellular [Ca^{2+}] (126). Hypoxia and metabolic inhibitors cause constriction only in the pulmonary circulation, whereas they cause vasodilation in most systemic vascular beds (10, 61, 103). Indeed, the response of even the proximal PAs to hypoxia is predominantly vasodilation (11, 20). The localization of the hypoxic response appears to result from diversity in the local expression of O_2-sensitive K^+ channels, with these voltage-gate K^+ channels (K_V) being predominantly functional in resistance (*vs* conduit) PAs (11), a finding recently confirmed by another group (112). Within the vasculature, hypoxia-sensitive whole cell K^+ current is specific to PASMC, and is not found in renal (93) or splanchnic (134) arterial SM. However, O_2 sensitive K^+ channels are found in all other O_2-sensitive tissues (26, 64, 86, 93, 120, 130, 134). Progress has been made recently in determining the molecular identity and regulatory pathways by which these channels respond to changes in Po_2.

In an interesting parallel to the carotid body, chronic hypoxia causes hypertrophy of PASMCs and diminishes the magnitude of the response to acute hypoxic ventilation. This is also true in PAs isolated from humans with chronic obstructive pulmonary disease (COPD). HPV was diminished in PAs from hypoxic patients (71), but was preserved in PAs from normoxic COPD patients (87). The magnitude of HPV *in vitro* was inversely related to the systemic PO_2 in these chronically patients (87). Thus both PAs and carotid bodies down-regulate their O_2 sensing functions in response to chronic hypoxia.

This chapter will specifically deal with the pathway of how O_2-sensitive K^+ channels and the network of mitochondria that permeate the vascular SMC interact to generate HPV. We will explore an increasingly accepted redox mechanism for HPV (3, 7, 10, 17, 72, 74), which may have relevance most O_2-sensitive tissues. This pathway involves PASMC mitochondria acting as redox sensors, producing activated oxygen species (AOS) in proportion to Po_2, which serve as diffusible mediators that modulate the activity of several O_2-sensitive K^+ channel (e.g., $K_V1.5$ and $K_V2.1$) (Fig. 4). These redox sensitive K^+ channels control tone through their effects on membrane potential and the L-type Ca^{2+} channel. Before delving into the interaction of all three factors as a "functional hypoxic-sensing unit", the characteristics of each component will be examined.

4. Mitochondria as Oxygen Sensors

The mitochondrion's role as the predominant site for O_2 consumption and ATP synthesis makes it an obvious candidate site for an O_2 sensor. The finding that inhibition of the mitochondrial electron transport chain (ETC) mimics hypoxia further supports this hypothesis (4). Indeed ETC inhibitors cause pulmonary vasoconstriction, systemic vasodilation and carotid body activation, a set of responses elicited by few other stimuli save hypoxia (4).

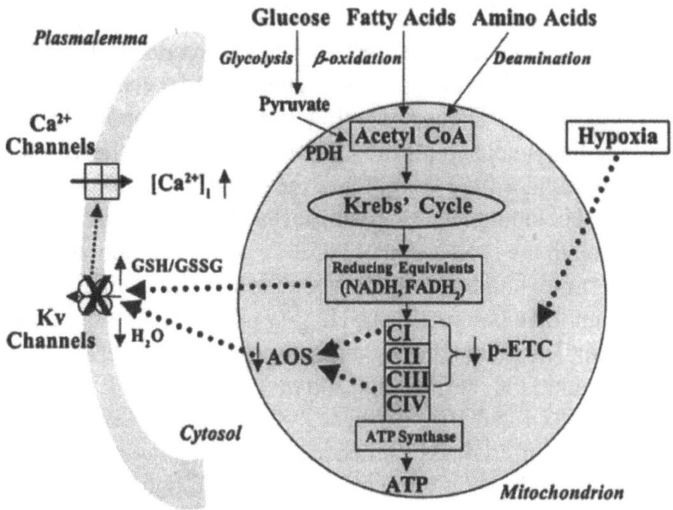

Figure 4: Mitochondria regulate K^+ channel function through the production of diffusible reducing equivalents and AOS in response to hypoxia. Hypoxia inhibits the proximal ETC (pETC), resulting in decreased production of AOS ($O_2^{\bullet-}$ and H_2O_2). The loss of tonic normoxic AOS reduces and inhibits K_V channels, depolarizing the PASMC and opening voltage-gated, L-type Ca^{2+} channels, causing $[Ca^{2+}]_i$ to rise and constriction to occur. PDH, pyruvate dehydogenase.

In 1981, Rounds and McMurtry (103) reported that certain inhibitors of ETC and oxidative phosphorylation (including azide, cyanide, antimycin A, and rotenone) mimicked the HPV in isolated blood-perfused lungs. Furthermore, the same mitochondrial inhibitors which block cytochrome c oxidase (cyanide or azide), stimulated hypoxia-mediated activation of the carotid body. These early studies suggested that the basis for this mimicry of HPV was a decrease in ATP production due to inhibition of oxidative phosphorylation. In support of this "energy hypothesis" inhibition of glycolysis also had similar effects as hypoxia or ETC inhibitors (79, 115). However, it appears that it is the mitochondrion's ability to alter cellular redox state and produce diffusible redox mediators, rather than its well-established role in producing ATP, that underlies its role as an O_2 sensor. Subsequent studies of HPV (induced by moderate hypoxia in perfused lungs) showed no association between HPV and depletion of ATP and adenylate charge (23). Although anoxia (Po_2 <10 mmHg) does cause ATP depletion, this results in pulmonary vasodilation, rather than constriction, in part by the activation of K_{ATP} channels (129). In the isolated rat lung, it has been reported that ATP and ATP/ADP ratios are preserved after exposure to an alveolar Po_2 of

7 mmHg, or to CO/O_2 ratios of 10/1 for up to 1 hr. Using ^{31}P-NMR, Buescher et al. showed no change in energy status in whole lungs during hypoxic ventilation when compared to its control normoxic counterpart (23). Moreover, other metabolic markers such as pH, phosphocreatinine, and phosphate are also conserved in PAs exposed to hypoxia (55). Thus, the theory that a change in energy status acts as an O_2 sensing mechanism seems unlikely. Teleologically, it would seem undesirable to wait until energy had been depleted before optimizing ventilation-perfusion matching (which could provide more O_2). Furthermore, the lungs consume little O_2 and, unlike in the heart, monitoring lung ATP levels provides little insight into the organ's function.

Another school of thought in O_2-sensing is that the mitochondria are involved in O_2 sensing through the production of AOS. In 1986, our laboratory proposed a link between mitochondrial AOS production, cellular redox status, K^+ conductance, membrane potential, and HPV (7). Subsequently, we found that inhibitors of complexes I and III, but not complex IV, cause pulmonary vasoconstriction, inhibit K^+ current (I_K), and prevent further HPV. Furthermore, inhibitors of complex I and III also cause systemic vasodilatation and do not inhibit I_K in systemic arterial SMCs, once again mimicking hypoxia (72). It is noteworthy that one of the putative mediators of HPV mimic hypoxia's opposing effects on the pulmonary and systemic vasculature (72). Further studies have shown that mitochondria generate superoxide anion ($O_2^{\bullet -}$) at complex I and III that is dismutated by mitochondrial specific manganese superoxide dismutase (MnSOD), generating the diffusible signal molecule hydrogen peroxide (H_2O_2). Several other groups have recently suggested that complex III may be more important than complex I (53, 125), although our group continues to find a role for both complexes in both the PA (72) and ductus arteriosus (74). The question remains as to how mitochondria are able to accomplish this task.

The O_2 sensing function of PASMC mitochondria is tied to the redox cascade within the ETC. Electrons from reduced nicotinamide adenine dinucleotide (NADH) and flavin adenine dinucleotide ($FADH_2$), are transferred down a redox gradient from potential of -0.35 for NADH/NAD^+ to +0.82 for O_2/H_2O (Fig. 5) (4). The terminal electron transfer to O_2 forms water. Four multicomponent megacomplexes accomplish the transfer of electrons. With each electron transferred down the ETC, a hydrogen ion is translocated across the inner mitochondrial membrane creating the very negative mitochondrial membrane potential ($\Delta\Psi_m$). It is the potential energy from this proton gradient that is utilized for ATP synthesis. Thus there is a link between electron donors, electron flux and production of AOS. It appears that complexes I and III are particularly important in O_2 sensing mechanisms because it is at these sites that AOS are generated (10, 19). The peroxides can then diffuse to the plasma membrane whilst it may be that superoxide anion can move through diisothiocyano-2,2 disulfonic acid stilbene (DIDS)- and voltage-sensitive anion channels, validating mitochondria as sources of cytosolic superoxide anion (39).

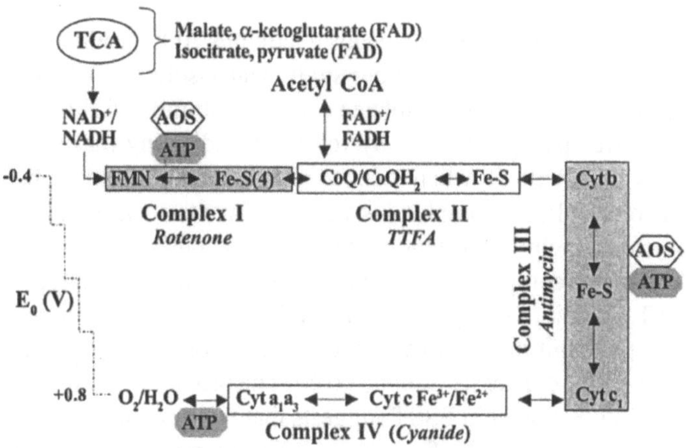

Figure 5: Sites of AOS generation in the mitochondrial ETC. Electrons flow down the ETC driven by a difference in the redox potential between -0.4 V in complex I and +0.8 V in complex IV. As the electrons flow towards their final target, O_2, they release electrochemical energy, which is used to synthesize ATP. Most of the "drop" in redox potential occurs in complexes I and III, sites that are responsible for most of the mitochondrial AOS production and ATP synthesis. Rotenone blocks complex I; thenoyltrifluoroacetone (TTFA) blocks complex II. Both agents inhibit electron entry to ubiquinone. Antimycin and cyanide are blockers of complexes III and IV, respectively.

4.1. Complex I (NADH Ubiquinone Oxidoreductase)

During oxidative phosphorylation, electrons pass through a series of membrane-bound multiprotein complexes that translocate protons across the membrane, resulting in a proton-motive force used by ATP synthase to make ATP (1). It couples the transfer of two electrons from NADH to ubiquinone to the translocation of four protons across the inner mitochondrial membrane. Complex I is the first of these electron transfer complexes and accounts for up to 40% of the pumped protons (50). In human mitochondria, complex I is also a major source of activated O_2 species. Complex I or the NADH-ubiquinone oxidoreductase is a macromolecular assembly of 43 subunits catalyzing electron transfer from NADH to ubiquinone through flavin mononucleotide and up to 7 iron-sulfur clusters (107). It is one of the most complex subunits known, with a molecular mass of about 1000 kDa (80). Thirty six of complex I's subunits are encoded by nuclear DNA whilst 7 are encoded by mitochondrial DNA. This includes several prosthetic groups including FMN, at least 7 Fe-S clusters, and five molecules of protein bound coenzyme Q; a natural acceptor of electrons. Complex I is the first site of oxidative phosphorylation but its function is still incompletely understood (21). It is hypothesized that Fe-S clusters might act as redox centers; however, the precise location within the cluster is still elusive.

Early experiments proved the involvement of complex I in AOS production. The addition of low concentrations of either NADH or NADPH, which feeds the electron to complex I at a slow rate, leads to copious AOS production, detected by lipid peroxidation (118). In another study, water-soluble CoQ homologues used as an electron acceptor from isolated complex I stimulated H_2O_2 production, which in turn was partly inhibited by rotenone, indicating that water-soluble quinones may react with O_2 when reduced at sites both prior and subsequent to the rotenone block (108). Complex I is very sensitive to a variety of structurally diverse toxicants, including rotenone, piericidin A, bullatacin, and pyridaben. Affinity labeling of complex I by rotenone analogues has shown recently that a 23 kDa PSST subunit of complex I is the likely binding site for these inhibitors and thus is the subunit that plays a key role in electron transfer by functionally coupling iron-sulfur cluster to quinone (108). The proximity of this binding site to a possible 4 Fe-S cluster in PSST suggests that this subunit is the site of final transfer of electrons, the Fe-S proteins to quinone and a likely site for semiquinone generation. There is evidence that the one-electron donor to oxygen in complex I is a non-physiological quinone reduction site. The cDNA of the human PSST protein was used to investigate tissue-specific expression and to localize the gene for this subunit to chromosome 19p13 (45). It is a hydrophilic site, which reduces several quinones to semiquinone, and since they are unstable, can reduce O_2 to $O_2^{\bullet-}$ (30). Thus, it may represent a potential site for $O_2^{\bullet-}$ generation. Indeed, duroquinone, a soluble analog of co-enzyme Q, constricts PAs and inhibits I_K in PASMCs (99).

In addition to direct studies for $O_2^{\bullet-}$ generation measurements, complex I deficiency (and its associated diseases) has been shown to generate AOS. Complex I deficiency displays a wide spectrum of phenotypes ranging from exercise intolerance to cataracts and development delay, which includes Leigh disease characterized by degeneration of basal ganglia and/or cardiomyopathy with or without cataract, and fatal infantile lactic acidosis (101). A more detailed investigation showed the correlation of increased levels of expression of MnSOD (a compensatory response) with a poor prognosis in complex I deficiency (91, 101). Excess AOS production is further complicated by damage to other systems such as mitochondrial aconitase, complex I, and succinate dehydrogenase which have functional Fe-S centers (site of $O_2^{\bullet-}$ reactivity) (34). This damage is pronounced in MnSOD knockout mice where $O_2^{\bullet-}$ produced in the mitochondria is not removed by the normal radical scavenger mechanisms (56).

4.2. Complex II and IV

Complex II is a flavoprotein dehydrogenase which uses succinate as an exogenous substrate. The main function of this complex is to transfer the electrons from $FADH_2$ (reduced ubiquinone) to FAD^+ (oxidized ubiquinol). Complex II is generally involved in electron flux from complex I to complex III. Similarly, complex IV, cytochrome oxidase, is composed of two cytochromes, a and a_3, and accepts electrons from reduced cytochrome c and passes them to

O_2. Thus, complexes II and IV are involved in shuttling electrons but play no role in generation of AOS. Inhibition of complex IV with cyanide does cause a transient pulmonary vasoconstriction, associated with increased AOS production (10). Recently we have demonstrated that while $\Delta\Psi_m$ does depolarize in response to cyanide in O_2-sensitive SMCs, these mitochondria are much less sensitive than those in cardiac cell lines, consistent with there being functionally different mitochondria in different cardiovascular tissues (72, 74).

4.3. Complex III (Ubiquinol-cytochrome *c* Oxidoreductase)

Complex III is responsible for taking reducing equivalents from complex II contained in ubiquinol and transferring them through reactions with cytochrome b, the Rieske Fe-S protein and cytochrome c_1, to the final electron acceptor cytochrome *c*. In the process, two species of semiquinone are generated. The Q-cycle mechanism proposed for the operation of the ubiquinol cytochrome c reductase outlines these two sites. Ubiquinol donates one electron to the Rieske Fe-S protein (a myxothiazol-inhibitor site) generating a semiquinone in proximity to the outer face of the inner membrane, which then reduces the first cytochrome b heme (b_L) (121). The second cytochrome b heme (b_H), situated closer to the matrix side of the membrane, accepts an electron from the first heme (b_L) and reduces ubiquinone to form ubisemiquinone and, with the passage of another electron, ubiquinol (the antimycin A-inhibitor site) (121). The transfer of the first electron is a relative low potential transfer with a redox potential of about +200 mV, however, the loss of the second electron from the resulting semiquinone anion to fully oxidized ubiquinol is much more energetic with a redox potential of about –200 mV. Thus, two ubiquinone molecules are oxidized at the outer site, transferring two electrons to cytochrome c (77).

The two inhibitors antimycin and myxothiazol, although they inhibit complex III electron transfer equally, have dramatically different effects on $O_2^{\bullet-}$ production. Blocking electron passage out of cytochrome b_H prevents the semiquinone from donating its electron and so, inhibition with antimycin, in some models, produces a more than tenfold increase in $O_2^{\bullet-}$ formation because it prevents ubisemiquinone formation at the cytosolic side of the inner mitochondrial membrane (57). However, in the pulmonary circulation we have found that antimycin, like rotenone, reduces normoxic generation of AOS (10, 19). Whether this reflects measurement errors or true differences in the effect of the inhibitors that relate to the type of mitochondria under study is unknown. Recent data suggest that there are significant differences in mitochondrial respiration and AOS production between lung and renal mitochondria (72).

However, the question remains about the relative importance and contribution of complexes I and III in O_2 sensing. First, let us consider the hemodynamic effects of ETC inhibitors (Fig. 6). Rotenone and antimycin, but not cyanide, cause an increase in PVR in isolated lungs, and a fall in renal vascular resistance, mimicking the effects of hypoxia. These opposing effects are also seen in isolated PA vs. RA rings (72). In the first systematic exploration of the theory that mitochondria are the pulmonary vascular O_2 sensors we showed that inhibiting complexes I and III (but not complex IV) caused PA constriction, inhibited subsequent HPV and like hypoxia reduced levels of AOS.

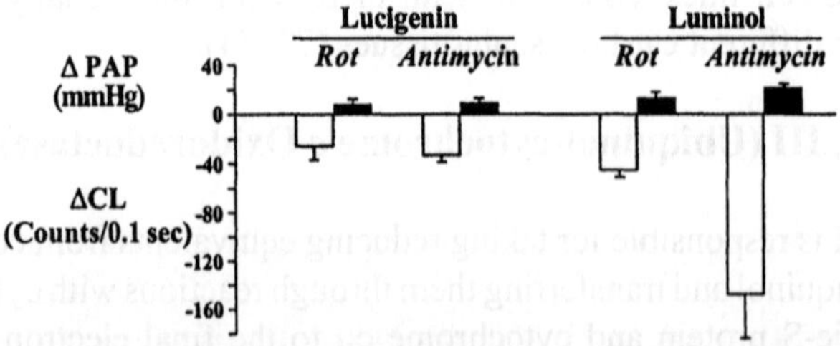

Figure 6: The effects of ETC inhibitors and hypoxia on AOS generation and PVR are concordant. In the Krebs'-albumin perfused isolated rat lungs, PA pressure reflects pulmonary vascular resistance because flow is held constant. Chemiluminescence, measured by two different methods (luminol and lucigenin enhancement), is decreased (open bars) whereas PA pressure is increased (solid bars) in response to rotenone and antimycin (Reproduced from Ref. 10).

Leach et al. reported that inhibition of complex I with rotenone resulted in inhibition of HPV which could not be reversed by bypassing complex I by providing succinate as a substrate for complex II (53). They found that succinate successfully reversed the rotenone-induced inhibition HPV. Moreover, inhibition of complex III abolished HPV. They concluded that complex III is the primary site for AOS production. However, studies published by our laboratory show otherwise (72, 74). Both rotenone and antimycin cause PA constriction of a similar magnitude to that elicited by hypoxia. Furthermore, these agents inhibit subsequent HPV (10). It is also noteworthy that chronic hypoxia decreases the constriction to both hypoxia and rotenone, while the pressor response to other vasoconstrictors is enhanced, suggesting again that hypoxia and certain ETC inhibitors share a common pathway (97). Cyanide does not mimic hypoxia in our hands (10), nor does it do so in studies by another group (125). We have also found that rotenone and antimycin, like hypoxia, decrease H_2O_2 levels regardless of the type of assays used (72). Thus, it provides evidence that perhaps both complexes I and III contribute to the AOS pool (72). Furthermore, a similar observation was made in DA where use of rotenone and antimycin A separately and in combination successfully relaxed O_2-preconstricted DA (74). In addition, these observations have been extended to the level of O_2-sensitive K^+ channel modulation and regulation of PA and DA tone. Thus, recent evidence suggests that both complexes I and III contribute to the AOS pool as suggested by our laboratory a decade ago. The difference in the conclusion reached by Leach et al. (53) and our laboratory might be due to the presence of PGF_2 and use of myxothiazol, a more proximal complex III inhibitor, in the perfusion system.

Moreover, use of rotenone in conjunction with succinate has been shown to generate reversed electron transport and methodological artifacts (114). Waypa's conclusion that the mitochondria function as O_2 sensors during hypoxia and that ROS generated in the proximal ETC act as second messengers to trigger calcium increases in PASMCs during acute hypoxia (125) mirrors our earlier conclusions (7, 10). However, our groups differ in two regards. We find AOS reduced in acute hypoxia (9, 13, 14, 97, 98) and we have evidence that complex I, as well as complex III, participate in the mechanism of HPV and O_2 responses in the human ductus arteriosus (74).

5. AOS as a Mediator

AOS from mitochondria and other cellular sources have been traditionally regarded as toxic by-products of metabolism with the potential to cause damage to lipids, proteins, and DNA. To protect against the potentially damaging effects of AOS, cells possess several isoforms of the antioxidant enzyme as superoxide dismutase (SOD; which reduces $O_2^{\bullet-}$ to H_2O_2), as well as catalase (which reduces H_2O_2 to H_2O and O_2). Thus, oxidative stress may be broadly defined as an imbalance between oxidant production and the antioxidant capacity of the cell to prevent oxidative injury. Oxidative stress has been implicated in a large number of human diseases including atherosclerosis, pulmonary fibrosis, cancer, neurodegenerative diseases and aging (29). Yet the relationship between oxidative stress and pathobiology of these diseases is not clear, largely due to a lack of understanding of the mechanisms by which AOS function in both physiological and disease states.

Accumulating evidence suggests that AOS are not always injurious by-products of cellular metabolism, but can also be important participants in cell signaling. The evidence can be found in bacteria as they use AOS to oxidize cysteine residues in oxidative stress-responsive protein (OxyR) to activate transcription (136). Similarly studies in *E. coli* showed that AOS was instrumental in oxidizing SoxR, triggering events which culminated in transcription of various factors including SOD (31). While bacteria use AOS as a physiological second messenger molecule, sea urchin uses H_2O_2 to form a protective envelope around the freshly fertilized oocyte thus enabling procreation in animals (110). In mammals, the role of AOS has been studied extensively and has been shown to be an integral second messenger molecule in events critical for immunomodulation, regeneration of tissues, maintaining vasoreactivity of blood vessels and neural conduction (119).

One of the primary AOS is $O_2^{\bullet-}$. Superoxide itself can be toxic, especially through inactivation of proteins that contain Fe-S centers such as aconitase, succinate dehydrogenase, and mitochondrial NADH-ubiquinone oxidoreductase (34). Fortunately, $O_2^{\bullet-}$ in aqueous solution is short-lived. This instability in aqueous solution is based on rapid dismutation of $O_2^{\bullet-}$ to H_2O_2, a reaction facilitated by higher concentrations of the protonated form of O_2^- ($HO_2^•$) in more acidic pH conditions (34). The stability is further reduced by enzymes that have evolved with the task of detoxifying oxygen free radicals, collectively named the superoxide dismutases. There are three of them in mammalian systems: a cytosolic CuZn superoxide (SOD1; CuZnSOD), an intramitochondrial manganese superoxide dismutase (SOD2; MnSOD) and extracellular CuZn superoxide dismutase (SOD3). Collectively, they catalyze the following reaction:

$$2H^+ + O_2^{•-} + O_2^{•-} \rightarrow H_2O_2 + O_2$$

The dismutation reaction has an overall rate constant of $5 \times 10^5\ M^{-1}\ s^{-1}$ at pH 7.0. SOD speeds up this reaction almost 10^4 fold ($K_d = 1.6 \times 10^9\ M^{-1}\ s^{-1}$) (34). Thus, in physiological systems, superoxides are converted to the stable H_2O_2. Because of its stability, H_2O_2 is used as a primary signaling molecule (24).

AOS have been implicated in HPV. Although, the precise mechanism behind AOS mediated HPV is not fully elucidated yet, the implication of AOS in HPV has been widely recognized. Though AOS has been an accepted signal transduction molecule of HPV, two schools of thoughts (outlined below) prevail on how the AOS levels and subsequent HPV is regulated in physiological system.

5.1. The Mitochondrial Redox Hypothesis

The redox theory of HPV proposes that there is a tonic, basal production of AOS in the PASMCs, most likely generated by mitochondria (7). The "normal" rate of generation of AOS causes physiological oxidation of K^+ channels in PASMCs, thus maintaining "normoxic vasodilation", thereby contributing to the low basal tone of pulmonary arteries. Under conditions of alveolar hypoxia, production of AOS falls in proportion to the inspired O_2 concentration (Fig. 7A). The result is the inhibition of normal AOS mediated dilatation and hence constriction (4). In support, our laboratory showed that a decrease in whole lung AOS production, as measured by chemiluminescence, precedes HPV (by seconds) (14). Non-redox dependent PA constrictors, such as angiotensin II, do not acutely alter AOS production. Furthermore, when antimycin A and rotenone were used to inhibit mitochondrial ETC, a similar decrease in measured chemiluminescence and increased PA pressure was observed (Fig. 7B) (10). Paky et al. also noticed inhibition of AOS production by hypoxia and antimycin A (82). However, these studies generated skepticism in some quarters because of concerns that lucigenin may itself trigger the redox cycle and purportedly has a propensity to measure cellular reductase activity independent from superoxide radicals. Furthermore, while the strength of these studies was the online measurement of AOS and its correlation with vascular tone, the chemiluminescence signal reflected the global superoxide anion generation by all pulmonary cells near the lung surface (potentially including neutrophils, airway epithelium and smooth muscle, endothelium, and vascular SMCs) (109). To address these concerns, we utilized three different radical detection methods (lucigenin-enhanced chemiluminescence, AmplexRed H_2O_2 assay, and 2'-7'-dichlorofluorescein (DCF)) for AOS measurement (72). We showed that in endothelium-denuded resistance PA rings, AOS levels were decreased during hypoxia as measured by these three different techniques. Furthermore, this decrease in AOS production was also noted in PA rings treated with proximal ETC inhibitors (72). This has been also supported by studies in other tissues and experimental models showing a similar decrease in AOS production during HPV (Fig. 7) (82, 97).

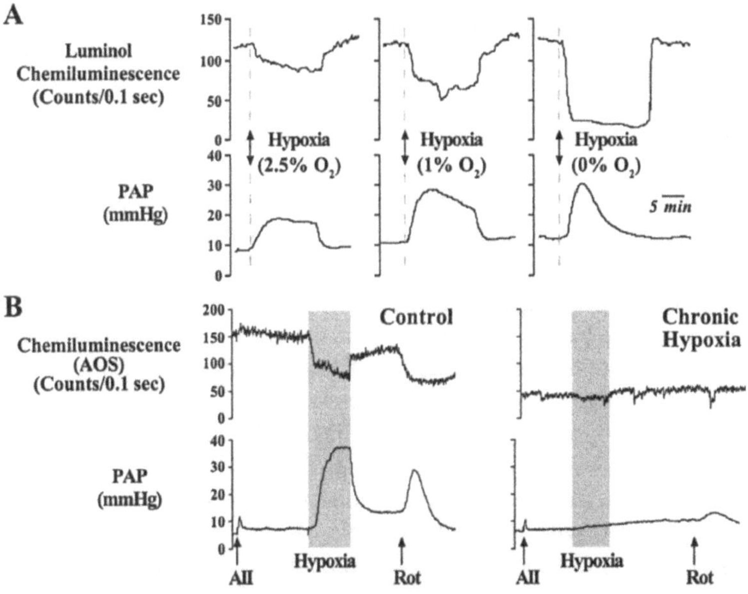

Figure 7. A: AOS production decreases with hypoxia, just prior to the onset of HPV. PA pressure and AOS production (luminol-enhanced chemiluminescence) were measured simultaneously in isolated rat lungs. The decrease in AOS production precedes the increase in PA pressure by seconds and occurs in a O_2 "dose dependent" manner. B: Chronic hypoxia reveals similarities between acute hypoxia and rotenone. Simultaneous luminol chemiluminescence and PA pressure measurements in a normoxic isolated rat lung show that rotenone constricts the PA and decreases AOS production in a manner similar to hypoxia. In lungs from a rat exposed to chronic hypoxia (0.45 atmospheres for ~3 weeks), HPV is suppressed while angiotensin II (AII) constriction is unchanged (Reproduced from Refs. 14, 97).

In the classical AOS response, ischemia-reperfusion, AOS are universally agreed to be reduced in the ischemic phase (analogous to hypoxia), and only increase with reperfusion (analogous to the termination of hypoxia) (42). Thus, it has been consistently shown that hypoxia or conditions mimicking hypoxia (proximal ETC inhibitors) decrease AOS production. In fact this hypothesis has been extended to another O_2-sensing tissue, i.e., the ductus arteriosus (DA) (74). Studies in our laboratory have shown that DASMCs responded in accordance to the mitochondrial redox hypothesis. In human DASMCs, O_2 causes constriction. We showed that, in response to increased ambient Po_2, H_2O_2 production was increased in DASMCs. This increase caused constriction of DASMCs, which was associated with an inhibition of plasmalemmal K^+ channels. Thus it appears that the response of the K^+ channels to H_2O_2 accounts for the opposing response of the PA and the DA to O_2. As in the PA, rotenone and antimycin mimic the effects of hypoxia on the DA (i.e., cause vasodilation) (74). Hence, it appears that during hypoxia, when ambient Po_2 is low, AOS production is decreased which subsequently translates into pulmonary vasoconstriction and dilatation of the ductus arteriosus.

5.2. Increased Mitochondrial AOS Production Hypothesis

In opposition to the theory that hypoxia results in a shift towards reduced state or decreased AOS levels, other laboratories have suggested a paradoxical increase in AOS generation during hypoxia. Marshall et al. found an increase in chemiluminescence during hypoxia in isolated PASMCs (66). A similar increase in oxidant signaling has also been reported in various hypoxic models. Recently, in parallel experiments in isolated perfused lungs and PASMCs (124), Waypa et al. showed that intracellular DCF was increased during hypoxic treatment , a response blocked by myxothiazol, a proximal ETC inhibitor. Based on these results the authors postulated that mitochondria indeed function as O_2 sensor, but that increased rather than decreased mitochondrial AOS production triggers HPV (124). They agreed with the proposed role of H_2O_2 as a probable signaling molecule. However, the use of DCF to measure AOS is controversial as previous studies suggest DCF as a suboptimal fluorescent probe to measure $O_2^{\bullet-}$ based radicals. DCF readily detects nitric oxide (in fact even more efficiently than it detects AOS) (95). In addition, DCF itself has been shown to be a source of H_2O_2 generation, further complicating and questioning the use of DCF as a fluorescent probe to detect O_2 radicals (102). Rota et al. evaluated the sensitivity and specificity of DCF and found that, depending on the chemical environment, DCF is quite effective in generating AOS and therefore should not be used as a reliable probe (102). Furthermore, previously published studies have shown a decrease in AOS production, as detected using DCF as a fluorescent marker (28). Collectively, these studies show that DCF may not be an optimal marker for measuring AOS levels. Teleologically, an increase in AOS levels in the face of hypoxia seems unlikely. Electrons accumulating during hypoxia in mitochondria would react with O_2 to produce $O_2^{\bullet-}$. However, under hypoxic conditions, it is difficult to conceptualize that $O_2^{\bullet-}$ production would increase as increased availability of electrons cannot be offset by decreased O_2 availability. Furthermore, even in the myocardium where ischemia-reperfusion studies have long been conducted, AOS production decreases during ischemia (58) and increased AOS is only seen during reperfusion (58, 137). It would seem that the AOS status in the lung during hypoxia is more analogous to that of the heart during ischemia-not the reperfusion or reoxygenation phase. With cessation of hypoxia AOS levels in the lung rise, as expected and as is consistent with the termination of ischemia in the heart (albeit without much of an overshoot). Therefore, the studies cumulatively suggest that, under hypoxic conditions, withdrawal of AOS is the mediator of vasoconstriction in pulmonary vascular beds.

6. K^+ Channel as an Effector

In 1986, we hypothesized that HPV related to inhibition of K^+ conductance (7). Hasunuma et al. found that the K_V channel blocker 4-aminopyridine (4-AP) caused pulmonary vasoconstriction that could, like HPV, be prevented by nifedipine in the isolated rat lung model (40). In 1992, studies done by Post et al. showed that hypoxia inhibit whole cell I_K in canine PASMCs and depolarize the membrane of canine PASMCs (93). This work initiated extensive research to evaluate the role of K^+ channels in HPV.

K^+ channels are proteins consisting of four transmembrane-bound pore-forming α subunits, often associated with four regulatory β subunits (2). The K^+ channel pore is conferred by the formation of tetramers of α subunits. The subunits determine the intrinsic conductance and voltage sensitivity of the channel. On the other hand, β-subunits are small subunits that associate with the α subunits and alter the channel's activation and inactivation kinetics (92). There are several types of K^+ channel α-subunits, including K_V channels, inward rectifier K^+ channels (K_{ir}), ATP-sensitive K^+ channels (K_{ATP}), and Ca^{2+}-activated K^+ channels (K_{Ca}) (92). Despite the existence of different kinds of K^+ channels, the K_V channels in particular have emerged as a participants in HPV, in large part because they control resting membrane potential and are redox-sensitive (opening when oxidized and closing when reduced) (97).

K_{ATP} channels are sensitive to changes in intracellular ATP and therefore are regulated by the cell's metabolic status (76). Because of this, K_{ATP} channels have been proposed as ideal candidates for HPV. However, it was quickly established that K_{ATP} channels do not contribute to HPV because glibenclamide, a K_{ATP} inhibitor, had no effect on pulmonary perfusion pressure in hypoxia (40). Similarly, the role of K_{ir} channels in HPV is less clear, as its presence in PASMC (as opposed to the endothelium where it is abundant) is still questionable (76). K_{Ca}, on the other hand, seem to act as a brake on HPV. Big conductance K_{Ca} (BK_{Ca}) channels control the arterial smooth muscle membrane in a negative feedback mechanism (76). At less negative or positive potentials, when the arterial bed is constricted and intracellular Ca^{2+} is increased, BK_{Ca} channels are activated, leading to K^+ efflux and restoration of membrane potential toward more negative values (hyperpolarization) (96). In cerebral arteries, hypoxia activates BK_{Ca} channels (37). Thus, although many different K^+ channels exist, the K_V channels are the only most likely candidates for HPV.

K_V channels are important determinants of membrane potential (E_m) of vascular SMCs. When K_V channels close and the tonic efflux of K^+ is decreased, the cell interior becomes relatively less negative (depolarized). At these potentials (i.e., -30 to -10 mV) the opening probability of L-type voltage-gated Ca^{2+} channels increases. This increases Ca^{2+} influx and activates Ca^{2+}-induced Ca^{2+} release, effectively increasing total Ca^{2+} levels inside the cell. This increase in cytosolic Ca^{2+} levels not only activates contraction via the actin-myosin apparatus but, in chronic settings, also increases the activation of immediate early genes to induce a proliferative response. Thus, regulation of K^+ channel activity and the subsequent regulation of Ca^{2+} is important to maintain vascular tone and the morphology of the PASMCs (75).

To date, nine families of K_V channel have been identified. These channels activate in a nonlinear fashion with depolarization and many are inhibited by 4-AP (100). Out of these, $K_V1.2$, $K_V1.5$, $K_V2.1$, $K_V3.1b$ and $K_V9.3$ are present in the pulmonary vasculature. One or more of them may form the PASMC O_2 sensitive K_V channel (27). The identification of the K_V channels involved in the E_m of PASMCs has been difficult, due to the lack of specific K_V channel blockers and to the fact that the whole-cell K^+ current is an ensemble of currents conducted by many types of K^+ channels. The fact that glyburide and charybdotoxin, inhibitors of K_{ATP} and BK_{Ca} channels, respectively, do not cause much increase in PVR whereas 4-AP causes similar constriction to hypoxia, suggests that K_V channels are key to HPV. Indeed the O_2 sensitive I_K in PASMCs is voltage sensitive (11, 16, 27, 132). To determine the molecular basis for the O_2-sensitive I_K, we "dissected" the whole cell current in rat PASMCs by making whole-cell patch clamp current recordings with or without antibodies directed against K_V channels in the intracellular solution (17). When anti-$K_V2.1$ antibody was added to patch pipette, the outward K^+ current was inhibited and the membrane depolarized, suggesting that $K_V2.1$ plays a role in setting the resting membrane potential in rat resistance PAs. When anti-$K_V1.5$ antibody was added to the patch pipette, whole cell K^+ currents significantly decreased and the ability of both hypoxia and 4-AP to increase Ca^{2+} in isolated PASMCs was attenuated. Thus, we showed that $K_V1.5$ and $K_V2.1$ are important components of whole-cell K^+ current in rat PASMCs. Moreover, it was found that although $K_V2.1$ sets the resting membrane potential, $K_V1.5$ is instrumental in mediating HPV (Fig. 8) (17). Subsequent studies have confirmed this, although it also suggests a role for $K_V1.2/1.5$ heterotetramers (44). Moreover, heteromeric channels comprised of $K_V2.1/K_V9.3$ have also been implicated in the mechanism of HPV (84), although, evidence of heteromeric channels existing *in vivo* is lacking (116). Recently it has been demonstrated that K_V channels are involved in O_2 sensing, although the specific populations involved have yet to be definitively established. Interestingly, $K_V1.5$ and $K_V2.1$ appear to be involved in the human DASMCs' response, although the sensitivity to H_2O_2 is reversed compared to the PA (74). In a model for the NEB cell (lung carcinoma line H146), acute hypoxia inhibits I_K and causes cell depolarization by inhibiting the tandem P domain family of K^+ channels, hTASK1 or hTASK3 (132). In NEB cells, K^+ channel inhibition results in serotonin secretion as a consequence of membrane depolarization and a rise in cytosolic Ca^{2+}(35). In the carotid body, it appears likely that several K^+ channels may be involved in O_2 sensing, including $K_V3.x$ and $K_V4.x$ channels (89, 105), and possibly TWIK or TASK, TASK-1, acid sensitive channels with four transmembrane segments and two pore domains (83).

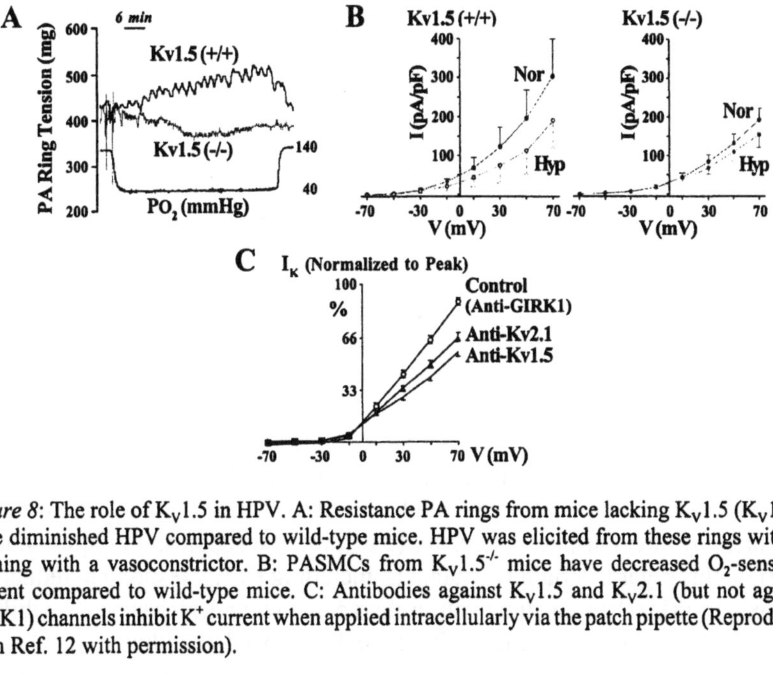

Figure 8: The role of $K_V1.5$ in HPV. A: Resistance PA rings from mice lacking $K_V1.5$ ($K_V1.5^{-/-}$) have diminished HPV compared to wild-type mice. HPV was elicited from these rings without priming with a vasoconstrictor. B: PASMCs from $K_V1.5^{-/-}$ mice have decreased O_2-sensitive current compared to wild-type mice. C: Antibodies against $K_V1.5$ and $K_V2.1$ (but not against GIRK1) channels inhibit K^+ current when applied intracellularly via the patch pipette (Reproduced from Ref. 12 with permission).

Exposure to hypoxia for 1-2 weeks elicits chronic hypoxic pulmonary hypertension (CH-PHT). Rodent CH-PHT is a relevant model for PHT that occurs in humans living at high altitude and in patients with chronic lung diseases. Paradoxically, acute hypoxic pulmonary vasoconstriction is blunted in CH-PHT, whilst the response to other vasoconstrictors is preserved or enhanced (71, 90). This is associated with inhibition of PASMC K^+ current (113) and a selective downregulation of mRNA expression of some, but not all, PASMC K^+ channels (e.g., $K_V1.5$, $K_V2.1$, $K_V4.3$ and $K_V9.3$) (97, 135). CH also selectively decreases function/expression of these channels in cultured PASMCs (18). Mice lacking $K_V1.5$ also have depressed HPV (16). Furthermore, anorexigens which have precipitated outbreaks of human PHT inhibit K^+ channels (11), including $K_V1.5$ (88). These data point strongly toward a role for $K_V1.5$, a slowly inactivating, 4-AP-sensitive K_V channel, in the mechanism of HPV. We recently utilized gene transfer to directly test the importance of $K_V1.5$ to the loss of HPV that occurs with chronic hypoxia. We hypothesized that restoration of $K_V1.5$ using gene transfer could be accomplished by nebulization of an adenoviral vector carrying the gene for $K_V1.5$ under a CMV promoter, thereby restoring $K_V1.5$ expression, reducing CH-PHT, and restoring HPV (94). In this recent study, we confirmed that administration of $K_V1.5$ to the pulmonary circulation via an aerosol is feasible and effective in eliciting transgene expression in resistance PASMCs. Furthermore, expression of cloned human PA $K_V1.5$ reduced PVR in experimental PHT. Furthermore, K_V gene therapy with an O_2-sensitive channel restored HPV and O_2 sensitive I_K in rats with established CH-PHT. This study supports a central role for $K_V1.5$ in the mechanism of HPV and is consistent with the hypothesis that a K^+ channel deficiency state is involved in the pathogenesis of PHT (5, 133).

The K^+ channel pore is formed by coordinated alignment of key amino acids in the S5-S6 region of each of the α-subunits. Some (but not all) K^+ channels are well suited to O_2 sensing by virtue of possessing key cysteine and methionine groups. Reduction or oxidation of these residues by a redox mediator, such as an AOS, could cause a conformational change in the channel and thus alter pore function (17). Studies with AOS have shown that they can regulate protein function by five possible ways: *i*) modification of proteins by oxidation of cysteine residues (e.g., S-glutathionylation), *ii*) formation of intra-molecular disulfide linkages (e.g., the oxidative stress-responsive protein OxyR), *iii*) protein dimerization by inter-molecular disulfide linkages (e.g., activation by various protein kinase), *iv*) di-tyrosine formation by H_2O_2-dependent reactions (e.g., $gp91^{phox}$ regulation of extracellular matrix), *v*) metal-catalyzed oxidation of protein by "Fenton-like" chemistry (e.g., protease action and ubiquitination) (119). In this regard, certain K^+ channels, including $K_V1.5$, respond to reduction and oxidation by changing their gating characteristics and open state probability. In the PASMCs, oxidants (e.g., H_2O_2, diamide oxidized glutathione) increase I_K whereas reducing agents (e.g., reduced glutathione and duroquinone) and agents that facilitate electron shuttling (e.g., the synthetic co-enzyme Q mimetic) inhibit I_K (99). In accordance with their electrophysiological effects, oxidants dilate the pulmonary circulation (mimicking O_2), while reducing agents mimic hypoxia (Fig. 9). Hypoxia and redox agents may alter the function of ion channels, specifically K^+ channels, directly (47) or by modulating levels of a diffusable redox mediator, such as an AOS. The existence of redox sensitive K^+ channels does not verify whether the electrophysiological effects of O_2 are direct or mediated via a remote sensor that produces a mediator which ultimately alters channel activity. However, research is warranted in this area to fully elucidate the precise mechanism behind AOS-mediated K_V inhibition.

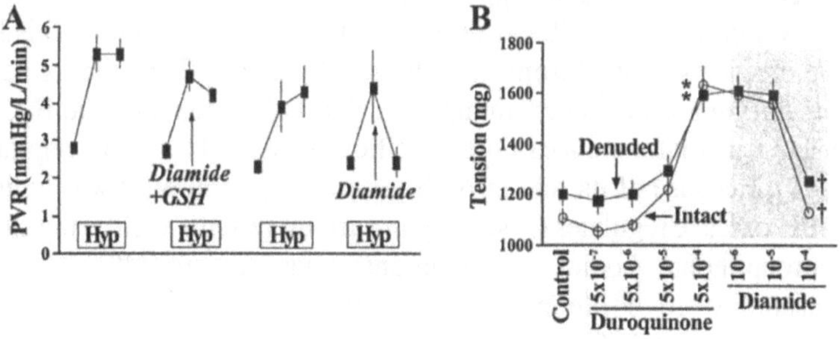

Figure 9: Redox control of PA tone. A: The sulfhydryl oxidant, diamide, reverses HPV in anesthetized dogs. Preincubation of diamide with reduced glutathione prevents this effect. B: Endothelium-denuded PA rings constrict in response to the electron shuttling agent duroquinone and this is reversed by diamide (Reproduced from Ref. 99 with permission).

Thus the K_V channels themselves are redox-sensitive, which leaves the question: to what signal are they responding. While we propose that this signal is mitochondria-derived H_2O_2, other hypotheses have been advanced: altering of a membrane-bound O_2-sensitive regulatory moiety that is adjacent or coupled to the channel subunits, activating mitochondrial NADPH oxidase and increasing $O_2^{\bullet -}$, inhibiting cytochrome P450, decreasing intracellular pH, releasing Ca^{2+} from intracellular stores and, recently, affecting directly the K_V channel protein (116). In this regard, Marshall and Marshall have suggested a gp91^{phox} containing NADPH oxidase is involved in HPV (making more AOS during hypoxia) (66). In contrast, we confirmed that HPV and O_2-sensitive I_K are preserved in mice lacking a functional NADPH oxidase (15).

7. Summary

In the redox theory of HPV, we propose the existence of a mitochondrial redox sensor in complexes I and III that is linked to several O_2-sensitive K^+ channels (e.g., $K_V1.5$) by diffusible redox mediators (e.g., H_2O_2). In hypoxia, a rapid and dynamic alteration in levels of AOS will reduce critical amino acids in certain Kv channels which would translate into a conformational change of the channel, decreasing its open probability. The resultant E_m depolarization increases intracellular Ca^{2+} by promoting Ca^{2+} influx. This increased Ca^{2+} would act on the actin-myosin apparatus thus precipitating HPV (Fig. 10).

One of the more viable alternatives to the mitochondrial origins of AOS is that a similar signal is derived from membrane-bound NADPH oxidase. The NADPH oxidase system is a multisubunit assembly consisting of a membrane-bound catalytic complex of the 91 kDa subunit of cytochrome oxidase ($gp91^{phox}$) and $p22^{phox}$, which altogether forms a flavo-cytochrome b_{558}. The cytosolic regulatory component consists of $p47^{phox}$, $p67^{phox}$ and other regulatory subunits such as the GTPase proteins Rac-1 or Rac-2 (49). Studies in neutrophils have shown that electrons derived from NADPH are shuttled to O_2 by the oxidase at a rapid rate during normoxia generating $O_2^{\bullet-}$. This $O_2^{\bullet-}$ is converted to H_2O_2 via SOD. The H_2O_2 mediator induces a site-specific oxidation of cytosolic regulatory proteins (41). Because of its role in AOS generation, initially, it was believed that NADPH oxidase regulates O_2 sensing in various systems such as the carotid body, neuroepithelial bodies and pulmonary vasculature (49). However, subsequent studies proved otherwise especially in patients suffering from chronic granulomatous disease (CGD). CGD is a genetic defect in one or more of the subunits of NADPH oxidase and presents as a broad spectrum infection of organ systems with abscess formation. Although NADPH oxidase system is disrupted, patients with CGD are still able to maintain HPV (128). Furthermore, mice-deficient in NADPH oxidase activity by mutation of the X-linked $gp91^{phox}$, showed that HPV was unaltered. In addition, studies with rotenone constriction also showed that hypoxic response was preserved in these NADPH oxidase-deficient mice. Thus, this form of NADPH oxidase does not contribute to HPV (15). However, it remains possible that a form of NADPH oxidase called novel oxidase (NOX) may have a role.

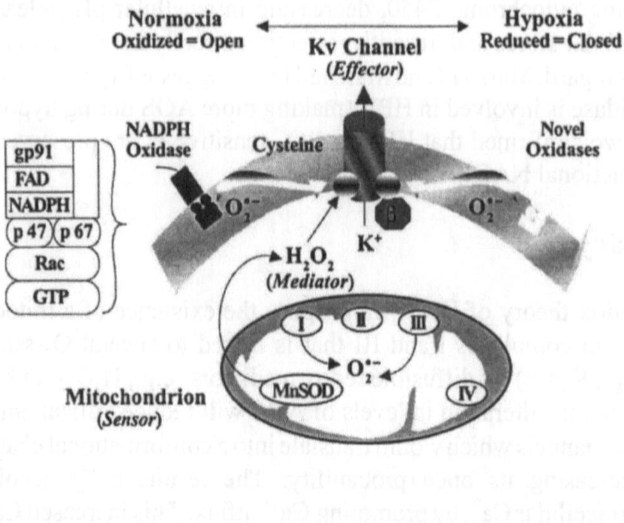

Figure 10: Alternative sources of AOS for the proposed redox mechanism for HPV. The redox theory of HPV suggests that AOS are produced in proportion to Po_2 by O_2 sensors, likely complexes I and III of the mitochondrial ETC, and possibly also cytochrome-based vascular oxidases. The changes in AOS production alter the gating and open probability of the effectors of HPV, the O_2-and redox-sensitive K_V channels such as $K_V1.5$ and $K_V2.1$.

Vascular SMCs also contain gp91^{phox} homologs, called NOX, which preferentially utilize NADPH versus NADH as substrate. NOXs are also inhibited by DPI and are potentially important sources of AOS production in systemic vascular SMCs (52). However, most studies of NOX have been performed in systemic arteries and have measured AOS production in response to vasoconstrictors (e.g., angiotensin II) or mitogens (e.g., PDGF). There is little evidence that NOX are involved in PASMC AOS production or HPV, since little AOS signal remains in the NADPH oxidase gp91^{phox} knockout mice. It is also noteworthy that angiotensin II does not cause an acute change in AOS production in mouse lungs at doses that cause vasoconstriction (14, 15). Perhaps the predominant source of AOS differs between Pas (where AOS may be signaling molecules serving to optimize O_2 uptake from the environment) and systemic arteries (where AOS may be involved in the pathogenesis of atherosclerosis and systemic hypertension).

8. Controversies

In 2003, five major controversies remain regarding the mechanism of HPV. *i*) The role of the endothelium and ET-1: There remains healthy debate as to the relative contribution of the endothelium to HPV. In particular, there is still debate about whether ET-1 is (81) or is not (104) essential for expression of HPV. It appears that ET-1, in some species, exaggerates HPV, particularly if hypoxia is sustained; this may occur by an endothelium-mediated inhibition of K_{ATP} channels (106). *ii*) The role of the intracellular Ca^{2+} stores: The contribution of intracellular Ca^{2+} pools to HPV is controversial, although there can be little debate that HPV is strongly inhibited by Ca^{2+} channel blockers. *iii*) The vascular O_2 sensor (mitochondria vs. NADPH oxidase): While there is growing evidence that mitochondria are the sensor for HPV, the precise complex involved remains controversial and a role for novel NADPH oxidases cannot be excluded. There is certainly evidence in some vascular beds that the NADPH oxidases may be the predominant source of AOS. Intrinsic to this controversy is the identity of the putative redox signaling mediator. In the case of NADPH oxidase, the proposed mediator is the superoxide radical whilst signaling from the mitochondria likely involves diffusible H_2O_2. *iv*) Effects of hypoxia on AOS production: Surprisingly, there also remains debate as to whether hypoxia increases or decreases AOS production. Part of the confusion here may relate to the means of achieving hypoxia (9), the severity and/or duration of hypoxia, and the important difference between hypoxia vs. reperfusion following hypoxia. *v*) The molecular identity of the O_2 sensitive K^+ channel(s): Finally, there is discussion about the molecular identities of the K_V channels that are involved in HPV, although there seems to be consensus around a relatively short list which includes $K_V1.2$, $K_V1.5$, $K_V2.1$, $K_V3.1b$ and $K_V9.3$.

These controversies can certainly be resolved in a collegial manner as scientists carefully examine the technical aspects of each others work that may underlie apparent discrepancies and more comprehensively employ genomics, proteomics, and integrative physiology, rather than relying on single inhibitors or single tissue preparations to reach conclusions. Finally, it appears virtually certain that many of the steps in HPV are shared with other specialized O_2 sensor pathways and thus careful attention should be paid to findings in the carotid body, neuroepithelial body, adrenomedullary cell and ductus arteriosus.

Acknowledgments

Drs. E.D. Michelakis and S.L. Archer are supported by the Alberta Heritage Foundation for Medical Research, the Canadian Foundation for Innovation and the Heart and Stroke Foundation of Canada and the Canadian Institutes for Health Research. Dr. Archer is the Heart and Stroke Foundation Chair for Research in Alberta.

References

1. Abrahams JP, Leslie AG, Lutter R, and Walker JE. Structure at 2.8 A resolution of F1-ATPase from bovine heart mitochondria. *Nature*. 1994; 370: 621-628.
2. Ackerman MJ and Clapham DE. Ion channels – basic science and clinical disease. *N. Engl. J. Med.* 1997; 336: 1575-1586.
3. Archer, S, McMurtry, IF, and Weir, EK. "Mechanisms of acute hypoxic and hyperoxic changes in pulmonary vascular reactivity." In *Pulmonary Vascular Physiology and Pathophysiology*, Weir EK, and Reeves JT , eds. New York, NY: Marcel Dekker Inc., 1988, p. 241-290.
4. Archer S and Michelakis E. The mechanism(s) of hypoxic pulmonary vasoconstriction: potassium channels, redox O_2 sensors, and controversies. *News Physiol. Sci.* 2002;17:131-137.
5. Archer S and Rich S. Primary pulmonary hypertension: a vascular biology and translational research "Work in progress". *Circulation*. 2000; 102: 2781-2791.
6. Archer S, Tolins JP, Raij L, and Weir EK. Hypoxic pulmonary vasoconstriction is enhanced by inhibition of the synthesis of an endothelium derived relaxing factor. *Biochem. Biophys. Res. Comm.* 1989; 164: 1198-1205.
7. Archer S, Will JA, and Weir EK. Redox status in the control of pulmonary vascular tone. *Herz.* 1986; 11: 127-141.
8. Archer S,Yankovich RD, Chesler E, and Weir EK. Comparative effects of nisoldipine, nifedipine and bepridil on experimental pulmonary hypertension. *J. Pharmacol. Exp. Ther.* 1985; 233: 12-17.
9. Archer SL, Hampl V, Nelson DP, Sidney E, Peterson DA, and Weir EK. Dithionite increases radical formation and decreases vasoconstriction in the lung. Evidence that dithionite does not mimic alveolar hypoxia. *Circ. Res.* 1995; 77: 174-181.
10. Archer SL, Huang J, Henry T, Peterson D, and Weir EK. A redox-based O_2 sensor in rat pulmonary vasculature. *Circ. Res.* 1993; 73: 1100-1112.
11. Archer SL , Huang JM, Reeve HL, Hampl V, Tolarová S, Michelakis E, and Weir EK. Differential distribution of electrophysiologically distinc myocytes in conduit and resistance arteries determines their response to nitric oxide and hypoxia. *Circ. Res.* 1996; 78: 431-442.

12. Archer SL, London B, Hampl V, Wu X, Nsair A, Puttagunta L, Hashimoto K, Waite RE, and Michelakis ED. Impairment of hypoxic pulmonary vasoconstriction in mice lacking the voltage-gated potassium channel Kv1.5. *Faseb J.* 2001; 15: 1801-1803.
13. Archer SL, Nelson D, and Weir EK. Characterization of radical production by xanthine / xanthine oxidase, in vitro and in rat lungs, using chemiluminescence. *J. Appl. Physiol.* 1989; 67: 1912-1921.
14. Archer SL, Nelson DP, and Weir EK. Simultaneous measurement of O_2 radicals and pulmonary vascular reactivity in rat lung. *J. Appl. Physiol.* 1989; 67: 1903-1911.
15. Archer SL, Reeve HL, Michelakis E, Puttagunta L, Waite R, Nelson DP, Dinauer MC, and Weir EK. O_2 sensing is preserved in mice lacking the gp91 phox subunit of NADPH oxidase. *Proc. Natl. Acad. Sci. USA.* 1999; 96: 7944-7949.
16. Archer, SL and Rusch, NJ, eds., *Potassium Channels in Cardiovascular Biology*. New York, NY: Kluwer Academic/Plenum Publishers, 2001.
17. Archer SL, Souil E, Dinh-Xuan AT, Schremmer B, Mercier JC, El Yaagoubi A, Nguyen-Huu L, Reeve HL, and Hampl V. Molecular identification of the role of voltage-gated K^+ channels, Kv1.5 and Kv2.1, in hypoxic pulmonary vasoconstriction and control of resting membrane potential in rat pulmonary artery myocytes. *J. Clin. Invest.* 1998; 101: 2319-2330.
18. Attali B, Wang N, Kolot A, Sobko A, Cherepanov V, and Soliven B. Characterization of delayed rectifier Kv channels in oligodendrocytes and progenitor cells. *J. Neurosci.* 1997; 17: 8234-8245.
19. Barja G. Mitochondrial oxygen radical generation and leak: sites of production in states 4 and 3, organ specificity, and relation to aging and longevity. *J. Bioenerg. Biomembr.* 1999; 31: 347-366.
20. Bennie RE, Packer CS, Powell DR, Jin N, and Rhoades RA. Biphasic contractile response of pulmonary artery to hypoxia. *Am. J. Physiol.* 1991; 261: L156-L163.
21. Brandt U. Proton-translocation by membrane-bound NADH:ubiquinone-oxireductase (complex I) through redox-gated ligand conduction. *Biochem.Biophys.Acta.* 1997;1318:79-91.
22. Brimioulle S, LeJeune P , and Naeije R. Effects of hypoxic pulmonary vasoconstriction on pulmonary gas exchange. *J. Appl. Physiol.* 1996; 81: 1535-1543.
23. Buescher PC, Pearse DB, Pillai RP, Litt MC, Mitchell MC, and Sylvester JT. Energy state and vasomotor tone in hypoxic pig lungs. *J. Appl. Physiol.* 1991; 70: 1874-1881.
24. Burke T and Wolin M. Hydrogen peroxide elicits pulmonary arterial relaxation and guanylate cyclase activation. *Am. J. Physiol.* 1987; 252: H721-H732.
25. Carlsson AJ, Bindslev L, and Hedenstierna G. Hypoxia-induced pulmonary vasoconstriction in the human lung. The effect of isoflurane anesthesia. *Anesthesiology.* 1987; 66: 312-316.
26. Conforti L and Millhorn DE. Selective inhibition of a slow-inactivating voltage-dependent K^+ channel in rat PC12 cells by hypoxia. *J. Physiol.* 1997; 502: 293-305.
27. Coppock EA, Martens JR, and Tamkun MM. Molecular basis of hypoxia-induced pulmonary vasoconstriction: role of voltage-gated K^+ channels. *Am. J. Physiol. Lung Cell. Mol. Physiol.* 2001; 281: L1-L12.
28. Corda S, Laplace C, Vicaut E, and Duranteau J. Rapid reactive oxygen species production by mitochondria in endothelial cells exposed to tumor necrosis factor-alpha is mediated by ceramide. *Am. J. Respir. Cell Mol. Biol.* 2001; 24: 762-768.
29. Cross CE, Halliwell B, Borish ET, Pryor WA, Ames BN, Saul RL, McCord JM, and Harman D. Oxygen radicals and human diseases. *Ann. Intern. Med.* 1987; 107: 526-545.
30. Degli Esposti M, Ngo A, McMullen GL, Ghelli A, Sparla F, Benelli B, Ratta M, and Linnane AW. The specificity of mitochondrial complex I for ubiquinones. *Biochem. J.* 1996; 313: 327-334.
31. Demple B and Amabile-Cuevas CF. Redox redux: the control of oxidative stress responses. *Cell.* 1991; 67: 837-839.
32. Dorrington KL, Clar C, Young JD, Jonas M, Tansley JG, and Robbins PA. Time course of the human pulmonary vascular response to 8 hours of isocapnic hypoxia. *Am. J. Physiol.* 1997; 273: H1126-H1134.

33. Franco-Obregon A, and Lopez-Barneo J. Differential oxygen sensitivity of calcium channels in rabbit smooth muscle cells of conduit and resistance pulmonary arteries. *J. Physiol.* 1996; 491: 511-518.
34. Fridovitch I. Superoxide and superoxide dismutase. *Annu. Rev. Biochem.* 1995; 64: 97-112.
35. Fu XW, Nurse CA, Wong V, and Cutz E. Hypoxia-induced secretion of serotonin from intact pulmonary neuroepithelial bodies in neonatal rabbit. *J. Physiol.* 2002; 539: 503-510.
36. Fu XM, Pan WD, Farragher SM, Wong V, and Cutz E. Neuroepithelial bodies in mammalian lung express functional serotonin type 3 receptor. *Am. J. Physiol. Lung Cell. Mol. Physiol.* 2001; 281: L931-L940.
37. Gebremedhin D, Bonnet P, Greene AS, England SK, Rusch NJ, Lombard JH, and Harder DR. Hypoxia increases the activity of Ca^{2+}-sensitive K^+ channels in cat cerebral arterial muscle cell membranes. *Pflügers Arch.* 1994; 428: 621-630.
38. Hales CA, Ahluwalia B, and Kazemi H. Strength of pulmonary vascular response to regional alveolar hypoxia. *J. Appl. Physiol.* 1975; 38: 1083-1087.
39. Han D, Antunes F, Canali R, Rettori D, and Cadenas E. Voltage-dependent anion channels control the release of superoxide anion from mitochondria to cytosol. *J. Biol. Chem.* 2002; 278: 5557-5563.
40. Hasunuma K, Rodman D , and McMurtry I. Effects of K^+ channel blockers on vascular tone in the perfused rat lung. *Ann. Rev. Respir. Dis.* 1991; 144: 884-887.
41. Henderson LM and Chappell JB. NADPH oxidase of neutrophils. *Biochim. Biophys. Acta.* 1996; 1273: 87-107.
42. Henry T, Archer SL, Nelson D, Weir EK, and From AH. Postischemic oxygen radical production varies with duration of ischemia. *Am. J. Physiol.* 1993; 264: H1478-H1484.
43. Hughes JD and Rubin LJ. Relation between mixed venous oxygen tension and pulmonary vascular tone during normoxic, hyperoxic and hypoxic ventilation in dogs. *Am. J. Cardiol.* 1984; 54: 1118-1123.
44. Hulme JT, Coppock EA, Felipe A, Martens JR, and Tamkun MM. Oxygen sensitivity of cloned voltage-gated K^+ channels expressed in the pulmonary vasculature. *Circ. Res.* 1999; 85: 489-497.
45. Hyslop SJ,Duncan AM, Pitkanen S, and Robinson BH. Assignment of the PSST subunit gene of human mitochondrial complex I to chromosome 19p13. *Genomics.* 1996; 37: 375-380.
46. Jensen KS, Micco AJ, Czartolomna J, Latham L, and Voelkel NF. Rapid onset of hypoxic vasoconstriction in isolated lungs. *J. Appl. Physiol.* 1992; 72: 2018-2023.
47. Jiang C, and Haddad GG. A direct mechanism for sensing low oxygen levels by central neurons. *Proc. Natl. Acad. Sci. USA.* 1994; 91: 7198-7201.
48. Johnson D and Georgieff MK. Pulmonary neuroendocrine cells. Their secretory products and their potential roles in health and chronic lung disease in infancy. *Ann. Rev. Respir.Dis.* 1989; 140: 1807-1812.
49. Jones RD, Hancock JT, and Morice AH. NADPH oxidase: a universal oxygen sensor? *Free Radic. Biol. Med.* 2000; 29: 416-424.
50. Kashani-Poor N, Zwicker K, Kerscher S, and Brandt U. A central functional role for the 49-kDa subunit within the catalytic core of mitochondrial complex I. *J. Biol. Chem.* 2001; 276: 24082-24087.
51. Kato M and Staub N. Response of small pulmonary arteries to unilobar alveolar hypoxia and hypercapnia. *Circ. Res.* 1966; 19: 426-440.
52. Lassegue B, Sorescu D, Szocs K, Yin QQ, Akers M, Zhang Y, Grant SL, Lambeth JD, and Griendling KK. Novel gp91^{phox} homologues in vascular smooth muscle cells : nox1 mediates angiotensin II-induced superoxide formation and redox- sensitive signaling pathways. *Circ. Res.* 2001; 88: 888-894.
53. Leach RM, Hill HM, Snetkov VA, Robertson TP, and Ward JPT. Divergent roles of glycolysis and the mitochondrial electron transport chain in hypoxic pulmonary vasoconstriction of the rat: identity of the hypoxic sensor. *J. Physiol.* 2001; 536: 211-224.
54. Leach RM, Robertson TP, Twort CH, and Ward JP. Hypoxic vasoconstriction in rat

pulmonary and mesenteric arteries. *Am. J. Physiol.* 1994; 266: L223-L231.
55. Leach RM, Sheehan DW, Chacko VP, and Sylvester JT. Energy state, pH, and vasomotor tone during hypoxia in precontracted pulmonary and femoral arteries. *Am. J. Physiol. Lung Cell. Mol. Physiol.* 2000; 278: L294-L304.
56. Li Y, Huang TT, Carlson EJ, Melov S, Ursell PC, Olson JL, Noble LJ, Yoshimura MP, Berger C, Chan PH, Wallace DC, and Epstein CJ.. Dilated cardiomyopathy and neonatal lethality in mutant mice lacking mangenese superoxide dismutase. *Nat. Genet.* 1995; 11:376-381.
57. Li Y, Zhu H, and Trush MA. Detection of mitochondria-derived reactive oxygen species production by chemilumigenic probes lucigenin and luminol. *Biochem. Biophys. Acta.* 1999; 1428: 1-12.
58. Liu P, Hock CE, Nagele R, and Wong PY. Formation of nitric oxide, superoxide, and peroxynitrite in myoardial ischemia-reperfusion injury in rats. *Am. J. Physiol.* 1997; 272: H2327-H2336.
59. Liu Q, Sham JSK, Shimoda LA, and Sylvester JT. Hypoxic constriction of porcine distal pulmonary arteries: endothelium and endothelin dependence. *Am. J. Physiol. Lung Cell. Mol. Physiol.* 2001; 280: L856-L865.
60. Lloyd TC Jr. PO_2-dependent pulmonary vasoconstriction caused by procaine. *J. Appl. Physiol.* 1966; 21: 1439-1442.
61. Lloyd TC Jr. Pulmonary vasoconstriction during histotoxic hypoxia. *J. Appl. Physiol.* 1965; 20: 488-490.
62. Lopez MG, Moro MA, Castillo CF, Artalejo CR, and Garcia AG. Variable, voltage-dependent, blocking effects of nitrendipine, verapamil, diltiazem, cinnarizine and cadmium on adrenomedullary secretion. *Br. J. Pharmacol.* 1989; 96: 725-731.
63. Lopez-Barneo J, Benot A, and Urena J., Oxygen sensing and the electrophysiology of arterial chemoreceptor cells. *News Physiol Sci.* 1993; 8: 191-195.
64. López-Barneo J, López-López JR, Ureña J, and González C. Chemotransduction in the carotid body: K^+ current modulated by Po_2 in type I chemoreceptor cells. *Science Wash. DC.* 1988; 242: 580-582.
65. Madden JA, Dawson CA, and Harder DR. Hypoxia-induced activation in small isolated pulmonary arteries from the cat. *J. Appl. Physiol.* 1985; 59: 113-118.
66. Marshall C, Mamary AJ, Verhoeven AJ, and Marshall BE. Pulmonary artery NADPH-oxidase is activated in hypoxic pulmonary vasoconstriction. *Am. J. Respir. Cell Mol. Biol.* 1996; 15: 633-644.
67. Marshall C and Marshall BE. Influence of perfusate PO_2 on hypoxic pulmonary vasoconstriction in rats. *Circ. Res.* 1983; 52: 691-696.
68. McCormack DG and Paterson NA. Loss of hypoxic pulmonary vasoconstriction in chronic pneumonia is not mediated by nitric oxide. *Am. J. Physiol.* 1993; 265: H1523-H1528.
69. McMurtry IF. BAY K 8644 potentiates and A23187 inhibits hypoxic vasoconstriction in rat lungs. *Am. J. Physiol.* 1985; 249: H741-H746.
70. McMurtry IF, Davidson AB, Reeves JT, and Grover RF. Inhibition of hypoxic pulmonary vasoconstriction by calcium antagonists in isolated rat lungs. *Circ. Res.* 1976; 38:99-104.
71. McMurtry IF, Petrun MD, and Reeves JT. Lungs from chronically hypoxic rats have decreased pressor response to acute hypoxia. *Am. J. Physiol.* 1978; 235: H104-H109.
72. Michelakis E, Hampl V, Nsair A, Wu X, Harry G, Haromy A, Gurtu R, and Archer SL. Diversity in mitochondrial function explains differences in vascular oxygen sensing. *Circ. Res.* 2002; 90: 1307-1315.
73. Michelakis E, Rebeyka I, Bateson J, Olley P, Puttagunta L, and Archer S. Voltage-gated potassium channels in human ductus arteriosus. *Lancet.* 2000; 356: 134-137.
74. Michelakis ED, Rebeyka I, Wu X, Nsair A, Thebaud B, Hashimoto K, Dyck JR, Haromy A, Harry G, Barr A, and Archer SL. O_2 sensing in the human ductus arteriosus: regulation of voltage-gated K^+ channels in smooth muscle cells by a mitochondrial redox sensor. *Circ. Res.* 2002; 91: 478-486.
75. Michelakis ED and Weir EK. The pathobiology of pulmonary hypertension. Smooth muscle

cells and ion channels. *Clin. Chest Med.* 2001; 22: 419-432.

76. Nelson MT and Quayle JM. Physiological roles and properties of potassium channels in arterial smooth muscle. *Am. J. Physiol.* 1995; 268: C799-C822.
77. Nicholls DG. Mitochondrial function and dysfunction in the cell: its relevance to aging and aging-related disease. *Int. J. Biochem. Cell Biol.* 2002; 34: 1372-1381.
78. Obeso A, Rocher A, Lopez-Lopez JR, and Gonzalez C. Intracellular Ca^{2+} deposits and catecholamine secretion by chemoreceptor cells of the rabbit carotid body. *Adv. Exp. Med. Biol.* 1996; 410: 279-284.
79. Ohe M, Mimata T, Haneda T, and Takishima T. Time course of pulmonary vasoconstriction with repeated hypoxia and glucose depletion. *Respir. Physiol.* 1986; 63: 177-186.
80. Ohnishi T. Iron-sulfur clusters/semiquinones in complex I. *Biochim. Biophys. Acta.* 1998; 1364: 186-206.
81. Oparil S, Chen SJ, Meng QC, Elton TS, Yano M, and Chen YF. Endothelin-A receptor antagonist prevents acute hypoxia-induced pulmonary hypertension in the rat. *Am. J. Physiol.* 1995; 12: L95-L100.
82. Paky A, Michael JR, Burke-Wolin TM, Wolin MS, and Gurtner GH. Endogenous production of superoxide by rabbit lungs: effects of hypoxia or metabolic inhibitors. *J. Appl. Physiol.* 1993; 74: 2868-2874.
83. Patel AJ and Honore E. Molecular physiology of oxygen-sensitive potassium channels. *Eur. Respir. J.* 2001; 18: 221-227.
84. Patel AJ, Lazdunski M, and Honore E. Kv2.1/Kv9.3, a novel ATP-dependent delayed-rectifier K^+ channel in oxygen-sensitive pulmonary artery myocytes. *EMBO. J.* 1997; 16: 6615-6625.
85. Pease RD, Benumof JL, and Trousdale FR. PAO_2 and PVO_2 interaction on hypoxic pulmonary vasoconstriction. *J. Appl. Physiol.* 1982; 53: 134-139.
86. Peers C. Hypoxic suppresion of K^+ currents in type I carotid body cells: selective effect on the Ca^{2+}-activated K^+ current. *Neurosci. Lett.* 1990; 119: 253-256.
87. Peinado VI, Santos S, Ramirez J, Roca J, Rodriguez-Roisin R, and Barbera JA. Response to hypoxia of pulmonary arteries in chronic obstructive pulmonary disease: an in vitro study. *Eur. Respir. J.* 2002; 20: 332-338.
88. Perchenet L, Hilfiger L, Mizrahi J, and Clement-Chomienne O. Effects of anorexinogen agents on cloned voltage-gated K^+ channel hKv1.5. *J. Pharmacol. Exp. Ther.* 2001; 298: 1108-1119.
89. Pérez-García MT, López-López JR, Riesco AM, Hoppe UC, Marbán E, González C, and Johns DC. Viral gene transfer of dominant-negative Kv4 construct suppresses an O_2-sensitive K^+ current in chemoreceptor cells. *J. Neurosci.* 2000; 20: 5689-5695.
90. Peterson DA, Reeve HL, Nelson D, Archer SL, and Weir EK. Triple-bonded unsaturated fatty acids are redox active compounds. *Lipids.* 2001; 36: 431-433.
91. Pitkanen S and Robinson BH. Mitochondrial complex I deficiency leads to increased production of superoxide radicals and induction of superoxide dismutase. *J. Clin. Invest.* 1996; 98: 345-351.
92. Pongs O. Molecular biology of voltage-dependent potassium channels. *Physiol. Rev.* 1992; 72: S62-S88.
93. Post JM, Hume JR, Archer SL, and Weir EK. Direct role for potassium channel inhibition in hypoxic pulmonary vasoconstriction. *Am. J. Physiol.* 1992; 262: C882-C890.
94. Pozeg Z, Michelakis ED, McMurtry S, Thebaud B, Wu X-C, Dyck JRB, Hashimoto K, Wang S, Moudgil R, Harry G, Sultanian R, Koshal A, and Archer SL. In vivo gene transfer of the O_2-sensitive potassium channel Kv1.5 reduces pulmonary hypertension and restores hypoxic pulmonary vasoconstriction in chronically hypoxic rats. *Circulation.* 2003; 107: 2037-2044.
95. Rao KM, Padmanabhan J, Kilby DL, Cohen HJ, Currie MS, and Weinberg JB. Flow cytometric analysis of nitric oxide production in human neutrophils using dichlorofluorescein diacetate in the presence of a calmodulin inhibitor. *J. Leukoc. Biol.* 1992; 51: 496-500.
96. Reeve HL, Archer S, and Weir EK. Ion channels in the pulmonary vasculature. *Pulm. Pharmacol. Ther.* 1997; 10: 243-252.

97. Reeve HL, Michelakis E, Nelson DP, Weir EK, and Archer SL. Alterations in a redox oxygen sensing mechanism in chronic hypoxia. *J. Appl. Physiol.* 2001; 90: 2249-2256.
98. Reeve HL, Tolarova S, Nelson DP, Archer S, and Weir EK. Redox control of oxygen sensing in the rabbit ductus arteriosus. *J. Physiol.* 2001; 533: 253-261.
99. Reeve HL, Weir EK, Nelson DP, Peterson DA, and Archer SL. Opposing effects of oxidants and antioxidants on K^+ channel activity and tone in rat vascular tissue. *Exp. Physiol.* 1995; 80: 825-834.
100. Rettig J, Heinemann SH, Wunder F, Lorra C, Parcej DN, Dolly JO, and Pongs O. Inactivation properties of voltage-gated K^+ channels altered by presence of β-subunit. *Nature.* 1994; 369: 289-294.
101. Robinson BH. Human complex I deficiency: clinical spectrum and involvement of oxygen free radicals in the pathogenicity of the defect. *Biochim. Biophys. Acta.* 1994; 1364: 271-276.
102. Rota C, Chignell CF, and Mason RP. Evidence for free radical formation during the oxidation of 2, 7-dichlorofluorecein to the fluorescent dye 2, 7-dichlorofluorescein by horseradish peroxidase: possible implications for oxidative stress measurements. *Free Radic. Biol. Med.*1999; 27: 873-881.
103. Rounds S and McMurtry IF. Inhibitors of oxidative ATP production cause transient vasoconstriction and block subsequent pressor responses in rat lungs. *Circ. Res.* 1981; 48: 393-400.
104. Salameh G, Karamsetty MR, Warburton RR, Klinger JR, Ou LC, and Hill NS. Differences in acute hypoxic pulmonary vasoresponsiveness between rat strains: role of endothelium. *J. Appl. Physiol.* 1999; 87: 356-62.
105. Sanchez D,Lopez-Lopez JR, Perez-Garcia MT, Sanz-Alfayate G, Obeso A, Ganfornina MD, and Gonzależ C. Molecular identification of Kvα subunits that contribute to the oxygen-sensitive K^+ current of chemoreceptor cells of the rabbit carotid body. *J. Physiol.* 2002; 542: 369-382.
106. Sato K, Morio Y, Morris KG, Rodman DM, and McMurtry IF. Mechanism of hypoxic pulmonary vasoconstriction involves ET_A receptor-mediated inhibition of K_{ATP} channel. *Am. J. Physiol. Lung Cell. Mol. Physiol.* 2000; 278: L434-L442.
107. Schuler F and Casida JE. The insecticide target in the PSST subunit of complex I. *Pest Manag. Sci.* 2001; 57: 932-940.
108. Schuler F,Yano T, Di Bernardo S, Yagi T, Yankovskaya V, Singer TP, and Casida JE. NADH-quinone oxidoreductase: PSST subunit couples electron transfer from iron-sulfur cluster N2 to quinone. *Proc. Natl. Acad. Sci. USA.* 1999; 96: 4149-4153.
109. Sham JSK. Hypoxic pulmonary vasoconstriction: ups and downs of reactive oxygen species. *Circ. Res.* 2002; 91: 649-651.
110. Shapiro BM. The control of oxidant stress at fertilization. *Science.* 1991; 252: 533-536.
111. Shirahata M, and Fitzgerald RS. Dependency of hypoxic chemotransduction in cat carotid body on voltage-gated calcium channels. *J. Appl. Physiol.* 1991; 71: 1062-1069.
112. Smirnov SV, Beck R, Tammaro P, Ishii T, and Aaronson PI. Electrophysiologically distinct smooth muscle cell subtypes in rat conduit and resistance pulmonary arteries. *J. Physiol.* 2002; 538: 867-878.
113. Smirnov SV, Robertson TP, Ward JPT, and Aaronson PI. Chronic hypoxia is associated with reduced delayed rectifier K^+ current in rat pulmonary artery muscle cells. *Am. J. Physiol.* 1994; 266: H365-H370.
114. St-Pierre J, Buckingham JA, Roebuck SJ, and Brand MD. Topology of superoxide production from different sites in the mitochondrial electron transport chain. *J. Biol. Chem.* 2002; 277: 44784-44790.
115. Stanbrook HS, and McMurtry IF. Inhibition of glycolysis potentiates hypoxic vasoconstriction in rat lungs. *J. Appl. Physiol.* 1983; 55: 1467-1473.
116. Sweeney M and Yuan JX-J. Hypoxic pulmonary vasoconstriction: role of voltage-gated potassium channels. *Respir. Res.* 2000; 1: 40-48.
117. Sylvester JT, Harabin AL, Peake MD, and Frank RS. Vasodilator and constrictor responses

to hypoxia in isolated pig lungs. *J. Appl. Physiol.* 1980; 49: 820-825.

118. Takeshige K and Minakami S. NADH and NADH-dependent formation of superoxide anions by bovine heart submitochondrial particles and NADH-ubiquinone reductase preparation. *Biochem. J.* 1979; 15: 129-135.

119. Thannickal VJ and Fanburg BL. Reactive oxygen species in cell signaling. *Am. J. Physiol. Lung Cell. Mol. Physiol.* 2000; 279: L1005-L1028.

120. Tristani-Firouzi M, Reeve HL, Tolarova S, Weir EK, and Archer SL. Oxygen-induced constriction of the rabbit ductus arteriosus occurs via inhibition of a 4-aminopyridine-sensitive potassium channel. *J. Clin. Invest.* 1996; 98: 1959-1965.

121. Trumpower BL. Energy transduction by coupling of proton translocation to electron transfer by the cytochrome bc1 complex. *J. Biol. Chem.* 1990; 265: 11409-11412.

122. Vadula MS, Kleinman JG, and Madden JA. Effect of hypoxia and norepinephrine on cytoplasmic free Ca^{2+} in pulmonary and cerebral arterial myocytes. *Am. J. Physiol.* 1993; 265: L591-L597.

123. Ward JPT and Aaronson PI. Mechanisms of hypoxic pulmonary vasoconstriction: can anyone be right? *Respir. Physiol.* 1999; 115: 261-271.

124. Waypa GB, Chandel NS, and Schumacker PT. Model for hypoxic pulmonary vasoconstriction involving mitochondrial oxygen sensing. *Circ. Res.* 2001; 88: 1259-1266.

125. Waypa GB, Marks JD, Mack MM, Boriboun C, Mungai PT, and Schumacker PT. Mitochondrial reactive oxygen species trigger calcium increases during hypoxia in pulmonary arterial myocytes. *Circ. Res.* 2002; 91: 719-726.

126. Weir EK and Archer SL. The mechanism of acute hypoxic pulmonary vasoconstriction: the tale of two channels. *FASEB J.* 1995; 9: 183-189.

127. Weksler B, Ng B, Lenert JT, and Burt ME. Isolated single lung perfusion in the rat: an experimental model. *J. Appl. Physiol.* 1993; 74: 2736-2739.

128. Wenger RH, Marti HH, Schuerer-Maly CC, Kvietikova I, Bauer C, Gassmann M, Maly FE. Hypoxic induction of gene expression in chronic granulomatous disease-derived B-cell lines: oxygen sensing is independent of the cytochrome b558-containing nicotinamide adenine dinucleotide phosphate oxidase. *Blood.* 1996; 87: 756-761.

129. Wiener CM, Dunn A, and Sylvester JT. ATP-dependent K^+ channels modulate vasoconstrictor responses to severe hypoxia in isolated ferret lungs. *J. Clin. Invest.* 1991; 88: 500-504.

130. Youngson C, Nurse C, Yeger H, and Cutz E. Oxygen sensing in airway chemoreceptors. *Nature.* 1993; 365: 153-155.

131. Yu AY, Shimoda LA, Iyer NV, Huso DL, Sun X, McWilliams R, Beaty T, Sham JSK, Wiener CM, Sylvester JT, and Semenza GL. Impaired physiological responses to chronic hypoxia in mice partially deficient for hypoxia-inducible factor 1α. *J. Clin. Invest.* 1999; 103: 691-696.

132. Yuan X-J. Voltage gated K^+ currents regulate resting membrane potential and $[Ca^{2+}]_i$ in pulmonary artery myocytes. *Circ. Res.* 1995; 77: 370-378.

133. Yuan X-J, Aldinger AM, Juhaszova M, Wang J, Conte JV Jr, Gaine SP, Orens JB, and Rubin LJ. Dysfunctional voltage-gated potassium channels in the pulmonary artery smooth muscle cells of patients with primary pulmonary hypertension. *Circulation.* 1998; 98: 1400-1406.

134. Yuan X-J, Goldman WF, Tod ML, Rubin LJ, and Blaustein MP. Hypoxia reduces potassium currents in cultured rat pulmonary but not mesenteric arterial myocytes. *Am. J. Physiol.* 1993; 264: L116-L123.

135. Yuan X-J, Wang J, Juhaszova M, Golovina VA, and Rubin LJ. Molecular basis and function of voltage-gated K^+ channels in pulmonary arterial smooth muscle cells. *Am. J. Physiol.* 1998; 274: L621-L635.

136. Zheng M, Aslund F, and Storz G. Activation of the OxyR transcription factor by reversible disulfide bond formation. *Science.* 1998; 279: 1718-1721.

137. Zweier JL, Flaherty JT, and Weisfeldt ML. Direct measurement of free radical generation following reperfusion of ischemic myocardium. *Proc. Natl. Acad. Sci. USA.* 1987; 84: 1404-1407.

Chapter 10

Hypoxic Regulation of K^+ Channel Expression and Function in Pulmonary Artery Smooth Muscle Cells

Hemal H. Patel, Carmelle V. Remillard, and Jason X.-J. Yuan
University of California, San Diego, California, U.S.A.

1. Introduction

The primary function of the pulmonary circulation is to maintain efficient gas exchange and provide adequate oxygenated blood to the body. The pulmonary vasculature is unlike systemic vascular beds in that pulmonary arteries are low-pressure, high-capacity vessels with thin distensible walls specifically adapted for gas exchange. Hypoxic pulmonary vasoconstriction (HPV) is specific to the pulmonary vasculature in that pulmonary arteries constrict in response to alveolar hypoxia to divert blood flow away from regions that are poorly ventilated by "sensing" the reduced alveolar oxygen tension (PO_2). However, in systemic vascular beds (e.g., coronary, cerebral, renal arteries), the physiological response to hypoxia is the exact opposite, causing vasodilation to increase blood flow and subsequent oxygen delivery.

HPV can actually be of benefit during acute hypoxic events (e.g., an acute asthma attack) by diverting blood from poorly aerated regions of the lung to distribute blood flow to regions where adequate gas exchange can occur. This provides an automatic and relatively immediate control system for matching perfusion with ventilation in order to maximize oxygenation of venous blood in the pulmonary artery. On the other hand, chronic hypoxic exposure (e.g., chronic obstructive pulmonary disease, congenital heart disease, or high altitude habitats) causes a prolonged hypoxic environment in the lungs. The prolonged hypoxia and/or protracted HPV result in profound pulmonary vascular remodeling and pulmonary hypertension, leading to possible right heart failure. It would appear that acute responses to hypoxia alter cell physiology by modifying the molecular and cellular machinery already present allowing for the immediacy of response; however, the chronic response would be expected to involve altered gene expression and modification of cell physiology at a more gross level.

The precise mechanism or mediator of HPV is currently unclear. This chapter will discuss the relevance of cytoplasmic Ca^{2+} and ion (e.g., K^+ and Ca^{2+}) channel function and expression on acute and chronic responses to hypoxia in the pulmonary arteries. The role of signaling microdomains as structured and organized modulators of cell physiology comprised of receptors, mediators (e.g., cyclases, reactive oxygen species, arachidonic acid metabolites), and effectors (e.g., ion channels) will be considered in both acute and chronic hypoxia. In the latter case, it is possibly accounting for remodeling via a redistribution or differential expression of microdomain components resulting in an altered phenotype of the pulmonary artery (PA) smooth muscle cell (SMC).

2. Role of Ca^{2+} in Pulmonary Vasoconstriction and PASMC Proliferation

Contraction of vascular smooth muscle permits active changes in diameter and wall tension of pulmonary blood vessels. An increase in cytoplasmic free Ca^{2+} concentration ($[Ca^{2+}]_{cyt}$) in PASMC is a major trigger for pulmonary vasoconstriction and an important stimulus for PASMC proliferation. When $[Ca^{2+}]_{cyt}$ rises, Ca^{2+} binds to calmodulin (CaM), and then activates myosin light chain kinase that phosphorylates the myosin light chain. The phosphorylation increases the myosin ATPase activity that hydrolyzes ATP to release energy. The subsequent cycling of the myosin cross-bridges produces displacement of the myosin filament in relation to the actin filament and causes contraction (96).

In isolated PA rings (Fig. 1A*a*), removal of extracellular Ca^{2+} almost abolished the sustained pulmonary vasoconstriction induced by high K^+ (which induces membrane depolarization by shifting the equilibrium potential for K^+) (Fig. 1A*b*) and markedly inhibited vasoconstriction induced by phenylephrine (PE, which activates α-adrenogic receptors in the plasma membrane) (Fig. 1A*c*). In the absence of extracellular Ca^{2+}, application of vasoactive substances, such as PE, to the PA rings induces a transient contraction followed by a small sustained contraction (Fig. 1A*c*). There results suggest that membrane depolarization-mediated pulmonary vasoconstriction completely depends on extracellular Ca^{2+}, while the agonist-mediated pulmonary vasoconstriction is composed of at least three components: a transient contraction and a small sustained contraction that are not dependent on extracellular Ca^{2+}, and a large sustained contraction that is dependent of extracellular Ca^{2+}. The extracellular Ca^{2+}-independent transient and sustained contractions are due to Ca^{2+} release from the intracellular stores to the cytosol and to Ca^{2+} sensitization of the contractile apparatus (or a Ca^{2+}-insensitive mechanism), respectively (57, 96). The Ca^{2+}-dependent sustained contraction is due to a rise in $[Ca^{2+}]_{cyt}$ as a result of Ca^{2+} influx through sarcolemmal Ca^{2+}-permeable channels.

Intracellular Ca^{2+} not only serves as a major trigger for vascular smooth muscle contraction (96), a rise in $[Ca^{2+}]_{cyt}$ is also essential for cell proliferation (8), migration (73), and gene expression (26). Ca^{2+} diffuses quickly between the cytosol and nucleus (3). In the cell cycle of mammalian cells, there are at least four Ca^{2+}/CaM-sensitive steps (Fig. 1B*a*): *a*) transition from G_0 (resting state) to G_1 phase (the beginning of DNA synthesis), *b*) transition from G_1 to S phase (an interphase during which replication of the nuclear DNA occurs), *c*) transition from G_2 to M phase (mitosis), and *d*) mitosis (M phase) (8). Maintenance of sufficient Ca^{2+} within the intracellular Ca^{2+} stores is required for cell growth; depletion of the SR Ca^{2+} store induces growth arrest (94) and may trigger apoptosis (32).

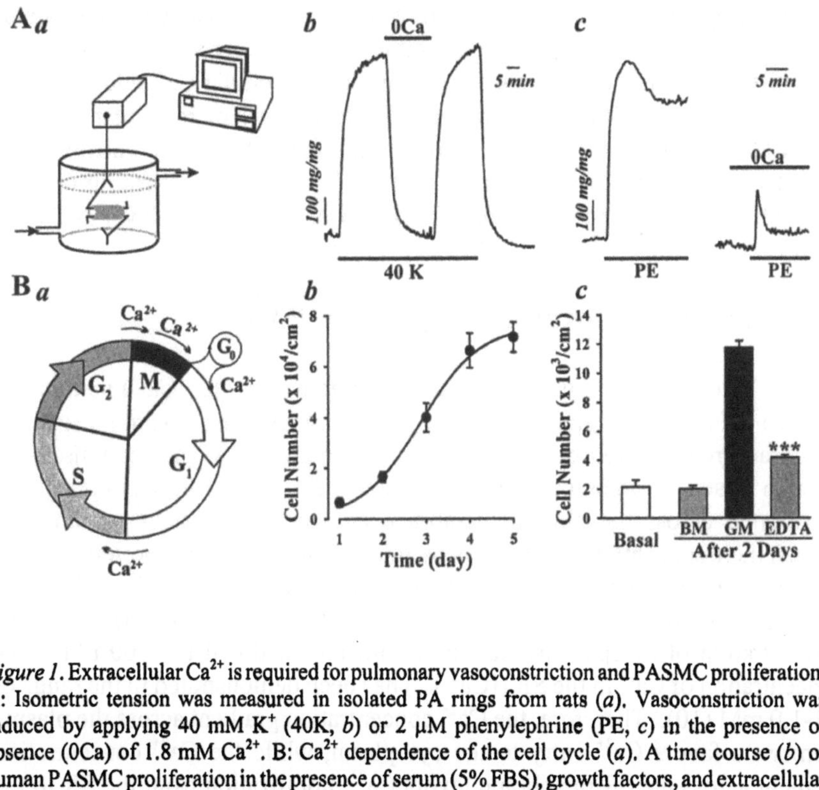

Figure 1. Extracellular Ca^{2+} is required for pulmonary vasoconstriction and PASMC proliferation. A: Isometric tension was measured in isolated PA rings from rats (*a*). Vasoconstriction was induced by applying 40 mM K^+ (40K, *b*) or 2 μM phenylephrine (PE, *c*) in the presence or absence (0Ca) of 1.8 mM Ca^{2+}. B: Ca^{2+} dependence of the cell cycle (*a*). A time course (*b*) of human PASMC proliferation in the presence of serum (5% FBS), growth factors, and extracellular Ca^{2+}. Chelation of extracellular Ca^{2+} with 2 mM EDTA inhibits PASMC proliferation (*c*). BM, smooth muscle basal medium; GM, serum- and growth factor-supplemented growth medium. ****P*<0.001 vs. GM (Reproduced from Refs. 28 and 52).

The Ca^{2+}/CaM-dependent propelling and progression of the cell cycle are regulated by both cytoplasmic and nuclear $[Ca^{2+}]$ (31). For example, a rise in $[Ca^{2+}]_{cyt}$ activates Ras and CREB, two transcription factors that are involved in cell proliferation (31). An increase in nuclear $[Ca^{2+}]$ activates AP-1, a family of transcription factors (e.g., c-Fos, c-Jun) that promote cell proliferation (8).

Furthermore, maintaining a sufficient Ca^{2+} in the sarcoplasmic (SR) or endoplasmic (ER) reticulum (an organelle in which protein and lipid synthesis and sorting take place) is also required for cell growth (94).

The Ca^{2+} dependence of cell growth can be exhibited by examining whether removal (or chelation) of extracellular Ca^{2+} or depletion of Ca^{2+} in the SR affects cell proliferation in the presence of serum and growth factors. As shown in Figure 1B, chelation of extracellular Ca^{2+} significantly inhibits human PASMC growth in media containing 5% fetal bovine serum and growth factors (epidermal growth factor, platelet-derived growth factor, endothelial growth factor, and/or insulin). Depletion of intracellularly-stored Ca^{2+} in the SR using thapsigargin or cyclopiazonic acid (CPA) further inhibits PASMC growth (28, 94). The synergistically inhibitory effects of removal of extracellular Ca^{2+} and depletion of stored Ca^{2+} on PASMC proliferation indicate that both extracellular Ca^{2+} and stored Ca^{2+} are required for cell growth, and suggest that Ca^{2+} influx is an important mechanism to maintain sufficient Ca^{2+} in the SR. These results provide compelling evidence that a constant Ca^{2+} influx and a sufficient level of Ca^{2+} in the SR are both required for PASMC to proliferate.

3. Regulation of Intracellular $[Ca^{2+}]$ in PASMC

The lipid membrane is impermeable to cations (e.g., K^+, Ca^{2+}) and anions (e.g., Cl^-). Flux of ions across either the plasma membrane or the intraorganelle membrane requires ion channels which selectively allow various ions to go through based on their concentration gradients. Furthermore, ions can be transported across the plasma membrane and the intraorganelle membrane against their concentration gradients by ion pumps (or ion transporters) (10).

The concentration of extracellular or intercellular Ca^{2+} ($[Ca^{2+}]_{out}$) is in the range of 1.6-2.0 mM (similar to the concentration in the plasma), while the concentration of cytoplasmic Ca^{2+} in an unstimulated cell is ~100 nM. The concentration of Ca^{2+} in the SR ($[Ca^{2+}]_{SR}$) has been demonstrated in the range of 0.1-1.0 mM (10). Therefore, the Ca^{2+} concentration gradient across the plasma membrane favors Ca^{2+} entry to the cytoplasm, while the Ca^{2+} concentration gradient across the SR membrane favors Ca^{2+} mobilization from the SR to the cytoplasm (Fig. 2A). The passive Ca^{2+} influx (from the extracellular milieu) and release (from the SR) to the cytoplasm are driven by the electrochemical gradient that causes $[Ca^{2+}]_{cyt}$ to rise without expenditure of energy.

In vascular smooth muscle cells, $[Ca^{2+}]_{cyt}$ can be increased by *a*) Ca^{2+} release from intracellular stores (mainly the SR) to the cytosol through Ca^{2+}-permeable channels, such as inositol 1,4,5-trisphosphate (IP_3) and ryanodine receptor-coupled Ca^{2+} release channels, in the SR membrane and *b*) Ca^{2+} influx from extracellular/intercellular site to the cytosol through Ca^{2+}-permeable channels in the plasma membrane (Fig. 2B). Increased $[Ca^{2+}]_{cyt}$ can be reduced to the resting level by Ca^{2+} extrusion from the cytoplasm to extracellular/ intercellular site via the Ca^{2+}-Mg^{2+} ATPase and Na^+/Ca^{2+} exchange in the plasma membrane and Ca^{2+} sequestration from the cytoplasm into the SR via the Ca^{2+}-Mg^{2+} ATPase in the SR membrane (SERCA) (Fig. 2B).

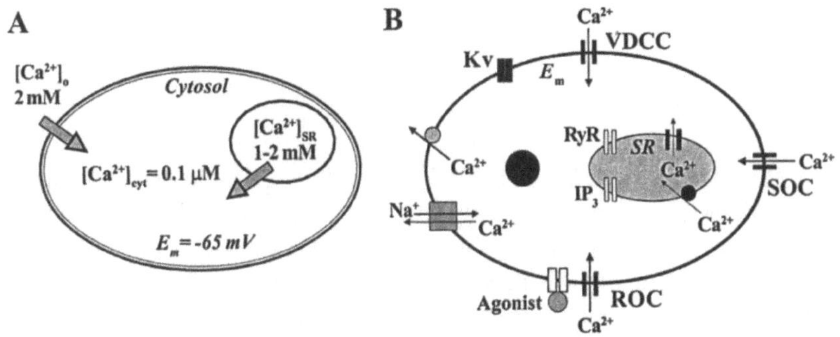

Figure 2. Schematic diagrams showing Ca^{2+} distribution and ion channels involved in regulating $[Ca^{2+}]_{cyt}$. A: Ca^{2+} is more concentrated in the extracellular milieu and in the sarcoplasmic reticulum (SR) than in the cytosol. B: Key ion channels involved in the regulation of $[Ca^{2+}]_{cyt}$ include the plasma membrane voltage-dependent Ca^{2+} channels (VDCC), store-operated Ca^{2+} channels (SOC), and receptor-operated Ca^{2+} channels (ROC). Energy-driven transporters such as the Ca^{2+}-ATPase and the Na^+-Ca^{2+} exchanger are primarily responsible for Ca^{2+} extrusion from the cytosol. On the SR membrane, ryanodine (RyR) and inositol 1,4,5-trisphosphate (IP_3) receptors act as Ca^{2+} release pathways from the SR while a Ca^{2+}-ATPase pump is responsible for Ca^{2+} re-uptake and replenishing of SR Ca^{2+} stores.

In vascular smooth muscle cells, the SR has been demonstrated as the dominant intracellular Ca^{2+} store (9). There are two functionally and spatially distinguishable SR: *a*) an IP_3-sensitive SR that is involved in Ca^{2+} mobilization during agonist stimulation and is sensitive to the SR Ca^{2+} pump inhibitors, cyclopiazonic acid (CPA) and thapsigargin (27); and *b*) a ryanodine-sensitive SR that is activated by caffeine and is responsible for Ca^{2+}-induced Ca^{2+} release (10). Both IP_3- and ryanodine-sensitive SR functionally exist in PASMC.

In pulmonary vascular smooth muscle cells, there are at least three functionally-distinct Ca^{2+}-permeable channels expressed in the plasma membrane (Fig. 2B): *a*) voltage-dependent Ca^{2+} channels (VDCC) that are opened by membrane depolarization (60), *b*) receptor-operated Ca^{2+} channels (ROC) that are opened by ligand binding with receptors (2); and *c*) store-operated Ca^{2+} channels (SOC) that are opened by depletion of Ca^{2+} from the intracellular Ca^{2+} stores (i.e., the SR) (67). By governing Ca^{2+} influx via VDCC, membrane potential (E_m) plays a critical role in regulating $[Ca^{2+}]_{cyt}$ (60) and excitation-contraction coupling in smooth muscle (96), as well as in regulating cell proliferation (78).

4. Regulation of Membrane Potential by K^+ Permeability

Based on the Nernst equation, the K^+ equilibrium potential (E_K) is determined by the transmembrane K^+ concentration gradient:

$$E_K = \frac{RT}{F} \ln \frac{[K]_o}{[K]_i}$$

where (RT/F) is 25.26 at 25°C; and $[K^+]_{out}$ and $[K^+]_{cyt}$ denote extracellular and cytoplasmic K^+ concentration, respectively. According to the Goldman-Hodgkin-Katz (GHK) equation, the reversal potential (E_{rev}) or zero-current potential is a function of the Na^+, K^+, Ca^{2+}, and Cl^- concentration gradient across the plasma membrane and the relative ion permeability (P):

$$E_{rev} = \frac{RT}{F} \ln \frac{P_K[K]_o + P_{Na}[Na]_o + P_{Cl}[Cl]_i}{P_K[K]_i + P_{Na}[Na]_i + P_{Cl}[Cl]_o}$$

where *o* and *i* refer to extra- and intra-cellular concentration, respectively; P_K, P_{Na}, and P_{Cl} denote the membrane permeability for K^+, Na^+, and Cl^-, respectively. Under resting conditions, the membrane permeability to K^+ (P_K) is much greater than the permeabilities to Na^+ (P_{Na}) and Cl^- (P_{Cl}). Thus, the resting E_m (close to E_{rev}) is mainly controlled by P_K and should be close to the E_K. However, the measured E_m in PASMC is often 20-35 mV apart from the E_K, this is because of the contribution of the Na^+ and Cl^- (as well as Ca^{2+}) permeabilities to the E_m.

4.1. Resting E_m is Regulated by K^+ Channel Activity

Resting E_m is primarily determined by K^+ permeability. A typical cell has about 30 times more K^+ (~140 mM) inside the cell compared to extracellular K^+ (~5 mM). Thus, when K^+ channels open, it is favorable for K^+ to move out of the cell according to its concentration gradient leading to hyperpolarization. There are at least five functionally distinguishable K^+ channels found in vascular smooth muscle cells (including PASMC): *a*) voltage-gated K^+ (Kv) channels, *b*) Ca^{2+}-activated K^+ (K_{Ca}) channels, *c*) inward-rectifier K^+ (K_{IR}) channels, *d*) ATP-sensitive K^+ (K_{ATP}) channels, and *e*) tandem pore two domain K^+ (K_T) channels (43, 61). E_m is controlled by a balance of outward currents (e.g., K^+ currents, Na^+/Ca^{2+} exchange currents, Na^+ pump currents) and inward currents (e.g., Na^+ currents, Ca^{2+} currents, Cl^- currents). In PASMC isolated from humans, rats, canines, and mice, activities of Kv (88, 95, 115), K_{ATP} (16), K_{Ca} (75), and K_T (29) channels all contribute to the regulation of E_m. Under resting conditions, inhibition of Kv channels by 4-aminopyridine (4-AP) causes membrane depolarization, induces Ca^{2+}-dependent action potentials, and increases $[Ca^{2+}]_{cyt}$ in PASMC (Fig. 3). These results suggest that activity of Kv channels plays an important role in regulating resting E_m in PASMC (115).

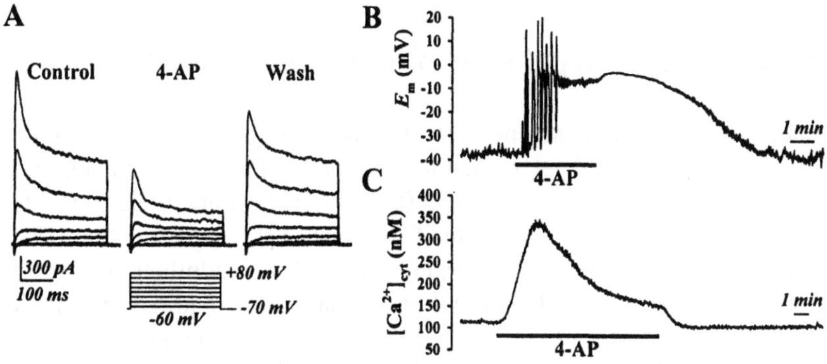

Figure 3. 4-Aminopyridine-sensitive Kv currents are involved in membrane depolarization and regulation of $[Ca^{2+}]_{cyt}$. A: Representative currents, elicited by depolarizing a human PASMC to test potentials ranging between -60 and +80 mV from a holding potential of -70 mV, before (Control), during (4-AP) and after (Wash) treatment with 5 mM 4-AP. E_m (B) and $[Ca^{2+}]_{cyt}$ (C) were measured in the current-clamp (I=0) mode and using fluorescence microscopy, respectively, before, during, and after application of 5 mM 4-AP (Reproduced from Refs. 37 and 78).

4.2. Molecular Composition of Kv Channels

Though current through each of the K^+ channels contributes to E_m, the major determinant of resting E_m in human and animal PASMC appears to be Kv channel activity (23, 115). At the molecular level, functional Kv channels are hetero- or homo-multimeric tetramers composed of two structurally distinct subunits: the pore forming α subunits and the cytoplasmic β subunits that regulate channel activity. Genes encoding α subunits for functional Kv channels are categorized into four subfamilies: *Shaker* (Kv1), *Shab* (Kv2), *Shaw* (Kv3), and *Shal* (Kv4). Recently, four additional subfamilies of inactive, or electrically silent, γ subunits (88, 129) were cloned (Kv5, Kv6, Kv8, and Kv9) and described in PASMC (19, 70, 79, 80, 105, 125). Though expression of the electrically silent γ subunits alone does not produce channel activity, coexpression of these subunits (e.g., Kv6 and Kv9) with functional α-subunits (e.g., Kv2.1 and Kv2.2) results in functionally complex channel activity (66, 70, 91, 129).

As for the regulatory β subunits, there are three subfamilies, Kvβ1, Kvβ2, and Kvβ3 (49). β subunits can confer fast inactivation on slow or non-inactivating delayed rectifier Kv channels (e.g., Kv1.2, Kv1.5) (86), determine the responses of the α subunits to protein kinases and hypoxia (48, 49, 69, 76), and block Kv channels by acting as open channel blockers. Interestingly, it has been reported that the Kvβ1 subunits have high homology to the NAD(P)H oxidoreductase superfamily (51), suggesting that the β subunits play a critical role in regulating α subunit activity by sensing and responding to changes in P_{O_2} (18, 69, 76), cellular redox status (86), and metabolism (e.g., ATP level) (70).

5. Acute Hypoxic Regulation of Kv Channel Function

Acute exposure to hypoxia increases pulmonary arterial pressure in rats (Fig. 4A) and causes a sustained yet reversible contraction of both isolated pulmonary arteries (Fig. 4B) and isolated PASMC (59, 117, 120). Functional removal of endothelium enhances HPV in isolated PA rings (120), suggesting that HPV is intrinsic to the individual PASMC independent of the endothelium, neuronal control, and circulating regulators. All of the machinery required to sense and respond to hypoxia and regulate pulmonary vascular tone is integrated within the individual PASMC (45, 46, 59, 120). HPV depends on external Ca^{2+} and is caused by a rise in $[Ca^{2+}]_{cyt}$ due, partially, to membrane depolarization-mediated opening of VDCC. This is supported by studies showing that the hypoxic contractile response is almost abolished by removal of extracellular Ca^{2+} (Fig. 4B) and the hypoxic pressor effect in intact rats and isolated perfused lungs is blunted by the VDCC blockers, verapamil and nifedipine (Fig. 4A) (52). In addition to opening VDCC, recent studies suggest that acute hypoxia could also raise $[Ca^{2+}]_{cyt}$ via capacitative Ca^{2+} entry (CCE) or, possibly, voltage-independent pathways (see Chapters 6 and 9) (89). HPV is a critical physiological process for maximal oxygenation of venous (or pulmonary arterial) blood, therefore, it must be assured by multiple redundant mechanisms or pathways.

5.1. Acute Hypoxia Induces Membrane Depolarization by Inhibiting Kv Channels

In rat PASMC, acute hypoxia reduces K^+ currents by inhibiting Kv channels (Fig. 4C*a*) (5, 63, 82, 83, 117). The inhibitory effect of hypoxia on Kv channels appears to be selective to PASMC; hypoxia has little effect on Kv currents ($I_{K(V)}$) in mesenteric artery smooth muscle cells (MASMC) isolated from the same rat (Fig. 4C*b*). The inability of acute hypoxia to reduce $I_{K(V)}$ in MASMC also explains why acute hypoxia has little effect on mesenteric vascular tone (120). The acute hypoxia-mediated inhibition of Kv channels causes a sustained membrane depolarization in PASMC (Fig. 4D) and elicits Ca^{2+}-dependent action potentials in some cells. The resultant increase in $[Ca^{2+}]_{cyt}$ (Fig. 4E) due to opening of voltage-dependent Ca^{2+} channels triggers PASMC contraction by activating CaM and myosin light chain kinase and eventually causes pulmonary vasoconstriction (96). Consistent with the effect in rat PASMC, acute hypoxia also reduces $I_{K(V)}$ in canine (82, 83) and rabbit (63) PASMC.

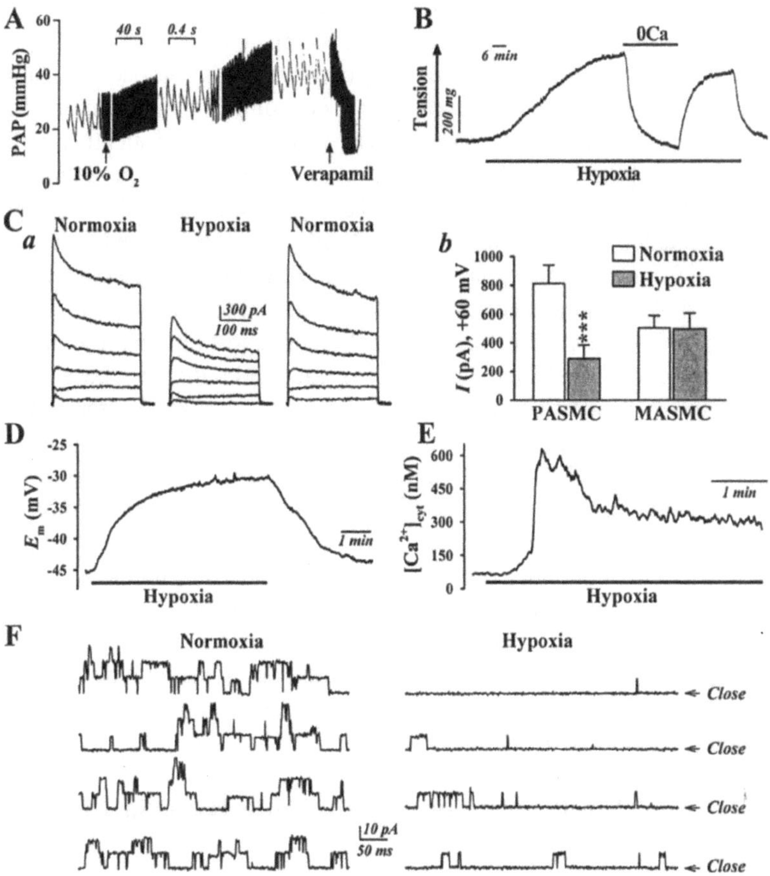

Figure 4. Hypoxia modulates Kv channel function, membrane potential, pulmonary vasomotor tone, and pulmonary arterial pressure (PAP). A: Decreasing Po_2 to 10% caused a gradual increase in PAP that was sensitive to treatment by verapamil, a Ca^{2+} channel blocker. B: Hypoxia causes vasoconstriction in an isolated small rat pulmonary artery ring. C: Hypoxia inhibits Kv currents (*a*) in PASMC but not in mesenteric arterial smooth muscle cells (MASMC, shaded bars, *b*). *** $P<0.001$ vs. normoxia. D and E: Change in E_m (D) and $[Ca^{2+}]_{cyt}$ (E) recorded upon decreasing Po_2 from 149 to 8 mmHg (Hypoxia). E: Hypoxia reduces the steady-state open probability of unitary K_{Ca} channels from ~1.10 to ~0.25 in cell-attached patches from isolated rat PASMC (Reproduced from Refs. 116-118).

In addition to blocking Kv channels, acute hypoxia also reduces K currents by inhibiting K_T (29, 63) and K_{Ca} channels (Fig. 4F) (42, 74). A voltage-insensitive, sustained K^+ current was observed in rabbit PASMC, which has been demonstrated to be generated by K^+ efflux through K_T channels (29, 63). Inhibition of the K_T channels would also contribute to initiating membrane depolarization and $[Ca^{2+}]_{cyt}$ rise in PASMC challenged with acute hypoxia (29).

A rise in $[Ca^{2+}]_{cyt}$ induced by membrane depolarization-mediated Ca^{2+} influx and/or agonist-induced Ca^{2+} release activates K_{Ca} channels in PASMC (74, 119). The resultant increase in whole-cell K_{Ca} currents limits vasoconstriction by inducing membrane hyperpolarization and concomitant vasodilation. Large conductance (200-250 pS) K_{Ca} channels in PASMC are synergistically regulated by cytoplasmic $[Ca^{2+}]$ and E_m. In other words, the acute hypoxia-mediated membrane depolarization and increase in $[Ca^{2+}]_{cyt}$ would markedly activate the large-conductance K_{Ca} channels and limit hypoxia-induced membrane depolarization and pulmonary vasoconstriction. However, the observed response of E_m (Fig. 4D) or pulmonary vascular tone (Fig. 4B) to hypoxia is a sustained depolarization or persistent vasoconstriction, suggesting that the negative feedback effect of K_{Ca} channel activation is inhibited during hypoxia. Indeed, acute hypoxia, in addition to inhibiting Kv and K_T channels for initiating membrane depolarization, also inhibits K_{Ca} channels in cell attached membrane patches of rat PASMC (Fig. 4F). Similar results have been reported in rabbit PASMC (42) and carotid body cells (110). Acute hypoxia-induced inhibition of K_{Ca} channels appears to be due to an intermediate released or produced during hypoxia, since the effect disappears in excised (inside-out) membrane patches (42, 82). However, chronic hypoxia-mediated inhibition of K_{Ca} channels appears to be caused by direct effect on the channel expression and function (74). Thus, hypoxia-induced inhibition of K_{Ca} channels, although it may not be involved in triggering membrane depolarization, certainly plays an important role in maintaining membrane depolarization.

5.2. Potential Mechanisms by Which Acute Hypoxia Inhibits Kv Channels

As mentioned above, a functional Kv channel is composed of four pore-forming α (and γ) subunits and four regulatory β subunits. In native cells including mammalian and human PASMC, functional Kv channels are believed to be mainly formed heterogenously by different α (and γ) and β subunits (17, 18). An attractive hypothesis is that the Kv channel (including the pore-forming α subunits, regulatory β subunits, and chaperone modulators) acts as both the sensor and effector of the response to acute hypoxia. In other words, hypoxia directly alters Kv channel functions by modulating amino acid residues that are sensitive to oxygen and redox status (44) without intermediates involved. The "direct-action" hypothesis appears to be supported by studies showing that acute hypoxia reduces the current generated by recombinant Kv channels expressed in heterologous transfection system (e.g., HEK-293, COS cells) (17, 69, 70). However, it has been reported that only certain α subunit homotetramers (e.g., Kv1.2, Kv3.2), α/α subunit heterotetromers (e.g., Kv1.2/Kv1.5, Kv1.2/Kv9.3), or α/β subunit heterotetromers (e.g., Kv1.2/Kvβ1.1, Kv1.5/Kvβ1.2), are affected by acute hypoxia (35, 64, 69, 70), although the homology of amino acid sequence among different Kv channel α subunits is very high (14). This suggests that acute hypoxia may also affect Kv channels by an "indirect-action" mechanism via, e.g., an intermediate or modulator (e.g., endothelin-1 or Ca^{2+}) produced or released during hypoxia (82, 93).

5.2.1. Role of Kv Channel β Subunits

The Kv channel β subunits have been demonstrated to serve as an oxygen sensor in hypoxia-induced inhibition of Kv channel activity (18, 76). If the regulatory β subunit senses low oxygen, it is likely that it can regulate the kinetics of channel opening and closing directly through physical interaction with the pore forming α subunits. The β subunit shares homology to NAD(P)H oxidase (51), possibly allowing it to sense oxygen tension and regulate the gating properties (e.g., conductance, voltage dependence, inactivation, pharmacological interaction with modulators, and binding activity with kinases) of the pore forming α subunit. In addition, certain β subunits have been described to possess oxygen-sensing ability (18, 70, 76). Once hypoxia ensues and low oxygen is sensed, it is likely that the channel protein adapts to the stress by closing, leading to depolarization, Ca^{2+} influx, vasoconstriction, and subsequent diversion of blood away from regions that have a poor oxygen saturation (17, 69).

5.2.2. Role of Redox Status and Oxygen Radicals

In addition, it has been suggested that other oxygen sensors in PASMC, such as the mitochondria or NADPH oxidase, may exist to use redox potential for regulating physiological properties. The notion that reactive oxygen species (ROS) function as signaling molecules is relatively new and based in a large part on the discovery of reactive species such as nitric oxide (NO) and hydrogen peroxide (H_2O_2) that have signaling potential. ROS have generally been regarded as destructive as they can contribute to protein and DNA damage; however, in many systems it has been suggested that low-level ROS generation can serve a signaling role (106, 107). First, it must be assumed that if a redox network exists for the regulation of channel activity as well as for cellular signaling, there must be a tonic generation of reactive species by the cell that maintains homeostasis and directs cellular activity. That is, the cell, through regulated and constant generation of ROS, maintains a sensory system that is required for maintaining normal activity. When stress (e.g., acute hypoxia or redox potential change) is introduced, this sensory system is perturbed so that the cell responds to the stimuli by changing membrane channel activity. Two possibilities exist: ROS directly affect channel function and expression or they modulate expression and function of proteins or molecules that secondarily affect channel activity.

Whether acute hypoxia mediates pulmonary vasoconstriction by decreasing or increasing ROS is still controversial (see Chapters 9, 15, 16, and 17). Since

O_2 is the substrate for the generation of ROS via many mechanisms (including NADPH oxidase, the electron transport chain of the mitochondria, NO synthase, or cyclooxygenase), it has been demonstrated that hypoxia reduces ROS production and changes redox status in PASMC (5). *In vitro* experiments have shown that ROS or oxidants (e.g., NO, diamide, H_2O_2) increase K^+ channel activity, whereas reducing agents (e.g., reduced form of glutathione, dithiothreitol, N-acetyl-L-cysteine) decrease K^+ currents in PASMC. Therefore, decreased tonic generation of ROS would inhibit K^+ channels,, induce membrane depolarization, and cause pulmonary vasoconstriction. However, acute hypoxia has also been demonstrated to increase the production of ROS (e.g., superoxide) and H_2O_2 by the mitochondria in rat pulmonary microvessel myocytes (106, 107). The increased mitochondrial H_2O_2 acts as a trigger for increasing $[Ca^{2+}]_{cyt}$ by mediating Ca^{2+} release or mobilization from intracellular stores (21, 107). Released Ca^{2+} from the SR could also serve as an inhibitor of Kv channels and cause membrane depolarization in PASMC (82).

5.2.3. Role of Cytochrome *c* and Bcl-2

Cytochrome *c* and Bcl-2 are both normally located in mitochondria. In PASMC, cytoplasmic application of cytochrome *c* increases Kv channel activity, whereas overexpression of Bcl-2 using an adenoviral vector decreases Kv channel activity (22, 81). It is still unknown whether acute or chronic hypoxia in PASMC inhibits the release or production of cytochrome *c* from mitochondria to the cytosol and enhances the translocation (or expression) of Bcl-2 to the plasma membrane. Based on the pro-apoptotic effect of cytochrome *c* and the anti-apoptotic effect of Bcl-2, as well as the involvement of K^+ channels in regulating apoptotic volume decrease (22, 38, 81), it is possible that cytochrome *c* and Bcl-2 may also participate in hypoxia-mediated inhibition of Kv channels in PASMC and in hypoxic pulmonary vasoconstriction (see below).

5.2.4. Role of Metabolic Inhibition

During normoxia, inhibition of oxidative phosphorylation and the citric acid cycle induces pulmonary vasoconstriction (90), while inhibition of glycolysis potentiates HPV (98). These observation suggest that alveolar hypoxia may reduce the phosphate potential [ATP/(ADP+Pi)], an indicator of the energy status, by decreasing oxidative phosphorylation in PASMC (90, 98). This hypoxia-induced change in energy status might subsequently decrease K conductance, induce membrane depolarization, increase Ca^{2+} influx, and cause pulmonary vasoconstriction. In isolated PASMC, the metabolic inhibitors, 2-deoxy-D-glucose (121), carbonyl cyanide-ρ-trifluoromethoxyphenyl-hydrazone (FCCP) (121), rotenone (5), and antimycin A (5), all mimic acute hypoxia to reduce $I_{K(V)}$. In contrast, intracellular application of ATP enhances $I_{K(V)}$ in native

PASMC (23) and cells transfected with Kv channels (70), suggesting that Kv channels are regulated by changes in ATP production. Although a global decrease of intracellular ATP content occurs slowly, hypoxia and metabolic inhibition may produce a rapid depletion of local or compartmentalized ATP contents (36) and subsequently reduce Kv channel activity.

5.2.5. Role of Cytochrome P-450 and NAD(P)H-Dependent Oxidoreductase

The cytochrome P-450 monooxygenase system (P-450), which is prevalent in lungs and in various vascular beds (30), is a family of heme-thiolate enzymes that catalyze oxidative reactions. The activity of P-450 requires a reducing agent, NADPH, as an electron donor and molecular (atmospheric) oxygen in a 1:1 stoichiometry. Activation of P-450 is oxygen dependent, and half-saturation of P-450 with oxygen occurs at a Po_2 of 20-100 Torr (56, 101), which is similar to the range of Po_2 that affects Kv channel activity and pulmonary vascular tone.

In PASMC, inhibition of P-450 by the structurally distinct inhibitors of P-450, the imidazole antimycotics (e.g., clotrimazole and miconazole) and the suicide substrate inhibitor, 1-aminobenzotriazole, decreased $I_{K(V)}$ and caused membrane depolarization (122). Induction of P-450 induced membrane hyperpolarization, while depletion of the enzyme mediated membrane depolarization (15). Taken together with the observations that carbon monoxide inhibits HPV (102) and hypoxia inhibits P-450 (30), these data suggest that P-450 or a heme-protein may function as an oxygen sensor in hypoxia-mediated regulation of Kv channel activity and in initiating HPV (56, 101).

P-450 is always attached to an NADPH-reductase which is required to reduce $NAD(P)^+$ to NAD(P)H. It is possible that the cytochrome P-450-NADPH oxidoreductase complex is localized either very closely with or attached physically to Kv channel α subunits, and that it functions as a regulatory modulator of the channel (similar to Kv channel β subunits). Under normoxic conditions, the complex requires and utilizes oxygen to produce an intermediate, namely an oxygen-dependent channel regulator (ODCR), which can be rapidly and efficiently delivered to the binding sites of the channel protein. The ODCR, constantly produced by the complex during normoxia, links the enzyme activity to the channel activity and keeps the channels open to maintain a negative E_m. During hypoxia, the production of ODCR is decreased because of inhibited activity of the cytochrome P-450-NADPH oxidoreductase complex and the channels are, therefore, closed. In this scenario, the oxidoreductase complex serves as the *oxygen sensor*, the ODCR acts like a *coupler*, and the Kv channel protein is considered the *effector* in the development of HPV. The observations supporting this hypothesis include: *a*) carbon monoxide inhibits HPV (102), while P-450 inhibitors decrease $I_{K(V)}$ and depolarize PASMC (123); *b*) NADPH-oxidase is activated during HPV, while the enzyme inhibitor, diphenyleneiodonium, decreases $I_{K(V)}$ (108) and blocks HPV (124); *c*) the

NADPH-P-450 oxidoreductase (and NOS) system is oxygen sensitive; *d*) NO reversibly activates Kv channels in PASMC (124) and causes membrane hyperpolarization and pulmonary vasodilation; and *e*) the P-450-derived epoxides (e.g., epoxyeicosatrienoic acid) activate K^+ channels (13, 33) and cause membrane hyperpolarization and pulmonary vasodilation (13, 102).

In addition to the cytochrome P-450-NADPH oxidoreductase, NO synthase (NOS) may also serve as an oxygen sensor given the fact that NOS is a P-450-type hemoprotein (12, 109). The potential candidates for ODCR (or the coupler) include oxygen radicals, epoxides (e.g, epoxyeicosatrienoic acid) and NO (P-450- or NOS-dependent metabolites), and cytochrome *c* or other metabolites from mitochondrial oxidative phosphorylation.

The localized change of redox potential may also be determined by localized distribution of NADPH-P-450 oxidoreductase (or NOS) that produces oxidants and reductants for regulating local redox potential. Consumption of NADPH also contributes to the regulation of local redox potentials. Thus, it is plausible that Kv channel activity is regulated by local redox potential that is controlled by local production of oxidants or reductants from attached or adjacent NADPH-P-450 oxidoreductase complexes.

6. Chronic Hypoxic Regulation of Kv Channel Expression

As mentioned earlier, the major physiological function of acute HPV is to maximize oxygenation of the venous blood in the pulmonary artery by optimizing the ventilation-perfusion ratio (see Chapter 1). However, persistent hypoxia (e.g., patients with airway obstructive diseases and congenital heart diseases) and chronic exposure to hypoxia (e.g., residents in high altitude) cause sustained pulmonary vasoconstriction and vascular remodeling characterized by significant medial (smooth muscle) hypertrophy, which lead to pulmonary hypertension.

In lungs from chronically hypoxic rats, the pressor response to acute hypoxia was impaired, whereas the pressor response to other vasoconstrictors (e.g., angiogensin II, prostaglandin $F_{2\alpha}$ and norepinephrine) was enhanced (54). This reduced pressor response to acute hypoxia might result from abnormalities in the mechanism that couples acute hypoxia to contraction of the pulmonary vascular smooth muscle. Based on these observations, McMurtry and his associates proposed almost two decades ago that chronic and acute hypoxia might share the same cellular and molecular mechanisms in mediating pulmonary vasoconstriction. Indeed, chronic hypoxia-mediated pulmonary vasoconstriction (and pulmonary vascular medial hypertrophy) may also be related to inhibition of Kv channels in PASMC (80, 95, 105).

The changes of ROS production, cellular redox status and cell metabolism (e.g., glycolysis and oxidative phosphorylation), and the increased regulatory effect of oxygen-sensitive regulatory molecules and proteins (e.g., Kv channel β subunits, P-450-NADPH oxidase-like oxygen-sensing complex) are phenomena common to both acute and hypoxic exposure. However, unlike acute hypoxia, chronic exposure to hypoxia causes structure changes in the pulmonary vasculature as an adaptive response to prolonged hypoxic stress. The major pathological changes involved in pulmonary vascular remodeling in patients with hypoxic cardiopulmonary diseases and in animals exposed to chronic hypoxia include arterial medial hypertrophy and muscularization of small arterioles due to enhanced proliferation and/or inhibited apoptosis of PASMC (84, 99).

A common hypothesis is that pulmonary vasoconstriction and PASMC proliferation use overlapping signaling processes that result in parallel intracellular events (e.g., a rise in cytoplasmic Ca^{2+}) causing pulmonary hypertension. As discussed earlier, an increase in $[Ca^{2+}]_{cyt}$ in PASMC is a major trigger for pulmonary vasoconstriction and an important stimulus for PASMC proliferation (78, 100, 113), and acute hypoxia-mediated inhibition of Kv channel function is one of the mechanisms involved in the development of acute HPV. Therefore, the hypoxia-induced decrease in Kv channel currents, membrane depolarization, and increase in $[Ca^{2+}]_{cyt}$ in PASMC may serve as a shared mechanism by acute and chronic hypoxia to cause pulmonary vasoconstriction and stimulate PASMC proliferation. Indeed, in addition to reducing K^+ channel activity, chronic hypoxia also mediates changes in the transcriptional and translational control of K^+ channel genes.

6.1. Chronic Hypoxia Downregulates Kv Channel Expression in PASMC

Under resting conditions, K^+ permeability is much greater than other ions (e.g., Na^+, Ca^{2+}, Cl^-) permeabilities. Therefore, resting E_m in PASMC is primarily determined by K^+ permeability which is mainly related to the whole-cell K^+ current (I_K). K^+ currents through K_V channels ($I_{K(V)}$), in particular, have been demonstrated to be predominately responsible for the maintenance of the E_m in PASMCs under resting conditions (61, 115). Whole-cell $I_{K(V)}$ at any given time is determined by the following equation:

$$I_{K(V)} = N \times i_{Kv} \times P_{open}$$

where N is the number of functional Kv channels expressed in the plasma membrane; i_{Kv} is the current through a single Kv channel; and P_{open} is the steady-state open probability of a Kv channel. When Kv channels close (i.e., i_{Kv} or P_{open} declines) and/or the number of Kv channels in the plasma membrane declines (i.e., N decreases) as expression of Kv channels is reduced, E_m becomes more depolarized as a result of decreased $I_{K(V)}$. In contrast, opening of Kv channels (i.e., a rise in i_{Kv} or P_{open}) and/or increasing expression of Kv channels (i.e., N

rises) cause membrane hyperpolarization (E_m becomes closer to the $E_K \approx$ -85 mV) as a result of increased $I_{K(V)}$. In other words, the mRNA and protein expression level of different Kv channel genes plays an important role in determining whole-cell $I_{K(V)}$ and, ultimately, in regulating E_m and $[Ca^{2+}]_{cyt}$ in PASMC.

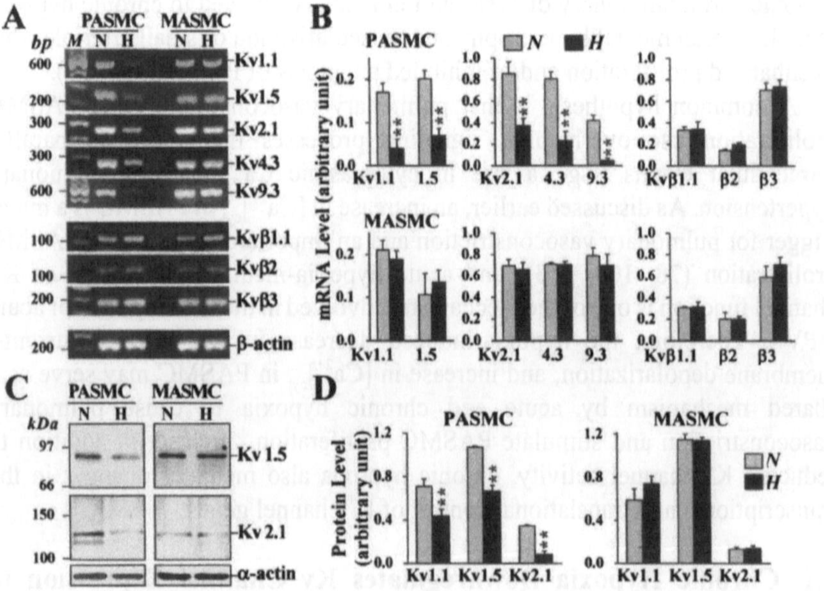

Figure 5. Hypoxia downregulates Kv channel expression in rat PASMC, but not in MASMC. A: PCR-amplified products for Kv channel α and β subunits in normoxia (N) and hypoxia (H). B: Summarized data (means±SE) showing mRNA levels of Kv channel α and β subunits normalized to β-actin transcript levels. ** $P<0.01$ vs. normoxia. C and D: Western blot analyses of Kv1.5 and 2.1 in normoxia and hypoxia. ** $P<0.01$ vs. normoxia (Reproduced from Ref. 80).

As demonstrated in Figure 5, chronic hypoxia downregulates the mRNA (Fig. 5A and B) and protein (Fig. 5C and D) expression of several pore-forming α subunits of Kv channels in rat PASMC, while no appreciable expression changes are seen in mesenteric artery smooth muscle cells (MASMC) (80, 105). These data indicate that: *i*) the chronic hypoxia-induced downregulation of Kv channels is specific or unique to PASMC (in comparison to systemic arterial smooth muscle cells), and *ii*) the inhibitory effect of chronic hypoxia only occurs on the pore-forming α subunits in PASMC but not on the regulatory β subunits (Fig. 5A and B). The decreased Kv channel α subunit expression would be expected to decrease the number of sarcolemmal Kv channels and decrease whole-cell $I_{K(V)}$ (because of decreased *N*). The unsuppressed regulatory β subunit expression would be expected to further limit the activity of existing pore-forming α subunits (because of decreased i_{Kv} and/or P_{open}), thereby further limiting Kv channel activity and the ability of the cell to maintain hyperpolarized.

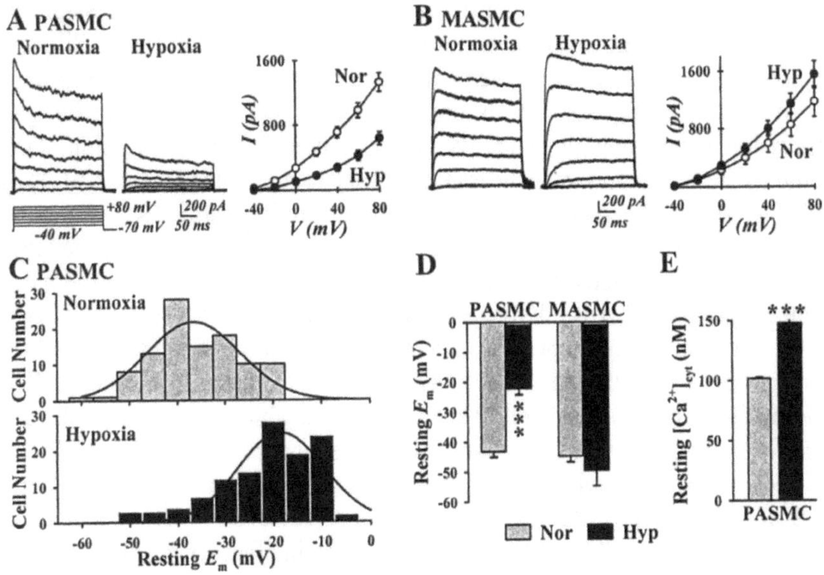

Figure 6. Hypoxia-induced regulation of $I_{K(V)}$, E_m, and $[Ca^{2+}]_{cyt}$ in PASMC. Representative and summarized data showing the effect of hypoxia on $I_{K(V)}$ in PASMC (A) and MASMC (B). C: Distribution histogram showing the range of resting E_m values recorded in PASMC under normoxic (top) and hypoxic (bottom) conditions. D: Summarized data (means±SE) showing resting E_m in PASMC and MASMC in normoxia (Nor) and hypoxia (Hyp). ***$P<0.001$ vs. normoxia. E: Resting $[Ca^{2+}]_{cyt}$ is increased in PASMC by 60-hr exposure to hypoxia. *** $P<0.001$ vs. normoxia (Reproduced from Ref. 80).

As shown in Figure 6, indeed, chronic hypoxia significantly reduced whole-cell $I_{K(V)}$, caused membrane depolarization, and increased $[Ca^{2+}]_{cyt}$ in PASMC, but had little effect on $I_{K(V)}$ and E_m in MASMC. The selective effect of chronic hypoxia on PASMC, in comparison to MASMC, indicates that pulmonary and systemic vascular smooth muscle cells are regulated by chronic hypoxia through different cellular and molecular mechanisms. The hypoxia-induced sustained membrane depolarization and increase in $[Ca^{2+}]_{cyt}$ in PASMC would trigger pulmonary vasoconstriction and contribute to the excessive PASMC proliferation or pulmonary arterial medial hypertrophy observed in lungs from chronically hypoxic animals as well as patients with hypoxic cardiopulmonary diseases. The sustained pulmonary vasoconstriction and vascular remodeling are two of the major causes for the development of pulmonary hypertension in humans and animals exposed to chronic hypoxia.

6.2. Promoter Expression and Regulation by Regulatory DNA Elements and Transcription Factors

The primary site of channel expression control is in the expression and function of the promoter of the genes encoding the channel proteins. Silencers are *cis*-acting regulatory DNA elements that downregulate gene transcription in a cell/tissue-specific fashion in order to either control cell-specific gene expression or prevent the negative consequences of overexpression. DNA elements that are important in the transcriptional regulation of Kv1.5, Kv1.4, and Kv3.1 genes have been identified (104). In the case of the cardiac Kv1.5 channel, a Kv1.5 repressor element (KRE) located in the 5' flanking region of the Kv1.5 gene, contains a dinucleotide repetitive element that is necessary for mediating silencer activity (58, 104). KRE selectively decreases the expression of Kv1.5 in cell lines that do not express Kv1.5 proteins. Deletion of the KRE repetitive element abolishes the silencing activity in transfected cells. In addition, a KRE binding factor (KBF) has recently been identified, which may regulate the cell-specific expression of Kv1.5 by abolishing the silencer effects of KRE. Whether the KRE or KBF is involved in regulating Kv1.5 transcription and expression in PASMC during hypoxia has not been evaluated. However, we can speculate that the expression of the KRE in the Kv1.5 gene (and in other Kv channel α subunit genes) may play an important role in the hypoxia-induced downregulation of Kv channel expression in PASMC. Alternatively, hypoxia may also exert its inhibitory effect on Kv channel α subunit expression by downregulating KBF expression, thereby allowing for silencing of Kv1.5 (and other Kv channels) gene expression and decreasing functional Kv channel availability.

It is generally accepted that changes in P_{O_2} can regulate gene expression via second messengers, protein kinases, and transcription factors such as activating protein-1 (AP-1), hypoxia-inducible factor-1 (HIF-1), nuclear factor-κB (NF-κB), nuclear factor interleukin-6 (NF-IL6), and early growth response-1 (EGR-1) (92). Transcription factors modulate channel protein expression by regulating genes that contain binding sites within their promoters (55). Two AP-1 family members, c-Fos and c-Jun, have been suggested as possible effectors of hypoxic gene regulation based on their ability to sense changes in redox potential and levels of cytoplasmic and nuclear $[Ca^{2+}]$, thereby modulating transcription and expression of the AP-1-responsive genes (1, 6). Under normoxic conditions, overexpression of c-*jun* in rat PASMC resulted in decreased Kv1.5 channel expression and activity, as well as increased Kv channel β subunit expression and increased current inactivation (113). The subsequent depolarization and Ca^{2+} increase enhanced cell proliferation. The divergent regulatory effects of c-Jun on Kv channel α (e.g., Kv1.5) and β (e.g., Kvβ2) subunit gene expression may be due to direct binding to the AP-1 binding sites of the channel gene promoter (as has been reported for the human MaxiK gene, h*slo*) (20) or to inducing expression of an intermediate that subsequently inhibits the channel expression.

We do have evidence that AP-1 binding activity is increased during chronic hypoxia (3% O_2, 72 hrs) in human pulmonary artery endothelial cells (25). Whether the AP-1 binding activity is regulated by chronic hypoxia in PASMC and whether the hypoxia-sensitive Kv channel genes all contain the AP-1 binding sites in their promoters remain unresolved.

While AP-1 binding may be involved more in cellular proliferation and the pulmonary remodeling induced by prolonged hypoxia, HIF-1 activation is intimately involved in hypoxia-induced pulmonary hypertension (111). The biological activity of HIF-1 is regulated by the expression and activity of the HIF-1α subunit. Under normoxic conditions, HIF-1 is ubiquitinated and rapidly degraded, thereby never allowed to form functional heterodimeric HIF-1 proteins. Under hypoxic conditions, HIF-1α ubiquitination is inhibited and HIF-1α dimerizes with HIF-1β subunits to form functional HIF-1 which is responsible for activation of a number of target genes including vascular endothelial growth factor (VEGF) and endothelin-1 (ET-1). The role of these growth factors and mitogens in HPV is discussed in the following section.

6.3. K^+ Channel Expression is Modulated by Growth Factors and Mitogens

As stated above, the increased activation of the transcription factors by chronic hypoxia can result in the increased transcriptional activation of genes encoding vasoactive agonists (e.g., ET-1), mitogens [e.g., VEGF, platelet-derived growth factor (PDGF), erythropoietin, and insulin-like growth factor 2 (IGF-2)], and several glycolytic enzymes (92). Many of these vasoactive agonists and growth factors are known to affect ion channel expression and function. For example, ET-1 can decrease the mRNA and protein expression of Kv channels, and the amplitude of $I_{K(V)}$ is decreased in proliferating PASMC in comparison to growth arrested cells (78). We have also shown that PDGF can stimulate rat PASMC proliferation by upregulating TRPC6, a gene putatively encoding for receptor-operated and/or store-operated Ca^{2+} channels (114). Therefore, the regulation of ion channel activity and expression by growth factors appears not to be limited to K^+ channels.

6.4. K^+ Channel Expression is Regulated by Auxiliary Regulatory Proteins

An emerging idea in the area of regulation of function and expression of Kv channels is the direct interaction of the pore-forming α subunits with K^+ channel-associated proteins (KChAPs) and K^+ channel-interacting proteins (KChIPs), regulators different from the Kv channel β subunits. These proteins have predominantly been described in the brain and heart and play an important role in regulating Kv channel function and expression in these tissues; little data exist regarding their role on channel expression in the pulmonary vasculature. KChAPs function mainly as chaperones (39, 40) for specific Kv channels (e.g. Kv1.3 and Kv4.3), leading to enhanced protein expression of the channels. However, this interaction is transient and does not affect channel activity.

Three KChIPs have been identified (KChIP1-3) that modulate $I_{K(V)}$. In terms of channel kinetics, KChIPs appear to work oppositely to the β subunit in that total current is modified, activation is slowed, and the chaperone proteins accelerate the channel recovery from inactivation (4). In addition, all KChIPs co-localize and co-immunoprecipitate with Kv channel α subunits, suggesting that they may be part of the channel complex (4). KChIPs have four EF-hand-like domains, each consisting of two perpendicular 10 to 12 residue α-helices with a 12-residue loop region, forming a single Ca^{2+}-binding site. The KChIP-induced modulation of Kv channel activity is sensitive to Ca^{2+} levels (72). This suggests that a possible mechanism by which the cell senses and responds to a transient rise in $[Ca^{2+}]_{cyt}$ would be the modulation of Kv channel activity by Ca^{2+} regulation of KChIPs or KChIP-Kv channel interaction.

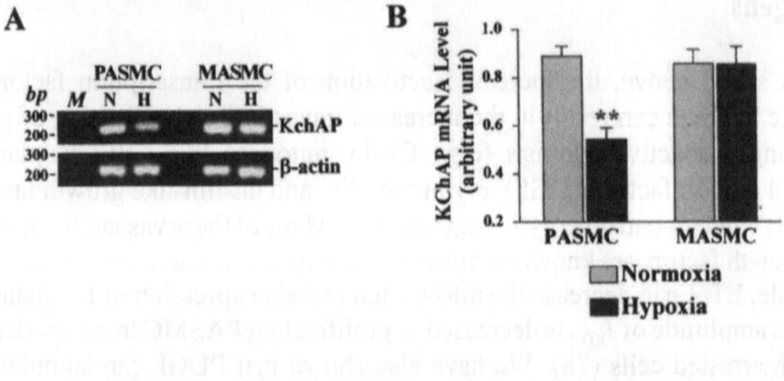

Figure 7. KChAP mRNA expression in PASMC and MASMC is differentially modulated by hypoxia. A: PCR-amplified products for KChAP and β-actin in PASMC and MASMC incubated under normoxic (N) and hypoxic (H) conditions. B: Summarized mRNA levels (normalized to β-actin control) from the data presented in A. ** $P<0.01$ vs. normoxia.

Studies in PASMC involving the characterization of KChAPs and KChIPs are limited with respect to hypoxia. Our preliminary data showed that chronic hypoxia downregulated KChAP in PASMC (Fig. 7). Inhibition of these intracellular regulators of Kv channels may result in decreased expression of Kv channels or inability of Kv channels to open. A chronically hypoxic environment would result in long-term effects on these proteins that then contribute to pulmonary vascular remodeling and subsequent pulmonary hypertension.

7. Role of Plasmalemmal K^+ Channels in Regulating Apoptosis in PASMC

The precise control of the balance between PASMC proliferation and apoptosis is important in maintaining the structural and functional integrity of the pulmonary vasculature. As was stated earlier, increased PASMC proliferation and/or decreased apoptosis, result in pulmonary vascular wall thickening that elevates pulmonary vascular resistance, a process that is partially responsible in the development of hypoxic pulmonary hypertension.

Apoptotic volume decrease or cell shrinkage is the early hallmarks of apoptosis (47, 112). Maintenance of a high concentration of cytoplasmic K^+ ($[K^+]_{cyt}$) is essential to the regulation of normal cell volume. Thus, apoptotic cell shrinkage may result partly from a decrease in $[K^+]_{cyt}$ due to increased K^+ efflux through opened K^+ channels (11, 41, 47, 112). In addition to its role in the control of cell volume, a high $[K^+]_{cyt}$ is required for suppression of caspases and nucleases (34), the final mediators of apoptosis (85, 103). Accordingly, activation of K^+ channels in the plasma membrane would induce apoptotic volume decrease and apoptosis by enhancing K^+ efflux or loss, whereas inhibition of K^+ channel activity would attenuate apoptotic volume decrease by maintaining sufficient $[K^+]$ in the cytoplasm to inhibit apoptosis.

In human PASMC, treatment with staurosporine, a potent inducer of apoptosis, increased $I_{K(V)}$ (Fig. 8A*a*) (37), caused cell shrinkage (Fig. 8A*b*), and increased the percentage of the cells undergoing apoptosis (Fig. 8A*c*). The time courses for staurosporine-mediated effects on $I_{K(V)}$, cell volume change and apoptosis were, however, quite different. The staurosporine-mediated increase in $I_{K(V)}$ took place at first followed by apoptotic cell shrinkage. The correlated increase in $I_{K(V)}$ and cell shrinkage far preceded the increase in the percentage of apoptotic cells (Fig. 8A*c*). Blockade of Kv channels with 4-aminopyridine inhibited staurosporine-induced increase in $I_{K(V)}$ (Fig. 8B*a*) and attenuated staurosporine-induced apoptosis (Fig. 8B*b*). These results suggest that the earliest event in apoptosis involves an increase in Kv channel activity which subsequently results in K^+ efflux and cell volume decrease, and ultimately causes apoptosis.

Bcl-2 is an antiapoptotic membrane protein which enhances cell survival by preventing cytochrome *c* release from the mitochondrial intermembrane space to the cytosol. In keeping with the known importance of transmembrane K^+ movement in modulating cell death, it is not surprising to observe that the overexpression of human Bcl-2 in rat PASMC downregulated the expression of Kv channel α subunits (particularly Kv1.1, Kv1.5, and Kv2.1) (Fig. 8C*a*) and decreased $I_{K(V)}$ (Fig. 8C*b*). Additionally, staurosporine-induced augmentation in $I_{K(V)}$ and apoptosis were both attenuated in the Bcl-2-transfected PASMC (Fig. 8C*c*) (22).

In summary, function and expression of K^+ channels play an important role in apoptosis in PASMC. Decreased activity of voltage-dependent K^+ channels due to decreased expression or function of K^+ channels during hypoxia leads to maintenance of a high $[K^+]_{cyt}$, which opposes apoptosis by preventing apoptotic cell shrinkage and by suppression of cytoplasmic caspases and nucleases. Therefore, decreased expression or function of Kv channels in PASMC may be an important anti-apoptotic mechanism and, hence, may significantly contribute to pulmonary vascular remodeling.

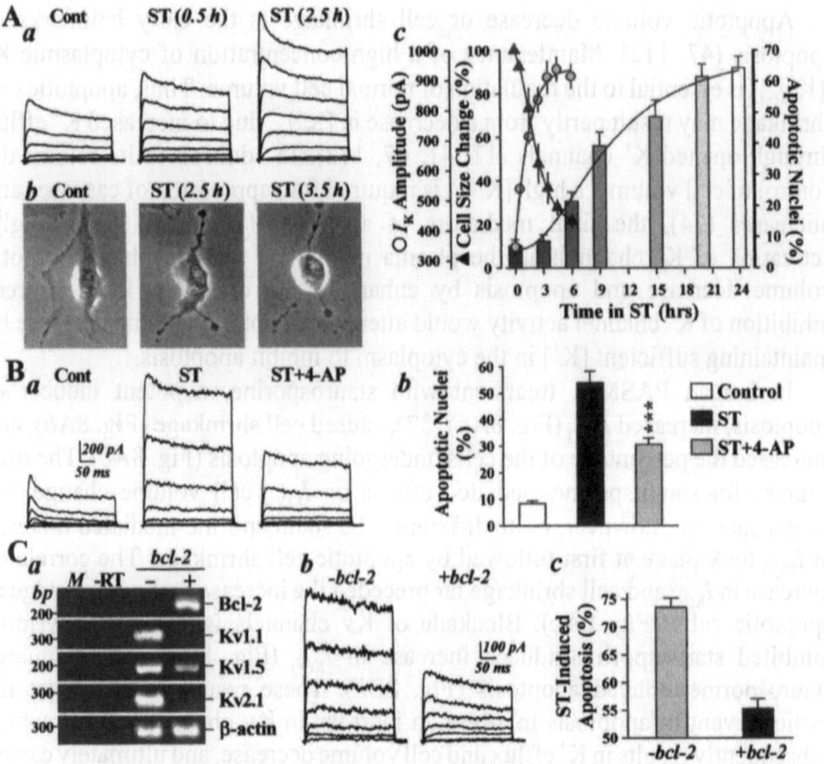

Figure 8. Kv channel activity in PASMC is modulated by pro- and anti-apoptotic agents. A: Staurosporine (ST), a pro-apoptotic agent, enhances $I_{K(V)}$ (*a*) and causes apoptotic cell shrinkage (*b*) in rat PASMC. Time-courses of the ST-induced $I_{K(V)}$ increase (*c*, circles), cell volume decrease (*c*, triangles), and the percentage of apoptotic cells (*c*, bars). Note that increased $I_{K(V)}$ precedes cell shrinkage and apoptosis. B: Inhibition of Kv channels with 4-AP (5 mM) attenuates ST-induced apoptosis in human PASMC. C: Overexpression (+*bcl-2*) of Bcl-2 downregulates mRNA expression of Kv1.1, Kv1.5, and Kv2.1 (*a*), decreases amplitude of $I_{K(V)}$ (*b*), and inhibits ST-induced apoptosis (*c*). ***P<0.001 vs. control (-*bcl-2*) (Reproduced from Refs. 23, 37, and 81).

8. Microdomains and Organized Signaling in Regulating Kv Channel Function in PASMC

An emerging idea in signal transduction posits the existence of spatially organized complexes of signaling molecules in microdomains (62, 65) of the plasma membrane. Recent studies have focused on the distribution of signaling molecules in caveolae, which are cholesterol and sphingolipid-enriched regions that can form distinct structural invaginations of the plasma membrane and are enriched of the protein caveolin (97). This notion of signaling microdomains is attractive in that it would account for a localization of receptors, signaling components, and effector molecules (e.g., ion channels) in a confined region to facilitate coordinated, precise, and rapid regulation of cell function.

Knock-out mice deficient in caveolin-1 protein, required in the structural formation of caveolae, phenotypically expresses pulmonary hypertension with marked right ventricular hypertrophy and appear to have unregulated NO generation (128). This suggests that the formation of caveolae is integral to the pulmonary vasculature and that stress such as chronic hypoxia may lead to malformation or lack of essential caveolae components. As this is done in a small rodent model, possible implication of caveolae in larger mammals may be different and it is debatable whether pulmonary hypertension is a manifestation of too little or too many caveolae in humans. Equally compelling and rational arguments can be made for both possibilities.

Though caveolae from cardiomyocytes have been show to be enriched in adrenergic receptors, adenylyl cyclase, and related signaling machinery, not much data exists for signaling complexes assembled in animal or human PASMC. It is likely that similar enrichment exists in PASMC, but it is equally likely that, because these cells contain machinery to respond to low Po_2, there may be enrichment of *a*) oxygen sensor(s) such as NADPH-P-450 oxidoreductase complex and/or a heme- and metal-containing motif; *b*) couplers (e.g., ODCR), signal transduction proteins, and/or modulators (e.g., KChAP, KChIP), and *c*) effectors such as Kv channels (e.g., Kv1.5) and Ca^{2+} channels (e.g., VDCC) in caveolae allowing for localized sensing and rapid response to modulate cell physiology (Fig. 9). It has been demonstrated that Kv1.5 specifically targets to caveolae, whereas depletion of cellular cholesterol and inhibition of sphingolipid synthesis alter Kv1.5 channel function (50).

It is possible that caveolae may contribute to regulation of Kv channels during acute and chronic hypoxia. Lipid rafts, of which caveolae are a subclass, are enriched in arachidonic acid (AA) and stresses (e.g., ischemia) can release AA (77, 87). Recent studies involving brain hypoxia (7) and heart ischemia (71) reveal through gene array analysis that a common response is the activation of 12-lipoxygenase, an enzyme that metabolizes AA and generates biologically-active metabolites. Other AA metabolizing enzymes are also induced and their tissue specific expression determines cellular responses to hypoxia. Epoxyeicosatrienoic acids (EETs) are cytochrome P450-derived AA metabolite and have the ability to regulate vascular tone. Though most EETs cause vasodilation through activation of K^+ channels, 14,15-EET activates Ca^{2+} channels and cause Ca^{2+} influx and induce vasoconstriction (24). Therefore, it is likely that there exists a delicate balance in PASMC between numerous factors that cause relaxation and contraction by modulating Kv channel activity, of which AA metabolites may be a particular regulatory component that can be easily exploited by hypoxia through closing of Kv channels and opening of Ca^{2+} channels. This would suggest that caveolae and their localized components contribute to the development of hypoxia-mediated Kv channel inhibition, pulmonary vasoconstriction, vascular remodeling, and pulmonary hypertension. Caveolae may thus serve as an organizing center for cellular signaling and the particular responses elicited by hypoxia may be a manifestation of the enriched signaling components in caveolae dependent on the remodeled state of the pulmonary vasculature.

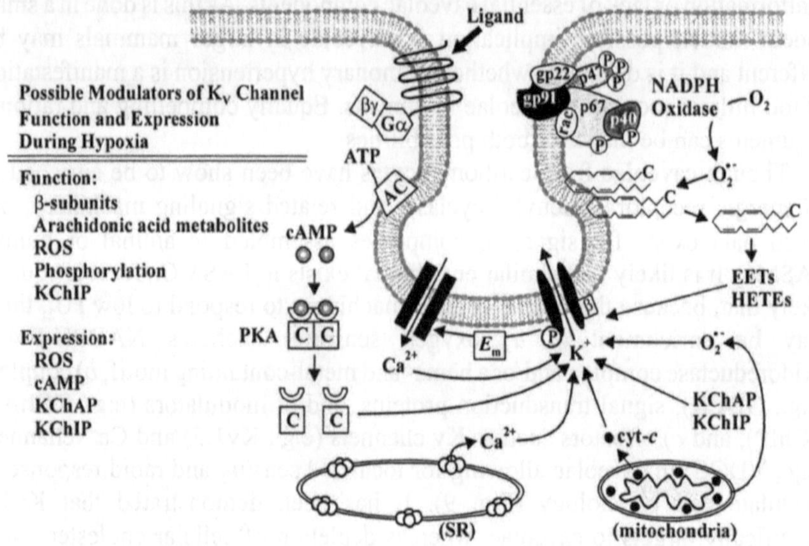

Figure 9. Schematic representation of the organization of ion channels, membrane receptors, and signal transduction proteins in a caveolae. Kv channels within the caveolae are modulated by protein kinase A (PKA), Ca^{2+} (from the SR and extracellular site), mitochondrial cyt-*c* and superoxide, EETs/HETEs, and NADPH oxidase-produced superoxide. NADPH oxidase is depicted as a multi-subunit enzyme complex comprised of gp91, gp22, p47, p67, p40, and rac proteins. AC, adenylate cyclase; EETs, epoxyeicosatrienoic acids; HETEs, hydroxyeicosatetraenoic acids, P, phosphorylation sites, KChIP, K^+ channel interacting protein; KChAP, K^+ channel associated proteins; ROS, reactive oxygen species.

9. Summary and the Road Ahead

HPV is a critical physiological mechanism to ensure maximal oxygenation of blood. Persistent HPV or pulmonary vascular remodeling during chronic hypoxia causes pulmonary hypertension that may lead to right heart failure. A rise in $[Ca^{2+}]_{cyt}$ in PASMC triggers pulmonary vasoconstriction and stimulates PASMC proliferation. Therefore, an increase in $[Ca^{2+}]_{cyt}$ in PASMC may serve as a critical step in acute HPV and in chronic hypoxia-mediated pulmonary vascular medial hypertrophy (Fig. 10). An important mechanism in elevating $[Ca^{2+}]_{cyt}$ is membrane depolarization-mediated Ca^{2+} influx through voltage-dependent Ca^{2+} channels. Given the fact that E_m is primarily determined by K^+ permeability and the resting E_m is regulated predominantly by whole-cell K^+ currents through Kv channels in PASMC, regulation of Kv channel function and expression has been demonstrated to be a mechanism by which hypoxia induces pulmonary vasoconstriction and stimulates PASMC proliferation.

Acute hypoxia may inhibits Kv channel function via multiple different pathways to assure the efficacy and the sensitivity of the response. Hypoxia may decrease whole-cell $I_{K(V)}$ by *a*) inhibiting oxidative phosphorylation, *b*) changing cellular redox status, *c*) altering ROS production, *d*) inhibiting the release of oxygen-dependent Kv channel regulators (e.g., NO, cytochrome *c*, H_2O_2, superoxide) and/or P-450-NADPH oxidoreductase metabolites (e.g., epoxides), *e*) inducing conformational changes of the channel α and/or β subunits via reducing or oxidizing disulfide bridges, and/or *f*) altering a membrane-delimited O_2-sensitive regulatory moiety that is adjacent or coupled to the channel protein. Chronic hypoxia may inhibit Kv channel activity by downregulating mRNA and protein expression and/or decreasing mRNA and protein stability of Kv channel α subunits. The underlying mechanisms involved in chronic hypoxia-induced inhibition of Kv channel expression include *a*) upregulation or downregulation of transcription factors (e.g., HIF-1, NF-κB, AP-1, and FixL/FixJ) and signal transduction proteins (e.g., P_{53}, P_{38}, MAP kinase, tyrosine kinase, PKC), *b*) metabolic inhibition (e.g., mitochondrial oxidative phosphorylation and glycolysis), *c*) mitochondrial production of ROS, *d*) inhibition of P-450-NADPH oxidase metabolites, and *v*) induction or inhibition of an intermediate that down- or up-regulating Kv channel α subunit gene expression (Fig. 10).

Vascular smooth muscle cells utilize a myriad of receptors, signaling molecules, and effector molecules (particularly ion channels) to maintain a delicate homeostatic balance. Controlled activity of PASMC with respect to growth, death, and function is dependent to a large part on the activity of functional Kv channels and Ca^{2+} channels. Activity of Kv channels is involved in *a*) modulating PASMC contractility via changes in E_m and $[Ca^{2+}]_{cyt}$, *b*) regulating PASMC growth through changes in cytoplasmic, nuclear and intracellularly-stored $[Ca^{2+}]$, and *c*) regulating PASMC apoptosis via controlling K^+ efflux and apoptotic cell shrinkage. Inhibition of Kv channel function (by acute hypoxia) and expression (by chronic hypoxia) decreases $I_{K(V)}$, induces E_m depolarization, promotes Ca^{2+} influx through VDCC, increases $[Ca^{2+}]_{cyt}$, causes pulmonary vasoconstriction, and enhances vascular medial hypertrophy by stimulating PASMC proliferation. Reduction of K^+ efflux due to inhibited Kv channels also inhibits apoptotic cell shrinkage and apoptosis in PASMC, and further contribute to the progression of pulmonary medial hypertrophy.

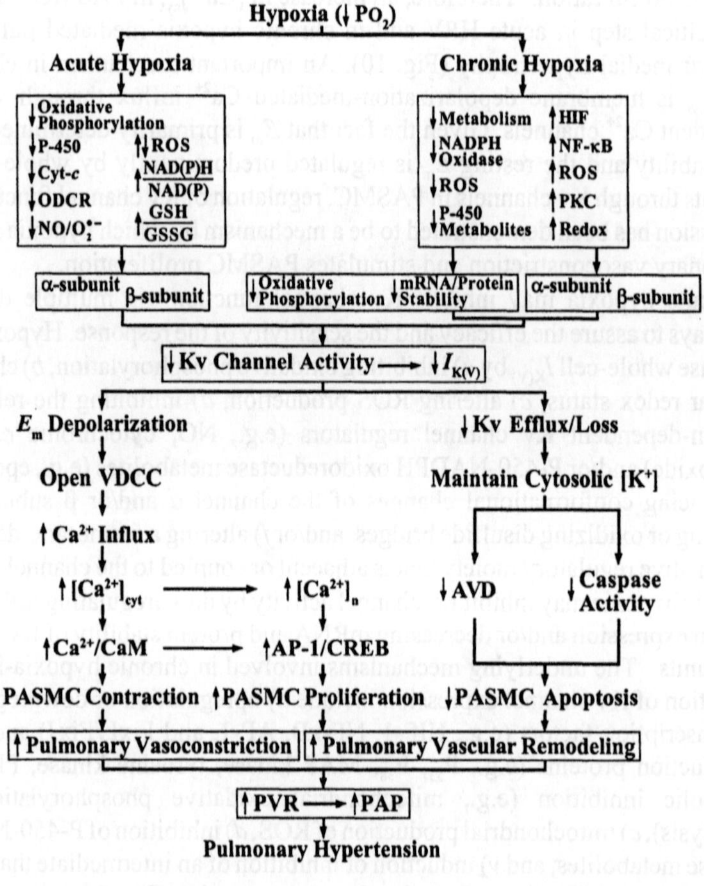

Figure 10. Flow chart illustrating how acute or chronic hypoxia causes pulmonary vasoconstriction and vascular remodeling through modulation of Kv channel activity, E_m, $[Ca^{2+}]_{cyt}$, as well as by regulating apoptosis. P-450, cytochrome P-450; ODCR, oxygen-dependent channel regulator; HIF hypoxia-inducible factor; NF-κB, nuclear factor-κB; PKC, protein kinase C; VDCC, voltage-dependent Ca^{2+} channel; CaM, calmodulin; AP-1, activating protein 1; CREB, cAMP-responsive element binding protein; AVD, apoptotic volume decrease; PVR, pulmonary vascular resistance; PAP, pulmonary arterial pressure.

It is possible that acute and chronic hypoxia are cellular responses along a continuum of vascular physiology ranging from contraction to pulmonary hypertension. It is that the former is an immediate response utilizing the complement of what is already present in a cell to respond to a stress with the latter being a more profound response requiring modification of the existing machinery to suit the cell to its new environment.

The progress of research in the area of HPV and hypoxia-induced pulmonary hypertension lies in an intricate understanding of this continuum and the steps that commit a cell to alter its response to hypoxia from the acute setting of utilizing existing components to the chronic setting of generating new components to create a new cellular environment. The targets of potential therapeutics lie in understanding the proteins and signals involved in pulmonary vascular remodeling. In this endeavor it will be essential to understand the organizing of signaling events in subcellular microdomains. The cell is not a random phenomenon; it is well organized. It is likely that this has ramifications for cellular signaling. The results will be the elucidation of signaling networks that integrate existing and new signaling pathways to begin to understand the complexity of a cellular response to a stimulus such as hypoxia.

Acknowledgment

This work was supported by grants from the National Heart, Lung, and Blood Institute of the National Institutes of Health (HL 64945, HL 54043, HL 66012, HL 69758, and HL 66941). We thank L.J. Rubin, O. Platoshyn, Y. Yu, S. Krick, D. Ekhterae, S. Zhang, E. Brevnova, I. Fantozzi, V.A. Golovina, C. Bailey, S. McDaniel, and J. Wang for their contributions to the work.

References

1. Abate C, Patel L, Rauscher FJ 3rd, Curran T. Redox regulation of fos and jun DNA-binding activity *in vitro*. *Science.* 1990; 24: 1157-1161.
2. Albert AP, and Large WA. Activation of store-operated channels by noradrenaline via protein kinase C in rabbit portal vein myocytes. *J. Physiol.* 2002; 544: 113-125.
3. Allbritton NL, Oancea E, Kugn MA, and Meyer T. Source of nuclear calcium signals. *Proc. Natl Acad. Sci. U.S.A.* 1994; 91: 12458-12462.
4. An WF, Bowlby MR, Betty M, Cao J, Ling H-P, Mendoza G, Hinson JW, Mattsson KI, Strassle BW, Trimmer JS, Rhodes KJ. Modulation of A-type potassium channels by a family of calcium sensors. *Nature.* 2000; 403: 553-556.
5. Archer SL, Huang J, Henry T, Peterson D, Weir EK. A redox-based O_2 sensor inorat pulmonary vasculature. *Circ. Res.* 1993; 73: 1100-1112.
6. Bannister AJ, Cook A, Kouzarides T. *In vitro* DNA binding activity of Fos/Jun and BZLF1 but not C/EBP is affected by redox changes. *Oncogene.* 6: 1243-1250, 1991.
7. Bernaudin M, Tang Y, Reilly M, Petit E, Sharp FR. Brain genomic response following hypoxia and re-oxygenation in the neonatal rat. *J. Biol. Chem.* 2002; 277: 39728-39738.
8. Berridge MJ. Calcium signalling and cell proliferation. *Bioessays.* 1995; 17: 491-500.
9. Berridge MJ. Inositol trisphosphate and calcium signalling. *Nature.* 1993; 361: 315-325.

10. Blaustein MP. Physiological effects of endogenous ouabain: control of intracellular Ca^{2+} stores and cell responsiveness. *Am. J. Physiol.* 1993; 264: C1367-C1387.
11. Bortner CD, Hughes FM Jr, and Cidlowski JA. A primary role for K^+ and Na^+ efflux in the activation of apoptosis. *J. Biol. Chem.* 1997; 272: 32436-32442.
12. Bredt DS, Hwang PM, Glatt CE, Lowenstein C, Reed RR, and Snyder SH. Cloned and expressed nitric oxide synthase structurally resembles cytochrome P-450 reductase. *Nature.* 1991; 351: 714-718.
13. Campbell WB, Gebremedhin D, Pratt PF, and Harder DR. Identification of epoxyeicosatrienoic acids as endothelium-derived hyperpolarizing factors. *Circ. Res.* 1996; 78: 415-423.
14. Chandy KG, and Gutman GA. Voltage-gated K^+ channels. In *Ligand- and Voltage-Gated Ion Channels*, North RA, ed. 1995. Boca Raton, FL: CRC, pp. 1-71.
15. Chen G, and Cheung DW. Modulation of endothelium-dependent hyperpolarization and relaxation to acetylcholine in rat mesenteric artery by cytochrome P450 enzyme activity. *Circ. Res.* 1996; 79: 827-833.
16. Clapp LH, Gurney AM, Standen NB, and Langton PD. Properties of the ATP-sensitive K^+ current activated by levcromakalim in isolated pulmonary arterial myocytes. *J. Membr. Biol.* 1994; 140: 205-213.
17. Coppock EA, Martens JR, and Tamkun MM. Molecular basis of hypoxia-induced pulmonary vasoconstriction: role of voltage-gated K^+ channels. *Am. J. Physiol. Lung Cell. Mol. Physiol.* 2001; 281: L1-L12.
18. Coppock EA, and Tamkun MM. Differential expression of K_V channel α- and β-subunits in the bovine pulmonary arterial circulation. *Am J. Physiol. Lung Cell. Mol. Physiol.* 2001; 281: L1350-L1360.
19. Davies AR, and Kozlowski RZ. Kv channel subunit expression in rat pulmonary arteries. *Lung.* 2001; 179: 147-161.
20. Dhulipala PDK, and Kotlikoff MI. Cloning and characterization of the promoters of the maxiK channel α and β subunits. *Biochim Biophys Acta.* 1444: 254-262, 1999.
21. Dipp M, Nye PCGm and Evans AM. Hypoxic release of calcium from the sarcoplasmic reticulum of pulmonary artery smooth muscle. *Am. J. Physiol. Lung Cell. Mol. Physiol.* 2001; 281: L318-L325.
22. Ekhterae D, Platoshyn O, Krick S, Yu Y, McDaniel SS, and Yuan JX-J. Bcl-2 decreases voltage-gated K^+ channel activity and enhances survival in vascular smooth muscle cells. *Am. J. Physiol. Cell Physiol.* 2001; 281: C157-C165.
23. Evans AM, Clapp LH, and Gurney Am. Augmentation by intracellular ATP of the delayed rectifier current idependently of the glibenclamide-sensitive K-current in rabbit arterial myocytes. *Br. J. Pharmacol.* 1994; 111: 972-974.
24. Fang X, Weintraub NL, Stoll LL, Spector AA. Epoxyeicosatrienoic acids increase intracellular calcium concentration in vascular smooth muscle cells. *Hypertension.* 1999; 34: 1242-1246.
25. Fantozzi I, Zhang S, Platoshyn O, Remillard CV, Cowling RT, and Yuan, JX-J. Hypoxia increases AP-1 binding activity by enhancing capacitative Ca^{2+} entry in human pulmonary artery endothelial cells. *Am. J. Physiol. Lung Cell. Mol. Physiol.* 2003; 10.1152/ajplung.00445.2002.
26. Ginty DD. Calcium regulation of gene expression: isn't that spatial? *Neuron.* 1997; 18: 183-186.
27. Golovina VA, and Blaustein MP. Spatially and functionally distinct Ca^{2+} stores in sarcoplasmic and endoplasmic reticulum. *Science.* 1997; 275: 1643-1648.
28. Golovina VA, Platoshyn O, Bailey CL, Wang J, Limsuwan A, Sweeney M, Rubin LJ, and Yuan JX-J. Upregulated TRP and enhanced capacitative Ca^{2+} entry in human pulmonary artery myocytes during proliferation. *Am. J. Physiol. Heart Circ. Physiol.* 2001; 280: H746-H755.

29. Gurney AM, Osipenko ON, MacMillan D, and Kempsill FEJ. Potassium channels underlying the resting potential of pulmonary artery smooth muscle cells. *Clin. Exp. Pharmacol. Physiol.* 2002; 29: 330-333.
30. Harder DR, Narayanan J, Birks EK, Liard JF, Imig JD, Lombard JH, Lange AR, and Roman RJ. Identification of a putative microvascular oxygen sensor. *Circ. Res.* 1996; 79: 54-61.
31. Hardingham GE, Chawla S, Johnson CM, and Bading H. Distinct functions of nuclear and cytoplasmic calcium in the control of gene expression. *Nature.* 1997; 385: 260-265.
32. He H, Lam M, McCormick TS, and Distelhorst CW. Maintenance of calcium homeostasis in the endoplasmic reticulum by Bcl-2. *J. Biol. Chem.* 1997; 138: 1219-1228.
33. Hu S, and Kim HS. Activation of K^+ channel in vascular smooth muscles by cytochrome P450 metabolites of arachidonic acid. *Eur. J. Pharmacol.* 1993; 230: 215-221.
34. Hughes FM.Jr, Bortner CD, Purdy GD, and Cidlowski JA. Intracellular K^+ suppresses the activation of apoptosis in lymphocytes. *J. Biol. Chem.* 1997; 272: 30567-30576.
35. Hulme JT, Coppock EA, Felipe A, Martens JR, and Tamkun MM. Oxygen sensitivity of cloned voltage-gated K^+ channels expressed in the pulmonary vasculature. *Circ. Res.* 1999; 85; 489-497.
36. Ishida Y, and Paul RJ. Effects of hypoxia on high-energy phosphagen content, energy metabolism and isometric force in guinea-pig taenia caeci. *J. Physiol.* 1990; 424: 41-56.
37. Krick S, Platoshyn O, McDaniel SS, Rubin LJ, and Yuan JX-J. Augmented K^+ currents and mitochondrial membrane depolarization in pulmonary artery myocyte apoptosis. *Am. J. Physiol. Lung Cell. Mol. Physiol.* 2001, 281: L887-L894.
38. Krick S, Platoshyn O, Sweeney M, Kim H, Yuan JX-J. Activation of K^+ channels induces apoptosis in vascular smooth muscle cells. *Am. J. Physiol. Cell Physiol.* 2001; 280: C970-C979.
39. Kuryshev YA, Gudz TI, Brown AM, Wible BA. KChAP as a chaperone for specific K^+ channels. *Am. J. Physiol. Cell Physiol.* 2000; 278: C931-C941.
40. Kuryshev YA, Wible BA, Gudz TI, Ramirez AN, Brown AM. KChAP/Kvβ1.2 interactions and their effects on cardiac Kv channel expression. *Am. J. Physiol. Cell Physiol.* 2001; 281: C290-299.
41. Lang F, Lepple-Wienhues A, Paulmichl M, Szabó I, Siemen D, and Gulbins E. Ion channels, cell volume, and apoptotic cell death. *Cell. Physiol. Biochem.* 1998; 8: 285-292.
42. Lee S, Park M, So I, and Earm YE. NADH and NAD modulates Ca^{2+}-activated K^+ channels in small pulmonary arterial smooth muscle cells of the rabbit. *Pflügers Arch.* 1994; 427: 378-380.
43. Lesage F, and Lazdunski M. Molecular and functional properties of two-pore-domain potassium channels. *Am. J. Physiol. Renal Physiol.* 2000; 279: F793-F801.
44. López-Barneo J. Oxygen-sensing by ion channel and the regulation of cellular functions. *Trends Neurosci.* 1996; 19: 435-440.
45. Madden JA, Ray DE, Keller PA, and Kleinman JG. Ion exchange activity in pulmonary artery smooth muscle cells: the response to hypoxia. *Am. J. Physiol. Lung Cell. Mol. Physiol.* 2001; 280: L264-L271.
46. Madden JA, Vadula KS, and Kurup VP. Effects of hypoxia and other vasoactive agents on pulmonary and cerebral artery smooth muscle cells. *Am. J. Physiol.* 1992; 263: L384-L393.
47. Maeno E, Ishizaki Y, Kanaseki T, Akihiro H, and Okada Y. Normotonic cell shrinkage because of disordered volume regulation is an early prerequisite to apoptosis. *Proc. Natl Acad. Sci. U.S.A.* 2000; 97: 9487-9492.
48. Martel J, Dupuis G, Deschênes P, and Payet MD. The sensitivity of the human Kv1.3 (hKv1.3) lymphocyte K^+ channel to regulation by PKA and PKC is partially lost in HEK 203 host cells. *J. Membr. Biol.* 1998; 161: 183-196.
49. Martens JR; Kwak Y-G, and Tamkun MM. Modulation of K_V channel α/β subunit interactions. *Trends Cardiovasc. Med.* 1999; 9: 253-258.
50. Martens JR, Sakamoto N, Sullivan SA, Grobaski TD, and Tamkun MM. Isoform-specific

localization of voltage-gated K^+ channels to distinct lipid raft populations. Targeting of Kv1.5 to caveolae. *J. Biol. Chem.* 2001; 276: 8409-8414.
51. McCormack T, McCormack K. Shaker K^+ channel β subunits belong to an NAD(P)H-dependent oxidoreductase superfamily. *Cell.* 1994; 79: 1133-1135.
52. McDaniel SS, Platoshyn O, Wang J, Yu Y, Sweeney M, Krick S, Rubin LJ, and Yuan JX-J. Capacitative Ca^{2+} entry in agonist-induced pulmonary vasoconstriction. *Am. J. Physiol. Lung Cell. Mol. Physiol.* 2001; 280: L870-L880.
53. McMurtry IF, Davidson AB, Reeves JT, Grover RF. Inhibition of hypoxic pulmonary vasoconstriction by calcium antagonists in isolated rat lungs. *Circ. Res.* 1976; 38: 99-104.
54. McMurtry IF, Petrun MD, and Reeves JT. Lungs from chronically hypoxic rats have decreased pressor response to acute hypoxia. *Am. J. Physiol.* 1978; 235: H104-H109.
55. Michiels C, Minet E. Michel G, Mottet D, Piret J-P, Raes M. HIF-1 and AP-1 cooperate to increase gene expression in hypoxia: Role of MAP kinases. *IUBMB Life.* 52: 49-53, 2001.
56. Miller MA, and Hales CA. Role of cytochrome P-450 in alveolar hypoxic pulmonary vasoconstriction in dogs. *J. Clin. Invest.* 1979; 64: 666-673.
57. Morgan KG. Calcium and vascular smooth muscle tone. *Am. J. Med.* 1987; 82: 9-15.
58. Mori Y, Folco E, and Koren G. GH3 cell-specific expression of Kv1.5 gene. Regulation by a silencer containing a dinucleotide repetitive element. *J. Biol. Chem.* 270: 27788-27796, 1995.
59. Murray TR, Chen L, Marshall BE, Macarak EJ. Hypoxic contraction of cultured pulmonary vascular smooth muscle cells. *Am. J. Respir. Cell. Mol. Biol.* 1990; 3: 457-465.
60. Nelson MT, Patlak JB, Worley JF, and Standen NB. Calcium channels, potassium channels, and voltage dependence of arterial smooth muscle tone. *Am. J. Physiol.* 1990; 259: C3-C18.
61. Nelson MT, and Quayle JM. Physiological roles and properties of potassium channels in arterial smooth muscle. *Am. J. Physiol.* 1995; 268: C799-C822.
62. Okamoto T, Schlegel A, Scherer PE, Lisanti MP. Caveolins, a family of scaffolding proteins for organizing "preassembled signaling complexes" at the plasma membrane. *J. Biol. Chem.* 1998; 273: 5419-5422.
63. Osipenko ON, Evans AM, Gurney AM. Regulation of the resting membrane potential of rabbit pulmonary artery myocytes by a low threshold, O_2-sensing potassium current. *Br. J. Pharmacol.* 1997; 120: 1461-1470.
64. Osipenko ON, Tate RJ, and Gurney AM. Potential role for Kv3.1b channels as oxygen sensors. *Circ. Res.* 2000; 86: 534-540.
65. Ostrom RS, Insel PA. Caveolar microdomains of the sarcolemma: compartmentation of signaling molecules comes of age [editorial]. *Circ. Res.* 1999; 84: 1110-1112.
66. Ottschytsch N, Raes A, Van Hoorick D, and Snyders DJ. Obligatory heterotetramerization of three previously uncharacterized Kv channel α subunits identified in the human genome. *Proc. Natl Acad. Sci. U.S.A.* 2002; 99: 7986-7991.
67. Parekh AB, and Penner R. Store depletion and calcium influx. *Physiol. Rev.* 1997; 77: 901-930.
68. Park MK, and Lee SH, Lee SJ, Ho W-K, and Earm YE. Different modulation of Ca-activated K channels by the intracellular redox potential in pulmonary and ear arterial smooth muscle cells of the rabbit. *Pflügers Arch.* 1995; 430: 308-314.
69. Patel AJ, and Honoré E. Molecular physiology of oxygen-sensitive potassium channels. *Eur. Respir. J.* 2001; 18: 221-227.
70. Patel AJ, Lazdunski M, and Honoré E. Kv2.1/Kv9.3, a novel ATP-dependent delayed-rectifier K^+ channel in oxygen-sensitive pulmonary artery myocytes. *EMBO J.* 1997; 16: 6615-6625.
71. Patel HH, Fryer RM, Gross ER, Bundey RA, Hsu AK, Isbell M, Eusebi LOV, Jensen RV, Gullans SR, Insel PA, Nithipatikom K, Gross GJ. 12-Lipoxygenase in opioid-induced delayed cardioprotection: gene array, mass spectrometric, and pharmacological analyses. *Circ. Res.* 2003; 92: 676-682.
72. Patel SP, Campbell DL, Strauss HC. Elucidating KChIP effects on Kv4.3 inactivation and

recovery kinetics with a minimal KChIP2 isoform. *J. Physiol.* 2002; 545: 5-11.

73. Pauly RR, Bilato C, Sollott SJ, Monticone R, Kelly PT, Lakatta EG, and Crow MT. Role of calcium/calmodulin-dependent protein kinase II in the regulation of vascular smooth muscle cell migration. *Circulation.* 1995; 91: 1107-1115.
74. Peng W, Hoidal JR, and Farrukh IS. Role of a novel K_{Ca} opener in regulating K^+ channels of hypoxic human pulmonary vascular cells. *Am. J. Respir. Cell. Mol. Biol.* 1999; 20: 737-745.
75. Peng W, Hoidal JR, Karwande SV, and Farrukh IS. Effect of chronic hypoxia on K^+ channels: regulation in human pulmonary vascular smooth muscle cells. *Am. J. Physiol.* 1997; 272: C1271-C1278.
76. Pérez-García MT, López-López JR, González C. Kvβ1.2 subunit coexpression in HEK293 cells confers O_2 sensitivity to Kv4.2 but not *Shaker* channels. *J Gen Physiol.* 1999; 113: 897-907.
77. Pike LJ, Han X, Chung K-N, Gross RW. Lipid rafts are enriched in arachidonic acid and plasmenylethanolamine and their composition is independent of caveolin-1 expression: a quantitative electrospray ionization/mass spectrometric analysis. *Biochemistry.* 2002; 41: 2075-2088.
78. Platoshyn O, Golovina VA, Bailey CL, Limsuwan A, Krick S, Juhaszova M, Seiden JE, Rubin LJ, and Yuan JX-J. Sustained membrane depolarization and pulmonary artery smooth muscle cell proliferation. *Am. J. Physiol. Cell Physiol.* 279: C1540-C1549, 2000.
79. Platoshyn O, Mandegar M, Yu Y, Golovina VA, Zhang S, Thistlethwaite PA, and Yuan JX-J. Functional ion channels in human pulmonary artery smooth muscle cells. *Biophys. J.* 82:249a, 2002.
80. Platoshyn O, Yu Y, Golovina VA, McDaniel SS, Krick S, Li L, Wang JY, Rubin LJ, and Yuan JX-J. Chronic hypoxia decreases K_V channel expression and function in pulmonary artery myocytes. *Am. J. Physiol. Lung Cell. Mol. Physiol.* 2001; 280: L801-L812.
81. Platoshyn O, Zhang S, McDaniel SS, and Yuan JX-J. Cytochrome *c* activates K^+ channels before inducing apoptosis. *Am. J. Physiol. Cell Physiol.* 2002; 283: C1298-C1305.
82. Post JM, Gelband CH, and Hume JR. $[Ca^{2+}]_i$ inhibition of K^+ channels in canine pulmonary artery. Novel mechanism for hypoxia-induced membrane depolarization. *Circ. Res.* 1995; 77: 131-139.
83. Post JM, Hume JR, Archer SL, Weir EK. Direct role for potassium channel inhibition in hypoxic pulmonary vasoconstriction. *Am J Physiol Cell Physiol.* 1992; 262: C882-C890.
84. Rabinovitch M. Pathobiology of pulmonary hypertension. Extracellular matrix. *Clin. Chest Med.* 2001; 22: 433-439.
85. Remillard CV, and Yuan JX-J. Activation of K^+ channels: an essential pathway in programmed cell death. *Am. J. Physiol. Lung Cell. Mol. Physiol.* 2003; in press..
86. Rettig J, Heinemann SH, Wunder F, Lorra C, Parcej DN, Dolly JO, Pongs O. Inactivation properties of voltage-gated K^+ channels altered by presence of β-subunit. *Nature.* 1994; 369: 289-294.
87. Revtyak GE, Buja LM, Chien KR, Campbell WB. Reduced arachidonate metabolism in ATP-depleted myocardial cells occurs early in cell injury. *Am. J. Physiol. Heart Circ. Physiol.* 1990; 259: H582-H591.
88. Robertson B. The real life of voltage-gated K^+ channels: more than model behaviour. *Trends Physiol. Sci.* 1997; 18:474-483.
89. Robertson TP, Hague D, Aaronson PI, Ward JPT. Voltage-independent calcium entry in hypoxic pulmonary vasoconstriction of intrapulmonary arteries of the rat. *J. Physiol.* 2000; 525: 669-680.
90. Rounds S, and McMurtry IF. Inhibitors of oxidative ATP production cause transient vasoconstriction and block subsequent pressor responses in rat lungs. *Circ. Res.* 1981; 48: 393-400.
91. Sano Y, Mochizuki S, Miyake A, Kitada C, Inamura K, Yokoi H, Nozawa K, Matsushime H, and Furuichi K. Molecular cloning and characterization of Kv6.3, a novel modulatory subunit

for voltage-gated K+ channel Kv2.1. *FEBS Lett.* 2002; 512: 230-234.
92. Semenza GL. Oxygen-regulated transcription factors and their role in pulmonary disease. *Respir. Res.* 1:159-162, 2000.
93. Shimoda LA, Sylvester JT, and Sham JSK. Inhibition of voltage-gated K^+ current in rat intrapulmonary arterial myocytes by endothelin-1. *Am. J. Physiol.* 1998; 274: L842-L853.
94. Short AD, Bian J, Ghosh TK, Waldron RT, Rybak SL, and Gill DL. Intracellular Ca^{2+} pool content is linked to control of cell growth. *Proc. Natl Acad. Sci. U.S.A.* 1993; 90: 4986-4990.
95. Smirnov SV, Robertson TP, Ward JPT, and Aaronson PI. Chronic hypoxia is associated with reduced delayed rectifier K^+ current in rat pulmonary artery muscle cells. *Am. J. Physiol.* 1994; 266: H365-H370.
96. Somlyo AP, and Somlyo AV. Signal transduction and regulation in smooth muscle. *Nature.* 1994; 372: 231-236.
97. Song KS, Scherer PE, Tang Z, Okamoto T, Li S, Chafel M, Chu C, Kohtz DS, Lisanti MP. Expression of caveolin-3 in skeletal, cardiac, and smooth muscle cells. Caveolin-3 is a component of the sarcolemma and co-fractionates with dystrophin and dystrophin-associated glycoproteins. *J. Biol. Chem.* 1996; 271: 15160-15165.
98. Stanbrook HS, and McMurtry IF. Inhibition of glycolysis potentiates hypoxic vasoconstriction in lungs. *J. Appl. Physiol.* 1983; 55: 1467-1473.
99. Stenmark KR, and Mecham RP. Cellular and molecular mechanisms of pulmonary vascular remodeling. *Annu. Rev. Physiol.* 1997: 59: 89-144.
100. Sweeney M, and Yuan JX-J. Hypoxic pulmonary vasoconstriction: role of voltage-gated potassium channels. *Respir. Res.* 2000; 1: 40-48.
101. Sylvester JT, and McGowan C. The effects of agents that bind to cytochrome P-450 on hypoxic pulmonary vasoconstriction. *Circ. Res.* 1978; 43: 429-437.
102. Tamayo L, López-López JR, Castañeda J, and González C. Carbon monoxide inhibits hypoxic pulmonary vasoconstriction in rats by a cGMP-independent mechanism. *Pflügers Arch.* 1997; 434: 698-704.
103. Thornberry NA, and Lazebnik Y. Caspases: enemies within. *Science.* 1998; 281: 1312-1316.
104. Valverde P, and Koren G. Purification and preliminary characterization of a cardiac Kv1.5 repressor element binding factor. *Circ. Res.* 1999; 84: 937-944.
105. Wang J, Juhaszova M, Rubin LJ, Yuan JX-J. Hypoxia inhibits gene expression of voltage-gated K^+ channel α subunits in pulmonary artery smooth muscle cells. *J. Clin. Invest.* 1997; 100: 2347-2353.
106. Waypa GB, Chandel NS, and Schumacker PT. Model for hypoxic pulmonary vasoconstriction involving mitochondrial oxygen sensing. *Circ. Res.* 2001; 88: 1259-1266.
107. Waypa GB, Marks JD, Mack MM, Boriboun C, Mungai PT, and Schumacker PT. Mitochondrial reactive oxygen species trigger calcium increases during hypoxia in pulmonary arterial myocytes. *Circ. Res.* 2002; 91: 719-726.
108. Weir EK, Wyatt CN, Reeve HL, Huang J, Archer SL, and Peers C. Diphenyleneiodonium inhibits both potassium and calcium currents in isolated pulmonary artery smooth muscle cells. *J. Appl. Physiol.* 1994; 76: 2611-2615.
109. White KA, and Marletta MA. Nitric oxide synthase is a cytochrome P-450 type hemoprotein. *Biochemistry.* 1992; 31: 6627-6631.
110. Wyatt CN, Wright C, Bee D, and Peers C. O_2-sensitive K^+ currents in carotid body chemoreceptor cells from normoxic and chronically hypoxic rats and their roles in hypoxic chemotransduction. *Proc. Natl Acad. Sci. U.S.A.* 1995; 92: 295-299.
111. Yu AY, Shimoda LA, Iyer NV, Huso DL, Sun X, McWilliams R, Beaty T, Sham JSK, Wiener CM, Sylvester JT, and Semenza GL. Impaired physiological responses to chronic hypoxia in mice partially deficient for hypoxia-inducible factor 1?. *J. Clin. Invest.* 1999; 103: 691-696.
112. Yu SP, and Choi DW. Ions, cell volume, and apoptosis. *Proc. Natl Acad. Sci. U.S.A.* 2000; 97: 9360-9362.
113. Yu Y, Platoshyn O, Zhang J, Krick S, Zhao Y, Rubin LJ, Rothman A, and Yuan JX-J. c-Jun decreases voltage-gated K^+ channel activity in pulmonary artery smooth muscle cells.

Circulation. 2001; 104: 1557-1563.

114. Yu Y, Sweeney M, Zhang S, Platoshyn O, Landsberg J, Rothman A, and Yuan JX-J. PDGF stimulates pulmonary vascular smooth muscle cell proliferation by upregulating TRPC6 expression. *Am. J. Physiol. Cell Physiol.* 2003; 284: C316-C330.

115. Yuan JX-J. Voltage-gated K^+ currents regulate resting membrane potential and $[Ca^{2+}]_i$ in pulmonary arterial myocytes. *Circ. Res.* 1995; 77: 370-378.

116. Yuan X-J. Mechanisms of hypoxic pulmonary vasoconstriction: The role of oxygen-sensing voltage-gated potassium channels. In *Oxygen Regulation of Ion Channels and Gene Expression*, López-Barneo J and Weir EK, eds. Armonk, NY: Futura Publishing Company, Inc., 1998, pp. 207-233.

117. Yuan X-J, Goldman WF, Tod ML, Rubin LJ, Blaustein MP. Hypoxia reduces potassium currents in cultured rat pulmonary but not mesenteric arterial myocytes. *Am. J. Physiol.* 1993; 264: L116-L123.

118. Yuan X-J, Salvaterra CG, Tod ML, Juhaszova M, Goldman WF, Rubin LJ, and Blaustein MP. The sodium gradient, potassium channels, and regulation of calcium in pulmonary and mesenteric arterial smooth muscles: effect of hypoxia. In *Ion Flux in Pulmonary Vascular Control*, Weir EK, Hume JR, and Reeves JT, eds. New York, NY: Plenum Publishers, 1993, pp. 205-222.

119. Yuan X-J, Sugiyama T, Goldman WF, Rubin LJ, and Blaustein MP. A mitochondrial uncoupler increases K_{Ca} currents but decreases K_V currents in pulmonary artery myocytes. *Am. J. Physiol.* 1996; 270: C321-C331.

120. Yuan X-J, Tod ML, Rubin LJ, Blaustein MP. Contrasting effects of hypoxia on tension in rat pulmonary and mesenteric arteries. *Am. J. Physiol. Heart Circ. Physiol.* 1990; 259: H281-H289.

121. Yuan X-J, Tod ML, Rubin LJ, and Blaustein MP. Deoxyglucose and reduced glutathione mimic effects of hypoxia on K^+ and Ca^{2+} conductances in pulmonary artery cells. *Am. J. Physiol.* 1994; 267: L52-L63.

122. Yuan X-J, Tod ML, Rubin LJ, and Blaustein MP. Hypoxic and metabolic regulation of voltage-gated K^+ channels in rat pulmonary artery smooth muscle cells. *Exp. Physiol.* 1995; 80: 803-813.

123. Yuan X-J, Tod ML, Rubin LJ, and Blaustein MP. Inhibition of cytochrome P-450 reduces voltage-gated K^+ currents in pulmonary arterial myocytes. *Am. J. Physiol.* 1995; 268: C259-C270.

124. Yuan X-J, Tod ML, Rubin LJ, and Blaustein MP. NO hyperpolarizes pulmonary artery smooth muscle cells and decreases the intracellular Ca^{2+} concentration by activating voltage-gated K^+ channels. *Proc. Natl Acad. Sci. U.S.A.* 1996; 93: 10489-10494.

125. Yuan X-J, Wang J, Juhaszova M, Golovina VA, and Rubin LJ. Molecular basis and function of voltage-gated K^+ channels in pulmonary artery smooth muscle cells. *Am. J. Physiol.* 1998; 274: L621-L635.

126. Zhang F, Carson RC, Zhang H, Gibson G, and Thomas HM 3rd. Pulmonary artery smooth muscle cell $[Ca^{2+}]_i$ and contraction: responses to diphenyleneiodonium and hypoxia. *Am. J. Physiol.* 1997; 273: L603-L611.

127. Zhang X, Wrzeszczynska MH, Horvath CM, and Darnell JE Jr. Interacting regions in Stat3 and c-Jun that participiate in cooperative transcriptional activation. *Mol. Cell. Biol.* 1999; 19: 7138-7146.

128. Zhao Y-Y, Liu Y, Stan R-V, Fan L, Gu Y, Dalton N, Chu P-H, Peterson K, Ross J, Jr., Chien KR. Defects in caveolin-1 cause dilated cardiomyopathy and pulmonary hypertension in knockout mice. *Proc. Natl. Acad. Sci. U.S.A.* 2002; 99: 11375-11380.

129. Zhu X-R, Netzer R, Bohlke K, Liu Q, and Pongs O. Structural and functional characterization of Kv6.2, a new γ-subunit of voltage-gated potassium channel. *Receptor Channels.* 1999; 6: 337-350.

Chapter 11

Transient Receptor Potential Channels and Capacitative Ca^{2+} Entry in Hypoxic Pulmonary Vasoconstriction

Alison M. Gurney and Lih-Chyuan Ng
University of Strathclyde, Glasgow, Scotland and University of Nevada, Reno, Nevada, U.S.A.

1. Introduction

Hypoxic pulmonary vasoconstriction (HPV) consists of a reversible, monophasic increase in pulmonary artery pressure that reaches a maximum within around 20 min. The response is retained in isolated pulmonary arteries, where hypoxia often induces an early transient contraction that peaks in around 10 min and a slower sustained contraction that develops over 30 min (46). Different mechanisms appear to underlie each phase as they can be separated on the basis of differential dependence on the endothelium, required only for the slower contraction, and sensitivity to pharmacological agents. Nevertheless, both phases are associated with a rise in the intracellular Ca^{2+} concentration ($[Ca^{2+}]_i$) of pulmonary artery smooth muscle cells (PASMCs) independent of the endothelium (31). This is brought about by the recruitment of cellular pathways that either release Ca^{2+} from intracellular stores in the sarcoplasmic reticulum (SR) or transport it into the cell from the extracellular medium. There is a great deal of controversy surrounding the underlying sources of Ca^{2+} and how the pathways are recruited. This chapter outlines the relationship between different Ca^{2+} generating pathways in the cell and discusses the evidence for their involvement in HPV. The main focus is on capacitative Ca^{2+} entry (CCE), which has recently attracted much interest due to the identification of transient receptor potential (TRP) proteins in pulmonary artery. These proteins represent a novel class of cation channels, some of which are implicated as mediators of CCE.

2. Capacitative Ca^{2+} Entry

At least three regulated pathways for Ca^{2+} entry contribute to the control of $[Ca^{2+}]_i$ in mammalian cells. These are voltage-gated, receptor-operated and capacitative Ca^{2+} entry. Voltage-gated Ca^{2+} entry is a major Ca^{2+} influx pathway in the smooth muscle cells of most blood vessels. The ion channel that mediates it is selectively permeable to Ca^{2+} and activated by membrane depolarization above a threshold of around -30 mV. The channel is selectively blocked by calcium antagonist drugs like nifedipine and diltiazem; their well established antihypertensive effects illustrate the important role of voltage-gated Ca^{2+} channels in regulating $[Ca^{2+}]_i$ and tone in vascular smooth muscle.

In most cells, Ca^{2+} entry is coupled to membrane receptors. Receptor activation often leads to membrane depolarization, which in turn stimulates voltage-dependent Ca^{2+} entry. There are, however, many examples of receptor activation stimulating Ca^{2+} influx into vascular smooth muscle cells without depolarizing the cell membrane (3). It can be brought about in several ways. For instance, receptors may be directly coupled to ion channels that gate the entry of Ca^{2+}. An example is the P2X receptor, where the binding of ATP causes a conformational change in the protein to open an ion channel that is intrinsic to the protein and permeable to Ca^{2+} and Na^{+}. G-protein coupled receptors, such as α-adrenergic or muscarinic receptors, can also activate voltage-independent Ca^{2+} entry (3), although in this case the receptor and ion channel are separate entities. These pathways are all examples of receptor-operated Ca^{2+} entry and can be identified as the component of agonist-induced Ca^{2+} influx that is insensitive to block by calcium antagonists. It is commonly seen as the component of agonist-induced contraction that requires the presence of Ca^{2+} in the extracellular medium, but is resistant to block by calcium antagonist drugs.

Many G-protein coupled receptors stimulate the release of Ca^{2+} from intracellular stores in the SR of vascular smooth muscle cells. The underlying mechanism involves stimulation of phospholipase C (PLC), which hydrolyzes phosphatidylinositol 4,5-bisphosphate (PIP_2) to release the Ca^{2+} signalling molecule, inositol 1,4,5-trisphosphate (IP_3). This triggers Ca^{2+} release via specific, IP_3-gated channels in the SR membrane, which are selectively permeable to Ca^{2+}. There is now overwhelming evidence that store-depletion, resulting from the release of Ca^{2+} from the SR, activates an additional Ca^{2+} influx pathway, known as CCE. CCE can be distinguished from voltage and receptor-operated Ca^{2+} entry, because it activates independently of receptor stimulation and at membrane voltages that inhibit the opening of voltage-gated Ca^{2+} channels. In fact, since the electrochemical driving force for Ca^{2+} entry is greater at negative membrane potentials, depolarization actually inhibits CCE. Although the mechanisms underlying the activation of CCE are still not understood, it has been established that it is store depletion itself that is responsible, and any procedure that causes store depletion will lead to its activation. Thus, in order to distinguish CCE from other Ca^{2+} entry pathways, experimental protocols often rely on procedures that deplete Ca^{2+} stores directly, by acting downstream of receptor binding. A common approach is to expose cells to thapsigargin or cyclopiazonic acid (CPA), which are selective inhibitors of the Ca^{2+} ATPase

(SERCA) in the SR membrane and prevent Ca^{2+} uptake by these organelles. In this way, Ca^{2+} leaking out of the store is not replenished. Another approach, for example when patch clamp recording, is to load cells with a high concentration of a Ca^{2+} buffer, such as EGTA or BAPTA. This also prevents uptake via SERCA by minimizing the cytosolic [Ca^{2+}] available to the pump.

3. Capacitative Ca^{2+} Entry in Pulmonary Artery Smooth Muscle Cells

By inhibiting SR Ca^{2+} uptake, SERCA inhibitors cause contraction of pulmonary artery smooth muscle (15, 27, 32). Some of the effects of CPA on rat pulmonary artery smooth muscle are illustrated in Figure 1.

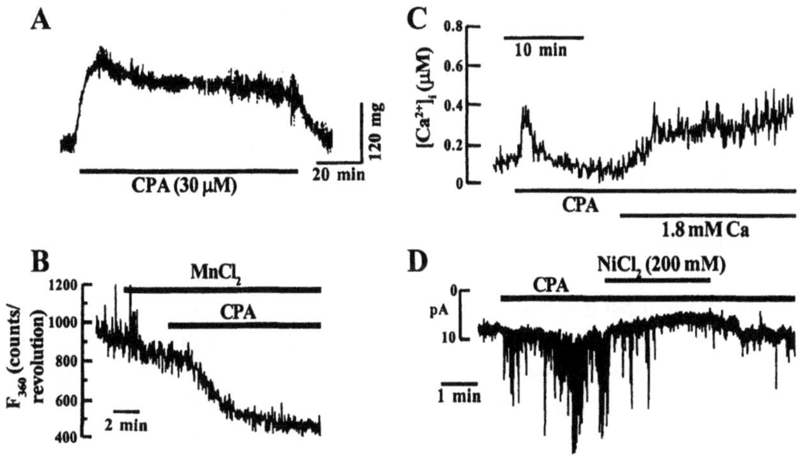

Figure 1. Store-depletion by SERCA inhibitors activates contraction, divalent cation influx and a Ni^{2+}-sensitive cation current. A: Contraction of intact pulmonary artery induced by CPA. B: In isolated myocytes from the same artery CPA induces a transient rise in $[Ca^{2+}]_i$ when applied in Ca^{2+}-free solution, due to Ca^{2+} release, but a sustained rise follows when Ca^{2+} is readmitted. C: The ability of CPA to accelerate quenching of fura-2 fluorescence by Mn^{2+} confirms that the sustained rise in $[Ca^{2+}]_i$ is due to CCE. D: Currents flowing through store-operated channels activated by CPA in physiological solution are small in amplitude, noisy, and abolished by low concentrations of Ni^{2+} (Modified from Ref. 27).

Contraction results from a rise in $[Ca^{2+}]_i$ brought about through several mechanisms. As in other cell types, the SERCA inhibitors prevent the SR from buffering Ca^{2+} as it enters the cells, resulting in the accumulation of Ca^{2+} in the cytosol. This can occur even in the absence of stimulated Ca^{2+} entry and is enhanced by the unopposed leak of Ca^{2+} from the SR, which eventually depletes the Ca^{2+} store. Invariably, however, SERCA inhibitors have been found to produce a sustained contraction of pulmonary arterial smooth muscle that is dependent upon the presence of extracellular Ca^{2+}, implying a need for Ca^{2+} entry (15, 27, 32). At least some of the Ca^{2+} may enter through voltage-operated Ca^{2+} channels, because nifedipine reduces the contractile response to SERCA inhibitors (27). These channels may be activated following the initial rise in $[Ca^{2+}]_i$ caused by SERCA inhibition, which would bring about the activation of Ca^{2+}-dependent Cl^- channels and consequently cause depolarization of the cell membrane. Other studies found no effect of L-type Ca^{2+} channel blockers on the contractile response (15), and the drugs were ineffective at blocking the rise in $[Ca^{2+}]_i$ produced by SERCA inhibitors in pulmonary artery myocytes (7, 27, 48). Thus, at least part of the contraction is mediated by Ca^{2+} entry through a pathway that is independent of voltage-gated Ca^{2+} channels. We found in rat pulmonary artery that nifedipine inhibited the CPA-induced contraction if applied within the first 30 min, but its effect then disappeared (15). This suggests that the initial effect of SERCA inhibitors is to release SR Ca^{2+}, activate Cl^- channels and stimulate voltage-dependent Ca^{2+} entry. But, as the stores become depleted, CCE emerges as the predominant pathway by which Ca^{2+} entry sustains contraction. This is further supported by the finding that both the sustained contraction and the Ca^{2+} influx pathway activated by CPA in rat pulmonary artery were 50% blocked by the cation channel blockers SKF965, Ni^{2+}, and Cd^{2+}, all at concentrations causing similar inhibition of a store-operated current in isolated myocytes (27).

Although the most direct support for CCE involvement in the contraction of pulmonary arteries derives from studies with SERCA inhibitors, evidence is emerging that CCE also contributes to the contraction brought about by receptor activation. Thus following store depletion in Ca^{2+}-free medium by continuous exposure to the α-adrenoceptor agonist, phenylephrine, the restoration of Ca^{2+} in the presence of the α-adrenoceptor antagonist, phentolamine, produced a contraction that was presumed to reflect CCE (24). The interpretation of this result is, however, critically dependent on the receptor-mediated events being abolished by phentolamine, because CCE and receptor-operated Ca^{2+} entry are not easily distinguished on pharmacological grounds. Contraction resistant to calcium antagonists has also been shown to result when extracellular Ca^{2+} is returned to vessels exposed briefly to agonist in Ca^{2+}-free medium to deplete the stores (15). Providing sufficient time was allowed for the receptor-activated events to cease, the contraction probably resulted from Ca^{2+} entry that was stimulated as a consequence of store depletion. Although CCE has been linked to pulmonary artery constriction, contractile responses to SERCA inhibitors are only a fraction of the amplitude of those generated by receptor agonists (27). Moreover, in the presence of physiological levels of Ca^{2+}, receptor-operated and voltage-gated Ca^{2+} influx may be sufficient to prevent store depletion and CCE activation during agonist-induced contraction (16, 21). Thus CCE is probably a relatively minor source of contractile Ca^{2+} under physiological conditions *in vivo*.

In support of this, we recently found that although low concentrations of La^{3+} had little effect on CPA-induced contraction, La^{3+} was a potent inhibitor of agonist-induced contraction (unpublished results). In cerebral arteries, CCE was found to mediate a substantial rise in smooth muscle $[Ca^{2+}]_i$ without evoking contraction, suggesting that $[Ca^{2+}]_i$ rose within a localized cell compartment that was inaccessible to the contractile proteins (9). It is likely therefore that CCE serves other functions within the cell. Replenishment of Ca^{2+} stores is likely to be its major role, but CCE has also been linked to the proliferation of PASMCs in culture (13, 37). Thus CCE could be important in the regulation of cell growth and the vascular remodeling that occurs in pulmonary hypertensive disease.

A direct pathway linking the extracellular medium to refilling of the Ca^{2+} store was initially suggested for vascular muscle in 1981 (4). Direct evidence for a store depletion-activated Ca^{2+} influx pathway in pulmonary artery smooth muscle appeared only in the last few years (7, 13, 27, 32, 48). Using fluorescent $[Ca^{2+}]_i$ indicators, these studies identified an increase in $[Ca^{2+}]_i$ brought about by store depletion, which required the presence of extracellular Ca^{2+}, was insensitive to calcium antagonists, inhibited by blockers of cation channels and inhibited by membrane depolarization. However, while thapsigargin and CPA activated this pathway in myocytes from rat and human pulmonary artery (7, 13, 27), they were an insufficient stimulus in the dog (48). The reason appears to be that dog pulmonary artery myocytes have separate IP_3-sensitive and ryanodine/caffeine-sensitive Ca^{2+} stores, both of which must be depleted in order to activate CCE.

4. Ca^{2+} Stores and Capacitative Ca^{2+} Entry in Hypoxic Pulmonary Vasoconstriction

Several strands of evidence support a key role for SR Ca^{2+} release in mediating the contractile response of pulmonary artery smooth muscle to hypoxia. Ca^{2+} store depletion with caffeine and/or ryanodine has been reported to blunt the hypoxia-induced rise in $[Ca^{2+}]_i$ in freshly isolated or cultured smooth muscle cells (33, 40) and to reduce the contraction of intact arteries to varying degrees (21, 22, 32). Other studies found that this treatment abolished hypoxic vasoconstriction (6, 11, 41). This led to the hypothesis that Ca^{2+} release via ryanodine receptor-channels is an early event in the cascade activated by hypoxia. The mechanism by which the cell senses a fall in O_2 tension and triggers the release of Ca^{2+} is still the subject of much debate. An attractive hypothesis invokes the diffusible messenger cyclic adenosine diphosphate-ribose (cADPR), which mobilizes Ca^{2+} from ryanodine-sensitive stores (8). It is based on the premise that a change in cellular redox state brought about by hypoxia increases the level of β-NADH, which inhibits cADPR hydrolase to promote cADPR formation from β-NAD^+. Inhibition of HPV by the cADPR antagonist, 8-bromo-cADPR, lends strong support for such a mechanism (8).

There are mixed reports on the effects of SERCA inhibitors on HPV. Thapsigargin and CPA inhibited the hypoxic contraction in rabbit pulmonary artery (41) and abolished it in the rat (11). In stark contrast, CPA and thapsigargin potentiated the contractile response to hypoxia in dog pulmonary artery (21). These apparently conflicting results may be reconciled by considering species differences in the functional properties of the SR Ca^{2+} stores. In canine myocytes, the caffeine- and ryanodine-sensitive store is structurally and functionally distinct from the IP_3-sensitive store and SERCA inhibitors deplete only the latter (21, 48). Thus if hypoxia only affects the ryanodine-sensitive store, SERCA inhibitors would not be expected to affect the response. The potentiation of HPV seen in dog vessels may be explained if some of the Ca^{2+} released by hypoxia is normally accumulated via SERCA into the IP_3-sensitive store. In contrast, the caffeine/ryanodine- and IP_3-sensitive Ca^{2+} stores display substantial functional overlap in rodent myocytes (27). So SERCA inhibitors (as well as caffeine and ryanodine) could deplete Ca^{2+} from both stores in these cells and consequently inhibit the Ca^{2+}-releasing effect of hypoxia.

Although Ca^{2+} release is important for HPV, it is not yet clear if the SR is the major source of contractile Ca^{2+}. This is because HPV and the associated rise in $[Ca^{2+}]_i$ are greatly diminished when Ca^{2+} is removed from the extracellular medium (6, 11, 22, 33, 41), implying that Ca^{2+} influx also contributes. It is still possible that hypoxia initially stimulates Ca^{2+} influx, which binds to the ryanodine receptor and triggers Ca^{2+}-induced Ca^{2+} release. Alternatively, hypoxia-induced Ca^{2+}release could trigger an increase in $[Ca^{2+}]_i$ by stimulating Ca^{2+} entry pathways. In support of this, nisoldipine reduced the contractile response to caffeine in canine pulmonary artery (21). Organic calcium antagonists have been widely found to blunt HPV and the associated rise in $[Ca^{2+}]_i$, both in isolated pulmonary arteries (22, 32, 33) and intact perfused lungs (25, 36). Interestingly, the potentiated contractile response to hypoxia observed in canine pulmonary artery in the presence of CPA was much less sensitive to nisoldipine than the control response to hypoxia in the absence of CPA (21). Since removing extracellular Ca^{2+} blocked this effect, hypoxic vasoconstriction was likely due to the recruitment of voltage-independent Ca^{2+} entry. This pathway resembled CCE in being blocked by SKF96365. Another recent study showed that voltage-independent Ca^{2+} entry contributes to HPV in rat pulmonary artery (32). The underlying pathway was permeable to Mn^{2+} and blocked by SKF96365 and low concentrations of La^{3+}, all of which are characteristic markers of non-selective cation channels. Although it has only recently attracted serious interest, evidence for a voltage-independent Ca^{2+} entry pathway in HPV has in fact been available for some time. In 1993, a Ca^{2+} entry pathway insensitive to nifedipine and verapamil was shown to underlie a sustained rise in $[Ca^{2+}]_i$ induced by hypoxia in cultured pulmonary arterial myocytes (33).

Although the voltage-independent Ca^{2+} entry pathway has been proposed to reflect CCE, this has yet to be confirmed. Its properties are equally compatible with cation channels mediating receptor-operated Ca^{2+} entry. Thus we found in rat pulmonary artery that La^{3+}, which was shown to inhibit HPV, is a poor inhibitor of CCE but a potent blocker of agonist-induced Ca^{2+} entry. Since agonist-induced pre-tone is required in most experiments to achieve a clear response to hypoxia, the use of currently available CCE inhibitors to investigate their role is problematic. The fact that pre-tone is needed implies that there is synergy between the pathways activated by agonist and hypoxia. Thus, by affecting this synergy, a drug interfering with receptor-mediated events could give the impression of inhibiting HPV without actually altering the events mediating HPV. More selective approaches for interfering with CCE will be needed to enable its contribution to HPV to be clearly elucidated.

5. Store Depletion-Activated Channels in Pulmonary Artery Smooth Muscle

Little is known yet about the ion channels mediating CCE in pulmonary artery smooth muscle. Store depletion-activated cation currents recorded from rat pulmonary artery myocytes (Fig. 1D) have amplitudes of only a few pA in the presence of physiological levels of Ca^{2+} (27). Store operated currents of similarly small amplitude have been recorded from rabbit and mouse arterial (5, 39) and rabbit venous (1, 20) smooth muscle cells. In contrast, large currents of several hundred pA were reported in cultured human pulmonary artery myocytes (14). This may reflect a phenotypic change from contractile to proliferative cells during culture, as CCE is enhanced during cell proliferation (14). There is general agreement from all these studies that cation-selective channels mediate the currents, although the degree of selectivity for Ca^{2+} over other cations may vary among different blood vessels. Thus the channel in rabbit portal vein shows a high degree of Ca^{2+} selectivity (1) whereas those in aorta and pulmonary artery discriminate poorly between different cations (27, 39). The store-operated current recorded from pulmonary artery myocytes reverses direction close to 0 mV in physiological conditions and is permeable to Ca^{2+}, Na^{+}, K^{+}, Cs^{+} and Mn^{2+} (27). Single channel currents have not yet been reported for store-operated channels (SOCs) in PASMCs, but channels reported for rabbit and mouse aorta and rabbit portal vein have conductances of around 3 pS and a relatively low open probability (1, 39). Once activated by store depletion the channels remain active when membrane patches are excised from the cell, but the channels cannot be activated be SERCA inhibitors applied after patch excision (2, 39). This is the behavior expected for a SOC. Store depletion may not be the only mechanism for activating these channels though, because noradrenaline could activate the same channels in excised outside-out patches via protein kinase C (2). This further complicates the distinction between SOC and ROC.

The pharmacology of store-depletion activated channels in vascular smooth muscle is poorly characterized. In rat pulmonary artery the CPA-induced current is blocked by low micromolar concentrations of Ni^{2+}, Cd^{2+} and SKF96365, but it is resistant to La^{3+}, which is effective only at 100 μM or higher (27). This pharmacological profile matches the pharmacology of CCE and the CPA-induced contraction reported for rat pulmonary artery in the same study, but is inconsistent with the high La^{3+} sensitivity reported by others for CCE in the same vessels (32). The reason for this discrepancy is unclear, but the latter study did not measure CCE directly. In general the pharmacology of SOCs is poorly defined. Although a number of drugs are widely used as blockers of SOCs, none of them are selective, especially when used at high concentrations. Thus it is unclear if the SOCs and CCE blocked by millimolar concentrations of Ni^{2+} and La^{3+} in some studies (14, 39) are the same as channels showing high sensitivity to these blockers.

6. Transient Receptor Potential Channels and Their Relationship to Capacitative Ca^{2+} Entry

The transient receptor potential (TRP) channel was first identified in *Drosophila* photoreceptors, where a mutation in the *trp* gene gives rise to a transient rather than sustained membrane response of the photoreceptor to light. This effect was associated with the loss of Ca^{2+} permeability, leading to the proposal that the *trp* gene encodes a Ca^{2+}-permeable ion channel (17). This was supported by strong sequence homology between the genes encoding the TRP protein and voltage-gated Ca^{2+} channels. Electrophysiological studies subsequently confirmed that both *trp* and the homologous *Drosophila* gene, *trpl*, form Ca^{2+}-permeable non-selective channels in heterologous expression systems.

In the mid 1990s, mammalian homologues of the *Drosophila trp* gene began to emerge and at least 19 genes are now known to encode human TRP proteins. All of these proteins consist of 6 putative membrane-spanning domains, domains 5 and 6 being linked by a short hydrophobic segment predicted to be the pore-forming region, with both the N and C termini located intracellularly (Fig. 2). Based on sequence homology/divergence, we now recognize three major subgroups of mammalian TRP proteins: TRPC, TRPV and TRPM. The TRPV nomenclature originates from the vanilloid receptor, which was the first identified member of the family. Of these channels, only the epithelial TRPV6 protein has been suggested to function as a SOC, although this is disputed. The TRPM nomenclature similarly originates from the first identified member, melastatin; none of these proteins have been implicated as a SOC. Several members of the TRPC family of channels have been suggested at some time to play a role in CCE. Thus, TRPC channels have recently received a great deal of interest in relation to CCE in vascular smooth muscle cells.

7. Properties of Transient Receptor Potential (TRPC) Channels

Seven members (TRPC1-C7) of the TRPC family have been identified. All are able to form cation channels in heterologous expression systems, except the human *trpc2*, which is a pseudo-gene. Structural homologies within the TRPC family and their relationship to the other mammalian TRP channels are illustrated in Figure 2A. Functional channels are thought to require the co-assembly of four TRPC subunits into a tetrameric complex (Fig. 2C). Table 1 lists some of the characteristic biophysical and pharmacological properties of the homomeric channels, measured from *in vitro* expression systems.

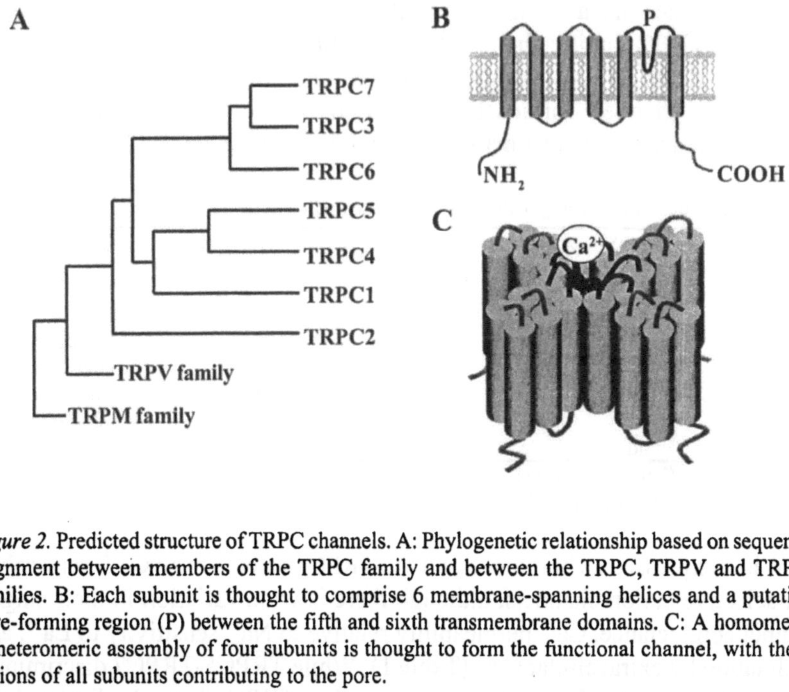

Figure 2. Predicted structure of TRPC channels. A: Phylogenetic relationship based on sequence alignment between members of the TRPC family and between the TRPC, TRPV and TRPM families. B: Each subunit is thought to comprise 6 membrane-spanning helices and a putative pore-forming region (P) between the fifth and sixth transmembrane domains. C: A homomeric or heteromeric assembly of four subunits is thought to form the functional channel, with the P regions of all subunits contributing to the pore.

TRPC proteins can also form heteromeric assemblies consisting of more than one subunit type. Clear rules governing the possible interactions within the TRPC family have emerged. TRPC2 does not interact with other TRPC proteins, while TRPC1, TRPC4 and TRPC5 can interact with each other, but not with other members of the family, and TRPC3, TRPC6 and TRPC7 can interact with each other, but not with other members of the family (12, 18). A similar distinction can be drawn in relation to the interactions of TRPC proteins with the *Drosophila* scaffolding protein INAD (identified from the Inactivation-No-After-Potential *Drosophila* mutant), which contains protein interaction sites known as PDZ domains and forms the backbone of a macromolecular signaling complex with TRP proteins. INAD can associate with TRPC1, TRPC4 and TRPC5, but not with TRPC3, TRPC6 or TRPC7 (12). The cloning of a human INAD-like protein (29) suggests that comparable complexes may be involved in the regulation of mammalian TRPC channels. In support of this, a PDZ domain in the Na^+/H^+ exchange regulatory factor (NHERF) binds to TRPC4 and TRPC5 as well as to PLC, suggesting that it could act as a scaffolding protein to bring these signaling molecules together (38). It would be interesting to know if NHERF can interact with other TRPC proteins, or if there are distinct scaffolding proteins for different TRPC complexes. Another distinction can be drawn in relation to diacylglycerol (DAG), which interacts directly with TRPC3 TRPC6 and TRPC7 to cause channel activation, but not with TRPC1, TRPC4 or TRPC5 (19, 28). These properties are all consistent with two functionally distinct subgroups within the TRPC family: the TRPC1/4/5 subfamily, which are most closely related in terms of evolutionary distance, and the TRPC3/6/7 subfamily. Heteromeric channels formed by interactions within these groups can have properties that are quite distinct from the homomeric channels (23, 35).

Table 1. Properties of Heterologously Expressed TRPC Channels

	γ (pS)	P_{Ca}/P_{Na}	IC_{50} for La^{3+} (μM)	Ca^{2+} Modulation
TRPC1	16[1]	1	5	inhibition
TRPC 2	n.d.	n.d.	n.d.	n.d.
TRPC 3	66	1.5	50	stimulation
TRPC4	41	1,7	>100	inhibition
TRPC 5	48-63	~10	>100	stimulation
TRPC 6	30	5	4	inhibition
TRPC 7	nd	5	~100	stimulation

[1]measured in Ca^{2+}-free conditions. γ, single channel conductance; P_{Ca}/P_{Na}, Ca^{2+} permeability relative to Na^+ (Data from Refs. 18, 20, 23, 26, 28, 30, 34, 43).

When expressed as homomers, all the TRPC proteins form Ca^{2+}-permeable, non-selective cation channels, although there is wide variation in their single-channel conductance, Ca^{2+} permeability relative to Na^+, sensitivity to La^{3+}, and modulation by extracellular Ca^{2+} (Table 1). While TRPC5-TRPC7 discriminate strongly between Ca^{2+} and Na^+ ions, the other TRPC channels display little selectivity for Ca^{2+} over Na^+ or other cations. As commonly found in other Ca^{2+}-permeable channels, Ca^{2+} passing through the pore can modulate TRPC channel activity. Inhibition of TRPC1 and TRPC4 is apparent at physiological (millimolar) levels of extracellular Ca^{2+}, but Ca^{2+} has been found to stimulate the activity of TRPC3, TRPC5 and TRPC7 channels. Lanthanides are often used as inhibitors of SOCs but, although La^{3+} blocks TRPC1 and TRPC3 at low micromolar concentrations, it is at least an order of magnitude less potent on TRPC6 and TRPC7 channels, and rather ineffective on TRPC4 and TRPC5 channels. These properties could all be helpful for the identification of particular TRPC channels underlying Ca^{2+} entry in vascular cells.

The mechanisms underlying the activation of TRPC channels are still controversial. TRPC1, TRPC3, TRPC4, TRPC5 and TRPC6 have all been implicated as SOCs at some time. However, much of the evidence is circumstantial and alternative mechanisms are likely. The evidence for and against a store-dependent mechanism of activation was reviewed recently in several articles (26, 43), so it is not covered here. It is however important to realize how controversial this area is when considering the potential role of TRPCs, acting as SOCs, in HPV. One of the problems may be that TRPC activation has so far been studied primarily in *in vitro* expression systems, which have generated conflicting results. This may reflect varying levels of protein expression, because expression levels were recently found to profoundly affect the mechanism by which TRPC3 channels are activated (42). Thus at relatively low levels of expression, TRPC3 was activated by store depletion, but at higher levels of expression store depletion was ineffective and the channel required receptor-coupled PLC to open. In general though, it is thought that TRPC3 and TRPC6 are usually activated directly by DAG, and that TRPC1, TRPC4, TRPC5 are more likely candidates for SOCs. TRPC7 channels have not been widely studied yet, but store depletion was recently shown to activate the human recombinant TRPC7 channel (30), but not the mouse homologue (28). It is therefore still an open question which TRPCs (if any) mediate store-depletion activated Ca^{2+} influx. The possibility remains that any one of TRPC1, TRPC3, TRPC4, TRPC5, TRPC 6 or TRPC7 could contribute, either as homomeric channels or, more likely as constituents of a heteromeric channel complex.

8. Transient Receptor Potential (TRPC) Channels in Pulmonary Arteries

The first evidence that TRPC channels might play a role in pulmonary artery was provided by Northern blot analyses, which demonstrated the presence of TRPC3, TRPC4, TRPC5 and TRPC7 mRNA in the lung. Reverse-transcription-PCR and antibody-labeling subsequently identified several TRPC isoforms in pulmonary vascular endothelial and smooth muscle cells. Vascular smooth muscle cells from a range of blood vessels have been found to contain TRPC subunits and there is typically co-expression of multiple TRPC species. Two studies on rat pulmonary artery smooth muscle agree that there is relatively abundant expression of TRPC1 and TRPC6 (24, 27), but they disagree on the levels of expression of other transcripts. Freshly isolated cells expressed significant amounts of TRPC3 and low levels of TRPC4 and TRPC5 (27), whereas cultured cells expressed relatively high levels of TRPC4 and less TRPC5 and TRPC2 (24). This may reflect changes in the expression pattern brought about during cell culture. The expression pattern differs again in canine PASMCs, where there is relatively abundant expression of TRPC4, significant levels of TRPC6 and TRPC7, but no TRPC1 or TRPC3 (44). The latter study also identified two splice variants of TRPC4 and 3 variants of TRPC7 that are expressed in pulmonary artery. The functional significance of this is not yet known.

Progress is, however, being made on identifying the functional roles of TRPC channels in vascular smooth muscle. It is becoming clear that TRPC6 is an essential component of the receptor-operated channel (ROC) in vascular muscle that is opened by α-adrenoceptor stimulation via PLC activation and the generation of DAG (20). This was demonstrated using antisense oligonucleotides, which abolished TRPC6 expression and inhibited the cation current activated by phenylephrine in rabbit portal vein myocytes, without affecting the current activated by thapsigargin. TRPC6 has also been implicated in myogenic tone, which is characteristic of resistance arteries (47). Again this was demonstrated using antisense oligonucleotides, introduced into small cerebellar arteries *in vitro* by reversible permeabilization. This greatly reduced TRPC6 expression and inhibited the smooth muscle depolarization and vasoconstriction caused by elevated transmural pressure, while also reducing the amplitude of a cation current in isolated smooth muscle cells. Similar approaches have helped to establish a likely role for the TRPC1 subunit in forming the SOC underlying CCE in pulmonary artery smooth muscle. In cultured human pulmonary artery smooth muscle cells treated with antisense oligonucleotides directed against TRPC1, both the mRNA and channel protein were down regulated while the amplitude of an ionic current and Ca^{2+} entry activated by CPA were inhibited, and cell proliferation was reduced (37). Consistent with this, TRPC1 expression is higher in proliferating cells than in growth-arrested cells (14).

Further evidence to support the involvement of TRPC1 in CCE was provided by an antibody directed against the pore region of the channel, which reduced the amplitude of store depletion-activated Ca^{2+} entry in cerebral arterioles (49). The antibody gave rise to membrane-localized immunostaining, confirming channel expression in the plasma membrane. Along with the finding that CCE can support contraction in pulmonary arterial muscle (27), these results are all consistent with a role for TRPC1 in mediating CCE-dependent contraction of pulmonary artery smooth muscle cells. The evidence is, however, circumstantial, but hopefully a direct test will soon be possible using similar antisense techniques. Although only TRPC1 has been clearly implicated in CCE in pulmonary artery smooth muscle cells, it could co-assemble with other subunits to form the functional CCE channel. As indicated above, both TRPC4 and TRPC5 are expressed in the same cells and either could combine with TRPC1. Since TRPC4 has been directly implicated as a component of the SOC in vascular endothelial cells (10), perhaps the most likely combination is TRPC1 with TRPC4.

9. Summary

Hypoxic pulmonary vasoconstriction involves a rise in $[Ca^{2+}]_i$ in PASMCs, brought about by the release of Ca^{2+} from intracellular stores in the SR and Ca^{2+} influx through voltage-gated and voltage-independent channels.

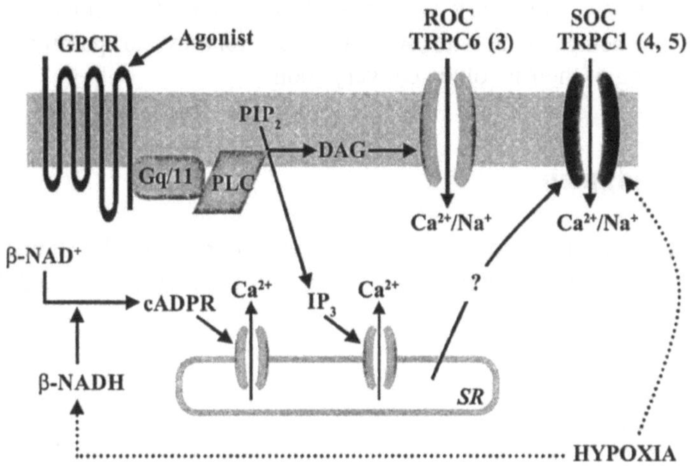

Figure 3. Simplified model depicting the relationship between G-protein coupled receptors (GPCR), SR Ca^{2+} stores and voltage-independent Ca^{2+} entry in pulmonary artery myocytes. Agonist binding to its GPCR causes it to interact with the $G_{q/11}$ protein, thereby stimulating PLC activity to generate DAG and IP_3 from PIP_2. DAG directly activates Ca^{2+} influx through non-selective ROCs, formed by TRPC6 proteins and possibly TRPC3, also found in these cells. IP_3 stimulates Ca^{2+} release from the SR via IP_3-gated Ca^{2+} channels to promote store depletion which, by an unknown mechanism, activates SOCs formed by TRPC1 (possibly co-assembled with TRPC4 or TRPC5). Hypoxia activates SOC activity and CCE by an unknown mechanism that could involve the elevation of β-NADH levels, enhanced synthesis of cADPR and SR Ca^{2+} release via ryanodine receptor-channels.

As summarized in Figure 3, there is a close relationship between the SR and Ca^{2+} entry pathways. Thus, in the presence of the pre-tone which is often required to study HPV, agonists acting via the $G_{q/11}$ protein would mobilize Ca^{2+} from the SR via IP_3 receptors and Ca^{2+} entry via ROCs. TRPC6 subunits are thought to form the ROC, possibly in a complex with other TRPC proteins. Hypoxia has a synergistic action, stimulating Ca^{2+} release from the SR via ryanodine receptor channels. The mechanism responsible for this effect is not yet known, but it may be mediated by the ryanodine receptor agonist cADPR. There is evidence that CCE is also stimulated by hypoxia and that this may provide the $[Ca^{2+}]_i$ needed to sustain the hypoxic contraction. This could be a direct effect of hypoxia, or it could result from excessive depletion of the SR Ca^{2+} store, brought about by the combined effect of agonist and hypoxia, causing Ca^{2+} release via synergistic pathways. Store depletion activates non-selective cation channels that are distinct from ROCs. These SOCs are probably composed of TRPC1 subunits, either as a homologous channel or in a heteromeric complex with TRPC4 or TRPC5. Since SOCs are permeable to Na^+ as well as Ca^{2+}, they could contribute to membrane depolarization, which would promote further Ca^{2+} entry via voltage-gated Ca^{2+} channels. Our knowledge of the roles of TRPC channels and CCE in hypoxic pulmonary vasoconstriction is still preliminary, but rapid progress is being made and we can look forward to gaining a better understanding of their involvement very soon.

Acknowledgments

Support was provided by the British Heart Foundation, the Biotechnology and Biological Sciences Research Council, the Wellcome Trust and Tenovus Scotland.

References

1. Albert AP and Large WA. A Ca^{2+}-permeable non-selective cation channel activated by depletion of internal Ca^{2+} stores in single rabbit portal vein myocytes. *J. Physiol.* 2002; 538: 717-728.
2. Albert AP and Large WA. Activation of store-operated channels by noradrenaline via protein kinase C in rabbit portal vein myocytes. *J. Physiol.* 2002; 544: 113-125.
3. Bolton TB. Mechanisms of action of transmitters and other substances on smooth muscle. *Physiol Rev.* 1979; 59: 606-718.
4. Casteels R and Droogmans G. Exchange characteristics of the noradrenaline-sensitive calcium store in vascular smooth muscle cells or rabbit ear artery. *J. Physiol.* 1981; 317: 263-279.
5. Curtis TM and Scholfield CN. Nifedipine blocks Ca^{2+} store refilling through a pathway not involving L-type Ca^{2+} channels in rabbit arteriolar smooth muscle. *J. Physiol.* 2001; 532: 609-623.
6. Dipp M, Nye PCG, and Evans AM. Hypoxic release of calcium from the sarcoplasmic reticulum of pulmonary artery smooth muscle. *Am. J. Physiol. Lung Cell. Mol. Physiol.* 2001; 281: L318-L325.
7. Doi S, Damron DS, Ogawa K, Tanaka S, Horibe M, and Murray PA. K^+ channel inhibition, calcium signaling, and vasomotor tone in canine pulmonary artery smooth muscle. *Am. J. Physiol. Lung Cell Mol. Physiol.* 2000; 279: L242-L251.
8. Evans AM and Dipp M. Hypoxic pulmonary vasoconstriction: cyclic adenosine diphosphate-ribose, smooth muscle Ca^{2+} stores and the endothelium. *Resp. Physiol. Neurobiol.* 2002; 132: 3-15.
9. Flemming R, Cheong A, Dedman AM, and Beech DJ. Discrete store-operated calcium influx into an intracellular compartment in rabbit arteriolar smooth muscle. *J. Physiol.* 2002; 543: 455-464.
10. Freichel M, Suh SH, and Pfeifer A. Lack of an endothelial store-operated Ca^{2+} current impairs agonist-dependent vasorelaxation in TRP4-/- mice. *Nat. Cell Biol.* 2001; 3: 121-127.
11. Gelband CH and Gelband H. Ca^{2+} release from intracellular stores is an initial step in hypoxic pulmonary vasoconstriction of rat pulmonary artery resistance vessels. *Circulation.* 1997; 96:

3647-3654.
12. Goel M, Sinkins WG, and Schilling WP. Selective association of TRPC channel subunits in rat brain synaptosomes. *J. Biol. Chem.* 2002; 277: 48303-48310.
13. Golovina VA. Cell proliferation is associated with enhanced capacitative Ca^{2+} entry in human arterial myocytes. *Am. J. Physiol.* 1999; 277: C343-C349.
14. Golovina VA, Platoshyn O, Bailey CL, Wang J, Limsuwan A, Sweeney M, Rubin LJ, and Yuan JX-J. Upregulated TRP and enhanced capacitative Ca^{2+} entry in human pulmonary artery myocytes during proliferation. *Am. J. Physiol. Heart Circ. Physiol.* 2001; 280: H746-H755.
15. Gonzalez De La Fuente P, Savineau JP, and Marthan R. Control of pulmonary vascular smooth muscle tone by sarcoplasmic reticulum Ca^{2+} pump blockers: thapsigargin and cyclopiazonic acid. *Pflügers Arch.* 1995; 429: 617-624.
16. Gurney AM and Allam M. Inhibition of calcium release from the sarcoplasmic reticulum of rabbit aorta by hydralazine. *Br. J. Pharmacol.* 1995; 114:38-244.
17. Hardie RC and Minke B. The trp gene is essential for a light-activated Ca^{2+} channel in *Drosophila* photoreceptors. *Neuron.* 1992; 8:43-651.
18. Hofmann T, Schaefer M, Schultz G, and Gudermann T. Subunit composition of mammalian transient receptor potential channels in living cells. *Proc. Natl. Acad. Sci. USA.* 2001; 99: 7461-7466.
19. Hofmann T, Obukhov AG, Schaefer M, Harteneck C, Gudermann T, and Schultz G. Direct activation of human TRPC6 and TRPC3 channels by diacylglycerol. *Nature.* 1999; 397: 259-263.
20. Inoue R, Okada T, Onoue H, Hara Y, Shimizu S, Naitoh S, Ito Y, and Mori Y. The transient receptor potential protein homologue TRP6 is the essential component of vascular α_1-adrenoceptor-activated Ca^{2+}-permeable cation channel. *Circ. Res.* 2001; 88:3 25-332.
21. Jabr RI, Toland H, Gelband CH, Wang XX, and Hume JR. Prominent role of intracellular Ca^{2+} release in hypoxic vasoconstriction of canine pulmonary artery. *Brit. J. Pharmacol.* 1997; 122: 12-30.
22. Jin N, Packer CS, and Rhoades RA. Pulmonary arterial hypoxic contraction: signal transduction. *Am. J. Physiol.* 1992; 263: L73-L78.
23. Lintschinger B, Balzer-Geldsetzer M, Baskaran T, Graier WF, Romanin C, Zhu MX, and Groschner K. Coassembly of Trp1 and Trp3 proteins generates diacylglycerol- and Ca^{2+}-sensitive cation channels. *J. Biol. Chem.* 2000; 275: 27799-27805.
24. McDaniel SS, Platoshyn O, Wang J, Yu Y, Sweeney M, Krick S, Rubin LJ, and Yuan JX-J. Capacitative Ca^{2+} entry in agonist-induced pulmonary vasoconstriction. *Am. J. Physiol. Lung Cell. Mol. Physiol.* 2001; 280: L870-L880.
25. McMurtry IF, Davidson AB, Reeves JT, and Grover RF. Inhibition of hypoxic pulmonary vasoconstriction by calcium antagonists in islated rat lungs. *Circ. Res.* 1976; 38: 99-104.
26. Minke B and Cook B. TRP channel proteins and signal transduction. *Physiol. Rev.* 2002; 82: 429-472.
27. Ng LC and Gurney AM. Store-operated channels mediate Ca^{2+} influx and contraction in rat pulmonary artery. *Circ. Res.* 2001; 89: 923-929.
28. Okada T, Inoue R, Yamazaki K, Maeda A, Kurosaki T, Yamakuni T, Tanaka I, Shimizu S, Ikenaka K, Imoto K, and Mori Y. Molecular and functional characterization of a novel mouse transient receptor potential protein homologue TRP7. Ca^{2+}-permeable cation channel that is constitutively activated and enhanced by stimulation of G protein-coupled receptor. *J. Biol. Chem.* 1999; 274: 27359-27370.
29. Philipp S and Flockerzi V. Molecular characterization of a novel human PDZ domain protein with homology to INAD from Drosophila melanogaster. *FEBS Lett.* 1997; 413: 243-248.
30. Riccio A, Mattei C, Kelsell RE, Medhurst AD, Calver AR, Randall AD, Davis JB, Benham CD, and Pangalos MN. Cloning and functional expression of human short TRP7, a candidate protein for store-operated Ca^{2+} influx. *J. Biol. Chem.* 2002; 277: 12302-12309.

31. Robertson TP, Aaronson PI, and Ward JPT. Ca^{2+}-sensitization during sustained hypoxic pulmonary vasoconstriction is endothelium-dependent. *Am. J. Physiol. Lung Cell. Mol. Physiol.* 2003; 284: L1121-L1126.
32. Robertson TP, Hague D, Aaronson PI, and Ward JPT. Voltage-independent calcium entry in hypoxic pulmonary vasoconstriction of intrapulmonary arteries of the rat. *J. Physiol. 2000*; 525: 669-680.
33. Salvaterra CG and Goldman WF. Acute hypoxia increases cytosolic calcium in cultured pulmonary arterial myocytes. *Am. J. Physiol.* 1993; 264: L323-L328.
34. Schaefer M, Plant TD, Obukhov AG, Hofmann T, Gudermann T, and Schultz G. Receptor-mediated regulation of the nonselective cation channels TRPC4 and TRPC5. *J. Biol. Chem.* 2000; 275: 17517-17526.
35. Strübing C, Krapivinsky G, Krapivinsky L, and Clapham DE. TRPC1 and TRPC5 form a novel cation channel in mammalian brain. *Neuron.* 2001; 29: 645-655.
36. Suggett AJ, Mohammed FH, and Barer GR. Angiotensin, hypoxia, verapamil and pulmonary vessels. *Clin. Exp. Pharmacol. Physiol.* 1980; 7: 263-274.
37. Sweeney M, Yu Y, Platoshyn O, Zhang S, McDaniel SS, and Yuan JX-J. Inhibition of endogenous TRP1 decreases capacitative Ca^{2+} entry and attenuates pulmonary artery smooth muscle cell proliferation. *Am. J. Physiol. Lung Cell. Mol. Physiol.* 2002; 283: L144-L155.
38. Tang Y, Tang J, Chen Z, Trost C, Flockerzi V, Li M, Ramesh V, and Zhu MX. Association of mammalian trp4 and phospholipase C isozymes with a PDZ domain-containing protein, NHERF. *J. Biol. Chem.* 2000; 275: 37559-37564.
39. Trepakova ES, Gericke M, Hirakawa Y, Weisbrod RM, Cohen RA, and Bolotina VM. Properties of a native cation channel activated by Ca^{2+} store depletion in vascular smooth muscle cells. *J. Biol. Chem.* 2001; 276: 7782-7790.
40. Vadula MS, Kleinman JG, and Madden JA. Effect of hypoxia and norepinephrine on cytoplasmic free Ca^{2+} in pulmonary and cerebral arterial myocytes. *Am. J. Physiol.* 1993; 265: L591-L597.
41. Vandier C, Delpech M, Rebocho M, and Bonnet P. Hypoxia enhances agonist-induced pulmonary arterial contraction by increasing calcium sequestration. *Am. J. Physiol.* 1997; 273: H1075-H1081.
42. Vazquez G, Wedel BJ, Trebak M, Bird GS, and Putney JW. Expression level of the canonical transient receptor potential (TRPC3) channel determines its mechanism of activation. *J. Biol. Chem.* 2003; 278: 21649-21654.
43. Vennekens R, Voets T, Bindels RJ, Droogmans G, and Nilius B. Current understanding of mammalian TRP homologues. *Cell Calcium.* 2002; 31: 253-264.
44. Walker RL, Hume JR, and Horowitz B. Differential expression and alternative splicing of TRP channel genes in smooth muscles. *Am. J. Physiol. Cell Physiol.* 2001; 280: C1184-C1192.
45. Wang Y-X, Zheng Y-M, Abdullaev I, and Kotlikoff MI. Metabolic inhibition with cyanide induces calcium release in pulmonary artery myocytes and *Xenopus* oocytes. *Am. J. Physiol. Cell Physiol.* 2003; 284: C378-C388.
46. Ward JPT and Aaronson PI. Mechanisms of hypoxic pulmonary vasoconstriction: can anyone be right? *Respir. Physiol.* 1999; 115: 261-271.
47. Welsh DG, Morielli AD, Nelson MT, and Brayden JE. Transient receptor potential channels regulate myogenic tone of resistance arteries. *Circ. Res.* 2002; 90: 248-250.
48. Wilson SM, Mason HS, Smith GD, Nicholson N, Johnston L, Janiak R, and Hume JR. Comparative capacitative calcium entry mechanisms in canine pulmonary and renal arterial smooth muscle cells. *J. Physiol.* 2002; 543: 917-931.
49. Xu SZ and Beech DJ. TrpC1 is a membrane-spanning subunit of store-operated Ca^{2+} channels in native vascular smooth muscle cells. *Circ. Res.* 2001; 88: 84-87.

IV. ROLE OF ENDOTHELIUM IN HYPOXIC PULMONARY VASOCONSTRICTION

Chapter 12

Endothelium-dependent Hypoxic Pulmonary Vasoconstriction

Jeremy P.T. Ward and Philip I. Aaronson
King's College London, London, United Kingdom

1. Introduction

Over the last decade, considerable progress has been made in unravelling the mechanisms that underlie the phenomenon of hypoxic pulmonary vasoconstriction (HPV), and it is becoming evident that HPV is a multifactorial process. There is strong evidence that the vascular smooth muscle (VSM) itself contains both the hypoxic sensor and the necessary effector pathways that lead to an increase in intracellular $[Ca^{2+}]$ ($[Ca^{2+}]_i$), though there is still controversy regarding the nature of the sensor and its downstream signal, and the mechanism(s) responsible for the rise in $[Ca^{2+}]_i$. Many of these issues are discussed elsewhere in this volume.

It has often been shown that hypoxia elicits a rise in $[Ca^{2+}]_i$ and an associated constriction in both isolated pulmonary VSM cells and arteries where the endothelium has been removed (11, 24, 43). However, the vasoconstriction in such cases is often either transient, or the hypoxic challenge relatively short (<15min) (reviewed in Ref. 39). Where the challenge is longer (> ~20 min), and particularly in small pulmonary arteries, there is strong evidence from numerous sources that removal of the endothelium substantially depresses or abolishes *sustained* HPV (7, 12, 17, 20-22). This *endothelium-dependent* HPV clearly involves additional mechanisms to those resident in the underlying VSM, and presumably requires the production of an endothelium-derived mediator.

The endothelium produces a wide range of vasoactive substances; there are therefore many potential pathways by which it could modulate HPV, including both vasoconstrictor and vasodilator mechanisms. However, it is generally accepted that hypoxia induces a rise in VSM $[Ca^{2+}]_i$ whether or not the endothelium is present. The question is therefore whether some product of the endothelium (or its lack) either potentiates this rise in $[Ca^{2+}]_i$ or, alternatively, modulates the relationship between $[Ca^{2+}]_i$ and force development. In addition, it needs to be determined whether this effect of the endothelium is specific to the pulmonary circulation, or whether it is a more generalized mechanism that only results in sustained vasoconstriction during hypoxia when combined with pulmonary-specific mechanisms in the underlying VSM.

2. Vasoconstrictor Versus Vasodilator Mechanisms

It is important to establish at the outset whether the endothelium-dependent component of HPV is due to an endothelium-derived vasoconstrictor, or, alternatively, to reduced activity of a vasodilator; both would lead to an increase in vascular tone. It has been suggested that HPV may be partly or wholly due to suppression of an endothelium-derived vasodilator influence such as NO or prostacyclin (2, 6, 35). If this were the case, it might be expected that removal of the endothelium or application of an eNOS blocker would mimic hypoxia and increase normoxic tension in pulmonary arteries, apparently suppressing HPV since the artery is already fully constricted. Except for a study on neonatal piglets (2), the large majority of studies do not support this contention, as endothelial-denudation is normally reported to cause a very minor increase in pulmonary artery basal tone if any and, as reported above, fairly consistently suppresses sustained HPV (17, 21, 22). Moreover, Gaine et al. (10) have shown that HPV can be restored in denuded arteries when tissue with an intact endothelium is included in the bath, though this did not reduce tension in normoxia. This does not however rule out an important modulatory influence of endothelium-derived vasorelaxant substances on HPV.

3. The Phenomenon of Pretone: A Role for the Endothelium?

There are several apparent qualitative differences that are commonly observed between studies on perfused lungs, intact animals, isolated arteries and VSM cells (see also Ref. 39). This is most likely at least partly due to differences in the extracellular milieu. Many factors modulate HPV, and of particular importance is the initial degree of vasoconstriction (or "pretone"). Pretone has been known for some time to have a strong potentiating or facilitatory effect on HPV, and in many cases (but not all), salt-perfused lungs, isolated pulmonary arteries and VSM cells only exhibit a really significant response to hypoxia in the presence of a small "priming" or facilitatory concentration of vasoconstrictor agonist, that is effectively synergistic with hypoxia (9, 21, 25, 32, 37, 40). It may therefore be difficult for the effects of any drug or inhibitor to be distinguished between those acting on specific mechanisms underlying HPV, and those acting non-specifically via an effect on this priming phenomenon, unless the latter can be accounted for (28).

In the context of this chapter, it is important to consider the potential source of priming agent(s), and to differentiate between agents that may be present whether or not there is hypoxia, and those that are released as a consequence of hypoxia and play a direct, specific, and critical role in the sustained HPV response. Several different types of vasoconstrictor can be used as priming agents, including prostaglandin $F_{2\alpha}$, angiotensin II, thromboxane analogues, endothelin 1 (ET-1) and α-adrenoceptor agonists, as well as high $[K^+]$ depolarization. Many vasoconstrictors are constitutively produced and could act as endogenous priming agents for HPV in intact animals and perfused lungs . These could easily be overlooked as the small amount of priming required for development of a hypoxic response may be insufficient to cause any significant vasoconstriction on its own (32), and agents affecting the primer therefore might not have any effect on pulmonary vascular resistance during normoxia, but may still have significant effects on HPV. The importance of taking the above into consideration when comparing between different preparations and methodologies cannot be over-emphasized.

Several vasoconstrictors, in particular ET-1, are produced by the endothelium, raising the possibility that the loss of sustained HPV after removal of the endothelium is due to loss of a priming agent rather than to a specific mediator released in response to hypoxia. However, there are two reasons why we consider this to be unlikely. The first is that increasing the concentration of priming agent tends to potentiate both the transient and sustained components of HPV in isolated artery preparations (25, 40), whereas removal of the endothelium only abolishes or depresses the sustained component (7, 20). This implies that priming facilitates the mechanisms underlying the hypoxia-induced rise in VSM $[Ca^{2+}]_i$, a concept consistent with some previous studies (32, 37). There is also good evidence that in sustained HPV, the endothelium promotes an increase in Ca^{2+} sensitivity of the contractile apparatus, but has no effect on VSM $[Ca^{2+}]_i$ (39); this important aspect is discussed in some detail below. This does not rule out the possibility that a single endothelium-derived mediator could be acting as both priming agent and mediator for sustained HPV.

4. The Endothelium, Ca^{2+} and Ca^{2+} Sensitization

Our studies in rat small pulmonary arteries have shown that the relationship between $[Ca^{2+}]_i$ and tension development during sustained HPV (20-60 min) is not straightforward. Whereas $[Ca^{2+}]_i$ (measured using Fura-2 simultaneously with tension) shows an initial transient rise followed by a raised but stable plateau, after an initial transient constriction, tension continues to rise progressively without any further change in $[Ca^{2+}]_i$ (Fig. 1) (20, 26, 28). Moreover, removal of the endothelium does not alter the hypoxia-induced rise in $[Ca^{2+}]_i$, even though the sustained component of HPV is abolished (39). This strongly suggests, as we originally proposed in 1995 (26), that the endothelium is releasing a factor that increases the Ca^{2+} sensitivity of the VSM contractile apparatus, and the concept of modulation of Ca^{2+} sensitivity as an important component of HPV is now gaining acceptance (1, 7, 38). The reader is also directed to the chapters in this volume on the critical role of Ca^{2+} sensitization in HPV, by Robertson and McMurtry (Chapters 7 and 24).

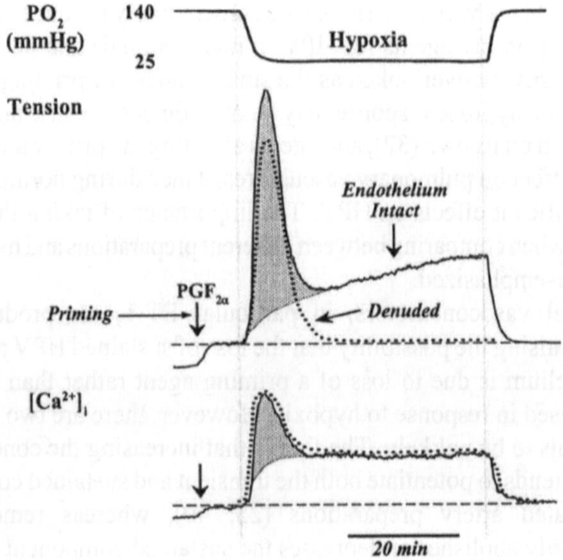

Figure 1: A typical experiment on a small intrapulmonary artery (~350 μm i.d.) from rat, showing simultaneous recordings of PO_2, tension, and $[Ca^{2+}]_i$ measured with Fura PE-3. The solid lines are from an experimental trace. $PGF_{2\alpha}$ was used as a priming agent, and the level of pretone is shown as a dashed line; on reoxygenation both tension and $[Ca^{2+}]_i$ fall back to the level of pretone. Hypoxia elicited a biphasic response, with a transient constriction and rise in $[Ca^{2+}]_i$ superimposed on a slowly developing but sustained vasoconstriction and a stable rise in $[Ca^{2+}]_i$ respectively. The transient component is absent in the present of 1 μM La^{3+} or with use of 20 mM K^+ as a priming agent; it is shown shaded in the figure. The effect of removal of the endothelium is shown as the dotted line; although the sustained component of tension is abolished, there is no effect on $[Ca^{2+}]_i$ (line offset for clarity).

Regulation of Ca^{2+} sensitivity is a recognized mechanism that contributes to the vasoconstrictor response to a variety of agonists in smooth muscle (34). Ca^{2+} sensitivity is determined by the balance between Ca^{2+}-calmodulin dependent phosphorylation of the myosin light chain (MLC) by myosin light chain kinase (MLCK), and its dephosphorylation by myosin phosphatase (SMPP-1M) (Fig. 2B) (34). Several protein kinases have been implicated in pathways leading to Ca^{2+} sensitization, including protein kinase C (PKC), Rho-associated kinases (ROK 1 and 2), and possibly MAP kinases.

4.1. Protein Kinase C

Activation of PKC isoforms has been long been recognized as leading to Ca^{2+}-independent vasoconstriction in smooth muscle (including pulmonary vascular smooth muscle), although it is now believed that conventional and novel PKC isoforms play little role in at least G-protein mediated regulation of Ca^{2+} sensitivity (34). Several studies in isolated perfused lungs have suggested that inhibition of PKC suppresses HPV and hypoxia-mediated pulmonary hypertension (3, 41). However, in isolated small pulmonary arteries, the wide spectrum PKC inhibitor, Ro31-8220, had no effect on endothelium-dependent HPV, although it completely abolished phorbol ester induced pulmonary vasoconstriction (26). Note that PKC inhibition may affect the action of endogenous priming agents (Fig. 2).

4.2. P^{38} MAP Kinase

Comparatively little work has been performed on the role of MAP kinases in HPV, though a recent study by Karamsetty et al. (16) shows that SB202190, an inhibitor of p^{38} MAP kinase, abolishes the sustained phase of HPV in isolated pulmonary arteries. Even though this effect appears to be relatively specific to the sustained component of HPV, it is not yet clear whether it relates to inhibition of the signal transduction pathways leading to Ca^{2+} sensitization, those linking the hypoxic sensor to Ca^{2+} mobilization in the vascular smooth muscle cells, or production of an endothelium-derived mediator in pulmonary arteries.

4.3. RhoA and Rho Associated Kinase

There has been considerable interest in the role of the monomeric G-protein RhoA and the inhibition of myosin phosphatase by its associated kinase, Rho kinase. It is now believed that this pathway is the major intermediate for regulation of Ca^{2+} sensitivity in smooth muscle (34). We have shown that an inhibitor of Rho kinase, Y-27632, selectively suppresses the sustained, but not transient, component of HPV in small pulmonary arteries, and abolishes the hypoxic pressor response in isolated perfused lungs at very low concentrations (27). Another study has also shown that Rho A and Rho kinase are both activated and are important for myosin light chain (MLC) phosphorylation during acute hypoxia-mediated pulmonary vasoconstriction (38). These results strongly suggest that RhoA and Rho kinase are a key component of the mechanisms underlying sustained and endothelium-dependent pulmonary vasoconstriction induced by acute hypoxia (Fig. 2B).

5. Endothelium-derived Mediators and HPV

Although there is strong evidence that the pulmonary vascular endothelium is important, possibly crucial, for development of sustained HPV, and that its action appears to be via an increase in vascular smooth muscle Ca^{2+} sensitivity mediated via RhoA/Rho kinase, the mediator responsible has not been identified (Fig. 2D). Moreover, the pulmonary endothelium also has a powerful modulatory influence on pulmonary vasomotor tone and vascular resistance via the action of numerous other mediators, and this may be modified during hypoxia. In the following section we briefly review the influence of key endothelium-derived mediators on HPV (Fig. 3), and discuss the possible identity of the putative mediator underlying the endothelium-dependent component of sustained HPV.

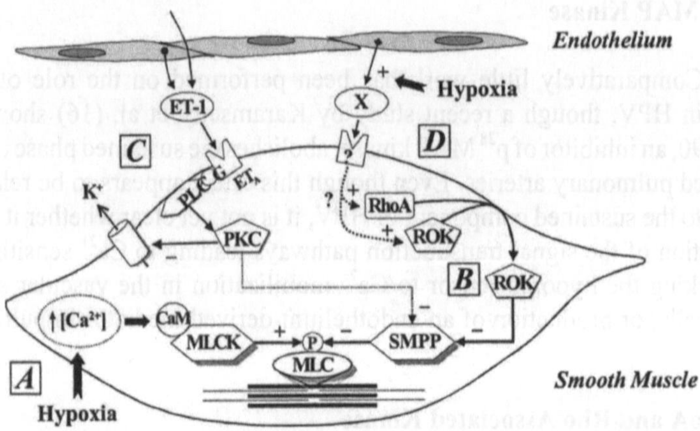

Figure 2: Schematic of potential pathways underlying sustained, endothelium-dependent HPV. **A**: Hypoxia causes an increase in VSM $[Ca^{2+}]_i$ via mechanisms independent of the endothelium; these are discussed elsewhere in this volume. **B**: Ca^{2+} sensitivity is regulated by the balance between phosphorylation of MLC by MLCK and its dephosphorylation by SMPP-1M; ROK inhibits SMPP via its regulatory subunit, thus enhancing MLC phosphorylation. PKC also inhibits SMPP-1M, but this effect is dominated by the Rho/Rho kinase pathway. **C**: ET-1 causes activation of PKC which, in turn, suppresses opening of voltage gated and other K^+ channels and causes depolarization; PKC also activates Ca^{2+} channels. Note that other priming agents also activate PKC and/or cause depolarization (see text). **D**: During sustained HPV an unidentified endothelium-derived mediator ("X") causes RhoA and ROK translocation to the membrane where ROK activates, leading to Ca^{2+} sensitization. PLC: phospholipase C; MLCK: myosin light chain kinase; ROK: Rho kinase; SMPP-1M: myosin phosphatase.

5.1. Cyclooxygenase Products

The cyclooxygenase pathway synthesizes a wide range of vasoactive products, with both vasoconstrictor and vasodilator properties. Numerous studies have failed to show that any vasoconstrictor prostanoid is involved in HPV, although prostacyclin may act as a physiological brake (1). It has been proposed that inhibition of prostacyclin production during hypoxia may form an important component of HPV (6), but inhibition of the cyclooxygenase pathway has little if any effect on HPV in isolated small arteries of rat or pig (21, 22).

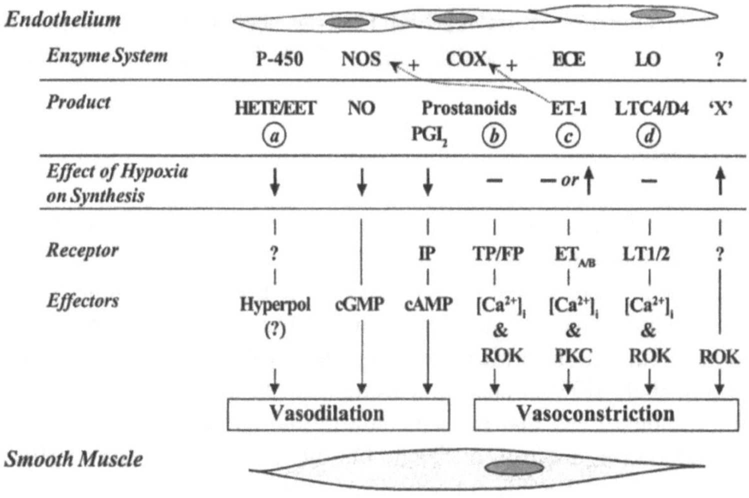

Figure 3: Pulmonary arterial endothelium-derived mediators affecting vascular tone. P450: Cytochrome P-450 monooxygenase; NOS: NO synthase; COX: cyclooxygenase; ECE: endothelin-converting enzyme; LO: lipoxygenase; LTC_4/D_4: cysteinyl leukotrienes C_4 and D_4; ROK: Rho kinase. Note a: P450 mono-oxygenase is expressed in both endothelium and VSM. Note b: Vasoconstrictor prostanoids are also produced by other cell types, and are increased in inflammation; thromboxane and $PGF_{2\alpha}$ have been implicated in the pulmonary vasculature. Note c: ET-1 also acts via ET_B receptors on the endothelium to activate NOS and COX; atypical ET_B receptors may promote constriction in VSM. The evidence that ET-1 production is increased in *acute* hypoxia is open to question, and there is strong evidence that it is not. Note d: Leukotriene production by the endothelium may only be significant in inflammation.

5.2. Lipoxygenase Products

The lipoxygenase pathway synthesizes a variety of inflammatory mediators from arachidonic acid, the most important in this context being the cysteinyl leukotrienes. These are powerful vasoconstrictors, and there is also evidence that they activate RhoA/Rho kinase (31). However, several studies have dismissed them as playing any role in HPV whatsoever (36, 41).

5.3. Cytochrome P-450 Products

The cytochrome P-450 mono-oxygenase system consists of a family of O_2-sensitive enzymes that are expressed in both pulmonary vascular smooth muslce and endothelium, and synthesize hydroxyeicosatetraenoic acids (HETEs) and cis-epoxyeicosatrienoic acids (EETs) (14). Both HETEs and EETs are normally regarded as pulmonary vasodilators (14), and there is little evidence for any major role for cytochrome P-450 products in HPV (42). Nevertheless, as hypoxia inhibits the production of 20-HETE it has been proposed that this effect could provide an important contribution to HPV by relieving a vasodilator influence (14).

5.4. Nitric Oxide

Endothelium-derived nitric oxide is an important modulator of pulmonary vascular tone and resistance. There is little if any evidence that it is involved in the mechanisms of HPV *per se*, but changes in its synthesis during hypoxia undoubtedly have a major modulatory influence in the intact animal and perfused lung. Inhibition of nitric oxide synthase (NOS) is almost universally found to potentiate HPV in such preparations, even though it may have little effect on pulmonary vascular resistance during normoxia (1). Results derived from isolated arteries are less consistent, but often there is little or no effect of NOS inhibition on HPV except when hypoxia is severe (1). The overall consensus is that production of NO by the lung decreases in moderate hypoxia (19), as might be expected because O_2 is a substrate for NOS. Nevertheless, it has been proposed that nitric oxide may act as a physiological brake on HPV (19), possibly at the level of the small pulmonary arteries and as a result of increased shear stress due to hypoxia-induced narrowing of these arteries (1).

5.5. Endothelin 1

ET-1 has been mooted as either the primary mediator, or an important priming factor, in HPV. The actions of this 21-amino acid polypeptide on the pulmonary vasculature are extraordinarily complex and context-dependent. It is generally held that ET-1 acts mainly via ET_A receptors on the pulmonary VSM to cause constriction, and via ET_B receptors on the endothelium to release nitric oxide and prostacyclin to cause relaxation (Fig. 3), with the latter effect predominating if ET-1 is applied in the presence of pre-existing tone (5). However this picture is complicated by the presence, at least in some species, of atypical ET_B receptors which mediate contraction (23), and by the reported ability of ET-1 to release vasoconstricting prostanoids (4).

The role of ET-1 in acute HPV remains controversial, since the substantial literature which supports its involvement is balanced by an equally persuasive body of work which denies it. These reports can be summarized by noting that selective ET_A or mixed ET_A/ET_B blockers inhibit HPV in some studies, but not in others of apparently similar quality (1). Recently, however, several papers have appeared which go some way towards resolving this debate.

Sato et al. (30) noted that studies supporting a role for ET-1 in HPV have mainly been carried out in intact animals, while those showing a lack of effect have typically been conducted in perfused lungs. They therefore examined the effects of ET-1 antagonists on HPV in conscious catheterized rats, as well as in isolated blood-perfused and salt-perfused rat lungs. In each case, blockade of the ET_A receptor greatly attenuated HPV. HPV was, however, restored in salt perfused lungs exposed to combined ET_A/ET_B blockade when hypoxia was applied in the presence of angiotensin II, and also when the lungs had been pretreated with the K_{ATP} channel antagonist glibenclamide. Based on the results of this elegant study, they proposed that ET-1 acts on pulmonary VSM to tonically inhibit K_{ATP} channels, thereby causing a small degree of ongoing depolarization which allows other processes stimulated by hypoxia to cause constriction (Fig. 2C). Angiotensin II could also play a similar role in 'enabling' HPV, thus explaining the observation of many investigators that HPV is greatly enhanced in isolated lung preparations if they are first primed with this drug (and see section 3 above). The work of Sato et al. (30) suggests, however, that ET-1 may be a more physiologically relevant priming agent.

A similar proposal was advanced by Sham et al. (32), who found that the very small contraction and rise in $[Ca^{2+}]_i$ induced by hypoxia in VSM cells isolated from distal porcine pulmonary arteries was greatly increased if cells were primed with a low concentration (0.1 pM) of ET-1, which itself had no effect. In a subsequent study of HPV in porcine small pulmonary arteries using a perfusion myograph (22), they observed that HPV was entirely abolished by removal of the endothelium. This effect was mimicked by the ET_A antagonist BQ-123, whereas blockade of cyclooxygenase with indomethacin had no effect, and inhibition of eNOS with L-NAME greatly enhanced HPV. Application of ET-1 (0.1 pM) to endothelium-denuded arteries was then able to restore the response to hypoxia. Again, ET-1 was suggested to be acting as a priming agent rather than specific mediator of HPV.

Johnson et al. (15) have very recently examined the effect of BQ-123 on pulmonary hemodynamics in non-sedated young human volunteers under normoxic and moderately hypoxic conditions (arterial $Po_2 = 100$ and 49 mmHg, respectively). BQ-123 caused a significant fall in pulmonary vascular resistance (PVR) under normoxic conditions, but did not cause any inhibition of the percentage increase in PVR elicited by hypoxia. Although these results support the concept that ET-1 supports a degree of basal tone in the human pulmonary vasculature, they argue against the proposal that this ET-1 mediated tone is an obligatory priming factor for HPV, at least in healthy young individuals. They do not, however, rule out the possibility that some degree of pretone contributed by other factors (including ET-1 acting via ET_B receptors) is required for the full expression of the hypoxic response in humans, or that ET-1 may play more of a role in conditions such as heart failure or primary pulmonary hypertension, in which its plasma levels and/or receptor expression may be elevated.

Possible subtleties which may complicate the role of ET-1 in HPV are evident in the report of Ambalavanan et al. (2), who studied HPV in chronically instrumented and sedated 3-4 week old piglets. They found that the ET_A receptor blockers EMD 122946 and BQ 610 partially suppressed the moderate hypoxia-elicited rise in PVR. The eNOS inhibitor L-NAME, however, caused a large increase in PVR, after which hypoxia had no further effect. The rise in PVR caused by L-NAME was not reversed by EMD 122946. Based on their results, the authors proposed that HPV in their model was being caused by a fall in NO release. They suggested that the partial reversal of HPV by EMD 122946 and BQ 610 was occurring because blockade of ET_A receptors was unmasking or enhancing an ET_B-mediated release of NO, rather than by preventing a direct ET_A-mediated constriction. One implication of this intriguing hypothesis, which is that combined ET_A/ET_B receptor antagonists should not be useful in preventing HPV, is however not supported by several other studies, in which non-selective antagonists ET-1 antagonists were effective in suppressing HPV (13).

On the face of it, these reports suggest that ET-1 is not the unique stimulus of HPV, at least in humans, rats, and piglets. Instead, ET-1 may be one of several priming factors which potentiate HPV non-specifically (see section 3) (39), depending on the extent to which it is contributing to basal pulmonary tone or excitation in a particular species and under the specific conditions obtaining at the time of study. It is clear, however, that the effects of ET-1 on the pulmonary vasculature are complex, and it is not unlikely that future studies will fuel rather than resolve the ongoing debate concerning its role in HPV.

5.6. Other Potential Mediators of Endothelium-dependent HPV

We originally suggested that the endothelium released a substance during sustained HPV that effectively altered the Ca^{2+}-tension relationship of the underlying VSM in 1995 (26). Since that time there has been a significant amount of work to support this hypothesis. So far however, no known endothelium-derived mediator has been shown to play an indispensable role in sustained HPV, and the identity of the mediator responsible for endothelium-dependent Ca^{2+} sensitization of the VSM during hypoxia remains unknown. Although ET-1 is the candidate of choice for many, and may indeed play an important role as an endogenous facilitator or priming agent for HPV (see above), blockade of all endothelin receptors does not even slightly diminish HPV

(7, 8, 10, 18). Therefore, there is clear evidence that the endothelium must release another vasoconstrictor mediator that is distinct from ET-1, and which is not a product of the cyclooxygenase or lipoxygenase pathways.

An important study in support of this hypothesis was published by Gaine et al. (10), who showed that removal of the endothelium from pig pulmonary artery rings abolished the slowly developing and sustained component of HPV, but that this component could be fully restored in the presence of pulmonary valve leaflets as a source of pulmonary endothelial cells. Neither indomethacin nor the ET_A antagonist BQ123 had any effect on either the initial sustained component of HPV before endothelium removal, nor that after restoration in the presence of the valve leaflets (10). This strongly suggests the action of an endothelium-derived factor that is not ET-1 or a cyclooxygenase product.

Further evidence for a specific mediator of sustained HPV and the associated Ca^{2+}-sensitization comes from our recent study on the properties of a factor derived from the effluent of hypoxic-perfused rat lungs (29). Lungs were perfused under normoxic and hypoxic conditions, and the factor extracted and partially purified. There are several distinct similarities between the effects of this factor and the sustained, endothelium-dependent component of HPV. The factor induced a slowly developing vasoconstriction in pulmonary arteries, which was insensitive to L-type channel blockade and not associated with a rise in $[Ca^{2+}]_i$; it also significantly potentiated the action of another vasoconstrictor. Both of these results imply that it induces an increase in Ca^{2+} sensitivity in the VSM. Importantly, the vasoconstriction induced by the factor was completely unaffected by either indomethacin or combined ET_A/ET_B blockade, strongly suggesting that it is neither a cyclooxygenase product nor ET-1. An interesting observation was that the factor has very little effect on mesenteric arteries, and that when similar experiments were performed with effluent from hypoxic perfused mesenteric beds no equivalent vasoconstrictor activity could be discovered (29). This raises the possibility that this factor not only has a specific action in the pulmonary circulation, but also is specifically produced there.

Evans and Dipp (8) have recently reported the results of a similar study using the effluent from perfused lungs, where they additionally showed that the vasoconstriction induced in endothelium-denuded pulmonary arteries by effluent from hypoxic lungs was suppressed by the Rho kinase inhibitor Y27632. Moreover, the Y27632 concentration-inhibition curve for this effect was almost identical to those observed for both sustained HPV in endothelium-intact pulmonary arteries and the hypoxic pressor response in perfused lungs; it is particularly noteworthy that Y27632 was considerably less effective in inhibiting the vasoconstrictor response to ET-1 (8). Taken together, these two reports strongly suggest that during hypoxia the lungs produce a mediator distinct from ET-1 which has very similar properties to those that would be predicted for the proposed endothelium-derived mediator of sustained HPV, in that both apparently act via a Rho kinase-mediated pathway to enhance the Ca^{2+} sensitivity of the VSM (Fig. 2D). It remains to be seen however whether the factor derived from the effluent of hypoxic perfused lungs is indeed produced by the endothelium. Moreover, attempts to identify the factor have proved to be technically very difficult.

6. Summary

There is a growing consensus that HPV is multi-factorial in mechanism, and, in particular, that for HPV to develop fully and be sustained there needs to be both a rise in VSM $[Ca^{2+}]_i$, initiated by mechanisms intrinsic to the VSM cell, and a concomitant increase in VSM Ca^{2+} sensitivity induced by an endothelium-derived mediator. The precise identity of the hypoxic sensor, the effector mechanisms leading to the rise in $[Ca^{2+}]_i$, and the endothelium-derived mediator remain however controversial. In terms of endothelium-dependent HPV, there is now fairly strong evidence that the RhoA/Rho kinase pathway is central to the hypoxia-induced Ca^{2+} sensitization, though other protein kinases may also be involved. Although this provides a potential therapeutic target for alleviation of both the detrimental effects of acute HPV in critically ill hypoxic patients, and pulmonary hypertension associated with chronic hypoxic lung disease, development of an antagonist to the putative endothelium-derived mediator would be a greater prize. However, the mediator must first be identified, and as yet no known endothelium-derived mediator has yet to be unequivocally demonstrated as essential for sustained HPV.

Acknowledgments

We are grateful to colleagues who have worked with us, notably Richard Leach, Vladimir Snetkov and Tom Robertson, and to the Wellcome Trust for supporting our research.

References

1. Aaronson PI, Robertson TP, and Ward JPT. Endothelium-derived mediators and hypoxic pulmonary vasoconstriction. *Respir. Physiol. Neurobiol.* 2002; 132: 107-120.
2. Ambalavanan N, Philips JB III, Bulger A, Oparil S, and Chen YF. Endothelin-A receptor blockade in porcine pulmonary hypertension. *Pediatr. Res.* 2002; 52: 913-921.
3. Barman SA. Effect of protein kinase C inhibition on hypoxic pulmonary vasoconstriction. *Am. J. Physiol. Lung Cell. Mol. Physiol.* 2001; 280: L888-L895.
4. Barnard JW, Barman SA, Adkins WK, Longenecker GL, and Taylor AE. Sustained effects of endothelin-1 on rabbit, dog, and rat pulmonary circulations. *Am. J. Physiol.* 1991; 261: H479-H486.
5. Cassin S, Kristova V, Davis T, Kadowitz P, and Gause G. Tone-dependent responses to

endothelin in the isolated perfused fetal sheep pulmonary circulation in situ. *J. Appl. Physiol.* 1991; 70: 1228-1234.

6. Demiryurek AT, Wadsworth RM, Kane KA, and Peacock AJ. The role of endothelium in hypoxic constriction of human pulmonary artery rings. *Am. Rev. Resp. Dis.* 1993; 147: 283-290.
7. Dipp M, Nye PC, and Evans AM. Hypoxic release of calcium from the sarcoplasmic reticulum of pulmonary artery smooth muscle.[comment]. *Am. J. Physiol. Lung Cell. Mol. Physiol.* 2001; 281: L318-L325.
8. Evans AM and Dipp M. Hypoxic pulmonary vasoconstriction: cyclic adenosine diphosphate-ribose, smooth muscle Ca^{2+} stores and the endothelium. *Respir. Physiol. Neurobiol.* 2002; 132: 3-15.
9. Fishman AP. Hypoxia on the pulmonary circulation. How and where it acts. *Circ. Res.* 1976; 38: 221-231.
10. Gaine SP, Hales MA, and Flavahan NA. Hypoxic pulmonary endothelial cells release a diffusible contractile factor distinct from endothelin. *Am. J. Physiol.* 1998; 274: L657-L664.
11. Gelband CH and Gelband H. Ca^{2+} release from intracellular stores is an initial step in hypoxic pulmonary vasoconstriction of rat pulmonary artery resistance vessels. *Circulation.* 1997; 96: 3647-3654.
12. Holden WE and McCall E. Hypoxia-induced contractions of porcine pulmonary artery strips depend on intact endothelium. *Exp. Lung Res.* 1984; 7: 101-112.
13. Holm P, Liska J, Clozel M, and Franco-Cereceda A. The endothelin antagonist bosentan: hemodynamic effects during normoxia and hypoxic pulmonary hypertension in pigs. *J. Thorac. Cardiovasc. Surg.* 1996; 112: 890-897.
14. Jacobs ER and Zeldin DC. The lung HETEs (and EETs) up. *Am. J. Physiol. Heart Circ. Physiol.* 2001; 280: H1-H10.
15. Johnson W, Nohria A, Garrett L, Fang JC, Igo J, Katai M, Ganz P, and Creager MA. Contribution of endothelin to pulmonary vascular tone under normoxic and hypoxic conditions. *Am. J. Physiol. Heart Circ. Physiol.* 2002; 283: H568-H575.
16. Karamsetty MR, Klinger JR, and Hill NS. Evidence for the role of p38 MAP kinase in hypoxia-induced pulmonary vasoconstriction. *Am. J. Physiol. Lung Cell. Mol. Physiol.* 2002; 283: L859-L866.
17. Kovitz KL, Aleskowitch TD, Sylvester JT, and Flavahan NA. Endothelium-derived contracting and relaxing factors contribute to hypoxic responses of pulmonary arteries. *Am. J. Physiol.* 1993; 265: H1139-48.
18. Lazor R, Feihl F, Waeber B, Kucera P, and Perret C. Endothelin-1 does not mediate the endothelium-dependent hypoxic contractions of small pulmonary arteries in rats. *Chest.* 1996; 110: 189-197.
19. Le Cras TD and McMurtry IF. Nitric oxide production in the hypoxic lung. *Am. J. Physiol. Lung Cell. Mol. Physiol.* 2001; 280: L575-L582.
20. Leach RM, Hill HS, Snetkov VA, Robertson TP, and Ward JPT. Divergent roles of glycolysis and the mitochondrial electron transport chain in hypoxic pulmonary vasoconstriction of the rat: Identity of the hypoxic sensor. *J. Physiol.* 2001; 536: 211-224.
21. Leach RM, Robertson TP, Twort CH, and Ward JPT. Hypoxic vasoconstriction in rat pulmonary and mesenteric arteries. *Am. J. Physiol.* 1994; 266: L223-L231.
22. Liu Q, Sham JSK, Shimoda LA, and Sylvester JT. Hypoxic constriction of porcine distal pulmonary arteries: endothelium and endothelin dependence. *Am. J. Physiol. Lung Cell. Mol. Physiol.* 2001; 280: L856-L865.
23. MacLean MR and McCulloch KM. Influence of applied tension and nitric oxide on responses to endothelins in rat pulmonary resistance arteries: effect of chronic hypoxia. *Br. J. Pharmacol.* 1998; 123: 991-999.
24. Madden JA, Vadula MS, and Kurup VP. Effects of hypoxia and other vasoactive agents on pulmonary and cerebral artery smooth muscle cells. *Am. J. Physiol.* 1992; 263: L384-L393.
25. Ozaki M, Marshall C, Amaki Y, and Marshall BE. Role of wall tension in hypoxic responses

of isolated rat pulmonary arteries. *Am. J. Physiol.* 1998; 19: L1069-L1077.

26. Robertson TP, Aaronson PI, and Ward JPT. Hypoxic vasoconstriction and intracellular Ca^{2+} in pulmonary arteries: Evidence for PKC-independent Ca^{2+} sensitization. *Am. J. Physiol.* 1995; 268: H301-H307.
27. Robertson TP, Dipp M, Ward JPT, Aaronson PI, and Evans AM. Inhibition of sustained hypoxic vasoconstriction by Y-27632 in isolated intrapulmonary arteries and perfused lung of the rat. *Br. J. Pharmacol.* 2000; 131: 5-9.
28. Robertson TP, Hague DE, Aaronson PI, and Ward JPT. Voltage-independent calcium entry in hypoxic pulmonary vasoconstriction of intrapulmonary arteries of the rat. *J. Physiol* 2000; 525: 669-680.
29. Robertson TP, Ward JPT, and Aaronson PI. Hypoxia induces the release of a pulmonary-selective, Ca^{2+}-sensitising, vasoconstrictor from the perfused rat lung. *Cardiovasc. Res.* 2001; 50: 145-150.
30. Sato K, Morio Y, Morris KG, Rodman DM, and McMurtry IF. Mechanism of hypoxic pulmonary vasoconstriction involves ET_A receptor-mediated inhibition of K_{ATP} channel. *Am. J. Physiol. Lung Cell. Mol. Physiol.* 2000; 278: L434-L442.
31. Setoguchi H, Nishimura J, Hirano K, Takahashi S, and Kanaide H. Leukotriene C_4 enhances the contraction of porcine tracheal smooth muscle through the activation of Y-27632, a rho kinase inhibitor, sensitive pathway. *Br. J. Pharmacol.* 2001; 132: 111-118.
32. Sham JSK, Crenshaw BR Jr., Deng L-H, Shimoda LA, and Sylvester JT. Effects of hypoxia in porcine pulmonary arterial myocytes: roles of K_V channel and endothelin-1. *Am. J. Physiol. Lung Cell. Mol. Physiol.* 2000; 279: L262-L272.
33. Shimoda LA, Sylvester JT, and Sham JSK. Inhibition of voltage-gated K^+ current in rat intrapulmonary arterial myocytes by endothelin-1. *Am. J. Physiol.* 1998; 274: L842-L853.
34. Somlyo AP, Wu X, Walker LA, and Somlyo AV. Pharmacomechanical coupling: the role of calcium, G-proteins, kinases and phosphatases. *Rev. Physiol. Biochem. Pharmacol.* 1999; 134: 201-234.
35. Terraz S, Baechtold F, Renard D, Barsi A, Rosselet A, Gnaegi A, Liaudet, L, Lazor R, Haefliger JA, Schaad N, Perret C, Kucera P, Markert M, and Feihl F. Hypoxic contraction of small pulmonary arteries from normal and endotoxemic rats: fundamental role of NO. *Am. J. Physiol.* 1999; 276: H1207-H1214.
36. Tseng CM, McGeady M, Privett T, Dunn A, and Sylvester JT. Does leukotriene C4 mediate hypoxic vasoconstriction in isolated ferret lungs? *J. Appl. Physiol.* 1990; 68: 253-259.
37. Turner JL and Kozlowski RZ. Relationship between membrane potential, delayed rectifier K^+ currents and hypoxia in rat pulmonary arterial myocytes. *Exp. Physiol.* 1997; 82: 629-645.
38. Wang Z, Jin N, Ganguli S, Swartz DR, Li L, and Rhoades RA. Rho-kinase activation is involved in hypoxia-induced pulmonary vasoconstriction. *Am. J. Resp. Cell. Mol. Biol.* 2001; 25: 628-635.
39. Ward JPT and Aaronson PI. Mechanisms of hypoxic pulmonary vasoconstriction: Can anyone be right? *Resp. Physiol.* 1999; 115: 261-271.
40. Ward JPT and Robertson TP. The role of the endothelium in hypoxic pulmonary vasoconstriction. *Exp. Physiol.* 1995; 80: 793-801.
41. Weissmann N, Seeger W, Conzen J, Kiss L, and Grimminger F. Effects of arachidonic acid metabolism on hypoxic vasoconstriction in rabbit lungs. *Eur. J. Pharmacol.* 1998; 356: 231-237.
42. Weissmann N, Voswinckel R, Hardebusch T, Rosseau S, Ghofrani HA, Schermuly R, Seeger W, and Grimminger F. Evidence for a role of protein kinase C in hypoxic pulmonary vasoconstriction. *Am. J. Physiol.* 1999; 20: L90-L95.
43. Yuan X-J, Tod ML, Rubin LJ, and Blaustein MP. Contrasting effects of hypoxia on tension in rat pulmonary and mesenteric arteries. *Am. J. Physiol.* 1990; 259: H281-H289.

V. MECHANISMS OF OXYGEN SENSING IN THE PULMONARY VASCULATURE

Chapter 13

Chemistry of Oxygen and Its Derivatives in the Lung

Lisa A. Palmer
University of Virginia, Charlottesville, Virginia, U.S.A.

1. Introduction

Oxygen is an abundant element in the world in which we live, making up 21% of the air we breath. It is necessary in the production of energy as the terminal electron acceptor in the mitochondrial electron transport chain. Blood transports oxygen and ensures its efficient delivery to all tissues of the body, with the subsequent removal of carbon dioxide. In the cell, changes in the amount of oxygen can serve as a molecular switch, turning on ion channels (20), altering the expression of factors which affect local and regional blood flow (10, 27, 28), and stabilizing transcription factors (15, 17).

Oxygen also exists and has influence in forms other than O_2, including the highly reactive oxygen species (ROS); superoxide, hydrogen peroxide, and the hydroxyl radical. ROS have been traditionally viewed as detrimental by-products of metabolism with the potential to cause damage to lipids, proteins, and DNA. A series of antioxidant systems exist to reduce the damage caused by the production of these ROS. These antioxidant systems can be grouped into two categories: *i*) enzymatic (superoxide dismutase, catalase, and glutathione peroxidases) and *ii*) soluble (vitamin E, vitamin C, β carotene). Imbalances between the production and neutralization of ROS can lead to diseases such as asthma, sarcoidosis, and chronic beryllium disease (6). However, the role of ROS may not be all negative, as some have been shown to have important physiologic function.

Cell signaling events have been associated with changes in oxygen concentration as well as the formation of ROS. Interactions between oxygen, hemoglobin, and nitric oxide have been implicated in regulating blood flow through the alteration of vascular tone by the formation of S-nitrosothiols (27, 28). Changes in the levels of oxygen have also been shown to regulate the activity of prolyl and asparagine hydroxylase, regulating the stability of the α subunit of the transcription factor hypoxia inducible factor-1 (13, 15, 17). Both increases and decreases in the levels of ROS have been reported to regulate ion channel activity in the vasculature thus regulating vascular tone to match ventilation to perfusion (32).

The concentration of oxygen and its derivatives play a role both in physiology as well pathology. In this chapter, we will review the chemistry and formation of ROS and the antioxidant machinery that limits the abundance of ROS. In addition, we will summarize the following relationships: *i*) ROS and oxygen sensing in hypoxic pulmonary vasoconstriction; *ii*) oxygen, nitric oxide and blood flow; and *iii*) oxygen regulation of the stability of the transcription factor, hypoxia-inducible factor-1 (HIF-1).

2. Oxygen and Reactive Oxygen Species

Oxygen is a necessity of life. It is essential for generating the energy needed to live by reducing molecular oxygen to water. During this process, oxygen undergoes a series of one-electron reductions in the mitochondrial electron transport chain to produce superoxide radical ($O_2^{\bullet-}$), hydrogen peroxide (H_2O_2), the hydroxyl radical (OH^-) and water (H_2O) (Equation 1).

$$O_2 \rightarrow O_2^{\bullet-} \rightarrow H_2O_2 \rightarrow HO^{\bullet} \rightarrow H_2O \qquad [1]$$

2.1. Superoxide ($O_2^{\bullet-}$)

The superoxide radical is produced from the one electron reduction of oxygen. The major source of this radical is from the oxygen reduction pathway in the electron transport chain of mitochondria and the endoplasmic reticulum. However, within the vascular wall, superoxides can be produced from NAD(P)H oxidase (2). The activated form of NAD(P)H oxidase is responsible for the single electron reduction of molecular oxygen to form superoxide (Equation 2).

$$NAD(P)H + 2O_2 \rightarrow NAD(P)^+ + H^+ + 2O_2^{\bullet-} \qquad [2]$$

A vascular isoform of this enzyme has been identified. It is constitutively active and is a major source of vascular superoxide production. Two cytochrome b_{558} subunits, $p22^{phox}$ and $gp91^{phox}$, of NADPH oxidase have been shown to be important for electron transport and the reduction of molecular oxygen to superoxide (2). Other sources of superoxide production include xanthine oxidase and nitric oxide synthase (NOS). Xanthine oxidase converts hypoxanthine to xanthine and produces superoxide as a by-product. NOS generates superoxide rather than NO under conditions in which L-arginine or tetrahydrobiopterin are limiting (26).

Vascular superoxide production sites include the endothelium, vascular smooth muscle cells and the fibroblasts within the adventitia (2). Unfortunately, the superoxide radical is dangerous to biological systems because it is easily converted to a more reactive oxygen species. This radical can generate the hydroxyl radical via a series of reactions known as the metal-catalyzed Haber-Weiss reaction or the superoxide radical-driven Fenton reaction (Equations 3, 4, and 5).

$$Fe^{3+} + O_2^{\bullet -} \rightarrow Fe^{2+} + O_2 \quad (\textit{Haber-Weiss Reaction}) \qquad [3]$$

$$Fe^{2+} + H_2O_2 \rightarrow Fe^{3+} + HO^{\bullet} + HO^{-} \quad (\textit{Fenton Reaction}) \qquad [4]$$

$$O_2^{\bullet -} + H_2O_2 \rightarrow O_2 + HO^{\bullet} + HO^{-} \qquad [5]$$

In addition, the superoxide radical can react with nitric oxide to give peroxynitrite ($ONOO^-$) (Equation 6). The products of both reactions (Equations 5 and 6) are very reactive.

$$NO + O_2^{\bullet -} \rightarrow ONOO^{-} \qquad [6]$$

Fortunately, the superoxide radical is removed *in vivo* by superoxide dismutase (SOD) (Equation 7).

$$2O_2^{\bullet -} + 2H^{+} \rightarrow H_2O_2 + O_2 \qquad [7]$$

2.2. Hydrogen Peroxide (H_2O_2)

Hydrogen peroxide (H_2O_2) has no unpaired electrons and is therefore not a radical. In biological systems, hydrogen peroxide is relatively stable, may be long-lived, can travel long distances, and readily passes through cell membranes. Hydrogen peroxide is produced from microsomes, mitochondria, and phagocytic cells mainly through the superoxide radical. Thus, any biological system that produces superoxide radical will produce hydrogen peroxide by nonenzymatic or superoxide dismutase-catalyzed dismutation (Equation 7). In addition, it can be directly produced by several oxidase systems including monamine oxidase, galactose oxidase, xanthine oxidase and amino acid oxidase (6). Hydrogen peroxide can be oxidized by eosinophil-specific peroxidase and neutrophil-specific peroxidase using halides as a cosubstrate to form potent oxidant hypohalous acids (HOX) and other reactive halogenating species (Equation 8).

$$H_2O_2 + X^{-} + H^{+} \rightarrow HOX + H_2O \; (X = Br^{-}, Cl^{-}) \qquad [8]$$

Hydrogen peroxide (H_2O_2) can inactivate enzymes through oxidation of essential thiol groups. More importantly, when hydrogen peroxide enters the cell it may react with Fe^{2+} or Cu^{1+} (Fenton reaction) to give the hydroxyl radical (Equation

4) (21). Dismutase and catalase contribute to the removal of H_2O_2 (Equation 9).

$$2H_2O_2 \rightarrow 2H_2O + O_2 \qquad [9]$$

2.3. Hydroxyl Radical ($OH^{\cdot}$)

Most of the damage produced by $O_2^{\cdot -}$ and H_2O_2 *in vivo* is due to the production of hydroxyl radicals via the metal catalyzed Haber-Weiss reaction or the superoxide radical-driven Fenton reaction (Equations 3-5). Alternately, generation of the hydroxyl radical may involve the one electron reduction of hypochlorite by $O_2^{\cdot -}$ or an Fe^{2+} complex (Equations 10 and 11).

$$HClO + O_2^{\cdot -} \rightarrow Cl^- + O_2 + HO^{\cdot} \qquad [10]$$

$$HClO + Fe^{2+} \rightarrow Cl^- + Fe^{3+} + HO^{\cdot} \qquad [11]$$

The hydroxyl radical has a very high oxidation potential and is highly reactive. Once formed *in vivo*, the hydroxyl radical may react rapidly with almost any molecule near its formation site. However, because of its high reactivity, it is short-lived, thus unable to react with molecules far away from its site of initiation.

3. Nitric Oxide and Reactive Nitrogen Species

Nitric oxide (NO) is a cell signaling molecule involved in a number of physiological and pathophysiological processes. NO is synthesized by a family of complex enzymes know as NOS. Three isoforms exist: bNOS (neuronal or Type I NOS), iNOS (inducible or Type II NOS) and eNOS (endothelial or Type III NOS). Each enzyme is a homodimer requiring three cosubstrates (L-arginine, NADPH, and O_2) and five cofactors (FAD, FMN, Ca^{2+}-calmodulin, heme and tetrahydrobiopterin) for activity.

$NO^{\cdot}$ is a free radical gas with a short half life under physiological conditions. NO chemistry and concentration determine its biological effect. The redox biochemistry of NO involves the action of three redox species: NO^+ (nitrosonium), NO^- (nitroxyl anion) and $NO^{\cdot}$ (the free radical gas), with each redox species having its own unique chemical properties. Together, NO can react with oxygen, superoxide anion, redox metals, and other radical species.

3.1. NO Reactions with Oxygen

NO does not rapidly react with most biological molecules. However, the effects of NO can be mediated through the formation of reactive nitrogen oxide species from reactions of NO with either O_2 or $O_2^{\bullet-}$ (34). Three types of reactions can occur: *i*) nitrosative reactions, adding the equivalent of NO^+ to an amino, thiol, or hydroxy aromatic group; *ii*) oxidative reactions, the removal of 1 or 2 electrons from a substrate or a hydroxylation reaction; and *iii*) nitration reactions, the addition of an equivalent of an NO_2^+ to an aromatic group.

NO_2, N_2O_3 and N_2O_4 can nitrosate sulfhydryl and tyrosine groups and oxidize substrates. In aqueous solution, N_2O_3 hydrolyzes to nitrite (Equations 12, 13, and 14) and is capable of the oxidation of redox-active complexes.

$$2NO^{\bullet} + O_2 \rightarrow 2NO_2 \qquad [12]$$

$$NO_2 + NO^{\bullet} \rightarrow N_2O_3 \qquad [13]$$

$$N_2O_3 + H_2O \rightarrow 2HNO_3 \qquad [14]$$

The primary reaction of N_2O_3 is nitrosation. Nitrosation of amines results in the formation of nitrosamines. Nitrosamines are carcinogenic and have deleterious consequences. Nitrosation of thiols results in the formation of nitrosothiols. On the other hand, nitrosation of thiols has been shown to have a variety of effects from inhibiting enzyme action to modifying signal transduction (34).

3.2. NO Reactions with Superoxide

The reaction of $NO^{\bullet}$ with superoxide results in the formation of peroxynitrite (^-ONOO) (Equation 15).

$$NO^{\bullet} + O_2^{\bullet-} \rightarrow {}^-OONO \qquad [15]$$

The formation of peroxynitrite is dependent on the concentrations of both NO and $O_2^{\bullet-}$. In healthy cells, the levels of $NO^{\bullet}$ are regulated by the activity of NOS and the presence of oxyhemoglobin. The levels of $O_2^{\bullet-}$ are regulated by the activity of superoxide dismutase. Generally, the formation of peroxynitrite is a detrimental reaction. Protonation of peroxynitrite leads to the formation of peroxynitrous acid (HOONO), which is thought to mediate most of the oxidation and nitration reactions mediated by peroxynitrite. However, in instances where NO or $O_2^{\bullet-}$ are in excess, reactions between peroxynitrite and either $NO^{\bullet}$ or $O_2^{\bullet-}$ can limit peroxynitrite activity (Equations 16 and 17) (34).

$$ONOO^- + NO^{\bullet} \rightarrow NO_2 + NO_2^- \qquad [16]$$

$$ONOO^- + O_2^{\bullet-} \rightarrow NO_2 + O_2 + NO^{\bullet} \qquad [17]$$

3.3. Interaction with Redox Metals

NO can interact directly with metal centers forming metal nitrosyl complexes (Equation 18). For example, the reaction of NO with proteins containing heme moieties such as guanylate cyclase, cytochrome P450 and NOS (34).

$$Fe^{2+} + NO \rightarrow Fe^{2+}\text{-}NO \qquad [18]$$

NO can also interact directly with metal-oxygen complexes like that seen with oxyhemoglobin (Equation 19) forming met hemoglobin and nitrate.

$$Hb(Fe\text{-}O_2) + NO \rightarrow \text{met } Hb(Fe^{3+}) + NO_3^- \qquad [19]$$

Lastly, NO can interact with high valent metal complexes formed from the oxidation of metal species or metal oxygen complexes by agents such as H_2O_2 (Equations 20 and 21).

$$Fe^{3+} + H_2O_2 \rightarrow Fe^{4+} = O + H_2O \qquad [20]$$

$$Fe^{4+} = O + NO \rightarrow Fe^{3+} + NO_2^- \qquad [21]$$

The addition of NO in this situation, results in the reduction of the hypervalent complex to a lower valent state, protecting tissues from peroxide-mediated damage.

3.4. NO Reactions with Radicals

$NO^{\bullet}$ has an unpaired electron, thus $NO^{\bullet}$ can react with other radical species. For instance, $NO^{\bullet}$ can react with the hydroxyl radical (Equation 22)

$$NO^{\bullet} + OH^- \rightarrow HNO_2 \qquad [22]$$

Alternatively, $NO^{\bullet}$ can react with NO_2 to form the reactive the nitrogen oxide species N_2O_3 (Equations 23 and 24).

$$NO_2 + NO^{\bullet} \rightarrow N_2O_3 \qquad [23]$$

$$N_2O_3 + H_2O \rightarrow 2HNO_3 \qquad [24]$$

Furthermore, alkyl, alkoxy, and alkylperoxide radicals have also been shown to react with $NO^{\cdot}$ (Equation 25) terminating lipid peroxidation chain reactions and protecting the cells against peroxide induced cytotoxicity.

$LOO^{\bullet} + NO \rightarrow ROONO$ [25]

4. Antioxidant Defense Mechanisms

Cellular antioxidant defenses can be separated into two types, enzymatic and non-enzymatic. Enzymatic defense systems include superoxide dismutase, glutathione peroxidases, and catalases. The non-enzymatic pathways include vitamin E, vitamin C and β carotene.

4.1. Enzymatic Defense Systems

4.1.1. Superoxide Dismutase (EC1.15.1.1)

Superoxide dismutases (SOD) are a family of metalloenzymes that catalyze the one-electron dismutation of $O_2^{\bullet-}$ to H_2O_2 (Equation 26).

$2O_2^{\bullet-} + 2H^+ \rightarrow H_2O_2 + O_2$ [26]

SOD provides the first line of enzymatic defense against intracellular free radical production by removing $O_2^{\bullet-}$. There are three types of SOD: copper/zinc SOD (Cu/Zn SOD, SOD1), manganese SOD (Mn SOD, SOD2), and extracellular SOD (EC-SOD, SOD 3).

Cu/Zn-SOD is a highly stable enzyme found in the cytosol and nucleus of all cell types (8). It is dimer composed of two 16 kDa subunits, each containing 1 copper and 1 zinc atom. The copper is essential for the enzyme's catalytic activity while the zinc stabilizes the protein's structure. In the normal rat lung, CuZn-SOD is expressed in alveolar epithelial cells, fibroblasts, and capillary endothelial cells (4). CuZn-SOD deficient mice develop normally to adulthood and show no apparent evidence of oxidative damage (30).

Mn-SOD is a homotetramer with a molecular weight of 88 kDa. Each subunit contains a manganese atom. Mn-SOD is made in the cytoplasm and is directed to the mitochondria by a signal peptide, were it dismutates $O_2^{\bullet-}$ generated by the respiratory chain. Mn-SOD has been localized to type II pneumocytes, alveolar macrophages and bronchial epithelium of the rat (4, 5) and bronchial epithelial cells of the human (7). Mn-SOD gene knockout mice all die within 10-21 days after birth from cardiomyopathy, neurodeneration, and metabolic acidosis (18, 19).

EC-SOD is a member of the CuZn-SOD family, and is produced in fibroblasts and glial cells and secreted into the extracellular fluid. EC-SOD mostly exists as a homotetramer glycoprotein with a molecular weight of 135 kDa. EC-SOD has an affinity for heparin, thus it exists in the vasculature bound to the surface of endothelial cells and to the extracellular matrix. It has been suggested that EC-SOD plays a role in modulating nitric oxide levels and nitric oxide bioavailability (24). Mice lacking the Sod3 gene have been generated and characterized. They appear to be normal when kept under normal laboratory conditions. However, when exposed to oxidative stress, survival is reduced with death due to pulmonary edema, consistent with the lung being the tissue with the highest amount of EC-SOD in mice (3). In the human lung, EC-SOD has been found in bronchial epithelial cells, alveolar type I cells, alveolar macrophages, chondrocytes, and endothelial cells (23).

4.1.2. Glutathione Peroxidase (EC 1.11.1.9)

Glutathione peroxidase (GP_X) is a homodimer with each 22 kDa subunit bound to a selenium ion. GP_X is located in both the cytosol and mitochondrial matrix in a 2:1 ratio. GP_X catalyzes the reduction of H_2O_2 and organic hydroperoxides to water and alcohol using glutathione (GSH) as the electron donor while GSH is oxidized to glutathione disulfide (GSSG) (Equations 27 and 28).

$$2GSH + H_2O_2 \rightarrow GSSG + 2H_2O \qquad [27]$$

$$2GSH + ROOH \rightarrow GSSG + H_2O \qquad [28]$$

GSSH is recycled to GSH through a reduction catalyzed by glutathione reductase in the presence of NADPH forming a redox cycle.

4.1.3. Catalase (EC 1.11.1.6)

Catalase (CAT) catalyzes the decomposition of H_2O_2 to H_2O (Equation 29).

$$2H_2O_2 \rightarrow 2H_2O + O_2 \qquad [29]$$

It shares this function with GP_X, but the substrate specificity and affinity and cellular location of these two antioxidant enzymes are different. CAT is a tetramer with a molecular weight of 240 kDa. CAT is mainly localized to peroxisomes, but mitochondria and other intracellular organelles such as endoplasmic reticulum may also contain some CAT activity. It is detectable in alveolar type II pneumocytes and macrophages (14). Heme (Fe^{3+}) is a required to be bound to the enzyme's active site for its catalytic functions. The primary function of CAT is to remove H_2O_2 produced in the peroxisomes due to enzymes such as flavoprotein dehydrogenase in the β-oxidation of fatty acids, urate oxidase, and the metabolism of D-amino acids.

4.2. Non-enzymatic Defense Systems

4.2.1. Vitamin E

Vitamin E (α tocopherol) is an important fat-soluable antioxidant. All cell membranes contain small levels of vitamin E with the majority concentrated in the inner mitochondrial membrane (21). Vitamin E quenches an electron from a free radical species and is converted to a vitamin E radical. The vitamin E radical can be reduced back to vitamin E via enzymatic or non-enzymatic processes by ascorbate or GSH.

4.2.2. Vitamin C

Vitamin C (ascorbate) is a water soluble vitamin present in the cytosol and extracellular fluid. It can interact directly with superoxide and hydroxyl radical. In addition, vitamin C can also donate an electron to the vitamin E radical. It is then oxidized to a semidehydroascorbate (SDA) radical. SDA is recycled by *i*) dihydrolipoate or *ii*) disproportionation reaction forming dehydroascorbate (DHA). In the presence of GSH, DHA reductase catalyzes the regeneration of ascorbate. Alternatively, SDA can be converted directly to ascorbate by SDA reductase.

4.2.3. β-Carotene

β-Carotene is a major carotenoid precursor of vitamin A. β-carotene inhibits lipid peroxidation initiated by oxygen or carbon centered free radicals.

5. ROS and Hypoxic Pulmonary Vasoconstriction

Pulmonary vasoconstriction in response to hypoxia (HPV) improves gas exchange efficiency by matching ventilation to perfusion. The main stimulus for HPV is alveolar hypoxia. The main site of HPV is the small pulmonary arteries (60-70 μm i.d.) (11). ROS are thought to play a role in signaling the changes in Po_2 in the pulmonary vasculature. Both increases (1) and decreases (31-33) in the formation of mitochondrial ROS have been proposed. The first model proposes that the decrease in available oxygen during hypoxia inhibits mitochondrial electron transport, resulting in an accumulation of reducing equivalents and a decrease in ROS production. Complex I (NADH: oxidoreductase) is thought to function as an oxygen sensor by maintaining basal levels of superoxide and hydrogen peroxide production during normoxia. In hypoxia, the level of ROS decreases and NADH levels increase, thus shifting the state of cytosolic redox toward more reduced levels (33). The second model suggests that a paradoxical increase in ROS generation occurs during hypoxia. Data do support this model: *i*) pharmacological antioxidants have been shown to block the hypoxic response in isolated buffered perfused lung suggesting an increase in ROS is required (32); *ii*) inhibition of CuZn-SOD to block hydrogen peroxide formation has been shown to abrogate HPV, suggesting that hydrogen peroxide is an important downstream signaling agent; *iii*) Inhibition of catalase augments the HPV response (22); and *iv*) hydrogen peroxide constricts the pulmonary circulation in normoxia (31). Taken together, this suggests that increased hydrogen peroxide is the signaling molecule involved in HPV. Unfortunately, experiments opposing both models have been reported, thus the role of ROS and the mechanism involved in sensing oxygen in HPV remains unresolved.

6. Oxygen, NO, Hemoglobin, and Blood Flow

Oxygen delivery to a tissue is determined by two factors: the oxygen content of the blood and the blood flow (28). The binding and release of oxygen is determined by the conformational state of hemoglobin (28). Systemically, blood flow is increased by hypoxia and decreased when oxygen supply exceeds demand (28). These responses are thought to be mediated by changes in the level of NO.

Hemoglobin is a tetramer composed to 2α and 2β subunits, with each subunit containing a heme. Each β subunit contains a reactive SH group at cysteine 93. NO has been shown to interact with both the heme and the thiol in hemoglobin (28). However, the binding of NO to the heme or the thiol of hemoglobin is dependent on the oxygen saturation of hemoglobin (10). There are two conformational states of hemoglobin: the oxygenated R state and the deoxygenated T state. In the oxygenated R state, S-nitrosylation of the reactive thiol at cysteine β93 is favored whereas, in the deoxygenated T-state, NO binds to the heme. One of the principle interactions between hemoglobin and the red blood cell (RBC) is through the chloride/bicarbonate anion exchange protein AE1 (25). Transnitrosation of the reactive thiol in AE1 allows the transfer of the NO group from the β chain of hemoglobin to that of the RBC membrane. Release of NO from this membrane by deoxygenation (R to T transition) results in vasodilatory activity. Thus, S-nitrosylated hemoglobin senses tissue Po_2 levels, and uses the allosteric transition from R to T to release NO and modify arteriolar tone thus matching oxygen demand to blood flow.

7. Oxygen, Hydroxylation, and Hypoxia-inducible Factor 1α

HIF-1 is a transcription factor whose activity is regulated by changes in the concentration of oxygen. This transcription factor is composed of two subunits, HIF-1α and HIF-1β. HIF-1α is the primary determinant of HIF-1 expression within the cell, whereas HIF-1β is constitutively expressed. Under normoxic conditions, HIF-1α is normally ubiquinated and targeted for destruction through interactions between the oxygen-dependent degradation (ODD) domain of HIF-1α and the tumor suppressor protein von Hippel Lindau (pVHL). In order for pVHL to interact with HIF-1α, proline 564 must be hydroxylated (13, 15). This critical event is mediated through a family of iron (II)-dependent prolyl hydroxylase enzymes that use oxygen as a substrate. Oxygen appears to be the rate limiting step for activity thus acting as a molecular switch. In addition, HIF-1α contains two transactivation domains responsible for recruiting transcriptional coactivators essential for gene expression. The C-terminal transactivation domain (C-TAD) operates independently of the ODD domain in recruiting the coactivator complexes p300/CBP under hypoxic conditions. Interestingly, the activity of C-TAD to recruit these coactivators is regulated by oxygen-dependent hydroxylation of asparagine (17); however the enzyme responsible for this modification has not, as of yet, been identified. It is interesting to note that both hydroxylation switches must be activated to fully induce HIF-1. One could speculate that multiple levels of regulation may allow for graded responses to small subtle changes in oxygen concentration.

8. Summary

Oxygen, reactive oxygen species, nitric oxide and reactive nitrogen species have been shown to be involved in regulating both physiological as well as pathological processes. The role oxygen and its derivatives in a living organism is dependent on the particular species formed, its concentration, location, and the presence of antioxidant machinery. Of course, oxygen is necessary in the production of energy. In addition, its derivatives play an important role in cellular signaling by acting as a molecular switch by altering ion channel activity, blood flow, and the stability of transcription factors.

References

1. Archer SL, Huang J, Henry T, Peterson D, an Weir EK. A redox-based O_2 sensor in rat pulmonary vasculature. *Circ. Res.* 1993; 73:1100-1112.
2. Berry C, Brosnan J, Fennell J, Hamilton CA, and Dominiczak AF. Oxidative stress and vascular damage in hypertension. *Curr. Opin. Nephrol. Hypertens.* 2001; 10:247-255.
3. Carlsson LM, Jonsson J, Edlund T, and Marklund SL. Mice lacking extracellular superoxide dismutase are more sensitive to hyperoxia. *Proc. Natl. Acad. Sci. USA.* 1995; 92: 6264-6268.
4. Chang LY, Kang BH, Slot JW Vincent R, and Crapo JD. Immunocytochemical localization of the sites of superoxide dismutase induction by hyperoxia in rat lungs. *Lab. Invest.* 1995; 73: 29-39.
5. Clyde BL, Chang LY, Auten RL, Ho YS, and Crapo JD. Distribution of manganese superoxide disumase mRNA in normal and hyperoxic rat lung. *Am. J. Respir. Cell Mol. Biol.* 1993; 8: 530-537.

6. Comhair SAA and Erzurum SC. Antioxidant Responses to oxidant-mediated lung diseases. *Am. J. Physiol. Lung Cell. Mol. Physiol.* 2001; 283: L246-L255.
7. Coursin DB, Cihla HP, Sempf J, Overley TD, and Oberley LW An immunohistochemical analysis of antioxidant and glutathione S-transferase enzyme levels in normal and neoplastic human lung. *Histol. Histopathol.* 1996; 11:851-860.
8. Crapo JD, Oury TD, Rabouille C, Slot JW, and Chang LY. Copper, zinc superoxide dismutase is primarily a cytosolic protein in human cells. *Proc. Natl. Acad. Sci. USA.* 1992; 89: 10405-10409.
9. Fridovich I. Fundamental Aspects of Reactive Oxygen Species, or what's the matter with oxygen? *Ann. NY Acad. Sci.* 1999; 893: 13-18.
10. Gow AJ and Stamler JS. Reactions between nitric oxide and hemoglobin under physiological conditions. *Nature* 1998; 391:169-173.
11. Gurney AM. Multiple sites of oxygen sensing and their contributions to hypoxic pulmonary vasoconstriction. *Respir. Physiol. Neurobiol.* 2002; 132:43-53.
12. Halliwell B and Gutteridge JMC. "Role of Free Radicals and Catalytic Metal Ions in Human Disease: an Overview." In *Methods in Enzymology* Volume 186 Oxygen Radicals in Biological Systems Part B Oxygen Radicals and Antioxidants, Packer L and Glazer AN eds. Academic Press Inc: 1990, pp1-88.
13. Ivan M, Kondo K, Yang H, Kim W, Valiando J, Ohh M., Salic A, Asara JM, Lane WS, and Kaelin WG. HIF1 Targeted for VHL Mediated destruction by Proline hydroxylation: Implications for O_2 sensing. *Science* 2001; 292: 464-467.
14. Kinnula VL, Crapo JD, and Raivio KO. Biology of disease: generation and disposal of reactive oxygen metabolites in the lung. *Lab. Invest.* 1995; 73: 3-19.
15. Jaakkola, P, Mole DR, Tian Y-M, Wilson MI, Gielbert J, Gaskell SJ, von Kreigsheim A, Hebestreit HF, Mukherji M, Schofield CJ, Maxwell PH, Puch CW, and Ratcliffe PJ. Targeting of HIFα to the von Hippel-Lindau ubiquitylation complex by O_2-regulated prolyl hydroxylation. *Science* 2001; 292: 468-472, 2001.
16. Jones RD, Hancock JT, and Morice AH. NADPH oxidase: a universal oxygen sensor? *Free Radic. Biol. Med.* 2000; 29: 416-424.
17. Lando D, Peet DJ, Whelan DA, Gorman JJ, and Whitelaw ML. Asparagine hydroxylation of the HIF Transactivation Domain: a hypoxic switch. *Science* 2002; 295: 858-861.
18. Lebowitz RM, Qhang H, Vogel H, Cartwright J Jr, Dionne L, Lu N, Huang S, and Matzuk MM. Neurodegeneration, myocardial injury, and perinatal death in mitochondrial superoxidase-deficient mice. *Proc. Natl. Acad. Sci. USA.* 1996; 93: 9782-9787.
19. Li Y, Huang TT, Carlson EJ, Melov S, Ursell PC, Olson JL, Noble LJ, Yoshimura MP, Berger C, Chan PH, Wallace DC, and Epstein CJ. Dilated cardiomyopathy and neonatal lethality in mutant mice lacking manganese superoxide dismutase. *Nat. Genet.* 1995; 11: 376-381.
20. López-Barneo J, Pardal R, and Ortega-Sáenz P. Cellular mechanisms of oxygen sensing. *Annu. Rev. Physiol.* 2001; 63: 259-87.
21. Matsuo M and Kaneko T. "The Chemistry of Reactive Oxygen Species and Related Free Radicals, " In *Free Radicals in Exercise and Aging*, Radak Z ed. Human Kinetics 2000, pp 1-34.
22. Monaco JA and Burke-Wolin T. NO and H_2O_2 mechanisms of guanylate cyclase activation in oxygen-dependent responses of rat pulmonary circulation. *Am. J. Physiol.* 1995; 268: L546-L550.
23. Oury TD, Chang L-Y, Marklund SL, Day BJ, and Crapo JD. Immunocytochemical localization of extracelular superoxide dismutase in human lung. *Lab. Invest.* 1994; 70: 889-897.
24. Oury TD, Day BJ, and Crapo JD. Extracellular superoxide dismutase; a regulator of nitric oxide bioavailability. *Lab. Invest.* 1996; 75: 617-636.
25. Pawloski JR, Hess DT, and Stamler JS. Export by red blood cells of nitric oxide bioactivity. *Nature* 2001; 409: 622-626.
26. Pou S, Pou WS, Bredt DS, Snyder SH, and Rosen GM. Generation of superoxide by purified

brain nitric oxide synthase. *J. Biol. Chem.* 1992; 267: 24173-24176.

27. Stamler J. Redox signaling: nitrosylation and related target interaction of nitric oxide. *Cell* 1994; 78:931-936.
28. Stamler JS, Jia L, Eu JP, McMahon TJ, Demchenko IT, Bonaventura J, Gernert K, and Piantadosi CA. Blood flow regulation by S-nitrosohemoglobin in the physiological oxygen gradient. *Science* 1997; 276: 2034-2037.
29. Squadrito GL and Pryor WA Oxidative chemistry of nitric oxide: The roles of superoxide, peroxynitrite and carbon dioxide. *Free Radic. Biol. Med.* 1998; 392-403.
30. Tsan MF. Superoxide dismutase and pulmonary oxygen toxicity: lessons from transgenic and knockout mice. *Int. J. Mol. Med.* 2001; 7: 13-19.
31. Waypa GB, Morton CA, Vincent PA, Mahoney JR, Johnston WK III, and Minnear FL. Oxidant-increased endothelial permeability: prevention with phosphodiesterase inhibition vs. cAMP production. *J. Appl. Physiol.* 2000; 88: 835-842.
32. Waypa GB, Chandel NS, and Schumacker PT. Model for hypoxic pulmonary vasoconstriction involving mitochondrial oxygen sensing. *Circ. Res.* 2001; 88: 1259-1266.
33. Waypa GB and Schumacker PT. O_2 sensing in hypoxic pulmonary vasoconstriction: the mitochondrial door re-opens. *Respir. Physiol. Neurobiol.* 2002; 132: 81-91.
34. Wink DA and Mitchell JB. Chemical biology of nitric oxide: Insights into regulatory, cytotoxic and cytoprotective mechanisms of nitric oxide. *Free Radic. Biol. Med.* 1998; 25: 434-456.

Chapter 14

Interaction of Oxidants With Pulmonary Vascular Signaling Systems

Sachin A. Gupte and Michael S. Wolin
New York Medical College, Valhalla, New York, U.S.A.

1. Introduction

Oxidants or free radicals were initially considered as highly reactive, intermediates or unstable species in chemical reactions. During the 1970's it was realized that free radicals were involved in "bacterial killing" by phagocytic cells and radiation therapy for cancer, and that xanthine oxidase and sites in the electron transport chain of mitochondria were biological sources of reactive oxygen species (ROS). Initially, ROS were considered as cytotoxic species, and implicated in mediating tissue injury by virtue of their reactivity with biological molecules. Evidence accumulated suggesting that elevated ROS rapidly react with cellular lipids, proteins, and DNA, and subsequently oxidants have now been associated with progression of various diseases, including hypertension, acute lung injury, acute respiratory distress syndrome (ARDS), stroke, renal failure, and ischemic heart disease. However, with the exception of consideration of a role for hydrogen peroxide (H_2O_2) in the mechanism of action of insulin, and in the regulation of soluble guanylate cyclase (sGC), the importance of ROS in biological regulation was not initially explored. After ROS, carbon-, and nitrogen (reactive nitrogen species; RNS)-centered radical species were detected in a variety of biological systems, an extensive study of their metabolic chemistry and physiological importance has recently begun (34-36).

Paradoxically, ROS are generated from molecular oxygen (O_2) in a manner that is of fundamental importance for cellular regulation. Multiple roles for ROS and RNS in regulating intracellular signaling pathways, and modulating cellular and organ function have recently been recognized. We are just beginning to appreciate their role as second messengers in regulating signaling pathways that potentially modulate vascular tone, carotid body function, thrombosis, inflammation, ischemia/reperfusion injury, receptor function, respiration, metabolism, growth, as well as, cell death (15, 24, 34, 35). A growing body of evidence also suggests that oxidant-elicited signaling plays a potentially important role as an O_2 sensor, which subsequently activates and/or inhibits sGC, and ion channel pathways that participate in promoting the genesis of pulmonary vasoconstriction and hypertension (33). Thus, evidence is evolving to establish fundamental signaling roles for ROS as primary regulators of intracellular Ca^{2+}, membrane potential control through K^+ channel regulation, phospholipid metabolism, and cAMP and cGMP mediated regulation in physiological and pathological conditions.

In this chapter, we will focus on the cellular sources of ROS and RNS involved in pulmonary vascular control, conditions during which ROS and RNS are elevated, and highlight the key cellular signaling systems regulated by ROS and RNS in normal and disease conditions. Furthermore, we will review the roles of ROS and RNS in modulating vascular smooth muscle cell (VSMC) signaling pathways, with an emphasis on pulmonary vascular signaling systems.

2. Oxidants Species and their Scavengers in the Pulmonary Vascular System

In the pulmonary artery, ROS are produced by all the major cell-types in the cytosol, mitochondria, peroxisomes, and extracellular region, which contribute to the local levels of each species. Although the production and metabolism of ROS are independent of each other and compartmentalized, transport of individual ROS across compartments is often possible. ROS and free radicals generally diffuse to all regions of the cell when their production exceeds the capacity of local scavenging systems. Depending upon their concentration in the milieu of a cell, they either trigger intracellular signaling or cause oxidative stress. For example, it was initially reported that escalated production of ROS participated in mediating acute lung injury caused by inflammatory stimuli, ischemia-reperfusion, and ARDS (6, 16). A potential role for ROS in pulmonary vascular signaling was subsequently proposed by Archer and Weir, who suggested that diminished ROS generation during hypoxia lead to closing of K^+ channels resulting in membrane depolarization and vasoconstriction (2). The detection of a H_2O_2-elicited relaxing mechanism in pulmonary artery (PA), which appeared to be mediated by stimulation of sGC through peroxide metabolism by catalase, was also suggested to be a mechanism suppressed by hypoxia (4). Recent studies have provided evidence that mitochondrial-derived ROS are significantly higher in the pulmonary artery smooth muscle cells (PASMC) as compared to the renal artery smooth muscle cells at basal conditions, and that these ROS can diffuse to the cytoplasm and cause vasodilation (19). While this study has shown that during hypoxia ROS generation is inhibited, others have demonstrated that ROS diffuse out into the cytoplasm during hypoxia and activate a cascade of signaling systems in PASMC leading to the hypoxic pulmonary vasoconstriction (HPV) (29). Thus, even though the actual sources of ROS in PASMC is uncertain, and it is unresolved whether ROS levels go up or down during hypoxia at physiological levels, the role of ROS in regulating intracellular signaling mechanisms is currently under scrutiny, and is of immense importance in vascular biological processes ranging from physiological responses to the alterations observed in pulmonary vascular diseases (33-35).

The production of ROS begins with the formation of superoxide anion ($O_2^{\bullet-}$) resulting from one-electron reduction of O_2 by various oxidases (Equation 1) that occur in biological systems:

$$O_2 - e \rightarrow O_2^{\bullet-} \quad [1]$$

Vascular tissue contains copper-zinc superoxide dismutase (CuZn-SOD), a mitochondrial manganese form of SOD (Mn-SOD), and an extracellular SOD (eSOD). In addition, metallothionein is expressed in pulmonary endothelium. These enzymes normally scavenge and detoxify $O_2^{\bullet-}$. In vascular tissue, the levels of $O_2^{\bullet-}$ are thought to be in the nanomolar range in the absence of SOD, and the presence of SOD is likely to lower $O_2^{\bullet-}$ concentrations into the picomolar range. However, under extreme conditions such as infection or inflammation, extracellular $O_2^{\bullet-}$ rises up to near-micromolar levels. At physiological pH, $O_2^{\bullet-}$ is both a free radical (because it contains an unpaired electron) and a negatively charged species (as a result of its pKa of 4.8). $O_2^{\bullet-}$ reacts with itself with a rate constant of 8×10^{-4} $mol^{-1}\cdot L\cdot s^{-1}$ to form H_2O_2 and O_2. SOD enzymes function to accelerate the removal of $O_2^{\bullet-}$ as a result of their rate constant of 2×10^{9} $mol^{-1}\cdot L\cdot s^{-1}$ for the reaction with $O_2^{\bullet-}$ (Equation 2):

$$O_2^{\bullet-} + 2H^+ \rightarrow O_2 + H_2O_2 \quad [2]$$

Peroxide is either derived from $O_2^{\bullet-}$ through Equation 2, or it is directly produced by certain oxidases through a two-electron reduction of oxygen (O_2). H_2O_2 is efficiently metabolized to H_2O and O_2 by catalase and only to H_2O by glutathione peroxidase. Intracellular levels of H_2O_2 are maintained at a concentration of ~1 nM. Although the actual local extracellular levels of H_2O_2 are not known with certainty, it is reported that levels are in the 100 nM range. H_2O_2 is a relatively stable species, and is not a free radical. Generally increased H_2O_2 readily reacts with transition metals like ferrous iron to form highly reactive hydroxyl radical-like species ($OH^{\bullet}$) through Haber-Weiss and Fenton type reactions (Equation 3 and 4):

$$O_2^{\bullet-} + H_2O_2 \rightarrow O_2 + HO^- + HO^{\bullet} \quad [3]$$

$$Fe^{3+} + O_2^{\bullet-} \rightarrow Fe^{2+} + O_2 \qquad Fe^{2+} + H_2O_2 \rightarrow Fe^{3+} + HO^- + HO^{\bullet} \quad [4]$$

The highly reactive nature of $OH^{\bullet}$ initiates lipid peroxidation of the cell membrane by initially creating lipid free radicals, as seen during ARDS (16). Formation of RNS through the interaction between nitric oxide (NO) and $O_2^{\bullet-}$ results in the production of additional reactive species. Normally, SOD prevents the reaction of $O_2^{\bullet-}$ and NO, as is shown in Equation 5. Because $O_2^{\bullet-}$ reacts with NO at a rate constant of 7×10^9 $mol^{-1} \cdot L \cdot s^{-1}$, which is three times the rate of its reaction with SOD, when the levels of NO increase into the nanomolar range it competes with SOD for $O_2^{\bullet-}$. This results in the production of peroxynitrite ($ONOO^-$; see Equation 5):

$$O_2^{\bullet-} + NO \rightarrow ONOO^- \qquad [5]$$

It is estimated that for every 10-fold increase in $O_2^{\bullet-}$ and NO, there is 100-fold increase in $ONOO^-$ formation, and $ONOO^-$ is a source of additional reactive free radical species such as nitrogen dioxide ($^{\bullet}NO_2$) (34).

3. Sources of ROS in Pulmonary Tissue

$O_2^{\bullet-}$ is formed in lungs and can be detected on the surface of pulmonary tissue and isolated PA under basal conditions. Pulmonary vascular tissue contains multiple oxidases whose expression and activity is highly regulated (Table 1). Pulmonary endothelial cells, PASMC, and macrophages generally have major $O_2^{\bullet-}$ producing enzymes; *vis-à-vis* NAD(P)H oxidase, which contain b_{558}-type cytochromes and subunits resembling the phagocytic cell NADPH oxidase system are present in the vascular tissue. For example, there is evidence for presence of $p22^{phox}$, $p47^{phox}$, $p67^{phox}$ and homologues of the NOX-2 ($gp91^{phox}$ subunit), currently termed NOX-1 and NOX-4 (10). In calf PASMC from small diameter PA, this enzyme is active under basal physiological conditions and there is evidence showing that it is activated under hypoxia (6, 33).

In contrast to the phagocytic oxidase, which uses NADPH for a co-factor, the PASMC generates significant levels of $O_2^{\bullet-}$ through the action of a microsomal-bound NADH oxidoreductase (22). Xanthine dehydrogenase, another NAD(P)H-dependent enzyme, is also present in the pulmonary endothelial cells, and is major source of $O_2^{\bullet-}$, during prolonged hypoxic conditions (6). Other proteins, like NO synthase (NOS), cytochrome P450, cyclooxygenase appear to be minor sources of $O_2^{\bullet-}$ under baseline physiological conditions. Previously discussed studies (2, 19, 29) suggest that the mitochondrial electron transport chain is an important Po_2-regulated source of ROS in pulmonary tissue. Many receptor-linked regulatory mechanisms are capable of activating $O_2^{\bullet-}$ production through the oxidases present in the vessel wall (See Table 1).

Table 1. The Sources of $O_2^{\bullet-}$ in Pulmonary Artery and Their Stimulators

Sources of $O_2^{\bullet-}$	Cell types	Stimulators
NAD(P)H oxidase-NOX1, NOX2, and NOX4	Endothelium, Vascular smooth muscle, fibroblasts	A-II, endothelin-1, TNFα, TxA_2, Overexpression of NOX and p22*phox* ΔPO_2
Xanthine oxidase	Endothelium	Hypoxia-reoxygenation
Mitochondria	All cell types	ΔPO_2, apoptotic stimuli?
Nitric oxide synthase	Endothelium, Mitochondria, Vascular smooth muscle	Endothelial activation, Loss of L-arginine, Deficiency of tetrahydrobiopterin
Cyclo-oxygenase	Endothelium	Arachidonic acid release, Increase in peroxide tone
Cytochrome P450	Endothelium	Endothelial activation?

4. Regulation of the Pulmonary Vascular Signaling System by ROS and RNS

Cells are normally protected from the cytotoxic effects of $O_2^{\bullet-}$ by SOD, but when elevated, $O_2^{\bullet-}$ can either be dismutated to produce H_2O_2, or it can interact with NO to produce $ONOO^-$. An over production of free radicals exert cytotoxic effects, but these oxidants species can also interact with cell-control mechanisms and potentially contribute to signaling processes. As shown in the Figure 1, certain signaling systems are very sensitive to low levels of $O_2^{\bullet-}$ and H_2O_2 normally produced by cells, and these mechanisms are likely to participate in physiological regulation. In addition, other signaling systems activated by conditions generating higher levels of oxidants are generally associated with the mechanisms contributing to pathophysiological processes and cellular responses to oxidative stress.

Mechanical changes can influence $O_2^{\bullet-}$ production. For example, alterations in the shear stress caused by increased blood flow promote endothelium-derived $O_2^{\bullet-}$ generation in the pulmonary circulation (6). Local levels of $O_2^{\bullet-}$ are also elevated during lung injury or distress conditions by an imbalance between its rate of formation and its rate of removal via SOD. When the concentrations of $O_2^{\bullet-}$ rise over the nanomolar concentration range, it directly interacts with catecholamines (including the neurotransmitter norepinephrine and the hormone epinephrine) and NO, which inactivates the bioactivity of these mediators. $O_2^{\bullet-}$ inhibits vascular relaxation to H_2O_2, which may be a result of $O_2^{\bullet-}$ inhibiting the activity of catalase. $O_2^{\bullet-}$ inactivates glutathione peroxidase by oxidizing selenium and aconitase by interacting with iron-sulfur center in the enzyme resulting in increased oxidant stress and an impairment of mitochondrial function. Increased levels of $O_2^{\bullet-}$ reductively release iron from the iron-sulfur centers and from ferritin, and the reaction of the released iron with other ROS and RNS result in the modulation of other signaling processes or in the initiation of tissue injury. Signaling systems most sensitive to changes in intracellular levels of H_2O_2 are linked to the metabolism of H_2O_2 by cyclooxygenase, catalase and GSH peroxidase. Accumulating evidences suggest that H_2O_2 increases the levels of oxidized GSH and promotes S-thiolation of proteins (disulfide formation of protein thiols with GSH) that relays intracellular signaling (6, 34-36). A new concept is that endothelium cytochrome P450-derived H_2O_2 may function to open K^+ channels and induce relaxation through hyperpolarization of the underlying VSM cells through gap junctions.

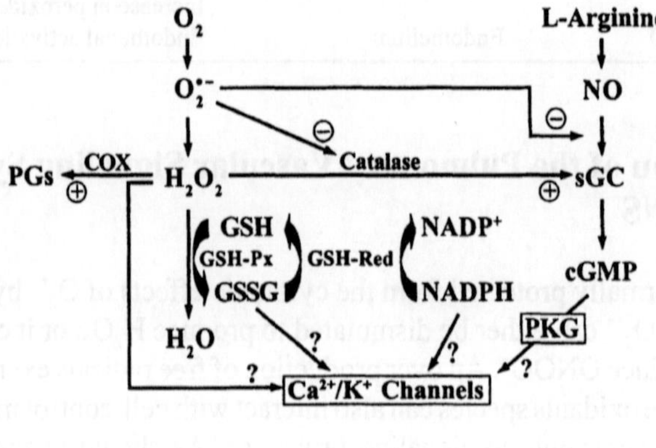

Figure 1. A schematic representation of the interactions of radicals, oxidants and redox changes with the signaling systems. Question marks indicate that the final pathway that changes the channel activity is still uncertain.

It is now well established that endothelium- and perhaps other sources such as smooth muscle- and mitochondrial-derived NO play pivotal roles in the regulation of pulmonary circulation (3, 17, 25). As such, NO is critically important in modulating pulmonary vascular function and inhibiting pulmonary vascular remodeling. NO induces relaxation of PA through activation of sGC-cGMP pathway and opening of big conductance calcium-activated K^+ channels (1). It has also been proposed that NO potentiates the Ca^{2+} uptake through sarcoplasmic reticulum (SR) Ca^{2+}-ATPase and promotes cGMP-independent relaxation of bovine PA under hypoxic conditions (21). In contrast to micromolar concentrations of H_2O_2 which stimulate the release of endothelium NOS-derived NO, $O_2^{\bullet}$- destroys NO, antagonizes endothelium-dependent relaxation, and has been reported to promote contraction of PA due to the formation of $ONOO^-$. High levels of $ONOO^-$ are cytotoxic, but low concentrations of $ONOO^-$ may participate in signaling-like mechanisms including PA relaxation, the inhibition of catalase and H_2O_2-elicited cGMP-mediated relaxation of PA, protein tyrosine nitration that may interfere with phosphorylation/dephosphorylation signaling pathways or alter protein functions, and based on studies in skeletal muscle it nitrosylates and inhibits the activity of SR Ca^{2+}-ATPase. Therefore, it is plausible that Po_2 governs the oxidative state of NO and modulates the action(s) of NO on the control of intracellular Ca^{2+} homeostasis, which ultimately regulates vasomotor tone and blood flow that participate in circulatory Po_2-elicited responses. Thus conditions that promote $ONOO^-$ formation could influence both acute blood flow responses linked to changes in Po_2, and the evolution of pulmonary vascular disease.

The sGC-derived second messenger cGMP is a key regulator of vascular relaxation. Both exogenously and endogenously produced NO and H_2O_2 promote relaxation through cGMP-mediated pathways. The H_2O_2 appears to activate sGC by two different pathways (Figure 1): *i*) H_2O_2 stimulates the production of NO by NOS in endothelium, and *ii*) vascular smooth muscle sGC activity is also increased by a mechanism involving the metabolism of H_2O_2 by catalase. Stimulation of sGC by the metabolism of peroxide by catalase occurs at high picomolar to low nanomolar concentrations of H_2O_2 and it promotes relaxation of the PA without changes in GSH redox (34). However, the specific pathways through which H_2O_2 and cGMP cause relaxation of pulmonary arteries remain to be defined. No one has yet studied in detail the role of ROS or RNS in modulating the protein kinase G pathway (cGMP-dependent protein kinase), although preliminary results have suggested that RNS and hypoxia may modulate protein kinase G activity in PASMC (25). Both $OH^{\bullet}$ and $O_2^{\bullet-}$ inhibit the stimulation of sGC by either NO or H_2O_2 (34). In fact endogenously $O_2^{\bullet-}$ generated originating from NADH oxidase inhibits sGC in bovine PA after inactivation of Cu-Zn SOD with diethyldithiocarbamate by impairing NO-elicited relaxation (34, 36). Thus, ROS have several mechanisms through which they can control the activity of sGC and relaxation of blood vessels mediated by cGMP.

Vascular phospholipases, including PLA_2, PLC, and PLD, are stimulated by ROS. Although the signaling pathways elicited by ROS, which control the activity of phospholipases, are not well understood, there is evidence that cytosolic PLA_2 is stimulated by H_2O_2 in endothelial cells via phosphorylation by tyrosine kinase-, mitogen-activated protein kinase (MAPK)-, and protein kinase C (PKC)-dependent pathways (34). Low levels of H_2O_2 and $ONOO^-$ stimulate the activities of lipoxygenase and cyclooxygenase enzymes, in contrast high levels of H_2O_2 have been suggested to inactivate cyclooxygenase and prostacyclin synthase, and increase thromboxane synthesis (13). The pulmonary vasoconstriction in pig and rabbit lungs caused by $O_2^{\bullet-}$ appears to be mediated by thromboxane, leukotrienes, and prostaglandins (6). Additionally, NO and $ONOO^-$ potentially interact with and regulate the activities of the heme-iron containing cytochrome P450, and the levels of 20-HETE production, which is a dilator of human PA that causes a prominent inhibition of vasoconstriction in lungs elicited by hypoxia (38).

Observations demonstrating the potential importance of oxidant regulation of ion transport mechanisms are slowly accumulating. It has been shown that elevated levels of H_2O_2 cause K^+ channel-dependent relaxation of VSM (19, 33, 37). Although the actual relationship between ROS or RNS and the function of cellular ion channels are not well characterized, evidence exists for the potential importance of several processes. A cysteine moiety in the α- or β-subunit of plasma membrane K_V channels have been shown to control the opening and closing of access to the channel by a "ball and chain" mechanism, thereby regulating channel activity. It has been shown that oxidation of the cysteine thiol group by ROS or RNS results in the opening of voltage-gated K^+ channels, namely $K_V1.1$, $K_V1.4$, $K_V1.5$, and $K_V2.1$, leading to hyperpolarization of the plasma membrane and relaxation of PA. Likewise, there are reports suggesting that Ca^{2+}-regulated K^+ channels mediate the vasodilation elicited by ROS in the rat pulmonary circulation (32, 33, 37). Oxidation of thiol molecules on the α_1-subunit of L-type Ca^{2+} channel protein by ROS or oxidizing agents also inactivates these channels by modulating voltage-gating and the influx of Ca^{2+} currents, which would result in relaxation of VSMC. Conversely, inward Ca^{2+} currents through non-specific cationic channels are increased due to oxidation of thiol groups on the channel proteins by free radicals. Other signaling mechanisms that are regulated by radical species, such as cAMP- and cGMP-dependent processes, control the function of ion channels in protein kinase-dependent or -independent manner. As $O_2^{\bullet-}$ modulates Na^+/H^+ antiport and Na^+/K^+ pump activity in cultured pulmonary artery endothelial cells (6), an alteration in ion pump activity could potentially regulate the ion channels on the plasma membranes of these cells. Furthermore, L-type Ca^{2+} channels are indirectly activated by phosphorylation of the channel protein through PKC, which is known to be stimulated by ROS, and this could cause an increase Ca^{2+} influx and VSMC dysfunction.

The mechanisms that control the uptake and/or release of SR Ca^{2+} are also sensitive to oxidants. $O_2^{\bullet-}$ appears to inhibit the breakdown of inositol 1,4,5-triphosphate and promote the release of Ca^{2+} from the SR of bovine VSMC. In contrast, although high levels of $O_2^{\bullet-}$ and H_2O_2 selectively disrupt SR Ca^{2+}-ATPase possibly through the irreversible oxidation of thiol groups or by directly attacking the ATP-binding sites, and there is evidence suggesting that $O_2^{\bullet-}$ augments the up-take of Ca^{2+} into the SR of bovine PASMC (6, 36). Thus, ROS induce different effects on the activity of cation channels, which produces diverse responses in VSMC. To add to this complexity arachidonic acid derived from phospholipid breakdown, prostaglandins derived from cyclooxygenase, and 15- and 20-HETE derived from cytochrome P450 following stimulation by ROS or RNS are also potent modulators of ion channel activity. As multiple vascular ion channels are potentially controlled by variety of redox- and oxidant-linked mechanisms, this could likely be the origin of diverse responses of different vascular beds to ROS or RNS.

Redox status of cytosolic NADP(H), NAD(H), and GSH are tightly controlled in vascular and non-vascular cells, and changes in the redox status of these systems are also potentially linked to controlling the expression of redox-regulated signaling mechanisms. It is currently thought that NADPH and GSH are major reducing systems in cells, and that the majority of cytosolic NADPH is maintained in its reduced form by the pentose phosphate pathway. Recent reports suggest that loss of control of NADPH/NADH/GSH redox potential, which is generally associated with ROS metabolism, is one of the important mediators of signaling systems in blood vessels. An example of this mediation are observations that the elevation of $NADP^+/NAD^+/GSSG$ in pulmonary and aorta VSMC opens K^+ channels and attenuates the release of Ca^{2+} from the SR associated with causing relaxation of PA and aorta (11). These redox changes may affect ion channel activities by GSSG-mediated S-thiolation of the channel protein or by direct interaction of $NADP^+/NAD^+$ with recently discovered nucleotide-binding sites on the $K_V1.4$ channel protein (23). Loss of NADPH also inactivates L-type Ca^{2+} currents in cardiac myocytes and suppresses myocardial contractility (12). In addition, NADPH-linked oxidoreductase systems are involved in the synthesis of NO and in preventing the oxidation of the heme and/or thiols on sGC, inactivating sGC and blocking cGMP-mediated NO-elicited relaxation of pulmonary vessels. NAD(H) redox has also been implicated in the control of $O_2^{\bullet-}$ production through its role as a substrate for NADH oxidase and interactions with several signaling pathways (34, 36). Thus, the redox status of NADP(H), NAD(H), and GSH is likely to have a major influence on the function of multiple oxidant-associated signaling mechanisms, which regulate vascular function. Imbalances in these redox systems are now emerging as factors, which have serious consequences on the signaling pathways that control vasomotor tone, and alterations in these systems maybe a formidable cause of dysfunction in various pulmonary vascular diseases.

Multiple components of protein phosphorylation cascades are altered by ROS and RNS. Modification of an essential thiol at the catalytic site of tyrosine-specific protein phosphatases by either ROS or RNS appears to inhibit the in phosphatase activities, and as a consequence, tyrosine kinase-related signaling pathways becomes activated (34). Thus, the stimulation of tyrosine phosphorylation by H_2O_2 activates most forms of PKC in a diacylglycerol-independent manner that leads to stimulation of numerous PKC-dependent vascular signaling pathways. ROS also activate the various MAPK systems that can potentially regulate vascular force generation, proliferation, and adaptive responses to injury (34). In pulmonary arteries, PKC has been reported to mediate vasoconstriction caused by increased $O_2^{\bullet-}$ (6). Interestingly, a recent study suggested that H_2O_2-induced activation of p42/p44 MAPK mediates the force generation of previously stretched PA (36). The p42/p44 MAPK pathway has been shown to stimulate a receptor-activated Ca^{2+}-independent contractile response in VSMC through the phosphorylation of caldesmon (7). Thus, ROS and RNS have multiple ways of interacting with processes that modulate the Ca^{2+}-sensitivity of the contractile apparatus, and augment phosphorylation of myosin light chains. The phosphorylation of proteins involved in the mitogenic processes, which are normally activated by receptor-regulated signaling systems, are also generally enhanced by ROS (24, 34). Some of the important interactions of ROS with signaling pathways that are potentially involved in regulating PA function during HPV are discussed in the following sections.

5. Role for ROS and RNS in Hypoxic Pulmonary Vasoconstriction and Pulmonary Hypertension

For over one hundred years, it has been known that during hypoxia or low Po_2 the systemic arteries dilate to supply more oxygen to the deprived organ, however, on the contrary the pulmonary arteries constrict to balance the ratio of perfusion to ventilation. This phenomenon is generally termed as acute hypoxic pulmonary vasoconstriction. If HPV is prolonged it leads to pulmonary hypertension (PH). PH is characterized by pulmonary arterial vasoconstriction and remodeling. Under physiological conditions, the pulmonary arterial circulation is a high-compliance, low-pressure, low-resistance system, which carries blood into pulmonary microcirculation for gas exchange. Pulmonary artery pressure varies with age, from early childhood to the sixth decade of life, where the upper limit of normal pulmonary arterial pressure is regarded as 20 mmHg. Based on this observation, PH generally is defined as a mean pulmonary artery pressure greater than 25 mmHg at rest or 30 mmHg during exercise. Details of PH have been discussed in other chapters; hence we will overview the roles of oxidants in regulating HPV and PH in the next section.

Although the mechanisms responsible for the opposite responses of systemic and pulmonary blood vessels to lowering of Po_2 are still not clearly understood, several working hypotheses have been proposed. Changes in the metabolism, loss of autacoids synthesis, and alterations in the ion channel activity has long been considered as a cause of HPV and development of PH (17, 33). As depicted in Figure 2, studies have suggested that generation in PASMC of oxidase- or mitochondrial-derived ROS also play a role in sensing O_2 tension and in the genesis of HPV (33). While the production of ROS in most arteries ceases during hypoxia or graded changes in Po_2 tension, a few studies have shown that mitochondrial-derived ROS generation detected by dichlorofluorescein diacetate method is paradoxically elevated in the PASMC, thus facilitating rise in the $[Ca^{2+}]_i$ levels, which elicits PA contraction (29, 30). In contrast, a recent study has demonstrated that basal high level of ROS in normoxic condition is significantly reduced in the PA during hypoxia, while it is elevated in the renal artery (19). As ROS levels in the cytosol appear to regulate the gating of K_V channels (32, 37), it has been proposed that loss of mitochondrial-derived ROS production during hypoxia blocks the outward $K_V1.5$ currents and causes an increase in $[Ca^{2+}]_i$ in PASMC thereby contracting the PA under low Po_2 conditions (19). Furthermore, inhibition of mitochondrial metabolism by dichloroacetate prevents and reverses PH by restoring expression and function of $K_V2.1$ channels (20). Inhibition of another glucose metabolic pathway, namely the pentose phosphate pathway, has also been shown to prevent HPV and PH by opening K_V channels (11).

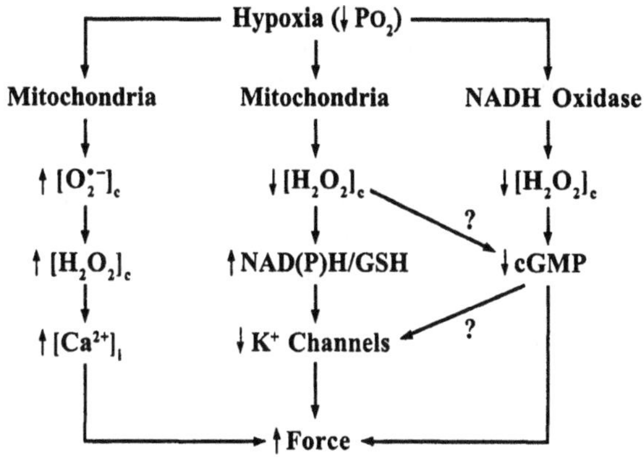

Figure 2. Scheme showing the three main proposed hypothesis for redox mechanisms involved in the development of acute hypoxic pulmonary vasoconstriction (HPV) that are currently under investigation. Although all three hypotheses propose a role for changes in cytosolic hydrogen peroxide $[H_2O_2]_c$ or superoxide $[O_2^{\bullet-}]_c$ derived from either mitochondria or cytosolic NAD(P)H oxidase, the downstream signaling pathways that affects force generation are not yet clearly understood.

Besides the regulation of K^+ channel activity, changes in Po_2 also inactivate the L-type Ca^{2+} channels in VSMC, which critically contributes to the local regulation of systemic circulation (9). In small PA, hypoxia elicits contraction by a mechanism which increases Ca^{2+} conductance (14). However, a direct evidence for Po_2 modulating voltage-gated Ca^{2+} channel activity in PASMC indicates longitudinal differences in Ca^{2+} channel density and O_2 sensitivity in myocytes from proximal and distal PA. For example, Ca^{2+} channels from proximal PA myocyte close and distal PA myocyte open to reduction in Po_2 (9). Although the role of NAD(P)H oxidase-derived ROS in Po_2 sensing and regulation of PA tone is uncertain, $O_2^{\bullet-}$ produced by non-phagocytic NAD(P)H oxidase during prolonged hypoxia appears to contribute to the constriction of PA in rabbit lungs (33). On the contrary, NADH oxidase-derived H_2O_2 has been shown to be increased in normoxia relative to hypoxia, and to cause PA relaxation through an increase in cGMP (4, 22, 35). Whereas, studies on gp91^{phox} knockout mouse indicate that the phagocytic NAD(P)H-oxidase does not appear to be an important regulator of pulmonary vascular function (33). The role of other NOX-type NAD(P)H oxidases have not yet been examined. Prolonged hypoxia also activates xanthine oxidase and this activation is a cause for increments in $O_2^{\bullet-}$ levels in the PA (6). Since ROS generation and PKC activation, the latter being stimulated by oxidants, have been reported to occur simultaneously during hypoxia (6), this implies that PKC may also be involved in initiating HPV through various mechanisms including activation of L-type Ca^{2+} channels.

As discussed above, redox changes and through their influence on ROS interactions with NO, the regulation of prostaglandins and endothelin-1 synthesis, the modulation of the activation states of K_V and Ca^{2+} channels, and the control of Ca^{2+} pump activity by ROS, are able to regulate processes including Ca^{2+} homeostasis, vascular tone and development of HPV. Loss of endothelial NO production in lungs and isolated PA has been associated with the development of HPV (17). Conditions that elevate the production of $O_2^{\bullet-}$, such as hypoxia, may inactivate NO in pulmonary vascular tissue associated with $ONOO^-$ formation and result in oxidant stress including decreased intracellular glutathione levels (25). Besides scavenging NO (which inhibits vascular remodeling), ROS and RNS may deactivate K^+ channels and activate hypoxia inducible factor (HIF) (28), both of which are well known to participate in cell proliferation and growth. Additionally, endothelin-1 stimulates PASMC proliferation through induction of $O_2^{\bullet-}$ under conditions that increase NAD(P)H oxidase activity (31). Therefore, ROS can further participate in offsetting cell proliferation and apoptosis processes, and promote the remodeling of VSM and progression of PH (Fig. 3).

6. Regulation of Pulmonary Vascular Remodeling by ROS and RNS During Pulmonary Hypertension

ROS and RNS are capable of mediating gene and protein expression through the regulation of transcript factors (e.g., nuclear factor-κB and activator protein-1) and many additional mechanisms. Although it is beyond the scope of this chapter to describe or discuss all the intracellular signaling mechanisms involved in triggering gene expression, examples of recent hypotheses are considered in this section and illustrated in Figure 3. Both $O_2^{\bullet-}$ and H_2O_2 activate tyrosine protein kinases, which subsequently stimulate MAP kinases and phospholipase Cγ, and ultimately alter gene expression mediated by the activation of several transcription factors in the nucleus. Other recent theories suggest that PO_2-sensitive proteins appear to contain specific regions called PAS domains. These regions are sites that potentially sense changes in PO_2 or ROS formation. ROS generation and lowering of Po_2 modifies the configuration of PAS domains, which facilitate dimerization of the proteins and their translocation to the nucleus where they up-regulate gene expression. For example, HIF, a PAS protein, senses lowering of Po_2 or ROS, and it subsequently translocates into the nucleus and induces expression of the VEGF gene in pulmonary tissue (27, 28). The absence of HIF and VEGF results in development of a respiratory distress syndrome that leads to premature death of neonatal mice, and impairs the proliferation of PASMC during hypoxia (5, 26). There is evidence that voltage-sensitive K^+ channels also possess the PAS domain and, therefore, are able to detect Po_2 changes (28). However, their roles in cell physiology during hypoxia have not yet been completely elucidated and needs further examination under conditions that promote PH.

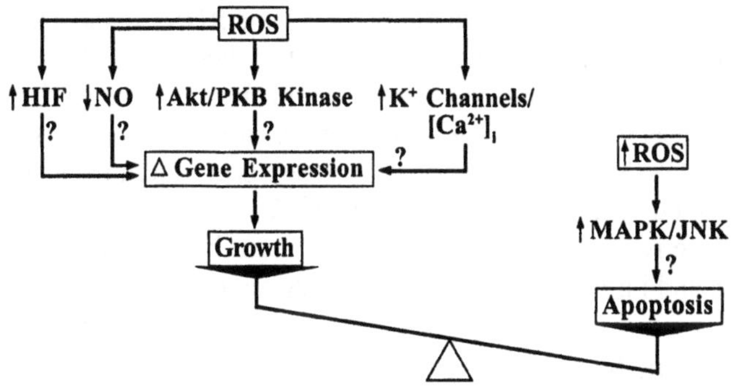

Figure 3. Roles for reactive oxygen species (ROS) and oxidant processes in poorly understood pathways that maintain the delicate balance between cell gene expression, growth, and apoptosis.

Other examples of up-regulated gene expression by redox or ROS signaling are: adhesion proteins, antioxidant enzymes, NOS, receptors, and many respiratory adaptations to hypoxic environments. ROS and RNS can also influence the growth and death of cells through a variety of mechanisms. Cellular release of ROS is now know to serve as an intercellular messenger to stimulate proliferation via mechanisms common to natural growth factors. For example, growth factors appear to activate Akt/protein kinase B pathways through oxidant mechanisms and initiate cell division. $O_2^{\bullet -}$ seems to mediate the downstream effects of Ras and Rac in non-phagocytic cells, and it contributes to the unchecked proliferation of Ras-transformed cells (15).

Recent evidence also indicates a role for $O_2^{\bullet -}$ and H_2O_2 in the control of VSM proliferation both *in vitro* and *in vivo* (10). As increased ROS production and PKC and MAPK are activated coincidentally in acute and chronically hypoxic pulmonary tissues, it might be appropriate to suggest a hypothesis that these pathways may converge to promote mitogenic activities leading to pulmonary vascular remodeling. Cytosolic Ca^{2+} concentration is another critically important factor that controls cell proliferation and cell cycle progression of pulmonary cells (18). Thus, there is a possibility that changes in intracellular Ca^{2+} levels initiated by ROS in PASMC in hypoxic conditions might be involved in promoting cell proliferation and remodeling during progressive PH. Oxidative stress has been shown to mediate hormone-induced hypertrophy, and under some circumstances to induce apoptosis. Studies have indicated that apoptosis contributes to postpartum arterial remodeling in the neonatal lamb. Furthermore, it has been reported that apoptosis of VSM cells is regulated by p53-dependent and -independent pathways (36). Although the mechanisms involved in the oxidant-induced apoptosis of VSM cells are still not clearly understood, it is quite possible that K^+ channels, which have been shown to promote apoptosis (18), could be one of the pathways that mediates apoptotic cell death of PASMC induced by ROS. Since overproduction of ROS and RNS triggers apoptotic cell death, ROS levels appear to be very critical in keeping a delicate balance between cell proliferation and apoptosis, a slight change in the ROS production may control vascular remodeling. For example, since K^+ channels, which are regulated by ROS (32), promote apoptosis under normal circumstances, changes in ROS production and down-regulation of K^+ channel or up-regulation of MAPK/JNK activities during hypoxia may in turn be a primary cause of cell proliferation and initiation of pulmonary artery remodeling leading to PH.

7. Concluding Remarks

ROS and RNS are emerging as important second messenger signaling molecules that participate in vital intracellular signaling pathways and cellular communication systems. Their active prominent interaction with NO and prostaglandins pathways, and ion channels and pumps that regulates a variety of signals in VSMC makes them potential mediators in HPV and attractive candidates in the pathogenesis of PH. A future direction is to precisely elucidate the pathways used by ROS to control in pulmonary vascular function, adaptation and pathology, in order to develop novel therapeutic approaches to treat PH.

References

1. Archer SL, Huang JM, Hampl V, Nelson DP, Shultz PJ, and Weir EK. Nitric oxide and cGMP cause vasorelaxation by activation of a charybdotoxin-sensitive K channel by cGMP-dependent protein kinase. *Proc. Natl. Acad. Sci. USA*. 1994; 91: 7583-7587.
2. Archer SL, Will JA, and Weir EK. Redox status in the control of pulmonary vascular tone. *Herz*. 1986; 11: 127-141.
3. Brophy, CM, Knoepp L, Xin J, and Pollock JS. Functional expression of NOS 1 in vascular smooth muscle. *Am. J. Physiol. Heart Circ. Physiol.* 2000; 278: H991-H997.
4. Burke-Wolin T and Wolin MS. H_2O_2 and cGMP may function as an O_2 sensor in the

pulmonary artery. *J. Appl. Physiol.* 1989; 66: 167-170.

5. Compernolle V, Brusselmans K, Acker T, Hoet P, Tjwa M, Beck H, Plaisance S, Dor Y, Keshet E, Lupu F, Nemery B, Dewerchin M, Van Veldhoven P, Plate K, Moons L, Collen D, and Carmeliet P. Loss of HIF-2α and inhibition of VEGF impair fetal lung maturation, whereas treatment with VEGF prevents fatal respiratory distress in premature mice. *Nat. Med.* 2002; 8: 702-710.
6. Demiryurek AT and Wadsworth RM. Superoxide in the pulmonary circulation. *Pharmacol. Ther.* 1999; 84: 355-365.
7. Dessy C, Kim I, Sougnez CL, Laporte R, and Morgan KG. A role for MAP kinase in differentiated smooth muscle contraction evoked by α-adrenoceptor stimulation. *Am. J. Physiol. Cell Physiol.* 1998; 275: C1081-C1086.
8. Franco-Obregon A, Urena J, and Lopez-Barneo J. Oxygen-sensitive calcium channels in vascular smooth muscle and their possible role in hypoxic arterial relaxation. *Proc. Natl. Acad. Sci. USA.* 1995; 92: 4715-4719.
9. Franco-Obregon A and Lopez-Barneo J. Differential oxygen sensitivity of calcium channels in rabbit smooth muscle cells of conduit and resistance pulmonary arteries. *J. Physiol.* 1996; 491: 511-518.
10. Griendling KK and Harrison DG. Dual role of reactive oxygen species in vascular growth. *Circ Res.* 1999; 85: 562-563.
11. Gupte SA, Li KX, Okada T, Sato K, and Oka M. Inhibitors of pentose phosphate pathway cause vasodilation: involvement of voltage-gated potassium channels. *J. Pharmacol. Exp. Ther.* 2002; 301: 299-305.
12. Gupte SA, Tateyama M, Okada T, Oka M, and Ochi R. Epiandrosterone, a metabolite of testosterone precursor, blocks L-type calcium channels of ventricular myocytes and inhibits myocardial contractility. *J. Mol. Cell. Cardiol.* 2002; 346:679-688.
13. Gurtner GH and Burke-Wolin T. Interactions of oxidant stress and vascular reactivity. *Am. J. Physiol.* 1991; 260: L207-L211.
14. Harder DR, Madden JA, and Dawson C. Hypoxic induction of Ca^{2+}-dependent action potentials in small pulmonary arteries of the cat. *J. Appl. Physiol.* 1985; 59: 1389-1393.
15. Irani K and Goldschmidt-Clermont PJ. Ras, superoxide and signal transduction. *Biochem. Pharmacol.* 1998; 55: 1339-1346.
16. Kinnula VL, Crapo JD, and Raivio KO. Generation and disposal of reactive oxygen metabolites in the lung. *Lab. Invest.* 1995; 73: 3-19.
17. Le Cras TD and McMurtry IF. Nitric oxide production in the hypoxic lung. *Am. J. Physiol. Lung Cell. Mol. Physiol.* 2001; 2804: L575-L582.
18. Mandegar M, Remillard CV, and Yuan JX-J. Ion channels in pulmonary arterial hypertension. *Prog. Cardiovasc. Dis.* 2002; 45: 81-114.
19. Michelakis ED, Hampl V, Nsair A, Wu XC, Harry G, Haromy A, Gurtu R, and Archer SL. Diversity in mitochondrial function explains differences in vascular oxygen sensing. *Circ. Res.* 2002; 90: 1307-1315.
20. Michelakis ED, McMurtry MS, Wu XC, Dyck JR, Moudgil R, Hopkins TA, Lopaschuk GD, Puttagunta L, Waite R, and Archer SL. Dichloroacetate, a metabolic modulator, prevents and reverses chronic hypoxic pulmonary hypertension in rats: role of increased expression and activity of voltage-gated potassium channels. *Circulation.* 2002; 1052: 244-250.
21. Mingone C, Gupte SA, Iesaki T, and Wolin MS. Hypoxia enhances a cGMP-independent nitric oxide relaxing mechanism in pulmonary artery. *Am. J. Physiol. Lung Cell. Mol. Physiol.* 2003; published ahead of print DOI:10.1152/ajplung.00362.2002
22. Mohazzab KM, and Wolin MS. Properties of a superoxide anion generating microsomal NADH-oxidoreductase, a potential pulmonary artery pO_2 sensor. *Am. J. Physiol.* 1994; 267: L823-L831.
23. Peri RB, Wible A, and Brown AM. Mutations in the Kvβ2 binding site for NADPH and their effects on Kv1.4. *J. Biol. Chem.* 2001; 276: 738-741.

24. Peterson D, and Weir EK. Redox signal transduction: reductive reasoning. *J. Lab. Clin. Med.* 2002; 140: 73-78.
25. Raj U and Shimoda L. Oxygen-dependent signaling in pulmonary vascular smooth muscle. *Am. J. Physiol. Lung Cell. Mol. Physiol.* 2002; 283:L671-L677.
26. Rose F, Grimminger F, Appel J, Heller M, Pies V, Weissmann N, Fink L, Schmidt S, Krick S, Camenisch G, Gassmann M, Seeger W, and Hanze J. Hypoxic pulmonary artery fibroblasts trigger proliferation of vascular smooth muscle cells: role of hypoxia-inducible transcription factors. *FASEB J.* 2002; 16: 1660-1661.
27. Semenza GL. Hypoxia-inducible factor 1: oxygen homeostasis and disease pathophysiology. *Trends Mol. Med.* 2001; 7: 345-350.
28. Taylor BL and Zhulin IB. PAS domains: internal sensors of oxygen, redox potential, and light. *Microbiol. Mol. Biol. Rev.* 1999; 63: 479-506.
29. Waypa GB, Chandel NS, and Schumacker PT. Model for hypoxic pulmonary vasoconstriction involving mitochondrial oxygen sensing. *Circ. Res.* 2001; 88: 1259-1266.
30. Waypa GB, Marks JD, Mack MM, Boriboun C, Mungai PT, and Schumacker PT. Mitochondrial reactive oxygen species trigger calcium increases during hypoxia in pulmonary arterial myocytes. *Circ. Res.* 2002; 91: 719-726.
31. Wedgwood S, Dettman RW, and Black S. M. ET-1 stimulates pulmonary arterial smooth muscle cell proliferation via induction of reactive oxygen species. *Am. J. Physiol. Lung Cell. Mol. Physiol.* 2001; 281: L1058-L1067.
32. Weir EK and Archer SL. The mechanism of acute hypoxic pulmonary vasoconstriction: the tale of two channels. *FASEB J.* 1995; 9: 183-189.
33. Weir EK, Hong Z, Porter VA, and Reeve HL. Redox signaling in oxygen sensing by vessels. *Respir. Physiol. Neurobiol.* 2002; 132: 121-130.
34. Wolin MS. Interactions of oxidants with vascular signaling systems. *Arterioscler. Thromb. Vasc. Biol.* 2000; 20: 1430-1442.
35. Wolin MS, Burke-Wolin TM, and Mohazzab-H KM. Roles of NAD(P)H oxidases and reactive oxygen species in vascular oxygen sensing mechanisms. *Respir. Physiol.* 1999; 115: 229-238.
36. Wolin MS, Gupte SA, and Oeckler RA. Superoxide in the vascular system. *J. Vasc. Res.* 2002; 39: 191-207.
37. Yuan JX-J. Oxygen-sensitive K^+ channel(s): where and what? *Am. J. Physiol. Lung Cell. Mol. Physiol.* 2001; 281: L1345-L1349.
38. Zhu D, Birks EK, Dawson CA, Patel M, Falck JR, Presberg K, Roman RJ, and Jacobs ER. Hypoxic pulmonary vasoconstriction is modified by P-450 metabolites. *Am. J. Physiol. Heart Circ. Physiol.* 2000; 279: H1526-H1533.

Chapter 15

Mitochondrial Oxygen Sensing in Hypoxic Pulmonary Vasoconstriction

Navdeep S. Chandel
Northwestern University Medical School, Chicago, Illinois, U.S.A.

1. Introduction

An acute decrease in alveolar oxygen (O_2) tension causes reversible hypoxic pulmonary vasoconstriction (HPV), an adaptive mechanism for diverting blood flow from poorly ventilated to better-ventilated regions of the lung to improve ventilation-perfusion matching. Significant advances have been made in determining the oxygen dependence of HPV in intact animals, isolated lungs, and pulmonary arterial rings. Recent studies using isolatedpulmonary arterial smooth muscle cells (PASMCs) indicate that hypoxia directly causes membrane depolarization, increased intracellular Ca^{2+} concentration ($[Ca^{2+}]_i$), and cell-shortening (30).

In contrast to the acute adaptive response to hypoxia, prolonged exposure to decreased oxygen tension results in pulmonary hypertension that is characterized by smooth muscle cell proliferation, intimal thickening, and extension of smooth muscle into previously non-muscular arterioles (38) The vascular remodeling that occurs during exposure to prolonged hypoxia is linked to a transcriptional increase in a variety of genes including endothelin-1 (ET-1) and serotonin transporter (5-HTT) (8, 21). Despite extensive characterization of the structural and functional changes that occur in pulmonary arteries in response to hypoxia, the precise cellular mechanism of O_2 sensing underlying cell contraction resulting in HPV or the increase in gene expression resulting in pulmonary arterial smooth muscle cell (PASMC) hypertrophy remain poorly understood. There are several proposed mechanisms for O_2 sensing, some of which are related to changes in reactive oxygen species (ROS), while others are independent of changes in ROS. This chapter focuses on the role of mitochondrial ROS as signaling molecules in regulating both HPV and pulmonary remodeling.

2. Hypoxia Augments Mitochondrial Reactive Oxygen Species

During mitochondrial respiration under normal oxygen conditions, O_2 is chemically reduced to water by the transfer of four electrons at cytochrome oxidase. The resulting free energy change is conserved in the form of ATP synthesis. It has been estimated that 2-3% of the O_2 consumed by mitochondria is incompletely reduced, yielding ROS (4). Univalent electron transfer to O_2 generates superoxide, a modestly stable free radical anion. Superoxide can potentially be generated at a number of different sites, including complex I, the ubisemiquinone site of complex III, and other electron transfer proteins (31). ROS do not appear to be generated by cytochrome oxidase itself, due to the high-affinity kinetic trapping of O_2 at the binuclear center, which occurs while the four electrons are sequentially transferred. The superoxide generated during the Q-cycle within the mitochondrial electron complex III (bc1 complex) is considered the main site of ROS generation within the electron transport chain (Figure 1). The Q-cycle involves the transfer of two electrons to ubiquinone from complex I or complex II resulting in the reduction of ubiquinone to ubiquinol. Subsequently, ubiquinol oxidation requires donation of two electrons: the first electron transfer is to the iron-sulfur centers and cytochrome c1 resulting in the oxidation of ubiquinol to ubisemiquinone. This reaction is inhibited by myxothiazol. The second electron is transferred to cytochrome b and results in the oxidation of ubisemiquinone to ubiquinone. This step is inhibited by antimycin A. The oxidation of ubisemiquinone to ubiquinone is the main site of ROS generation during the Q-cycle. The availability of O_2, the reduction state of the electron carriers, and the mitochondrial membrane potential determine the ability of electron transport chain to generate superoxide.

Recent data suggest that hypoxia increases the generation of ROS within mitochondria. Using fluorescent dyes to detect an oxidative signal, we initially studied the oxidation of 2', 7'-dichlorofluorescin (DCFH) in cardiomyocytes and Hep3B cells under controlled O_2 conditions (5, 7). The cells were studied in a flow-through chamber maintained at 37°C on an inverted microscope. The perfusate was bubbled with different O_2 concentrations in a water-jacketed equilibration column mounted above the microscope stage and was delivered to the chamber via a short length of tubing. The reduced diacetate form of the dye DCFH was continually present in the medium (5 μM). Oxidation of the dye within the cell yields the fluorescent compound 2',7'-dichlorofluorescein (DCF), which was detected with a 12-bit digital cooled charge-coupled device camera. In the absence of dye, no fluorescence can be detected. After addition of dye to the medium, evidence of cell fluorescence is detected within a few minutes. Dye oxidized outside of the cell, or dye oxidized within the cell that leaks out, is carried away in the perfusate flow. Under steady-state conditions, cellular fluorescence reflects a balance between the rate of oxidation of DCFH in the cell and the rate at which oxidized dye escapes and is carried away. Paradoxically, oxidation of DCFH increased as the O_2 concentration was decreased from normoxic levels (16% O_2), with minimal effect seen at 8% and a maximal effects observed at 1% O_2. Within minutes of decreasing the O_2 concentration, intracellular fluorescence increased, reflecting an increase in the rate of dye oxidation. On return to normoxia, fluorescence decreased as the rate of escape of oxidized dye exceeded the rate of dye oxidation in the cell. The increase in DCFH oxidation can be prevented by ebselen, a synthetic glutathione peroxidase, and by diethyldithiocarbamate (DDC), a cytosolic inhibitor of Cu,Zn superoxide dismutase (SOD). Ebselen and DDC prevent the cytosolic formation of hydrogen peroxide (H_2O_2) indicating that the increase in DCFH oxidation during hypoxia is primarily occurring by H_2O_2 production.

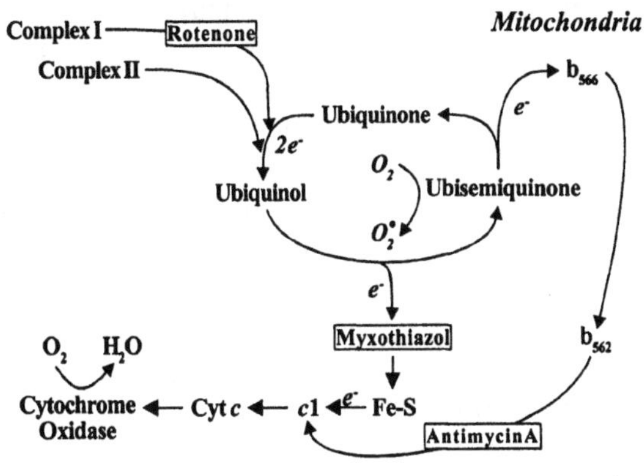

Figure 1. In the electron transport chain, ubisemiquinone appears to be a major site of superoxide generation because of its predisposition for univalent electron transfer to O_2. Electron transport inhibition with rotenone (complex I) or myxothiazol limits the generation of superoxide by attenuating the formation of ubisemiquinone whereas antimycin A, an inhibitor of complex III, augments superoxide generation by increasing the lifetime of ubisemiquinone.

The potential role of mitochondria as a source of ROS generation has been examined by oxidation of DCFH in the presence of electron transport chain inhibitors (Fig. 1). Both rotenone or myxothiazol prevent the increase in oxidation of DCFH during hypoxia. Rotenone inhibits complex I while myxothiazol blocks electron transfer to the Rieske iron-sulfur center within complex III. In contrast, antimycin A, which inhibits the oxidation of cytochrome b_{562} at complex III, augmented the increase in the oxidation of DCFH. Thus, inhibitors that block electron transfer upstream from that site (rotenone, myxothiazol) tend to prevent the formation of ubisemiquinone and thereby diminish ROS generation. Mitochondrial inhibitors that act at more downstream sites (antimycin, cyanide) tend to augment ROS generation by increasing the generation of ubisemiquinone. Further evidence of mitochondria as a source of oxidant production during hypoxia comes from the observation that 4'-diisothiocyanatostilbene-2,2'-disulphonic acid (DIDS) abolishes the oxidation of DCFH during hypoxia. DIDS blocks anion channels in both inner and outer mitochondrial membranes, thereby preventing the leakage of superoxide from mitochondria into the cytosol.

The pharmacological inhibition of the electron transport chain has been confirmed using ρ^0 cells, a functional mitochondria-deficient cell line. Rapidly-dividing cells can be mutated into a ρ^0 state by incubation with ethidium bromide, which inhibits the replication of mitochondrial DNA that is required for critical subunits of certain mitochondrial electron transport complexes and for part of the F_0F_1 ATP synthase. The ρ^0 cells are therefore incapable of supporting mitochondrial respiration or oxidative phosphorylation. Survival and growth of ρ^0 cells requires glycolytically-derived ATP. Because ρ^0 cells lack mitochondrial electron transport, they cannot generate mitochondrial ROS. In ρ^0 Hep3B cells, hypoxia failed to stimulate oxidative signaling, as evidenced by an absence of DCFH oxidation. Based on studies using both mitochondrial inhibitors and ρ^0 cells, we conclude that hypoxia stimulates superoxide production at the ubisemiquinone site. Subsequently, superoxide enters the cytosol by anion channels where it is converted into H_2O_2 by cytosolic Cu,Zn superoxide dismutase.

After our initial observations in HEP3B cells and cardiac myocytes on DCFH oxidation, Gillespie and colleagues demonstrated that primary cultures of rat main PASMCs increased DCFH oxidation during hypoxia (19). Hypoxia-induced DCF fluorescence was attenuated by the addition of the antioxidants dimethylthiourea and catalase. Recently, we also demonstrated that rat PASMCs acutely exposed to hypoxia exhibit a marked increase in intracellular DCF fluorescence suggestive of an increase in ROS (34). Myxothiazol attenuated the increase in DCFH oxidation during hypoxia, implicating the mitochondrion as a major source of oxidant production in PASMCs under hypoxic conditions. Thus, a variety of mammalian cells including PASMCs exhibit an increase in ROS during hypoxia.

3. Mitochondrial Reactive Oxygen Species are Required and Sufficient to Elicit Hypoxic Pulmonary Vasoconstriction

HPV occurs mainly in the small distal pulmonary arteries. Recent studies have indicated that acute hypoxia selectively inhibits voltage-gated K^+ channels (K_V) (3). This results in membrane depolarization, opening of L-type voltage-dependent Ca^{2+} channels, and elevation of $[Ca^{2+}]_i$. An increase in $[Ca^{2+}]_i$ causes calmodulin-mediated activation of myosin light chain kinase, actin-myosin interaction, and contraction. Voltage-gated Ca^{2+} channels provide a major influx pathway for the increase in $[Ca^{2+}]_i$. An alternative model proposes that the initial release of Ca^{2+} occurs from sarcoplasmic reticulum (SR) (10, 25). In this model, hypoxia causes the release of Ca^{2+} from intracellular stores, causing an inhibition of K_V channels, membrane depolarization, and opening of L-type voltage-gated Ca^{2+} channels. Clearly, an elevation in $[Ca^{2+}]_i$, irrespective of the source, is responsible for the contraction. How decreases in oxygen tension are detected and coupled to an increase in Ca^{2+} is not fully understood.

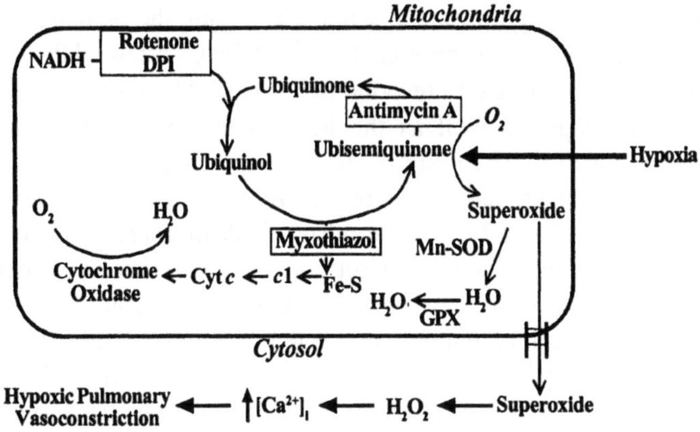

Figure 2. Hypoxia increases the generation of mitochondrial superoxide at the ubisemiquinone site. Subsequently, the superoxide escapes mitochondria into the cytosol via anion channels. In the cytosol, superoxide is converted to hydrogen peroxide, which is sufficient and required for the hypoxia-induced $[Ca^{2+}]_i$ increases in PA myocytes

We have proposed that mitochondria serve as oxygen sensors by increasing superoxide production within complex III of the electron transport chain during hypoxia in PASMCs (34, 35). The increase levels in mitochondrial superoxide production results in efflux of this anion into the cytosol through anion channels localized on both inner and outer mitochondrial membranes. Subsequently, the superoxide is converted to hydrogen peroxide which is sufficient and required to elevate $[Ca^{2+}]_i$ (Fig. 2). Primary evidence for this model comes from parallel experiments in isolated perfused lungs and PASMCs from rats. Pulmonary vasoconstriction in isolated lungs or PASMC contraction during hypoxia is blocked by myxothiazol, diphenyleneiodonium (DPI), or rotenone. All three mitochondrial inhibitors block electron transport chain upstream from the ubisemiquinone site of superoxide production. None of these pharmacological electron transport agents affect the vasoconstriction induced by U46619 under normoxia. DPI, rotenone, and myxothiazol also attenuate the increase in $[Ca^{2+}]_i$ in response to hypoxia without affecting the rise in $[Ca^{2+}]_i$ due to angiotensin II during normoxia. Antimycin A, which inhibits mitochondrial electron transport chain downstream from the ubisemiquinone site of superoxide production, maintains HPV and has no effect on hypoxia-induced calcium signaling. The mitochondrial anion channel blocker DIDS abolishes pulmonary vasoconstriction in isolated lungs and contraction in PASMCs during hypoxia without affecting the response to U46619. Ebselen, a synthetic glutathione peroxidase, abolishes pulmonary vasoconstriction in isolated lungs and contraction in PASMCs during hypoxia without affecting the response to U46619. H_2O_2 is sufficient to invoke contraction in PASMCs and to increase in calcium during normoxia. Adenovirus transfection of PASMCs with catalase prevents the increase in calcium during hypoxia or in H_2O_2 during normoxia, without affecting the increase in calcium due to angiotensin II during normoxia. The ρ^0 PASMCs retained the contractile response to U46619 but failed to respond to hypoxia. These results suggest that HPV in isolated lungs, as well as the increase in calcium and subsequent PASMC contraction during hypoxia, occur independently of mitochondrial ATP production and are triggered by increased release of superoxide from the ubisemiquinone site within the mitochondrial electron transport chain.

Our findings that mitochondria function as the O_2 sensor upstream of the increase in calcium in triggering HPV has been corroborated by Ward and colleagues (20). They found that rotenone and myxothiazol attenuated hypoxia-induced increases in $[Ca^{2+}]_i$ and subsequent vasoconstriction. By contrast, inhibition of complex IV with cyanide augmented the hypoxic response but had no effect on $[Ca^{2+}]_i$. Interestingly, in pulmonary arteries that had been treated with rotenone, Ward and colleagues were able to restore HPV by using succinate to shuttle electrons into complex III via complex II, effectively bypassing the rotenone inhibition of complex I. However our observations have been challenged by Archer and colleagues who have reported a decrease in mitochondrial ROS generation. According to their hypothesis, hypoxia reduces mitochondrial electron transport thereby changing the ratio of redox couples (i.e., GSSG/GSH and NAD^+/NADH) toward a more reduced state resulting in a decrease in ROS generation. This leads to an inhibition of K_V channels, membrane depolarization, Ca^{2+} influx via L-type Ca^{2+} channels, increased $[Ca^{2+}]_i$, and vasoconstriction (23) (see Chapter 9). Thus the major controversy in O_2 sensing in PASMC lies in whether ROS levels increase or decrease during hypoxia.

4. Mitochondria Regulate Hypoxic Stabilization of Hypoxia-inducible Factor-1

Cells respond to hypoxia by activating a multitude of responses designed to prevent cells from reaching 0% oxygen. The best-characterized cellular response is the activation of the transcription factor HIF-1. HIF-1 is a dimeric transcription factor composed of HIF-1α and HIF-1β subunits. HIF-1 is a transcriptional activator that is required for the upregulation expression of genes encoding for vascular endothelial growth factor, erythropoietin, glycolytic enzymes, and endothelin-1 in response to low oxygen concentration (27). HIF-1 plays important roles in normal development, physiological responses to hypoxia, and the pathophysiology of common human diseases. In mice, complete HIF-1α deficiency results in embryonic lethality at mid-gestation because of cardiac and vascular malformations (17). Mice that are partially HIF-1α deficient due to the loss of one allele (heterozygous) develop normally. However, when these mice are subjected to long-term hypoxia (10% O_2 for 3 weeks), they have impaired hypoxia-induced pulmonary hypertension as indicated by a diminished medial wall hypertrophy in small pulmonary arterioles (37). Furthermore, chronic hypoxia increases cell capacitance, reduces K_V current density, and depolarizes PASMCs, and these effects are reduced or absent in PASMCs from mice partially deficient for HIF-1 (28). The mechanism underlying hypoxic activation of HIF-1 is important for understanding the pathology of diseases such as pulmonary hypertension.

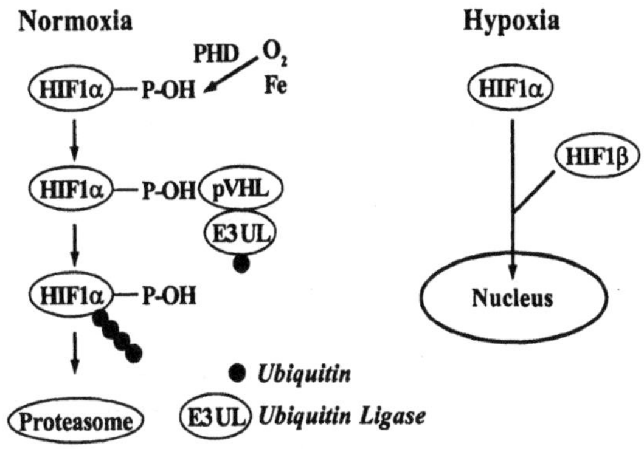

Figure 3. Prolyl hydroxylases (PHD) catalyze hydroxylation of proline residues in HIF-1α under normal oxygen conditions. Proline hydroxylation requires oxygen as a substrate and iron as a cofactor. Prolyl hydroxylation is recognized by pVHL which targets HIF-1α for ubiquitination and subsequent protein degradation by the 26S proteasome. Under hypoxia, HIF-1α is not hydroxylated thereby HIF-1α protein interacts with HIF-1β

HIF-1 is a heterodimer of two basic helix loop-helix/PAS proteins, HIF-1α and the aryl hydrocarbon nuclear trans-locator (ARNT or HIF-1β) (33). ARNT protein levels are constitutively expressed and not significantly affected by oxygen. In contrast, HIF-1α protein is present only in hypoxic cells. During normoxia (21% O_2), HIF-1α is polyubiquitinated by an E3 ubiquitin ligase complex that contains the von Hippel-Lindau tumor suppressor protein (pVHL), elongin B, elongin C, Cul2, and Rbx1 (22, 24). The binding of pVHL to the oxygen-dependent degradation (ODD) domain is located in the central region of HIF-1α. pVHL binding to HIF-1α is dependent on the hydroxylation of proline residues within HIF-1α (Fig. 3) (16, 18). This hydroxylated prolyl residue forms two critical hydrogen bonds with pVHL side chains present within the β domain.

This constitutes the pVHL substrate recognition unit. The enzymatic hydroxylation reaction is inherently oxygen-dependent since the oxygen atom of the hydroxy group is derived from molecular oxygen. In addition, prolyl hydroxylation requires 2-oxoglutarate and iron as cofactors. 2-oxoglutarate is required because the hydroxylation reaction is coupled to the decarboxylation of 2-oxoglutarate to succinate, which accepts the remaining oxygen atom. In mammalian cells, HIF prolyl hydroxylation is carried out by one of three homologs of *C. elegans* Egl-9 (EGLN1, EGLN2, and EGLN3; also called PHD2, PHD1, and PHD3, respectively, or HPH-2, HPH-3, and HPH-1, respectively) (9). Presumably, the prolyl hydroxylated mediated degradation of HIF-1α protein is suppressed under hypoxic conditions ranging from 0-5% O_2.

A fundamental question for understanding HIF-1α regulation involves the mechanism by which cells sense the lack of oxygen and initiate a signaling cascade that results in the stabilization of HIF-1α protein. Early progress in understanding molecular mechanisms underlying mammalian oxygen sensing came from the observation that erythropoietin mRNA can be induced under normoxic conditions in the human hepatoma Hep3B cell line by incubation with transition metals such as cobalt and iron chelators such as desferrioxamine (DFO) (12). This led to the proposal that a rapidly turning over heme protein capable of interacting with O_2 is a putative oxygen sensor. However, studies using heme synthesis inhibitors failed to show any effect on the hypoxic activation of HIF-1, suggesting that rapidly turning over heme proteins are not involved in hypoxia sensing (29). Subsequently, the NADPH oxidase was proposed as a possible oxygen sensor by decreasing reactive oxygen species (ROS) generation during hypoxia (1). The decrease in ROS triggers the stabilization of HIF-1. However, this model is confounded by the observation that DPI, a wide-ranging inhibitor of flavoprotein-containing enzymes including NAD(P)H oxidase, does not trigger HIF-1 stabilization during normoxia. Rather, DPI inhibits the hypoxic induction of HIF-1 dependent genes (11). We propose a model in which the increased generation of reactive oxygen species at complex III of the mitochondrial electron transport chain serves as the oxygen sensor for HIF-1α protein stabilization during hypoxia. In support of this model, hypoxia increases ROS generation, HIF-1α protein accumulation, and the expression of a luciferase reporter construct under the control of a hypoxic response element in wild-type cells, but not in cells depleted of their mitochondrial DNA (ρ^0 cells) (5, 6). Furthermore, catalase overexpression abolishes the luciferase expression in response to hypoxia. Hydrogen peroxide can stabilize HIF-1α protein levels and activate luciferase expression under normoxic conditions in both wild-type and ρ^0 cells. Thus, ROS are both required and sufficient to trigger HIF-1 activation. The site of ROS generation during hypoxia is localized to complex III within the mitochondrial electron transport chain. Mitochondrial complex I inhibitors, such as rotenone, that prevent electron flux upstream of complex III, ablate ROS generation during hypoxia and subsequently, the hypoxic stabilization of HIF-1α protein stabilization. These results have been corroborated by Lamanna and colleagues who have demonstrated that the neurotoxin 1-methyl-4-phenyl-1,2,3,6-tetrahydropyridine (MPTP), a complex I inhibitor, prevents the hypoxic stabilization of HIF-1α protein in PC12 cells (2). These investigators have also shown that hypoxic stabilization of HIF-1α protein is severely reduced in human xenomitochondrial cybrids harboring a partial (40%) complex I deficiency. Further evidence that mitochondria regulate HIF-1 activation comes from Stratford and colleagues, who demonstrated that the complex IV inhibitor cyanide is sufficient to activate HIF-1 dependent transcription in wild-type Chinese hamster ovary (CHO) cells and HT1080 cells under normoxic conditions (36). Cyanide inhibits electron transport chain downstream of complex III, thus eliciting an increase in ROS generation.

Recently, Ratcliffe and colleagues have challenged the role of mitochondria as a potential oxygen sensor. These investigators have demonstrated that ρ^0 cells are able to stabilize HIF-1α protein levels at oxygen concentration of 0.1% O_2 (32). They have proposed that the prolyl hydroxylases are oxygen sensors that regulate hypoxic stabilization of HIF-1α protein. This model is based on the observation that prolyl hydroxylases require molecular oxygen and iron to catalyze the hydroxylation of proline residues within HIF-1α. In the absence of oxygen or iron, HIF-1α would not undergo proline hydroxylation and subsequent pVHL mediated ubiquitin-targeted degradation. The hydroxylation inhibition due to anoxia or iron chelation would indicate prolyl hydroxylases as the sensor. However, it is not known if prolyl hydroxylase would intrinsically be inhibited at oxygen concentration as low as 1-2%, where HIF-1 is activated. We recently investigated the response to ρ^0 cells to both hypoxia and anoxia because Ratcliffe and colleagues used conditions close to anoxia in examining their hypoxic response to ρ^0 cells. Our results demonstrate that the stabilization of HIF-1α protein at 1-2% O_2 did not occur in ρ^0 cells. However, ρ^0 cells were able to stabilize HIF-1α protein at 0% O_2 or in the presence of an iron chelator under normoxic conditions (26). This observation is consistent with the requirement of proline hydroxylation as a mechanism for HIF-1α protein degradation under normal oxygen conditions. In the absence of oxygen, hydroxylation of proline residues within HIF-1α by prolyl hydroxylases cannot occur and intracellular signaling events are not required for the stabilization of HIF-1α protein. Thus, prolyl hydroxylases would effectively serve directly as the oxygen sensors during anoxia or during iron chelation under normoxia. Furthermore, if prolyl hydroxylase is, by itself, the oxygen sensor for both hypoxia- and anoxia-induced HIF-1 then there would be no signaling required upstream of prolyl hydroxylase, i.e., kinases/ROS upstream of prolyl hydroxylase. However, other investigators have shown that hypoxia (1% O_2) stimulates Rac1 activity, and Rac1 is required for the hypoxic stabilization of HIF-1α protein (15). Both the hypoxic activation of Rac1 and the stabilization of HIF-1α protein were abolished by the complex I inhibitor rotenone. These results indicate that Rac1 is downstream of mitochondrial signaling. Moreover, mitochondrial dependent oxidant signaling has been shown to regulate HIF-1α protein accumulation following exposure to TNFα (14). Non-mitochondrial dependent oxidant signaling has also been shown to stabilize HIF-1α protein under normoxia. For example, thrombin or angiotensin II stabilizes HIF-1α under normoxia through an increase in ROS generation from non-mitochondrial sources (13). Further support for the idea that hypoxic signaling is distinct from anoxia or iron chelation comes from the observation that DPI, an inhibitor of a wide range of flavoproteins including complex I, prevents stabilization of HIF-1α protein and HIF-1 target genes at oxygen levels of 1% (11). However, DPI fails to affect stabilization of HIF-1 in response to the iron chelator desferrioxamine (DFO). This observation is consistent with the notion that iron chelators or lack of oxygen directly inhibit prolyl hydroxylase activity due to substrate limitations and stabilize HIF-1α protein (Fig. 3). Interestingly, DPI can prevent a variety of other hypoxic responses such as pulmonary vasoconstriction and carotid body nerve firing (1). We speculate that the ultimate target of the oxidant dependent signaling pathway originating from mitochondria during hypoxia or non-mitochondrial sources such as angiotensin II during normoxia is to inhibit proline hydroxylation (Fig. 4).

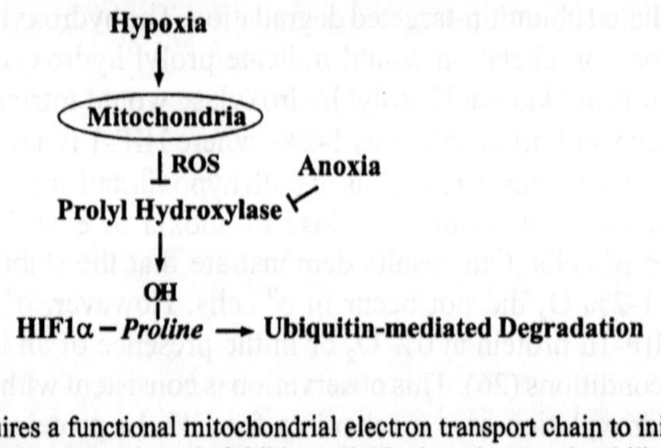

Figure 4. Hypoxia requires a functional mitochondrial electron transport chain to initiate oxidant dependent signaling that ultimately inhibits prolyl hydroxylases and stabilizes HIF-1α. By contrast, in the absence of oxygen (anoxia), proline hydroxylation cannot occur because oxygen is a required substrate for hydroxylation. Thus, anoxia would directly inhibit prolyl hydroxylases thereby stabilizing HIF-1α.

5. Conclusions

Currently, there are several proposed mechanisms of oxygen sensing in pulmonary artery smooth muscle cells. Most of the models are related to either increases or decreases in reactive oxygen species. Discrepancies among investigators are likely related to differences in preparations and detection methods of detecting ROS. Furthermore, the source of ROS generation has been proposed as either the mitochondrial electron transport chain or NADPH oxidase. Our own observations suggest that a similar mitochondrial O_2 sensing mechanism may be responsible for HPV and for HIF-1-mediated transcriptional activation during hypoxia. The HIF-1 mediated transcription is likely to be important for pulmonary remodeling during chronic exposure to hypoxia. The next challenge is to develop other genetic means to perturb mitochondrial function and arrive at similar conclusions that were provided by ρ^0 cells.

Acknowledgments

Supported by NIH grant GM60472-03 (NSC).

References

1. Acker H. Mechanisms and meaning of cellular oxygen sensing in the organism. *Respir. Physiol.* 1994; 95: 1-10.
2. Agani FH, Pichiule P, Chavez JC, and LaManna JC. The role of mitochondria in the regulation of hypoxia-inducible factor 1 expression during hypoxia. *J. Biol. Chem.* 2000; 275: 35863-35867.
3. Archer SL, Souil E, Dinh-Xuan AT, Schremmer B, Mercier JC, El Yaagoubi A, Nguyen-Huu L, Reeve HL, and Hampl V. Molecular identification of the role of voltage-gated K^+ channels, Kv1.5 and Kv2.1, in hypoxic pulmonary vasoconstriction and control of resting membrane potential in rat pulmonary artery myocytes. *J. Clin. Invest.* 1998; 101: 2319-2330.
4. Chance B and Williams GR. Respiratory enzymes in oxidative phosphorylation. *J. Biol. Chem.* 1955; 217: 383-393.
5. Chandel NS, Maltepe E, Goldwasser E, Mathieu CE, Simon MC, and Schumacker PT. Mitochondrial reactive oxygen species trigger hypoxia-induced transcription. *Proc. Natl. Acad. Sci. USA.* 1998; 95: 5015-5019.
6. Chandel NS, McClintock DS, Feliciano SE, Wood TM, Melendez JA, Rodriguez AM, and Schumacker PT. Reactive oxygen species generated at mitochondrial complex III stabilize hypoxia-inducible factor-1α during hypoxia: a mechanism of O_2 sensing. *J. Biol. Chem.* 2000; 275: 25130-25138.
7. Duranteau, J, Chandel NS, Kulisz A, Shao Z, and Schumacker PT. Intracellular signaling by reactive oxygen species during hypoxia in cardiomyocytes. *J. Biol. Chem.* 1998; 273: 11619-11624.
8. Eddahibi S, Raffestin B, Hamon M, and Adnot S. Is the serotonin transporter involved in the pathogenesis of pulmonary hypertension? *J. Lab. Clin. Med.* 2002; 139: 194-201.
9. Epstein AC, Gleadle JM, McNeill LA, Hewitson KS, O'Rourke J, Mole DR, Mukherji M, Metzen E, Wilson MI, Dhanda A, Tian YM, Masson N, Hamilton DL, Jaakkola P, Barstead R, Hodgkin J, Maxwell PH, Pugh CW, Schofield CJ, and Ratcliffe PJ. *C. elegans* EGL-9 and mammalian homologs define a family of dioxygenases that regulate HIF by prolyl hydroxylation. *Cell* 2001; 107: 43-54.
10. Gelband CH and Gelband H. Ca^{2+} release from intracellular stores is an initial step in hypoxic pulmonary vasoconstriction of rat pulmonary artery resistance vessels. *Circulation.* 1997; 96: 3647–3654.

11. Gleadle JM, Ebert BL, and Ratcliffe PJ. Diphenylene iodonium inhibits the induction of erythropoietin and other mammalian genes by hypoxia. Implications for the mechanism of oxygen sensing. *Eur. J. Biochem.* 1995; 234:92-99.
12. Goldberg MA, Dunning SP, and Bunn HF. Regulation of the erythropoietin gene: evidence that the oxygen sensor is a heme protein. *Science.* 1988; 242: 1412-1415.
13. Görlach A, Diebold I, Schini-Kerth VB, Berchner-Pfannschmidt U, Roth U, Brandes RP, Kietzmann T, and Busse R. Thrombin activates the hypoxia-inducible factor-1 signaling pathway in vascular smooth muscle cells: Role of the $p22^{phox}$-containing NADPH oxidase. *Circ. Res.* 2001; 89: 47-54.
14. Haddad JJ and Land SC. A non-hypoxic, ROS-sensitive pathway mediates TNF-α-dependent regulation of HIF-1α. *FEBS Lett.* 2001; 505: 269-274.
15. Hirota K and Semenza GL. Rac1 activity is required for the activation of hypoxia-inducible factor 1. *J. Biol. Chem.* 2001; 276: 21166-21172.
16. Ivan M, Kondo K, Yang H, Kim W, Valiando J, Ohh M, Salic A, Asara JM, Lane WS, and Kaelin WG Jr. HIFα targeted for VHL-mediated destruction by proline hydroxylation: implications for O_2 sensing. *Science* 2001; 292: 464-468.
17. Iyer NV, Kotch LE, Agani F, Leung SW, Laughner E, Wenger RH, Gassmann M, Gearhart JD, Lawler AM, Yu AY, and Semenza GL. Cellular and developmental control of O_2 homeostasis by hypoxia-inducible factor 1α. *Genes Dev.* 1998; 12: 149-162.
18. Jaakkola P, Mole DR, Tian Y-M, Wilson MI, Gielbert J, Gaskell SJ, von Kriegsheim A, Hebestreit HF, Mukherji M, Schofield CJ, Maxwell PH, Pugh CW, and Ratcliffe PJ. Targeting of HIF-α to the von Hippel-Lindau ubiquitylation complex by O_2-regulated prolyl hydroxylation. *Science* 2001; 292: 468-472.
19. Killilea DW, Hester R, Balczon R, Babal P, and Gillespie MN. Free radical production in hypoxic pulmonary artery smooth muscle cells. *Am. J. Physiol. Lung Cell. Mol. Physiol.* 2000; 279: L408-412.
20. Leach RM, Hill HM, Snetkov VA, Robertson TP, and Ward JPT. Divergent roles of glycolysis and the mitochondrial electron transport chain in hypoxic pulmonary vasoconstriction of the rat: identity of the hypoxic sensor. *J. Physiol.* 2001; 536: 211–224.
21. MacLean MR. Endothelin-1: a mediator of pulmonary hypertension? *Pulm. Pharmacol. Ther.* 1998; 11: 125-132.
22. Maxwell PH, Wiesener MS, Chang GW, Clifford SC, Vaux EC, Cockman ME, Wykoff CC, Pugh CW, Maher ER, and Ratcliffe PJ. The tumour suppressor protein VHL targets hypoxia-inducible factors for oxygen-dependent proteolysis. *Nature* 1999; 399: 271-275.
23. Michelakis ED, Hampl V, Nsair A, Wu X, Harry G, Haromy A, Gurtu R, and Archer SL. Diversity in mitochondrial function explains differences in vascular oxygen sensing. *Circ. Res.* 2002; 90: 1307-1315.
24. Ohh M, Park CW, Ivan M, Hoffman MA, Kim TY, Huang LE, Pavletich N, Chau V, and Kaelin WG. Ubiquitination of hypoxia-inducible factor requires direct binding to the β-domain of the von Hippel-Lindau protein. *Nat. Cell Biol.* 2000; 2: 423-427.
25. Robertson TP, Hague D, Aaronson PI, and Ward JPT. Voltage-independent calcium entry in hypoxic pulmonary vasoconstriction of intrapulmonary arteries of the rat. *J. Physiol.* 2000; 525: 669-680.
26. Schroedl C, McClintock DS, Budinger GRS, and Chandel NS. Hypoxic but not anoxic stabilization of HIF-1α requires mitochondrial reactive oxygen species. *Am. J. Physiol. Lung Cell. Mol. Physiol.* 2002; 283: L922-L931.
27. Semenza GL. Regulation of mammalian O_2 homeostasis by hypoxia-inducible factor 1. *Annu. Rev. Cell Dev. Biol.* 1999; 15: 551-578.
28. Shimoda LA, Manalo DJ, Sham JSK, Semenza GL, and Sylvester JT. Partial HIF-1α deficiency impairs pulmonary arterial myocyte electrophysiological responses to hypoxia. *Am. J. Physiol. Lung Cell. Mol. Physiol.* 2001; 281: L202-L208.
29. Srinivas V, Zhu X, Salceda S, Nakamura R, and Caro J. Hypoxia-inducible factor 1α (HIF-

1α) is a non-heme iron protein. Implications for oxygen sensing. *J. Biol. Chem.* 1998; 273: 18019-18022.

30. Sylvester JT, Sham JSK, Shimoda LA, and Liu Q. "Cellular mechanisms of acute hypoxic pulmonary vasoconstriction." In *Respiratory-Circulatory Interactions in Health and Disease,* Scharf SM, Pinsky MR, and Magder S, eds. New York, NY: Marcel Dekker; 2001, pp. 351-359.
31. Turrens JF, Alexandre A, and Lehninger AL. Ubisemiquinone is the electron donor for superoxide formation by complex III of heart mitochondria. *Arch. Biochem. Biophys.* 1985; 237: 408-414.
32. Vaux EC, Metzen E, Yeates KM, and Ratcliffe PJ. Regulation of hypoxia-inducible factor is preserved in the absence of a functioning mitochondrial respiratory chain. *Blood* 2001; 98: 296-302.
33. Wang GL, Jiang B-H, Rue EA, and Semenza GL. Hypoxia-inducible factor 1 is a basic-helix-loop-helix-PAS heterodimer regulated by cellular O_2 tension. *Proc. Natl. Acad. Sci. USA.* 1995; 92: 5510-5514.
34. Waypa GB, Chandel NS, and Schumacker PT. Model for hypoxic pulmonary vasoconstriction involving mitochondrial oxygen sensing. *Circ. Res.* 2001; 88: 1259-1266.
35. Waypa GB, Marks JD, Mack MM, Boriboun C, Mungai PT, and Schumacker PT. Mitochondrial reactive oxygen species trigger calcium increases during hypoxia in pulmonary arterial myocytes. *Circ. Res.* 2002; 91: 719-726.
36. Williams KJ, Telfer BA, Airley RE, Peters HP, Sheridan MR, van der Kogel AJ, Harris AL, and Stratford IJ. A protective role for HIF-1 in response to redox manipulation and glucose deprivation: implications for tumorigenesis. *Oncogene* 2002; 21: 282-290.
37. Yu AY, Shimoda LA, Iyer NV, Huso DL, Sun X, McWilliams R, Beaty T, Sham JS, Wiener CM, Sylvester JT, and Semenza GL. Impaired physiological responses to chronic hypoxia in mice partially deficient for hypoxia-inducible factor 1α. *J. Clin. Invest.* 1999; 103: 691-696.
38. Yuan X-J and Rubin LJ. "Pathology of pulmonary hypertension." In *Respiratory-Circulatory Interactions in Health and Disease*, Scharf SM, Pinsky MR, and Magder S, eds. New York, NY: Marcel Dekker; 2001, pp. 447-477.

Chapter 16

Redox Oxygen Sensing in Hypoxic Pulmonary Vasoconstriction

Andrea Olschewski and E. Kenneth Weir
Justus-Liebig-University, Giessen, Germany and University of Minnesota, Minneapolis, Minnesota, U.S.A.

1. Introduction

There are two phases in the response of the pulmonary vasculature to low O_2-tension. The acute, initial transient vasoconstrictor phase (phase 1) is independent of the endothelium. The slowly developing increase in vascular tone (phase 2) appears to be dependent on the presence of endothelium and extracellular Ca^{2+}. Pulmonary vasoconstrictive responsiveness to acute hypoxia has been demonstrated not only in whole animals (68, 74) and in isolated lungs (73, 75), but also in isolated pulmonary arteries (33, 36, 40), as well as in endothelium-denuded pulmonary artery rings (34, 81) and in single isolated smooth muscle cells from resistance pulmonary arteries (41), indicating that pulmonary artery smooth muscle cells (PASMCs) themselves respond to hypoxia, thus representing both sensor and effector cells with respect to this fundamental stimulus.

It has been proposed that the initial phase 1 is mediated by the inhibition of one or several K^+ channels which control the resting membrane potential (E_m), leading to cell depolarization, opening of voltage-gated Ca^{2+} channels, entry of Ca^{2+} into the cells and thus myocyte contraction (5, 9, 52, 59, 72, 79). In addition to Ca^{2+} influx, Ca^{2+} stores in the cell also release Ca^{2+} during physiological signaling of acute hypoxia. Hypoxia has also been found to induce Ca^{2+} release from intracellular stores, such as the ryanodine receptor (RyR)-sensitive and the inositol 1,4,5-trisphosphate (IP_3) receptor (IP_3R)-sensitive Ca^{2+} stores, suggesting that it acts through multiple pathways to induce HPV (22, 24, 62, 76).

Many hypotheses have been put forth to explain O_2-sensing in PASMCs, some of which are related to reactive oxygen species (ROS). O_2-sensing could occur through enzymatic production of ROS that then alter the redox state of signaling molecules and the function of effectors.

2. The Effect of Hypoxia on Cellular Redox Status

2.1. Sources of ROS Production

Cellular oxygen sensing is a highly conserved mechanism in evolution, most likely developed by primitive organisms with the first appearance of oxygen approximately 3 billion years ago (21, 50). It has long been recognized that ROS such as superoxide anion ($O_2^{\bullet-}$), hydrogen peroxide (H_2O_2) and hydroxyl radical ($OH^{\bullet}$) are produced in aerobic cells as by-products of oxidative metabolism, either during mitochondrial electron transport or by several oxidoreductases and the metal-catalyzed oxidation of metabolites.

Stimulated production of ROS by a multicomponent nicotinamide adenine dinucleotide phosphate reduced (NAD(P)H) oxidase was first described in phagocytic cells like neutrophils and macrophages due to the transient consumption of O_2 ("the respiratory burst"). During the last decade, production of ROS by NAD(P)H oxidase or by the NADH isoforms has also been demonstrated in a variety of cells in O_2-sensitive tissues, including neuroepithelial body (NEB) cells (78), PASMCs (42, 46, 47), endothelial cells (85), and carotid bodies (38). A central role for an NAD(P)H oxidase in O_2 sensing in lung NEB cells was recently examined by Fu et al. (30) who found that K^+ channels of NEB cells in lung slices recorded from a knockout mouse, lacking the gp91^{phox} subunit, were insensitive to acute hypoxia. They concluded that during normoxia NAD(P)H oxidase usually generates ROS that keep the K^+ channels open. During hypoxia the decrease in ROS would allow the channels to close. In bovine pulmonary arteries the changes in O_2 tension (Po_2) may regulate vascular relaxation through changes in ROS (77). Wolin et al. paper reported that H_2O_2 production increased in normoxia, relative to hypoxia, and caused PA SMC relaxation through an increase in cGMP. Diphenylene iodonium (DPI), the non-selective NAD(P)H oxidase inhibitor, would be expected to reduce ROS and thus resemble hypoxia. In isolated and cultured NEB cells DPI does in fact inhibit K^+ current as would be predicted if H_2O_2 produced by NAD(P)H oxidase promotes the K^+ current seen in normoxia (69). DPI did not mimic the hypoxia response when given during normoxia, although this could be due to inhibition of Ca^{2+} channels. In the study of Archer et al. (7), the gp91^{phox} knockout mouse has normal HPV and hypoxia still inhibits K^+ channels in their PASMCs, suggesting that NAD(P)H oxidase containing gp91^{phox}, is not required for HPV (Fig. 1). Consequently, the role of NAD(P)H oxidase in oxygen sensing is still uncertain.

Mitochondria were proposed over 30 years ago as potential O_2 sensors in the carotid body (45). Under hypoxic conditions, depletion of high-energy phosphates, a shift toward the reduced form of redox couples, or cytochromes with an unusually low affinity for Po_2 could act as a sensor. There are several studies providing further support for involvement of mitochondria in O_2-sensing in Hep 3B cells (15, 16, 17), cardiomyocytes (12, 13, 17) and in human ductus arteriosus (44). Complementary to these findings, Waypa et al. have recently shown that proximal but not distal inhibitors, acting at electron transport chain (ETC) in mitochondria, suppressed HPV without affecting the response to other vasoconstrictors. They proposed that mitochondrial complex III is important in HPV, increasing ROS production under hypoxia and H_2O_2 is probably the signal molecule (70, 71).

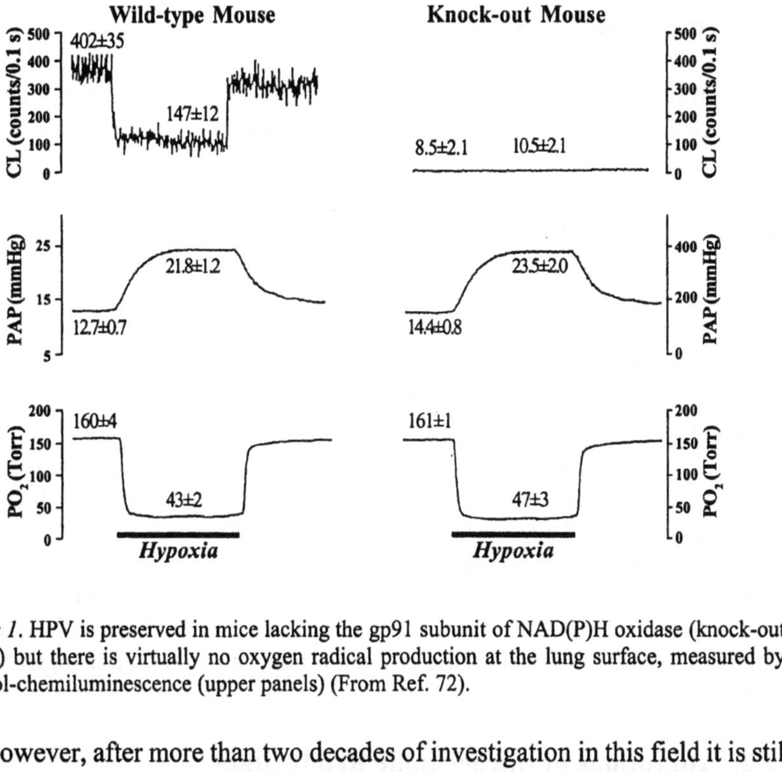

Figure 1. HPV is preserved in mice lacking the gp91 subunit of NAD(P)H oxidase (knock-out mouse) but there is virtually no oxygen radical production at the lung surface, measured by luminol-chemiluminescence (upper panels) (From Ref. 72).

However, after more than two decades of investigation in this field it is still disputed whether ROS go up or down during acute exposure of the lung to hypoxia at physiological levels. Several studies have shown that acute hypoxia decreases production and tissue levels of ROS. In a small cell lung carcinoma cell line (H-146), hypoxia has been shown to reduce ROS production measured by DCF fluorescence (51). In addition, a decrease in superoxide anion production in neutrophils was reported as measured by cytochrome *c* reduction (31). Similar observations were made using chemiluminescence techniques, with luminol or lucigenin enhancement, during hypoxia in the rat lung (6), mouse lung (7), rabbit lung (53) and rabbit ductus arteriosus (60). Finally, Michelakis et al. recently provided supportive evidence for decreased ROS release in denuded resistance pulmonary artery rings during hypoxia using three different detection methods, DCF fluorescence, lucigenin-enhanced chemiluminescence and AmplexRed H_2O_2 assay (43). Similarly, an increase in Po_2 increased ROS production in the human ductus arteriosus, as demonstrated by the same group (44). In contrast, measurements with either DCF fluorescence or lucigenin-enhanced chemiluminescence in cultured PASMCs all showed increases in ROS production during hypoxia (37, 42, 70). This discrepancy may relate to differences in preparation methods or experimental techniques used. In addition, questions arise concerning the precise subcellular locations of ROS production.

2.2. Cytosolic Redox Status

Regardless of the direction of the change in the level of ROS, ROS signaling can be divided into two general mechanisms of action: *a*) direct effect of ROS and *b*) alterations in intracellular redox state.

2.2.1. Direct Effect of ROS

ROS can alter protein structure and function by modifying critical amino acid residues by removing or donating electrons. The reduction or oxidation of the sulfhydryl groups of amino acids, such as cysteine, have emerged as a molecular mechanism behind many cellular processes, including DNA transcription, ion channel modification, and regulation of cytosolic free Ca^{2+} concentration ($[Ca^{2+}]_{cyt}$). Oxidative modifications of critical amino acids within the functional domain of proteins (notably voltagc-gated K^+ channels) may occur in several ways. The best described of such modifications involves cysteine residues. The sulfhydryl group (-SH) of a single cysteine residue may be oxidized to form sulfenic (-SOH), sulfinic (-SO2H), sulfonic (-SO3H), or S-glutathionylated (-SSG) derivatives. Such alterations caused by ROS may alter the activity of K^+ channels, changing channel gating or open probability and link changes in Po_2 to vascular tone.

2.2.2. Alterations in Intracellular Redox State

The effects of ROS in signalling have often been attributed to a "shift" in the redox potential of the cells. In comparison with the extracellular environment, the cytosol is normally maintained under strong "reducing" conditions. Schafer and Buettner (63) suggested recently that the glutathione disulfide (GSSG)/glutathione (GSG) ratio can serve as a good indicator of the cellular redox state. This ratio results primarily from a combination of the rates of H_2O_2 removal by GSH (reduced glutathione) peroxidase and GSSG reduction by GSH reductase, regulating the GSH concentration. Hypoxia increases the ratios of the reduced to the oxidized forms of the cytosolic redox pairs, such as glutathione

(GSH/GSSG) and/or NADH/NAD, NADPH/NADP (4, 18, 64) and thus shifts the cells to a more reduced state. The redox couples within the cytosol could also affect and modify channel activity, as will be extensively reviewed in the following part of the chapter.

3. Redox Modulation of Ion Channels

Redox modulation of ion channel activity seems to be an important regulatory mechanism under physiological conditions for several vasomotor functions, including regulation of cell E_m and vascular tone, activation of transcription factors required for the expression of genes, apoptosis and necrosis, and cellular protective mechanisms against ischemic or hypoxic insults.

Several studies have provided evidence that O_2-sensing in different tissues is mediated by effector K^+ channels with cytosolic redox as a sensor. G-protein-coupled inwardly rectifiying K^+ (K_{IR}) channels, present in excitable and endocrine tissues, have been shown to be activated by the reducing agent dithiothreitol (DTT) (84). For DTT-dependent activation, a cysteine-residue located in the N-terminal cytoplasmic domain of the channel seems to be essential. Interestingly, when this cysteine-residue is mutated, DTT-dependent activation is abolished, but receptor-mediated channel activation is not affected, suggesting that intracellular redox potential acts in a receptor-independent manner. It is possible that G-protein signaling may involve redox changes (57). Opening of G-protein-regulated-K_{IR} channels can slow cardiac pacemaking, shut down secretion or inhibit neuronal firing in the central nervous system and thus can protect cells against ischemic or hypoxic insults.

In the lung NEB cells and the H-146 cell line, the O_2-sensitive K^+ currents are augmented by H_2O_2 or by activation of the NADPH oxidase with phorbol esters (51, 69). In the HERG channel (which generates the cardiac K_{IR} currents) activation is accelerated by H_2O_2 (10). Consequently oxidation tends to increase the potassium current and reduction to decrease it.

The large conductance Ca^{2+}-activated K^+ (K_{Ca}) channels in different tissues have also been reported to be hypoxia and redox sensitive. In this case, the reducing agent GSH reversibly activates the neuronal K_{Ca} channel by increasing the activation rate constant and decreasing the deactivation rate constant (20). In myocytes of the taenia caeci, DTT and GSH significantly increases and the oxidizing agent thimerosal decreases the open probability of K_{Ca} channels in excised patches (39). In contrast, the activity of large conductance K_{Ca} channels in CA1 pyramidal neurons is markedly increased by the oxidizing agent 5,5'-dithio-bis-(2-nitrobenzoic acid) (DTNB). DTT has no effect on channel activity but can reverse DTNB-induced enhancement (32). The different behavior of these channels in response to redox changes could be explained by the heterogeneity of the K_{Ca} channel in different tissues.

3.1. Redox Control of Pulmonary Vascular Ion Channels

The E_m of arterial smooth muscle cells is an important regulator of arterial tone and hence arterial diameter. These cells have steady or slowly changing resting E_m around -65 to -50 mV *in vitro*, close to the predicted equilibrium potential for K^+ ions (E_K is approximately -85 mV with physiological extracellular K^+, ~5 mM) (49). The opening of K^+ channels in the cell membrane increases K^+ efflux, which causes membrane hyperpolarization. This closes voltage-dependent Ca^{2+} channels, decreasing Ca^{2+} entry and leading to pulmonary vasodilatation. Conversely, inhibition of K^+ channels activity causes membrane depolarization, leads to cell contraction and vasoconstriction. Therefore, alterations in K^+ channel activity play an important role in the development of hypoxic pulmonary vasoconstriction (72). It has been proposed that the reduction in Po_2 from normoxic to hypoxic levels, which causes reversible hypoxic pulmonary vasoconstriction, shifts the ratio of redox couples toward a more reduced state in PASMCs, leading to inhibition of K^+ current, membrane depolarization, Ca^{2+} influx via L-type Ca^{2+} channels, an increase in $[Ca^{2+}]_{cyt}$ and pulmonary vasoconstriction.

Three classes of K^+ channels have been identified in PASMCs: voltage-gated K^+ (Kv) channels (25, 58, 95), Ca^{2+}-activated K^+ (K_{Ca}) channels (3, 56) and ATP-sensitive K^+ (K_{ATP}) channels (49). The K^+ channels which control E_m in PASMCs, and inhibition of which initiates HPV, conduct an outward current which activates at potentials less negative then -50 mV, is slowly inactivating and blocked by the Kv channel blocker 4-aminopyridine (4-AP) but not by inhibitors of K_{Ca} or K_{ATP} channels (5, 8, 79). The large conductance K_{Ca} channels are activated by intracellular Ca^{2+} and membrane depolarization. The opening of K_{Ca} channels hyperpolarizes the cell membrane and thus prevents excessive vasoconstriction. K_{ATP} channels are activated after the depletion of intracellular ATP level for example, by hypoxic vasoconstriction associated with anoxia or after exposure to metabolic poisons. Therefore, the activity of K_{ATP} channels could bear an important role for the regulation of basal vascular tone under severe pathological conditions.

The existence of redox modulation in K_{Ca} channel activity has been shown in studies carried out in PASMCs. Reducing agents, such as DTT, reduced glutathione, and reduced nicotinamide adenine dinucleotide (NADH) decreased, and oxidizing agents, such as DTNB, oxidized glutathione and NAD, increased the K_{Ca} channel activity in PASMCs. The increased activity due to oxidizing agents was diminished by applying reducing agents (55). In contrast, in isolated PASMCs from large pulmonary arteries the K_{Ca} channel activity was unaffected by NAD and GSSG or NADH and GSH (66). This suggests that the change in the intracellular redox state, which would be expected during acute hypoxia, does not alter the activity of K_{Ca} channel in PASMCs from large conduit pulmonary arteries.

A variety of redox-active agents have been studied in PASMCs, showing that exogenous reducing agents mimic the effect of hypoxia on several types of O_2-sensitive K^+ channels, E_m and $[Ca^{2+}]_{cyt}$. Yuan (82) provided support for the role of a redox-based O_2-sensing system in PASMCs. Reduced glutathione caused an inhibition in the whole-cell K^+ current and significant membrane depolarization. In the presence of GSH, hypoxia had no further effect on K^+ current or resting E_m, suggesting that both hypoxia and GSH block the same K^+ channels. In addition, oxidized glutathione (GSSG), co-enzyme Q_{10} and duroquinone, on the other hand, increase whole cell K^+ current and hyperpolarize the resting E_m (61, 72). Park et al. showed, that DTT partially blocks Kv current in PASMCs and accelerates the inactivation kinetics, but does not affect steady-state activation and inactivation (54). On the contrary, the oxidizing agent 2,2'-dithio-bis(5-nitropyridine) (DTNBP) increases Kv current and accelerates activation kinetics (54). The effect of DTT and DTNB on whole-cell K^+ current and resting E_m of adult rat PASMCs is shown in Figure 2. Under normoxic conditions superfusion of PA SMCs with experimental solution containing DTT reduced K^+ current and depolarized the resting E_m. DTNB showed a time-dependent membrane hyperpolarization of these cells under hypoxic conditions, which is consistent with the opening of K^+ channels.

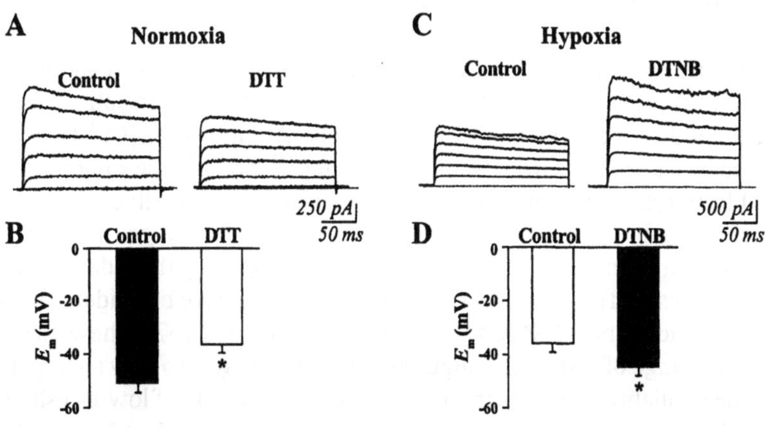

Figure 2. Effect of the reducing agent dithiothreitol (DTT) and oxidizing agent dithionitrobenzoic-acid (DTNB) on whole-cell outward K^+ current (I_K) and resting E_m of adult rat PASMCs. Currents were evoked from a holding potential of -70 mV to +50 mV in 20-mV steps. A: Representative traces demonstrate I_K under normoxic conditions (left) and after treatment with 3 mM DTT (right). B: DTT depolarizes the resting E_m in normoxia (n=5). C: The figure shows the augmenting effect of DTNB on I_K during hypoxia. D: DTNB causes PASMC hyperpolarization under hypoxic conditions (n=4). Data are mean±SE. * $P<0.05$ vs. control.

The studies presented here support the hypothesis that K^+ channels and in particular Kv channels are a target for redox active agents. These results suggest that pulmonary vascular tone can be modulated by electron-transfer agents at the level of K^+ channels. In addition, there appear to be similarities between the effects of hypoxia and reducing agents in PASMCs. Shifting the cellular redox status to a more reduced state, in both cases, causes pulmonary vasoconstriction by inhibition of K^+ channels and subsequent E_m depolarization.

4. Redox Modulation of Cytosolic Ca^{2+} Concentration

An increase of $[Ca^{2+}]_{cyt}$ is necessary to elicit vascular smooth muscle cell contraction. When $[Ca^{2+}]_{cyt}$ rises in the smooth muscle cell, Ca^{2+} binds to calmodulin. The Ca^{2+}/calmodulin complex activates myosin light chain kinase, which promotes myosin light chain phosphorylation and actin/myosin interaction. The actin/myosin crossbridges underlie smooth muscle contraction, prompting vasoconstriction. When calcium signaling is stimulated in smooth muscle cells, Ca^{2+} enters the cytoplasm from one of two general sources: it enters the cell across the plasmalemmal Ca^{2+} channels or it is released from intracellular Ca^{2+} stores, such as endoplasmic or sarcoplasmic reticulum (SR). These processes may occur either simultaneously or sequentially.

4.1. Redox Control of Ca^{2+} Channels

There are three main pathways by which extracellular Ca^{2+} can enter the PASMCs: *a*) voltage-dependent Ca^{2+} channels, *b*) receptor-activated Ca^{2+} channels, and *c*) store-operated Ca^{2+} channels. Additionally, $[Ca^{2+}]_{cyt}$ can be increased by Ca^{2+} release from two intracellular Ca^{2+} pools in the smooth muscle cells: *i*) from ryanodine-sensitive stores and *ii*) from IP_3-sensitive stores.

A growing body of work indicates that changes in oxygen-tension may initiate calcium signaling of smooth muscle cells by shifting the redox status. T-type (low-threshold) voltage-dependent Ca^{2+}-channels have been identified in a few tissues, including PASMCs (48) and sensory neurons (67). These channels act over the range of resting E_m, suggesting that they may also play an important role in the regulation of vascular tone. The redox-sensitivity of low-threshold T-type Ca^{2+}-channels was recently examined by Todorovic et al. (67), who found that the reducing agent DTT selectively augmented Ca^{2+}-current and the current was accelerated in the presence of DTT. The effects of DTT on T-type Ca^{2+}-currents were mimicked by the endogenous reducing agent L-cysteine. Similarly to DTT, L-cysteine also increased rates of activation and inactivation of the current. In contrast, the oxidizing agent DTNB blocked T-type Ca^{2+} channels at higher concentrations and, at lower concentrations DTNB fully reversed the effects of DTT on current kinetics, peak current and the time course of recovery. It is possible that other thiol-containing compounds like L-homocysteine or glutathione could also be endogenous redox modulators of T-type Ca^{2+} channels and that these agents could work in concert to upregulate this channel function. Recombinant T-type Ca^{2+} channel current flowing through the α1H subunit, expressed in HEK 293 cells, was found to be enhanced by the reducing agent glutathione (GSH) and inhibited by oxidized glutathione (GSSG) (27). In contrast, T-type Ca^{2+} current through the α1G subunit was unaffected by GSH, indicating that different members of the T-type Ca^{2+} channel family are differently regulated by redox agents and perhaps by hypoxia.

L-type voltage-dependent Ca^{2+} channels from vascular smooth cells have also been shown to be oxygen sensitive (29). Modulation by reducing and oxidizing agents of both native and recombinant L-type Ca^{2+} channels has been investigated by several groups, and differing effects are apparent. Campbell et al. demonstrated that DTT caused inhibition of native L-type Ca^{2+} current in ventricular myocytes and DTNB enhanced it (14). On the contrary, investigations by Fearon et al. showed that DTT was without effect on the recombinant L-type Ca^{2+} channel current which was not previously pretreated by the oxidizing agents thimerosal or by *p*-chloromercuribenzene sulphonic acid (PCMBS) (26). Using the stably transfected α1-subunit of L-type Ca^{2+} channels from PASMCs, it was found that the poorly membrane-permeable oxidizing agent thimerosal inhibited the whole-cell Ca^{2+} current. This effect was fully reversed by DTT, indicating that the inhibition of the current was dependent on oxidation of an extracellularly accessible cysteine residue (19).

4.2. Redox Regulation of Ca^{2+} Release from Sarcoplasmic Reticulum

Several studies on the redox regulation of the SR Ca^{2+}-release channel complex, consisting of the RyR, the FK506 binding protein and other associated proteins and molecules such as junctin, triadin and calmodulin, have been conducted on skeletal and cardiac muscle cells. Feng et al. have provided direct evidence that ryanodine channel activity follows transmembrane redox potential (28). The molecular basis of O_2-responsiveness in these cells was identified to involve 6-8 of 50 RyR thiols whose redox state is dynamically controlled, as described by Eu et al. (23). Oxygen exposure of native SR leads to activation of the RyR-coupled Ca^{2+} release channel due to the oxidation of 6 RyR thiols per subunit and enhanced ryanodine binding to RyR across the spectrum of physiological Ca^{2+} concentrations. From their studies on SR vesicles, Abramson and Salama (1, 2) proposed a model for redox regulation of the channel that involves three different sulfhydryl groups which exist in close proximity and react through thiol oxidation and thiol-disulfide interchange to open or close the channel. Biochemical evidence suggests that the SR Ca^{2+}-release channels in cardiac myocytes can also be regulated by the redox potential of the cells.

Boraso and Williams demonstrated that H_2O_2 increased the open probability of the Ca^{2+}-release channel in sheep and the channel activity was suppressed by DTT (Fig. 3) (11). Although they reported the involvement of H_2O_2 in the process of oxidative modification of the RyR, it is not clear if H_2O_2 itself reacted with RyR or that some product of H_2O_2, probably the hydroxyl radical is the true reactive agent. The effect of H_2O_2 and DTT on $[Ca^{2+}]_{cyt}$ in cardiomyocytes is consistent with our observations on the tension of the ductus arteriosus (DA) (60).

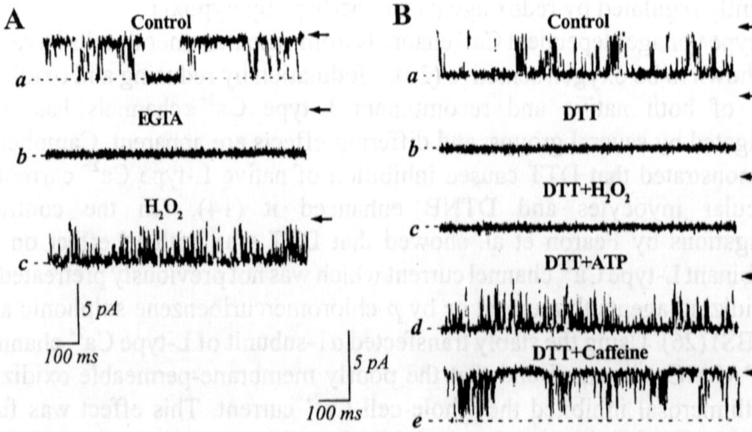

Figure 3. Modification of the gating of the cardiac SR Ca^{2+}-release channel by H_2O_2 and DTT. A: Representative current fluctuation in control (*a*) and after addition of 12 mM EGTA (*b*). 5 mM H_2O_2, applied to the cytosolic side of membrane, reopened the channel (*c*). B: Single channel recordings before (*a*) and after addition of DTT (*b*). Under these conditions H_2O_2 was not able to reopen channels (*c*). Application of ATP (*d*) and caffeine (*e*) increased the open probability of channels. Cardiac SR Ca^{2+}-release channels were activated by 10 μM Ca^{2+}. Holding potential for single-channel recording was 0 mV. Dashed lines, closed-channel level; arrows, open-channel level (From Ref. 11).

H_2O_2 constricted and DTT dilated DA rings consistently. The opposite effects of H_2O_2 and DTT on DA rings would indicate a potential mechanism for normoxia-induced vasoconstriction and hypoxia-induced vasodilatation in DA. Suzuki et al. showed that H_2O_2 enhanced Ca^{2+} release from SR (65). The augmentation of Ca^{2+} release requires the thiol reductant DTT or glutathione (GSH), suggesting that thiol-disulfide exchange reactions regulate Ca^{2+} release from the SR by converting the disulfide structure to another thiol through reduction (by GSH or DTT) and oxidation (H_2O_2) reactions. This is consistent with the proposal by Abramson and Salama in skeletal muscle that thiol-disulfide interchange reactions within the Ca^{2+}-release channel molecule convert the open/closed states of the channel (1, 2). Consistent with this observation is the observation that thiol reductants DTT and GSH suppress SR Ca^{2+}-release channel activity (11, 83). The results shown that redox modification of RyR channels affect Ca^{2+} signaling in cardiac and skeletal muscle. In both cases, oxidation produces a significant increase in Ca^{2+} release. Reversible activation of cardiac RyR channels by oxidation could be relevant in the heart, especially when there is an increase in free radical production such as in ischemia/reperfusion situations. Sustained or uncontrolled oxidation of RyR channels may elicit an imbalance in Ca^{2+} homeostasis, which could trigger apoptosis.

Wilson et al. presented a possible new mechanism for acute hypoxia-sensing in PASMCs involving Ca^{2+} release from ryanodine-sensitive SR stores. These observations suggest that acute hypoxia increases α-NADH levels, which then increase the net amount of cyclic ADP-ribose (cADPR) synthetized from β-NAD^+ by cADP-ribosyl cyclase, and simultaneously inhibit cADP degradation. The increased level of cADPR promotes Ca^{2+} release from RyR and elicits vasoconstriction (76).

The influence of ROS on Ca^{2+} release through IP_3R-coupled channels has been examined in smooth muscle cells. Kaplin et al. (35) investigated redox modulation of Ca^{2+} release in purified IP_3 receptors reconstituted in lipid vesicles. Reduced nicotinamide adenine dinucleotide (NADH) increased IP_3-mediated Ca^{2+} flux. $[Ca^{2+}]_{cyt}$ was markedly and specifically increased by direct intracellular injection of NADH, suggesting that direct regulation of IP_3 by NADH could be responsible for elevated $[Ca^{2+}]_{cyt}$ occurring in the earliest phase of hypoxia.

These findings suggest that in the mechanisms that are necessary for sensing and responding acutely to hypoxia, both intracellular and extracellular Ca^{2+} are important. However, the nature of the redox-related protein modification is in question and the molecular details are not known. It seems likely that sensitive thiol-groups within the channel complexes are an essential component of a transmembrane redox sensor and may be involved in mediating changes in Ca^{2+}-signaling during changes in Po_2.

Acknowledgments

Dr. E. K. Weir is supported by grants from the VA and NIH (HL 65322). Dr. A. Olschewski is supported by the Deutsche Forschungsgemeinschaft (DFG: Ol 127/1-1) and by the Award of the Justus-Liebig-University Gießen. We thank B. Agari for excellent technical assistance.

References

1. Abramson JJ, and Salama G. Sulfhydryl oxidation and Ca^{2+} release from sarcoplasmic reticulum. *Mol. Cell. Biochem.* 1988; 82 :81-84.
2. Abramson JJ, and Salama G. Critical sulfhydryls regulate calcium release from sarcoplasmic reticulum. *J. Bioenerg. Biomembr.* 1989; 21:283-294.
3. Albarwani S, Robertson BE, Nye PC, and Kozlowski RZ. Biophysical properties of Ca^{2+}- and Mg-ATP-activated K^+ channels in pulmonary arterial smooth muscle cells isolated from the

rat. *Pflügers Arch.* 1994; 428:446-454.

4. Archer SL, Huang J, Henry T, Peterson D, and Weir EK. A redox-based O_2 sensor in rat pulmonary vasculature. *Circ. Res.* 1993; 73:1100-1112.
5. Archer SL, Huang JM, Reeve HL, Hampl V, Tolarova S, Michelakis E, and Weir EK. Differential distribution of electrophysiologically distinct myocytes in conduit and resistance arteries determines their response to nitric oxide and hypoxia. *Circ. Res.* 1996; 78:431-442.
6. Archer SL, Nelson DP, and Weir EK. Detection of activated O_2 species in vitro and in rat lungs by chemiluminescence. *J. Appl. Physiol.* 1989; 67:1912-1921.
7. Archer SL, Reeve HL, Michelakis E, Puttagunta L, Waite R, Nelson DP, Dinauer MC, and Weir EK. O_2 sensing is preserved in mice lacking the gp91 phox subunit of NADPH oxidase. *Proc. Natl. Acad. Sci. U. S. A.* 1999; 96:7944-7949.
8. Archer SL, Souil E, Dinh-Xuan AT, Schremmer B, Mercier JC, El Yaagoubi A, Nguyen-Huu L, Reeve HL, and Hampl V. Molecular identification of the role of voltage-gated K^+ channels, Kv1.5 and Kv2.1, in hypoxic pulmonary vasoconstriction and control of resting membrane potential in rat pulmonary artery myocytes. *J. Clin. Invest.* 1998; 101:2319-2330.
9. Archer SL, Will JA, and Weir EK. Redox status in the control of pulmonary vascular tone. *Herz.* 1986; 11:127-141.
10. Berube J, Caouette D, and Daleau P. Hydrogen peroxide modifies the kinetics of HERG channel expressed in a mammalian cell line. *J. Pharmacol. Exp. Ther.* 2001; 297:96-102.
11. Boraso A and Williams AJ. Modification of the gating of the cardiac sarcoplasmic reticulum Ca^{2+}-release channel by H_2O_2 and dithiothreitol. *Am. J. Physiol.* 1994; 267:H1010-H1016.
12. Budinger GR, Chandel NS, Shao ZH, Li CQ, Mehmed A, Becker LB, and Schumacker PT. Cellular energy utilization and supply during hypoxia in embryonic cardiac myocytes. *Am. J. Physiol.* 1996; 270:L44-L53.
13. Budinger GR, Duranteau J, Chandel NS, and Schumacker PT. Hibernation during hypoxia in cardiomyocytes. Role of mitochondria as the O_2 sensor. *J. Biol. Chem.* 1998; 273:3330-3336.
14. Campbell DL, Stamler JS, and Strauss HC. Redox modulation of L-type calcium channels in ferret ventricular myocytes. Dual mechanism regulation by nitric oxide and S-nitrosothiols. *J. Gen. Physiol.* 1996; 108:277-293.
15. Chandel NS, Maltepe E, Goldwasser E, Mathieu CE, Simon MC, and Schumacker PT. Mitochondrial reactive oxygen species trigger hypoxia-induced transcription. *Proc. Natl. Acad. Sci. U. S. A.* 1998; 95:11715-11720.
16. Chandel NS, McClintock DS, Feliciano CE, Wood TM, Melendez JA, Rodriguez AM, and Schumacker PT. Reactive oxygen species generated at mitochondrial complex III stabilize hypoxia-inducible factor-1α during hypoxia: a mechanism of O_2 sensing. *J. Biol. Chem.* 2000; 275:25130-25138.
17. Chandel NS, and Schumacker PT. Cellular oxygen sensing by mitochondria: old questions, new insight. *J. Appl. Physiol.* 2000; 88:1880-1889.
18. Chander A, Dhariwal KR, Viswanathan R, and Venkitasubramanian TA. Pyridine nucleotides in lung and liver of hypoxic rats. *Life Sci.* 1980; 26:1935-1945.
19. Chiamvimonvat N, O'Rourke B, Kamp TJ, Kallen RG, Hofmann F, Flockerzi V, and Marban E. Functional consequences of sulfhydryl modification in the pore-forming subunits of cardiovascular Ca^{2+} and Na^+ channels. *Circ. Res.* 1995; 76:325-334.
20. Chung S, Jung W, Uhm DY, Ha TS, and Park CS. Glutathione potentiates cloned rat brain large conductance Ca^{2+}-activated K^+ channels (r*Slo*). *Neurosci. Lett.* 2002; 318:9-12.
21. Des Marais DJ. Earth's early biosphere. *Gravit. Space Biol. Bull.* 1998; 11:23-30.
22. Dipp M, Nye PC, and Evans AM. Hypoxic release of calcium from the sarcoplasmic reticulum of pulmonary artery smooth muscle. *Am. J. Physiol. Lung Cell. Mol. Physiol.* 2001; 281:L318-L325.
23. Eu JP, Sun J, Xu L, Stamler JS, and Meissner G. The skeletal muscle calcium release channel: coupled O_2 sensor and NO signaling functions. *Cell.* 2000; 102:499-509.
24. Evans AM and Dipp M. Hypoxic pulmonary vasoconstriction: cyclic adenosine diphosphate-ribose, smooth muscle Ca^{2+} stores and the endothelium. *Respir. Physiol. Neurobiol.* 2002;

132:3-15.

25. Evans AM, Osipenko ON, and Gurney AM. Properties of a novel K^+ current that is active at resting potential in rabbit pulmonary artery smooth muscle cells. *J. Physiol.* 1996; 496:407-420.
26. Fearon IM, Palmer AC, Balmforth AJ, Ball SG, Varadi G, and Peers C. Modulation of recombinant human cardiac L-type Ca^{2+} channel α_{1C} subunits by redox agents and hypoxia. *J. Physiol.* 1999; 514:629-637.
27. Fearon IM, Randall AD, Perez-Reyes E, and Peers C. Modulation of recombinant T-type Ca^{2+} channels by hypoxia and glutathione. *Pflügers Arch.* 2000; 441:181-188.
28. Feng W, Liu G, Allen PD, and Pessah IN. Transmembrane redox sensor of ryanodine receptor complex. *J. Biol. Chem.* 2000; 275:35902-35907.
29. Franco-Obregon A and Lopez-Barneo J. Low PO_2 inhibits calcium channel activity in arterial smooth muscle cells. *Am. J. Physiol.* 1996; 271:H2290-H2299.
30. Fu XW, Wang D, Nurse CA, Dinauer MC, and Cutz E. NADPH oxidase is an O_2 sensor in airway chemoreceptors: evidence from K^+ current modulation in wild-type and oxidase-deficient mice. *Proc. Natl. Acad. Sci. U. S. A.* 2000; 97:4374-4379.
31. Gabig TG, Bearman SI, and Babior BM. Effects of oxygen tension and pH on the respiratory burst of human neutrophils. *Blood.* 1979; 53:1133-1139.
32. Gong L, Gao TM, Huang H, and Tong Z. Redox modulation of large conductance calcium-activated potassium channels in CA1 pyramidal neurons from adult rat hippocampus. *Neurosci. Lett.* 2000; 286-191-194.
33. Harder DR, Madden JA, and Dawson C. Hypoxic induction of Ca^{2+}-dependent action potentials in small pulmonary arteries of the cat. *J. Appl. Physiol.* 1985; 59:1389-1393.
34. Jin N, Packer CS, and Rhoades RA. Pulmonary arterial hypoxic contraction: signal transduction. *Am. J. Physiol.* 1992; 263:L73-L78.
35. Kaplin AI, Snyder SH, and Linden DJ. Reduced nicotinamide adenine dinucleotide-selective stimulation of inositol 1,4,5-trisphosphate receptors mediates hypoxic mobilization of calcium. *J. Neurosci.* 1996; 16:2002-2011.
36. Kato M, and Staub NC. Response of small pulmonary arteries to unilobar hypoxia and hypercapnia. *Circ. Res.* 1966; 19:426-440.
37. Killilea DW, Hester R, Balczon R, Babal P, and Gillespie MN. Free radical production in hypoxic pulmonary artery smooth muscle cells. *Am. J. Physiol.* 2000; 279:L408-L412.
38. Kummer W and Acker H. Immunohistochemical demonstration of four subunits of neutrophil NAD(P)H oxidase in type I cells of carotid body. *J. Appl. Physiol.* 1995; 78:1904-1909.
39. Lang RJ, Harvey JR, McPhee GJ, and Klemm MF. Nitric oxide and thiol reagent modulation of Ca^{2+}-activated K^+ (BK_{Ca}) channels in myocytes of the guinea-pig taenia caeci. *J. Physiol.* 2000; 525:363-376.
40. Madden JA, Dawson CA, and Harder DR. Hypoxia-induced activation in small isolated pulmonary arteries from the cat. *J. Appl. Physiol.* 1985; 59:113-118.
41. Madden JA, Vadula MS, and Kurup VP. Effects of hypoxia and other vasoactive agents on pulmonary and cerebral artery smooth muscle cells. *Am. J. Physiol.* 1992; 263:L384-L393.
42. Marshall C, Mamary AJ, Verhoeven AJ, and Marshall BE. Pulmonary artery NADPH-oxidase is activated in hypoxic pulmonary vasoconstriction. *Am. J. Respir. Cell Mol. Biol.* 1996; 15:633-644.
43. Michelakis ED, Hampl V, Nsair A, Wu X, Harry G, Haromy A, Gurtu R, and Archer SL. Diversity in mitochondrial function explains differences in vascular oxygen sensing. *Clrc. Res.* 2002; 90:1307-1315.
44. Michelakis ED, Rebeyka I, Wu XC, Nsair A, Thébaud B, Hashimoto K, Dyck JRB, Haromy A, Harry G, Barr A, and Archer SL. O_2 sensing in the human ductus arteriosus - Regulation of voltage-gated K^+ channels in smooth muscle cells by a mitochondrial redox sensor. *Circ. Res.* 2002; 91:478-486.
45. Mills E, and Jobsis FF. Mitochondrial respiratory chain of carotid body and chemoreceptor response to changes in oxygen tension. *J. Neurophysiol.* 1972; 35:405-428.

46. Mohazzab KM, Fayngersh RP, Kaminski PM, and Wolin MS. Potential role of NADH oxidoreductase-derived reactive O_2 species in calf pulmonary arterial PO_2-elicited responses. *Am. J. Physiol.* 1995; 269:L637-L644.
47. Mohazzab KM, and Wolin MS. Properties of a superoxide anion-generating microsomal NADH oxidoreductase, a potential pulmonary artery PO_2 sensor. *Am. J. Physiol.* 1994; 267:L823-L831.
48. Muramatsu M, Tyler RC, Rodman DM, and McMurtry IF. Possible role of T-type Ca^{2+} channels in L-NNA vasoconstriction of hypertensive rat lungs. *Am. J. Physiol.* 1997; 272:H2616-H2621.
49. Nelson MT, and Quayle JM. Physiological roles and properties of potassium channels in arterial smooth muscle. *Am. J. Physiol.* 1995; 268:C799-C822.
50. Nunn FJ. Evolution of the atmosphere. *Proc. Geol. Assoc.* 1998; 109:1-13.
51. O'Kelly I, Lewis A, Peers C, and Kemp PJ. O_2 sensing by airway chemoreceptor-derived cells. Protein kinase c activation reveals functional evidence for involvement of NADPH oxidase. *J. Biol. Chem.* 2000; 275:7684-7692.
52. Osipenko ON, Evans AM, and Gurney AM. Regulation of the resting potential of rabbit pulmonary artery myocytes by a low threshold, O_2-sensing potassium current. *Br. J. Pharmacol.* 1997; 120:1461-1470.
53. Paky A, Michael JR, Burke-Wolin TM, Wolin MS, and Gurtner GH. Endogenous production of superoxide by rabbit lungs: effects of hypoxia or metabolic inhibitors. *J. Appl. Physiol.* 1993; 74:2868-2874.
54. Park MK, Bae YM, Lee SH, Ho WK, and Earm YE. Modulation of voltage-dependent K^+ channel by redox potential in pulmonary and ear arterial smooth muscle cells of the rabbit. *Pflügers Arch.* 1997; 434:764-771.
55. Park MK, Lee SH, Lee SJ, Ho WK, and Earm YE. Different modulation of Ca-activated K channels by the intracellular redox potential in pulmonary and ear arterial smooth muscle cells of the rabbit. *Pflügers Arch.* 1995; 430:308-314.
56. Peng W, Hoidal JR, and Farrukh IS. Role of a novel KCa opener in regulating K^+ channels of hypoxic human pulmonary vascular cells. *Am. J. Respir. Cell Mol. Biol.* 1999; 20:737-745.
57. Peterson DA, Peterson DC, Reeve HL, Archer SL, and Weir EK. GTP (γS) and GDP (βS) as electron donors: new wine in old bottles. *Life Sci.* 1999; 65:1135-1140.
58. Post JM, Gelband CH, and Hume JR. $[Ca^{2+}]_i$ inhibition of K^+ channels in canine pulmonary artery. Novel mechanism for hypoxia-induced membrane depolarization. *Circ. Res.* 1995; 77:131-139.
59. Post JM, Hume JR, Archer SL, and Weir EK. Direct role for potassium channel inhibition in hypoxic pulmonary vasoconstriction. *Am. J. Physiol.* 1992; 262:C882-C890.
60. Reeve HL, Tolarova S, Nelson DP, Archer S, and Weir EK. Redox control of oxygen sensing in the rabbit ductus arteriosus. *J. Physiol.* 2001; 533:253-261.
61. Reeve HL, Weir EK, Nelson DP, Peterson DA, and Archer SL. Opposing effects of oxidants and antioxidants on K^+ channel activity and tone in rat vascular tissue. *Exp. Physiol.* 1995; 80:825-834.
62. Salvaterra CG, and Goldman WF. Acute hypoxia increases cytosolic calcium in cultured pulmonary arterial myocytes. *Am. J. Physiol.* 1993; 264:L323-L328.
63. Schafer FQ, and Buettner GR. Redox environment of the cell as viewed through the redox state of the glutathione disulfide/glutathione couple. *Free Radic. Biol. Med.* 2001; 30:1191-1212.
64. Shigemori K, Ishizaki T, Matsukawa S, Sakai A, Nakai T, and Miyabo S. Adenine nucleotides via activation of ATP-sensitive K^+ channels modulate hypoxic response in rat pulmonary arteries. *Am. J. Physiol.* 1996; 270:L803-L809.
65. Suzuki YJ, Cleemann L, Abernethy DR, and Morad M. Glutathione is a cofactor for H_2O_2-mediated stimulation of Ca^{2+}-induced Ca^{2+} release in cardiac myocytes. *Free Radic. Biol. Med.* 1998; 24:318-325.
66. Thuringer D, and Findlay I. Contrasting effects of intracellular redox couples on the

regulation of maxi-K channels in isolated myocytes from rabbit pulmonary artery. *J. Physiol.* 1997; 500:583-592.

67. Todorovic SM, Jevtovic-Todorovic V, Meyenburg A, Mennerick S, Perez-Reyes E, Romano C, Olney JW, and Zorumski CF. Redox modulation of T-type calcium channels in rat peripheral nociceptors. *Neuron.* 2001; 31:75-85.
68. Vejlstrup NG, and Dorrington KL. Intense slow hypoxic pulmonary vasoconstriction in gas-filled and liquid-filled lungs: an in vivo study in the rabbit. *Acta Physiol. Scand.* 1993; 148:305-313.
69. Wang D, Youngson C, Wong V, Yeger H, Dinauer MC, Vega-Saenz ME, Rudy B, and Cutz E. NADPH-oxidase and a hydrogen peroxide-sensitive K^+ channel may function as an oxygen sensor complex in airway chemoreceptors and small cell lung carcinoma cell lines. *Proc. Natl. Acad. Sci. U. S. A.* 1996; 93:13182-13187.
70. Waypa GB, Chandel NS, and Schumacker PT. Model for hypoxic pulmonary vasoconstriction involving mitochondrial oxygen sensing. *Circ. Res.* 2001; 88:1259-1266.
71. Waypa GB, Marks JD, Mack MM, Boriboun C, Mungai PT, and Schumacker PT. Mitochondrial reactive oxygen species trigger calcium increases during hypoxia in pulmonary arterial myocytes. *Circ. Res.* 2002; 91:719-726.
72. Weir EK, and Archer SL. The mechanism of acute hypoxic pulmonary vasoconstriction: the tale of two channels. *FASEB J.* 1995; 9:183-189.
73. Weissmann N, Grimminger F, Walmrath D, and Seeger W. Hypoxic vasoconstriction in buffer-perfused rabbit lungs. *Respir. Physiol.* 1995; 100:159-169.
74. Welling KL, Sanchez R, Ravn JB, Larsen B, and Amtorp O. Effect of prolonged alveolar hypoxia on pulmonary arterial pressure and segmental vascular resistance. *J. Appl. Physiol.* 1993; 75:1194-1200.
75. Wiener CM, and Sylvester JT. Effects of glucose on hypoxic vasoconstriction in isolated ferret lungs. *J. Appl. Physiol.* 1991; 70:439-446.
76. Wilson HL, Dipp M, Thomas JM, Lad C, Galione A, and Evans AM. Adp-ribosyl cyclase and cyclic ADP-ribose hydrolase act as a redox sensor: a primary role for cyclic ADP-ribose in hypoxic pulmonary vasoconstriction. *J. Biol. Chem.* 2001; 276:11180-11188.
77. Wolin MS, Burke-Wolin TM, and Mohazzab H. Roles for NAD(P)H oxidases and reactive oxygen species in vascular oxygen sensing mechanisms. *Respir. Physiol.* 1999; 115:229-238.
78. Youngson C, Nurse C, Yeger H, and Cutz E. Oxygen sensing in airway chemoreceptors. *Nature* 1993; 365:153-155.
79. Yuan X-J. Voltage-gated K^+ currents regulate resting membrane potential and $[Ca^{2+}]_i$ in pulmonary arterial myocytes. *Circ. Res.* 1995; 77:370-378.
80. Yuan X-J, Goldman WF, Tod ML, Rubin LJ, and Blaustein MP. Hypoxia reduces potassium currents in cultured rat pulmonary but not mesenteric arterial myocytes. *Am. J. Physiol.* 1993; 264:L116-L123.
81. Yuan X-J, Tod ML, Rubin LJ, and Blaustein MP. Contrasting effects of hypoxia on tension in rat pulmonary and mesenteric arteries. *Am. J. Physiol.* 1990; 259:H281-H289.
82. Yuan X-J, Tod ML, Rubin LJ, and Blaustein MP. Deoxyglucose and reduced glutathione mimic effects of hypoxia on K^+ and Ca^{2+} conductances in pulmonary artery cells. *Am. J. Physiol.* 1994; 267:L52-L63.
83. Zaidi NF, Lagenaur CF, Abramson JJ, Pessah I, and Salama G. Reactive disulfides trigger Ca^{2+} release from sarcoplasmic reticulum via an oxidation reaction. *J. Biol. Chem.* 1989; 264:21725-21736.
84. Zeidner G, Sadja R, and Reuveny E. Redox-dependent gating of G protein-coupled inwardly rectifying K^+ channels. *J. Biol. Chem.* 2001; 276:35564-35570.
85. Zulueta JJ, Yu FS, Hertig IA, Thannickal VJ, and Hassoun PM. Release of hydrogen peroxide in response to hypoxia-reoxygenation: role of an NAD(P)H oxidase-like enzyme in endothelial cell plasma membrane. *Am. J. Respir. Cell Mol. Biol.* 1995; 12:41-49.

Chapter 17

Mitochondrial Diversity in the Vasculature: Implications for Vascular Oxygen Sensing

Sean McMurtry and Evangelos D. Michelakis
University of Alberta, Edmonton, Canada

1. Introduction

In recent years, it has become clear that mitochondria play a central role in several aspects of animal physiology and pathophysiology (18). Although the main role of mitochondria is to produce ATP, additional roles have recently been recognized. Mitochondria take up Ca^{2+} and play a key role in cellular Ca^{2+} homeostasis; the high capacity of mitochondrial Ca^{2+} uptake provides a means of matching energy demand with ATP production, since several mitochondrial enzymes are activated by Ca^{2+} (18). Mitochondria have also been proposed as important O_2 sensors in a variety of O_2-sensitive tissues, including the pulmonary arteries, the ductus arteriosus, and the carotid body (3). The role of mitochondria as O_2 sensors (see Chapter 9) allows them to match energy (or O_2) demand with O_2 delivery in the body, either by optimizing ventilation-perfusion matching (via hypoxic pulmonary vasoconstriction) or O_2 delivery (via activation of the brain respiration center by the carotid body). An important factor in the mitochondria-based O_2 sensing pathways is the activated oxygen species (AOS) produced in the mitochondrial electron transport chain (ETC). Some AOS, like hydrogen peroxide (H_2O_2), have a long effective diffusion radius and can leak to the cytosol and membrane and activate second messengers and ion channels that control vascular tone (3). Furthermore, mitochondria are recognized as a key regulator of apoptosis (22), a critical factor in the development or regression of pulmonary vascular remodeling (15, 28). By being involved in the regulation of both *vascular tone* and *vascular remodeling*, mitochondria might be more important in vascular disease than currently recognized and will undoubtedly be excellent targets for drug development.

There is definitive evidence that mitochondrial diversity exists among different organs or tissues. In addition to summarizing the existing evidence for such diversity, this chapter introduces the concept of mitochondrial diversity in the vasculature. We discuss recently published evidence that mitochondria are different between the vascular smooth muscle cells (SMC) in pulmonary versus systemic arteries. Such diversity might at least in part explain the still unresolved "mystery" of why the pulmonary arteries (PA) constrict to hypoxia while the systemic arteries dilate or why the pulmonary circulatory system is a "low pressure" circulation compared to the systemic circulatory system. Furthermore, this diversity might provide clues for the development of drugs that will target the pulmonary but not the systemic arteries, a much-desired feature of the ideal candidate treatment for pulmonary arterial hypertension.

2. Mitochondrial Diversity: Evidence from Mitochondrial Diseases

Mitochondrial abnormalities have been implicated in the pathogenesis of many genetic diseases and several degenerative diseases like Parkinson's (50). In most genetic mitochondrial diseases there is a very specific mutation in the mitochondrial or nuclear DNA encoding a specific mitochondrial protein. For example, Leber's hereditary optical neuropathy (LHON), a highly selective degeneration of the optic nerve, presents with acute blindness in mid-life. A specific mutation in the mitochondrial gene (ND6) encoding for a subunit of NADH dehydrogenase has been identified (51). This results in a substantial decrease in the activity of complex I of the ETC. Complex I is a critical part of the mitochondrial ETC and thus of cellular respiration. However, the phenotype of the LHON is surprisingly restricted to the optic nerve.

In Friedrich's ataxia, an autosomal recessive disease that results in cerebellar ataxia, peripheral neuropathy, and hypertrophic cardiomyopathy, the mutant protein, frataxin, is targeted to the mitochondrial inner membrane and functions to transport iron out of the mitochondrion (47). With the loss of this protein, iron accumulates in the mitochondrial matrix, stimulating the conversion of H_2O_2 to $OH^{\cdot}$ by the Fenton reaction. This inactivates the mitochondrial Fe-S center enzymes (complexes I, II, III, and aconitase), which in turn reduces mitochondrial energy production (47). The phenotype of this ataxia is restricted to the central nervous system (CNS) and the heart.

These observations suggest the presence of mitochondrial diversity among tissues, where the role or the relative importance of a mitochondrial protein or pathway varies from one tissue to another; disease is expressed only in tissues in which the affected mitochondrial proteins are of critical importance.

3. Diversity of Mitochondrial among Organs: Evidence from Metabolic Pathways

It has been well-known that some organ-specific metabolic pathways represent good examples of mitochondrial diversity (33).

3.1. Liver mitochondria

Synthesis of carbamoyl-phosphate and citrulline, first two steps of urea synthesis, is located only within liver mitochondria. A liver-specific ornithine-citrulline antiporter is responsible for the import of ornithine into, and the export of citrulline out of the liver mitochondria (33). Gluconeogenesis is a liver- and kidney-specific pathway. The formation of oxaloacetate occurs via pyruvate carboxylase in the mitochondrial matrix and plays a central role in controlling gluconeogenesis (24). Another peculiarity of liver mitochondria is the generation of the ketone bodies acetoactate and b-hydroxybutyrate from acetyl-CoA. b-hydroxy-butyrate dehydrogenase, which converts acetoacetate to b-hydroxy-butyrate using mitochondrial NADH, is liver-specific mitochondrial enzyme and can therefore be used to indicate mitochondrial NAD-redox state in liver (54).

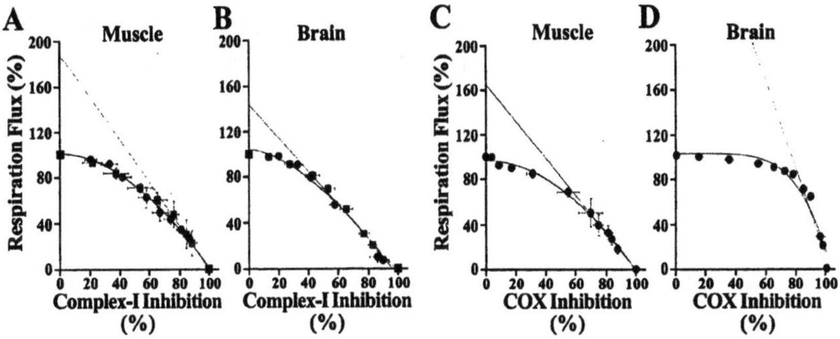

Figure 1. Different respiration rates and activities of ETC complex I and IV in muscle versus brain mitochondria. The inhibition plots are constructed from the relative ADP-stimulated O_2 consumption rates plotted against the percentage decrease in NADH:coenzyme Q1 reductase activity (complex I) at identical concentrations of amytal (•) and rotenone (■) (A and B) or plotted against the percentage decrease in COX activity (complex IV) at identical KCN concentrations (C and D). The intersection of the regression line with the ordinate gives the reserve capacity of the enzyme investigated (From Ref. 33).

3.2. Kidney Mitochondria

Mitochondrial glutaminase I, catalyzes the kidney-specific desamination of glutamine to glutamate and ammonium ions (30).

3.3. Sperm Mitochondria

An intriguing property of sperm mitochondria is the existence of a lactic dehydrogenase, which is in redox equilibrium with the mitochondrial NAD-system (11, 39). This presumably allows a very effective utilization of lactic acid, which is present in high quantities in the fluids of the female genital tract.

This diversity of mitochondrial function among organs is not restricted only within specific metabolic pathways due to the presence or absence of a specific enzyme. The overall function of mitochondria, as measured by respiration (i.e., O_2 consumption) from different organs is often different, suggesting differences in the expression or the activity of ETC complexes. In Figure 1, differences in the respiration between brain and muscle mitochondria are shown. The difference in the activities of complex I and IV (cytochrome *c* oxidase, COX) is shown by the response to complex I and IV inhibitors (rotenone and cyanide), while the experimental conditions are the same.

Differential expression of complex I has been described in the brain (43) and significant diversity has also been described in complex IV (COX) (33). Isoforms have been found for several COX subunits and are expressed in a tissue-specific and developmental manner (Fig. 2) (33). For example, for the isoform of subunit IV, high IV-2 expression was observed in adult lung and lower expression in all other tissues investigated, including fetal lung, implying that there may be effects of O_2 on ETC subunit gene expression (26) (Fig. 2).

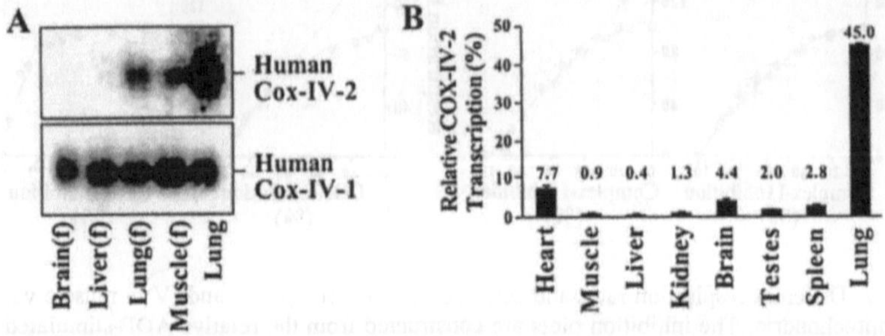

Figure 2. Diversity of cytochrome c oxidase isoforms mRNA expression in human and rat tissues. A: Northern blot analysis showing mRNA expression of cox-IV-1 and cox-IV-2 in human fetal (f) and adult tissues. The developmental stage of fetal tissues is 20 (brain), 24 (liver), 37 (lung), and 28 (muscle) weeks. B: Quantitative PCR analysis of mRNA levels of the cytochrome c oxidase subunit IV-2 isoform in rat tissues (From Ref. 26).

Diversity in the expression of the mitochondrial enzyme pyruvate dehydrogenase kinase (PDK) isoforms has also been described (10). PDK-2 is differentially expressed in the lungs versus the heart or systemic smooth muscle. Thus the PDK-2 inhibitor dichloroacetate (lower K_m for PDK-2 than for PDK 1-3-4) (10) may have selective effects on the pulmonary circulation. This is in agreement with the finding that dichloroacetate reverses pulmonary hypertension (perhaps via redox effects on K^+ channel expression and function) without affecting the left ventricle or lowering systemic blood pressure (38). This is an example of taking advantage of mitochondrial diversity to design therapeutic strategies for vascular disease that are selective and specific.

4. Diversity of Mitochondria Within Cells

The presence of two distinct intracellular populations of mitochondria within both skeletal and cardiac muscles has been well known (42, 53). Subsarcolemmal mitochondria are dispersed beneath the muscle cell membrane while intermyofibrillar mitochondria penetrate the myofibrillar apparatus. The advantage of two distinct, specialized mitochondrial populations may relate to their location with respect to metabolite production within the cell, and to O_2 gradients and blood-borne metabolites in relation to capillaries. The biochemical differences between these two populations emerge when the mitochondria are studied *in vitro* (42, 53). In general, intermyofibrillar mitochondria have a higher VO_2, a greater capacity to oxidize fatty acids, and higher rates of protein import. These differences are not due to differential damaging of the two populations of mitochondria during centrifugation in the isolation process. Recent studies in intact cells have shown that functional and morphological heterogeneities occur in vivo in many cell types (14).

5. Mitochondria and the Vasculature

Although vascular mitochondria are emerging as leading candidates for O_2 sensors, their role in vascular function in health and disease is unexplored. Mitochondria are likely to be important regulators of vascular function since they are involved in important mechanisms of vascular tone control and vascular remodeling (e.g., regulation of important apoptosis pathways).

5.1. Mitochondria-derived AOS Regulate Vascular Tone

Mitochondria are major sources of AOS, primarily superoxide and H_2O_2. Up to 5% of the molecular O_2 consumed during mitochondrial respiration is reduced univalently to superoxide and most if not all of the H_2O_2 produced in mitochondria arises from dismutation of superoxide (18). The dismutation is mediated by a mitochondria-specific superoxide dismutase (SOD), manganese SOD (MnSOD). At high levels, AOS can be detrimental for the cell mostly due to their potential for DNA damage. It has been proposed that diseased mitochondria that produce high levels of AOS can initiate apoptosis in an effort to prevent the survival of cells carrying damaged/mutated DNA (18). However, at low levels, mitochondria-derived AOS can mediate important cellular functions. For example, mitochondrial-derived superoxide initiates the stabilization of hypoxia-inducible factor1α (13), which in turn induces the expression of several genes important in hypoxia, such as erythropoietin. Mitochondria-derived H_2O_2 (9), which has a much longer effective diffusion radius than superoxide (55), can diffuse to the cytoplasm and activate soluble guanylate cyclase (12, 56) or sarcolemmal K^+ channels in vascular smooth muscle (6, 25, 52), thus causing vasodilation.

5.2. The Production of AOS is Closely Linked to Mitochondrial Respiration and Membrane Potential

Substrates (glucose, fatty acids, etc.) enter the tricarboxylic acid cycle (TCA) promoting reduction of NAD to NADH and FAD to $FADH_2$ (see Chapter 9). As these are re-oxidized, they supply electrons to the ETC (complexes I to IV). These electrons are transferred through the enzyme complexes to O_2, producing H_2O. During this process, protons are translocated out of the inner mitochondrial membrane, generating a potential across the inner membrane between -150 to - 200 mV ($\Delta\Psi_m$). Therefore, the $\Delta\Psi_m$ reflects the level of the respiratory activity in the mitochondria (18). ATP synthesis takes place at a different site, the ATP synthase (complex V), which contains a proton channel (F_0) and an ATP synthase (F_1). This enzyme is driven by the downhill movement of protons through the F_0 channel, and phosphorylates ADP to produce ATP. In other words, the enzyme uses the stored energy of $\Delta\Psi_m$ to synthesize ATP, thus coupling respiration to ATP synthesis. A decreased proton gradient (decreased $\Delta\Psi_m$) would result in decreased activity of the ATP synthase (18). Decreased activity in any ETC complex would disrupt the normal flow of electrons and the associated pumping of protons, thereby decreasing $\Delta\Psi_m$. When $\Delta\Psi_m$ decreases significantly, the ATP synthase reverses function and starts ATP hydrolysis in order to pump protons out of the matrix and restore $\Delta\Psi_m$. This is critical for the mitochondrial survival since the proton gradient confers an osmotic gradient in addition to the electrical gradient, and thus decreased $\Delta\Psi_m$ results in mitochondrial swelling. This is not surprising if one thinks of the "relative independence" of the mitochondria, which are thought to be the descendants of a bacterial endosymbiont (18).

The source of AOS production in mitochondria is complexes I and III (18). In some tissues, the primary source of AOS is complex I (7, 27) while in others, it is complex III (13, 49). The ETC complexes consist of many subunits. For example, complex I consists of 42 subunits, 7 of which are encoded by mitochondrial DNA (mtDNA) (18). Inhibition of the complex proximal to the site of AOS production site would cause a decrease in AOS production. Inhibition of the complex distal to this site would disrupt the flow of electron down the respiratory chain, diverting them to react with O_2 and generate AOS. Mutations in several subunits of complex I have been described in many human muscular and neurodegenerative diseases that are associated with increased neuronal AOS production, including Parkinson's disease (18). Fibroblast mitochondria from patients with complex I and complex III deficiency show increased superoxide production, but not mitochondria from patients with complex IV deficiency (45). Barrientos and Moraes used a genetic model (40%

complex I-inhibited human-ape xenomitochondrial hybrids) and a drug-induced model (0-100% complex I-inhibited cells using different concentrations of rotenone) and showed a direct quantitative correlation between the level of complex I function, cell respiration, AOS production and $\Delta\Psi_m$. Inhibition of complex I results in collapse of the proton gradient (decreased $\Delta\Psi_m$) and therefore decrease in the mitochondrial activity including decreased complex I activity, respiration, and $\Delta\Psi_m$, and increased superoxide production (7).

There are 3 forms of SOD: CuSOD found primarily in the cytoplasm, extracellular SOD present in the vascular wall, and MnSOD (55). Unlike the other two, MnSOD undergoes significant modulation. Dynamic induction of MnSOD occurs through a redox-sensitive mediator (44). Many patients with complex I (but not complex IV) deficiency have increased MnSOD levels (45). This is a defensive mechanism of the mitochondrion to the stress induced by the increased superoxide production that the patients display. It is possible therefore that differences in a cell's redox status might result in different levels of expressed MnSOD.

5.3. Apoptosis

Apoptosis or programmed cell death is a process through which multicellular organisms dispose of cells efficiently (8). This energy-dependent process is distinct from the more conventional necrotic death. Apoptosis occurs without loss of integrity of the plasma membrane, and so avoids inciting inflammatory responses due to spillage of cellular contents, as occurs with necrosis (8). Apoptotic cells are cleared *in vivo* by phagocytosis, with the dying cells literally being eaten alive, and the resulting corpses being digested so that the constituent components are recycled back into the body. The regulation of apoptosis can be simplified into two major pathways: receptor-mediated (5) and mitochondria-dependent (22, 59).

Mitochondria-dependent apoptosis, is initiated by the opening of the mitochondrial transition pore (MTP), a mega channel through which inducers of apoptosis including cytochrome *c* (a caspase activator), Smac (second mitochondria-derived activator of caspases, also known as DIABLO) and apoptosis inducing factor (AIF, a nuclease activator) leak to the cytoplasm (22, 59). The MTP is formed at the contact of inner and outer mitochondria membrane and is composed of the adenine nucleotide translocator (ANT, located on the inner membrane, which is otherwise impermeable) and the voltage-dependent anion channel (VDAC, on the outer membrane site). Cyclophilin D (which is inhibited by cyclosporine A) is also a part of the pore on the mitochondrial matrix (59). A multitude of factors control the opening of the MTP, including $\Delta\Psi_m$, Ca^{2+}, pH, and redox signals, as well as anti- and pro-apoptotic members of the Bcl-2 family (59). Mitochondrial depolarization is associated with increased AOS production (as discussed above), and opening of the MTP and leakage of cytochrome *c*. After it leaks into the cytoplasm, cytochrome *c* forms a complex with Apaf-1 and procaspase-9. The result is activation of caspase-9 which, in turn, activates caspase-3 and several other caspases to complete the process of apoptosis (22).

An intriguing relation between the sarcolemmal K^+ channel activity (which regulates intracellular K^+ levels) and apoptosis has recently been shown in PASMC (31, 32, 46). Downregulation of K^+ channels in PASMC from patients with pulmonary arterial hypertension (PAH) contributes to apoptosis suppression by maintaining a high level of intracellular K^+, and to the vascular remodeling that characterizes PAH (57). The interplay between mitochondria and the K^+ channels in the plasma membrane (that also exhibit significant diversity of expression within the pulmonary arterial tree (2)) will undoubtedly provide new insights in the pathogenesis and treatment of vascular disease.

In summary, vascular SMC mitochondria, either by the production of AOS or by the opening of the MTP can regulate vascular tone and apoptosis, respectively. There is now evidence that induction of apoptosis might be a critical event in the regression of vascular remodeling (especially medial hypertrophy) that occurs in the treatment of vascular disease like pulmonary hypertension (15). Therefore, in addition to their role in vascular O_2 sensing, mitochondria might be more important than recognized in both normal vascular function and vascular disease.

6. A Comprehensive Study of Vascular Mitochondria

Both the production of AOS and the opening of the MTP are directly related to mitochondrial respiration and mitochondrial $\Delta\Psi_m$. Therefore, measuring AOS production (using chemiluminescence, AmplexRed and DCF assays), respiration (with classic polarographic methods) and $\Delta\Psi_m$ (using $\Delta\Psi_m$-sensitive dyes and dynamic multiphoton confocal imaging, which allows for measurement of $\Delta\Psi_m$ *in situ* in living cells) allows for a functional study of mitochondria. Along with perfused organs, tissue baths and isolated vascular SMC electrophysiology, these methods can be used for a comprehensive study of the role of mitochondria in vascular function.

We used these methods and addressed the hypothesis that mitochondrial diversity exists within the vasculature and specifically between the pulmonary and systemic vascular beds (37). We speculated that such diversity might explain why the pulmonary and systemic circulations respond in opposing ways to hypoxia. In the oxygen sensing model described by Moudgil et al. (see Chapter 9), PASMC mitochondria (the sensors) respond to changes in O_2 tension by the production of AOS (the mediator) that tonically regulate the function of PASMC voltage-gated K^+ channels (K_V, the effector) (3). Since K_V channels are present in both the pulmonary and renal circulations (19, 21, 36) we speculated that the opposing responses to hypoxia might be due to differential production of AOS between the PA and systemic vascular SMC in response to hypoxia. Furthermore potential differences in the tonic production of AOS at baseline (normoxia) might even help explain why the baseline pressure of the pulmonary circulation is lower compared to the systemic circulation.

This hypothesis was based on the following previously described observations. First, the mechanism of O_2 sensing lies within the vascular SMC and although modulated by endothelial or circulating factors, it occurs independent of them. Madden et al. have shown that pulmonary SMC contract and have increased intracellular Ca^{2+} levels in response to hypoxia, while cerebral vascular SMC relax and have decreased Ca^{2+} levels (35). Second, the only class of drugs that can mimic hypoxia in a variety of O_2-sensitive tissues (PA, ductus arteriosus, carotid body) is the group of inhibitors of the proximal ETC, i.e., rotenone and antimycin-A which block the AOS-producing complexes I and III, respectively (3). Both hypoxia and these inhibitors cause PASMC contraction, decreased AOS levels, and inhibition of K^+ currents (1, 4).

Rotenone mimics hypoxia constricting the pulmonary while dilating the renal circulation. Rotenone (5 μM) and physiological hypoxia (Po_2, ~40 mmHg) both constrict pulmonary arteries (PA) whereas they dilate the renal arteries (RA) (Fig. 3A). As shown in Figure 3A, the lungs and kidneys of the rat are perfused in series with the same perfusate and while the flow is kept constant with pumps. This difference is independent of nitric oxide and prostaglandins since the experiments were performed in the presence of inhibitors of the endothelial nitric oxide synthase, and cyclooxygenase. In contrast, both vascular beds constrict to angiotensin II and 4-aminopyridine (4-AP, 5 mM), a K_V channel blocker.

Further evidence that the opposing response to hypoxia and rotenone is intrinsic to the SMC and independent of the endothelium, comes from tissue bath experiments. In endothelium-denuded vessels, rotenone and hypoxia causes PA constriction and RA dilation, while 4-AP constricts both artery types (37).

6.1. Rotenone Mimics Hypoxic Inhibition of K^+ Current in PASMC

Whole-cell patch clamping in freshly isolated SMC from resistance PA and RAs shows that rotenone inhibits outward K^+ current (I_K) in PA while it activates I_K in RASMC (Fig. 3B). The effects of both hypoxia and rotenone are rapid, occurring within 5 minutes. These data are in agreement with earlier work showing that metabolic inhibitors decrease PASMC but not systemic (mesenteric) vascular SMC current (58).

The facts that most of the O_2-sensitive channels are present in both the PA and RA, and that 4-AP constricts both vessels (19, 21, 36), suggest that the opposing responses to hypoxia are likely not due to a differential expression of K_V channels. Rather, the opposing responses to hypoxia might be due to differences in the O_2 sensor (i.e., the mitochondria) that perhaps results in the differential regulation of the effector of O_2 sensing (i.e., AOS).

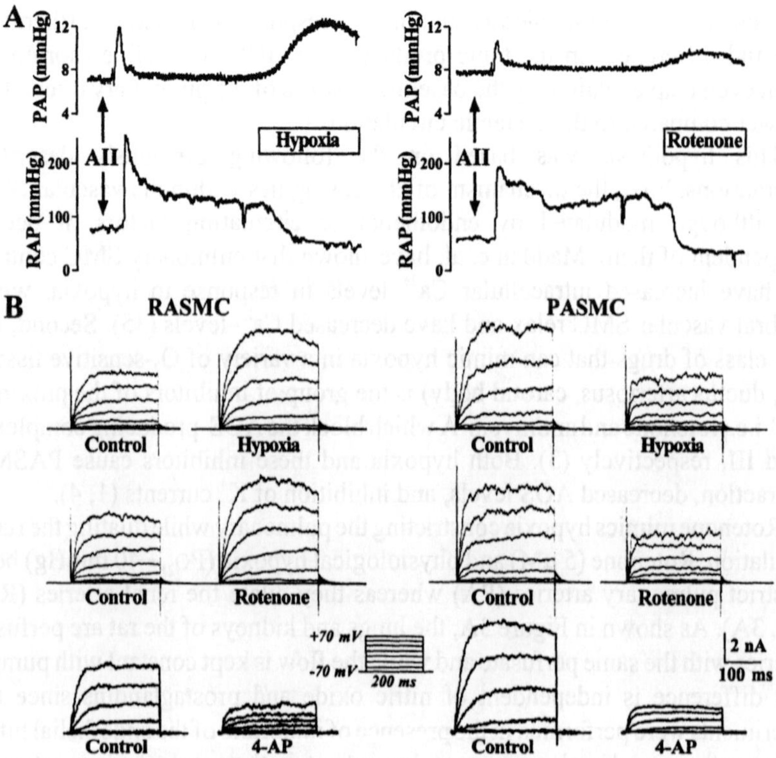

Figure 3. Divergent effects of rotenone and hypoxia on renal (RAP) versus pulmonary (PAP) arterial pressure (A) and on K^+ currents (I_K) in PASMC and MASMC (B). A: A representative trace and mean data from the isolated perfused lung-kidney model are shown. Rotenone (5 µM) and hypoxia (PO_2, 40 mmHg) increase PAP and decrease RAP in the presence of L-NAME and meclofenamate. 4-AP (5 mM) and angiotensin II (AII, 0.1 µg/50 µl bolus) increase both RAP and PAP. B: Hypoxia and rotenone increase whole-cell I_K in PASMC but decrease I_K in RASMC. In contrast, 4-AP inhibits I_K in both PASMC and RASMC (From Ref. 37).

6.2. Differential Regulation of Mitochondrial AOS and H_2O_2 Production between Pulmonary and Renal Arteries

We measured the baseline levels of AOS production (during normoxia) as well as the effects of hypoxia and ETC inhibitors in isolated, denuded, resistance PA and RA rings using three independent methods:

a). Lucigenin-enhanced chemiluminescence, which preferentially detects superoxide anion levels (34), shows that baseline AOS levels are significantly higher in the PA compared to the RA, even after incubation with 1 µM diphenyliodonium (DPI), an inhibitor of NAD(P)H oxidase (16) (Fig. 4). Rotenone's effects on the production of AOS parallel those of hypoxia. They both decrease AOS production in isolated PA rings, confirming previous studies on whole lung preparations (1). In contrast, hypoxia and rotenone increased AOS production in the RA (Fig. 4), confirming studies in other systemic arteries, such as coronary arteries (29).

b). The AmplexRed assay and the DCF (2',7'-dichlorofluorescin diacetate) fluorescence assay, which are specific for H_2O_2 show that, concordant with the chemiluminescence data, the production of H_2O_2 is higher in the PA versus the RA at baseline. Once again, proximal ETC inhibitors (rotenone and antimycin A, 50 μM) mimic hypoxic effect and decrease the production of H_2O_2 in the PA. However, neither hypoxia nor the proximal ETC blockers alter the production of H_2O_2 in the RA (Fig. 4). Cyanide (1 μM), a distal ETC complex IV blocker, does not affect H_2O_2 production in the PA and RA. Interestingly, we observed a significant auto-fluorescence of the aqueous sodium cyanide solution itself at a higher dose (10 μM) which prevented us from using this dose (37).

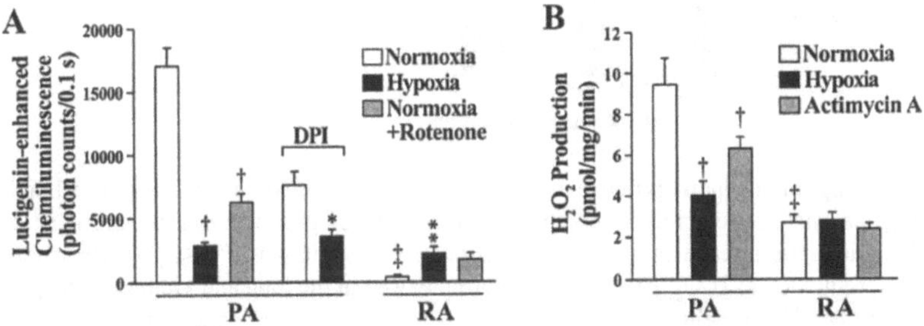

Figure 4. Differences of mitochondria-derived activated oxygen species (AOS) production in pulmonary (PA) and renal (RA) arteries. A: Lucigenin-enhanced chemiluminescence at normoxic baseline is higher in PA than in RA rings without endothelium (‡ $P<0.001$) in the absence or presence of the NAD(P)H oxidase inhibitor DPI (* $P<0.05$). Rotenone and hypoxia (Po_2, 40 mmHg) decrease AOS in PA (†$P<0.01$) and increase AOS in RA (** $P<0.05$). B: PA produces more H_2O_2 than RA (‡ $P<0.01$). Hypoxia and antimycin A decrease H_2O_2 production in PA († $P<0.05$) but not in RA (From Ref. 37).

To further show that the differences of the AOS production are intrinsic to the mitochondria, we isolated mitochondria from lungs and kidneys. The production of AOS and H_2O_2 from isolated mitochondria was measured with lucigenin-enhanced chemiluminescence and AmplexRed. Once again, the lung mitochondria produced more AOS and H_2O_2 than the kidney mitochondria and, while hypoxia and proximal ETC inhibitors decrease AOS and H_2O_2 in the lung, they do not in the kidney (Fig. 5A). In addition, in agreement with the vessel data, the complex IV ETC blocker cyanide (CN^-) does not alter AOS production, confirming previous reports (1) (Fig. 5A). These data are also consistent with complexes I and III, but not IV, being the major sites of AOS production within the ETC (18).

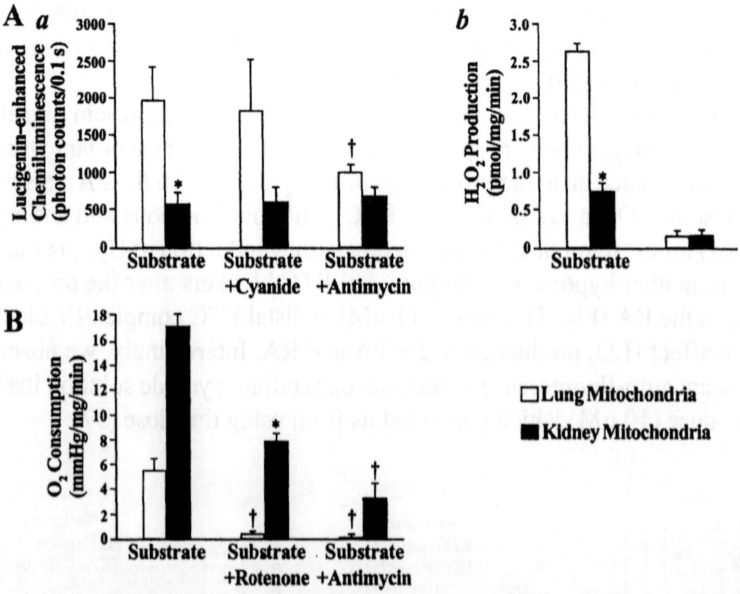

Figure 5. Lung mitochondria produce more AOS and are less active than kidney mitochondria. A: Lucigenin-enhanced chemiluminescence (*a*) indicating AOS production in the lung and kidney mitochondria in the absence or presence of cyanide (CN) or antimycin (* $P<0.05$). The catalase-sensitive production of H_2O_2 (*b*) measured by AmplexRed in the lung and kidney mitochondria (*$P<0.05$). B: The baseline respiration is greater in kidney mitochondria than in lung mitochondria. Rotenone and antimycin A inhibit respiration in both organs († $P<0.01$, * $P<0.05$). Glutamate (10 mM) and succinate (2.5 mM) were used as substrates (From Ref. 37).

6.3. The RA and Kidney Mitochondria Have Higher Respiratory Rates and More Hyperpolarized Membrane Potentials Than Those in the PASMC and Lungs

Why would the two groups of mitochondria have different levels of AOS production? Since the AOS production is linked with mitochondrial respiration, we studied O_2 consumption in both lung and kidney mitochondria. We show that lung mitochondria have significantly lower rates of respiration, compared to kidney mitochondria (Fig. 5B). Rotenone and antimycin A inhibit respiration in both lung and kidney mitochondria. Care was taken to study the same amount of mitochondria for both tissues, based on the protein content. Because of the tissue characteristics of lungs versus kidneys, the mitochondria isolation process could theoretically have resulted in preferential damage of one preparation versus the other, complicating any comparison. We therefore used two different assays to ensure that the quality of the mitochondria is similar in both preparations. Both preparations have almost identical respiratory control ratios and lysed/intact NADH-supported respiration ratios (17, 41). We isolated organ mitochondria because it is impossible to isolate adequate mitochondria from the small resistance rat PA and RA. Interestingly the data are in agreement with the measurements in isolated vessels, perhaps suggesting that the redox and metabolic environments of each organ parallel those of their vasculature.

We did however study vascular mitochondria in situ in freshly isolated PASMC and RASMC. As discussed above, the AOS production correlates with $\Delta\Psi_m$ and respiration, in that lower respiration is associated with depolarized $\Delta\Psi_m$ and higher AOS levels. Using two different $\Delta\Psi_m$-sensitive dyes (JC-1 and TMRM), we showed that mitochondria from RASMC have significantly higher $\Delta\Psi_m$ than mitochondria from PASMC under identical loading and imaging conditions. Extensive filamentous networks of mitochondria throughout the cytoplasm are shown in the representative pictures in Figure 6A. When the SMC were superfused with a hypoxic solution, the PASMC mitochondria hyperpolarized whereas the RASMC mitochondria depolarized (Fig. 6B).

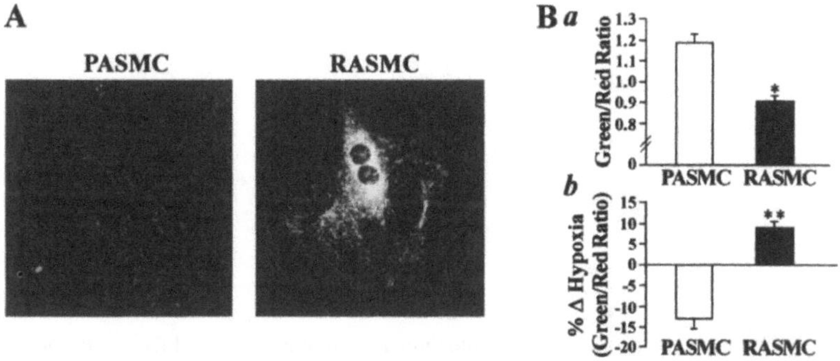

Figure 6. Mitochondrial membrane potential ($\Delta\Psi_m$) is more negative in RASMC than in PASMC. A: JC-1-loaded PASMC and RASMC. B: Mean data (as the ratio of the relative fluorescent units in the green and red channels) showing baseline $\Delta\Psi_m$ in PASMC and RASMC (*a*). Hypoxia hyperpolarizes PASMC mitochondria and depolarizes RASMC mitochondria (*b*). **P*<0.05, ***P*<0.0001 vs. PASMC (From Ref. 37).

7. The PA Mitochondria Express Less Complex I and III Subunits but More MnSOD than the RA Mitochondria

The depolarization in mitochondria is likely a direct result of the decreased respiration, since decreased activity of the ETC would result in less H^+ pumped out of the inner membrane while, at the same time, the electrons are not flowing down the ETC and are readily available to react with molecular O_2 to generate more AOS. Although we did not study ETC complexes' activity, we compared the expression of specific subunits of complexes I and III and showed that, in the PA, there is, as expected, less of both complexes I and III. At the same time, MnSOD protein expression is much higher in the PA than the RA, while MnSOD mRNA expression is higher in RA than in PA (Fig. 7). This is in agreement with our previous observations since MnSOD is known to be dynamically regulated and it is upregulated when the superoxide levels are high, as it occurs in the PA. It also explains the results shown in Figure 4 where, while the superoxide levels increase with hypoxia in the kidney mitochondria, the H_2O_2 levels do not. This might be because there is inadequate level of MnSOD expression that is required for the conversion of superoxide to H_2O_2.

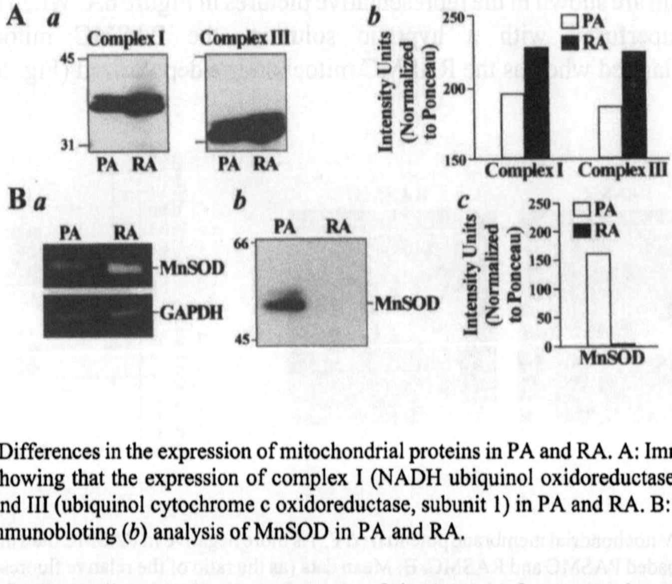

Figure 7. Differences in the expression of mitochondrial proteins in PA and RA. A: Immunoblot analysis showing that the expression of complex I (NADH ubiquinol oxidoreductase, 39 kDa subunit) and III (ubiquinol cytochrome c oxidoreductase, subunit 1) in PA and RA. B: RT-PCR (*a*) and immunobloting (*b*) analysis of MnSOD in PA and RA.

The findings discussed support many of the known features of vascular O_2 sensing. The high level of H_2O_2 production at baseline in the PA could explain the relatively dilated (low pressure) state of the pulmonary circulation in comparison to its systemic counterpart. Tonically produced H_2O_2 from the proximal ETC can diffuse to the cytoplasm and cause K^+ channel activation, PASMC hyperpolarization, decreased opening of voltage-gated Ca^{2+} channels, decreased $[Ca^{2+}]_i$ levels, and vasodilation. During acute hypoxia, the tonic production of this *mitochondrial-derived vasodilator* (H_2O_2) decreases, thus promoting I_K inhibition and vasoconstriction, i.e., hypoxic pulmonary vasoconstriction. In contrast, a hypoxia-induced increase in AOS (Fig. 4A) would increase I_K (20) in renal arterial smooth muscle cells and promote RA vasodilation (Fig. 3A).

8. Unanswered Questions

a) Based on knowledge derived from human and animal disease models, are PASMC a case of "congenital proximal ETC deficiency"? The difference between PA and RA mitochondria is analogous to the comparison of mitochondria from healthy subjects versus patients with ETC complex I deficiency. Fibroblasts from complex I deficient patients have increased mitochondria-derived AOS production compared with controls (7, 45). In addition, their fibroblasts exhibit impaired mitochondrial function, as reflected by depolarized $\Delta\Psi_m$ (7, 45). Furthermore, superoxide-induced MnSOD is upregulated in these patients (45). The normoxic PASMC, like the fibroblasts from patients with complex I deficiency, have (compared to the kidney) decreased levels of complex I (Fig. 7), decreased mitochondrial respiration (Fig. 5B), and depolarized $\Delta\Psi_m$ (Fig. 6). Lung mitochondria also make more superoxide and H_2O_2 than the kidney mitochondria (Fig. 5A). The PA's higher oxidative stress is associated with an apparently homeostatic induction of MnSOD expression (Fig. 7).

The relative decrease in the expression of proximal ETC complexes in the PA, compared to the RA, might be under genetic control, but since some of the ETC subunits are encoded in the mitochondria and some in the nucleus, it is not known whether this is a "genetic" difference, based on nuclear or mitochondrial DNA. Several transcription factors tightly regulate the coordination of the expression of nuclear and mitochondrial genes in order to form the very complex ETC multiple subunit complexes (40). Many respiratory genes have consensus regulatory sequences for the transcription factors nuclear respiratory factors (NRF-1 and 2). NRF-1 sites also exist in the promoters of regulators of mtDNA replication, particularly mitochondrial transcription factor A (Tfam) (48). AOS and intracellular Ca^{2+} are also important modulators of these transcription factors, although more research in needed in this field (40). The tight regulation and coordination of mitochondrial genes by the nucleus perhaps reflects their endosymbiotic origin; mitochondria are thought to have arisen 1.5 billion years ago from a symbiotic association between a glycolytic proto-eucaryotic cell and an oxidative bacterium.

b) Are the differences in PASMC and RASMC mitochondrial function a result of the difference redox and O_2 environments of the lungs and kidneys? On the other hand, significant ETC inhibition might result from the higher AOS levels in the PASMC. This is possible because the proximal ETC complexes are rich in thiols, making these complexes redox-sensitive. In other words, the differences in the function of mitochondria between the PASMC and RASMC could be a result of the differences in the redox-environment of the two organs. As a result of the higher levels of O_2 that the resistant pulmonary circulation is exposed to compared with the systemic circulation, the lung resistance circulation resides in a more oxidized redox state. This is supported by the significantly higher levels of expressed reduced glutathione (GSH) in the lung versus the kidney (37).

Furthermore, as discussed earlier, changes in O_2 levels might regulate the expression of ETC complexes. As shown in Figure 2, the expression of complex IV-2 dramatically increases in the lung at the transition from the fetal (low O_2) to the postnatal (high O_2) life.

9. Conclusion

The concept of mitochondrial diversity in the vasculature is new and requires further study as it is likely relevant to other aspects of vascular function. In addition to its role in vascular O_2-sensing the diversity concept may have implications for apoptosis, vascular wall remodeling or ischemia-reperfusion injury. It remains to be shown whether this mitochondrial diversity is due to a genetic difference intrinsic to the mitochondria or a result of the different redox environment that the PA are exposed to, compared to the RA.

Acknowledgments

This work is supported by a Bristol Mayer Squibs Fellowship, the Clinician Investigator Program at the University of Alberta, CIHR, AHFMR, CFI, and Alberta Heart & Stroke Foundation.

References

1. Archer SL, Huang J, Henry T, Peterson D, and Weir EK. A redox based oxygen sensor in rat pulmonary vasculature. *Circ. Res.* 1993; 73: 1100-1112.
2. Archer SL, Huang JMC, Reeve HL, Hampl V, Tolarová S, Michelakis ED, and Weir EK. Differential distribution of electrophysiologically distinct myocytes in conduit and resistance arteries determines their response to nitric oxide and hypoxia. *Circ. Res.* 1996; 78: 431-442.
3. Archer S and Michelakis E. The mechanism(s) of hypoxic pulmonary vasoconstriction: potassium channels, redox O_2 sensors, and controversies. *News Physiol. Sci.* 2002; 17: 131-137.
4. Archer SL, Nelson DP, and Weir EK. Simultaneous measurement of oxygen radicals and pulmonary vascular reactivity in the isolated rat lung. *J. Appl. Physiol.* 1989; 67: 1903-1911.
5. Ashcroft FM. Adenosine 5'-triphosphate-sensitive potassium channels. *Annu. Rev. Neurosci.* 1988; 11: 97-118.
6. Barlow RS and White RE. Hydrogen peroxide relaxes porcine coronary arteries by stimulating BK_{Ca} channel activity. *Am. J. Physiol.* 1998; 275: H1283-H1289.
7. Barrientos A and Moraes C. Titrating the effects of mitochondrial complex I impairment in the cell physiology. *J. Biol. Chem.* 1999; 274: 16188-16197.
8. Bennett MR. Apoptosis in the cardiovascular system. *Heart.* 2002; 87: 480-487.
9. Boveris A and Chance B. The mitochondrial generation of hydrogen peroxide. *Biochem. J.* 1973; 134: 707-716.

10. Bowker-Kinley MM, Davis WI, Wu P, Harris RA, and Popov KM. Evidence for existence of tissue-specific regulation of the mammalian pyruvate dehydrogenase complex. *Biochem. J.* 1998; 329: 191-196.
11. Brooks DE and Mann T. Relation between the oxidation state of nicotinamide-adenine dinucleotide and the metabolism of spermatozoa. *Biochem. J.* 1972; 129: 1023-1034.
12. Burke T and Wolin M. Hydrogen peroxide elicits pulmonary arterial relaxation and guanylate cyclase activation. *Am. J. Physiol.* 1987; 252: H721-H732.
13. Chandel NS, McClintock DS, Feliciano CE, Wood TM, Melendez JA, Rodriguez AM, and Schumacker PT. Reactive oxygen species generated at mitochondrial complex III stabilize hypoxia-inducible factor-1α during hypoxia: a mechanism of O_2 sensing. *J. Biol. Chem.* 2000; 275: 25130-25138.
14. Collins TJ, Berridge MJ, Lipp P, and Bootman MD. Mitochondria are morphologically and functionally heterogeneous within cells. *EMBO J.* 2002; 21: 1616-1627.
15. Cowan KN, Heilbut A, Humpl T, Lam C, Ito S, and Rabinovitch M. Complete reversal of fatal pulmonary hypertension in rats by a serine elastase inhibitor. *Nat. Med.* 2000; 6: 698-702.
16. Cross A and Jones O. The effect of the inhibitor diphenylene iodonium on the superoxide generating system of neutrophils. *Biochem. J.* 1986; 237: 111-116.
17. Darley-Usmar VM, Rickwood D, and Wolson MT, *Mitochondria: A Practical Approach.* Oxford: IRL Press, 1987.
18. Duchen MR. Contributions of mitochondria to animal physiology: from homeostatic sensor to calcium signalling and cell death. *J. Physiol.* 1999; 516: 1-17.
19. Fergus DJ, Martens JR, and England SK. Kv channel subunits that contribute to voltage-gated K^+ current in renal vascular smooth muscle. *Pflügers Arch.* 2003; 445: 697-704.
20. Gebremedhin D, Bonnet P, Greene AS, England SK, Rusch NJ, Lombard JH, and Harder DR. Hypoxia increases the activity of Ca^{2+}-sensitive K^+ channels in cat cerebral arterial muscle cell membranes. *Pflügers Arch.* 1994; 428: 621-630.
21. Gordienko DV, Clausen C, and Goligorsky MS. Ionic currents and endothelin signaling in smooth muscle cells from rat renal resistance arteries. *Am. J. Physiol.* 1994; 266: F325-F341.
22. Green DR and Reed JC. Mitochondria and apoptosis. *Science.* 1998; 281: 1309-1312.
23. Green L and Smith T. The use of digitalis in patients with pulmonary disease. *Ann. Intern. Med.* 1977; 87: 459-465.
24. Groen AK, van Roermund CW, Vervoorn RC, and Tager JM. Control of gluconeogenesis in rat liver cells. Flux control coefficients of the enzymes in the gluconeogenic pathway in the absence and presence of glucagon. *Biochem. J.* 1986; 237: 379-389.
25. Hayabuchi Y, Nakaya Y, Matsuoka S, and Kuroda Y. Hydrogen peroxide-induced vascular relaxation in porcine coronary arteries is mediated by Ca^{2+}-activated K^+ channels. *Heart Vessels.* 1998; 13: 9-17.
26. Huttemann M, Kadenbach B, and Grossman LI. Mammalian subunit IV isoforms of cytochrome c oxidase. *Gene.* 2001; 267: 111-123.
27. Ide T, Tsutsui H, Kinugawa S, Utsumi H, Kang D, Hattori N, Uchida K, Arimura K, Egashira K, and Takeshita A. Mitochondrial electron transport complex I is a potential source of oxygen free radicals in the failing myocardium. *Circ. Res.* 1999; 85: 357-363.
28. Jeffery TK and Morrell NW. Molecular and cellular basis of pulmonary vascular remodeling in pulmonary hypertension. *Prog. Cardiovasc. Dis.* 2002; 45: 173-202.
29. Kaminski PM and Wolin MS. Hypoxia increases superoxide anion production from bovine coronary microvessels, but not cardiac myocytes, via increased xanthine oxidase. *Microcirculation.* 1994; 1: 231-236.
30. Kovacevic Z, McGivan JD, and Chappell JB. Conditions for activity of glutaminase in kidney mitochondria. *Biochem. J.* 1970; 118: 265-274.
31. Krick S, Platoshyn O, Sweeney M, Kim H, and Yuan JX-J. Activation of K^+ channels induces apoptosis in vascular smooth muscle cells. *Am. J. Physiol. Cell Physiol.* 2001; 280: C970-C979.

32. Krick S, Platoshyn O, Sweeney M, McDaniel SS, Zhang S, Rubin LJ, and Yuan JX-J. Nitric oxide induces apoptosis by activating K^+ channels in pulmonary vascular smooth muscle cells. *Am. J. Physiol. Heart Circ. Physiol.* 2002; 282: H184-H193.
33. Kunz WS. Different metabolic properties of mitochondrial oxidative phosphorylation in different cell types - important implications for mitochondrial cytopathies. *Exp. Physiol.* 2003; 88: 149-154.
34. Li Y, Zhu H, Kuppusamy P, Roubaud V, Zweier J, and Trush M. Validation of lucigenin (bis-N-methylacridinium) as a chemilumigenic probe for detecting superoxide anion radical production by enzymatic and cellular systems. *J. Biol. Chem.* 1998; 273: 2015-2023.
35. Madden J, Vadula M, and Kurup V. Effects of hypoxia and other vasoactive agents on pulmonary and cerebral artery smooth muscle cells. *Am. J. Physiol.* 1992; 263: L384-L393.
36. Martens JR and Gelband CH. Alterations in rat interlobar artery membrane potential and K^+ channels in genetic and nongenetic hypertension. *Circ. Res.* 1996; 79: 295-301.
37. Michelakis ED, Hampl V, Nsair A, Wu X, Harry G, Haromy A, Gurtu R, and Archer SL. Diversity in mitochondrial function explains differences in vascular oxygen sensing. *Circ. Res.* 2002; 90: 1307-1315.
38. Michelakis ED, McMurtry MS, Wu XC, Dyck JR, Moudgil R, Hopkins TA, Lopaschuk GD, Puttagunta L, Waite R, and Archer SL. Dichloroacetate, a metabolic modulator, prevents and reverses chronic hypoxic pulmonary hypertension in rats: role of increased expression and activity of voltage-gated potassium channels. *Circulation.* 2002; 105: 244-250.
39. Milkowski AL and Lardy HA. Factors affecting the redox state of bovine epididymal spermatozoa. *Arch. Biochem. Biophys.* 1977; 181: 270-277.
40. Moyes CD and Hood DA. Origins and consequences of mitochondrial variation in vertebrate muscle. *Annu. Rev. Physiol.* 2003; 65: 177-201.
41. Nedergaard J and Cannon B. Overview-Preparation and properties of mitochondria from different sources. *Meth. Enzymol.*. 1979; 55: 3-33.
42. Palmer JW, Tandler B, and Hoppel CL. Biochemical properties of subsarcolemmal and interfibrillar mitochondria isolated from rat cardiac muscle. *J. Biol. Chem.* 1977; 252: 8731-8739.
43. Pettus EH, Betarbet R, Cottrell B, Wallace DC, Madyastha V, and Greenamyre JT. Immunocytochemical characterization of the mitochondrially encoded ND1 subunit of complex I (NADH: ubiquinone oxidoreductase) in rat brain. *J. Neurochem.* 2000; 75: 383-392.
44. Pitkanen S, Raha S, and Robinson BH. Diagnosis of complex I deficiency in patients with lactic acidemia using skin fibroblast cultures. *Biochem. Mol. Med.* 1996; 59: 134-137.
45. Pitkanen S and Robinson BH. Mitochondrial complex I deficiency leads to increased production of superoxide radicals and induction of superoxide dismutase. *J. Clin. Invest.* 1996; 98: 345-351.
46. Platoshyn O, Zhang S, McDaniel SS, and Yuan JX-J. Cytochrome *c* activates K^+ channels before inducing apoptosis. *Am. J. Physiol. Cell Physiol.* 2002; 283: C1298-C1305.
47. Rotig A, de Lonlay P, Chretien D, Foury F, Koenig M, Sidi D, Munnich A, and Rustin P. Aconitase and mitochondrial iron-sulphur protein deficiency in Friedreich ataxia. *Nat. Genet.* 1997; 17: 215-217.
48. Scarpulla RC. Nuclear control of respiratory chain expression in mammalian cells. *J. Bioenerg. Biomembr.* 1997; 29: 109-119.
49. Vanden Hoek TL, Becker LB, Shao Z, Li C, and Schumacker PT. Reactive oxygen species released from mitochondria during brief hypoxia induce preconditioning in cardiomyocytes. *J. Biol. Chem.* 1998; 273: 18092-18098.
50. Wallace DC. Mitochondrial diseases in man and mouse. *Science.* 1999; 283: 1482-1488.
51. Wallace DC, Singh G, Lott MT, Hodge JA, Schurr TG, Lezza AM, Elsas LJ 2nd, and Nikoskelainen EK. Mitochondrial DNA mutation associated with Leber's hereditary optic neuropathy. *Science.* 1988; 242: 1427-1430.
52. Wang D, Youngson C, Wong V, Yeger H, Dinauer MC, Vega-Saenz Miera E, Rudy B, and

Cutz E. NADPH-oxidase and a hydrogen peroxide-sensitive K^+ channel may function as an oxygen sensor complex in airway chemoreceptors and small cell lung carcinoma cell lines. *Proc. Natl. Acad. Sci. USA*. 1996; 93: 13182-13187.

53. Weinstein ES, Benson DW, and Fry DE. Subpopulations of human heart mitochondria. *J. Surg. Res*. 1986; 40: 495-498.
54. Williamson DH, Lund P, and Krebs HA. The redox state of free nicotinamide-adenine dinucleotide in the cytoplasm and mitochondria of rat liver. *Biochem. J.* 1967; 103: 514-527.
55. Wolin MS. Interactions of oxidants with vascular signaling systems. *Arterioscler. Thromb. Vasc. Biol.* 2000; 20: 1430-1442.
56. Wolin M and Burke T. Hydrogen peroxide elicits activation of bovine pulmoary arterial soluble guanylate cyclase by a mechanism associated with its metabolism by catalase. *Biochem. Biophys. Res. Comm.* 1987; 143: 20-25.
57. Yuan X-J, Aldinger A, Orens J, Conte J, and Rubin L. Dysfunctional voltage-gated potassium channels in the pulmonary artery smooth muscle cells of patients with primary pulmonary hypertension. *Circulation*. 1996; 94: 1-49.
58. Yuan X-J, Tod ML, Rubin LJ, and Blaustein MP. Deoxyglucose and reduced glutathione mimic effects of hypoxia on K^+ and Ca^{2+} conductances in pulmonary artery cells. *Am. J. Physiol.* 1994; 267: L52-L63.
59. Zamzami N and Kroemer G. The mitochondrion in apoptosis: how Pandora's box opens. *Nat. Rev. Mol. Cell Biol.* 2001; 2: 67-71.

Chapter 18

Hypoxia, Cell Metabolism, and cADPR Accumulation

A. Mark Evans
University of St. Andrews, St Andrews, Fife, United Kingdom

1. Introduction

Pulmonary arteries feed the lung in mammals and carry the entire output from the right side of the heart. In common with arteries that feed the gills of cyclostomes such as the Pacific hagfish and sea lamprey (64), and the skin of *Xenopus* (57), pulmonary arteries constrict in response to hypoxia and thereby regulate blood supply to the primary sites of gaseous exchange (81). In mammals this process has been termed hypoxic pulmonary vasoconstriction (HPV), and serves to aid ventilation perfusion matching (see Chapter 6). It is important to note, however, that a variety of O_2-sensing cells are present in the peripheral systems of mammals in addition to pulmonary artery smooth muscle and endothelial cells, respectively. These include carotid and aortic body cells, neuro-epithelial bodies and erythropoeitin (EPO)-producing cells, respectively.

Glomus cells within the carotid and aortic bodies monitor arterial Po_2, in addition to plasma pH, and are activated by physiological hypoxia. Carotid bodies are located at the bifurcation of the common carotid artery, and aortic bodies are located at the aortic arch. Stimulation of these peripheral chemo-receptors activates sensory afferent fibres that terminate in the dorsal group of respiratory neurons in the nucleus tractus solitarius of the medulla. Activation of these fibres increases ventilatory and vasomotor outflow from the solitary tract, thereby optimizing O_2 supply to the tissues (30).

Neuroepithelial bodies form highly specialized cell structures within the lung parenchyma and airway epithelia (50, 51). They are primarily located at the points of bifurcation within the airways of the lung, are activated by airway hypoxia (28, 92), and may integrate with the central nervous system via vagal afferents (3, 51, 82). Thus, neuroepithelial bodies, by releasing transmitters, may in turn trigger a reflex increase in ventilation to compensate for a relative O_2 deficit in the airways (28, 92).

EPO-secreting cells are mainly located in the kidney, although extrarenal cells are located in, for example, the liver. EPO cells also monitor arterial Po_2, and increase secretion of EPO in response to hypoxia. EPO then activates erythropoiesis at the bone marrow, leading to an increase in circulating erythrocytes. Thus, the O_2 carrying capacity of the blood is increased in line with demand (31, 42, 73).

Together, the O_2-sensing tissues derived, in part, from each of the aforementioned cell types, function to maintain O_2 supply to the body within an acceptable physiological range. While their respective roles and the precise nature of their response to hypoxia may differ, it seems likely that each cell type may share a common sensor that may couple O_2-dependent changes in cell metabolism to Ca^{2+} signaling pathways that effect the end response.

It seems equally likely, however, that the metabolic sensor(s) in question plays an integral role in the regulation of cellular activity in general, with the precise nature of the signaling cascade tailored to suit the function of a given cell type. The acute sensitivity of O_2-sensing cells to relatively small changes in Po_2 may be determined, therefore, by the "metabolic setting" of these cells within their respective physiological range of Po_2.

Within the present chapter, I will discuss the possible mechanisms by which hypoxia may alter cellular metabolism in pulmonary artery smooth muscle, and thereby promote acute HPV. The wider implications of metabolic regulation of cell signaling in O_2-sensing cells in general will then be considered.

2. Hypoxic Pulmonary Vasoconstriction

As mentioned previously, pulmonary arteries constrict in response to hypoxia and thereby regulate blood supply to the primary sites of gaseous exchange (81). Thus, HPV aids ventilation perfusion matching in the lung. In isolated pulmonary arteries, HPV is biphasic (Fig. 1A). Thus, hypoxia induces an initial transient constriction, which is then followed by a slow tonic constriction. It is apparent that HPV is a multifactorial process, with at least three discrete components of constriction (Fig. 1A) (22, 26). Our findings suggest that an initial transient (component 1) and maintained plateau (component 2) constriction are respectively promoted by direct effects of hypoxia on pulmonary artery smooth muscle (22, 23, 90). These components appear to be driven by two discrete mechanisms of Ca^{2+} release from the sarcoplasmic reticulum (SR). Full progression of HPV is then ensured by release of an endothelium-derived vasoconstrictor (component 3) (22, 26; see Chapters 6 and 12).

3. Mobilization of SR Ca^{2+} Stores by Hypoxia in Pulmonary Artery Smooth Muscle

Our findings are consistent with the view that hypoxia mobilizes smooth muscle SR Ca^{2+} stores by at least two discrete mechanisms, and thereby initiates

HPV. The initial transient constriction of isolated pulmonary arteries by hypoxia (component 1), appears to be mediated by metabolic inhibition of an SR Ca^{2+} ATPase (SERCA) subtype that serves the peripheral SR proximal to the plasma membrane, or by an as yet unidentified mechanism. In marked contrast, we have provided strong evidence to support our assertion that maintained smooth muscle constriction by hypoxia (component 2) is mediated by sustained Ca^{2+} release from a central SR compartment. And that this results from, at least in part, activation of ryanodine receptors (RyRs) by cyclic adenosine diphosphate-ribose (cADPR), a Ca^{2+} mobilizing messenger derived from β-NAD^+ (55).

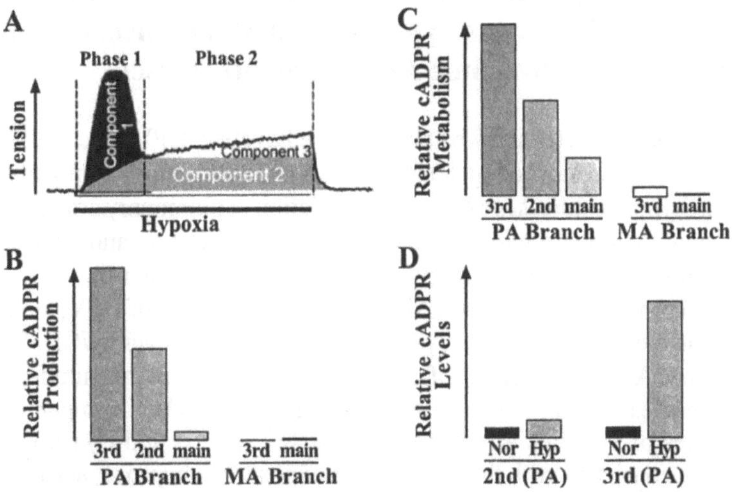

Figure 1. Three components of HPV, the distribution of the enzyme activities for cADPR production and metabolism in pulmonary versus systemic arteries, and increased cADPR accumulation in pulmonary arteries by hypoxia. A: Three components of constriction by hypoxia of isolated pulmonary arteries; 1, cADPR-independent SR Ca^{2+} release in the smooth muscle; 2, cADPR-dependent SR Ca^{2+} release in the smooth muscle; 3, release of an endothelium-derived vasoconstrictor. B and C: Synthesis of cADPR from β-NAD^+ (B) and metabolism of cADPR (C) in smooth muscle homogenates from a series of pulmonary (PA) and mesenteric (MA) arteries. D: Relative cADPR levels in 2nd and 3rd order branches of the pulmonary artery, respectively, during normoxia (Nor) and hypoxia (Hyp).

Our findings therefore suggest that cADPR accumulation by hypoxia represents one of a number of diverging processes that may or may not be of primary importance to the regulation of a given type of O_2-sensing cell. This is clear from the finding that cADPR does not appear to play a primary role in mediating the initial transient constriction of pulmonary artery smooth muscle by hypoxia, or in mediating vasoconstrictor release by hypoxia from the pulmonary artery endothelium (see Chapter 6 and 12). Thus, regulation by hypoxia of cADPR accumulation likely represents a point of bifurcation in the proposed O_2-sensitive signaling cascade within pulmonary artery smooth muscle. We will proceed, therefore, with the premise that a common metabolic signaling pathway exists in all cells, and that this may promote, among other events, cADPR accumulation by hypoxia in a manner dependent on the "metabolic setting" of a given cell. If this is the case, identification of the precise mechanism by which hypoxia regulates cADPR synthesis will lead us to the identification of the "primary metabolic sensor" and "primary effector", respectively, in O_2-sensing cells.

4. Functional Significance of the Differential Distribution of the Enzyme Activities for the Synthesis and Metabolism of cADPR in Pulmonary and Systemic Artery Smooth Muscle

As mentioned previously, we have identified a role in HPV for cADPR, a β-NAD^+ metabolite and an endogenous regulator of RyRs. Our experiments have shown that hypoxia increases cADPR accumulation in pulmonary artery smooth muscle, and thereby initiates maintained SR Ca^{2+} release and consequent constriction via the activation of RyRs (22, 55, 90; see Chapter 6).

Consistent with the view that cADPR acts as a mediator of HPV, we found the enzyme activities for the synthesis (ADP-ribosyl cyclase/CD38) and metabolism (cADPR hydrolase) of cADPR to be an order of magnitude higher in homogenates of pulmonary artery smooth muscle when compared with homogenates of systemic artery smooth muscle (Fig. 1B and C) (90). These findings are consistent with the idea that, on exposure to hypoxia, activation of the primary metabolic sensor will lead to a greater level of cADPR accumulation in pulmonary artery smooth muscle in general, than in systemic artery smooth muscle. As a consequence the mobilization of SR Ca^{2+} stores by cADPR will be greater in pulmonary when compared with systemic artery smooth muscle. The differential distribution of the enzyme activities for cADPR synthesis and metabolism could therefore underpin, at least in part, pulmonary artery constriction (81) and systemic artery dilation (74) by hypoxia, respectively (see Chapter 6). This may be determined by the absolute level of the enzyme activities for cADPR accumulation and/or their spatial localization within pulmonary and systemic artery smooth muscle, respectively. Such findings would be consistent with my proposal that the metabolic sensor(s) and primary effector(s) are likely common to all cells, with cADPR accumulation offering, in part, the pulmonary selectivity required of a mediator of HPV.

We have also identified a differential distribution with respect to the enzyme activities for cADPR synthesis and metabolism within the pulmonary arterial tree itself. In short, the level of these activities was inversely related to pulmonary artery diameter (Fig. 1B and C) (90), as indeed is the magnitude of pulmonary artery constriction by hypoxia (45). Thus, metabolic activation of these enzyme activities may also provide an amplification of cADPR accumulation and thereby SR Ca^{2+} release in near-resistance-sized pulmonary arteries when compared with that observed in the larger conduit arteries. We may thereby account for the increased magnitude of vasoconstriction by hypoxia in distal segments of the pulmonary arterial tree. Consistent with this proposal, we found the increase in cADPR levels by hypoxia to be 2 fold in smooth muscle homogenates from 2nd order branches of the pulmonary arterial tree, and 10 fold in those from 3rd order branches (Fig. 1D) (90).

5. Metabolic Regulation of cADPR Accumulation

Despite extensive investigations, we have yet to determine the precise mechanism by which hypoxia regulates cADPR accumulation. It seems likely that a change in the metabolic state of pulmonary artery smooth muscle will mediate this effect via one or a group of primary "metabolic sensor." In subsequent sections of this chapter, therefore, I will review the information currently available to us, in an effort to determine the likely identity of the "metabolic sensor(s)" and "metabolic effector(s)" involved.

When considering the possible mechanisms by which changes in cell metabolism may regulate cell activity, a logical place to start would be with a general consideration of the process of cellular O_2 utilization and energy production (ATP) under normoxic conditions, with particular reference, as required, to arterial smooth muscle.

As we have all been informed from cradle to laboratory, there are two primary pathways to consider in this respect, namely glycolysis and oxidative phosphorylation by mitochondria, respectively. The activity of each pathway is inextricably linked, inversely with respect to anaerobic metabolism, due to homeostatic mechanisms that seek to maintain the desired cellular energy state.

6. Glycolysis

Glycolysis describes the first step in the process by which cells utilize glucose to provide for their energy needs. Glucose uptake by cellular glucose transporters (GLUT), is followed by a sequence of enzymatic reactions that constitute glycolysis (Fig. 2). The initial steps in glycolysis are catalyzed by hexokinase and phosphofructokinase, respectively, each of which utilize a molecule of ATP to provide the necessary energy for completion of their role in glucose metabolism. The next step is driven by glyceraldehyde 3-phosphate dehydrogenase and results in the reduction of 2 molecules of β-NAD^+ to yield 2 molecules of β-NADH, which either enter mitochondria via the malate aspartate shuttle or are utilized by lactate dehydrogenase. The final steps in the process, driven by phosphoglycerate kinase and pyruvate kinase respectively, produce 2 molecules of ATP each and generate the final product of glycolysis, pyruvate. Thus, glycolysis yields a total of 2 molecules of ATP. Clearly, glycolysis is an intensive process that provides for only limited ATP production per cycle. Significantly, however, vascular smooth muscle exhibits a greater dependence on glycolysis than other cell types. In fact, at least 30% of the ATP requirement of smooth muscle may be supplied by glycolysis during normoxia, and this can account for as much as 90% of glucose utilization (38, 68).

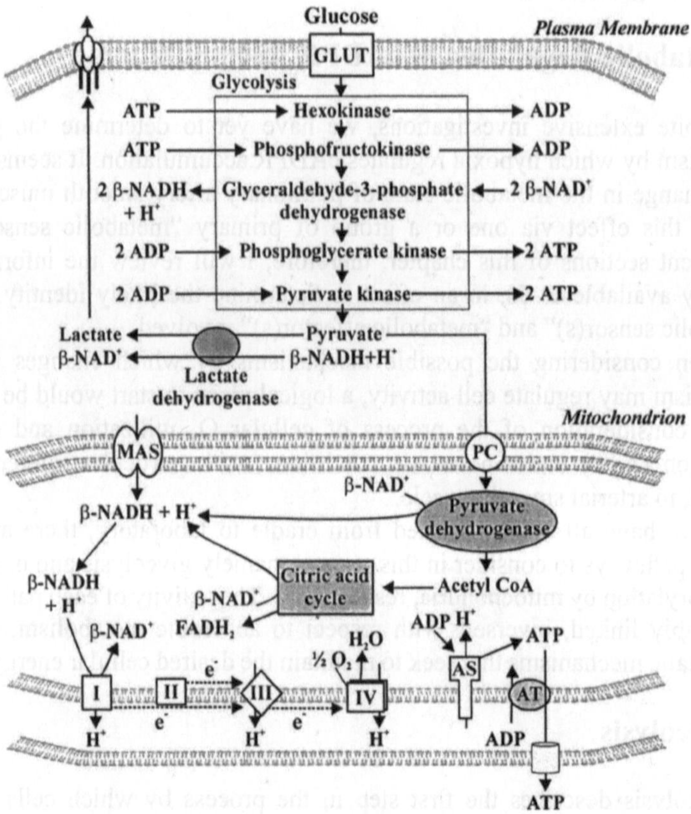

Figure 2. Glycolysis and oxidative-phosphorylation by mitochondria. GLUT, glucose transporter; MAS, malate aspartate shuttle; PC, pyruvate carrier; AS, ATP synthetase; AT, ATP translocase; I, II, III and IV, each complex of the mitochondrial electron transport chain; e^-, electron transport from complex I and II to complex III via coenzyme Q/ubiquinone, and from complex III to complex IV via cytochrome *c*, respectively.

An important consideration with respect to O_2-sensing is that this high level of glycolysis may lead to the generation of substantial amounts of lactate, which must then be removed from the cell (see below). This is because pyruvate derived from glycolysis has two possible fates. It can enter mitochondria via the pyruvate carrier, where it is dehydrogenated to yield acetyl CoA for the Krebs cycle. Alternatively, pyruvate, when surplus to mitochondrial requirements, may be broken down by lactate dehydrogenase to form lactate.

7. Oxidative Phosphorylation by Mitochondria

Oxidative phosphorylation by mitochondria is driven, in part, by acetyl CoA, which is derived from pyruvate (Fig. 2). Acetyl CoA enters the Krebs cycle where cycles of substrate dehydrogenation lead to the reduction of β-NAD$^+$ to β-NADH. Together with β-NADH derived from glycolysis, this provides electrons and protons for the electron transport chain (ETC) on oxidation of β-NADH at complex I. In addition, $FADH_2$, produced by succinate dehydrogenase, provides electrons and protons for the ETC upon oxidation at complex II. Transfer of electrons then occurs between complexes I-IV in the inner mitochondrial membrane. The final step in this process is electron transfer to O_2 by complex IV and the associated reduction of O_2 to H_2O. According to Mitchell's chemiosmotic theory (61), electron transfer drives proton pumps at complex I, III and IV, respectively, resulting in a proton gradient across the inner mitochondrial membrane. The proton gradient drives proton influx back into the mitochondria via proton (F_0) channels associated with ATP synthase. Subsequent activation of ATP synthetase leads to the production of ATP from ADP:

$$\beta\text{-NADH} + H^+ + 3ADP + 3P_i + 0.5O_2 \rightarrow \beta\text{-NAD}^+ + 3ATP + H_2O$$

Oxidative phosphorylation by mitochondria is a far more efficient method of ATP production when compared to glycolysis. The generation of 38 ATP molecules per cycle is testimony to this fact. However, it is important to note that arterial smooth muscle cells contain a relatively small number of mitochondria. This may be why vascular smooth muscle exhibits a relatively high dependency on glycolysis under aerobic conditions (38, 68). This relative shift in dependency from oxidative phosphorylation to glycolysis in vascular smooth muscle not only suggests that ATP production by glycolysis is important under normoxic conditions, but that glycolysis may afford the increased tolerance of vascular smooth muscle to prolonged hypoxia when compared to other cell types. In systemic arteries this has been linked to the need to maintain adequate blood supply to the tissues (38, 68). However, this may also point to a primary role for glycolysis in maintaining the energy state of pulmonary artery smooth muscle during hypoxia and, therefore, to the importance of glycolysis to the development of HPV.

8. ATP Mobility

ATP has limited mobility within cells, and thus tends to be concentrated around the sites of production, with the sites of production being close to primary sites of utilization. However, ATP translocation is enhanced, indirectly, by the phosphocreatine shuttle. A high-energy bond is transferred from ATP to creatine at the site of ATP production to yield phosphocreatine. The greater mobility of phosphocreatine allows for the transfer of the high-energy bond to ADP at relatively distant sites of utilization. This reversible process is known as the creatine kinase reaction (16), and is catalyzed by creatine kinase:

$$ATP + Creatine \leftrightarrow Phosphocreatine + ADP + H^+$$

The mere fact that the phosphocreatine shuttle is required to transport ATP from the primary sites of production to distant sites of utilization, tells us one thing. In short, on exposure to hypoxia it is likely that there will be a relative ATP deficit between those cellular compartments that are some distance from the sites of ATP production, when compared to those cellular compartments within which ATP production is concentrated. Not surprisingly, therefore, there is evidence to suggest that both ATP production and utilization are compartmentalized within smooth muscle (12, 38, 68). Furthermore, we cannot rule out the possibility that O_2-sensing cells may preferentially supply glucose to certain cellular compartments when exposed to hypoxia.

9. Compartmentalization of Energy Production

It has been suggested that membrane bound glycolytic enzymes restrict glycolysis to the plasma membrane, and that changes in glycolysis may primarily modulate membrane associated mechanisms (38, 68). However, recent studies have also demonstrated that there are two glycolytic pathways within vascular smooth muscle cells, one serving the subplasmalemmal space whilst the other serves the center of the cell (12). The relative dependence of vascular smooth muscle on ATP production by glycolysis suggests that these divergent glycolytic pathways may play an important role in determining the regulation by hypoxia of distinct processes in different cellular compartments. Such compartmentalization may be facilitated by the limited mobility of ATP.

There is also evidence to suggest that mitochondria may be localized in particular cellular compartments, and that there may be a particularly close association between mitochondria and the SR (20, 17, 33). Furthermore, it has been suggested that multiple mitochondrial subtypes may serve different cellular compartments, raising the possibility that each subtype may support discrete cellular functions (17, 25, 60). Indirect support for this viewpoint may also be derived from the finding that mitochondrial DNA in cells which exhibit a high dependence on glycolysis may differ from that found in other cell types (4).

10. Effects of Hypoxia on the Cellular Energy State

Logically, on exposure to hypoxia the primary point at which one would expect cellular O_2 deficit to impact is at the mitochondria, and more specifically at the point of oxidative phosphorylation. This is clear from the fact that the terminal cytochrome oxidase has a high affinity for O_2, and is, therefore, sensitive to even moderate hypoxia. Herein, the sequence of events that seek to maintain homeostasis with respect to the cellular energy state are relatively well characterized. We should be aware, however, that although oxidative phosphorylation by mitochondria is inhibited by hypoxia, it likely continues to contribute to ATP supply i.e. whilst O_2 is available oxidative phosphorylation will occur during hypoxia, albeit at a reduced rate.

10.1. Increased Glucose Uptake and Acceleration of Glycolysis

Investigations into the effect of hypoxia on skeletal muscle have demonstrated that glucose uptake is increased and glycolysis accelerated (14). There is also evidence to suggest that active glucose uptake is facilitated and glycolysis accelerated by hypoxia in pulmonary artery smooth muscle and/or endothelial cells (52, 54). This is not surprising, since glycolysis is rate-determined by glucose uptake in the lung (70). In bladder smooth muscle, it was reported that at least 60% of ATP production may be derived from glycolysis during hypoxia (87). Furthermore, it would appear that ATP production by glycolysis and other supporting mechanisms, may be sufficient to compensate for a fall in ATP production by oxidative phosphorylation, and sufficient to support maximal contraction (87). Consistent with this viewpoint, we have demonstrated that constriction by K^+ (i.e., depolarization-induced Ca^{2+} influx via voltage-gated Ca^{2+} channels) remains unaffected during hypoxia (22, 23). Indirect support for a primary role for glycolysis in HPV may also be derived from the finding that increased glucose levels may augment HPV, whilst their reduction or direct inhibition of glycolysis, respectively, may attenuate HPV (88, 93). In addition, it has been noted that hypoxia increases glucose uptake into systemic arteries to a lesser extent than in pulmonary arteries (52, 54).

Investigations on O_2-sensing cells have also provided indirect support for the view that hypoxia accelerates anaerobic glycolysis. Thus, hypoxia has been shown to increase β-NADH levels in all O_2-sensing cells from which measures have been taken (5, 11, 79, 92). This is probably a consequence of both inhibition of oxidative phosphorylation by mitochondria and acceleration of glycolysis, and likely reflects an increase in mitochondrial and cytoplasmic β-

NADH, respectively. As a result, we may observe reduced β-NADH oxidation and pyruvate utilization by mitochondria, allied to saturation of lactate dehydrogenase by pyruvate and/or limitations associated with lactate excretion.

$$\text{Pyruvate} + \beta\text{-NADH} + H^+ \leftrightarrow \text{Lactate } \beta\text{-NAD}^+$$

Although it is clear that glycolysis is accelerated and that glycolytic ATP generation may be sufficient to support maximal smooth muscle contraction when combined with a reduced rate of mitochondrial oxidative phosphorylation, it is important to note that other processes contribute to the maintenance of the cellular energy state.

10.2. The Adenylate Kinase Reaction

During hypoxia, a fall in ATP production by oxidative phosphorylation will also be compensated for by the conversion of 2 molecules of ADP to ATP. This is known as the adenylate kinase reaction:

$$\text{ADP} + \text{ADP} \rightarrow \text{ATP} + \text{AMP}$$

Significantly, for every molecule of ATP generated, the cell produces one molecule of AMP (see below).

11. The "Metabolic Setting" in O_2-sensing Cells

The acute sensitivity of O_2-sensing cells to relatively small changes in Po_2, is likely determined by their respective "metabolic setting" (e.g., basal glycolytic and mitochondrial activities, and the levels of metabolic intermediates such as ATP, ADP, AMP, β-NAD$^+$, β-NADH). Given their variety in both form and function, therefore, it is important to note that while the primary elements of the signaling cascade may be common to all O_2-sensing cells, the physiological range of normoxic and hypoxic Po_2, respectively, experienced by a given cell type will differ. As a consequence, the metabolic setting in each cell type may also differ at a given Po_2. Thus, when considering the regulation of a given cell type by hypoxia, we must ensure that the precise physiological range of Po_2 experienced by a particular cell type is adhered to. Furthermore, when studying the modulation by hypoxia of mitochondrial function and glycolysis, respectively, we must consider the impact on the metabolic setting of any physiological or pharmacological manoeuvre, before we draw conclusions as to the nature of the O_2-sensor or the mechanism of O_2-sensitive signaling.

12. Metabolic Regulation of Ca^{2+} Signaling and cADPR Synthesis in Pulmonary Artery Smooth Muscle

Given that cells may contain different mitochondrial subtypes it is possible that mitochondria, or a subset of mitochondria, within O_2-sensing cells may be particularly sensitive to hypoxia due to a relatively high affinity of their terminal cytochrome for O_2, and/or their initial "metabolic setting". Furthermore, different mitochondrial subtypes may serve different cellular compartments in O_2-sensing cells, and may contribute to the differential regulation of these compartments in response to hypoxia. Subsequent acceleration of two compartmentalized glycolytic pathways, one serving the subplasmalemmal space and one the center of the cell, may offer even greater compartmentalization with respect to the regulation of cell activity by hypoxia. Compartmentalization of energy supply may be further enhanced by limited ATP mobility and by regulation of energy transfer by the phosphocreatine shuttle. One or more of these factors may underpin component 1 of HPV in isolated arteries. Thus a fall in ATP levels could be restricted to a cellular compartment within which an SR Ca^{2+} ATPase subtype serves peripheral SR proximal to the plasma membrane, leading to a net increase in Ca^{2+} leak from the SR (see Chapter 6). Indirect support for this viewpoint may be derived from the finding that glycolysis can be tightly coupled to SR Ca^{2+} ATPase activity (9), and from the proposal that a general and maintained fall in cytoplasmic ATP may not be observed in pulmonary artery smooth muscle (54). The fact that constriction by K^+ remains unaffected during hypoxia also adds weight to this proposal (22, 23). It should be noted, however, that this consideration is not consistent with the tenor of this chapter. In short, it is unlikely that a fall in ATP levels *per se* triggers cADPR accumulation. A role for reduced ATP supply during component 1 of HPV would thus appear inconsistent with my proposal that a common metabolic sensor(s) regulates O_2-sensitive signaling in all cell types. It is possible, therefore, that an as yet unidentified effector may be involved in triggering component 1 of HPV.

The precise mechanism(s) of signal transduction here on in remains obscure. Investigations on the metabolic regulation of O_2-sensing cells have led to two divergent hypotheses regarding the mechanism of signal transduction activated by hypoxia, irrespective of the effector system under investigation (e.g. inhibition of K^+ channels, mobilization of intracellular Ca^{2+}, regulation of transcription factors). On the one hand it has been suggested that O_2-sensitivity may be conferred by a reduction in the cellular redox state, and consequent reduction of key elements of the cell's signaling mechanisms by reduced cellular redox couples (5, 11, 89). On the other, it has been suggested that the effects of hypoxia may be mediated by a paradoxical increase in the generation of reactive oxygen species (ROS) by mitochondria as O_2 availability falls (52, 84).

13. Does the Cellular Redox State Play a Role in the Regulation of cADPR Accumulation in Pulmonary Artery Smooth Muscle?

We were drawn to investigate the possibility that cADPR may be a mediator of HPV for two reasons. As mentioned previously, cADPR had been proposed as an endogenous regulator of RyRs (29, 55), and we had obtained evidence of a role for smooth muscle SR Ca^{2+} release via RyRs in maintained HPV (23). An additional attraction, however, was the fact that cADPR is a β-NAD^+ metabolite. This was because hypoxia had been shown to increase β-NADH levels in all O_2-sensing cells studied to date (5, 11, 79, 92). When taken together, these findings suggested that cADPR synthesis itself may, in some way, be sensitive to changes in the metabolic state of pulmonary artery smooth muscle, and that it may thereby play a role in HPV.

As discussed previously, a fall in PO_2 likely inhibits oxidative phosphorylation due to the high affinity of the terminal cytochrome oxidase for O_2. Consequent reduction in the rate of β-NADH oxidation by mitochondria and acceleration of anaerobic glycolysis, likely results in increased accumulation of β-NADH in the cytoplasm and mitochondria. Consistent with this view, chemiluminescence studies (5) and direct measurement (79) have demonstrated that hypoxia reduces the cellular redox state in pulmonary artery smooth muscle, in line with classical theory (15). It was therefore suggested that a general reduction in the cellular redox state may mediate signaling by hypoxia in pulmonary arteries, and a variety of redox couples were considered to play a role, namely GSH, β-NADPH, and β-NADH, respectively (86).

Subsequent investigations have, however, brought into question the idea that inhibition of β-NADPH oxidase by hypoxia, and subsequent β-NADPH accumulation represent the primary O_2-sensing pathway. In fact, Marshall et al. (58) demonstrated that the activity of NADPH oxidase may actually be increased by hypoxia in pulmonary artery smooth muscle. In addition, in knock-out mice lacking the neutrophil NADPH oxidase, HPV remained unaffected (7). Moreover, direct extraction and measurement of β-NADPH and β-$NADP^+$, respectively, in pulmonary artery smooth muscle has revealed that the β-NADPH: β-$NADP^+$ ratio is high under normoxic conditions and remains fairly constant in the presence of hypoxia (79). Given the above, it may not be surprising to discover that extremely high concentrations (10 mM) of β-NADPH are without effect on cADPR synthesis from a fixed concentration of β-NAD^+ in pulmonary artery smooth muscle homogenates (unpublished observation).

In contrast, paired measurement of β-NAD^+ and β-NADH levels, respectively, in extracts from pulmonary arteries yielded quite different findings. β-NAD^+ levels were found to be high (>1 mM) under normoxic conditions whilst β-NADH levels were very low (<80 μM). Most significantly, however, moderate hypoxia (~30 Torr) increased β-NADH levels at least 5 fold (79). This is consistent, therefore, with the proposal that hypoxia, by inhibiting β-NADH oxidation and oxidative phosphorylation by mitochondria, and accelerating anaerobic glycolysis, leads to an increase in cytoplasmic β-NADH concentration. The increase in cytoplasmic β-NADH is likely allied to saturation of lactate dehydrogenase and/or limitations with respect to the removal of lactate from the cell, as mentioned previously. Consistent with this view, Leach et al. (52) measured an increase in NAD(P)H autofluorescence by hypoxia before and after block of β-NADH oxidation by the mitochondrial electron transport chain.

Initially, and perhaps naively, we considered the possibility that β-NADH may be a better substrate for cADPR synthesis than β-NAD$^+$. This proposal could not have been further from the truth, in that we found β-NADH to be a poor substrate for cADPR synthesis. In fact, when applied to pulmonary artery smooth muscle homogenates, 25mM β-NADH produced a total cADPR yield that amounted to no more than 30% of that derived from 2.5 mM β-NAD$^+$ (Fig. 3A) (90). We therefore considered the possibility that the β-NADH : β-NAD$^+$ ratio may be a significant factor, rather than the absolute level of either component. Thus, we investigated the concentration-dependent effects of β-NADH on cADPR production from a fixed concentration of β-NAD$^+$. Surprisingly, β-NADH induced a concentration-dependent and synergistic increase in cADPR production from β-NAD$^+$ (Fig. 3B) (90). Moreover, the range over which a change in the β-NADH:β-NAD$^+$ ratio augmented cADPR accumulation was equivalent to the ratio change predicted from direct measurement of β-NAD$^+$ and β-NADH in extracts from pulmonary arteries during normoxia and moderate hypoxia (79, 90). In short, β-NADH may shift the curve for cADPR production from β-NAD$^+$ to the left and raise maximal cADPR accumulation (Fig. 3B, inset).

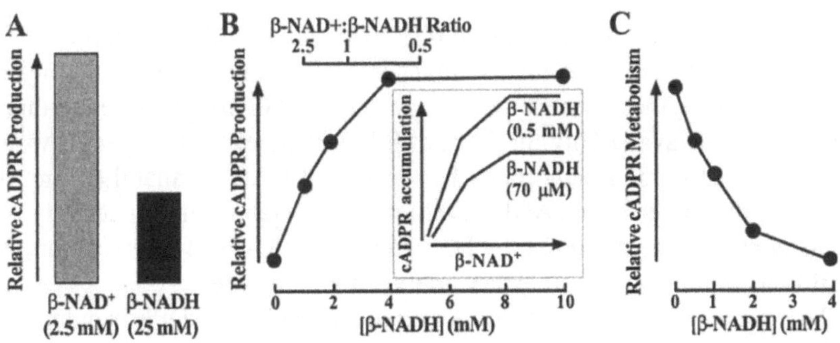

Figure 3. Regulation of cADPR accumulation by β-NADH. A: Relative cADPR accumulation from 2.5 mM β-NAD$^+$ and 25 mM β-NADH, respectively, in pulmonary artery smooth muscle homogenates. B: cADPR accumulation from 2.5 mM β-NAD$^+$ versus β-NADH concentration in pulmonary artery smooth muscle homogenates. Inset: predicted effect of increased β-NADH concentration on cADPR accumulation versus β-NAD$^+$ concentration. C: cADPR metabolism in pulmonary artery smooth muscle homogenates versus β-NADH concentration.

Considering our findings with respect to the facilitation of cADPR synthesis by β-NADH in a little more detail, however, we can gather further significant information. Firstly, we demonstrated that β-NADH inhibited cADPR metabolism in pulmonary artery smooth muscle homogenates in a concentration-dependent manner, equivalent in range to that by which β-NADH augmented cADPR synthesis. However, the reduction in cADPR metabolism was not sufficient to account for the total increase in cADPR synthesis by β-NADH (Fig. 3C). This is clear from the fact that 2mM β-NADH increased cADPR synthesis from 2.5 mM β-NAD^+ by ~12 nmoles mg $protein^{-1}$ hr^{-1}, but only reduced cADPR metabolism by ~3 nmoles mg $protein^{-1}$ hr^{-1} (90). Most significantly, however, our findings show that the maximal increase in cADPR accumulation by β-NADH from a fixed concentration of β-NAD^+ was approximately 2 fold. This is despite the fact that the level of substrate available was sufficient to promote almost maximal cADPR synthesis from β-NAD^+ in the absence of β-NADH. In contrast, the maximal increase in smooth muscle cADPR content by hypoxia was 10 fold, in equal quantities of pulmonary artery smooth muscle. Clearly, the latter measurements were taken under conditions where the increase in cellular β-NADH would lead to a consequent reduction in substrate (β-NAD^+) availability for cADPR synthesis, whilst the *in vitro* measurements did not allow for a fall in β-NAD^+ levels as β-NADH was increased. In short, it seems unlikely that β-NADH acts as the primary mediator of cADPR accumulation by hypoxia.

13.1 A Po_2 Window for cADPR Accumulation by Hypoxia

Further consideration of our developing model for regulation of cADPR synthesis by hypoxia in pulmonary artery smooth muscle, offered yet more insights into the process of HPV. As mentioned above, a rise in β-NADH levels will result in a consequent fall in β-NAD^+ availability, the substrate for cADPR synthesis. This would be exacerbated if re-oxidation of β-NADH to β-NAD^+ is tightly coupled to the maintenance of anaerobic glycolysis. Our model would predict, therefore, that severe hypoxia may trigger a fall in cADPR levels due to limited substrate availability. In short, as β-NADH rises, cADPR levels would begin to fall at a point determined by the Po_2, β-NAD^+ availability, and the kinetics of the enzymes for cADPR synthesis and metabolism, respectively. We would therefore expect to observe a Po_2 window within which hypoxia promotes cADPR accumulation in pulmonary artery smooth muscle.

To examine this hypothesis, we measured the levels of cADPR in pulmonary artery smooth muscle from third order branches of the pulmonary arterial tree under normoxic, hypoxic and near anoxic conditions. As mentioned previously, we observed a 10-fold increase in cADPR levels by hypoxia (16-21 Torr) when compared to levels measured under normoxia (155-160 Torr) (Fig. 4A). Consistent with the proposal that there may be a Po_2 window for cADPR accumulation by hypoxia, however, we measured a fall in cADPR levels under near anoxic conditions (5-8 Torr), to a level not significantly different from that observed in the presence of normoxia (Fig. 4A) (24). Furthermore, the PO_2 window for cADPR accumulation by hypoxia, was associated with a concomitant rise in β-NADH levels that was inversely related to the PO_2 of the bath solution between normoxia (150-160 Torr) and near anoxia (5-8 Torr) (Fig. 4B) (Evans, unpublished data). We can conclude, therefore, that cADPR accumulation by hypoxia may be limited by β-NAD^+ availability. Furthermore, we have shown that the fall in cADPR levels by anoxia was associated with a failure of HPV in intact arteries (see Chapter 6). This further strengthens our proposal that cADPR-dependent SR Ca^{2+} release in pulmonary artery smooth muscle is a primary mediator of maintained HPV.

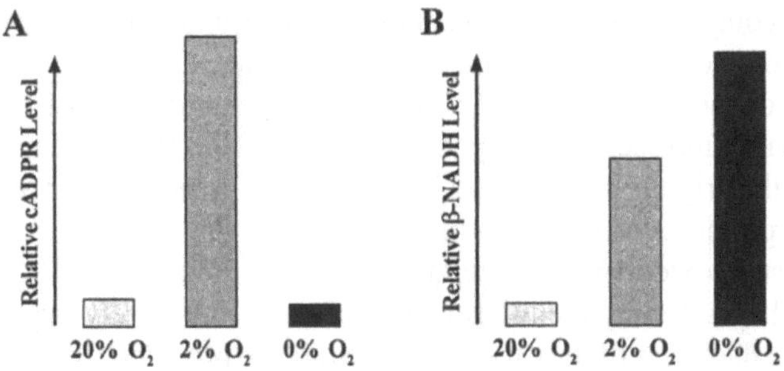

Figure 4. PO_2 window for cADPR accumulation. Relative cADPR (A) and β-NADH (B) levels in pulmonary artery smooth muscle during normoxia (20% O_2; 155-160 Torr), hypoxia (2% O_2; 16-21 Torr) and anoxia (0% O_2; 5-8 Torr), respectively.

In summary, increased β-NADH may effect a rise in cADPR synthesis and concomitant reduction in cADPR metabolism within a PO_2 window determined by substrate (β-NAD^+) availability. However, our findings also suggest that another mechanism may be required to mediate the extent of cADPR accumulation by hypoxia observed in pulmonary artery smooth muscle, and that this mechanism and not β-NADH may perhaps represent the primary effector in O_2-sensing cells.

14. Is cADPR Accumulation by Hypoxia Mediated by the Production of Reactive Oxygen Species by the Mitochondrial Electron Transport Chain?

Recent investigations on HPV have suggested that complex III of the electron transport chain may promote HPV. In short, it has been proposed that hypoxia inhibits oxidative phosphorylation, leading to a consequent increase in ROS at complex III in the electron transport chain, that is presumably driven by β-NADH and $FADH_2$ oxidation by complex I and complex II, respectively (52, 84, 85). More recently, however, it has been argued that the site of ROS production may be complex II and not complex III (67).

Irrespective of the exact site of ROS production, the possibility that a paradoxical increase in superoxide production by mitochondria may promote HPV, and therefore cADPR accumulation in pulmonary artery smooth muscle has been raised (52, 85). This proposal gains direct support from the finding that superoxide can increase cADPR synthesis by directly activating ADP-ribosyl cyclase (48, 62). Furthermore, Leach et al. (52, 53) suggest that it is an increase in superoxide and not an increase in β-NADH that promotes cADPR accumulation. This is based on their observation that HPV may be blocked by inhibitors of complex I of the electron transport chain, and subsequently restored by addition of succinate, i.e., by facilitating $FADH_2$ oxidation by complex II. Under each of these experimental conditions they report that the increase in NAD(P)H autofluorescence by hypoxia in intact arteries remained unaltered, a finding that they rightly cite as being inconsistent with the idea that an increase in β-NADH promotes cADPR accumulation.

Clearly, superoxide can, under certain conditions, promote cADPR synthesis. However, when assessing the information before us, it is vitally important that we recognize the limits of the experimental evidence provided by a given study before we draw any conclusion as to the likely mechanisms involved. Herein lies the problem. Leach et al. (52) relied upon pre-constriction by prostaglandin $F_{2\alpha}$ ($PGF_{2\alpha}$) in order to elicit constriction of pulmonary arteries by hypoxia. Furthermore, under the conditions of their experiments they have been unable to record maintained cADPR-dependent constriction by hypoxia in arteries without endothelium (2). In short, they have yet to observe the component of HPV driven by cADPR-dependent SR Ca^{2+} release by hypoxia. In this respect, interpretation of the findings of Waypa et al. (84) in the rat lung is also problematic. In these studies too, pre-constriction was required to procure HPV. Given this fact, it is impossible to draw any conclusions as to the mechanisms by which hypoxia mediates, by a direct action on the smooth muscle, maintained smooth muscle constriction in pulmonary arteries. These studies could therefore be seen as providing strong evidence against the idea that hypoxia promotes cADPR accumulation by increasing superoxide production at complex II/III of the mitochondrial electron transport chain. In short, the proposed increase in superoxide and/or H_2O_2 by hypoxia alone cannot explain cADPR accumulation by hypoxia in pulmonary artery smooth muscle. Support for this viewpoint may be derived from a recent investigation on carotid body glomus cells, which provided contrary findings with respect to the effects of mitochondrial inhibitors. In short, Ortega-Saenz et al. (65) conclude that the response of glomus cells to hypoxia is not linked to mitochondria in a simple way, and does not appear to be mediated by ROS production at complex II, or III of the ETC. This view is supported by the fact that mitochondrial inhibitors and ROS scavengers, respectively, do not alter the function of O_2-sensing cells in a consistent, nor reproducible manner (32). These contrary findings underline the need to monitor the "metabolic setting" of O_2-sensing cells under all experimental conditions, and the need to strictly adhere to the physiological range of Po_2 for a given cell type. This is highlighted by the Po_2 window for cADPR accumulation by hypoxia. In short, a level of hypoxia too moderate to induce HPV in the absence of preconstriction may be insufficient to provide the metabolic stress (i.e., activation of the primary effector) required for significant cADPR accumulation in the smooth muscle. On the other hand, severe hypoxia could induce a marked increase in the β-NADH:β-NAD^+ ratio without cADPR accumulation, due to a reduction in substrate (β-NAD^+) availability. Either way it may be possible to observe increased superoxide and/or β-NADH formation, respectively, in pulmonary artery smooth muscle in the absence of cADPR-dependent, maintained constriction by hypoxia.

In addition to their studies on the whole lung, however, Waypa et al. (85) presented data from cultured pulmonary artery smooth muscle cells. These data are consistent with the idea that an increase in ROS production, measured by dichlorofluorescein fluorescence, at complex III in the ETC triggers an increase in cytoplasmic Ca^{2+} concentration. It is notable, however, that the increase in the Fura-2 fluorescence ratio reported is small and non-uniform in individual cells, and does not appear to be associated with cell contraction. Furthermore, one glance at the literature informs us that investigations from a variety of laboratories have provided contrary data (32). For example, in pulmonary arteries a decrease in ROS has been measured by lucigenin chemiluminescence (5, 6, 8), consistent with the classical view that a fall in Po_2 results in a consequent fall in ROS (15). Not surprisingly, therefore, it has been stated that "the direct measurements of ROS are so demanding that they generate data supporting opposing views" (32, 77, 78). It is also important to note that cultured cells are known to lack many of the antioxidants normally present in cells *in vivo*, and are probably unrepresentative of wild type cells (34). Therefore, studies on cultured cells may not be consistent with those carried out on acutely isolated cells, isolated arteries, or whole lungs, respectively. In this respect it should not be forgotten that the investigations of Leach et al. (52) and Waypa et al. (84) on the effects of mitochondrial inhibitors on HPV in isolated pulmonary arteries and intact lungs, respectively, did not offer parallel measures of ROS levels. Furthermore, previous studies on pulmonary artery endothelial cells have shown a decrease in H_2O_2 production by hypoxia (91). This is significant, as the studies of Leach et al. (52) and Waypa et al. (84) may have been biased towards investigation of the effects of hypoxia on vasoconstrictor release from the pulmonary artery endothelium (see Chapter 6).

When considering the idea that cell signaling by hypoxia is mediated by ROS, however, I have most difficulty in explaining the fact that: *a*) Hyperoxia increases cellular ROS in a variety of preparations, including the rat lung and mitochondria derived from the lung (27, 41, 69, 76) and *b*) Hyperoxia but not hypoxia, increases the functional expression of antioxidants (18, 19, 40, 80). This is because short-term (hours) exposure to hyperoxia has little effect on resting tone of pulmonary arteries, and does not effect a change in pulmonary vascular resistance or the distribution of blood flow in the lung (36, 44). Allied to this, investigations on carotid body chemoreceptors have demonstrated that they are activated by hypoxia and inhibited by hyperoxia; the latter response being attributed to increased ROS (1,83). Furthermore, others have demonstrated that H_2O_2 decreases chemosensory discharge by the carotid body (66). Clearly, if increased cytoplasmic ROS acted as the primary effector of cell signaling by hypoxia and a common signaling pathway was present in all types of O_2-sensing cell, then one would expect hypoxia and hyperoxia, respectively, to have similar effects. This is clearly not the case.

Consideration of ROS metabolism also questions a role for ROS in signaling by hypoxia. Superoxide dismutases rapidly convert superoxide to H_2O_2, which is subsequently reduced by glutathione peroxidase. Thus, H_2O_2 + 2GSH yields H_2O + GSSG. GSH may then be regenerated by an NADPH glutathione reductase. Given these facts, an increase in cytoplasmic superoxide by hypoxia may be expected to elicit a consequent reduction in GSH and/or β-NADPH levels in pulmonary artery smooth muscle. In marked contrast, however, hypoxia has been shown to increase cellular GSH in pulmonary arteries and pulmonary artery endothelial cells (8, 21, 89), and hypoxia does not appear to alter β-NADPH levels in pulmonary artery smooth muscle (79). Furthermore, hyperoxia and not hypoxia increases glutathione uptake into endothelial cells (21). These findings are also inconsistent, therefore, with the idea that an increase in cytoplasmic superoxide and/or H_2O_2 act as primary mediators of HPV. There is also indirect evidence against a role for cytoplasmic ROS in increasing cADPR accumulation by hypoxia in pulmonary arteries. In short, superoxide has been shown to activate RyRs (46) and cADPR synthesis (48, 62), respectively, over a similar concentration range. One might expect SR Ca^{2+} release by hypoxia, therefore, to promote a degree of maintained pulmonary vasoconstriction in the presence of the cADPR antagonist 8-bromo-cADPR, because 8-bromo-cADPR does not block RyR activation *per se* (10, 22). In contrast, however, we have shown that 8-bromo-cADPR abolishes cADPR-dependent activation of RyRs in pulmonary artery smooth muscle and HPV (10, 22).

14.1. Intramitochondrial Regulation of cADPR Synthesis in the Cytoplasm

One possible explanation for contrary findings with respect to the role of ROS in HPV could, however, lie in the site of production and effect, respectively. Previous investigations have demonstrated that ADP-ribosyl cyclase/CD38 activities may be present in the plasma membrane, SR membrane and mitochondrial membrane, respectively (56, 90). Furthermore, there is evidence to suggest that CD38 may be anchored to the mitochondrial outer membrane by proteins distinct from those that may anchor it to the plasma membrane (56). It is possible, therefore, that a buildup of mitochondrial superoxide may promote cADPR production in the cytoplasm. This is clear from the fact that enzyme activities of membrane associated CD38 are generally thought to be conferred by the extracellular/ extra-organellar domain, whilst the intra-organellar domain may offer a site for regulation (55). Thus, mitochondrial superoxide and/or H_2O_2 could directly modulate cADPR synthesis and thereby RyR activation due to the close association between mitochondria and the SR (17, 20, 33). To confirm this proposal, however, future investigations must demonstrate that cADPR synthesis occurs in the mitochondrial fraction of pulmonary artery smooth muscle homogenates free from any microsomal contamination. Furthermore, if cADPR synthesis does occur in a pure mitochondrial fraction, we must establish that the relatively small number of mitochondria in pulmonary artery smooth muscle is sufficient to support the observed level of cADPR accumulation by hypoxia. We must also account for the fact that hyperoxia and not hypoxia, increases the functional expression of antioxidants (18, 19, 40, 80).

15. Is AMP-activated Protein Kinase the Metabolic Sensor and Primary Effector in O_2-sensing Cells?

Given that hypoxia likely triggers an increase in cellular AMP levels, it seems possible that the AMP-activated protein kinase (AMPK) may play a role in regulating the activity of O_2-sensing cells. AMPK is a multi-substrate serine/threonine kinase, which acts as a metabolic sensor and modulates many aspects of cell metabolism in eukaryotes (37,47). AMPK effectively detects metabolic stress through changes in the local ATP:AMP ratio, and may be activated by vigorous exercise, nutrient starvation and hypoxia (37,47). AMPK is an α,β,γ heterotrimer, and its activity is increased 2-4 fold by the association of AMP with an allosteric site. However, the AMPK activity is also regulated by an upstream kinase, namely AMPK kinase (39). AMPK kinase is also activated by AMP and regulates AMPK by phosphorylation of Thr172 within the activation loop of the AMPK α subunit (37, 47). In response to hypoxia, AMPK acts to maintain ATP levels with the minimum amount of O_2 utilization. AMPK does so by inducing the expression and subsequent translocation of GLUT to the plasma membrane (43, 49, 75), by accelerating glycolysis via phosphorylation of phosphofructokinase (59), and by inhibiting creatine kinase (71). These properties are consistent with the likely effects of hypoxia on pulmonary artery smooth muscle cell metabolism. AMPK kinase and AMPK, respectively, may therefore act as the "primary metabolic sensors" and "primary effectors" in O_2-sensing cells, with the activity of AMPK kinase/AMPK determined by the "metabolic setting" under normoxic and hypoxic conditions, respectively.

Given that AMPK targets an array of proteins including ion channels (35), it is possible that AMP-dependent activation of AMPK kinase and AMPK, respectively, in pulmonary artery smooth muscle may also regulate Ca^{2+} signaling. Thus, phosphorylation of ADP-ribosyl cyclase by AMPK could elicit a consequent increase in cADPR-dependent Ca^{2+} signaling by hypoxia. Furthermore, AMPK activation could also, by a direct or indirect action, inhibit potassium channel activity in the plasma membrane and SERCA activity in the peripheral SR proximal to the plasma membrane.

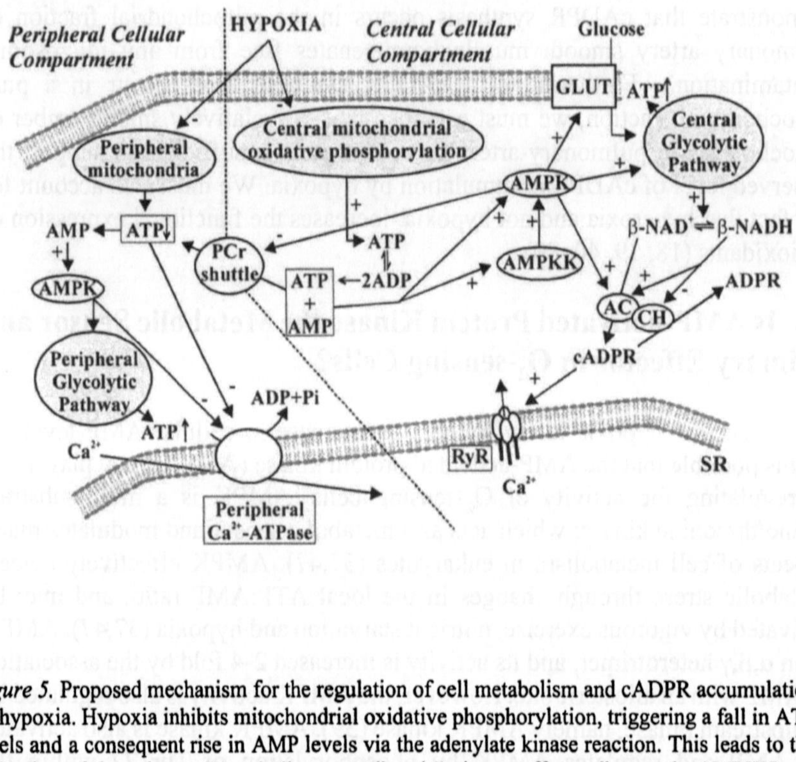

Figure 5. Proposed mechanism for the regulation of cell metabolism and cADPR accumulation by hypoxia. Hypoxia inhibits mitochondrial oxidative phosphorylation, triggering a fall in ATP levels and a consequent rise in AMP levels via the adenylate kinase reaction. This leads to the activation of the primary "metabolic sensors" and "primary effectors", namely AMPKK and AMPK, respectively. AMPK then accelerates glucose transport, glycolysis, and inhibits creatine kinase in an effort to maintain ATP levels at the primary sites of utilization. In addition, however, AMPK may activate ADP-ribosyl cyclase and cADPR accumulation, and inhibit SERCA activity in the peripheral SR proximal to the plasma membrane. AMPK, AMP activated kinase; AMPKK, AMP kinase kinase; GLUT, glucose transporter; AC, ADP-ribosyl cyclase; CH, cADPR hydrolase; ADPR, ADP-ribose; Pcr, phosphocreatine.

16 . Possible Role for cADPR Signaling by Hypoxia in Other O_2-sensing Cells

It is important to note that cADPR formation by hypoxia is a point of bifurcation in the O_2-sensitive signaling cascade, which may be initiated by AMPK kinase and AMPK. This is clear from our finding that cADPR does not play a primary role in the regulation of endothelial cell function by hypoxia (22, 90). Thus, cADPR may not be a primary mediator in all O_2-sensing cells. It should be noted, however, that recent studies have suggested that cADPR may play a modulatory role with respect to the regulation of exocytosis in both endocrine and exocrine cells (13, 63). It is quite possible, therefore, that cADPR may play a similar modulatory role in exocytosis by hypoxia in EPO-secreting cells, carotid body glomus cells, neuroepithelial bodies, and pulmonary artery endothelial cells, respectively.

17. Summary

In summary, it is clear that hypoxia promotes pulmonary artery constriction via the activation of a "metabolic sensor(s)" and a "primary effector(s)". The "primary effector" may then modulate a variety of other cell functions associated with HPV, including cADPR accumulation. Hypoxia likely triggers these events by inhibiting oxidative phosphorylation by mitochondria, leading to an increase in cytoplasmic AMP levels as the adenylate kinase reaction seeks to maintain ATP levels. I propose that a build up of AMP leads to the activation of the primary "metabolic sensors/primary effectors" in O_2-sensing cells, namely AMPK kinase and AMPK, respectively, and that AMPK, in addition to promoting glucose uptake and anaerobic glycolysis, may activate ADP-ribosyl cyclase and cADPR accumulation by hypoxia. Subsequently, increased β-NADH formation by anaerobic glycolysis may augment cADPR accumulation. As the severity of hypoxia increases, however, β-NADH levels may increase still further, leading to a consequent fall in cADPR levels due to reduced substrate availability (β-NAD^+), and ultimately to the failure of maintained HPV (Fig. 5). While, I cannot rule out the possibility that a paradoxical increase in mitochondrial ROS by hypoxia may also promote cADPR accumulation, this seems unlikely.

Acknowledgment

I would like to thank the Wellcome Trust for funding the research described in this chapter.

References

1. Alcayaga J, Iturriaga R, and Zapata P. Time structure, temporal correlation and coherence of chemosensory impulses propagated through both carotid nerves of the cat. *Bio. Res.* 1997; 30:125-133.
2. Aaronson PI, Robertson TP, and Ward JPT. Endothelium-derived mediators and hypoxic pulmonary vasoconstriction. *Resp. Physiol. Neurobiol.* 2002;132:107-120.
3. Adriaensen D, and Scheuermann D. Neuroendocrine cells and nerves of the lung. *Anatomical Record.* 1993;236:70-85.
4. Annex BH, and Williams RS. Mitochondrial DNA structure and expression in specialized subtypes of mammalian striated muscle. *Mol. Cell Biol.*, 1990; 10:5671-5678.
5. Archer SL, Huang J, Henry T, Peterson D, and Weir EK. A redox-based O_2 sensor in rat pulmonary vasculature. *Circ. Res.* 1993;73:1100-1112.
6. Archer S, Nelson D, and Weir E. Simultaneous measurement of oxygen radicals and pulmonary vascular reactivity in isolated rat lung. *J. Appl. Physiol.* 1989; 67:1903-1911.
7. Archer S, Reeve H, Michelakis E, Puttagunta L, Waite R, Nelson D, Dinauer M, and Weir E. Oxygen-sensing is preserved in mice lacking the gp91 phox subunit of NADPH oxidase. *Proc. Nat. Acad. Sci. U.S.A.* 1999; 96:7944-7949.
8. Archer SL, Will JA, and Weir EK. Redox status in the control of pulmonary vascular tone. *Herz* 1986; 11:127-141.
9. Boehme E, Ventura-Claier R, Mateo P, Lechene P, and Veksler V. Glycolysis supports calcium uptake by the sarcoplasmic reticulum in skinned ventricular fibers of mice deficient in mitochondrial and cytosolic creatine kinase. *J. Mol. Cell Cardiol.* 2000; 32:891-902.
10. Boittin F-X, Dipp M, Kinnear NP, Galione A, and Evans AM. Vasodilation by the calcium mobilising messenger cyclic ADP-ribose. *J. Biol. Chem.* 2003; 278: 9602-9608.
11. Biscoe TJ, and Duchen MR. Responses of type I cells dissociated from the rabbit carotid body to hypoxia. *J. Physiol.* 1990; 428: 39-59.
12. Baron JT, Gu L, and Parrillo JE. NADH/NAD redox state of cytoplasmic glycolytic compartments in vascular smooth muscle. *Am. J. Physiol.* 2000; 279: H2872-H2878.
13. Cancela JM, Gerasimenko OV, Gerasimenko JV, Galione A, Tepiken AV, and Peterson OH. Two different but converging messenger pathways to intracellular calcium release: the roles of nicotinic acid adenine dinucleotide phosphate, cyclic ADP-ribose and inositol trisphosphate. *EMBO J.* 2000; 19:2549-2557.
14. Cartee GD, Dounen AG, Ramlai T, Klip A, and Holloszky JO. Stimulation of glucose transport in skeletal muscle by hypoxia. *J. Appl. Physiol.* 1991; 70:1593-1600.
15. Chance B, Sies H, and Boveris A. Hydroperoxide metabolism in mammalian organs. *Physiol. Rev.* 1979;59:527-605.
16. Clarke JF. The creatine kinase system in smooth muscle. *Mol. Cell Biochem.* 1994;133:221-232.
17. Collins TJ, Berridge MJ, Lipp P, and Bootman MD. Mitochondria are morpphologically and functionally heterogeneous within cells. *EMBO J.* 2002; 21:1616-1627.
18. Comhair SA, and Erzurum SC. Antioxidant responses to oxidant-mediated lung diseases. *Am. J. Phyiol.* 2002; 283:L246-L255.
19. Coursin DB, Cihlaa HP, Will JA, and McCreary JL. Adaption to chronic hypoxia. Biochemical effects and the response to subsequent lethal hyperoxia. *Am. Rev. Respir. Dis.* 1987; 135:1002-1006.
20. Dalen H. An ultrastructural study of the hypertophied human papillary muscle cell with special emphasis on specific staining patterns, mitochondrial projections and association between mitochondria and SR. *Virchows Arch. A .Pathol . Anat . Histopathol.* 1989; 414:187-189.
21. Deneke SM, Steiger V, and Fanburg BL. Effect of hyperoxia on glutathione levels and glutamic acid uptake in endoothelial cells. *J. Appl. Physiol.* 1987; 63:1966-1971.

22. Dipp M, and Evans AM. cADPR is the primary trigger for hypoxic pulmonary vasoconstriction in the rat lung *in-situ*. *Circ. Res.* 2001; 89:77-83.
23. Dipp M, Nye PCG, and Evans AM. Hypoxia induces sustained sarcoplasmic reticulum calcium release in rabbit pulmonary artery smooth muscle in the absence of calcium influx. *Am. J. Physiol.* 2001; 281:L318-L325.
24. Dipp M, Thomas JM, Galione A, and Evans AM. A Po_2 window for smooth muscle cADPR accumulation and constriction by hypoxia in rabbit pulmonary arteries. *J. Physiol.* 2003; 547 P:72C.
25. Duchen MR. Mitochondria and calcium in cell physiology and pathophysiology. *Cell Calcium.* 2000;28:339-348.
26. Evans AM, Dipp M. Hypoxic pulmonary vasoconstriction: cyclic adenosine diphosphate-ribose, smooth muscle calcium stores and the endothelium. *Resp. Physiol. Neurobiol.* 2002; 132:3-15.
27. Freeman BA, Topolosky MK, and Crapo JD. Hyperoxia increases oxygen radical production in rat lung homogenates. *Archiv. Biochem. Biophys.* 1982; 216:477-484.
28. Fu XW, Nurse CA, Wang YT, and Cutz E. ective modulation of membrane cuurrents by hypoxia in intact airway chemoreceptors from neonatal rabbit. *J. Physiol.* 1999; 514:139-150.
29. Galione A, Lee HC, and Busa WB. Ca^{2+}-induced Ca^{2+} release in sea urchin egg homogenates: modulation by cyclic ADP-ribose. *Science.* 1991; 253:1143-1146.
30. Gonzalez C, Almaraz L, Obeso A, and Rigaul R. Carotid body chemoreceptors: from natural stimuli to sensory discharges. *Physiol. Rev.* 1994; 74:829-898.
31. Gonzalez C. "Sensitivity to physiologic hypoxia." In *Oxygen regulation of ion channels and gene expression,* Lopez Barneo J, and Weir EK, eds. Armonk, NY: Futura Publishing Co Inc., 1998, pp.321-336.
32. Gonzalez C, Sanz-Alfayate G, Agapito T, Gomez-Nino A, Rocher A, and Obeso A. Significance of ROS in oxygen sensing in cell systems with sensitivity to physiological hypoxia. *Respiratory Physiol. Neurobiol.* 2002; 132:17-41.
33. Hajonczky G, Csordas G, Madesh M, and Pacher P. The machinery of local calcium signalling between sarco-endoplasmic reticulum and mitochondria. *J. Physiol.* 2000; 529:69-81.
34. Halliwell B, and Gutteridge JMC. Free radicals and antiooxidant in the year, a historical look to the future. *Ann. New York Acad. Sci.* 2000; 899:136-147.
35. Hallows KR, McCane JE, Kemp BE, Witters LA, and Foskett JK. Regulation of channel gating by AMP-activated protein kinase modulates cystic fibrosis transmembrane conductance regulator activity in lung submucosa cells. *J. Biol. Chem.* 2003; 278:998-1004.
36. Hambreaeus-Jonzon K, Bindslev L, Mellgard AJ, and Hedenstierna G. Hypoxic pulmonary vasoconstriction in human lungs. A stimulus response study. *Anaesthesiology* 1997; 86:308-315.
37. Hardie DG, Carling D, and Carlson M. The AMP-activated/SNF1 protein kinase subfamily: Metabolic sensors of the eukaryotic cell. *Annu. Rev. Biochem.* 1998; 67:821-855.
38. Hardin CD, Allen TJ, and Paul RJ. "Metabolism and energetics of vascular smooth muscle." In *Physiology and pathophysiology of the Heart,* Sperelakis N, ed. Morwell, MA: Kluwer Academic, pp. 571-595.
39. Hawley SA, Selbert MA, Goldstein EG, Edleman AM, Carling D, and Hardie DG. 5'-AMP activates AMP-activated protein kinase cascade, and Ca^{2+}/calmodulin activates the calmodulin-dependent protein kinase 1 cascade via three independent mechanisms. *J. Biol. Chem.* 1995; 270:27186-27191.
40. Ho YS, Dey MS, and Crapo JD. Antioxidant enzyme expression in rat lungs during hyperoxia. *Am. J. Physiol.* 1996; 270:L810-L818.
41. Jamieson D, Chance B, Cadenas E, and Boveris A. The relation between free radical production and hyperoxia. *Annu. Rev. Physiol.* 1986; 48:703-719.
42. Jelkmann W. Erythropoeitin: structure, control of production and function. *Physiol. Rev.* 1992; 72:449-489.

43. Jessen N, Pold R, Buhl ES, Jensen LS, Schmitz O, and Lund S. Effects of AICAR and exercise on insulin-stimulated glucose uptake, signalling and GLUT-4 content in rat muscles. *J. Appl. Physiol.* 2003; 94:1373-1379.
44. Jones R, Zapol WM, and Reid L. Pulmonary artery remodelling and pulmonary hypertension after exposure to hyperoxia for 7 days. A morphogenic and haemodynamic study. *Am. J. Physiol.* 1984; 117:273-285.
45. Kato M, and Staub NC. Response of small pulmonary arteries to unilobar hypoxia and hypercapnia. *Circ. Res.* 1966; 19: 426-440.
46. Kawakami M, and Okabe E. Superoxide anion readical triggered calcium release from cardiac sarcoplasmic reticulum through ryanodine receptor calcium channel. *Mol. Pharm.* 1998; 53:497-503.
47. Kemp BE, Mitchelhill KI, Stapleton D, Michell BJ, Chen ZP, and Witters LA. Dealing with energy demand: the AMP-activated protein kinase. TIBS. 1999; 24:22-25.
48. Kumasaka Shoji H, and Okabe E. Novel mechanisms involved in superoxide anion radical-triggered calcium release from cardiac sarcoplasmic reticulum linked to cyclic ADP-ribose stimulation. *Antioxid. Redox Signal.* 1999; 1:55-69.
49. Kurth-Kraczek EJ, Hirshman MF, Goodyear LJ, and Winder WW. 5'AMP-activated protein kinase activation causes GLUT4 translocation in skeletal muscle. *Diabetes.* 1999; 48:1667-1671.
50. Lauweryns JM, Cokelaere J, and Theunynck P. Serotonin producing neuroepithelial bodies in rabbit respiratory mucosa. *Science.* 1973; 27:410-413.
51. Lauweryns JM, and Van Lommel A. Unltrastructure of nerve endings and synaptic junctions in rabbit pulmonary neuroepithelial bodies: A single and serial section study. *J. Anat.* 1987; 151:65-83.
52. Leach RM, Hill HS, Snetkov VA, Robertson TP, and Ward JPT. Divergent roles of glycolysis and the mitochondrial electron transport chain in hypoxic pulmonary vasoconstriction of the rat. Identity of the hypoxic sensor. *J. Physiol.* 2001; 536:211-224.
53. Leach RM, Hill HS, Snetkov VA, and Ward JPT. Hypoxia, energy state and pulmonary vasomotor tone. *Respir. Physiol. Neurobiol.* 2002; 132:55-67.
54. Leach RM, Sheehan DW, Chacko VP, and Sylvester JT. Energy state, pH, and vasomotor tone during hypoxia in precontracted pulmonary and femoral arteries. *Am. J. Physiol.* 2000; 278:L294-L304.
55. Lee HC. Physiological functions of cyclic ADP-ribose and NAADP as calcium messengers. *Annu. Rev. Pharmacol. Toxicol.* 2001; 41:317-345.
56. Liang M, Chini EN, Cheng J, and Dousa TP. Synthesis of NAADP and cADPR in mitochondria. *Arch. Biochem. Biophys.* 1999; 371:317-325.
57. Malvin GM, and Walker BR. Sites and ionic mechanisms of hypoxic pulmonary vasoconstriction in frog skin. *Am. J. Physiol. Regul. Integr. Comp. Physiol.* 2001; 280:R1308-R1314.
58. Marshall BE, Marshall C, Frasch F, and Hanson CW. Pulmonary artery NADPH oxidase is activated in hypoxic pulmonary vasoconstriction. *Am. J. Respir. Cell Mol. Biol.* 1996; 15:633-644.
59. Marsin AS, Bouzin C, Bertrand L, and Hue L. Phosphorylation and activation of heart PFK-2 by AMPK has a role in the stimulation of glycolysis during ischaemia. *Curr. Biol.* 2000; 19:1247-1255.
60. McKenna MC, Stevenson JH, Huang X, Tildon JT, Zeilke CL, and Hopkins IB. Mitochondrial malic enzyme activity is much higher in mitochondria from cortical terminals compared with mitochondria from primary culture of cortical neurons or cerebellar granule cells. *Neurochem. Int.* 2000; 38:451-459.
61. Mitchell P, and Moyle J. Chemiosmotic hypothesis of oxidative phosphorylation. *Nature* 1967; 213:137-139.
62. Okabe E, Tsujimoto Y, and Koboyashi Y. Calmodulin and cyclic ADP-ribose interaction in calcium signalling related to cardiac sarcoplasmic reticulum: superoxide anion radical-

triggered calcium release. *Antioxid. Redox Signal.* 2000; 1:47-54.
63. Okamoto H. The CD38 ADP-ribose signalling system in insulin secretion. *Mol. Cell Biochem.* 1999; 193:115-118.
64. Olson KR, Russell MJ, and Forster ME. Hypoxic pulmonary vasoconstriction of cyclostome systemic vessels: the antecedent of hypoxic pulmonary vasoconstriction? *Am. J. Physiol. Regul. Integr. Comp. Physiol.* 2001; 280:R198-R206.
65. Ortega-Saenz P, Pardai R, Garcia-Fernandez M, and Lopez-Barneo J. Rotenone selectively occludes sensitivity to hypoxia in rat carotid body glomus cells. *J. Physiol.* 2003; 548:789-800.
66. Osani S, Mokashi A, Rosanov C, Buerk DG, and Lahiri S. Potential role of H_2O_2 in chemoreception in the rat carotid body. *J. Auton. Nerv. Syst.* 1997; 63:39-45.
67. Paddenberg R, Ishaq B, Goldenberg A, Faulhammer P, Rose F, Weissmann N, Braun-Dullaeus RC, and Kummer W. Essential role of complex II of the respiratory chain in hypoxia-induced ROS generation in the pulmonary vasculature. *Am. J. Physiol.* 2003; 284:L710-L719.
68. Paul R. Smooth muscle energetics. *Annu. Rev. Physiol.* 1989; 51:331-349.
69. Paky A, Micahel JR, Burke-Wolin TM, Wolin MS, and Gurtner GH. Endogenous production of superoxide by rabbit lungs: effects of hypoxia or metabolic inhibitors. *J. Appl. Physiol.* 1993; 74:2868-2874.
70. Perez-Diaz J, Martin-Requero A, Ayuso-Parrila MS, and Parilla R. Metabolic features of isolated rat lung cells. I. Factors controlling glucose utilisation. *Am. J. Physiol.* 1977; 232:E394-E400.
71. Ponticos M, Lu QL, Morgan JE, Morgan JE, Hardie DG, Partridge TA, and Carling D. Dual regulation of the AMP-activated protein kinase provides a novel mechanism for the control of creatine kinase in skeletal muscle. *EMBO* 1998; 17:1688-1699.
72. Reeve H, Michelakis E, Nelson D, and Weir E. Alterations in redox oxygen sensing mechanism in chronic hypoxia. *J. Appl. Physiol.* 2001; 90:2249-2256.
73. Richalet JP. Oxygen sensors in the organism: examples of regulation under altitude hypoxia in mammals. *Comp. Biochem. Physiol. A Physiol.* 1997; 118:9-14.
74. Roy CS, and Sherrington CS. The regulation of the blood supply of the brain. *J. Physiol.* 1890; 11:85.
75. Russell RR, Bergeron R, Shulman GI, and Young LH. Translocation of myocardial GLUT-4 and increased glucose uptake through activation of AMPK by AICAR. *Am. J. Physiol.* 1999; 277:H643-H649.
76. Sanders SP, Zweier JL, Kuppusamy P, Harrison SJ, Bassett DJ, Gabrielson EW, and Sylvester JT. Hyperoxic sheep pulmonary microvascular endothelial cells generate free radicals via mitochondrial electron transport. *J. Clin. Invest.* 1993; 91:46-52.
77. Semenza GL. Perspectives in oxygen sensing. *Cell.* 1999; 98:281-284.
78. Semenza GL. Chairman's summary: mechanisms of oxygen sensing. *Adv. Exp. Med. Biol.* 2000; 475;303-310.
79. Shigemori K, Ishizaki T, Matsukawa S, Sakai A, Nakai T, and Miyabo S. Adenine nucelotides via activation of ATP-sensitive K^+ channels modulate hypoxic response in rat pulmonary arteries. *Am. J. Physiol.* 1996; 270:L803-L809.
80. Simon LM, Liu J, Theodore J, and Robin ED. Effect of hyperoxia, hypoxia and maturation on superoxide dismutase activity in isolated alveolar macrophages. *Am. Rev. Respir. Dis.* 1977; 115:279-284.
81. von Euler US, and Liljestrand G. Observations on the pulmonary arterial blood pressure in the cat. *Acta Physiol. Scand.* 1946; 12:301-320.
82. Von Lommel A, Lauweryns JM, De Leyn P, Wouters P, Shreinemakers H, and Lerut T. Pulmonary neuroepthelial bodies in neonatal and adult dogs: Histochemistry, ultrastructure and effects of unilateral lung denervation. *Lung* 1995; 173;13-23.
83. Wang HY, and Fitzgerald RS. Muscarinic modulation of hypoxia-induced release of catecholamines from the cat carotid body. *Brain Res.* 2002; 927:122-137.

84. Waypa GB, Chandel NS, and Schumacker PT. Model for hypoxic pulmonary vasoconstriction involving mitochondrial oxygen sensing. *Circ. Res.* 2001; 88:1259-1266.
85. Waypa GB, Marks JD, Mack, MM, Borboun C, Mungai PT, and Schumacker PT. Mitochondrial reactive oxygen species trigger calcium increases during hypoxia in pulmonary artery myocytes. *Circ. Res.* 2002; 91:719-726.
86. Weir E, and Archer S. The mechanism of acute hypoxic pulmonary vasoconstriction: the tale of two channels. *FASEB.* 1995; 9:183-189.
87. Wendt IR. Effects of substrate and hypoxia on smooth muscle metabolism and contraction. *Am. J. Physiol.* 1989; 256:C719-C727.
88. Wiener CM, and Sylvester JT. Effects of glucose on hypoxic pulmonary vasoconstriction in isolated ferret lungs. *J. Appl. Physiol.* 1993;74:2426-2431.
89. White R, Jackson J, McMurtry I, and Repine J. Hypoxia increases glutathione redox cycle and protects lungs against oxidants. *J. Appl. Physiol.* 1988; 65:2607-2616.
90. Wilson HL, Dipp M, Thomas JM, Lad C, Galione A, and Evans AM. ADP-ribosyl cyclase and cyclic ADP-ribose hydrolase act as a redox sensor: a primary role for cADPR in hypoxic pulmonary vasoconstriction. *J. Biol. Chem.* 2001; 276:11180-11188.
91. Yang W, and Block ER. Effect of hypoxia and reoxygenation on the formation and release of reactive oxygen species by porcine pulmonary artery endothelial cells. *J. Cell Physiol.* 1995; 164:414-423.
92. Youngson C, Nurse C, Yeger H, and Katz E. Oxygen sensing in airway chemoreceptors. *Nature.* 1993; 356:153-155.
93. Zhao Y, Packer CS, and Rhoades RA. The vein utilizes different sources of energy than the artery during hypoxic pulmonary vasoconstriction. *Exp. Lung Res.* 1996; 22:51-63.

VI. OXYGEN-SENSING MECHANISMS IN OTHER ORGANS AND TISSUES

Chapter 19

Involvement of Intracellular Reactive Oxygen Species in the Control of Gene Expression by Oxygen

Agnes Görlach, Helmut Acker, and Thomas Kietzmann
Experimentelle Kinderkardiologie, Deutsches Herzzentrum München, München; MPI für molekulare Physiologie, Dortmund; and Institut für Biochemie und Molekulare Zellbiologie, Göttingen, Germany

1. Introduction

Adaptation to changes in the ambient O_2 tension (PO_2) is essentially required for the adequate energy supply of humans and all other aerobic living organisms. In mammals, for instance, the respiratory and cardiovascular systems allow the provision and appropriate distribution of O_2 to serve as the terminal electron acceptor during the phosphorylation of adenosine diphosphate (ADP) into adenosine triphosphate (ATP). This so called oxidative phosphorylation is the major biochemical reaction for the generation of energy present in form of ATP. The process of extracting O_2 from the environment, and its distribution not only for oxidative phosphorylation but also as a substrate for other biochemical reactions, has been conserved throughout evolution by the development of advanced systems which tightly maintain O_2 homeostasis, i.e., keep the O_2 concentrations even in a single cell within a narrow physiological range.

This is important since, although the physiological mean arterial PO_2 (dissolved free O_2 concentration) is 74-104 mmHg (104-146 µmol/l) (36, 79, 80), differences in the vascularization and in the tissue diffusion properties may result in a heterogeneous PO_2 distribution within a single organ, leading to regions where the PO_2 is only approximately 7 mmHg. Moreover, it is conceivable that cells adjacent to an arterial inflow have other metabolic capacities or electrical activities than cells located at the venous end. The heterogenous PO_2 distribution patterns can become deleterious under pathophysiological conditions. This is influenced further by other factors such as the hemoglobin concentration in erythrocytes, which enable the O_2-carrying capacity of the blood. A low hematocrit with normal heart and lung function will result in anemia and thus in a lowered O_2-carrying capacity. In contrast, hypoxemia may occur also under arterial normoxia but reduced hemoglobin-bound oxygen content. In the case that heart function and hematocrit are normal but lung function is impaired, arterial blood may be both hypoxic and hypoxemic (79). Thus an oxygen sensing system with an elaborate sequence of adaptive mechanisms is required in order to maintain O_2 homeostasis, which is essential for survival, adequate organ function, and to avoid diseases like anemia, infarction, tumor development or oxygen induced retinopathy (Fig. 1).

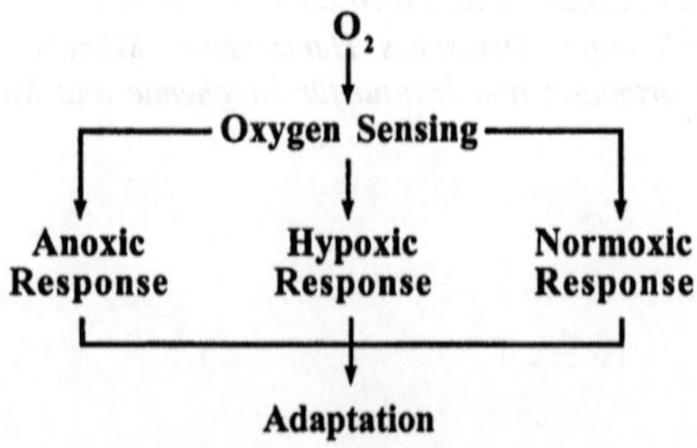

Figure 1. Oxygen responses. Aerobic living organisms need an O_2 sensing system which allows the cells to adapt to changes in the ambient Po_2 via generation of anoxic, hypoxic or normoxic responses. Adaptation mechanisms may include respiratory responses, changes in protein synthesis or modulation of channel activities and gene expression.

Increasing evidence has shown that almost all cells are able to adapt to changes in Po_2 by using cellular O_2-sensing systems which have to control short-term reactions within seconds or minutes by modification of channel or enzyme activities, as well as the long-term adaptation of cellular functions via regulation of gene expression. An O_2-sensing system that affects gene expression should consist of the sensor proper from which the O_2 signal is transmitted to a regulator proper, which should possess DNA or RNA binding ability to modulate gene activity. Along the signaling cascade between the O_2 sensor and the regulatory transcription factor(s), several chemical modifications (e.g., phosphorylation, hydroxylation and oxidation reactions), might be involved (Fig. 2). Within this article we will try to summarize the current knowledge of the O_2 sensing process which leads to a modulation of gene activity.

2. Cellular Response to Hypoxia

So far, the best-studied adaptation of mammalian cells to variations in the oxygen availability is the response to hypoxia. Once exposed to low oxygen, chemoreceptor cells of the carotid body initiate a ventilatory response which should compensate the decreased O_2 supply in the lung. Furthermore, the vasculature in the lung contracts, whereas the peripheral vasculature dilates, resulting in more efficient perfusion and gas exchange. Enhanced expression of erythropoietin (EPO) in the kidney and, to some extent, in the liver results in increased erythropoiesis in the bone marrow and in an elevation in the blood's O_2-carrying capacity.

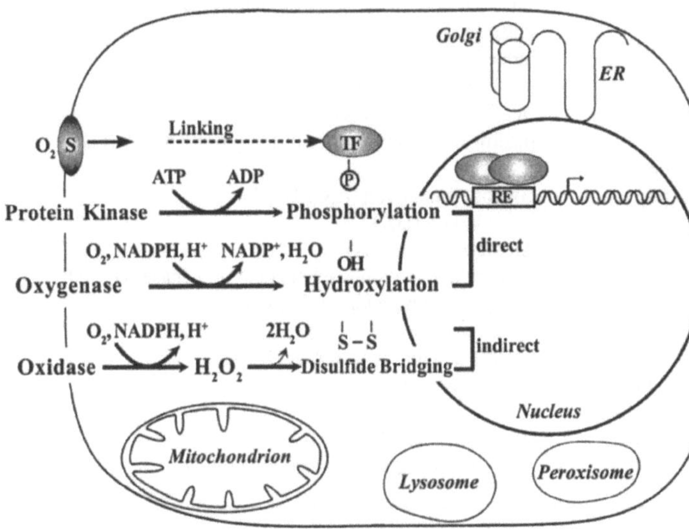

Figure 2. Models of an O_2 sensing system. An O_2-sensing system which modulates gene activity should ideally consist of an O_2 sensor (S) which should possess enzymatic activity and may be ideally situated at the cell membrane. From the O_2 sensor a signal is transmitted to a regulatory transcription factor (TF) which should possess DNA or RNA binding ability to modulate gene activity. The transmission of the O_2 signal can involve several chemical modifications of the regulator either mediated directly by the sensor or indirectly via the generation of a second messenger such as H_2O_2. In the case that the sensor contains a kinase or oxygenase activity, the regulatory TF will be phosphorylated or hydroxylated. If the sensor contains an oxidase, oxygen intermediates such as H_2O_2 may be generated as second messengers leading then to the modification of the regulatory TF. The chemical modifications of the regulatory TF would then modify its binding affinity to response elements (RE) in the promoters or regulatory sites of certain genes, thus leading to a modification in their transcription.

Additionally, various types of organs and tissues increase the expression of vascular endothelial growth factor (VEGF) and its receptors to promote angiogenesis, thereby improving blood supply to tissue. Moreover, increased expression of glycolytic enzymes elevates the efficiency of anaerobic ATP generation. Besides these more general responses evolved to enhance O_2 supply and to maintain energy provision, many other physiological and pathophysiological processes are known to be regulated by O_2. This is illustrated by the increasing number of genes encoding transporters, enzymes, hormones and growth factors which are expressed differentially (i.e., higher or lower) under conditions of low Po_2 (hypoxia) compared to normoxia (Table 1).

Table 1. Examples of Processes and Genes Activated by Hypoxia

Processes	Hypoxia-induced gene	Ref.
Hematopoesis	Erythropoietin (EPO)	29
Angiogenesis	Vascular endothelial growth factor (VEGF)	30
Fibrinolysis	Plasminogen activator inhibitor-1 (PAI-1)	55
Inflammation	Inducible nitric oxide synthase (iNOS)	73
Growth and	p53 Tumor-suppressor protein	34
Differentiation	Bcl-2 apoptosis-preventing protein	92
	Insulin-like growth factor binding protein-1 (IGFBP-1)	98
Carbohydrate metabolism		
Glucose transport	Glucose transporter-1 (GLUT-1)	28
Glycolysis	Glucokinase (GK)	19,24,25,54,89,90
	Aldolase A (ALD-A)	
	Glyceraldehyde 3-phosphate dehydrogenase (GAPDH)	
	Phosphoglycerate kinase 1 (PGK1)	
	Pyruvate kinase M (PK_M)	
	Lactate dehydrogenase A (LDH-A)	
NH_3-amino acid metabolism	L-Arginine transporter	67
Xenobiotic metabolism		
Heme metabolism	Heme oxygenase-1 (HO-1)	64
H_2O- and electrolyte regulation	Angiotensin converting enzyme (ACE)	58

A special case of the response to hypoxia is the so called anoxic response, i.e., the response to a Po_2 very close to 0 mmHg. This response is best studied in lower vertebrates such as turtles which can live for months under water without ventilation because of their drastically reduced protein synthesis, and thus O_2 consumption, which allows the survival of cellular functions under extreme conditions (42). Similarly, a reduction of protein synthesis was also observed after stroke in the peri-infarct area (penumbra) of the brain, facilitating cell survival by oxygen sensing (91).

3. The Nature of the Oxygen Sensor(s)

Identifying the nature of the O_2 sensor in mammalian cells has stimulated research for many years. Evidence has now been collected that a non-respiratory chain heme protein (29, 57), a part of the mitochondrial electron transport chain (2, 14), or a proline hydroxylase modifying the transcription factor hypoxia-inducible factor 1α (HIF-1α) (10, 22, 62) may act as oxygen sensor(s).

3.1. Role of a NADPH Oxidase as a Non-respiratory Chain Heme Protein in O_2 Sensing

Based on experiments in rat carotid body preparations and multicellular spheroids derived from the hepatoma cell line HepG2 as well as molecular studies in different liver-derived cell types, it was suggested that the oxygen sensor may be a heme protein (29). Spectrophotometrical analyses of carotid body preparations and HepG2 cells suggested the presence of a non-mitochondrial, b-type cytochrome with absorbance maxima at 559 and 427 nm similar to the cytochrome b558 of the NADPH oxidase from neutrophils (1, 33).

Although initially described as a neutrophil-specific enzyme, various NADPH oxidases have now been identified in several non-phagocytic cells including carotid body cells, airway chemoreceptor cells, HepG2 cells, endothelial cells and smooth muscle cells (26, 31, 33, 37, 61, 102). In neutrophils, the cytochrome b558-containing NADPH oxidase generates superoxide anion radicals ($O_2^{\bullet -}$) which can then be converted into H_2O_2 or $OH^{\bullet}$, the latter two often referred to as reactive oxygen species (ROS) (35). Whereas in neutrophils the generation of high amounts of ROS in the respiratory burst is essential for host defense, NADPH oxidases expressed in non-phagocytic cells appear to generate lower amounts of ROS which may play an important function in ROS-dependent signaling processes (6). Although neutrophils show decreased generation of superoxide anion radicals under hypoxic conditions, the leukocyte respiratory burst NADPH oxidase did not appear to be directly involved in oxygen sensing (105). However, the role of a "low output" NADPH oxidase isoenzyme as O_2 sensor was supported by the findings that the flavin oxidase inhibitor diphenylene iodonium (DPI) inhibited the hypoxia-dependent expression of the vascular VEGF, lactate dehydrogenase A (LDHA) and glucose transporter 1 (GLUT-1) in several cell lines (27).

Furthermore, in mice lacking the $gp91^{phox}$ subunit of the (high output) neutrophil NADPH oxidase oxygen sensing of the pulmonary airway chemoreceptors was disrupted (25) whereas hypoxia-dependent electrophysiological responses in the carotid body and pulmonary artery smooth muscle cells remained intact (4), indicating that different tissue and cell types may contain different types of oxygen sensors or oxygen sensing NADPH oxidases. Interestingly, in mice lacking the NADPH oxidase subunit p47phox, the ventilatory and chemoreceptor responses to hypoxia were elevated whereas EPO production of the kidneys remained unchanged (85).

3.2. Role of Cytochrome c Oxidase as an Electron Transport Chain Heme Protein in O_2 Sensing

Since a decrease in cytochrome c oxidase enzyme activity can be observed at Po_2 levels below 2 to 3 mm Hg, it was proposed that this enzyme acts as an O_2 sensor (13). Studies in embryonic cardiomyocytes showed that, under decreasing Po_2 (15% to 1% O_2), an increased oxidation of the non-fluorescent 2',7-dichlorofluorescin (DCFH) to the fluorescent DCF, an indicator of ROS presence, occurred (17). Paradoxically, upon return to normoxia, DCF fluorescence decreased again (17). Similar to cardiomyocytes, DCF fluorescence was also increased in response to hypoxia in Hep3B cells (13). The complex I respiratory chain inhibitor rotenone and the complex III inhibitor myxothiazol also abolished the hypoxia-dependent EPO, VEGF, and PGK-1 mRNA induction in Hep3B cells (13). Thus, the authors hypothesized that hypoxia is sensed by the cytochrome c oxidase, which then reduces its V_{max}, thereby enhancing the half-life of reduced electron carriers upstream of cytochrome a3 such as ubisemiquinone anions. The ubisemiquinone anion can then transfer an electron to molecular O_2 yielding $O_2^{\bullet -}$ radicals which are then dismutated to the signal molecule H_2O_2 (14).

Interestingly, in pulmonary arteries, hypoxia and the proximal respiratory chain inhibitors rotenone and antimycin A decreased the generation of ROS as was measured with three different techniques (lucigenin-based chemiluminescence, the AmplexRed H_2O_2 assay kit and DCF fluorescence), supporting the findings of many other laboratories which showed decreased ROS formation under hypoxia. However, in renal arteries, hypoxia and rotenone increased ROS production. These authors attributed their divergent observations to differences in the mitochondria between pulmonary and renal arteries (74).

However, the relevance of the mitochondrial respiratory chain as an ubiquitous oxygen sensor was questioned by several observations. In chinese hamster ovary cell lines (such as Gal32 and CCLI 6B-2) with defects in genes encoding different components of the electron transport chain, and in Hep3B cells lacking mitochondrial DNA (Hep3Bρ0), hypoxia-dependent activation of the transcription factor HIF-1α remained intact when compared to the wild type cells. Furthermore, H_2O_2 levels decreased under hypoxia in Hep3Bρ0 as well as in control cells further suggesting that the source of H_2O_2 was not located in the mitochondria (99).

3.3. Role of Proline and Asparagine Hydroxylases in O_2 Sensing

Recently, it has been demonstrated that a family of newly identified prolyl and asparaginyl hydroxylases play a more direct role in O_2 sensing mechanisms leading to the modulation of gene activity. Since the action of these enzymes is dependent on the presence of molecular O_2, iron, oxoglutarate and ascorbic acid it is likely that they might act as proximate direct O_2 sensors. The major substrate of these new prolyl hydroxylases is the O_2-dependent degradation domain (ODD) of the transcription factor HIF-1α (10, 22) whereas the asparaginyl hydroxylase appears to target the C-terminal transactivation domain (63) (see below). However, the expression of the prolyl hydroxylases is dependent on the oxygen tension (22) and the catalytic hydroxylation process exerted by the hydroxylases appears to require a radical cycling system (108), supporting the requirement of an upstream oxygen sensor such as a heme protein (see above) and an important role of ROS in oxygen signaling.

4. The Nature of the Oxygen-sensitive Regulators (Transcription Factors)

Several transcription factors, such as activator protein 1 (AP-1), early growth response protein-1 (EGR-1) (107, 108), CAAT enhancer binding protein beta (C/EBPβ/NF-IL-6), nuclear factor κB (NFκB) (60), and the hypoxia-inducible factor-1 (HIF-1), were found to be involved in the modulation of gene expression by O_2. Among these regulatory proteins HIF-1 appears to play a central role and is the best studied so far (86-88, 104).

HIF-1 is a dimer of HIF-1α and HIF-1β (ARNT) both belonging to the basic helix-loop-helix (bHLH) PAS (Per-ARNT-Sim) transcription factor family. Two other HIF α-subunits (HIF-2α/EPAS/HRF/HLF and HIF-3α) as well as two other ARNT isoforms (ARNT2 and ARNT3/BMAL-1/MOP-3) have been identified. This may then give rise to the formation of several HIF-isoforms composed of different HIF α-subunits and ARNT isoforms (103). Although other HIF isoforms appear to exist, HIF-1 is considered to be the major regulator of physiologically important genes such as those encoding EPO, VEGF, LDH and PAI-1. Furthermore, HIF-1 also seems to be required for carotid body neural activity and ventilatory adaptation to chronic hypoxia since carotid bodies from *hif1a*$^{+/-}$ mice responded to cyanide but not to hypoxia, whereas chronic hypoxia resulted in a diminished ventilatory response to subsequent exposure to acute hypoxia in *hif1a*$^{+/-}$ mice (59).

Oxygen sensitivity of the HIF-1 complex is conferred only by the HIF-1α protein via regulation of its protein stability and coactivator recruitment. Both, regulation of protein stability and coactivator recruitment involved hydroxylation reactions carried out by specific prolyl and asparaginyl hydroxylases (see above and below).

Moreover, HIF-1α was found to be modified by redox factor-1 (Ref-1) (44) and to interact with coactivators such as CBP/p300, TIE-2 and SRC-1 in a redox-dependent manner (12, 21), suggesting a role for ROS in the oxygen signaling pathway.

5. Coupling Between O_2 Sensors and Regulators by Chemical Modifications

5.1. Oxygen Sensing and HIF-1α Phosphorylation

The first evidence indicating that phosphorylation reactions are involved in the regulation of HIF-1 activity were obtained by electrophoretic mobility shift assay (EMSA) which showed that treatment of nuclear extracts from hypoxic Hep3B cells with phosphatase disrupts the HIF-1/DNA complex (101). Furthermore, application of the serine/threonine kinase inhibitor 2-aminopurine, the tyrosine kinase inhibitor genistein, as well as the serine threonine phosphatase inhibitor NaF inhibited HIF-1-DNA-binding activity and HIF-1 stabilization under hypoxic conditions (100). These data suggested that phosphorylation, as well as dephosphorylation, could both be involved in HIF-1 activation. Several kinases were shown to be activated under hypoxia in different cell types and are therefore possible candidates for HIF-1 activation. This is the case for some of the mitogen-activated protein kinase (MAPK) family members and for phosphatidylinositol-3 kinase (PI3-K) and protein kinase B (PKB, also known as AKT) (66, 71, 81, 82, 94).

Several reports demonstrated that the pathway involving PI3-K, which generates phosphatidylinositol-3,4,5-phosphate [PI(3,4,5)P_3], which then regulates the activity or subcellular localization of a variety of signaling molecules such as phosphatidiylinositol-dependent kinase (PDK) and PKB, plays an important role in activating the HIF pathway. The co-transfection of a HIF-1 reporter plasmid either with dominant-negative vectors for PI3-K or for PKB impaired the activation of HIF-1 as well as VEGF gene transcription in hypoxic NIH3T3R cells (71). Likewise, overexpression of PKB enhanced HIF-1α levels and stimulated HIF-1-dependent PAI-1 expression as well as EPO-HRE-regulated reporter gene activity in primary rat hepatocytes and HepG2 cells (56).

However, in contrast to growth factors, hormones and coagulation factors including PDGF, angiotensin II, insulin and thrombin as well as to H_2O_2 which activate PKB within minutes, enhanced phosphorylation of PKB has been observed after hours of exposure to hypoxia (8). Interestingly, these agonists were also able to stimulate the HIF pathway in a ROS- and PI3-K/PKB-dependent manner independent from the O_2 tension (68, 75, 84, 97). This suggested a cross talk between these agonist-dependent pathways and the O_2 signaling cascade leading to the activation of HIF. This assumption was further supported by observations that MAP kinases (including p38 MAP kinase and/or p42/44 MAP kinase) also contribute to the activation of the HIF pathway. Overexpression of the p38 upstream kinases MKK3 and MKK6 resulted in enhanced HIF-1α levels and stimulated HIF-1-dependent PAI-1 expression as well as EPO-HRE-regulated reporter gene activity under normoxic and hypoxic conditions (52) whereas inhibition of p38 MAP kinase prevented thrombin-induced HIF activation in a ROS-dependent manner (32). On the other hand, p42/44 MAP kinase has been shown to directly phosphorylate HIF-1α in vitro under normoxic conditions (81) but overexpression of p42/44 MAP kinase did not alter HIF-1α protein levels and did not stimulate HIF-1-dependent PAI-1 expression or EPO-HRE-regulated reporter gene activity (52). Moreover, in contrast to p38 MAP kinase, p42/44 MAP kinase was not activated by H_2O_2 and did not contribute to thrombin-induced activation of the HIF cascade (32) but stimulated angiotensin II-dependent activation of HIF (82). The reasons for these discrepancies are not clear at the moment but may relate to cell type specific differences in kinase expression and sensitivity to hypoxia and ROS.

5.2. Oxygen Sensing and HIF-1α Hydroxylation

The amount of the HIF-1α protein is rapidly increased when cells are exposed to hypoxia. In contrast, under normoxic conditions, the HIF-1α protein is remarkably unstable suggesting that the formation of the active HIF-1 dimer depends mainly on hypoxia-induced stabilization of HIF-1α. Two transactivation domains (TAD) were identified within HIF-α which confer the sensitivity towards O_2. Both domains are 100% conserved in human, mouse and rat HIF-1α. The N-terminal TAD (TAD-N) is present within a region (amino acids 401-603) which was found to be critically involved in the O_2-dependent destabilization of the HIF-1α protein via proteasomal degradation (45, 84) and was thus named the oxygen-dependent degradation domain (ODD) (45).

In normoxia, HIF α-subunit destabilization is mediated by O_2-dependent hydroxylation of at least two proline residues within the ODD (47, 68, 97). This process allows binding of the von Hippel-Lindau tumor suppressor protein (pVHL) (70, 75, 78, 97). pVHL is found in a multiprotein complex with elongins B/C, Cul2, and Rbx1 forming an E3 ubiquitin ligase complex called VEC. This modular enzyme then initiates degradation by the ubiquitin-proteasome pathway (46, 48, 49).

The C-terminal TAD (TAD-C, ranging from amino acid 776-826), and the TAD-N have been shown to interact with coactivators such as CREB-binding protein (CBP)/p300) (3, 18, 21), the steroid receptor coactivator-1 (SRC-1), transcription intermediary factor-2 (TIF-2) and redox factor-1 (Ref-1) (12), thereby facilitating enhanced transcriptional activity. A new enzyme named factor inhibiting HIF (FIH) (69) prevents the recruitment of the coactivator CBP/p300 by hydroxylating an asparaginyl residue in the TAD-C (63). FIH, like the HIF prolyl hydroxylases, is an O_2-, oxoglutarate-, iron-. and ascorbate-dependent enzyme and may thus also be considered as a putative O_2 sensor (39, 63, 65, 72).

5.3. ROS as Messengers in the Oxygen Sensing Cascade

Whenever oxygen is not completely reduced to water within the organism or a cell,, oxygen intermediates such as $O_2^{\bullet-}$, H_2O_2, and $OH^{\bullet}$ can be generated. Besides their ability to exert oxidative stress resulting in the damage of membranes, the oxidation of proteins and the mutation of DNA, ROS are also important determinants for the normal growth and metabolism in a variety of cells. The latter has become evident for the last decade and pointed to a rather widespread and exciting role of H_2O_2 and ROS as second messengers for various signals in bacteria, plants and mammalian cells (16, 83, 93). Since the production of ROS increases proportionally with the O_2 tension they are likely to be involved in the modification of transcription factors modulating gene activity in response to the ambient Po_2. The role of ROS as O_2 messengers has been supported by the finding that, similar to a typical response to hypoxia, treatment of healthy human volunteers with the antioxidant N-acetyl-cysteine (NAC) enhances the hypoxic ventilatory response (HVR) and blood erythropoietin concentration (40). Thus, NAC or its biochemical derivates, cysteine and glutathione, appear to mimic hypoxia. Furthermore, the proposal that H_2O_2 and derived ROS such as $OH^{\bullet}$ can serve as mediators of the O_2 signal was based on findings in HepG2 cells and carotid body preparations showing the presence of a low output oxidase which may be an NADPH oxidase isoform able to convert oxygen to superoxide and thus act as an oxygen sensor (33). Subsequently, Po_2-dependent $OH^{\bullet}$ production has been demonstrated in hepatoma cells and primary hepatocytes implicating that hypoxia is associated with decreased $OH^{\bullet}$ levels. The response to hypoxia can be mimicked under normoxia by the application of the $OH^{\bullet}$ radical scavenger dimethyl thiurea (DMTU) which reduces $OH^{\bullet}$ levels (20, 50) (Fig. 3). In line with these observations were findings by several laboratories demonstrating decreased amounts of ROS under hypoxic conditions in the lung (4).

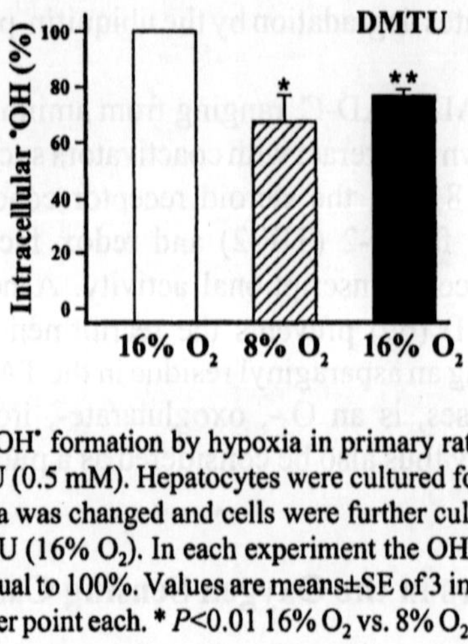

Figure 3. Reduction of $OH^{\bullet}$ formation by hypoxia in primary rat hepatocyte cultures. Mimicry of venous Po_2 by DMTU (0.5 mM). Hepatocytes were cultured for 24 hrs under normoxia (16% O_2). At 24 hrs, the media was changed and cells were further cultured for 2 hrs under 8% O_2 or in the presence of DMTU (16% O_2). In each experiment the $OH^{\bullet}$ level measured in the control under 16% O_2 was set equal to 100%. Values are means±SE of 3 independent culture experiments with 32 measurements per point each. * $P<0.01$ 16% O_2 vs. 8% O_2, ** $P<0.01$ control vs. DMTU.

5.4. Oxygen Sensing and Modification of HIF-1α by ROS

The concept that H_2O_2 and other ROS play an important role in oxygen sensing was further supported by findings demonstrating that application of H_2O_2 repressed the hypoxia- and HIF-1-dependent EPO production in HepG2 cells (89) and decreased the upregulation of tyrosine hydroxylase expression in PC12 cells by hypoxia. Addition of H_2O_2 prevented the downregulation of the glucagon-dependent PCK mRNA expression by hypoxia in primary hepatocytes (51). These findings implicate that addition of H_2O_2 prevents the hypoxic response and restores the normoxic response. This assumption was further supported by findings demonstrating that addition of H_2O_2 to cells grown under hypoxia resulted in the destabilization of the HIF-1α protein in Hep3B cells (44) and the HIF-2α protein in HeLa cells (106). Furthermore, treatment with the antioxidants pyrrolidine dithiocarbamate (PDTC) and NAC increased HIF-1α levels in alveolar type II epithelial cells (38). The redox processes modifying both HIF-1α and HIF-2α appeared to predominantly affect the C-terminal transactivation domain (TAD-C). Within this domain, the cysteine 800 of HIF-1α and the cysteine 848 of HIF-2α seem to be critical for transactivation, and the oxidation/reduction state of these cysteines is dependent on the presence of Ref-1 (12, 21, 44, 97). The regulation of HIF-1 DNA binding activity by ROS appeared to be evolutionary conserved since *Drosophila* HIF-D was also shown to be sensitive to redox modifications (7, 76, 77).

Whereas superoxide anion radicals are less likely to act as a second messenger since they are not freely diffusible, their dismutation product H_2O_2 is more suitable to function as a second messenger in O_2-sensing. Due to its freely diffusible non-charged character it can participate in two- and one-electron transfer reactions in the cells. Usually H_2O_2 is degraded by glutathione peroxidase in the cytosol and mitochondria or by catalase in peroxisomes or non-enzymatically converted into hydroxyl anions and hydroxyl radicals ($OH^{\bullet}$) in the presence of Fe^{2+} in a Fenton reaction:

$$H_2O_2 + Fe^{2+} \rightarrow Fe^{3+} + OH^- + OH^{\bullet}$$

Thus, a Fenton reaction adjacent to proteins or even transcription factors may affect Fe-S clusters or cysteine residues within regulatory proteins. It has been demonstrated that this H_2O_2-degrading reaction takes place in a perinuclear compartment and is involved in O_2 modulated gene expression (20, 50) as well as in carotid body nervous discharge (15). Thus, it appears likely that the redox-sensitive HIF-1α may be a target within an O_2 sensing system involving ROS and a Fenton reaction. However, it remains open in which cellular compartment the Fenton reaction takes place and whether transcription factors regulating the O_2-dependent gene expression or ion channels triggering nervous activity are located in this compartment.

5.5. Measurement of Intracellular Reactive Oxygen Species

Most of the chemicals commonly used to detect ROS may interact with a variety of radicals. Therefore, it appears to be difficult to use an indicator reacting with $OH^{\bullet}$ in more or less specific way and to allow the detection and subcellular localization of the Fenton reaction. This problem was solved by the use of the non-fluorescent dye dihydrorhodamine 123 (DHR 123) which can be converted by $OH^{\bullet}$ into fluorescent rhodamine 123 (RH 123) (20, 50, 53).

Thereby, the OH$^•$ rearrange the π-system of DHR 123, yielding fluorescent RH 123 (20). Furthermore, specific experimental conditions are required to measure the Fenton reaction in response to hypoxia such as: *i*) a non-phototoxic irradiation of the cells to avoid ROS generation during fluorescence excitation; *ii*) a minimal intracellular DHR 123 deposition to minimize secondary RH 123 dye distribution via diffusion and channel transport (103); *iii*) the start of kinetic measurements under conditions where ROS levels are expected, like hypoxia, since DHR 123 to RH 123 conversion is irreversible; *iv*) the maintenance of physiological conditions of tissue culture during the measurements to minimize cell stress; and *v*) an independent proof of intracellular ROS turnover.

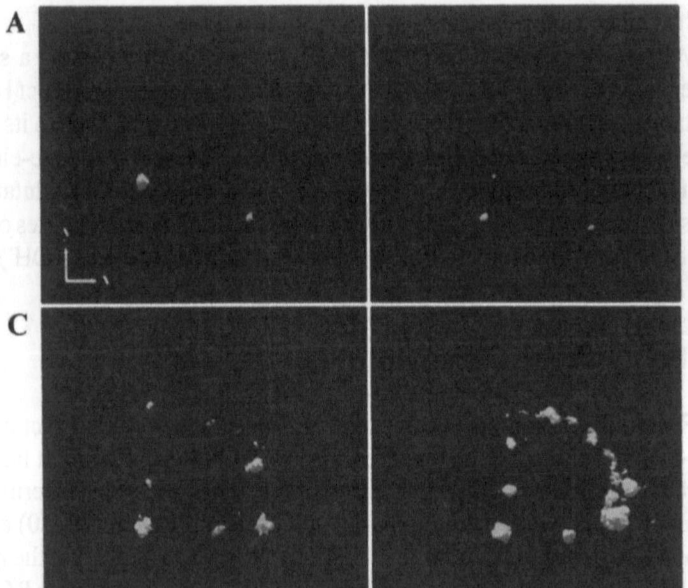

Figure 4. Oxygen-dependent generation of OH$^•$. One single HepG2 cell cultured in hypoxia for 60 mins was treated for 5 mins with 30 μM DHR123 before rhodamine 123 fluorescence (in white) was visualized by 2P-CLSM (A). Then the same cell was exposed to normoxia and imaged again (C). White perinuclear spots indicate an increased oxdation of DHR 123 to RH 123 under normoxia due to an enhanced Fenton reaction mediated OH· generation. The same HepG2 cell under hypoxia (B) and normoxia (D) was challenged with light of a wavelength range between 400 and 500 nm for 5 seconds to mediate OH$^•$ generation by photoreduction. Normoxia shows a significant higher OH$^•$ generation mediated by photoreduction than hypoxia. Dimensions of the X, Y, Z axis are given in μm.

These criteria can be fulfilled by using two photon confocal laser scanning microscopy which is not phototoxic due to infrared light (43). RH 123 fluorescence was registered by a photo multiplier, digitized and visualized. The signal-to-noise ratio was determined and images were deconvolved using the Maximum Likelihood Estimation (MLE) method. The data were reconstructed with the Application Visualization System (Waltham). Calculation of isosurfaces was performed using a marching cube algorithm (96). In our measurements, a 5-min incubation of cells (with 30 μM DHR 123) kept under physiological conditions in a microscope tissue culture chamber enabling observation at 37 °C under a variable gas atmosphere gave an optimal dye deposit which was fully convertible to fluorescent RH 123 only under normoxia in combination with 5-sec blue light illumination. This drastic ROS increase under normoxia was contrasted by the nearly-missing illumination reaction under hypoxia, which renders short term blue light illumination as an ideal prove for intracellular ROS turnover (Fig. 4). Under these conditions hypoxia was accompanied by a decrease in ROS levels, in agreement with earlier studies (23).

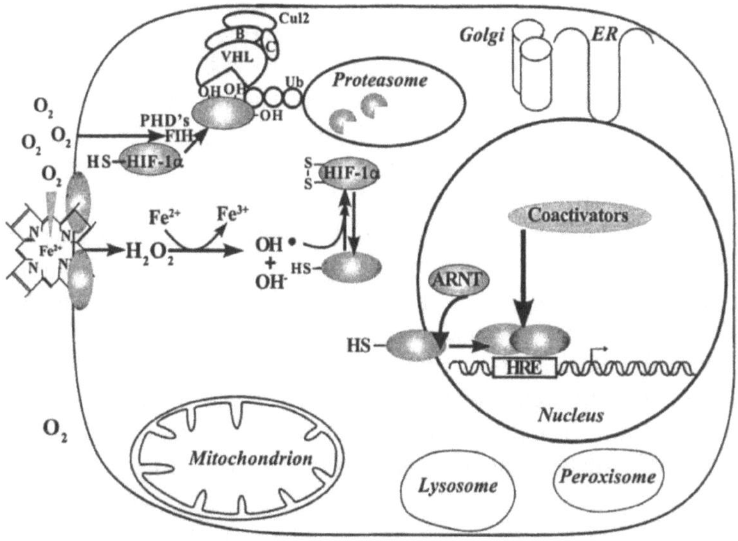

Figure 5. A Fenton reaction in transmission of the O_2 signal and regulation of O_2 modulated genes. PO_2 is sensed via a heme protein producing H_2O_2 below the threshold exerting oxidative stress. H_2O_2 is then diffused to the close vicinity of the nucleus due to the Fenton reaction, yielding OH•. Under a high PO_2, OH• oxidizes SH-groups (e.g., the cystein 800 in HIF-1α) and shifts the balance to the oxidized state. Furthermore, O_2 which escapes binding by the heme protein may be additionally used by the asparagine (FIH) and proline hydroxylases (PHD'S) which modify HIF-1α directly to inhibit cofactor recruitment and to mediate VHL binding to promote proteasomal degradation, respectively. In hypoxia, reduced ROS levels and decreased HIF-1α hydroxylation lead to nuclear translocation of HIF-1α, its dimerization with HIF-1β (ARNT), recruitment of cofactors and binding to hypoxia response elements (HRE). In addition, crosstalk between different signaling cascades activated by high or low ROS levels may also influence the stability of HIF-1β, thus allowing to fine tune HIF-1-dependent target gene expression.

However, other reports described enhanced ROS levels in hypoxia which decreased by returning to normoxia although the ROS-induced dye oxidation should be irreversible (13, 14). These findings may be explained by intensity changes of the oxidized dye, but do not appear to be due to changes in ROS.

Our findings suggest that the Fenton reaction primarily occurs in the cytoplasm. Since only 0.15% of the electron flow during mitochondrial respiration gives rise to the formation of ROS, this rather small amount of ROS is unlikely to escape to the cytoplasm due to the effective mitochondrial ROS scavenging systems including manganese superoxide dismutase and glutathione peroxidase (95). Our findings thus underline the importance of cytoplasmic, but not mitochondrial ROS as second messengers in O_2 sensing.

6. Conclusion

While the nature of the O_2 sensor(s) is not known in detail, a number of investigations support a model where a heme-containing specific "low output" NADPH oxidase isoform or a cytochrome b-type NAD(P)H oxidoreductase (109) have been proposed as candidate oxygen sensors. This complex leads to the formation of H_2O_2 in dependence of the PO_2 which, in the presence of Fe^{2+}, is converted in a Fenton reaction. The generated ROS in form of hydroxyl radicals ($OH^{\cdot}$) may prevent activation of HIF-1 and HIF-1-dependent signaling by providing an oxidizing environment. Furthermore, O_2 which escapes binding by the heme protein may be additionally used by the asparagine and proline hydroxylases which modify HIF-1α directly to inhibit cofactor recruitment and to mediate proteasomal degradation, respectively. By reducing O_2 the ROS levels, activation of the HIF-1 signaling cascade may be allowed. In addition, crosstalk between different signaling cascades activated by high or low ROS levels may also influence the stability and/or activity of HIF-1α, thus allowing to fine tune HIF-1-dependent target gene expression (Fig. 5).

Acknowledgments

Supported by grants from the Deutsche Forschungsgemeinschaft (GO 709/4-1 to AG, SFB 402 Teilprojekt A1 and GRK 335 to TK), Silicon Graphics (SGI), Nikon, Newport, and BMBF (13N7447/5 to HA).

References

1. Acker H. Mechanisms and meaning of cellular oxygen sensing in the organism. *Respir. Physiol.* 1994; 95: 1-10.
2. Agani FH, Pichiule P, Chavez JC, and LaManna JC. The role of mitochondria in the

regulation of hypoxia-inducible factor 1 expression during hypoxia. *J. Biol. Chem.* 2000; 275: 35863-35867.
3. Arany Z, Huang LE, Eckner R, Bhattacharya S, Jiang C, Goldberg MA, Bunn HF, and Livingston DM. An essential role for p300/CBP in the cellular response to hypoxia. *Proc. Natl. Acad. Sci. USA.* 1996; 93: 12969-12973.
4. Archer SL, Huang J, Henry T, Peterson D, and Weir EK. A redox-based O_2 sensor in rat pulmonary vasculature. *Circ. Res.* 1993; 73:1100-1112.
5. Archer SL, Reeve HL, Michelakis E, Puttagunta L, Waite R, Nelson DP, Dinauer MC, and Weir EK. O_2 sensing is preserved in mice lacking the gp91 phox subunit of NADPH oxidase. *Proc. Natl. Acad. Sci. USA.* 1999; 96: 7944-7949.
6. Babior BM. NADPH oxidase: an update. *Blood.*1999; 93: 1464-1476.
7. Bacon NC, Wappner P, O'Rourke JF, Bartlett SM, Shilo B, Pugh CW, and Ratcliffe PJ. Regulation of the *Drosophila* bHLH-PAS protein Sima by hypoxia: functional evidence for homology with mammalian HIF-1α. *Biochem. Biophys. Res. Commun.* 1998; 249: 811-816.
8. Beitner-Johnson D, Rust RT, Hsieh TC, and Millhorn DE. Hypoxia activates Akt and induces phosphorylation of GSK-3 in PC12 cells. *Cell Signal.* 2001; 13: 23-27.
9. Bestvater F, Spiess E, Stobrawa G, Hacker M, Feurer T, Porwol T, Berchner-Pfannschmidt V, Wotzlaw C, and Acker H. Two-photon fluorescence absorption and emission spectra of dyes relevant for cell imaging. *J. Microsc.* 2002; 208: 108-115.
10. Bruick RK and McKnight SL. A conserved family of prolyl-4-hydroxylases that modify HIF. *Science.* 2001; 294: 1337-1340
11. Bruick RK and McKnight SL. Transcription. Oxygen sensing gets a second wind. *Science.* 2002; 295: 807-808.
12. Carrero P, Okamoto K, Coumailleau P, O'Brien S, Tanaka H, and Poellinger L. Redox-regulated recruitment of the transcriptional coactivators CREB-binding protein and SRC-1 to hypoxia-inducible factor 1α. *Mol. Cell. Biol.* 2000; 20: 402-415.
13. Chandel NS, Maltepe E, Goldwasser E, Mathieu CE, Simon MC, and Schumacker PT. Mitochondrial reactive oxygen species trigger hypoxia-induced transcription. *Proc. Natl. Acad. Sci. USA.* 1998; 95: 11715-11720.
14. Chandel NS, McClintock DS, Feliciano CE, Wood TM, Melendez JA, Rodriguez AM, and Schumacker PT. Reactive oxygen species generated at mitochondrial complex III stabilize hypoxia-inducible factor-1α during hypoxia: a mechanism of O_2 sensing. *J. Biol. Chem.* 2000; 275: 25130-25138.
15. Daudu PA, Roy A, Rozanov C, Mokashi A, and Lahiri S. Extra- and intracellular free iron and the carotid body responses. *Respir. Physiol. Neurobiol.* 2002; 130:21-31.
16. Demple B. Study of redox-regulated transcription factors in prokaryotes. *Methods.* 1997; 11: 267-278.
17. Duranteau J, Chandel NS, Kulisz A, Shao Z, and Schumacker PT. Intracellular signaling by reactive oxygen species during hypoxia in cardiomyocytes. *J. Biol. Chem.* 1998; 273: 11619-11624.
18. Ebert BL and Bunn HF. Regulation of transcription by hypoxia requires a multiprotein complex that includes hypoxia-inducible factor 1, an adjacent transcription factor, and p300/CREB binding protein. *Mol. Cell. Biol.* 1998; 18:4089-4096,
19. Ebert BL, Gleadle JM, O'Rourke JF, Bartlett SM, Poulton J, and Ratcliffe PJ. Isoenzyme-specific regulation of genes involved in energy metabolism by hypoxia: similarities with the regulation of erythropoietin. *Biochem. J.* 1996; 313:809-814.
20. Ehleben W, Porwol T, Fandrey J, Kummer W, and Acker H. Cobalt and desferrioxamine reveal crucial members of the oxygen sensing pathway in HepG2 cells. *Kidney Int.* 1997; 51: 483-491.
21. Ema M, Hirota K, Mimura J, Abe H, Yodoi J, Sogawa K, Poellinger L, and Fujii-Kuriyama Y. Molecular mechanisms of transcription activation by HLF and HIF1α in response to hypoxia: their stabilization and redox signal-induced interaction with CBP/p300. *EMBO J.*

1999; 18: 1905-1914.

22. Epstein AC, Gleadle JM, McNeill LA, Hewitson KS, O'Rourke J, Mole DR, Mukherji M, Metzen E, Wilson MI, Dhanda A, Tian YM, Masson N, Hamilton DL, Jaakkola P, Barstead R, Hodgkin J, Maxwell PH, Pugh CW, Schofield CJ, and Ratcliffe PJ. *C. elegans* EGL-9 and mammalian homologs define a family of dioxygenases that regulate HIF by prolyl hydroxylation. *Cell.* 2001; 107: 43-54.
23. Fandrey J and Genius J. Reactive oxygen species as regulators of oxygen dependent gene expression. *Adv. Exp. Med. Biol.* 2000; 475: 153-159.
24. Firth JD, Ebert BL, Pugh CW, and Ratcliffe PJ. Oxygen-regulated control elements in the phosphoglycerate kinase 1 and lactate dehydrogenase A genes: similarities with the erythropoietin 3' enhancer. *Proc. Natl. Acad. Sci. USA.* 1994; 91: 6496-6500.
25. Fu XW, Wang D, Nurse CA, Dinauer MC, and Cutz E. NADPH oxidase is an O_2 sensor in airway chemoreceptors: Evidence from K^+ current modulation in wild-type and oxidase-deficient mice. *Proc. Natl. Acad. Sci. USA*. 2000; 97: 4374-4379.
26. Fukui T, Lassegue B, Kai H, Alexander RW, and Griendling KK. Cytochrome b-558 alpha-subunit cloning and expression in rat aortic smooth muscle cells. *Biochem. Biophys. Acta.* 1995; 1231: 215-219.
27. Gleadle JM, Ebert BL, and Ratcliffe PJ. Diphenylene iodonium inhibits the induction of erythropoietin and other mammalian genes by hypoxia. Implications for the mechanism of oxygen sensing. *Eur. J. Biochem.* 1995; 234: 92-99.
28. Gleadle JM and Ratcliffe PJ. Induction of hypoxia-inducible factor-1, erythropoietin, vascular endothelial growth factor, and glucose transporter-1 by hypoxia: evidence against a regulatory role for Src kinase. *Blood.* 1997; 89: 503-509.
29. Goldberg MA, Dunning SP, and Bunn HF. Regulation of the erythropoietin gene: evidence that the oxygen sensor is a heme protein. *Science.* 1988; 242: 1412-1415.
30. Goldberg MA and Schneider TJ. Similarities between the oxygen-sensing mechanisms regulating the expression of vascular endothelial growth factor and erythropoietin. *J. Biol. Chem.* 1994; 269: 4355-4359.
31. Görlach A, Brandes RP, Nguyen K, Amidi M, Dehghani F, and Busse R. A gp91phox containing NADPH oxidase selectively expressed in endothelial cells is a major source of oxygen radical generation in the arterial wall. *Circ. Res.* 2000; 87: 26-32.
32. Görlach A, Diebold I, Schini-Kerth VB, Berchner-Pfannschmidt U, Roth U, Brandes RP, Kietzmann T, and Busse R. Thrombin activates the hypoxia-inducible factor-1 signaling pathway in vascular smooth muscle cells: Role of the $p22^{phox}$-containing NADPH oxidase. *Circ. Res.* 2001; 89:47-54.
33. Görlach A, Holtermann G, Jelkmann W, Hancock JT, Jones SA, Jones OT, and Acker H. Photometric characteristics of haem proteins in erythropoietin-producing hepatoma cells (HepG2). *Biochem. J.* 1993; 290: 771-776.
34. Graeber TG, Peterson JF, Tsai M, Monica K, Fornace-AJ J, and Giaccia AJ. Hypoxia induces accumulation of p53 protein, but activation of a G1-phase checkpoint by low-oxygen conditions is independent of p53 status. *Mol. Cell. Biol.* 1994; 14: 6264-6277.
35. Graven KK, Troxler RF, Kornfeld H, Panchenko MV, and Farber HW. Regulation of endothelial cell glyceraldehyde-3-phosphate dehydrogenase expression by hypoxia. *J. Biol. Chem.* 1994; 269: 24446-24453.
36. Greger R and Bleich M. "Normal values for physiological parameters." In *Comprehensive Human Physiology*, Greger R and Windhorst U, eds. Berlin-Heidelberg: Springer Verlag, 1996, pp. 2427-2447.
37. Griendling KK and Harrison DG. Dual role of reactive oxygen species in vascular growth. *Circ. Res.* 1999; 85: 562-563
38. Haddad JJ, Olver RE, and Land SC. Antioxidant/pro-oxidant equilibrium regulates HIF-1α and NF-κ B redox sensitivity. Evidence for inhibition by glutathione oxidation in alveolar epithelial cells. *J. Biol. Chem.* 2000; 275: 21130-21139.

39. Hewitson KS, McNeill LA, Riordan MV, Tian YM, Bullock AN, Welford RW Elkins JM, Oldham NJ, Bhattacharya S, Gleadle JM, Ratcliffe PJ, Pugh CW, and Schofield CJ. Hypoxia-inducible factor (HIF) asparagine hydroxylase is identical to factor inhibiting HIF (FIH) and is related to the cupin structural family. *J. Biol. Chem.* 2002; 277: 26351-26355.
40. Hildebrandt W, Alexander S, Bartsch P, and Droge W. Effect of N-acetyl-cysteine on the hypoxic ventilatory response and erythropoietin production: linkage between plasma thiol redox state and O_2 chemosensitivity. *Blood.* 2002; 99: 1552-1555.
41. Hirsch-Ernst KI, Kietzmann T, Ziemann C, Jungermann K, and Kahl GF. Physiological oxygen tensions modulate expression of the mdr1b multidrug-resistance gene in primary rat hepatocyte cultures. *Biochem. J.* 2000; 350: 443-451.
42. Hochachka PW and Lutz PL. Mechanism, origin, and evolution of anoxia tolerance in animals. *Comp. Biochem. Physiol. B Biochem. Mol. Biol.* 2001; 130: 435-459.
43. Hockberger PE, Skimina TA, Centonze VE, Lavin C, Chu S, Dadras S, Reddy JK, and White JG. Activation of flavin-containing oxidases underlies light-induced production of H_2O_2 in mammalian cells. *Proc. Natl. Acad. Sci. USA.* 1999; 96: 6255-6260.
44. Huang LE, Arany Z, Livingston DM, and Bunn HF. Activation of hypoxia-inducible transcription factor depends primarily upon redox-sensitive stabilization of its α subunit. *J. Biol. Chem.* 1996; 271: 32253-32259.
45. Huang LE, Gu J, Schau M, and Bunn HF. Regulation of hypoxia-inducible factor 1α is mediated by an O_2-dependent degradation domain via the ubiquitin-proteasome pathway. *Proc. Natl. Acad. Sci. USA.* 1998; 95: 7987-7992.
46. Ivan M, Kondo K, Yang H, Kim W, Valiando J, Ohh M, Salic A, Asara JM, Lane WS, and Kaelin WG Jr. HIFα targeted for VHL-mediated destruction by proline hydroxylation: implications for O_2 sensing. *Science.* 2001; 292: 464-468.
47. Jaakkola P, Mole DR, Tian YM, Wilson MI, Gielbert J, Gaskell SJ, Kriegsheim Av, Hebestreit HF, Mukherji M, Schofield CJ, Maxwell PH, Pugh CW, and Ratcliffe PJ. Targeting of HIF-α to the von Hippel-Lindau ubiquitylation complex by O_2-regulated prolyl hydroxylation. *Science.* 2001; 292: 468-472.
48. Kaelin WGJ and Maher ER. The VHL tumour-suppressor gene paradigm. *Trends Genet.* 1998; 14: 423-426.
49. Kamura T, Sato S, Haque D, Liu L, Kaelin-WG J, Conaway RC, and Conaway JW. The Elongin BC complex interacts with the conserved SOCS-box motif present in members of the SOCS, ras, WD-40 repeat, and ankyrin repeat families. *Genes Dev.* 1998; 12: 3872-3881.
50. Kietzmann T, Fandrey J, and Acker H. Oxygen radicals as messengers in oxygen-dependent gene expression. *News Physiol. Sci.* 2000; 15: 202-208.
51. Kietzmann T, Freimann S, Bratke J, and Jungermann K. Regulation of the gluconeogenic phosphoenolpyruvate carboxykinase and glycolytic aldolase A gene expression by O_2 in rat hepatocyte cultures. Involvement of hydrogen peroxide as mediator in the response to O_2. *FEBS Lett.* 1996; 388: 228-232.
52. Kietzmann T, Jungermann K, and Görlach A. Regulation of the hypoxia-dependent plasminogen activator inhibitor 1 expression by MAP kinases in HepG2 cells. *Thromb. Haemost.* 2003; In press.
53. Kietzmann T, Porwol T, Zierold K, Jungermann K, and Acker H. Involvement of a local Fenton reaction in the reciprocal modulation by O_2 of the glucagon-dependent activation of the phosphoenolpyruvate carboxykinase gene and the insulin-dependent activation of the glucokinase gene in rat hepatocytes. *Biochem. J.* 1998; 335: 425-432.
54. Kietzmann T, Roth U, Freimann S, and Jungermann K. Arterial oxygen partial pressures reduce the insulin-dependent induction of the perivenously located glucokinase in rat hepatocyte cultures: mimicry of arterial oxygen pressures by H_2O_2. *Biochem. J.* 1997; 321: 17-20.
55. Kietzmann T, Roth U, and Jungermann K. Induction of the plasminogen activator inhibitor-1 gene expression by mild hypoxia via a hypoxia response element binding the hypoxia

inducible factor-1 in rat hepatocytes. *Blood.* 1999; 94: 4177-4185.

56. Kietzmann T, Samoylenko A, Roth U, and Jungermann K. Hypoxia-inducible factor-1 and hypoxia response elements mediate the induction of plasminogen activator inhibitor-1 gene expression by insulin in primary rat hepatocytes. *Blood.* 2003; 101: 907-914.
57. Kietzmann T, Schmidt H, Unthan FK, Probst I, and Jungermann K. A ferro-heme protein senses oxygen levels, which modulate the glucagon-dependent activation of the phosphoenolpyruvate carboxykinase gene in rat hepatocyte cultures. *Biochem. Biophys. Res. Commun.* 1993; 195: 792-798.
58. King SJ, Booyse FM, Lin PH, Traylor M, Narkates AJ, and Oparil S. Hypoxia stimulates endothelial cell angiotensin-converting enzyme antigen synthesis. *Am. J. Physiol.* 1989; 256: C1231-C1238.
59. Kline DD, Peng YJ, Manalo DJ, Semenza GL, and Prabhakar NR. Defective carotid body function and impaired ventilatory responses to chronic hypoxia in mice partially deficient for hypoxia-inducible factor 1α. *Proc. Natl. Acad. Sci. USA.* 2002; 99: 821-826.
60. Koong AC, Chen EY, and Giaccia AJ. Hypoxia causes the activation of nuclear factor kappa B through the phosphorylation of I kappa B alpha on tyrosine residues. *Cancer Res.* 1994; 54: 1425-1430.
61. Lambeth JD. Nox/Duox family of nicotinamide adenine dinucleotide (phosphate) oxidases. *Curr. Opin .Hematol.* 2002; 9: 11-17.
62. Lando D, Peet DJ, Gorman JJ, Whelan DA, Whitelaw ML, and Bruick RK. FIH-1 is an asparaginyl hydroxylase enzyme that regulates the transcriptional activity of hypoxia-inducible factor. *Genes Dev.* 2002; 16: 1466-1471.
63. Lando D, Peet DJ, Whelan DA, Gorman JJ, and Whitelaw ML. Asparagine hydroxylation of the HIF transactivation domain a hypoxic switch. *Science.* 2002; 295: 858-861.
64. Lee PJ, Jiang B-H, Chin BY, Iyer NV, Alam J, Semenza GL, and Choi AMK. Hypoxia-inducible factor-1 mediates transcriptional activation of the heme oxygenase-1 gene in response to hypoxia. *J. Biol. Chem.* 1997; 272: 5375-5381.
65. Lee C, Kim SJ, Jeong DG, Lee SM, and Ryu SE. Structure of human FIH-1 reveals a unique active site pocket and interaction sites for HIF-1 and VHL. *J. Biol. Chem.* 2003; 278: 7558-7563.
66. Lee E, Yim S, Lee S-K, and Park H. Two transactivation domains of hypoxia-inducible factor-1α regulated by the MEK-1/p42/p44 MAPK pathway. *Mol. Cells.* 2002; 14: 9-15.
67. Louis CA, Reichner JS, Henry WL Jr, Mastrofrancesco B, Gotoh T, Mori M, and Albina JE. Distinct arginase isoforms expressed in primary and transformed macrophages: regulation by oxygen tension. *Am. J. Physiol.* 1998; 274: R775-R782.
68. Masson N, Willam C, Maxwell PH, Pugh CW, and Ratcliffe PJ. Independent function of two destruction domains in hypoxia-inducible factor-α chains activated by prolyl hydroxylation. *EMBO J.* 2001; 20: 5197-5206.
69. Mahon PC, Hirota K, and Semenza GL. FIH-1: a novel protein that interacts with HIF-1α and VHL to mediate repression of HIF-1 transcriptional activity. *Genes Dev.* 2001; 15: 2675-2686.
70. Maxwell PH, Wiesener MS, Chang GW, Clifford SC, Vaux EC, Cockman ME, Wykoff CC, Pugh CW, Maher ER, and Ratcliffe PJ. The tumour suppressor protein VHL targets hypoxia-inducible factors for oxygen-dependent proteolysis. *Nature.* 1999; 399: 271-275.
71. Mazure NM, Chen EY, Laderoute KR, and Giaccia AJ. Induction of vascular endothelial growth factor by hypoxia is modulated by a phosphatidylinositol 3-kinase/Akt signaling pathway in Ha-ras-transformed cells through a hypoxia inducible factor-1 transcriptional element. *Blood.* 1997; 90: 3322-3331.
72. McNeill LA, Hewitson KS, Claridge TD, Seibel JF, Horsfall LE, and Schofield CJ. Hypoxia-inducible factor asparaginyl hydroxylase (FIH-1) catalyses hydroxylation at the β-carbon of asparagine-803. *Biochem. J.* 2002; 367: 571-575.
73. Melillo G, Musso T, Sica A, Taylor LS, Cox GW, and Varesio L. A hypoxia-responsive element mediates a novel pathway of activation of the inducible nitric oxide synthase

promoter. *J. Exp. Med.* 1995; 182: 1683-1693.
74. Michelakis ED, Hampl V, Nsair A, Wu XC, Harry G, Haromy A, Gurty R, and Archer SL. Diversity in mitochondrial function explains differences in vascular oxygen sensing. *Circ. Res.* 2002; 90: 1307-1315.
75. Min JH, Yang H, Ivan M, Gertler F, Kaelin WG, Jr., and Pavletich NP. Structure of an HIF-1 α-pVHL complex: hydroxyproline recognition in signaling. *Science* 2002; 296: 1886-1889.
76. Nagao M, Ebert BL, Ratcliffe PJ, and Pugh CW. Drosophila melanogaster SL2 cells contain a hypoxically inducible DNA binding complex which recognises mammalian HIF-binding sites. *FEBS Lett.* 1996; 387: 161-166.
77. Nambu JR, Chen W, Hu S, and Crews ST. The *Drosophila melanogaster* similar bHLH-PAS gene encodes a protein related to human hypoxia-inducible factor 1 α and Drosophila single-minded. *Gene.* 1996; 172: 249-254.
78. Ohh M, Park CW, Ivan M, Hoffman MA, Kim TY, Huang LE, Pavletich N, Chau V, and Kaelin WG. Ubiquitination of hypoxia-inducible factor requires direct binding to the β-domain of the von Hippel-Lindau protein. *Nat. Cell Biol.* 2000; 2: 423-427.
79. Piiper J. "Oxygen supply and energy metabolism." In *Comprehensive Human Physiology*, Greger R and Windhorst U, eds. Berlin-Heidelberg: Springer Verlag, 1996, pp. 2063-2069.
80. Piiper J. "Respiratory gas transport and acid-base equilibrium in blood." In *Comprehensive Human Physiology*, Greger R and Windhorst U, eds. Berlin-Heidelberg: Springer Verlag, 1996, pp. 2051-2062.
81. Richard DE, Berra E, Gothie E, Roux D, and Pouyssegur J. p42/p44 mitogen-activated protein kinases phosphorylate hypoxia-inducible factor 1 α (HIF-1 α) and enhance the transcriptional activity of HIF-1. *J. Biol. Chem.* 1999; 274: 32631-32637.
82. Richard DE, Berra E, and Pouyssegur J. Nonhypoxic pathway mediates the induction of hypoxia-inducible factor 1α in vascular smooth muscle cells. *J. Biol. Chem.* 2000; 275: 26765-26771.
83. Rosner JL and Storz G. Regulation of bacterial responses to oxidative stress. *Curr. Top .Cell. Regul.* 1997; 35: 163-77.
84. Salceda S and Caro J. Hypoxia-inducible factor 1α (HIF-1α) protein is rapidly degraded by the ubiquitin-proteasome system under normoxic conditions. Its stabilization by hypoxia depends on redox-induced changes. *J. Biol. Chem.* 1997; 272: 22642-22647.
85. Sanders KA, Sundar KM, He L, Dinger B, Fidone S, and Hoidal JR. Role of components of the phagocytic NADPH oxidase in oxygen sensing. *J. Appl. Physiol.* 2002; 93: 1357-1364.
86. Semenza GL. HIF-1: mediator of physiological and pathophysiological responses to hypoxia. *J. Appl. Physiol.* 2000; 88: 1474-1480.
87. Semenza GL. Hypoxia-inducible factor 1: control of oxygen homeostasis in health and disease. *Pediatr. Res.* 2001; 49: 614-617.
88. Semenza GL. Signal transduction to hypoxia-inducible factor 1. *Biochem. Pharmacol.* 2002; 64: 993-998.
89. Semenza GL, Jiang BH, Leung SW, Passantino R, Concordet JP, Maire P, and Giallonyo A. Hypoxia response elements in the aldolase A, enolase 1, and lactate dehydrogenase A gene promoters contain essential binding sites for hypoxia-inducible factor 1. *J. Biol. Chem* .1996; 271: 32529-32537.
90. Semenza GL, Roth PH, Fang HM, and Wang GL. Transcriptional regulation of genes encoding glycolytic enzymes by hypoxia-inducible factor 1. *J. Biol. Chem.* 1994; 269: 23757-23763.
91. Sharp FR, Lu A, Tang Y, and Millhorn DE. Multiple molecular penumbras after focal cerebral ischemia. *J. Cereb. Blood Flow Metab.* 2000; 20: 1011-1032.
92. Shimizu S, Eguchi Y, Kosaka H, Kamiike W, Matsuda H, and Tsujimoto Y. Prevention of hypoxia-induced cell death by Bcl-2 and Bcl-xL. *Nature.* 1995; 374: 811-813.
93. Shirasu K, Nakajima H, Rajasekhar VK, Dixon RA, and Lamb C. Salicylic acid potentiates an agonist-dependent gain control that amplifies pathogen signals in the activation of defense

mechanisms. *Plant Cell.* 1997; 9: 261-270.

94. Sodhi A, Montaner S, Miyazaki H, and Gutkind JS. MAPK and Akt act cooperatively but independently on hypoxia inducible factor-1α in *ras*V12 upregulation of VEGF. *Biochem. Biophys. Res. Commun.* 2001; 287: 292-300.
95. St-Pierre J, Buckingham JA, Roebuck SJ, and Brand MD. Topology of superoxide production from different sites in the mitochondrial electron transport chain. *J. Biol. Chem.* 2003; 277: 44784-44790.
96. Strohmaier AR, Porwol T, Acker H, and Spiess E. Three-dimensional organization of microtubules in tumor cells studied by confocal laser scanning microscopy and computer-assisted deconvolution and image reconstruction. *Cells Tissues Organs.* 2000; 167: 1-8.
97. Tanimoto K, Makino Y, Pereira T, and Poellinger L. Mechanism of regulation of the hypoxia-inducible factor-1α by the von Hippel-Lindau tumor suppressor protein. *EMBO J.* 2000; 19: 4298-4309.
98. Tazuke SI, Mazure NM, Sugawara J, Carland G, Faessen GH, Suen LF, Irwin JC, Powell DR, Giaccia AJ, and Giudice LC. Hypoxia stimulates insulin-like growth factor binding protein 1 (IGFBP-1) gene expression in HepG2 cells: a possible model for IGFBP-1 expression in fetal hypoxia. *Proc. Natl. Acad. Sci. USA.* 1998; 95: 10188-10193.
99. Vaux EC, Metzen E, Yeates KM, and Ratcliffe PJ. Regulation of hypoxia-inducible factor is preserved in the absence of a functioning mitochondrial respiratory chain. *Blood.* 2001; 98: 296-302.
100. Wang GL, Jiang BH, and Semenza GL. Effect of protein kinase and phosphatase inhibitors on expression of hypoxia-inducible factor 1. *Biochem. Biophys. Res. Commun* .1995; 216: 669-675.
101. Wang GL and Semenza GL. Characterization of hypoxia-inducible factor 1 and regulation of DNA binding activity by hypoxia. *J. Biol. Chem.* 1993; 268: 21513-21518.
102. Wang D, Youngson C, Wong V, Yeger H, Dinauer MC, Vega-Saenz ME, Rudy B, and Cutz E. NADPH-oxidase and a hydrogen peroxide-sensitive K^+ channel may function as an oxygen sensor complex in airway chemoreceptors and small cell lung carcinoma cell lines. *Proc. Natl. Acad. Sci. USA.* 1996; 93: 13182-13187.
103. Wenger RH. Cellular adaptation to hypoxia: O_2-sensing protein hydroxylases, hypoxia-inducible transcription factors, and O_2-regulated gene expression. *FASEB J.* 2002; 16: 1151-1162.
104. Wenger RH and Gassmann M. Oxygen(es) and the hypoxia-inducible factor-1. *Biol. Chem.* 1997; 378: 609-616.
105. Wenger RH, Marti HH, Schuerer MC, Kvietikova I, Bauer C, Gassmann M, and Maly FE.Hypoxic induction of gene expression in chronic granulomatous disease-derived B-cell lines: oxygen sensing is independent of the cytochrome b558-containing nicotinamide adenine dinucleotide phosphate oxidase. *Blood.* 1996; 87: 756-761.
106. Wiesener MS, Turley H, Allen WE, Willam C, Eckardt KU, Talks KL, Wood SM, Gatter KC, Harris AL, Pugh CW, Ratcliffe PJ, and Maxwell PH. Induction of endothelial PAS domain protein-1 by hypoxia: characterization and comparison with hypoxia-inducible factor-1α. *Blood.* 1998; 92: 2260-2268.
107. Yan SF, Lu J, Zou YS, Soh WJ, Cohen DM, Buttrick PM, Cooper DR, Steinberg SF, Mackman N, Pinsky DJ, and Stern DM. Hypoxia-associated induction of early growth response-1 gene expression. *J. Biol. Chem.* 1999; 274: 15030-15040.
108. Yan SF, Zou YS, Mendelsohn M, Gao Y, Naka Y, Du YS, Pinsky D, and Stern D. Nuclear factor interleukin 6 motifs mediate tissue-specific gene transcription in hypoxia. *J. Biol. Chem.* 1997; 272: 4287-4294.
109. Zhu H, Qiu H, Yoon HW, Huang S, and Bunn HF. Identification of a cytochrome b-type NAD(P)H oxidoreductase ubiquitously expressed in human cells. *Proc. Natl. Acad. Sci. USA.* 1999; 96: 14742-14747.

Chapter 20

Oxygen Sensing, Oxygen-sensitive Ion Channels and Mitochondrial Function in Arterial Chemoreceptors

José López-Barneo, Patricia Ortega-Sáenz, Maria García-Fernández and Ricardo Pardal
Universidad de Sevilla, Seville, Spain

1. Introduction

Oxygen sensing is a fundamental biological process necessary for the adaptation of living organisms to variable habitats and physiologic situations. In mammals, acute hypoxia triggers fast respiratory and cardiovascular counter-regulatory adjustments to ensure sufficient O_2 supply to the most critical organs such as the brain or the heart. Reduction of blood O_2 tension (Po_2) is sensed by the arterial chemoreceptors, which generate afferent chemosensory discharges that activate the brain to evoke hyperventilation and sympathetic activation (for a review see Ref. 22). The main arterial chemoreceptor is the carotid body, a minute bilateral organ located in the bifurcation of the carotid artery which contains neurosecretory glomus, or type I, cells and the less numerous substentacular, or type II, cells. Glomus cells, the O_2-sensitive elements in the carotid body, are electrically excitable (9, 20) and have O_2-sensitive potassium channels in their membranes (3, 7, 20, 33, 36). It is broadly accepted that inhibition of these channels by low Po_2 is a major O_2-dependent event leading to membrane depolarization, external calcium influx and activation of neurotransmitter release, which, in turn, stimulates the afferent sensory fibers. This model of chemotransduction, suggested by the electrophysiological experiments, has been confirmed in single Fura 2-loaded cells by monitoring cytosolic $[Ca^{2+}]$ and quantal catecholamine secretion (4, 5, 21, 26, 30, 39).

2. O_2-sensitive K^+ Channels and Stimulus-secretion Coupling in Glomus Cells

The typical macroscopic ionic currents recorded from dispersed rabbit carotid body cells maintained in normoxic conditions ($Po_2 \approx 150$ mmHg) are illustrated in Figure 1A. On depolarization the cells generate an inward current, mainly mediated by Ca^{2+} channels, followed by an outward K^+ current. The large inward tail current at the end of the pulse reflects the influx of Ca^{2+} through the channels and their closing time course. The outward K^+ current is selectively and reversibly reduced in amplitude upon exposure to the same external solution but with a $Po_2 \approx 30$ mmHg. Figure 1B shows single-channel recordings obtained from an excised membrane patch containing one O_2-sensitive ion channel whose peak open probability was reduced to almost 50% on switching from normoxia to hypoxia (11, 12).

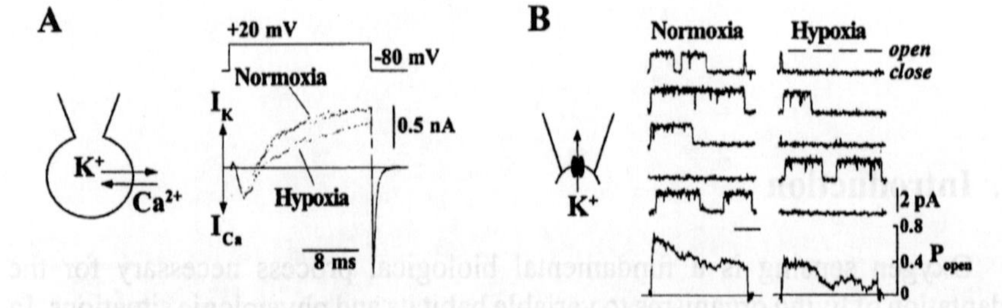

Figure 1. Major O_2 dependent electrophysiological properties of glomus cells. A: Macroscopic inward and outward currents of glomus cells and reversible inhibition of the outward current by hypoxia ($Po_2 \approx 20$ mmHg). Control and recovery traces in normoxia ($Po_2 = 150$ mmHg) are shown superimposed. In this experiment TTX was added to block the Na^+ conductance. B: Single-channel recordings from an excised membrane patch containing at most one open O_2-sensitive K^+ channel. Depolarizing pulses were applied from -80 to +20 mV. Ensemble averages indicating the single-channel open probability in normoxia and hypoxia are from 15 and 22 successive recordings, respectively (modified from Refs. 11, 26).

Reduction of K^+ conductance in hypoxia is expected to produce an increase in glomus cell excitability leading to Ca^{2+} influx and secretion. This has been shown to occur in dispersed glomus cells loaded with the Ca^{2+} indicator Fura2-AM in which single cell secretion was monitored by amperometry (5, 26, 39). Note in Figure 2A that, upon exposure to hypoxia, the increase in cytosolic $[Ca^{2+}]$ is paralleled by the appearance of spike-like quantal events corresponding to the release of dopamine (the most abundant catecholamine in glomus cells) from individual vesicles. The current model of glomus cell activation by hypoxia is shown diagrammatically in Figure 2B. Sensing of low Po_2 is done through inhibition of the K^+ conductance (G_K, 1), which leads to increase of action potential firing frequency (2), Ca^{2+} influx (3) and secretion (4) of the transmitters that activate the afferent fibers of the sinus nerve (5). However, the K^+ channel type modulated by Po_2 appears to change among the various species (22), and inhibition (26) and potentiation (37) of the Ca^{2+} current by hypoxia in rabbit glomus cells have also been reported.

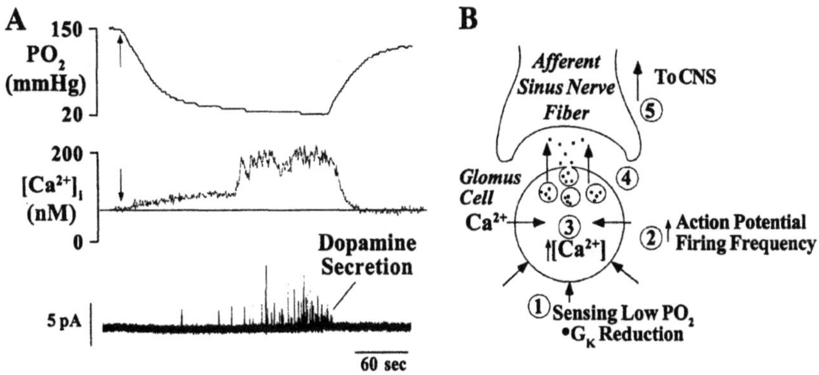

Figure 2. Secretory response of single glomus cells to low Po_2. A: Parallel changes of Po_2, $[Ca^{2+}]_i$, and dopamine secretion in a glomus cell in response to hypoxia. Cytosolic $[Ca^{2+}]$ was measured by microfluorimetry with Fura-2. Dopamine release was monitored by amperometry with a 6 μm polarized carbon fiber electrode and quantal secretory events appeared as spike like activity representing the fusion of individual secretory vesicles. B: Schematic diagram of the major steps in a model of sensory transduction involving O_2-sensitive ion channels in glomus cells. 1) O_2 sensing by K^+ channels. 2) Hypoxic reduction of the K^+ conductance (G_K) leads to an increase in the action potential frequency and Ca^{2+} influx. 3) Rise of cytosolic $[Ca^{2+}]$. 4) Transmitter release. 5) Activation of the afferent fibers of the sinus nerve (Modified from Refs. 19, 26, 39).

3. O_2-Sensitivity of Glomus Cells in Carotid Body Thin Slices and Secretory Responses to K^+ Channel Blockers

The model of acute O_2 sensing based on the regulation of membrane K^+ channels described in the carotid body has been demonstrated to operate in other neurosecretory systems, such as cells in the lung neuroepithelial bodies (43), chromaffin cells of the adrenal medulla (38), or PC-12 cells (44). There are, however, controversies on whether O_2-sensitive membrane electrical events in glomus cells are directly involved in the chemosensing process. The major argument supporting this notion is that inhibitors of the potassium current, like tetraethylammonium (TEA), 4-aminopyridine (4-AP), or charybdotoxin (CTX), do not enhance either action potential firing frequency of the afferent sensory fibers or secretory activity in the whole carotid body preparations used in these experiments (8, 16, 29). Although it was shown in a study that CTX can depolarize dialyzed rat glomus cells (42) it has been reported that neither TEA nor 4-AP influence the membrane potential of the same cells (3). To investigate the reasons for the discrepancies among laboratories using dispersed rat glomus cells and the contradictions between findings in isolated cells and the whole organ, we have developed a carotid body slice preparation to study the O_2-sensitivity of glomus cells in the best possible physiological conditions (30, 31). We also attempted to obtain a preparation of glomus cells with consistent properties since cellular O_2-sensitivity is a labile phenomenon easily destroyed by uncontrolled variables during the enzymatic treatment and mechanical disruption of the tissue (32).

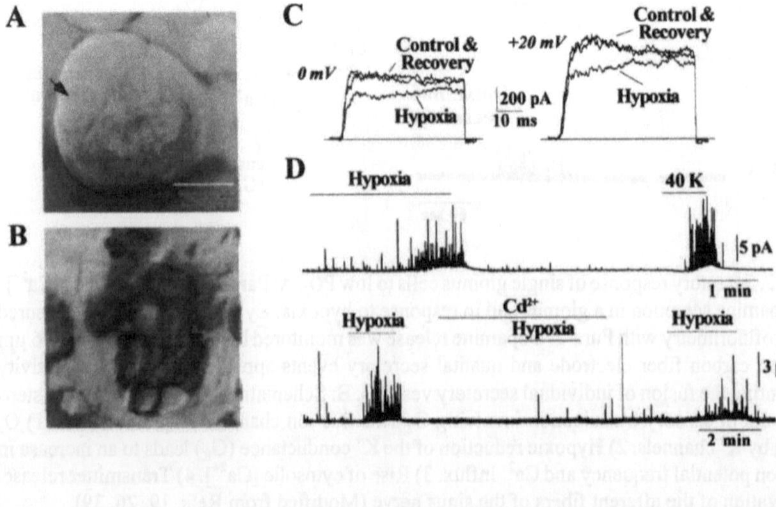

Figure 3. Responses to hypoxia of carotid body glomus cells in slices. A: Morphological appearance of a typical glomerulus within a cultured rat carotid body thin slice maintained in culture for 72 h. A well-defined single cell is indicated by the arrow. B: Carotid body slice immunostained with antibodies against tyrosine hydroxylase. The carotid body was fixed, then sliced and stained. Note the typical appearance of glomus cells with large nuclei and a thin layer of stained cytoplasm. The organization of type I cells in glomeruli is similar to that seen in fresh slices. C: Superimposed K^+ currents from a glomus cell elicited by depolarizing pulses from −80 mV to 0 (left) or +20 mV (right) in the three experimental conditions (control, hypoxia and recovery). Note the reversible reduction of the current by low PO_2 (≈20 mmHg). D: Top, Amperometric recordings from an O_2-sensitive glomus cell illustrating the increase of secretory activity elicited by hypoxia (PO_2 ≈20 mmHg). Note also the typical response of the cell to high extracellular K^+. Bottom, Secretory activity recorded from an O_2 sensitive glomus cell to illustrate the reversible abolishment of the response to hypoxia during the blockade of Ca^{2+} channels by addition of 0.2 mM cadmium to the extracellular solution (Modified from Refs. 30, 32).

The procedures followed to make carotid body slices are described in Pardal et al. (30) and Pardal and López-Barneo (32). In healthy slices, clusters of glomus cells were clearly distinguished from the surrounding tissue (Fig. 3A). These clusters, of similar appearance to the glomeruli described in histological preparations of the carotid body, contained numerous ovoid cells of ≈10 to 12 μm of diameter. The similarity between fresh and fixed carotid body preparations can be appreciated by observing slices immunostained with antibodies against tyrosine hydroxylase (TH), where glomus cells appear in clusters with intensely stained thin cytoplasmic layers and big clear nuclei (Fig. 3B). For the experiments, a slice was transferred to a recording chamber mounted on the stage of an upright microscope, where it was continuously perfused by gravity (flow

1 to 2 ml/min) with a solution containing (in mM) 117 NaCl, 4.5 KCl, 23 $NaHCO_3$, 1 $MgCl_2$, 2.5 $CaCl_2$, and 10 glucose. The recording electrodes (either patch pipette or amperometric carbon fiber) were placed adjacent to a well-identified cell within a glomerulus, such as the one indicated by the arrow in Figure 3A. In experiments designed to test the effect of hypoxia, the control (normoxic) solution was bubbled with a gas mixture of 5% CO_2, 20% O_2, and 75% N_2 ($Po_2 \approx 150$ mmHg), and the hypoxic solution with 5% CO_2, and 95% N_2 (Po_2 in the chamber≈20 mmHg). After switching from normoxia to hypoxia, complete equilibration of the new solution in the chamber required between 1 and 2 min.

As described in dispersed rabbit (11, 13, 20, 34) and rat (23, 33, 36) glomus cells the amplitude of macroscopic voltage-dependent currents recorded from cells in rat carotid body slices is reduced when exposed to low Po_2 (Fig. 3C). However, reversible reduction of K^+ current amplitude by hypoxia is a response seen less consistently in our slices than the increase of secretory activity monitored by amperometry. This could mean that in patch clamped rat glomus cells the O_2-sensing mechanism is altered and the sensitivity to low Po_2 decreased possibly due to the intracellular dialysis. It is also possible that, besides voltage-gated K^+ channels, other conductances, mediated by voltage-gated Ca^{2+} (37) or K^+-selective leaky (3) channels, not studied so far in slices, also contribute to mediate the low Po_2-induced secretory response. In slices with well-defined glomeruli, low Po_2 induces consistently a progressive increase in the frequency of secretory events (Fig. 3D), reaching values of 48.6±19 spikes/min (n=24 cells). At the peak of the response to hypoxia the secretory events normally fuse into a broad concentration envelope that quickly declines after switching to the control, normoxic, solution. As expected from electrically excitable cells, all the glomus cells that respond to hypoxia are also activated by solutions with high external K^+ (Fig. 3D, top). Interestingly, glomus cells unresponsive to hypoxia but activated by depolarization with high external K^+ are occasionally observed. The neurosecretory response to hypoxia of rat glomus cells in slices is completely abolished by the addition of the voltage-dependent calcium channel blocker cadmium (Fig. 3D, bottom) or the removal of extracellular calcium with EGTA (30). These observations confirm previous data on dispersed rabbit carotid body cells (26, 39).

The major contributors to the voltage-gated O_2-sensitive macroscopic K^+ currents in rat glomus cells are the Ca^{2+}-dependent maxi-K^+ channels (23, 33, 42). Because these channels are blocked by TEA or iberiotoxin (IbTX), we have studied whether, like hypoxia, these agents induce Ca^{2+} entry and secretion in glomus cells. In most cells studied (33 of 34 cells), application of 5 mM TEA to the bath solution elicits an increase in the secretory activity similar to that triggered by hypoxia (Fig. 4A). In response to the K^+ channel blocker the frequency of secretory events (42±17 spikes/min, n=6 cells) is similar to that obtained in low Po_2 (Student's *t* test, $P<0.05$). The average quantal charge of events induced by TEA is 43±30 fC (n=275 spikes in 6 cells). This value and the distribution of quantal events are also similar to those estimated with events elicited by hypoxia (43±26 fC; n=576 spikes in 14 cells), suggesting that both stimuli can trigger the release of the same type of secretory vesicle (Fig. 4B). The effect of TEA can be observed even in quiescent cells, without any measurable spontaneous quantal release, as well as in O_2-insensitive glomus cells. We have also tested the effect of IbTX, a selective blocker of Ca^{2+}- and voltage-activated maxi K^+ channels (10). Figure 4C illustrates the increase of secretory activity in a glomus cell exposed to 200 nM IbTX. The response is similar to the ones obtained with TEA or hypoxia, although the recovery phase seems to be somewhat longer possibly due to slower wash-out of IbTX. These observations indicate that direct blockade of the K^+ channels with TEA or IbTX can elicit secretion from rat glomus cells in the slices.

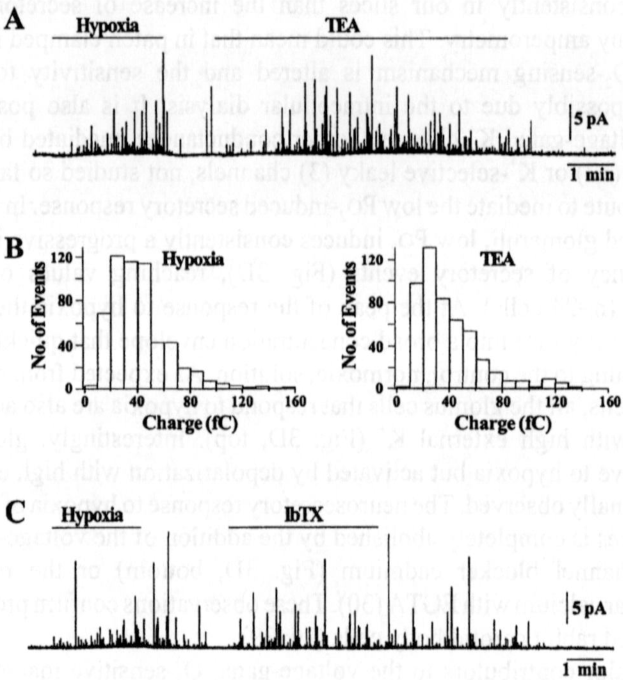

Figure 4. Secretory responses of intact glomus cells to K^+ channel blockers. A: Amperometric recording from a glomus cell illustrating the similar effects elicited by low PO_2 and the application of 5 mM TEA to the external solution. B: Frequency histograms of the quantal charge of events elicited by hypoxia and TEA. Note that the parameters of the distributions are the same in the two experimental conditions. C: Secretory activity induced in a glomus cell by hypoxia and 200 nM IbTX (Modified from Ref. 30).

4. Mitochondrial Function and Responsiveness to Hypoxia of Glomus Cells

Despite the progress in the understanding of glomus cell electrophysiology and its responses to hypoxia, how O_2 is sensed in the carotid body remains unknown. It has been reported that modulation of some K^+ channels by Po_2 in glomus cells is maintained in excised patches, thus suggesting that O_2 sensing depends on membrane-delimited mechanisms (11, 18, 19, 35). On the other hand, mitochondria have traditionally been considered as the site for glomus cell O_2 sensing by several investigators because these organelle consume most of the cellular O_2 and similarly to hypoxia, inhibitors of the electron transport chain (ETC) or mitochondrial uncouplers increase the afferent activity of the carotid body sinus nerve (24, 27). Besides in the carotid body, mitochondria have also been postulated to participate in O_2 sensing in other acutely responding systems, such as pulmonary vascular smooth muscle (1, 17, 41) or chromaffin cells (15, 25). We have investigated in our carotid body slice preparation whether sensitivity of intact glomus cells to hypoxia is altered by mitochondrial dysfunction (28). We have tested the effect of inhibition at complex I with rotenone, at complex II with thenoyltrifluoroacetone (TTFA), at complex III with myxothiazol or antimycin A (respectively proximal and distal inhibitors of this complex) and at complex IV with cyanide. These agents were used in a broad range of concentrations, however the lowest concentrations used were at least 5 to 10 times the reported K_{50} values (6, 40).

The typical secretory response to hypoxia ($Po_2 \approx 20$ mmHg) of a glomus cell in the carotid body slice is illustrated in Figure 5A. As shown in previous figures (see also Refs. 30, 31), low Po_2 induces spike-like quantal events corresponding to catecholamine release from individual vesicles. For these type of experiments we calculated the cumulative secretion signal (lower trace in Fig. 5A), a value of electric charge obtained by the sum of the time integral of successive amperometric events that is proportional to the number of catecholamine molecules oxidized. Thus, the secretion rate, in femtocoulombs per min (fC/min), is given by the amount of charge transferred to the recording electrode during 60 seconds once the solutions are equilibrated in the recording chamber. We have observed that the ETC inhibitors induce in one to three minutes an exocytotic response from glomus cells (Fig. 5B-F; see also Fig. 6). The difference in the secretagogue potency of applied ETC inhibitors is not very marked when they are used at concentrations above saturation. The mean area of individual quantal events (in fC, mean±sd) triggered by the ETC inhibitors (rotenone: 40±18, n=245 spikes in 5 cells; myxothiazol: 42±30, n=102 spikes in 7 cells; antimycin A: 40±22, n=132 spikes in 5 cells; cyanide: 48±33, n=116 spikes in 7 cells) is not significantly different ($p>0.1$) from the average values estimated with events evoked by hypoxia (43±26, n=576 spikes in 14 cells, see Fig. 4 above). These data indicate that hypoxia and the ETC inhibitors induce the release of a common type of catecholaminergic vesicle. Secretion evoked by all the ETC inhibitors can be completely abolished by blockade of membrane Ca^{2+} channels with Cd^{2+} (Fig. 5B-F). Only the secretory response induced by concentrations of cyanide in the millimolar range is partially maintained in the presence of 0.3 mM extracellular Cd^{2+}, thus suggesting Ca^{2+} release from intracellular stores (2). These data indicate that, as described in cells exposed to hypoxia (Fig. 3) (5, 30, 39), activation of carotid body glomus cells by mitochondrial ETC inhibitors largely depends on extracellular Ca^{2+} influx through channels of the plasma membrane (28).

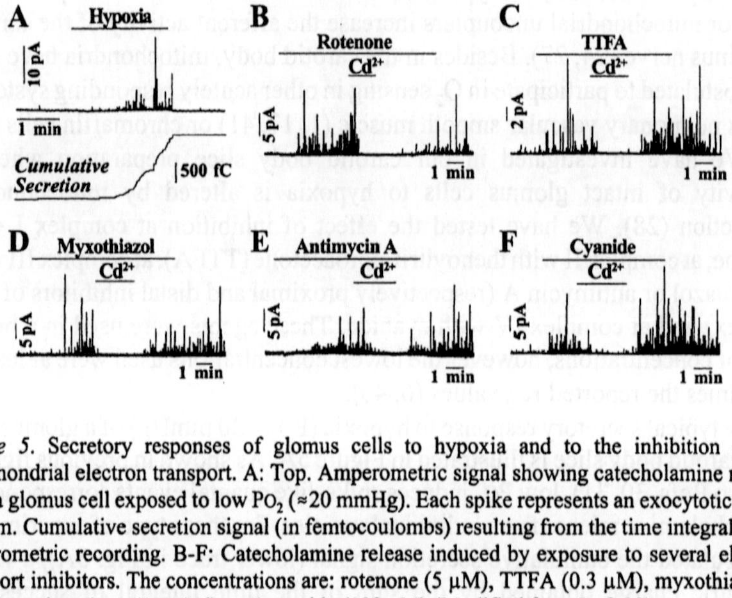

Figure 5. Secretory responses of glomus cells to hypoxia and to the inhibition of the mitochondrial electron transport. A: Top. Amperometric signal showing catecholamine release from a glomus cell exposed to low Po_2 (≈20 mmHg). Each spike represents an exocytotic event. Bottom. Cumulative secretion signal (in femtocoulombs) resulting from the time integral of the amperometric recording. B-F: Catecholamine release induced by exposure to several electron transport inhibitors. The concentrations are: rotenone (5 μM), TTFA (0.3 μM), myxothiazol (1 μg/ml), antimycin A (1 μg/ml), and cyanide (100 μM) (Modified from Ref. 28).

The interaction between hypoxia and the mitochondrial electron flow has been studied in cells exposed to low Po_2 before and during application of ETC inhibitors acting at either proximal or distal mitochondrial complexes. The rationale behind these experiments is that if hypoxia exerts its effect through alteration of the mitochondrial electron flow, preincubation with saturating concentrations of ETC blockers would prevent any further effect of low Po_2. In contrast, the effects of hypoxia and ETC inhibition would be additive, at least partially, if they were acting through separate pathways. Figure 6 (A and B) illustrates that when O_2-responsive glomus cells are treated with cyanide or antimycin A, the concomitant exposure to low Po_2 elicits further increase in the secretory activity. In each case the amperometric recordings are shown on the left and right panels are the cumulative secretion signals recorded during hypoxia in the presence of the ETC inhibitors. The secretion rates measured immediately before exposure to hypoxia are illustrated diagrammatically by the slopes of the cumulative secretion signals. The amperometric recordings show that hypoxic responsiveness was preserved in cells treated with the ETC inhibitors. In parallel, the cumulative secretion traces clearly illustrate that in the presence of ETC inhibitors hypoxia induces a reversible increase in the slope of the signals. Similar results have been obtained in experiments performed with myxothiazol, or TTFA (28).

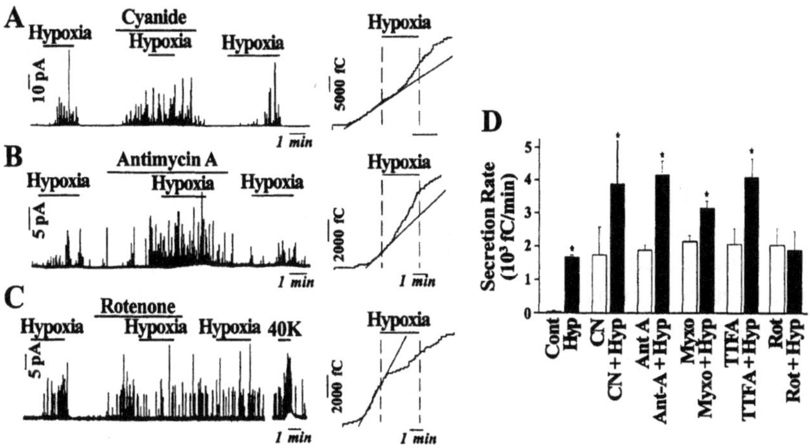

Figure 6. Secretory responses of glomus cells exposed concomitantly to hypoxia (Po_2, 20 mmHg) and to blockade of the mitochondrial electron transport. A-C: Amperometric recordings (left panels). The concentration of drugs are: cyanide (100 μM), antimycin A (1 μg/ml) and rotenone (5 μM). Right panels, Cumulative secretion signals before, during and after the exposure to hypoxia in the presence of the ETC inhibitors. The straight lines represent the slopes (secretion rates) of the cumulative secretion signals immediately before the exposure to hypoxia. D: Average secretion rate measured in cells in various experimental conditions. Secretion rate in the ordinate is expressed in fC/min (mean±SE). Experimental conditions: Control (Po_2, 150 mmHg, 75±15 fC/min, n=17 cells) and hypoxia (1710±65 fC/min, n=17 cells). Cyanide (CN, 0.1 μM, 1771±842 fC/min, n=4 cells) and CN plus hypoxia (3932±1339 fC/min, n=4 cells). Antimycin A (0.1-1 μg/ml, 1910±151 fC/min, n=13 cells) and antimycin A plus hypoxia (4201±421 fC/min, n=7 cells). Myxothiazol (0.1-1 μg/ml, 2167±199 fC/min, n=6 cells) and myxothiazol plus hypoxia (3188±240 fC/min, n=6 cells). TTFA (0.1-0.3 μM; 2093±488 fC/min, n=5 cells) and TTFA plus hypoxia (4134±587 fC/min, n=5 cells). Rotenone (0.1-5 μM, 2058±550 fC/min, n=14 cells), rotenone plus hypoxia (1915±552 fC/min, n=12 cells). Asterisks indicate statistically significant difference ($P<0.05$) between each pair of samples (Modified from Ref. 28).

An exception among the mitochondrial inhibitors tested is rotenone, a flavoprotein inhibitor that blocks mitochondrial complex I. Figure 6C shows that, as other ETC inhibitors, rotenone elicits secretion from the cells, however previous exposure to rotenone abolishes any further increase of secretion by hypoxia. In cells treated with rotenone the secretory response to depolarization with high potassium is unaltered (Fig. 6C). The average secretion rates measured in several cells exposed to hypoxia and the ETC inhibitors are given in Figure 6D. This summary plot shows that inhibition at various sites along the ETC with saturating concentrations of mitochondrial inhibitors induces a secretory activity in glomus cells of a magnitude comparable to that evoked by low Po_2 ($\approx$20 mmHg). With the exception of rotenone, the effects of hypoxia and ETC inhibitors are additive, thus suggesting that they might act through separate signaling pathways. Selective occlusion of hypoxia sensitivity by rotenone has been observed in all the cells studied with concentrations of the drug at 0.1-5 μM (28). The lowest concentration used in these experiments (0.1 μM) can produce full blockade of complex I (6, 40) or saturation of rotenone binding sites (14).

5. Summary

In this chapter, we present the current model of carotid body oxygen sensing based on the inhibition by hypoxia of O_2-sensitive K^+ channels. The basic mechanisms described in carotid body glomus cells also operate in other neurosecretory systems acutely responding to low Po_2 (22). Besides the studies in dispersed glomus cells, we describe the properties of cells in carotid body thin slices. This preparation has allowed us to study the responses of glomus cells to hypoxia and K^+ channel blockers in almost optimal physiologic conditions (30, 31, 32). As described in dispersed cells (23, 33, 42), voltage- and Ca^{2+}-dependent outward currents, inhibited by TEA and IbTX, are reduced in amplitude upon exposure to low Po_2. Glomus cells in the slices consistently exhibit a secretory response to hypoxia, which can be easily monitored by the amperometric detection of catecholamines with a polarized carbon-fiber electrode. Transmitter release induced by physiologic levels of low Po_2 ($\approx$ 20-30 mmHg) is completely dependent on extracellular calcium influx. This observation further supports the view that glomus cells work as O_2-sensitive presynaptic-like elements, in which external calcium entry after low Po_2-induced depolarization constitutes the principal event leading to transmitter release and activation of the afferent sensory fibers (4, 21, 30, 39). Exposure of cells in the slices to K^+ channel blockers, like TEA or IbTX, induces a secretory activity resembling the effect of low Po_2. These results suggest that although leak K^+-selective channels (3) have also a role in the responsiveness of rat glomus cells to hypoxia, the direct blockade of the O_2-sensitive voltage-dependent K^+ currents, which in these cells are those inhibited by TEA and IbTX, is sufficient to induce secretion.

The work on carotid body slices demonstrates the importance of voltage-gated K^+ channels as effector molecules in the acute response of glomus cells to hypoxia. Finally, we report that inhibitors of proximal and distal complexes of the mitochondrial ETC elicit, as hypoxia, a powerful external Ca^{2+}-dependent secretory response in intact rat carotid body glomus cells. Cellular sensitivity to hypoxia is maintained after blockade of the mitochondrial ETC, thus suggesting that mitochondrial electron flow is not linked in a simple way to acute regulation of glomus cell activity by changes of O_2 tension. Hypoxia and mitochondrial inhibitors, acting through separate pathways, converge to raise cytosolic $[Ca^{2+}]$, which triggers secretion. However, we have identified rotenone as highly selective and specific inhibitor of the responsiveness to hypoxia (28). Therefore, this drug and its derivatives could be used as tools to pursue investigation on the molecular characterization and location of the O_2 sensing mechanism in arterial chemoreceptor cells.

Acknowledgments

Research was supported by grants from the Spanish Ministry of Science and Technology and the Andalusian Government. M. G-F. is a fellow of the FPU program of the Spanish Ministry of Education. J. L.-B. received the "Ayuda a la investigación 2000" of the Juan March Foundation.

References

1. Archer SL, Huang J, Henry T, Peterson D, and Weir EK. A redox-based O_2 sensor in rat pulmonary vasculature. *Circ. Res.* 1993; 73: 1100-1112.
2. Biscoe TJ and Duchen MR. Responses of type I cells dissociated from the rabbit carotid body to hypoxia. *J. Physiol.* 1990; 42: 39-59.
3. Buckler KJ. A novel oxygen-sensitive potassium current in rat carotid body type I cells. *J. Physiol.* 1997; 498: 649-662.
4. Buckler KJ and Vaughan-Jones RD. Effects of hypoxia on membrane potential and intracellular calcium in rat neonatal carotid body type I cells. *J. Physiol.* 1994; 476: 423-428.
5. Carpenter E, Hatton CJ, and Peers C. Effects of hypoxia and dithionite on catecholamine release from isolated type I cells of the rat carotid body. *J. Physiol.* 2000; 523: 719-729.
6. Degli Esposti M. Inhibitors of the NADH-ubiquinone reductase: an overview. *Biochem. Biophys. Acta.* 1998; 1364: 222-235.
7. Delpiano MA and Hescheler J. Evidence for a Po_2-sensitive K^+ channel in the type-I cell of the rabbit carotid body. *FEBS Lett.* 1989; 249: 195-198.
8. Doyle TP and Donnelly DF. Effect of Na^+ and K^+ channel blockade on baseline and anoxia induced catecholamine release from rat carotid body. *J. Appl. Physiol.* 1994; 77: 2606-2611.
9. Duchen MR, Caddy KWT, Kirby GC, Patterson DL, Ponte J, and Biscoe TJ. Biophysical studies of the cellular elements of the rabbit carotid body. *Neuroscience.* 1988; 26: 291-311.
10. Gálvez A, Gimenez-Gallego G, Reuben JP, Roy-Contancin L, Feigenbaum P, Kaczorowski GJ, and García ML. Purification and characterization of a unique, potent, peptidyl probe for the high conductance calcium-activated potassium channel from venom of the scorpion *Buthus tamulus. J. Biol. Chem.* 1990; 265: 11083-11090.
11. Ganfornina MD and López-Barneo J. Single K^+ channels in membrane patches of arterial chemoreceptor cells are modulated by O_2 tension. *Proc. Natl. Acad. Sci. USA.* 1991; 88: 2927-2930.
12. Ganfornina MD and López-Barneo J. Potassium channel types in arterial chemoreceptor cells

and their selective modulation by oxygen. *J. Gen. Physiol.* 1992; 100: 401-426.

13. Hescheler J, Delpiano MA, Acker H, and Pietruschka F. Ionic currents on type-I cells of the rabbit carotid body measured by voltage-clamp experiments and the effect of hypoxia. *Brain Res.* 1989; 486: 79-88.
14. Higgins DS and Greenamyre JT. [^{3}H]Dihydrorotenone binding to NADH: Ubiquinone reductase (complex I) of the electron transport chain: An autoradiographic study. *J. Neurosci.* 1996; 16: 3807-3816.
15. Inoue M, Fujishiro N, Imanaga I, and Sakamoto Y. Role of ATP decrease in secretion induced by mitochondrial dysfunction in guinea-pig adrenal chromaffin cells. *J. Physiol.* 2002; 539: 145-155.
16. Lahiri S, Roy A, Rozanov C, and Mokashi A. K^+-current modulated by Po_2 in type I cells in rat carotid body is not a chemosensor. *Brain Res.* 1998; 794: 162-165.
17. Leach RM, Hill HM, Snetkov VA, Robertson TP, and Ward JPT. Divergent roles of glycolysis and the mitochondrial electron transport chain in hypoxic pulmonary vasoconstriction of the rat: identity of the hypoxic sensor. *J. Physiol.* 2001; 536: 211-224.
18. Lewis A, Peers C, Ashford ML, and Kemp PJ. Hypoxia inhibits human recombinant large conductance, Ca^{2+}-activated K^+ (maxi-K) channels by a mechanism which is membrane delimited and Ca^{2+} sensitive. *J. Physiol.* 2002; 540: 771-780.
19. López-Barneo J. Oxygen-sensitive ion channels: how ubiquitous are they? *Trends Neurosci.* 1994; 17: 133-135.
20. López-Barneo J, López-López JR, Ureña J, and González C. Chemotransduction in the carotid body: K^+ current modulated by Po_2 in type I chemoreceptor cells. *Science.* 1988; 242: 580-582.
21. López-Barneo J, Benot AR, and Ureña J. Oxygen sensing and the electrophysiology of arterial chemoreceptor cells. *News Physiol. Sci.* 1993; 8: 191-195.
22. López-Barneo J, Pardal R, and Ortega-Sáenz P. Cellular mechanisms of oxygen sensing. *Annu. Rev. Physiol.* 2001; 63: 259-287.
23. López-López JR, González C, and Pérez-García MT. Properties of ionic currents from isolated adult rat carotid body chemoreceptor cells: effect of hypoxia. *J. Physiol.* 1997; 499: 429-441.
24. Mills E, and Jöbsis FF. Mitochondrial respiratory chain of carotid body and chemoreceptor response to changes in oxygen tension. *J. Neurophysiol.* 1972; 35: 405-428.
25. Mojet MH, Mills E, and Duchen MR. Hypoxia-induced catecholamine secretion in isolated newborn rat adrenal chromaffin cells is mimicked by inhibition of mitochondrial respiration. *J. Physiol.* 1997; 504: 175-189.
26. Montoro RJ, Ureña J, Fernández-Chacón R, Álvarez de Toledo G, and López-Barneo J. Oxygen sensing by ion channels and chemotransduction in single glomus cells. *J. Gen. Physiol.* 1996; 107: 133-143.
27. Mulligan E, Lahiri S, and Storey BT. Carotid body O_2 chemoreception and mitochondrial oxydative phosphorylation. *J. Appl. Physiol.* 1981; 51: 438-446.
28. Ortega-Sáenz P, Pardal R, García-Fernández M, and López-Barneo J. Rotenone selectively occludes sensitivity to hypoxia in rat carotid body glomus cells. *J. Physiol.* 2003; 548: 789-900.
29. Osanai S, Buerk DG, Mokashi A, Chugh DK, and Lahiri S. Cat carotid body chemosensory discharge (in vitro) is insensitive to charybdotoxin. *Brain Res.* 1997; 747: 324-327.
30. Pardal R, Ludewig U, García-Hirschfeld J, and López-Barneo J. Secretory responses of intact glomus cells in thin slices of rat carotid body to hypoxia and tetraethylammonium. *Proc. Natl. Acad. Sci. USA.* 2000; 97: 2361-2366.
31. Pardal R and López-Barneo J. Low glucose-sensing cells in the carotid body. *Nature Neurosci.* 2002; 5: 197-198.
32. Pardal R and López-Barneo J. Carotid body thin slices: responses of glomus cells to hypoxia and K^+-channel blockers. *Respir. Physiol. Neurobiol.* 2002; 132: 69-79.
33. Peers C. Hypoxic suppression of K^+ currents in type I carotid body cells: selective effect on the Ca^{2+}-activated K^+ current. *Neurosci. Lett.* 1990; 119: 253-256.

34. Pérez-García T, López-López JR, Riesco AM, Hoppe UC, Marbán E, González C, and Johns DC. Viral gene transfer of dominant negative Kv4 construct suppresses an O_2 sensitive K^+ current in chemoreceptor cells. *J. Neurosci.* 2000; 20: 5689-5695.
35. Riesco-Fagundo AM, Pérez-García MT, González C, and López-López JR. O_2 modulates large conductance Ca^{2+}-dependent K^+ channels of rat chemoreceptor cells by a membrane-restricted and CO-sensitive mechanism. *Circ. Re.s* 2001; 89: 430-436.
36. Stea A and Nurse CA. Whole-cell and perforated-patch recordings from O_2-sensitive rat carotid body cells grown in short- and long-term culture. *Pflügers Arch.* 1991; 418: 93-101.
37. Summers BA, Overholt JL, and Prabhakar NR. Augmentation of L-type calcium current by hypoxia in rabbit carotid body glomus cells: evidence for a PKC-sensitive pathway. *J. Neurophysiol.* 2000; 84: 1636-1644.
38. Thompson RJ and Nurse CA. Anoxia differentially modulates multiple K^+ currents and depolarizes neonatal rat adrenal chromaffin cells. *J. Physiol.* 1998; 512: 421-434.
39. Ureña J, Fernández-Chacón R, Benot AR, Álvarez de Toledo G, and López-Barneo J. Hypoxia induces voltage-dependent Ca^{2+} entry and quantal dopamine secretion in carotid body glomus cells. *Proc. Natl. Acad. Sci. USA.* 1994; 91: 10208-10211.
40. Vaux EC, Metzen E, Yeates KM, and Ratcliffe PJ. Regulation of hypoxia inducible factor is preserved in the absence of a functioning mitochondrial respiratory chain. *Blood.* 2001; 98: 296-302.
41. Waypa GB, Chandel NS, and Schumacker PT. Model for hypoxic pulmonary vasconstriction involving mitochondrial oxygen sensing. *Circ. Res.* 2001; 88: 1259-1266.
42. Wyatt CN and Peers C. Ca^{2+}-activated K^+ channels in isolated type I cells of the neonatal rat carotid body. *J. Physiol.*1995; 483: 559-565.
43. Youngson C, Nurse C, Yeger H, and Cutz E. Oxygen sensing in airway chemoreceptors. *Nature.* 1993; 365: 153-155.
44. Zhu WH, Conforti L, Czyzyk-Krzesk MF, and Millhorn DE. Membrane depolarization in PC-12 cells during hypoxia is regulated by an O_2-sensitive K^+ current. *Am. J. Physiol.* 1996; 271, C658-C665.

Chapter 21

Oxygen Sensing by Adrenomedullary Chromaffin Cells

Roger J. Thompson and Colin A. Nurse
University of Colorado, Denver, Colorado, U.S.A. and McMaster University, Hamilton, Ontario, Canada

1. Introduction

The adrenal gland is composed of two separate functional regions: The outer cortex is principally responsible for corticosteroid synthesis and secretion. The inner medulla, containing the adrenomedullary chromaffin cells (AMC), synthesizes and secretes catecholamines (i.e., adrenaline, noradrenaline and dopamine), and is perhaps best known for its contribution to the 'fight-or-flight' response. In mature animals, metabolic or physiological stress increases activity in the sympathetic nervous system leading to acetylcholine (ACh) release from splanchnic nerve endings, which innervate AMC. Released ACh activates nicotinic receptors on AMC, resulting in secretion of catecholamines (CA) into the blood. Circulating CA have well-defined roles in the fight-or-flight response of adult animals (Fig. 1), ensuring that adequate blood flow to vital organs (heart and lung) is maintained.

In species that are relatively immature at birth, such as rat and man, sympathetic innervation to several target organs, including the splanchnic projections to the adrenal medulla, are non-functional at birth (25). This suggests that CA secretion from immature adrenal glands is not under nervous regulation. However, despite the lack of neurogenic control of adrenal CA secretion in neonatal animals, several studies have suggested that AMC secrete CA during physiological stressors (e.g., hypoxia), and that the released CA are vital for survival of the neonate during birth and subsequent hypoxic events.

As early as 1961, the oxygen sensitivity of the adrenal medulla was recognized. Comline and Silver (5) demonstrated that asphyxiated fetal sheep had elevated plasma CA and depleted adrenal CA even in the absence of mature sympathetic innervation. In the 1980s, Seidler and Slotkin described a similar non-neurogenic regulation of CA release in the neonatal rat, by demonstrating that prior to maturation of innervation to the adrenal gland, 1 hr of inspired hypoxia (5% O_2) depleted adrenal CA. CA release was associated with lung fluid absorption and initiation of surfactant secretion, clamping of the fetal heart rate, and brown fat mobilization (Fig. 1) (25-27). The hypoxia-induced CA release was not dependent upon activation of sympathetic nerves because block of nicotinic receptors with chlorisondamine did not prevent the hypoxia-induced CA surge (25). Seidler and Slotkin (28) also demonstrated that these non-neurogenic responses to hypoxia were absent in adult animals, but that three weeks after splanchnic nerve transection in these animals, the non-neurogenic hypoxia-induced CA secretion returned. Taken together, these studies suggested that AMC may sense hypoxia prior to the maturation of sympathetic innervation and that this O_2-sensitivity is lost in mature animals.

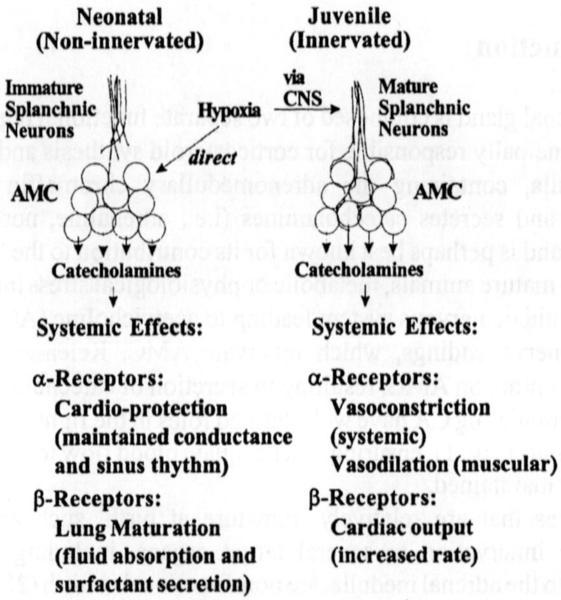

Figure 1. Comparison of the role of hypoxia-induced secretion of catecholamines from adrenomedullary chromaffin cells (AMC) of different ages. Neonatal [postnatal (P) 1-2 day old] AMC are sensitive to hypoxia and secrete catecholamines in the absence of functional splanchnic innervation. Catecholamines released into the circulation promote neonatal survival during hypoxia via initiation of surfactant secretion in the lung and maintenance of the heart's conduction characteristics. In contrast, juvenile (P13-20) AMC may not respond directly to hypoxia (though some adult cells may), but release catecholamines during splanchnic nerve activation via the central nervous system (CNS). These differences in physiological responses can be considered adaptations of the neonate that contribute to physiological changes associated with the transition to extrauterine life.

2. Catecholamine Release From Neonatal AMC Promotes Survival During Hypoxia

Birth is associated with both fetal hypoxia and hypercapnia due to intermittent occlusions of the umbilical cord (15). In mature animals, physiologic adjustments to hypoxia are initiated and maintained by several systems, including ventilatory reflex activation via the carotid body and hypoxic pulmonary vasoconstriction. The transition from fetal to extra-uterine life requires that the lungs be cleared of fluid and that surfactant be secreted to enable proper lung expansion and gas exchange across the alveoli. It is known that CA play a pivotal role in these processes through the activation of β-adrenergic receptors in the lung (34) and α-adrenergic receptors in the heart (27).

Seidler and Slotkin (25) demonstrated that CA derived from the adrenal gland are important for lung maturation after birth, and are vital for neonatal survival during hypoxia. Administration of the β_2-blocker ICI-118551 compromised the survival of neonatal rats [postnatal (P) day 1-2] during hypoxia, but did not affect mortality in more mature animals (P14) that had further developed lung function (25). Whereas adrenalectomy dramatically compromised the ability of the neonatal rat to survive hypoxia (5% O_2 for 1 hr), interference of CA release from sympathetic nerve endings did not (25). This suggests a vital role for adrenal-derived CA in the hypoxic tolerance of P1-2 rats. Interestingly, hypoxia in the presence of the cardiac-specific β_1-receptor blocker atenolol did not affect neonatal mortality, suggesting that cardiac β-receptors are not involved in tolerance of neonates to hypoxia (27).

The principal cardioprotective effect of adrenal CA is maintenance of heart rate and conduction characteristics (Fig. 1) (27). Administration of the α-receptor blocker phenoxybenzamine (PBX) to 1 day-old rat pups during normoxia (21% inspired O_2) did not alter cardiac function. However, PBX applied concomitantly with hypoxia caused a marked decline in heart rate, slowing of sinus rhythm, atrioventricular block, and cardiac failure (27, 28). Although mature animals express few cardiac α-receptors relative to β- receptors, neonatal animals predominantly express the α-type (27). As the neonate matures, cardiac α receptors are replaced with the β subtype and the cardioprotective effects of adrenal-derived CA disappears (27). The hypoxia-induced secretion of adrenal CA, and the presence of several physiological mechanisms that utilize circulating CA to promote development and survival of neonates can be considered adaptations of the adult stress response that are designed to promote neonatal survival during and after birth.

3. The Oxygen Sensing Mechanism of Chromaffin Cells

The work described above led us to hypothesize that neonatal (P1-2) 'noninnervated' AMC function as direct sensors of PO_2, and that this O_2-sensing mechanism is absent in juvenile (P14) cells, which are functionally innervated. To test this hypothesis, we isolated AMC from both age groups, maintained them in short-term cell culture (1-4 days) and determined whether or not they were hypoxia-sensitive using patch clamp recording of whole-cell currents and membrane potential. CA secretion was also measured by HPLC with electrochemical detection.

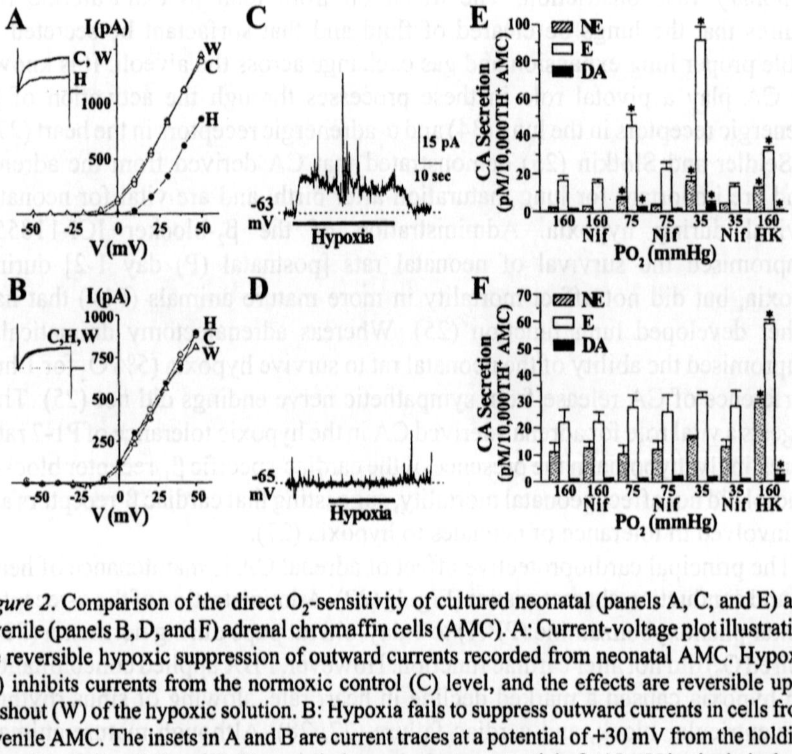

Figure 2. Comparison of the direct O_2-sensitivity of cultured neonatal (panels A, C, and E) and juvenile (panels B, D, and F) adrenal chromaffin cells (AMC). A: Current-voltage plot illustrating the reversible hypoxic suppression of outward currents recorded from neonatal AMC. Hypoxia (H) inhibits currents from the normoxic control (C) level, and the effects are reversible upon washout (W) of the hypoxic solution. B: Hypoxia fails to suppress outward currents in cells from juvenile AMC. The insets in A and B are current traces at a potential of +30 mV from the holding potential of -60 mV. C and D: Hypoxia induces a receptor potential of ~ 15 mV in singly isolated neonatal but not juvenile AMC. E and F: Hypoxia induces catecholamine secretion in cultures of neonatal but not juvenile AMC and secretion is inhibited by the L-type Ca^{2+} channel blocker nifedipine (Nif). "NE", norepinephrine; "E", epinephrine; "DA", dopamine; "HK", high K^+. Panels C-F are reproduced with permission from Ref. 33.

3.1. Development of O_2 Sensing in AMC

Similar to the mechanism of O_2-sensing by the glomus cells of the carotid body (18) and pulmonary arterial myocytes (2), exposure of neonatal AMC to hypoxia (PO_2 ~ 5 mmHg) reversibly inhibited outward currents and induced a membrane depolarization (receptor potential) of ~ 15 mV (Fig. 2A, C, and E) (31,

33). Additionally, 1 hr of hypoxia stimulated CA secretion in cultures of neonatal AMC by ~6 × basal, and secretion of all three CA (dopamine, norepinephrine and epinephrine) was enhanced. Hypoxia-induced CA secretion was dependent on extracellular Ca^{2+} and attenuated by the L-type Ca^{2+} channel blocker nifedipine (10 μM) (33). In a parallel study, Mojet et al. (21) demonstrated that the electron transport chain (ETC) inhibitor cyanide (CN; 2.5 mM) mimicked the effects of hypoxia on CA secretion, as detected by carbon fiber electrodes placed adjacent to isolated neonatal AMC. Interestingly, in both studies responses to hypoxia were absent in juvenile AMC (Fig. 2B, D, F).

In contrast to the experiments summarized above, several reports have suggested that adult AMC may express O_2-sensing mechanisms. In these studies AMC were maintained in either long-term (7 days) (20) or short-term (1-2 days) (16) culture, or studied in an adrenal slice preparation (30). Hypoxia (~5% O_2) suppressed K^+ currents, induced membrane depolarization, increased intracellular Ca^{2+}, and enhanced CA secretion in ~50% of cultured adult AMC (16, 20). Additionally, AMC in slices of adult adrenals may exhibit a hypoxia-induced increase in intracellular Ca^{2+} of similar magnitude to that observed in cells from neonatal slices (30). Interestingly, 10 nM ryanodine, which releases Ca^{2+} from the endoplasmic reticulum, prolonged the hypoxia-induced Ca^{2+} rise in adult but not neonatal cells, suggesting that Ca^{2+}-induced Ca^{2+} release may contribute to the sustained Ca^{2+} rise is adult cells (30). The presence of O_2-sensitive responses in adult AMC is in contrast with our previous report where only 1/27 juvenile cells responded significantly to hypoxia as determined by K^+ current inhibition (33). Unfortunately, it was not determined if hypoxia inhibited outward currents or caused membrane depolarization of adult AMC in adrenal slices.

How can the presence of O_2-sensing in adult AMC be reconciled with its absence in juvenile cells? It seems reasonable to suggest that in long-term culture experiments, there was a return of the non-neurogenic O_2-sensing mechanism, similar to the effects of denervating adult AMC *in vivo* (28). However, this point is more difficult to reconcile in the studies that used cells maintained in short-term culture or in slice preparations. One possible explanation is that there is a quantitative difference in the number of juvenile and adult AMC that respond to hypoxia. Such a difference could arise if the number of functional synapses between splanchnic nerve endings and AMC are culled during maturation, so that there are fewer in the adult than juvenile adrenal gland. Indeed, it has been reported that innervation of the adrenal gland develops postnatally (11). One further potential explanation for the observed O_2-sensitivity of adult AMC in slices is the potentiation of the hypoxic responses due to hyperoxic exposure. It is routine in slice preparations to bubble tissue sections in 95-100% O_2 for ~1 hr, and exposure to hyperoxia has been reported to augment the hypoxic ventilatory response of rats in a nitric oxide (NO)-dependent manner (10). Thus, it is conceivable that exposure to hyperoxia alters the sensitivity of AMC to hypoxia.

3.2. Identification and Characterization of the O_2-sensitive Ion Channels in AMC

Ligand-gated nicotinic ACh receptors mediate CA secretion from AMC in mature animals via splanchnic nerve stimulation. AMC also express several types of voltage-gated ion channels that are intimately involved in the regulation of CA secretion, and that are candidates for the hypoxia-sensitive ion channels. In addition to K^+ channels (see below), AMC express a TTX-sensitive Na^+ channel and three types of voltage-dependent Ca^{2+} channels: the dihydropyridine-sensitive L-type, the conotoxin-sensitive N-type, and agatoxin-sensitive P-type Ca^{2+} channels (4). Interestingly, the L-type channel appears most efficiently coupled to depolarization-induced CA secretion in adult AMC (4), and the hypoxia-induced CA secretion from neonatal AMC could be blocked by the L-type channel blocker nifedipine (1, 20, 33).

Several voltage-dependent K^+ channels are found in AMC, including delayed rectifier (K_V) and large conductance Ca^{2+}-dependent (BK) K^+ channels. Several of these classes of ion channels are known to be regulated by hypoxia in various cell types (18). For example, BK channels are thought to be a major component of the O_2-sensitive K^+ (KO_2) current in carotid body glomus cells (23) and the delayed rectifier K^+ channels, $K_V1.5$ and $K_V2.1$ are thought to mediate the hypoxia-sensitive outward currents in pulmonary myocytes (3). In order to identify the types of K^+ currents that mediate the hypoxia-sensitive outward currents in AMC, we tested the ability of inhibitors of different classes of K^+ channels to block the hypoxia-induced suppression of outward currents in neonatal AMC (31). The major component (~65%) of the KO_2 current (I_{KO2}) was blocked by removal of extracellular Ca^{2+} or by 50-100 nM iberiotoxin (IbTx), suggesting that BK channels are a major contributor. Thus the remaining ~35 % of I_{KO2} was attributable mainly to delayed rectifier K_V channels which are sensitive to 20 mM tetraethylammonium (TEA) (31). Additionally, AMC express a pinacidil-activated and glibenclamide-sensitive, ATP-sensitive K^+ current (K_{ATP}) that is augmented by hypoxia (31). This current is reminiscent of the one carried by O_2-sensitive channels in neurons of the substantia nigra, which are thought to limit membrane depolarization during hypoxia (13).

What is the molecular identity of the O_2-sensitive K_V channels in AMC? Recent work from our laboratory raises the possibility that these channels are K_V 1.2/K_V1.5 heteromultimers, though this requires validation. Both $K_V1.2$ and $K_V1.5$ subunits appear to be expressed in neonatal AMC based on immunocytochemistry, and the heteromultimer likely comprises one of the O_2-sensitive K^+ channels expressed by the immortalized chromaffin cell line, i.e., v-myc, adrenal-derived HNK1$^+$ (MAH) cells, which are derived from embryonic AMC (8). These K_V channel subunits are also potential molecular components of the O_2-sensitive outward current in pulmonary myocytes (3, 29) and PC12 cells ($K_V1.2$ only) (6). The molecular characterization of the O_2-sensitive K^+

channels in neonatal AMC need to be confirmed using expression systems and mutagenesis of the candidate proteins.

It should be noted that none of the O_2-sensitive channels described above were responsible for generation of the receptor potential in neonatal AMC, although K_{ATP} channels could modulate its magnitude. The hypoxia-induced membrane depolarization persisted in the presence IbTx, Cd^{2+}, TEA, and 4-aminopyridine (31). Though the resting membrane potential of AMC hyperpolarized in a Na^+- and Ca^{2+}-free solution, the receptor potential remained constant. This argues against a role for cation selective channels in generating the receptor potential in rat AMC, as has been suggested for adult guinea pig cells (12). Interestingly, glibenclamide augmented the magnitude of the receptor potential in our studies, suggesting that K_{ATP} channels may play a similar protective role to limit the extent of membrane depolarization during hypoxia in AMC as in central neurons. Consistent with this notion, cromakalim, an activator of K_{ATP} channels, reversed the hypoxia-induced stimulation of CA secretion in adult AMC maintained in long-term culture (20). There is recent evidence that the receptor potential may be mediated by small-conductance Ca^{2+}-dependent K^+ channels (SK) because apamin, an inhibitor of SK channels, can depolarize AMC and block the receptor potential (14, 16, 22).

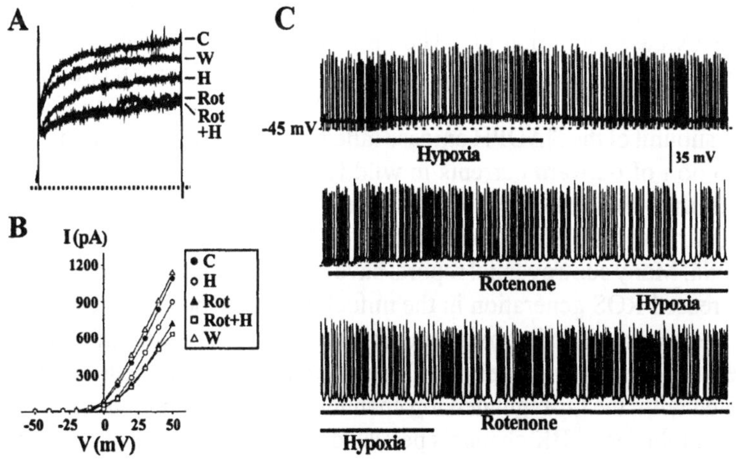

Figure 3. The mitochondrial electron transport chain complex I inhibitor, rotenone, mimics and attenuates the effects of hypoxia on neonatal adrenomedullary chromaffin cells. A: Nystatin perforated patch clamp recordings (step to +30 mV from –60 mV) from a neonatal AMC exposed to control (C), hypoxia (H), 10 μM rotenone (Rot), hypoxia+rotenone (Rot+H), and washout (W). Both hypoxia and rotenone inhibited outward currents and the combined effects of rotenone and hypoxia did not exceed the suppression seen by rotenone alone. B: Current-voltage plots for the cell currents shown in A. C: Current-clamp recording from a single AMC in a small cluster of ~ 10 cells. Both spontaneous action potential generation and a receptor potential of ~ 12 mV are visible. All current-clamp recordings from the same cell are 500 ms in duration and illustrate that 10 μM rotenone depolarizes the cell and blocks the hypoxia-induced receptor potential.

The hypoxic inhibition of voltage-dependent K^+ channels suggests that hypoxia may induce CA secretion via modulation of the action potential waveform. We recorded membrane potential in neonatal rat and mouse AMC from singly-isolated and small clusters of cells. Interestingly, single isolated cells were often quiescent and fired action potentials superimposed on the receptor potential (Fig. 2C) (33), whereas those in small clusters (>8 cells) frequently fired spontaneous rhythmic action potentials at ~1 Hz (Fig. 3C) (32). Exposure to hypoxia did not significantly modulate the frequency of action potentials in clustered AMC, but caused a reversible broadening of the spike duration that was primarily associated with a prolongation of the decay phase, as would be expected if K^+ channels were inhibited (Fig. 3C) (32). However, we also observed a slight increase in the rise time of the action potential that could arise from inhibition of a conductance that is active at the resting membrane potential. Candidates for these channels include non-selective cation conductances (12) and SK channels (14, 16, 22).

4. Investigations of the O_2 Sensor in AMC

Several models have been developed to identify the O_2-sensor and signal transduction pathways in various O_2-sensitive cells. In one of the earlier models, the signaling pathway involved reduced superoxide ($O_2^{\bullet -}$) generation via NADPH oxidase inhibition (7). The best evidence that NADPH oxidase can function as the O_2 sensor comes from studies on airway chemoreceptors, i.e., the pulmonary neuroepithelial bodies (NEBs). NEBs deficient in the O_2-binding gp91^{phox} subunit of the NADPH oxidase failed to respond to hypoxia as measured by inhibition of outward currents in wild type cells (9). An alternative model proposes that changes in reactive oxygen species (ROS) generated by the mitochondrial electron transport chain act as the key signal mediating the effect of hypoxia. Many controversial reports have appeared suggesting both increased and decreased ROS generation in the mitochondria as the link between the O_2-sensor and activation of hypoxia inducible factor 1α (35) and/or inhibition of K^+ channels (19). In other models, an O_2 sensor that is a component of, or closely associated with, the O_2-regulated ion channel itself has been proposed (18). The hypoxic inhibition of BK channels persisted in excised patches (17), suggesting that K^+ channel inhibition by hypoxia may in some cases involve a membrane-delimited mechanism. However, since mitochondria have been reported to be associated with excised membrane patches (24), a role for these organelles as the O_2 sensor still remains viable.

4.1. Is NADPH Oxidase the O_2 Sensor in AMC?

To investigate if NADPH oxidase functions as the O_2-sensor in neonatal

AMC we utilized the gp91^{phox} knockout mouse model and monitored outward K^+ currents, membrane depolarization, and CA secretion following exposure to hypoxia. Wild type neonatal mouse AMC appeared to express similar O_2-sensitive properties as the rat; thus, hypoxia caused K^+ channel inhibition, depolarization of the resting potential (i.e., receptor potential), action potential broadening and enhanced CA secretion (32). Interestingly, all aspects of the O_2-sensing mechanism persisted in neonatal AMC from the oxidase-deficient or knockout mice, indicating that the NADPH oxidase is unlikely to be an important O_2 sensor in AMC. It is important to note that the possibility still remains that an O_2-binding gp91^{phox}-like protein is expressed in AMC, or alternatively, the cell may compensate for the gp91^{phox} disruption, such that O_2 sensing is maintained. The simplest interpretation of the data is that the NADPH oxidase is unlikely to be an important O_2 sensor in AMC because: *i*) pulmonary NEBs from the same gp91^{phox}-deficient animals lack O_2-sensitivity, indicating this protein complex can function as the O_2-sensor in some tissues, and *ii*) pharmacological inhibition of the ETC (see below) can mimic and attenuate the effects of hypoxia on K^+ currents, whereas loss of NADPH oxidase function can not.

4.2. Electron Transport Chain Inhibition May Act as the O_2 Sensor in AMC

Several specific inhibitors of each of the four complexes of the ETC have been extensively characterized and can be utilized to investigate whether or not the O_2-sensor is consistent with a mitochondrial location. Figure 3 (A and B) show that rotenone (a complex I inhibitor) can inhibit outward currents in neonatal AMC. It should be noted that hypoxia plus rotenone caused no further suppression of outward current, suggesting convergence of the two pathways (22). Figure 3C shows a typical response to hypoxia (~5 mmHg) of a spontaneously active AMC, recorded from a small cell cluster. Both hypoxia and rotenone induced membrane depolarizations that were not additive (Fig. 3C and D), and were associated with action potential broadening (Fig. 4). This suggests that hypoxia and block of electron transport at complex I utilize a common mechanism for ion channel inhibition, and tentatively localizes the O_2 sensor of AMC to mitochondria. Interestingly, we did not observe any significant effect of 5 mM cyanide (a complex IV inhibitor) on neonatal AMC, with the exception of 1 of 5 cells tested where CN induced membrane hyperpolarization (Fig. 4) (22). This latter observation contrasts with that of Mojet et al. (21) who showed that CN raised intracellular Ca^{2+} and activated CA secretion from AMC, presumably mimicking hypoxia. One possibility to explain this discrepancy is that CN is acting at other O_2-dependent cellular systems or that it activates secretion by a different pathway to hypoxia.

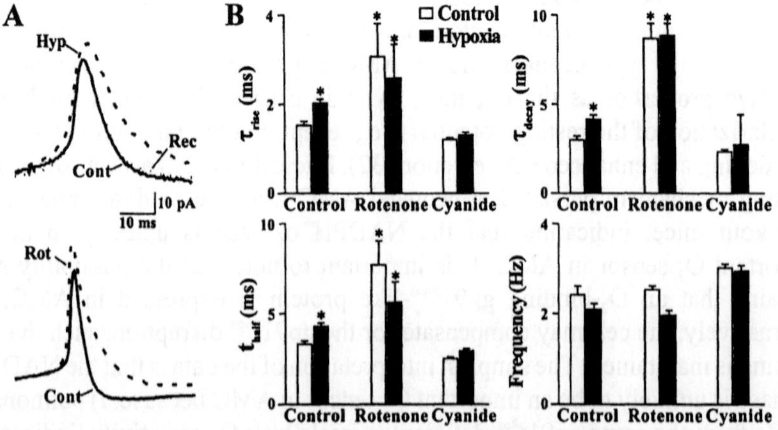

Figure 4. Hypoxia and rotenone prolong action potential duration in neonatal AMC. A: Hypoxia (Hyp) and rotenone (Rot) reversibly broadened action potentials recorded from spontaneously active neonatal chromaffin cells. For comparison purposes and to eliminate the effects of hypoxia-induced membrane depolarization, action potentials were adjusted to start at the same resting potential. "Rec", recovery/washout. B: Effects of hypoxia (n=90 spikes from 4 cells), rotenone (n=50 spikes from 3 cells), and cyanide (n=30 spikes from 2 cells) on action potential duration and spike frequency. Hypoxia and rotenone (5 μM), but not cyanide (2.5 mM), caused a slight increase in the rise time of the action potential (τ_{rise}), significantly prolonged the decay phase (τ_{decay}) and half width of the action potential (t_{half}). Hypoxia and rotenone, but not cyanide caused a slight but non-significant decrease in action potential frequency. Note that the effects of hypoxia were attenuated or abolished in the presence of rotenone, but not cyanide. All records were obtained using the nystatin perforated patch configuration of whole-cell recording in I=0 mode. * $P<0.05$ vs. normoxic control.

4.3. Reactive Oxygen Species May Function as the Second Messenger During Hypoxic Chemoreception

If hypoxia is sensed by the mitochondria, how is the signal transduced to plasmalemmal K^+ channels? Our data suggest that hypoxia *decreases* ROS generation from the ETC at a site upstream of the classical O_2 and CN binding site of cytochrome c oxidase (complex IV) and downstream of the rotenone binding site in complex I. We tested if alterations of ROS could account for K^+ channel inhibition during hypoxia. Exogenous H_2O_2, an O_2-signaling pathway second messenger, reversed the effect of hypoxia on K^+ currents, and the ROS scavenger, N-acetylcysteine, mimicked and attenuated the hypoxic inhibition of K^+ currents (Thompson and Nurse, unpublished data). Taken together, these data suggest that hypoxia decreases ROS generation by the ETC of AMC. A similar mechanism, involving inhibition of mitochondrial ROS generation has been proposed to explain the O_2-sensitivity of pulmonary myocytes (19).

5. A Model of O_2 Sensing in Neonatal AMC

We propose the following model for the O_2 signaling pathway in AMC (Fig. 5). During hypoxia, O_2 availability is decreased and ROS generation at a critical site of the proximal ETC is reduced. The key ROS may be H_2O_2, which is generated in the mitochondria by the dismutation of $O_2^{\bullet-}$, is freely diffusible, and can be long-lived depending on catalase levels in the cell. The combined inhibition of BK, SK, and K_V channels leads to or enhances membrane depolarization and action potential broadening, increasing Ca^{2+} influx through nifedipine-sensitive L-type Ca^{2+} channels and CA secretion. Although controversial, it appears that the ability of AMC to directly respond to hypoxia disappears along a time course similar to the maturation of sympathetic innervation to the adrenal gland.

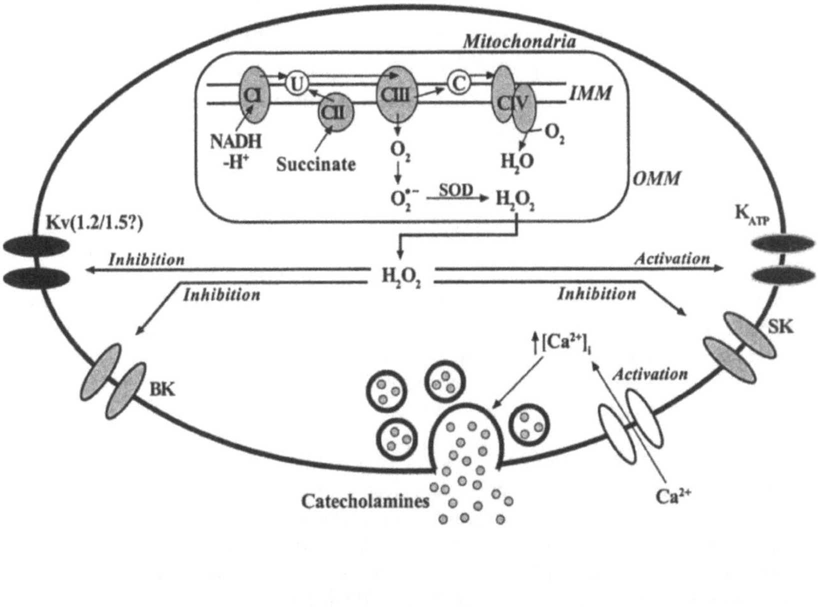

Figure 5. A working model of the O_2-sensing mechanism of neonatal AMC. Hypoxia is detected by the electron transport chain (ETC), a component of the inner mitochondrial membrane (IMM), reducing the production of superoxide radical ($O_2^{\bullet-}$) from the ETC. $O_2^{\bullet-}$ is rapidly dismutated to H_2O_2 by the mitochondrial enzyme superoxide dismutase (SOD) and crosses the outer mitochondrial membrane (OMM). The overall decrease in cytoplasmic H_2O_2 modulates plasma membrane ion channels. It is proposed that SK channels are inhibited and K_{ATP} channels activated, resulting in a receptor potential. K_{ATP} channels appear to limit the magnitude of the receptor potential. Additionally, in spontaneously active AMC, H_2O_2 is hypothesized to inhibit large-conductance Ca^{2+}-activated (BK) and delayed-rectifier (K_V1.2/1.5) channels, thereby broadening the action potentials. The combined receptor potential and modulation of the action potential waveform opens L-type Ca^{2+} channels, causing Ca^{2+} influx and catecholamine exocytosis.

Clearly, this model will have to be tested more rigorously using excised patches from AMC to determine modulation of the channels by putative second messengers (e.g., H_2O_2) and with knockout models that lack key ETC proteins. The fact that the O_2-sensitivity of these cells appears to be intimately associated with innervation (25, 33) means that this naturally-occurring developmental change in phenotype may be used as a powerful tool for elucidating the molecular identity of the sensor. Future work should focus on determining how the change in O_2-sensing phenotype develops and what are the cellular and molecular mechanisms that contribute to it.

Acknowledgments

The authors greatly acknowledge the technical assistance of Cathy Vollmer. We also thank S. Farragher, I. Samjoo, and Drs. A. Jackson, I. Fearon. L. Doering and E. Cutz for their contributions to the experiments described in this chapter. This work was supported by operating grants to CAN from the Heart and Stroke Foundation of Ontario and Natural Sciences and Research Engineering Council of Canada.

References

1. Adams M, Simonetta G, and McMillen IC. The non-neurogenic catecholamine response of the fetal adrenal to hypoxia is dependent on activation of voltage sensitive Ca^{2+} channels. *Brain Res. Devel. Brain Res.* 1996; 94: 182-189.
2. Archer S and Michelakis E. The mechanism(s) of hypoxic pulmonary vasoconstriction: potassium channels, redox O_2 sensors, and controversies. *News Physiol. Sci.* 2002; 17: 131-137.
3. Archer SL, Souil E, Dinh-Xuan AT, Schremmer B, Mercier J-C, Yaagoubi AE, Nguyen-Huu L, Reeve HL, and Hampl V. Molecular identification of the role of voltage-gated K^+ channels, Kv1.5 and Kv2.1, in hypoxic pulmonary vasoconstriction and control of resting membrane potential in rat pulmonary artery myocytes. *J. Clin. Invest.* 1998;101: 2319-2330.
4. Artalejo C, Adams M, and Fox A. Three types of Ca^{2+} channels trigger secretion with different efficacies in chromaffin cells. *Nature.* 1994; 367: 72-76.
5. Comline RS and Silver M. The release of adrenaline and noradrenaline from the adrenal glands of the foetal sheep. *J. Physiol.* 1961; 156: 424-444.
6. Conforti L, Bodi I, Nisbit J, and Millhorn D. O_2-sensitive K^+ channels: role of the Kv1.2-subunit in mediating the hypoxic response. *J. Physiol.* 2000; 524: 783-793.
7. Cross A, Henderson L, Jones O, Delpiano M, Hetschel J, and Acker H. Involvement of and NAD(P)H oxidase as pO_2 sensor protein in the rat carotid body. *Biochem. J.* 1990; 272: 743-747.
8. Fearon IM, Thompson RJ, Samjoo I, Vollmer C, Doering L, and Nurse C. O_2-sensitive K^+ channels in rat adrenal-derived MAH cells. *J. Physiol.* 2002; 545: 807-818.
9. Fu X, Wang D, Nurse C, Dinauer M, and Cutz E. NADPH oxidase is an O_2 sensor in airway chemoreceptors: evidence from K^+ current modulation in wild-type and oxidase-deficient mice. *Proc. Nat. Acad. Sci. USA.* 2000; 97: 4374-4379.

10. Gozal D. Potentiation of hypoxic ventalitory response by hyperoxia in the conscious rat: putative role of nitric oxide. *J. Appl. Physiol.* 1998; 85: 129-132.
11. Holgert H, Dagerlind A, Hokfelt T, and Lagercrantz H. Neuronal markers, peptides and enzymes in nerves and chromaffin cells in the rat adrenal medulla during postnatal development. *Brain Res. Devel. Brain Res.*1994; 83: 35-52.
12. Inoue M, Fujishiro N, and Imanaga I. Na^+ pump inhibition and non-selective cation channel activation by cyanide and anoxia in guinea-pig chromaffin cells. *J. Physiol.* 1999; 519: 385-396.
13. Jiang C and Haddad G. Short periods of hypoxia activate a K^+ current in central neurons. *Brain Res.* 1993; 614: 352-356.
14. Keating DJ, Rychkov GY, and Roberts ML. Oxygen sensitivity in the sheep adrenal medulla: role of SK channels. *Am. J. Physiol Cell Physiol.* 2001; 281: C1434-C1431
15. Lagercrantz H and Bistoletti P. Catecholamine release in the newborn infant at birth. *Pediatr. Res.* 1977; 8: 889-893.
16. Lee J, Lim W, Eun S-Y, Kim SJ, and Kim J. Inhibition of apamin-sensitive K^+ current by hypoxia in adult rat adrenal chromaffin cells. *Pflügers Arch.* 2000; 439: 700-704.
17. Lewis A, Peers C, Ashford M, and Kemp P. Hypoxia inhibits human recombinant large conductance, Ca^{2+}-activated K^+ (maxi-K) channels by a mechanism which is membrane delimited and Ca^{2+} sensitive. *J. Physiol.* 2002; 540: 771-780.
18. López-Barneo J, Pardal R, and Ortega-Sáenz P. Cellular mechanism of oxygen sensing. *Annu. Rev. Physiol.* 2001; 63: 259-287.
19. Michelakis ED, Hampl V, Nsair A, Wu XC, Harry G, Haromy A, Gurtu R, Archer SL. Diversity in mitochondrial function explains differences in vascular oxygen sensing. *Circ. Res.* 2002; 90: 1307-1315.
20. Mochizuki-Oda N, Takeuchi Y, Matsumura K, Oosawa Y, and Watanabe Y. Hypoxia-induced catecholamine release and intracellular Ca^{2+} increase via suppression of K^+ channels in cultured rat adrenal chromaffin cells. *J. Neurochem.* 1997; 69: 377-387.
21. Mojet M, Mills E, and Duchen MR. Hypoxia-induced catecholamine secretion in isolated newborn rat adrenal chromaffin cells is mimicked by inhibition of mitochondrial respiration. *J. Physiol.* 1997; 504: 175-189.
22. Nurse CA, Fearon IM, Jackson A, and Thompson RJ. "Oxygen sensing by neonatal adrenal chromaffin cells: A role for mitochondria?" In *Oxygen Sensing: Responses and Adaptations to Hypoxia,* Lahiri S, Semenza G, and Prabhakar N, eds. New York, NY: Marcel Dekker Inc., 2003, pp. 603-618.
23. Peers C. Hypoxic suppression of K^+ currents in type I carotid body cells: selective effect on the Ca^{2+}-activated K^+ current. *Neurosci. Lett.* 1990; 119: 253-256.
24. Rustenbeck I, Dickel C, Herrmann C, and Grimmsmann T. Mitochondria present in excised patches from pancreatic B-cells may form microcompartments with ATP-dependent potassium channels. *Biosci. Reports.* 1999; 19:89-98.
25. Seidler FJ and Slotkin T. Adrenomedullary function in the neonatal rat: responses to acute hypoxia. *J Physiol.* 1985; 385: 1-16.
26. Seidler FJ and Slotkin TA. Ontogeny of adrenomedullary responses to hypoxia and hypoglycemia: role of splanchnic innervation. *Brain Res. Bul.* 1986; 16: 11-14.
27. Seidler FJ, Brown K, Smith PG, and Slotkin TA. Toxic effects of hypoxia on neonatal cardiac function in the rat: á-adrenergic mechanisms. *Toxicol. Lett.* 1987; 37: 79-84.
28. Slotkin TA and Seidler FJ. Adrenomedullary catecholamine release in the fetus and newborn: secretory mechanisms and their role in stress and survival. *J. Devel. Physiol.* 1988; 10:1-16.
29. Sweeney M and Yuan JX-J. Hypoxic pulmonary vasoconstriction: role of voltage-gated potassium channels. *Resp. Physiol.* 2000; 1: 40-48.
30. Takeuchi Y, Mochizuki-Oda N, Yamada H, Kurokawa K, and Watanabe Y. Nonneurogenic hypoxia sensitivity in rat adrenal slices. *Biochem. Biophys. Res. Comm.* 2001; 289: 51-56.
31. Thompson RJ and Nurse CA. Anoxia differentially modulates multiple K^+ currents and depolarizes neonatal rat adrenal chromaffin cells. *J. Physiol.* 1998; 512: 421-434.

32. Thompson RJ, Farragher SM, Cutz E, and Nurse CA. Developmental regulation of O_2 sensing in neonatal adrenal chromaffin cells from wild-type and NADPH-oxidase-deficient mice. *Pflugers Arch.* 2002; 444: 539-548.
33. Thompson RJ, Jackson A, and Nurse CA. Developmental loss of hypoxic chemosensitivity in rat adrenomedullary chromaffin cells. *J. Physiol.* 1997; 498: 503-510.
34. Walters DV and Olver RE. The role of catecholamines in lung liquid absorption at birth. *Pediatr. Res.* 1978; 12: 239-242.
35. Waypa G, Chandel N, and Schumacker P. Model for hypoxic pulmonary vasoconstriction involving mitochondrial oxygen sensing. *Circ. Res.* 2001; 88: 1259-1266.

Chapter 22

Oxygen-sensitive Ion Channels in Pheochromocytoma (PC12) Cells

Laura Conforti and David E. Millhorn
University of Cincinnati, Cincinnati, Ohio, U.S.A.

1. Introduction

The ability to sense changes in O_2 availability is conserved among various cell types (21). In particular, chemosensitive cells such as carotid body type I cells and pulmonary artery smooth muscle cells respond to hypoxia with a characteristic sequence of events: inhibition of O_2-sensitive K^+ (Ko_2) currents, membrane depolarization, and changes in intracellular Ca^{2+} concentration ($[Ca^{2+}]_i$) (21). $[Ca^{2+}]_i$ is important to regulate cell function (e.g., contraction and neurotransmitter release) and gene regulation. Inhibition of K^+ channels is thus a very important event that links hypoxia to cell function. Although much progress has been made in identifying the different steps of the O_2-sensing process, the molecular mechanisms underneath are not fully understood. A major drawback in obtaining a more comprehensive understanding of the cellular response to hypoxia has been the lack of a suitable cell line that will provide a continuously replicating supply of chemosensitive cells that exhibit many characteristics of their normal counterparts. We have established that pheochromocytoma (PC12) cells respond to hypoxia in a manner that is reminiscent of O_2 sensitive cells *in vivo*. This cell line has been since utilized as a model system to study the biophysical and molecular mechanisms by which cells respond to reduced O_2 tension. In the current chapter, we will discuss the characteristics of this cell line. In particular, we will focus on the properties and functional role of the O_2-regulated K^+ channels expressed in PC12 cells.

2. Pheochromocytoma (PC12) Cells

The PC12 clonal cells were originally derived from rat adrenal medullary tumors (12). These cells synthesize catecholamines (in particular dopamine and norepinephrine) and release them in response to a variety of stimuli (12).

Although this cell line has been widely used as a model system for neurobiological and neurochemical studies, in the past ten years, PC12 cells have proven to be an important tool in better understanding the different aspects of O_2-sensing. The interest in PC12 cells as possible chemosensitive cell line arose from the many similarities that exist between PC12 and the carotid body (CB) type I cells. The type I (glomus) cells are the O_2-sensing cells of the CB and their function is to transmit information concerning the arterial O_2 tension to primary sensory afferent terminals by release of neurotransmitters (21). Both PC12 cells and CB type I cells have neural crest origin; both demonstrate chromaffin cell-like characteristics and synthesize catecholamines, especially dopamine, as their major neurotransmitter. Importantly, CB type I cells respond to hypoxia by releasing catecholamines and upregulating tyrosine hydroxylase (TH, the rate limiting enzyme in the production of dopamine) activity and expression (4, 9). Millhorn and colleagues first demonstrated that gene regulation and signal transduction pathways involved in the regulation of TH production are similarly influenced by hypoxia in PC12 cells (27). Prolonged exposure of PC12 cells to hypoxia causes an increase in TH gene expression in PC12 cells just like in CB type I cells (9, 10). This increase in TH gene level upon hypoxia is due to both elevated gene transcription rate and prolonged mRNA stability (10). These observations suggested that PC12 cells represent a very useful model system for detailed study of O_2 signal transduction. Further confirmation of their chemosensitive phenotype came from studying the effects of acute hypoxia on PC12 cells: exposure to hypoxia triggers the same sequence of early O_2-sensing events observed in CB and other chemosensitive cells.

3. Oxygen Sensitivity of PC12 Cells

An initial critical step in the process of O_2-sensing common to chemosensitive cells is membrane depolarization upon exposure to hypoxia (21, 22). This event is essential for activating voltage-sensitive Ca^{2+} channels and producing the increase in $[Ca^{2+}]_i$ necessary to trigger many physiological responses such as constriction (in pulmonary artery) and neurotransmitter release (in CB). Like other oxygen-chemosensitive cells, PC12 cells respond to acute exposure to hypoxia with membrane depolarization. Whole-cell current-clamp studies of membrane potential show that hypoxia causes membrane depolarization in PC12 cells (Fig. 1A) and that the degree of membrane depolarization is proportional to the severity of the hypoxic stimulus and is independent of external space Ca^{2+} (31). The membrane depolarization is necessary to activate voltage-dependent Ca^{2+} channels, thereby increasing cytosolic Ca^{2+}. Indeed, Fura-2 experiments indicated that hypoxia induces a 2-3 fold increase in $[Ca^{2+}]_i$ in PC12 cells (Fig. 1B). A similar effect of hypoxia on Ca^{2+} homeostasis has been reported for CB type I cells and pulmonary artery smooth muscle cells (21, 22, 25). Ultimately, these responses lead to transmitter release. Indeed, amperometric measurements indicated that hypoxia evokes dopamine and norepinephrine release in PC12 cells (19). The hypoxia-induced exocytosis occurs via depolarization, leading to Ca^{2+} influx primarily via N-type Ca^{2+} channels (30). Overall, a primary response to acute hypoxia in PC12 cells is facilitation of release of neurotransmitters such as dopamine (DA), norepinephrine (NE) and adenosine (18, 19, 30, 31). Although the role of these transmitters in transduction of the hypoxic stimulus is still controversial, recent findings show that DA and adenosine exert a feedback control on cellular excitability and function in PC12 cells during hypoxia via stimulation of D_2 and A_2 receptors, respectively (18, 32).

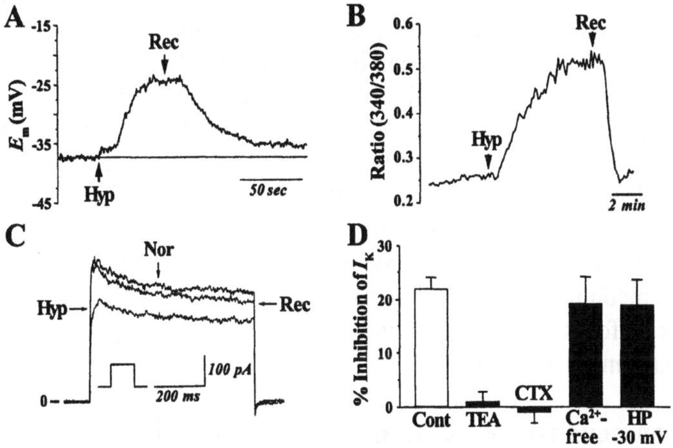

Figure 1. O_2-sensing mechanisms triggered by acute hypoxia in PC12 cells. A: Acute hypoxia induces cell depolarization. Membrane potential (E_m) was recorded in current-clamp mode. B: Acute hypoxia increases $[Ca^{2+}]_i$. Arrows indicate point of introduction of hypoxia (Hyp; $Po_2<10$ mmHg) and return to normoxic conditions (Rec; $Po_2 \approx 150$ mmHg) in panels A and B. C: Acute hypoxia inhibits a Ko_2 current. Superimposed current traces were recorded in normoxia (Nor), after steady-state inhibition by hypoxia (Hyp) and after returning to normoxia (Rec). D: The effect of hypoxia on the outward K^+ current, elicited by a 800-ms test pulse of +50 mV, was determined under control conditions with a holding potential (HP) of -90 mV (open bar, standard bath solution with 2 mM $CaCl_2$ and Ca^{2+}-free pipette solution, n=11), in the presence of 5 mM TEA (n=4) or 20 nM CTX (n=7), in Ca^{2+}-free external medium (n=5) or using a HP of -30 mV (n=5).

All the above series of events necessary to produce the functional response to hypoxia are triggered by the initial inhibition of a K^+ conductance (20). PC12 cells, like other chemosensitive cells, express an Ko_2 channel that is inhibited by hypoxia (21). Exposure to hypoxia ($Po_2<10$ mmHg) induces inhibition of an outward slow-inactivating voltage-dependent K^+ current, and this effect is reversible upon returning to normoxia (Po_2=150 mmHg) (Fig. 1C). The magnitude of hypoxia-induced inhibition of the K current depends on the severity of hypoxia. Perfusion with progressively lower Po_2 reduces the K current in a step-wise fashion (31). The Ko_2 current in PC12 cells is present at the voltage range of their resting potential (*ca.* -35 to -45 mV), thus its inhibition results in membrane depolarization (31). Detailed whole-cell voltage-clamp studies have indicated that the Ko_2 current in PC12 cells is a slowly-inactivating voltage-dependent K^+ current which is blocked by tetraethylammonium (TEA), a blocker of voltage-dependent K (K_V) channels (31). Taylor and Peers (30) have confirmed that inhibition of a TEA-sensitive K^+ conductance is indeed responsible for the hypoxia-evoked depolarization and consequent exocytosis in PC12 cells. Figure 1D illustrates the characteristics of the Ko_2 current in PC12 cells by showing the relative inhibition of the K^+ current induced by hypoxia under various experimental conditions. In control conditions, hypoxia inhibits the K^+ current by approximately 20%. Exposure of cells to 5 mM external TEA results in loss of the hypoxia-induced inhibition of the K current. Furthermore, the Ko_2 current in PC12 cells is inhibited by charybdotoxin, a potent blocker of K_V channels (in particular $K_V1.2$ and $K_V1.3$) and Ca^{2+}-activated K (K_{Ca}) channels. Although K_{Ca} channels are expressed in PC12 cells, they do not appear to be O_2-sensitive. Indeed, the Ko_2 current is not sensitive to Ca^{2+}. Furthermore, the Ko_2 current is not sensitive to the holding voltage. In fact, the same percentage inhibition of the K^+ current by hypoxia was observed in experiments performed in Ca^{2+}-free medium and in experiments where the holding potential was kept at -30 mV. Therefore the Ko_2 current in PC12 cells does not appear to be either a K_{Ca} or a transient K^+ current.

4. Oxygen-sensitive K^+ Channels in PC12 Cells

The outward Ko_2 current that is selectively and reversibly inhibited by hypoxia in PC12 cells is carried by a K_V channel. Ko_2 currents have been characterized in other chemosensitive cells. In most chemosensitive cells such as carotid body type I cells and rat pulmonary artery smooth muscle cells , the Ko_2 current is voltage-dependent; however, in rat carotid body and in neuroepithelial body, a background K^+ current was also recently proposed to constitute an O_2-sensitive K^+ conductance (1, 3, 11, 14). At least four types of K^+ channels are expressed in PC12 cells, distinguished by their unitary current conductances and current-voltage relationships: a small conductance K^+ channel (Ksm; 14 pS), a Ca^{2+}-activated K^+ (K_{Ca}) channel (102 pS) and two K^+ channels with similar conductance (20 pS; Fig. 2A-B) (7, 15). These last two channels differ in their time-dependent inactivation: one is a slowly-inactivating channel (Kdr), while the other belongs to the family of fast transient K^+ channels (Ktr). Patch clamp studies revealed that hypoxia selectively inhibits only the 20 pS slowly-inactivating delayed-rectifier type of K^+ channel in PC12 cells (7). Figure 2C shows the single-channel properties of the 20 pS Ko_2 channel recorded in a cell-attached patch in PC12 cells and the effect of hypoxia on the channel activity. Single-channels currents were elicited by step-pulse depolarization in normoxia and 2 min after hypoxia. The inhibitory effect of hypoxia was apparent in the decreased amplitude of the ensemble-averaged currents due to a decrease in open probability (P_o) of the channels, with no change in conductance (Fig. 2C, bottom panels) (7). The effect of hypoxia on the Ko_2 channel in PC12 cells as well as CB type I cells and other recombinant Ko_2 channels is manifested by a similar mechanism involving a decrease in channel P_o, a slowing of activation kinetics, and little effect on channel closing (7, 11, 23). Interestingly, the sensitivity to hypoxia of the Ko_2 channel is maintained in inside-out patch configuration both in PC12 cells and CB type I cells, suggesting that cytosolic soluble factors might not be required for the hypoxic response (7, 11).

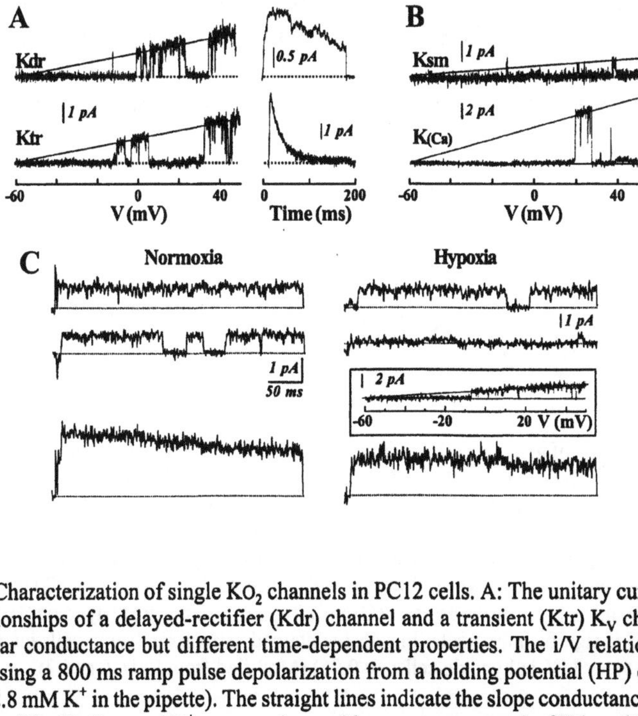

Figure 2. Characterization of single Ko_2 channels in PC12 cells. A: The unitary current-voltage (i/V) relationships of a delayed-rectifier (Kdr) channel and a transient (Ktr) K_V channel which have similar conductance but different time-dependent properties. The i/V relationships were obtained using a 800 ms ramp pulse depolarization from a holding potential (HP) of -60 mV to +50 mV (2.8 mM K^+ in the pipette). The straight lines indicate the slope conductance for Kdr (19 pS) and Ktr (20 pS). Outward K^+ currents (ensemble-average currents) of Kdr and Ktr channels, elicited by a 180 ms step pulse from -60 to +50 mV in the same patch, respectively, are shown in right panels. B: The i/V relationships of the small conductance (Ksm) K_V channel and a large conductance K_{Ca} channel show that the slope conductances are 15 and 102 pS for Ksm and K_{Ca} channels, respectively. C: Hypoxia inhibits the activity of the 20-pS Kdr channel. Top panels show representative currents, elicited by 180 ms step depolarizing pulses from -60 to +50 mV before (normoxia) and after 2-min exposure to hypoxia (10% O_2). The corresponding ensemble-averaged currents (from 100 consecutive traces) are shown in bottom panels. Dashed lines represent the zero current. The slope conductance, measured by ramp depolarization, is shown in the inset (Modified from Refs. 6 and 7).

5. Molecular Properties of Ko_2 Channels in PC12 Cells

Further molecular biological experiments have indicated that the Ko_2 channel in PC12 cells belongs to the K_V1 subfamily (6, 7). K_V channels are tetrameric arrangements of 4 separate pore-forming α subunits and auxiliary β subunits (5). The genes that encode functional K_V α subunits are classified in 4 major subfamilies: K_V1-K_V4. New subfamilies ($K_V5.1$-K_V10) are recently added. The molecular composition of Ko_2 channels is still under investigation. The K_V α subunits implicated in forming Ko_2 channels in PA are: $K_V2.1$ (homomultimer or heteromultimer in combination with the electrically silent K_V 9.3 α subunit), $K_V1.5$, $K_V1.2$, $K_V3.1b$ (21). $K_V4.1$ and $K_V4.3$ have been proposed in rabbit CB (26). We have identified the $K_V1.2$ α subunit as an important component of the Ko_2 channel in PC12 cells (6, 7). The O_2-sensitivity of $K_V1.2$ homomultimers was also confirmed in recombinant studies in L cells or *Xenopus* oocytes (6, 16).

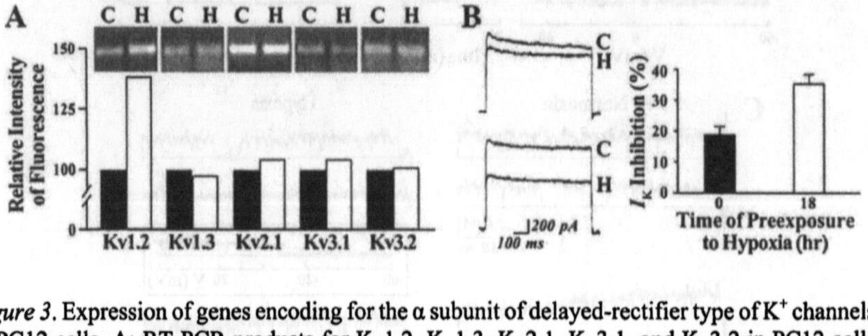

Figure 3. Expression of genes encoding for the α subunit of delayed-rectifier type of K^+ channels in PC12 cells. A: RT-PCR products for $K_V1.2$, $K_V1.3$, $K_V2.1$, $K_V3.1$, and $K_V3.2$ in PC12 cells maintained for 18 hrs in a normoxic (C) or hypoxic (H, 10% O_2)) incubator. The relative intensity (in arbitrary units) of each band, averaged from 7 experiments for $K_V1.2$ and 4 experiments for other K_V channels, is shown in the bar graph. B: O_2-sensitivity of PC12 cells after prolonged exposure to hypoxia (18 hr, 10% O_2). Whole-cell K^+ currents (I_K) elicited by a 800 ms test pulse from -70 to +50 mV were measured before (C) and after 1-min exposure to hypoxia (H). Hypoxic inhibition of I_K is significantly greater in cells maintained in hypoxia than in cells maintained in normoxia (19 % vs 35% in hypoxic cells; $P<0.05$) (Modified from Refs. 6 and 7).

PC12 express different genes that may encode the α subunits of delayed rectifier types of K^+ channels: $K_V1.2$, $K_V1.3$, K_VTo identify the K_V α subunits that form the Ko_2 channel in PC12 cells, we studied the regulation of K_V gene expression by chronic hypoxia. It is known that O_2-sensitive tissues (e.g., the carotid bodies) enhance their chemosensitivity in the process of adaptation to chronic hypoxia (29). Regulation of the Ko_2 channel expression by chronic hypoxia has also been shown to occur in other cell types (2, 8, 25). We found that expression of the $K_V1.2$, but not other K_V genes, was increased by ~40% after prolonged exposure to hypoxia (Fig. 3A). The increased expression of the

$K_V1.2$ gene correlated with an enhanced acute response to hypoxia in those cells pre-exposed to 10% O_2 for 18 hrs (Fig. 3B) (7). These data provided the first evidence that the Ko_2 channel in PC12 cells is a homo- or heterotetramer of $K_V1.2$ subunit, as the pharmacological and electrophysiological studies described above had suggested. Indeed, it was shown previously that expression of $K_V1.2$ gene in a mammalian cell line leads to a delayed rectifier K^+ channel of 18 pS conductance (13). The conductance and gating properties of this channel are very similar to those of the Ko_2 channel we identified in PC12 cells. Expression of other K_V α genes is associated with K^+ channels of much different conductance (e.g., 27 pS for $K_V3.1$) (13).

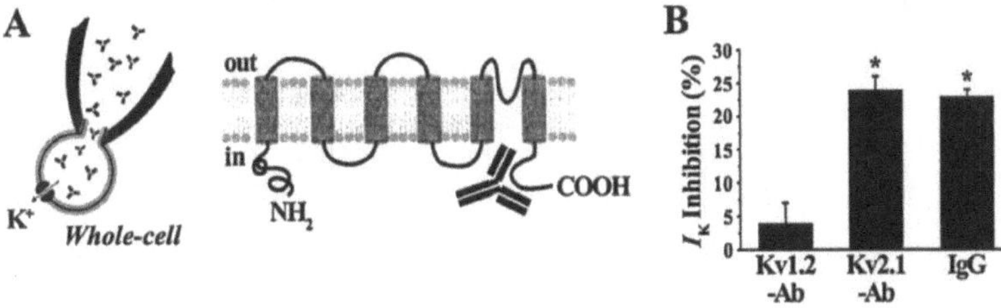

Figure 4. Effect of acute hypoxia on K^+ currents after selectively blocking $K_V1.2$ and $K_V2.1$ channels with specific antibodies. A. Schematic diagram showing how the antibodies (Ab) are delivered to the cytosol through the pipette (left), and block K^+ efflux by binding to the C-terminal epitope of the channel polypeptide (right). B: Hypoxia-induced inhibition of K^+ currents (I_K) in cells dialyzed with an anti-$K_V1.2$ antibody (Ab), an anti-$K_V2.1$ Ab, or an irrelevant antibody (IgG). * $P \leq 0.001$ vs. $K_V1.2$ Ab (Modified from Refs. 6 and 7).

In addition to $K_V1.2$, PC12 cells also express a $K_V2.1$ α subunit. $K_V2.1$ channels have been proposed as possible Ko_2 channels in PA (2, 24). Although the $K_V2.1$ gene is not regulated by chronic hypoxia in PC12 cells, it is still possible that a functional Ko_2 channel formed by the $K_V2.1$ subunit might be also expressed. It is known that K_V channel regulation can post-transcriptionally occur, independently on mRNA levels. In PC12 cells $K_V2.1$ protein levels are increased by exposure to nerve growth factor, with no change in steady-state levels of $K_V2.1$ mRNA (28). The molecular nature of the functional Ko_2 channel/s in PC12 cells was established by selective block of the channel activity with the corresponding antibody (Ab) (6). The anti-K_V antibodies used are prepared against the C-terminal part of their corresponding K_V channel protein. The anti-K_V Ab dialyzes into the cell through the patch pipette, binds to the corresponding antigenic portion of the K_Vv α subunit and blocks, possibly by steric hindrance, the flow of K^+ ions through the channel pore (Fig. 4A). Using antibodies against the $K_V1.2$ polypeptide to block $K_V1.2$ channels and by comparing the obtained results with similar experiments performed in the presence of anti-$K_V2.1$ Ab, we have confirmed that the Ko_2 channel in PC12 cells is composed of $K_V1.2$ α subunits (Fig. 4B). The specificity of the anti-$K_V1.2$

antibody was established by immunohistochemical and western blot experiments. We have also established the feasibility of using the anti-$K_V1.2$ antibody to selectively block the K^+ current carried by $K_V1.2$ channels (6). The use of K_V channel antibodies as experimental tools to assess the molecular nature of the Ko_2 channels has been done by others and elegantly applied recently by Sanchez and colleagues (2, 26). Dialysis of PC12 cells with specific antibodies against the Kv1.2 α subunit prevented the hypoxia-induced inhibition of voltage-activated K^+ currents. Cells dialyzed with anti-$K_V2.1$ Ab maintained their response to hypoxia, as shown in Fig. 4B. This important finding confirmed that, in PC12 cells, a functional $K_V1.2$ α subunit is necessary for the response of the Ko_2 channel to hypoxia and that the $K_V2.1$ channels are not O_2-sensitive.

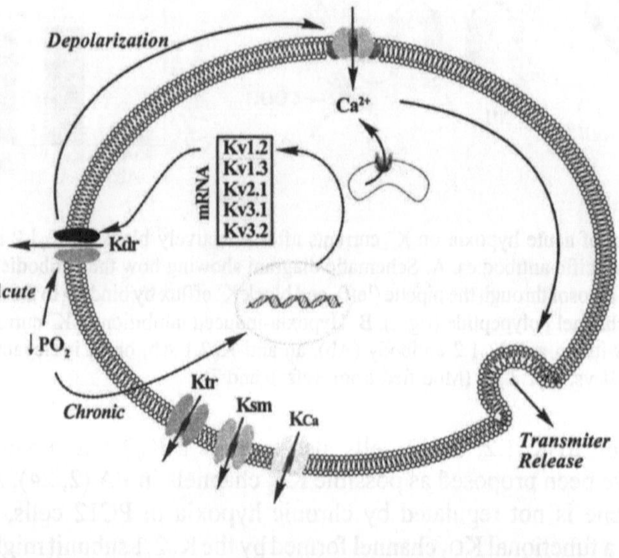

Figure 5. A model for hypoxic modulation of Ko_2 channels in PC12 cells and its functional consequences. PC12 cells express at least four types of voltage-dependent K^+ channels: delayed rectifier (Kdr), transient (Ktr), small conductance (Ksm) and Ca^{2+}-activated (K_{Ca}) K^+ channels. Exposure to acute hypoxia selectively inhibits the Kdr channel. The subsequent decrease in K^+ currents through Kdr channels results in membrane depolarization, activation of voltage-dependent Ca^{2+} channels, increase in $[Ca^{2+}]_i$, and transmitter release. Kdr channels are encoded by different genes. Chronic hypoxia increases the expression of O_2-sensitive $K_V1.2$ channels.

6. Summary

The limitations residing in using fresh chemosensitive cells to study O_2-sensing mechanisms (i.e., small number of O_2-sensitive cells present and cell heterogeneity) have been greatly overcome by the availability of immortalized cellular models such as PC12 cells and, more recently, H146, a model of neuroepithelial cells (17, 27). Overall, the physiological response to reduced O_2

tension in PC12 cells is similar to the response in other O_2-sensitive tissues. PC12 cells, like other chemosensitive cells, express voltage-dependent Ko_2 channels (Fig. 5). We have gathered various evidences indicating that the Ko_2 channel in PC12 cells belongs to the K_V1 subfamily of K_V channels and is formed by the $K_V1.2$ α subunit. Hypoxia exerts two effects on Ko_2 channels: an acute effect occurring within minutes consists of inhibition of the channel activity, and a chronic effect occurring after prolonged exposure to hypoxia which is associated with regulation of the channel itself (21). In PC12 cells acute hypoxia inhibits the Ko_2 channel and this inhibition is responsible for the cascade of events that culminate in neurotransmitter release: membrane depolarization, activation of voltage-gated Ca^{2+} channels and increase in cytosolic Ca^{2+}. Chronic hypoxia selectively upregulates expression of the $K_V1.2$ α gene and its corresponding functional Ko_2 channel. Hence, various finding have confirmed the chemosensitive phenotype of PC12 cells so that this cell line will continue to provide a useful model to study the signaling mechanisms that underlie the O_2 sensitivity of these K^+ channels and other proteins important in the process of O_2 sensing.

Acknowledgments

This work is supported by grants from the American Heart Association (#0030091N to LC) and grant numbers HL3831 and HL55929 to DEM.

References

1. Archer SL, Huang JMC, Reeve HL, Hampl V, Tolarová S, Michelakis E, and Weir EK. Differential distribution of electrophysiologically distinct myocytes in conduit and resistance arteries determines their response to nitric oxide and hypoxia. *Circ. Res.* 1996; 78: 431-442.
2. Archer SL, Souil E, Dinh-Xuan AT, Schremmer B, Mercier JC, El Yaagoubi A, Nguyen-Huu L, Reeve HL, and Hampl V. Molecular identification of the role of voltage-gated K^+ channels, Kv1.5 and Kv2.1, in hypoxic pulmonary vasoconstriction and control of resting membrane potential in rat pulmonary artery myocytes. *J. Clin. Invest.* 1998; 101: 2319-2330.
3. Buckler KJ, Williams BA, and Honore E. An oxygen-, acid- and anaesthetic-sensitive TASK-like background potassium channel in rat arterial chemoreceptor cells. *J. Physiol.* 2000; 525: 135-142.
4. Bunn HF and Poyton RO. Oxygen sensing and molecular adaptation to hypoxia. *Physiol. Rev.* 1996; 76: 839-885.
5. Coetzee WA, Amarillo Y, Chiu J, Chow A, Lau D, McCormack T, Moreno H, Nadal MS, Ozaita A, Pountney D, Saganich M, Vega-Saenz de Miera E, and Rudy B. Molecular diversity of K^+ channels. *Ann. N.Y. Acad. Sci.* 1999; 868: 233-285.
6. Conforti L, Bodi I, Nisbet JW, and Millhorn DE. O_2-sensitive K^+ channels: role of the $K_V1.2$-α subunit in mediating the hypoxic response. *J. Physiol.* 2000; 524: 783-793.
7. Conforti L and Millhorn DE. Selective inhibition of a slow-inactivating voltage-dependent K^+ channel in rat PC12 cells by hypoxia. *J. Physiol.* 1997; 502: 293-305.

8. Conforti L, Petrovic M, Mohammad D, Lee S, Ma Q, Barone S, and Filpovich AH. Hypoxia regulates expression and activity of Kv1.3 channels in T lymphocytes: a possible role in T cell proliferation. *J. Immunol.* 2003; 170: 695-702.
9. Czyzyk-Krzeska MF, Bayliss DA, Lawson EE, and Millhorn DE. Regulation of tyrosine hydroxylase gene expression in the rat carotid body by hypoxia. *J. Neurochem.* 1992; 58: 1538-1546.
10. Czyzyk-Krzeska MF, Furnari BA, Lawson EE, and Millhorn DE. Hypoxia increases rate of transcription and stability of tyrosine hydroxylase mRNA in pheochromocytoma (PC12) cells. *J. Biol. Chem.* 1994; 269: 760-764.
11. Ganfornina MD and López-Barneo J. Potassium channel types in arterial chemoreceptor cells and their selective modulation by oxygen. *J. Gen. Physiol.* 1992; 100: 401-426.
12. Greene LA and Tischler AS. Establishment of a noradrenergic clonal line of rat adrenal pheochromocytoma cells which respond to nerve growth factor. *Proc. Natl. Acad. Sci. USA.* 1976; 73: 2424-2428.
13. Grissmer S, Nguyen AN, Aiyar J, Hanson DC, Mather RJ, Gutman GA, Karmilowicz MJ, Auperin DD, and Chandy KG. Pharmacological characterization of five cloned voltage-gated K^+ channels, types Kv1.1, 1.2, 1.3, 1.5, and 3.1, stably expressed in mammalian cell lines. *Mol. Pharmacol.* 1994; 45: 1227-1234.
14. Hartness ME, Lewis A, Searle GJ. O'Kelly I, Peers C, and Kemp PJ. Combined antisense and pharmacological approaches implicate hTASK as an airway O_2 sensing K^+ channel. *J. Biol. Chem.* 2001; 276: 26499-26508.
15. Hoshi T and Aldrich RW. Voltage-dependent K^+ currents and underlying single K^+ channels in pheochromocytoma cells. *J. Gen. Physiol.* 1988; 91: 73-106.
16. Hulme JT, Coppock EA, Felipe A, Martens JR, and Tamkun M. Oxygen sensitivity of cloned voltage-gated K^+ channels expressed in the pulmonary vasculature. *Circ. Res.* 1999; 85: 489-497.
17. Kemp PJ, Searle GJ, Hartness ME, Lewis A, Miller P, Williams S, Wootton P, Adriaensen D, and Peers C. Acute oxygen sensing in cellular models: Relevance to the physiology of pulmonary neuroepithelial and carotid bodies. *Anat. Rec.* 2003; 270: 41-50.
18. Kobayashi S, Conforti L, Pun RY, and Millhorn DE. Adenosine modulates hypoxia-induced responses in rat PC12 cells via the A_{2A} receptor. *J. Physiol.* 1998; 508: 95-107.
19. Kumar GK, Overholt JL, Bright GR, Hui KY, Lu H, Gratzi M, and Prabhakar NR. Release of dopamine and norepinephrine by hypoxia from PC-12 cells. *Am. J. Physiol.* 1998; 274: C1592-C1600.
20. López-Barneo J. Oxygen-sensing by ion channels and the regulation of cellular functions. *Trends Neurosci.* 1996; 19: 435-440.
21. López-Barneo J, Pardal R, and Ortega-Sáenz P. Cellular mechanism of oxygen sensing. *Annu. Rev. Physiol.* 2001; 63: 259-287.
22. Olschewski A, Hong Z, Nelson DP, and Weir EK. Graded response of K^+ current, membrane potential, and $[Ca^{2+}]_i$ to hypoxia in pulmonary arterial smooth muscle. *Am. J. Physiol. Lung Cell. Mol. Physiol.* 2002; 283: L1143-L1150.
23. Osipenko ON, Tate RJ, and Gurney AM. Potential role for Kv3.1b channels as oxygen sensors. *Circ. Res.* 2000; 86: 534-540.
24. Patel AJ, Lazdunski M, and Honore E. Kv2.1/Kv9.3, a novel ATP-dependent delayed-rectifier K^+ channel in oxygen-sensitive pulmonary artery myocytes. *EMBO J.* 1997; 16: 6615-6625.
25. Platoshyn O, Golovina VA, Bailey CL, Limsuwan A, Krick S, Juhaszova M, Seiden JE, Rubin LJ, and Yuan JX-J. Sustained membrane depolarization and pulmonary artery smooth muscle cell proliferation. *Am. J. Physiol. Cell Physiol.* 2000; 279: C1540-C1549.
26. Sanchez D, López-López JR, Pérez-García MT, Sanz-Alfayate G, Obeso A, Ganfornina MD, and Gonzalez C. Molecular identification of Kv α subunits that contribute to the oxygen-sensitive K^+ current of chemoreceptor cells of the rabbit carotid body. *J. Physiol.* 2002; 542: 369-382.
27. Seta KA, Spicer Z, Yuan Y, Lu G, and Millhorn DE. Responding to hypoxia: lessons from

a model cell line. *Sci. STKE.* 2002; 2002: RE11.
28. Sharma N, D'Arcangelo G, Kleinlaus A, Halegoua S, and Trimmer JS. Nerve growth factor regulates the abundance and distribution of K^+ channels in PC12 cells. *J. Cell. Biol.* 1993; 123: 1835-1843.
29. Stea A, Jackson A, Macintyre L, and Nurse CA. Long-term modulation of inward currents in O_2 chemoreceptors by chronic hypoxia and cyclic AMP in vitro. *J. Neurosci.* 1995; 15: 2192-2202.
30. Taylor SC and Peers C. Hypoxia evokes catecholamine secretion from rat pheochromocytoma PC-12 cells. *Biochem. Biophys. Res. Commun.* 1998; 248: 13-17.
31. Zhu WH, Conforti L, Czyzyk-Krzeska MF, and Millhorn DE. Membrane depolarization in PC-12 cells during hypoxia is regulated by an O_2-sensitive K^+ current. *Am. J. Physiol.* 1996; 271: C658-C665.
32. Zhu WH, Conforti L, and Millhorn DE. Expression of dopamine D_2 receptor in PC-12 cells and regulation of membrane conductances by dopamine. *Am. J. Physiol.* 1997; 273: C1143-C1150.

VII. PATHOLOGY AND MECHANISMS OF HYPOXIA-INDUCED PULMONARY HYPERTENSION

Chapter 23

Pulmonary Vascular Remodeling in Hypoxic Pulmonary Hypertension

Marlene Rabinovitch
Stanford University School of Medicine, Stanford California, U.S.A.

1. Introduction

Chronic hypoxia causes structural changes in the pulmonary arteries. These features are further influenced by inequalities of flow and by loss of tissue when there is associated lung disease. This chapter will address the clinical settings in which hypoxia leads to pulmonary vascular disease and cor pulmonale, the nature of the vascular abnormalities observed and the cellular and molecular basis for their development, progression and potential to regress.

2. Clinical Settings in Which Hypoxia Causes Pulmonary Vascular Disease

2.1. High-altitude Hypoxia

In persons living at high altitude, there is chronic elevation of pulmonary artery pressure, only a small portion of which is reversible with administration of oxygen. The peripheral arteries are more muscular than normal and have a decreased lumen diameter. The severity of pulmonary hypertension is variable and almost always improves on return to sea level, at least at rest. In response to exercise, it may increase markedly, suggesting limited functional reserve, owing to persistent structural abnormalities (17).

2.2. Upper Airway Obstruction

Severe upper airway obstruction from a variety of causes may be complicated by the development of pulmonary hypertension. These include, obstructive sleep apnea which may be associated with the Pickwickian syndrome. Obesity, through the increased work of breathing and CO_2 production, stresses the respiratory control system and, hypoventilation causes further CO_2 retention and hypoxia, Damage to the respiratory center is rarely a primary disorder or can be cause of secondary to trauma or other neurologic disease. Although removal of the airway obstruction often results in a prompt return to normal pulmonary artery pressure and resolution of the heart failure these symptoms can persist for some time due to slow regression of hypoxia-induced structural changes in the pulmonary vascular bed.

2.3. Lung Disease

Lung disease is the most common cause of hypoxia-induced pulmonary vascular disease. It is the result of hypoxic vasoconstriction compounded by the adverse effects of long-standing polycythemia (microthrombi) and elevated pulmonary venous pressure secondary to dysfunction of the hypoxemic left ventricle. Treatment of the lung disease will decrease the level of pulmonary artery pressure and hence allow some regression of the structural changes. Pulmonary vascular disease is rare with obstructive lung disorders, such as asthma, and more common with parenchymal disorders associated with loss of tissue, such as cystic fibrosis (41). Restrictive lung disease also may be associated with pulmonary hypertension. These include diffuse interstitial fibrosis, radiation fibrosis and chemotherapy toxicity, and infiltrative lung tumors. Neuromuscular disorders affecting the chest wall, (e.g., Duchenne muscular dystrophy, and Werdnig-Hoffman disease, also can cause pulmonary vascular remodeling, as can diseases affecting the vertebrae and rib cage, such as severe scoliosis (13).

2.4. Persistent Pulmonary Hypertension

Perhaps best studied, is the hypoxia causing persistent pulmonary hypertension in the newborn, where the sustained vasoconstriction of the fetal circulation is maintained in the perinatal period. Resolution of the hypoxia allows for dilation of the pulmonary circulation and is generally associated with prompt return to normal hemodynamics.

3. Structural Changes in the Pulmonary Arteries

3.1. Clinical Studies

In clinical tissue from patients living at high altitude or with severe lung disease and associated pulmonary hypertension, the structural features in the pulmonary circulation include extension of muscle into arteries that are peripheral and normally non-muscular, medial hypertrophy of the muscular arteries as well as adventitial thickening and loss in the number of distal arteries relative to alveoli and in association with loss of alveoli.

3.2. Experimental Studies

A variety of studies in animals have been undertaken to address the mechanisms leading to the evolution of the pulmonary vascular disease in association with hypoxia. It has been noted that at altitude severe pulmonary hypertension may in fact represent a maladaptive response related to the amount of intrinsic muscle in the vessel wall. The llama, develops little acute or chronic pulmonary hypertension and has the least muscle pulmonary circulation of all species tested (49). In the rat, the development of structural changes and sustained pulmonary hypertension appears to occur after 3 days of chronic hypoxia (air at half atmospheric pressure, PO_2 = approximately 40 mm Hg). Over the ensuing 2 weeks, pulmonary artery pressure progressively doubles and is associated with progressive extension of muscle into peripheral normally nonmuscular arteries, medial hypertrophy of normally muscular arteries, and reduction in arterial number, all proceeding in parallel, and right ventricular hypertrophy is evident. The hemodynamic and structural response progress slowly thereafter, suggesting an adaptation. The structural changes induced by hypoxia appear to be less severe in mature female experimental animals, and those of both sexes exposed to hypoxia through the period of development may show less potential for regression (39) (29). Regression of smooth-muscle hypertrophy may be accompanied by a relative increase in the amount of elastin and collagen (30); so the vessel, although less muscular, nonetheless may be abnormally noncompliant. Collagen increases in hypertensive pulmonary arteries, and stretch may be the stimulus (5); inhibition of collagen synthesis decreases chronic hypoxic pulmonary hypertension and vascular changes (34). Loss of small arteries may not, however, be recovered, explaining why the pulmonary vascular bed may have limited functional reserve.

3.3. Adventitial Remodeling: Elastin and Collagen

Stenmark et al. (43) took newborn calves to a simulated high altitude of 4,300 m and observed severe pulmonary hypertension with right-to-left shunting owing to the rapid development of suprasystemic levels of pulmonary artery pressure. There was striking medial hypertrophy and remarkable proliferation of a dense adventitial sheath that, in large vessels, was sometimes seen to exhibit neovascularization (Fig. 1). Hypoxia induced adventitial fibroblast proliferation has been related to protein kinase Cζ (11). The contribution of the extensive vaso vasorum of the adventitia has been recently investigated (12). Stem cells have been identified in these vessels which have the capacity to differentiate into both endothelial as well as smooth muscle cells and may play a critical role in the structural remodeling that is associated with hypoxia (12). Studies have shown striking synthesis of elastin in the pulmonary arteries of these neonatal calves (36). There may be a role for IGF-1, as increased expression of this growth factor stimulates elastin synthesis in cultured vascular cells (14), as does TGF-ß (25). Studies in the piglet by Allen and Haworth (1) showed that when the animal is subjected to hypoxia from birth, the fetal medial musculature does not regress, but connective tissue synthesis is not stimulated. When hypoxic exposure begins at 3 days of age, however, that is, after the fall in pulmonary vascular resistance and the regression of the fetal musculature, neosynthesis of elastin is apparent.

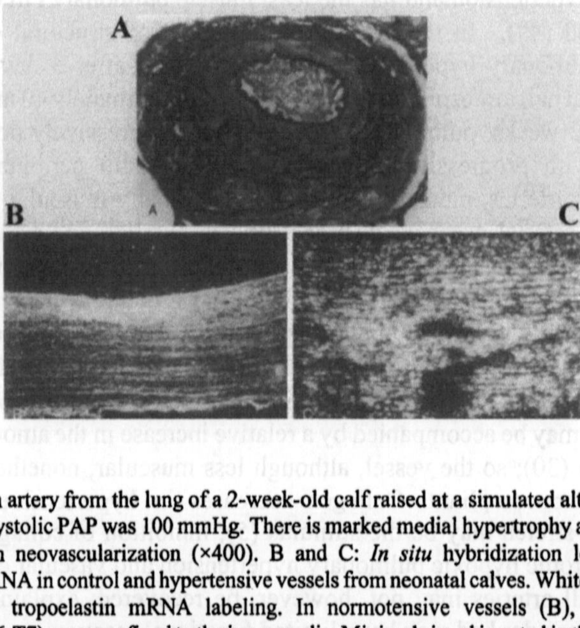

Figure 1. A: An artery from the lung of a 2-week-old calf raised at a simulated altitude of 4,300 m from birth. Systolic PAP was 100 mmHg. There is marked medial hypertrophy and adventitial thickening with neovascularization (×400). B and C: *In situ* hybridization localization of tropoelastin mRNA in control and hypertensive vessels from neonatal calves. White staining over areas indicates tropoelastin mRNA labeling. In normotensive vessels (B), labeled cells (^{35}S-labeled T66-T7) were confined to the inner media. Minimal signal is noted in the outer vessel wall. In vessels from hypertensive animals (14 days of hypoxia) (C), intense autoradiographic signal was observed throughout the media, albeit in a patchy distribution (Modified from Ref. 36).

4. Relating Acute Vasoconstriction and Vasoactive Mediators to Hypoxic Remodeling

A variety of studies attempted to show how acute vasoconstriction or a direct hypoxic 'injury' initiates the structural changes observed in the pulmonary arteries. Angiotensin II appears to be unimportant as a hypoxic vasoconstrictor but nonetheless is critical to the mechanism responsible for inducing vascular disease because of its biological properties in stimulating vascular smooth muscle cell hypertrophy and proliferation (33). Chronic hypoxia is associated with an increase in angiotensin I and II receptors (9). On the other hand, we showed that chronic infusion of angiotensin II by releasing prostacyclin (PGI_2) is actually protective of the pulmonary hypertension and vascular changes that develop during chronic hypoxia (40). Fetal and neonatal lamb pulmonary artery smooth muscle cells in culture show a decrease in the production of PGI_2 in response to acute hypoxia (38), as do bovine endothelial cells (27).

Although a variety of vasoactive blockers will lower pressure, endothelin (ET) receptor blockade appears to be most promising as a selective strategy in reversing acute hypoxic pulmonary vasoconstriction (HPV) as well as the remodeling associated with chronic hypoxia. For example, an ET-A receptor blocker has proven most effective in reducing the structural abnormalities associated with hypoxia-induced pulmonary hypertension in piglets (2). These studies have provided convincing evidence that the high ET levels are causally related to HPV. In addition, hypoxia inhibits ET-B receptor-mediated NO synthesis (NOS) (42). Impaired NOS and vasodilatation in chronic hypoxia is also associated with impaired cGMP-dependent mechanisms (6). To this end protection against the structural changes associated with chronic hypoxia has been achieved with sildenafil, related in part to the induction of natiuretic peptides (57). Agents also shown to be effective in the treatment of chronic hypoxic pulmonary hypertension in rats, include inhibitors of 5-lipoxygenase activating protein (FLAP) (51) as well as of cyclic 3'-5' GMP-specific phosphodiesterase (10) and continuous inhalation of NO (22). Recently much attention has focused on the NOS inhibitor asymetric dimethylarginine, ADMA and reduced concentrations of its metabolite dimethylarginine dimethylaminohydrolase has been observed (3). Additional strategies have been aimed at preventing acute HPV by activating K^+ channels. However, recent gene therapy studies replacing the Kv1.5 channel have restored HPV while at the same time attenuating the remodeling associated with chronic hypoxia, specifically the medial hypertrophy of the muscular arteries (35).

5. Relating Genetics to Hypoxic Pulmonary Vascular Disease

Studies in transgenic mice suggest that genetic factors might modulate the response to chronic hypoxia. For example, in the absence of hemoxygenase 1, there is reduced production of CO and its associated vasodilatory effects (54). PGI_2 synthetase overexpression is protective against the hemodynamic and vascular changes of pulmonary hypertension. Serotonin has been implicated either in the increased vasoreactivity of the Fawn hooded rat and there is attenuated severity of disease in mice lacking the serotonin transporter gene or serotonin receptors (15, 19). We also showed that increased endogenous expression of serine elastase inhibitors will effectively reduce chronic hypoxia-induced pulmonary hypertension (56) as will exogenous administration of elastase inhibitors (28).

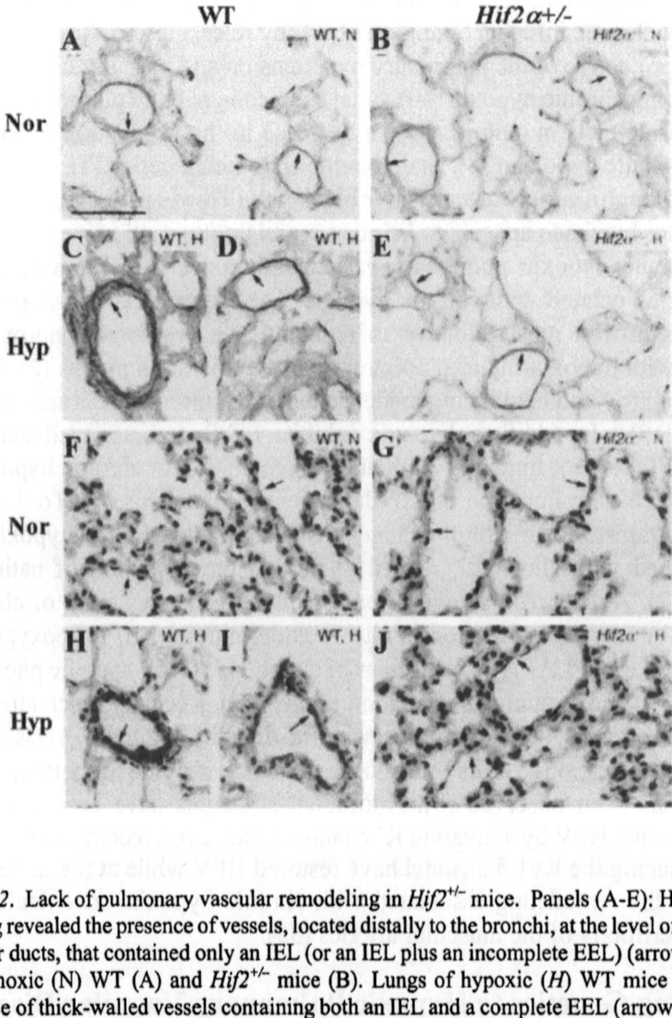

Figure 2. Lack of pulmonary vascular remodeling in *Hif2*$^{+/-}$ mice. Panels (A-E): Hart's elastin staining revealed the presence of vessels, located distally to the bronchi, at the level of alveoli and alveolar ducts, that contained only an IEL (or an IEL plus an incomplete EEL) (arrows) in lungs of normoxic (N) WT (A) and *Hif2*$^{+/-}$ mice (B). Lungs of hypoxic (*H*) WT mice showed the presence of thick-walled vessels containing both an IEL and a complete EEL (arrows; C and D), whereas no hypoxia-induced vascular remodeling occurred in *Hif2*$^{+/-}$ mice (arrows; E). F-J: SMC α-actin staining shows the presence of partially muscularized peripheral vessels (arrows) in lungs of normoxic WT (F) and *Hif2*$^{+/-}$ mice (G). Chronic hypoxia caused pulmonary vascular remodeling in WT mice, as revealed by the presence of fully muscularized vessels (arrows; H and I), but not in *Hif2*$^{+/-}$ mice (arrows, J). Bar = 50 μm in all panels (Modified from Ref. 7).

Both vascular endothelial growth factor (VEGF)-A and VEGF-B overexpressing mice show attenuation of chronic hypoxic pulmonary vascular disease (26). The protective mechanism of VEGF-B is unknown. Paradoxically, VEGF-B deficient mice, also show attenuation of chronic hypoxic pulmonary hypertension and pulmonary medial thickening (52). In VEGF-A overexpressing mice the protective mechanism is related to induction of nitric oxide synthase

(NOS). This is supported by the fact that in the NOS deficient mouse alveolar and associated vascular growth is impaired in hypoxia (4). Consistent with this, blockade of the VEGF receptor in newborn rat pups causes alveolar deficiency and pulmonary artery hypertrophy and pulmonary hypertension (24). In fact the combination of VEGF receptor blockade and hypoxia causes a very severe form of pulmonary hypertension and associated vascular changes (e.g., obliteration of small vessels in which there is evidence of intravascular endothelial cells) (44). The mechanism is thought to be related to apoptosis of endothelial cells and then the emergence of a resistant population of cells. In fact a caspase inhibitor which protects against apoptosis as well as with a bradykinin antagonist prevented these obliterative lesions whereas the pulmonary arterial medial hypertrophy persisted.

Carbon monoxide aggravates the effects of hypobaric hypoxia as reflected by an increase in the number of muscularized peripheral pulmonary arteries and in a more severe increase in pulmonary vascular resistance. Overexpression of hemoxygenase-1 protects mice against the inflammation and structural remodeling of chronic hypoxia (31). Heterozygous deletion of the transcription factor, hypoxia-inducible factor (HIF)-1α, is associated with delayed development of polycythemia, right ventricular hypertrophy and pulmonary hypertension in chronically hypoxic rats (55). More recently the heterozygous deletion of the homologue, HIF-2α, was shown to be associated with protection against pulmonary hypertension in association with suppression in endothelin-1 and plasma catecholamine levels (Fig. 2) (7).

6. Growth Factors and Hypoxia-Induced Structural Remodeling of Pulmonary Arteries

Studies by Thompson et al. (45) focused on the efficacy of growth inhibitors in preventing vascular disease. Heparin infusion decreases the severity of hypoxia-induced vascular changes, presumably by decreasing smooth muscle hyperplasia. Sulfonate groups in the heparin molecule have been shown to play a critical role in this growth inhibitory effect (16) which is influenced by the Na^+/H^+ antiporter (37). Other studies have also shown that calcitonin gene related peptide administered by gene transfer not only attenuates HPV and remodeling but enhances the effects of phosphodiesterase inhibitors (8).

7. Proteolytic Activity and Hypoxic Pulmonary Vascular Disease

While some studies have shown that increased activity of metalloproteinases is prevalent during the regression of vascular disease (48) but inhibition of metalloproteinases appears to aggravate experimental pulmonary hypertension in rats (50). Our group has focused on the specific contribution of

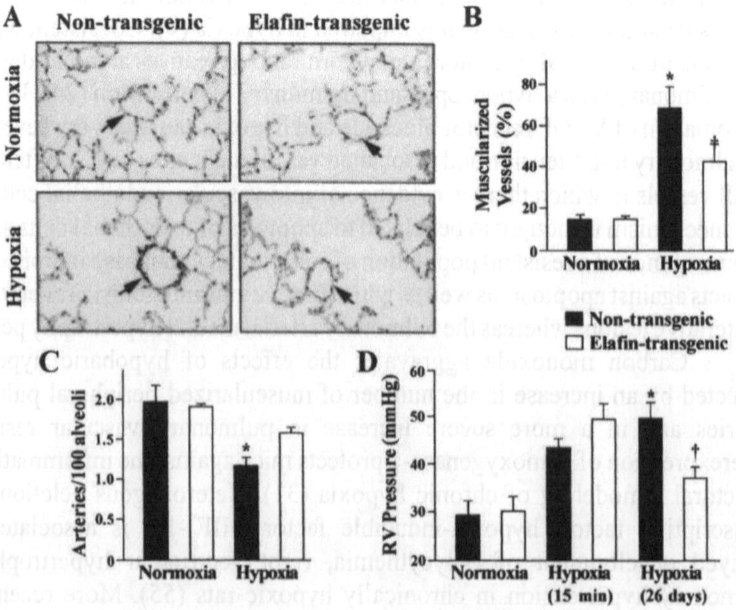

Figure 3. A and B: Muscularization of distal pulmonary arteries in mice hypoxic for 26 days. Lung sections were immunostained for SMC α-actin before hematoxylin staining. A: Fully muscularized artery associated with alveolar duct of a nontransgenic hypoxic mouse. Arrows denote alveolar duct-associated arteries. B: Morphometric analyses of fully plus partially muscularized pulmonary arteries (15-50 μm external diameter) associated with alveolar ducts. **P*<0.004 vs normoxia. †*P*=0.035 vs nontransgenic hypoxia. C: Loss of distal pulmonary arteries in mice subjected to chronic hypoxia for 26 days. Lung sections were stained with van Gieson elastin stain. Mean±SE of arteries per 100 alveoli are presented. **P*<0.02 vs normoxia. †*P*<0.02 vs nontransgenic hypoxia. D: Increase in RV pressure in mice subjected to acute (10% O_2 for 15 min) or chronic hypoxia (26 days). Mean±SE. **P*<0.01 vs normoxic, nontransgenic acute hypoxia. †*P*<0.004 vs nontransgenic chronic hypoxia (Modified from Ref. 56).

serine elastase activity to the progression and regression of pulmonary vascular disease. We showed that there is heightened activity of a serine elastase 2 days after exposing rats to chronic hypobaric hypoxia and that inhibition of elastase by infusion of elastase inhibitors will greatly ameliorate the severity of pulmonary hypertension as well as the associated structural abnormalities which include extension of muscle into normally non-muscular peripheral arteries, medial hypertrophy of muscular arteries and reduction in the number of small peripheral arteries (28). Similar repression of disease is observed in a transgenic mouse in which there overexpression of the naturally occurring serine elastase inhibitor, elafin, is targeted to the vasculature under the regulation of the preproendothelin promoter (Fig. 3) (56). In unpublished studies we showed that regression of hypoxia-induced vascular disease in rats, which occurs upon return to room air, is accompanied by smooth muscle cell apoptosis. We reported that induction of smooth muscle cell apoptosis via elastase inhibition can reverse the fatal form of pulmonary hypertension induced by monocrotaline.

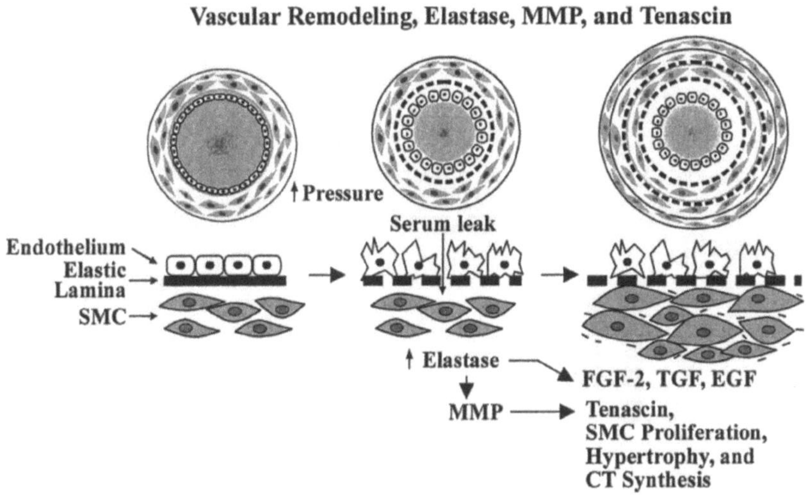

Figure 4. Schema showing how an elastase can induce changes in matrix proteins resulting in smooth muscle cell proliferation and migration.

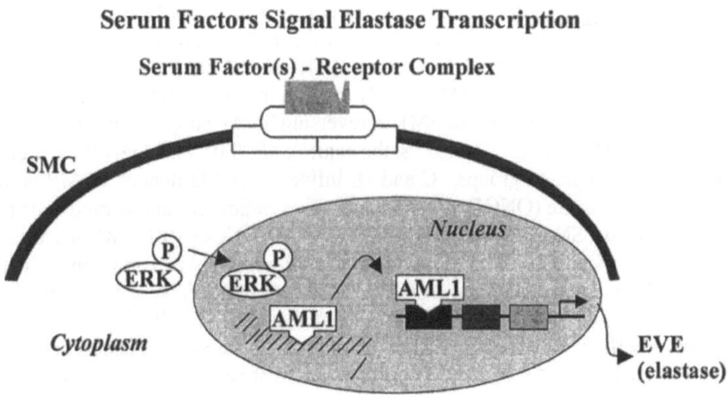

Figure 5. Schema showing how serum factors signal elastin gene transcription. Serum factors including apoA1 induce MAP kinase activity, leading to nuclear translocation of AML1, and increase in elastin gene transcription.

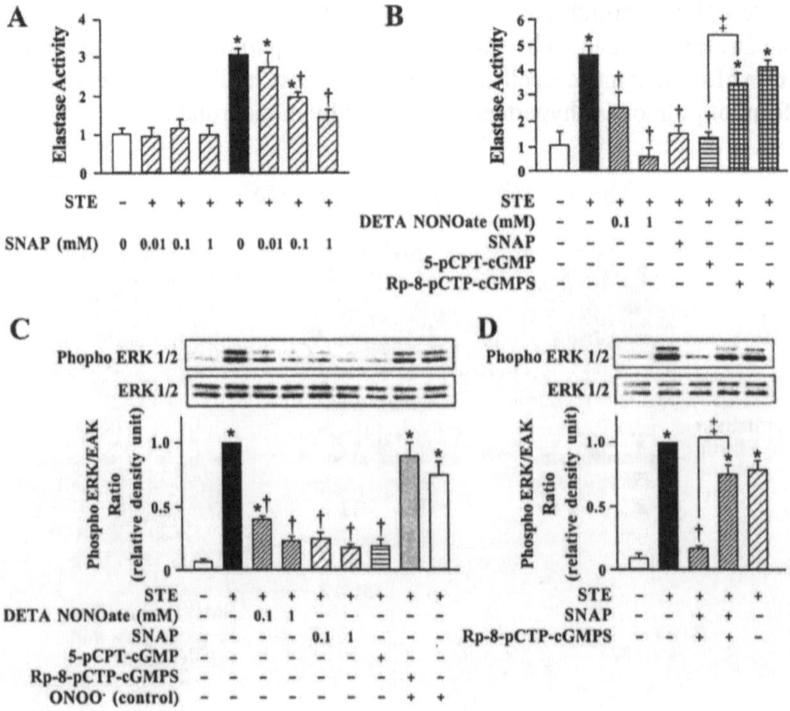

Figure 6. Influence of NO donors and a cGMP mimetic and inhibitor on SMC elastase activity. Elastase activity was evaluated by measured solubilized [^{3}H]-elastin in the culture medium after 24 hrs of incubation. A: Serum-starved SMCs were maintained serum free or incubated with serum-treated elastin (STE) after pretreatment with the NO donor SNAP (0.1-1 mM) for 30 min. B: Comparison of SMC elastase activity under control conditions or stimulated by serum-treated elastin (STE) after pretreatment with NO donors DETA NONOate (0.1-1 mM), SNAP (1 mM), and the cGMP mimetic 8-pCPT-cGMP (1 mM) for 30 min. The effect of pretreatment with the PKG inhibitor Rp-8-pCPT-cGMP (20 μM) on reversing SNAP suppression of elastase activity was also evaluated. Means±SE. **P*<0.05 vs. the control; †*P*<0.05 vs. STE with no pretreatment; ‡*P*<0.05 between designated groups. C and D: Influence of NO donors, cGMP mimetic and inhibitor, and peroxynitrite (ONOO$^-$) on ERK phosphorylation. Serum-starved cells pretreated with DETA NONOate, SNAP (0.1-1 mM), 8-pCPT-cGMP, ONOO$^-$ (0.1 mM), and an inactivated negative control for ONOO$^-$ (0.1 mM) (C) or retreated with SNAP (1 mM) in the presence or absence of coadministration with Rp-8-pCPT-cGMP (20 μM) (D) were stimulated by STE or control elastin for 5 min. Cell lysates (10 μg) were analyzed by SDS-PAGE and immunoblotted with phospho-specific or -nonspecific ERK antibodies. Means ± SE (n=3 experiments). **P*< 0.05 vs. the control elastin; †*P*< 0.05 vs. STE with no pretreatment; ‡*P*<0.05 between designated groups (Modified from Ref. 32).

In our laboratory we are currently investigating a transgenic mouse in which the response to chronic hypoxia produces more severe pulmonary hypertension that does not regress upon return to room air. The most striking feature is the failure of new growth of peripheral arteries. The mechanisms involved are being by correlating this phenotype with the genotype as determined by a micro-array approach.

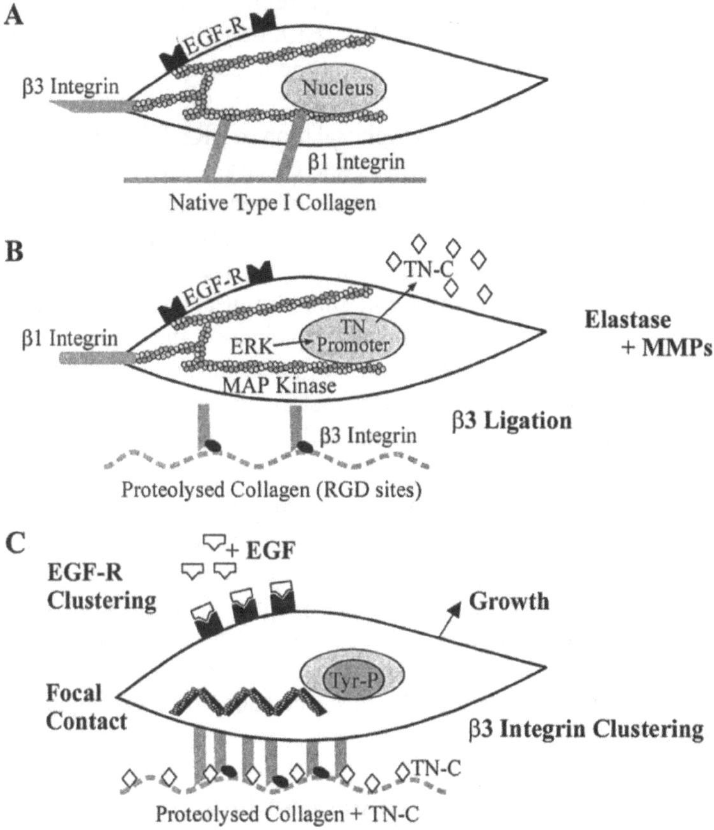

Figure 7. Hypothetical model for the regulation and function of Tenascin-C (TN-C) in vascular SMC. A: Vascular SMC attach and spread over native type-I collagen using β1 integrins. Under serum free conditions, the cells withdraw from the cell cycle and become quiescent. B: Degradation of native type I collagen by matrix metalloproteinases (MMPs) leads to exposure of cryptic RGD sites that preferentially bind β3 subunit-containing integrins. In turn, occupancy and activation of β3 integrins signals the production of TN-C. C: Incorporation of multivalent TN-C protein into the underlying substrate leads to further aggregation and activation of β3-containing integrins (αβ3), and to the accumulation of tyrosine-phosphorylated (Tyr-P) signaling molecules and actin into a focal adhesion complex. Even in the absence of the EGF ligand, the TN-C-dependent reorganization of the cytoskeleton leads to clustering of actin-associated EGF-Rs.

Cell culture studies in our laboratory have been carried out to explain how elastase might be regulated in pulmonary artery smooth muscle cells and how the activity of this enzyme may lead to pulmonary vascular disease (Fig 4).

We have shown that plasma factors that could be present in high concentration if the subendothelium when here is endothelial perturbation, can induce production of the elastase enzyme by smooth muscle cells. The mechanism requires activity of mitogen activated protein (MAP) kinases specifically extracellular regulated kinase 1 and 2 (ERK1/2) (20,21,46). This leads to increased expression and DNA binding of the transcription factor AML1 (32) which is required for the activity of elastase (Fig. 5) (53). Studies by our group showed that NO donors repress phosphorylation of ERK1/2 and the consequent expression and DNA binding of AML1 and elastase activity (Fig. 6) (32). This suggests a mechanism whereby NO may influence not only the vasoactive response to chronic hypoxia but also the structural remodeling. Increased production of elastase causes release of growth factors such as FGF-2 in a mitogenically active form resulting in smooth muscle cell proliferation (47). These growth factors can activate metalloproteinases, and augmentation of the proteolytic cascade leads to upregulation of tenascin C which in turn amplifies the proliferative response to growth factors by inducing changes in the cytoskeleton which lead to the clustering and facilitate transactivation of growth factor receptors (Fig. 7).

We reason that breakdown of elastin and other extracellular matrix protein by elastase could in this way cause abnormal proliferation of medial smooth muscle cells causing medial hypertrophy and of pericytes causing muscularization of abnormally muscular peripheral arteries. Loss of small arteries may also be a function of breakdown of the extracellular matrix and basement membrane of small vessels causing endothelial cell apoptosis. Inhibition of elastase causes smooth muscle cell apoptosis leading to regression of medial hypertrophy.

8. New Directions in Understanding Hypoxia Induced Structural Remodeling

The current use of transgenic mice with specific genes overexpressed or deleted will continue to provide new insights into the pathogenesis of hypoxia induced pulmonary vascular disease. Microarray technology as well as proteomic approaches should reveal genes and gene products that are associated with the phenotypes observed (18). Cell biology pathways will identify novel pathways to intervene in reversing structural abnormalities.

References

1. Allen K and Haworth SG. Impaired adaptation of intrapulmonary arteries to intrauterine life in the newborn pig exposed to hypoxia: an ultrastructural study. *Fed. Proc.* 1986; 45:879a.
2. Ambalavanan N, Philips JB, 3rd, Bulger A, Oparil S, and Chen YF. Endothelin-A receptor

blockade in porcine pulmonary hypertension. *Pediatr. Res.* 2002; 52:913-921.
3. Arrigoni FI, Vallance P, Haworth SG, and Leiper JM. Metabolism of asymmetric dimethylarginines is regulated in the lung developmentally and with pulmonary hypertension induced by hypobaric hypoxia. *Circulation* 2003; 107:1195-1201.
4. Balasubramaniam V, Tang JR, Maxey A, Plopper CG, and Abman SH. Mild hypoxia impairs alveolarization in the endothelial nitric oxide synthase-deficient mouse. *Am. J. Physiol. Lung Cell. Mol. Physiol.* 2003; 284:L964-971.
5. Belik J, Keeley FW, Baldwin F, and Rabinovitch M. Pulmonary hypertension and vascular remodeling in fetal sheep. *Am. J. Physiol.* 1994; 266:H2303-H2309.
6. Berkenbosch JW, Baribeau J, and Perrault T. Decreased synthesis and vasodilation to nitric oxide in piglets with hypoxia-induced pulmonary hypertension. *Am. J. Physiol. Lung Cell. Mol. Physiol.* 2000; 278: L276-L283.
7. Brusselmans K, Compernolle V, Tjwa M, Wiesener MS, Maxwell PH, Collen D, and Carmeliet P. Heterozygous deficiency of hypoxia-inducible factor-2alpha protects mice against pulmonary hypertension and right ventricular dysfunction during prolonged hypoxia. *J. Clin. Invest.* 2003; 111: 1519-1527.
8. Champion HC, Bivalacqua TJ, Toyoda K, Heistad DD, Hyman AL, and Kadowitz PJ. In vivo gene transfer of prepro-calcitonin gene-related peptide to the lung attenuates chronic hypoxia-induced pulmonary hypertension in the mouse. *Circulation* 1000; 101:923-930.
9. Chassagne C, Eddahibi S, Adamy C, Rideau D, Marotte F, Dubois-Rande JL, Adnot S, Samuel JL, and Teiger E. Modulation of angiotensin II receptor expression during development and regression of hypoxic pulmonary hypertension. *Am. J. Respir. Cell Mol. Biol.* 2000; 22:323-332.
10. Cohen AH, Hanson K, Morris K, Fouty B, McMurty IF, Clarke W, and Rodman DM. Inhibition of cyclic 3'-5'-guanosine monophosphate-specific phosphodiesterase selectively vasodilates the pulmonary circulation in chronically hypoxic rats. *J. Clin. Invest.* 1996; 97:172-179.
11. Das M, Dempsey EC, Bouchey D, Reyland ME, and Stenmark KR. Chronic hypoxia induces exaggerated growth responses in pulmonary artery adventitial fibroblasts: Potential contribution of specific protein kinase c isozymes. *Am. J. Respir. Cell Mol. Biol.* 2000; 22: 15-25.
12. Davie NJ, Crossno JT, Frid MG, Hofmeister SE, Reeves JT, Hyde DM, Carpenter TC, Brunetti JA, McNiece IK, and Stenmark KR. Hypoxia-induced pulmonary artery adventitial remodeling and neovascularization: potential contribution of circulating progenitor cells (R1). *Am. J. Physiol. Lung. Cell. Mol. Physiol.* 2003 (in press).
13. Davies G and Reid L. Effect of scoliosis on growth of alveoli and pulmonary arteries and on right ventricle. *Arch. Dis. Child.* 1971; 46: 623-632.
14. Dempsey EC, Stenmark KR, McMurtry IF, O'Brien RF, Voelkel NF, and Badesch DB. Insulin-like growth factor I and protein kinase C activation stimulate pulmonary artery smooth muscle cell proliferation through separate but synergistic pathways. *J. Cell. Physiol.* 1990; 144: 159-165.
15. Eddahibi S, Humbert M, Fadel E, Raffestin B, Darmon M, Capron F, Simmoneau G, Dartevelle P, Hamon M, and Adnot S. Serotonin transporter overexpression is responsible for pulmonary artery smooth muscle hyperplasia in primary pulmonary hypertension. *J. Clin. Invest.* 2001; 108: 1141-1150.
16. Garg HG, Thompson BT, and Hales CA. Structural determinants of antiproliferative activity of heparin on pulmonary artery smooth muscle cells. *Am. J. Physiol. Lung Cell. Mol. Physiol.* 2000; 279: L779-789.
17. Grover RF, Vogel JH, Voigt GC, and Blount SGJ. Reversal of high altitude pulmonary hypertension. *Am. J. Cardiol.* 1966; 18: 928-932.
18. Hoshikawa Y, Nana-Sinkam P, Moore MD, Sotto-Santiago S, Phang T, Keith RL, Morris KG, Kondo T, Tuder RM, Voelkel NF, and Geraci MW. Hypoxia induces different genes in the lungs of rats compared with mice. *Physiol. Genomics* 2003; 12: 209-219.

19. Keegan A, Morecroft I, Smillie D, Hicks MN, and MacLean MR. Contribution of the 5-HT_{1B} receptor to hypoxia-induced pulmonary hypertension: converging evidence using 5-HT_{1B}-receptor knockout mice and the 5-$HT_{1B/1D}$-receptor antagonist GR127935. *Circ. Res.* 2001; 89: 1231-1239.
20. Kobayashi J and Rabinovitch M. Elastin-bound serum factor and endothelial cell factor induce pulmolnary artery smooth muscle cell elastolytic activity. *Circulation* 1994; 90(4, part II): I-417.
21. Kobayashi J, Wigle D, Childs T, Zhu L, Keeley FW, and Rabinovitch M. Serum-induced vascular smooth muscle cell elastolytic activity through tyrosine kinase intracellular signalling. *J. Cell. Physiol.* 1994; 160: 121-131.
22. Kouyoumdjian C, Adnot S, Levame M, Eddahibi S, Bousbaa H, and Raffestin B. Continuous inhalation of nitric oxide protects against development of pulmonary hypertension in chronically hypoxic rats. *J. Clin. Invest.* 1994; 94: 578-584.
23. Launay JM, Herve P, Peoc'h K, Tournois C, Callebert J, Nebigil CG, Etienne N, Drouet L, Humbert M, Simonneau G, and Maroteaux L. Function of the serotonin 5-hydroxytryptamine 2B receptor in pulmonary hypertension. *Nat. Med.* 2002; 8: 1129-1135.
24. Le Cras TD, Markham NE, Tuder RM, Voelkel NF, and Abman SH. Treatment of newborn rats with a VEGF receptor inhibitor causes pulmonary hypertension and abnormal lung structure. *Am. J. Physiol. Lung Cell. Mol. Physiol.* 2002; 283: L555-562.
25. Liu JM and Davidson JM. The elastogenic effect of recombinant transforming growth factor-beta on porcine aortic smooth muscle cells. *Biochem. Biophys. Res. Commun.* 1988; 154: 895-901.
26. Louzier V, Raffestin B, Leroux A, Branellec D, Caillaud JM, Levame M, Eddahibi S, and Adnot S. Role of VEGF-B in the lung during development of chronic hypoxic pulmonary hypertension. *Am. J. Physiol. Lung Cell. Mol. Physiol.* 2003; 284: L926-937.
27. Madden MC, Vender RL, and Friedman M. Effect of hypoxia on prostacyclin production in cultured pulmonary artery endothelium. *Prostaglandins* 1986; 31: 1049-1062.
28. Maruyama K, Ye CL, Woo M, Venkatacharya H, Lines LD, Silver MM, and Rabinovitch M. Chronic hypoxic pulmonary hypertension in rats and increased elastolytic activity. *Am. J. Physiol.* 1991; 261: H1716-1726.
29. McMurtry IF, Frith CH, and Will DH. Cardiopulmonary responses of male and female swine to simulated high altitude. *J. Appl. Physiol.* 1973; 35: 459-462.
30. Meyrick B and Reid L. Endothelial and subintimal changes in rat hilar pulmonary artery during recovery from hypoxia. *Lab. Invest.* 1980; 42: 603-615.
31. Minamino T, Christou H, Hsieh CM, Liu Y, Dhawan V, Abraham NG, Perrella MA, Mitsialis SA, and Kourembanas S. Targeted expression of heme oxygenase-1 prevents the pulmonary inflammatory and vascular responses to hypoxia. *Proc. Natl. Acad. Sci. U.S.A.* 2001; 98: 8798-8803.
32. Mitani Y, Zaidi SHE, Dufourcq P, Thompson K, and Rabinovitch M. Nitric oxide reduces vascular smooth muscle cell elastase activity through cGMP-mediated suppression of ERK phosphorylation and AML1B nuclear partitioning. *FASEB J.* 2000; 14: 805-814.
33. Morrell NW, Morris KG, and Stenmark KR. Role of angiotensin converting enzyme and angiotensin II in the development of hypoxic pulmonary hypertension. *Am. J. Physiol.* 1995; 269: H1186-H1194.
34. Poiani GJ, Tozzi CA, Choe JK, Yohn SE, and Riley DJ. An antifibrotic agent reduces blood pressure in established pulmonary hypertension in the rat. *J. Appl. Physiol.* 1990; 68: 1542-1546.
35. Pozeg ZI, Michelakis ED, McMurtry MS, Thebaud B, Wu XC, Dyck JR, Hashimoto K, Wang S, Moudgil R, Harry G, Sultanian R, Koshal A, and Archer SL. In vivo gene transfer of the O_2-sensitive potassium channel Kv1.5 reduces pulmonary hypertension and restores hypoxic pulmonary vasoconstriction in chronically hypoxic rats. *Circulation* 2003; 107: 2037-2044.
36. Prosser IW, Stenmark KR, Suthar M, Crouch EC, Mecham RP, and Parks WC. Regional heterogeneity of elastin and collagen gene expression in intralobar arteries in response to

hypoxic pulmonary hypertension as demonstrated by *in situ* hybridization. *Am. J. Pathol.* 1989; 135: 1073-1088.

37. Quinn DA, Dahlberg CG, Bonventre JP, Scheid CR, Honeyman T, Joseph PM, Thompson BT, and Hales CA. The role of Na^+/H^+ exchange and growth factors in pulmonary artery smooth muscle cell proliferation. *Am. J. Respir. Cell Mol. Biol.* 1996; 14: 139-145.
38. Rabinovitch M, Boudreau N, Vella G, Coceani F, and Olley PM. Oxygen-related prostaglandin synthesis in ductus arteriosus and other vascular cells. *Pediatr. Res.* 1989; 26: 330-335.
39. Rabinovitch M, Gamble WJ, Miettinen OS, and Reid L. Age and sex influence on pulmonary hypertension of chronic hypoxia on recovery. *Am. J. Physiol.* 1981; 240: H62-H72.
40. Rabinovitch M, Mullen M, Rosenberg H, Maruyama K, O'Brodovich H, and Olley P. Angiotensin II prevents hypoxic pulmonary hypertension and vascular changes in rats. *Am. J. Physiol.* 1988; 254: H500-H508.
41. Ryland D and Reid L. The pulmonary circulation in cystic fibrosis. *Thorax* 1975; 30: 285-308.
42. Sato K, Rodman DM, and McMurtry IF. Hypoxia inhibits increased ETB receptor-mediated NO synthesis in hypertensive rat lungs. *Am. J. Physiol.* 1999; 276: L571-581.
43. Stenmark KR, Fasules J, Hyde DM, Voelkel NF, Henson J, Tucker A, Wilson H, and Reeves JT. Severe pulmonary hypertension and arterial adventitial changes in newborn calves at 4,300 m. *J. Appl. Physiol.* 1987; 62: 821-830.
44. Tarasevicienе-Stewart L, Gera L, Hirth P, Voelkel NF, Tuder RM, and Stewart JM. A bradykinin antagonist and a caspase inhibitor prevent severe pulmonary hypertension in a rat model. *Can. J. Physiol. Pharmacol.* 2002; 80: 269-274.
45. Thompson BT, Spence CR, Janssens SP, Joseph PM, and Hales CA. Inhibition of hypoxic pulmonary hypertension by heparins of differing in vitro antiproliferative potency. *Am. J. Respir. Crit. Care. Med.* 1994; 149: 1512-1517.
46. Thompson K, Kobayashi J, Childs T, Wigle D, and Rabinovitch M. Endothelial and serum factors which include apolipoprotein A1 tether elastin to smooth muscle cells inducing serine elastase activity via tyrosine kinase-mediated transcription and translation. *J. Cell Physiol.* 1998; 174: 78-89.
47. Thompson K and Rabinovitch M. Exogenous leukocyte and endogenous elastases can mediate mitogenic activity in pulmonary artery smooth muscle cells by release of extracellular-matrix bound basic fibroblast growth factor. *J. Cell Physiol.* 1996; 166: 495-505.
48. Tozzi CA, Wilson FJ, Yu SY, and Riley DJ. Vascular connective tissue is rapidly degraded during early regression of pulmonary hypertension. *Chest* 1991; 99: 41S-42S.
49. Tucker A, McMurtry IF, Reeves JT, Alexander AF, Will DH, and Grover RF. Lung vascular smooth muscle as a determinant of pulmonary hypertension at high altitude. *Am. J. Physiol.* 1975; 228: 762-767.
50. Vieillard-Baron A, Frisdal E, Eddahibi S, Deprez I, Baker AH, Newby AC, Berger P, Levame M, Raffestin B, Adnot S, and d'Ortho MP. Inhibition of matrix metalloproteinases by lung TIMP-1 gene transfer or doxycycline aggravates pulmonary hypertension in rats. *Circ. Res.* 2000; 87: 418-425.
51. Voelkel NF, Tuder RM, Wade K, Hoper M, Lepley RA, Goulet JL, Koller BH, and Fitzpatrick F. Inhibition of 5-lipoxygenase-activating protein (FLAP) reduces pulmonary vascular reactivity and pulmonary hypertension in hypoxic rats. *J. Clin. Invest.* 1996; 97: 2491-2498.
52. Wanstall JC, Gambino A, Jeffery TK, Cahill MM, Bellomo D, Hayward NK, and Kay GF. Vascular endothelial growth factor-B-deficient mice show impaired development of hypoxic pulmonary hypertension. *Cardiovasc. Res.* 2002; 55: 361-368.
53. Wigle DA, Thompson KE, Yablonsky S, Zaidi SHE, Coulber C, Jones PL, and Rabinovitch M. AML1-like transcription factor induces serine elastase activity in ovine pulmonary artery smooth muscle cells. *Circ. Res.* 1998; 83(3):252-263.
54. Yet SF, Perrella MA, Layne MD, Hsieh CM, Maemura K, Kobzik L, Wiesel P, Christou H, Kourembanas S, and Lee ME. Hypoxia induces severe right ventricular dilatation and

infarction in heme oxygenase-1 null mice. *J. Clin. Invest.* 1999; 103: R23-R29.

55. Yu A, Shimoda LA, Iyer NV, Huso DL, Sun X, McWilliams R, Beaty T, Sham JS, Wiener CM, and Sylvester JT. Impaired physiological responses to chronic hypoxia in mice partially deficient for hypoxia-inducible factor-1-alpha. *J. Clin. Invest.* 1999; 103: 691-696.
56. Zaidi SHE, You X-M, Ciura S, Husain M, and Rabinovitch M. Overexpression of the serine elastase inhibitor elafin protects transgenic mice from hypoxic pulmonary hypertension. *Circulation* 2002; 105: 516-521.
57. Zhao L, Mason NA, Strange JW, Walker H, and Wilkins MR. Beneficial effects of phosphodiesterase 5 inhibition in pulmonary hypertension are influenced by natriuretic Peptide activity. *Circulation* 2003; 107: 234-237.

Chapter 24

Rho/Rho-kinase Signaling in Hypoxic Pulmonary Hypertension

Ivan F. McMurtry, Natalie R. Bauer, Sarah A. Gebb, Karen A. Fagan, Tetsutaro Nagaoka, Masahiko Oka, and Tom P. Robertson
University of Colorado Health Sciences Center, Denver, Colorado and University of Georgia, Athens, Georgia, U.S.A.

1. Introduction

Hypoxic pulmonary hypertension (HPH) contributes to the morbidity and mortality of adult and pediatric patients with various lung and heart diseases. The pathogenesis of HPH comprises sustained vasoconstriction, and structural remodeling of the pulmonary arteries that is characterized by medial and adventitial thickening of muscular arteries and muscularization of the normally non-muscular arterioles (28). The vasoconstriction involves direct hypoxic activation of vascular smooth muscle and endothelial cells, increased activity of vasoconstrictors such as endothelin-1 (ET-1) and serotonin (5-HT), and deficient activity of vasodilators such as nitric oxide (NO) and prostacyclin (PGI_2). The hypoxic activation and mediator imbalance are also implicated in the complex pathobiology of pulmonary arterial wall thickening that includes medial smooth muscle and adventitial fibroblast cell growth and migration, and deposition of extracellular matrix proteins.

The role of intracellular signaling via the small GTPase RhoA and its downstream effector Rho-associated kinase (Rho/Rho-kinase signaling) in the pathogenesis of HPH is unknown. However, recent advances in the cell biology and pathophysiology of this signal transduction pathway in the systemic circulation suggest that it may play important roles in both the sustained vasoconstriction and arterial remodeling of HPH. The importance of Rho/Rho-kinase-mediated Ca^{2+} sensitization of vascular smooth muscle cell (VSMC) contraction in acute hypoxic pulmonary vasoconstriction (HPV) is reviewed in Chapter 7 of this book. The objective here is to provide an overview of Rho/Rho-kinase signaling and to detail the emerging evidence that it is also involved in the pulmonary vascular response to chronic hypoxia (Fig. 1).

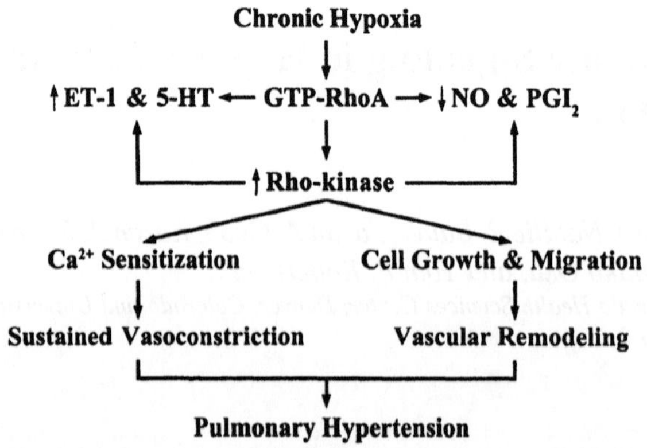

Figure 1. Overview of the postulated role of Rho/Rho-kinase signaling in the pathogenesis of HPH. Chronic hypoxia and/or the associated increased activity of vasoconstrictors such as ET-1 and 5-HT, and decreased activity of vasodilators such as NO and PGI_2, activate RhoA which then stimulates Rho-kinase. Rho-kinase contributes to pulmonary hypertension by: mediating Ca^{2+} sensitization of arterial smooth muscle cell contraction and sustained pulmonary vasoconstriction, promoting vascular smooth muscle and fibroblast cell growth and migration and vascular remodeling, and regulating the expression of genes involved in increased activity of vasoconstrictors and deficient production of vasodilators.

2. Rho/Rho-kinase Signaling

RhoA GTPase (RhoA) is a member of the Rho (Ras homologous) family of small GTP-binding proteins that includes Rac, Cdc42, and numerous other members (3, 51, 54). These GTPases transduce signals from extracellular stimuli to intracellular target proteins (effectors) that regulate several complex cellular processes, including reorganization of the actin cytoskeleton, cell contraction, adhesion, migration, gene expression, and growth. RhoA cycles between an inactive GDP-bound and active GTP-bound form that is targeted to cell membranes where it stimulates its downstream effectors. RhoA activity is regulated, in ways that are not yet well understood, by three classes of proteins. It is inhibited by guanosine nucleotide disassociation inhibitors (e.g., Rho GDIα), activated by guanosine nucleotide exchange factors (e.g., p115-Rho, PDZ-Rho, and LARG GEFs), and inactivated by GTPase-activating proteins (e.g., Graf and p190-Rho GAP). Many cellular responses to activation of G protein-coupled receptors (GPCR) are mediated via $G\alpha_{12/13}$, $G\alpha_q$, or Gq subunit-dependent stimulation of Rho GEFs and activation of RhoA (51). GPCR-independent signaling to RhoA also occurs and can be mediated by receptor and non-receptor tyrosine kinases and by recruitment of other GTPases (51, 54). Several vasoconstrictors, including ET-1, 5-HT, and thromboxane, activate RhoA in VSMC (19, 52). There is indirect evidence that hypoxia also activates RhoA in VSMC (50, 66) and endothelial cells (63).

Rho-kinase, a serine/threonine protein kinase, is a well-characterized RhoA effector (3, 51, 54). There are two isoforms of Rho-kinase, Rho-associated coiled coil-forming protein kinase (ROCK) I and II (also named ROKβ and ROKα, respectively), with few known functional differences between them. Other RhoA effectors include mDia1/2, protein kinase N (PKN) (also called PKC-related kinase 1, PRK1), PRK2, PIP5-kinase, citron kinase, rhophilin, rhotekin, phospholipase D, and Kv1.2. Many targets of Rho-kinase have been identified, including the myosin-binding subunit (MBS) of myosin light chain (MLC) phosphatase, myosin phosphatase inhibitor protein CPI-17, MLC, calponin, LIM-kinase, adducin, ERM family proteins, and intermediate filaments.

3. Rho/Rho-kinase in Ca^{2+} Sensitization and Pulmonary Vasoconstriction

VSMC tone is regulated by phosphorylation (causing contraction) and dephosphorylation (causing relaxation) of the 20-kDa regulatory MLC. Phosphorylation of MLC is catalyzed by Ca^{2+}/calmodulin-dependent MLC kinase (MLCK) and dephosphorylation by Ca^{2+}-independent MLC phosphatase (MLCP) that is targeted to myosin by its regulatory MBS. Thus, the balance in activities of MLCK and MLCP regulates contraction. At a given level of cytosolic Ca^{2+}, the activity of both enzymes can be modulated by second messenger-mediated pathways to change MLC phosphorylation and force, i.e., to change the Ca^{2+} sensitivity of contraction. There are multiple mechanisms of Ca^{2+} sensitization, but two major pathways lead to inhibition of MLCP and increased phosphorylation of MLC: one by Rho-kinase-mediated phosphorylation of the MBS, and the other by PKC-mediated phosphorylation and activation of the 17-kDa MLCP-inhibitor protein CPI-17 (61). CPI-17 is also phosphorylated by Rho-kinase, and recent studies suggest this pathway is a more physiologically significant mechanism of Ca^{2+} sensitization of vasoconstriction than is the Rho-kinase-mediated phosphorylation of MBS (43). The opposing roles of MLCK and MLCP in regulation of MLC phosphorylation and contraction, and the pathway for promotion of contraction by Rho-kinase-mediated inhibition of MLCP, are illustrated in Figure 2.

Because Ca^{2+} sensitization can augment vasoconstriction, it follows that Ca^{2+} *desensitization* can cause vasodilation. In addition to inducing VSMC relaxation by decreasing cytosolic [Ca^{2+}] and MLCK activity, the NO → soluble guanylate cyclase → cGMP → cGMP-dependent protein kinase (cGK) pathway also decreases Ca^{2+} sensitivity (4). cGMP-induced Ca^{2+} desensitization has been attributed to cGK-mediated phosphorylation and inhibition of RhoA, and a resultant increase in MLCP activity. However, NO/cGMP-induced vasorelaxation appears to involve only a transient increase in MLCP activity, and sustained relaxation may be due to other cGMP-mediated mechanisms that are not related to dephosphorylation of MLC, such as phosphorylation of telokin or HSP20 (13). Vasodilation by cAMP similarly involves diverse mechanisms (4), including inhibition of Rho/Rho-kinase signaling and increased activity of MLCP (51, 54).

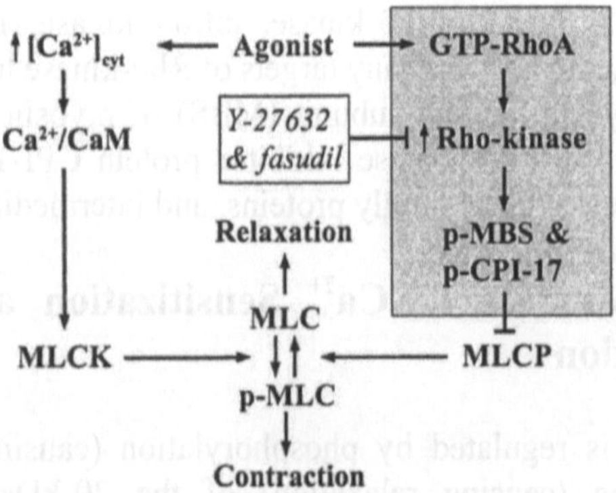

Figure 2. Diagram of the relationship between Ca^{2+} signaling and Rho/Rho-kinase-mediated Ca^{2+} sensitization in the regulation of VSMC tone. It is generally believed that increased cytosolic [Ca^{2+}] and Ca^{2+}/calmodulin (CaM)-dependent activation of MLCK initiate MLC phosphorylation and contraction. However, it is becoming increasingly clear that sustained contraction, even in face of decreasing cytosolic [Ca^{2+}] is due to activation of RhoA and Rho-kinase-mediated inhibition of MLCP which maintains, or even promotes, MLC phosphorylation. The shaded area illustrates the signaling pathway by which GTP-RhoA-dependent stimulation of Rho-kinase leads to phosphorylation of the myosin binding subunit (p-MBS) and/or the inhibitory protein CPI-17 (p-CPI-17) that then inhibits MLCP and promotes MLC phosphorylation and contraction. The Rho-kinase blockers Y-27632 and fasudil inhibit this pathway leading to increased MLCP activity, dephosphorylation of MLC, and relaxation.

Several agonists elicit Ca^{2+} sensitization-associated contractions in pulmonary arteries that are attenuated by the Rho-kinase inhibitor Y-27632 (8, 26, 27). As reviewed in Chapter 7, Rho/Rho-kinase-mediated Ca^{2+} sensitization also plays a major role in the mechanism of HPV. Y-27632 both prevents and reverses HPV in rat pulmonary arteries and perfused lungs, and it has been suggested that activation of Rho-kinase is essential to the sustained hypoxic response (50). Acute hypoxia causes an early and sustained phosphorylation of MLC in rat pulmonary VSMC in culture, and the sustained response is inhibited by the *Clostridium botulinum* toxin C3 exoenzyme (which ADP-ribosylates and inactivates Rho) and by Y-27632 (66). It was suggested that HPV is initiated by Ca^{2+}-induced stimulation of MLCK, and the sustained response is due to Rho/Rho-kinase-mediated inhibition of MLCP. While Rho/Rho-kinase signaling clearly regulates acute pulmonary vasoreactivity, the role of Rho/Rho-kinase-mediated Ca^{2+} sensitization in the increased vascular tone and vasoreactivity of hypoxic hypertensive lungs is uncertain. As discussed below in Sections 9 and

10, we have begun to investigate if and by what mechanisms Rho/Rho-kinase-mediated Ca^{2+} sensitization of vasoconstriction is increased in hypoxic hypertensive pulmonary arteries.

4. Rho/Rho-kinase in Vascular Smooth Muscle Cell Growth

Thrombin-induced DNA synthesis in rat aortic VSMC is associated with activation of RhoA and inhibited by C3 exoenzyme and Y-27632 (55). The activation of RhoA apparently augments the response to activated Ras by reducing the expression of the cyclin-dependent kinase (Cdk) inhibitor $p27^{Kip1}$ via activation of phosphatidylinositol 3-kinase (55). In human aortic and saphenous vein VSMC, RhoA mediates platelet derived growth factor (PDGF)-induced DNA synthesis by downregulating $p27^{Kip1}$ and stimulating Cdk-2 and the hyperphosphorylation of retinoblastoma protein (32). In this case, activation of RhoA is apparently necessary and sufficient for serum- and PDGF-induced DNA synthesis. Angiotensin II activates RhoA in rat aortic VSMC, and C3 exoenzyme and Y-27632 inhibit angiotensin II-induced protein synthesis (68). Similarly, C3 exoenzyme and Y-27632 inhibit stretch-induced activation of extracellular signal-regulated kinase (ERK) and stimulation of DNA synthesis in rat VSMC (45). Inhibition of protein geranylgeranylation, which prevents the membrane targeting and activation of RhoA, inhibits serum-induced proliferation and promotes apoptosis of rat pulmonary microvascular smooth muscle cells (62). However, the contribution of Rho/Rho-kinase signaling to growth factor/vasoconstrictor-induced proliferation, hypertrophy, and migration of pulmonary VSMC remains to be defined. Preliminary results in our laboratory indicate that Y-27632 inhibits 5-HT- but not PDGF-induced DNA synthesis in primary cultures of rat main pulmonary VSMC (unpublished data). Much remains to be learned about if and how Rho/Rho-kinase signaling interacts with numerous other signaling molecules and pathways to regulate pulmonary vascular cell growth and migration in HPH.

5. Rho/Rho-kinase in Gene Expression

Rho/Rho-kinase signaling regulates the expression of several genes relevant to control of vascular tone and remodeling. In endothelial cells, Rho/Rho-kinase signaling inhibits expression of eNOS (12, 31, 63) but upregulates that of pre-proET-1 (21). Rho/Rho-kinase also inhibits expression of iNOS in cytokine-stimulated VSMC (40). The mechanical induction of ET_B receptors in rat aortic VSMC is mediated partly by Rho-kinase (5), and the expression of the angiotensin II type 1 receptor is dependent on RhoA but not inhibited by Y-27632 (24). While Rho/Rho-kinase signaling inhibits cyclooxygenase (COX)-2 expression and subsequent PGI_2 production in VSMC (10), RhoA activation induces COX-2 expression in fibroblasts (60). Rho/Rho-kinase activation is also important in SRF-dependent transcription of VSMC marker genes and cell differentiation (20). It is unknown if Rho/Rho-kinase signaling plays a role in any of the changes in gene expression that occur in HPH. However, it can be speculated that stimulation of this signal transduction pathway in hypoxic hypertensive lungs might limit the upregulation and activity of eNOS, and perhaps of other vasodilators, and promote the upregulation of ET-1 and other vasoconstrictors.

6. Rho/Rho-kinase in Systemic Vascular Diseases

Rho/Rho-kinase signaling is implicated in the pathogenesis of various systemic vascular diseases (58). The Rho-kinase inhibitors Y-27632 and fasudil cause greater acute systemic vasodilation in hypertensive rats and patients than in their normotensive counterparts, implying an increased contribution of Rho-kinase-mediated Ca^{2+} sensitization to the regulation of systemic vascular tone in hypertension. Treatment with Rho-kinase inhibitors also prevents coronary artery medial hypertrophy and perivascular fibrosis in hypertensive rats (29, 39). 5-HT-induced vasospasm of IL-1β-treated coronary arteries is associated with hyperphosphorylation of MBS and MLC, and Rho-kinase inhibition prevents both the vasospasm and hyperphosphorylation (59). Fasudil reduces acetylcholine-induced coronary vasospasm in patients with vasospastic angina (36). In vivo gene transfer of dominant-negative Rho-kinase reverses IL-1β-induced coronary arteriosclerosis (38), and Rho-kinase inhibitors suppress neointimal formation after balloon injury (56).

In contrast to the considerable evidence for its involvement in systemic vascular diseases, there is little information on the importance of Rho/Rho-kinase signaling in the pathogenesis of pulmonary hypertension. Recent studies in our laboratory (16, 37, 42, 47) and others (1, 18, 22, 44) indicate that Rho/Rho-kinase signaling plays a role in both HPH and monocrotaline-induced pulmonary hypertension. However, a great deal remains to be learned about how, when, and where in the hypertensive pulmonary vasculature this pathway is activated and exactly what role(s) it plays, and by what mechanisms, in mediating increased vascular tone and remodeling.

7. Role of Increased ET-1 and 5-HT in Hypoxic Pulmonary Hypertension

Although several different vasoconstrictors are implicated in the complex pathophysiology of HPH, it is evident that ET-1 and 5-HT play major roles in the vasoconstriction and vascular remodeling (6, 11, 28). ET-1 is a pulmonary vasoconstrictor and co-mitogen, and either selective ET_A or mixed ET_A/ET_B

receptor blockade attenuates HPH in rats. ET-1-induced vasoconstriction is mediated through activation of smooth muscle cell ET_A and ET_B receptors, and is moderated by endothelial cell ET_B receptor-mediated production of NO and PGI_2. ET-1, endothelin converting enzyme, and ET_A and ET_B receptors are upregulated in hypertensive rat lungs, and ET_A receptor-mediated vasoconstriction is increased in distal hypertensive pulmonary arteries of hypoxic rats. Interestingly, ET-1-induced contraction of hypoxia-induced hypertensive pulmonary VSMC is largely independent of changes in cytosolic $[Ca^{2+}]$, and might be due primarily to Ca^{2+} sensitization (57). The major role of the ET_B receptor in HPH appears to be protective (25, 53). In addition to vasoconstriction, ET-1 can also stimulate proliferation of pulmonary VSMC from several species via activation of ET_A or both ET_A and ET_B receptors (9). Although ET-1 activates RhoA in VSMC (19, 52), the interactions between ET-1 and Rho/Rho-kinase signaling in the pathogenesis of HPH are unknown. It will be important to determine if Rho/Rho-kinase signaling contributes to endogenous ET-1-induced vasoconstriction and VSMC growth, as well as the upregulation of ET-1 and ET_A and ET_B receptors, in HPH.

5-HT is involved in both the vasoconstriction and vascular remodeling of HPH (11, 35). 5-HT-induced pulmonary vasoconstriction is mediated primarily by the $5\text{-}HT_{2A}$ receptor in normotensive arteries, but in hypertensive arteries, $5\text{-}HT_{1B}$ receptor expression is increased, and an augmented vasoconstriction is mediated by both $5\text{-}HT_{2A}$ and $5\text{-}HT_{1B}$ receptors. HPH is attenuated in rats by the $5\text{-}HT_{1B/1D}$ receptor antagonist GR-127935, and in $5\text{-}HT_{1B}$ (30) and $5\text{-}HT_{2B}$ (33) knockout mice. 5-HT-induced proliferation and hypertrophy of pulmonary VSMC depend on 5-HT transport into the cell, a process enhanced by hypoxia-induced upregulation of the 5-HT transporter (5-HTT) (11). 5-HT-induced growth of bovine pulmonary VSMC is associated with tyrosine phosphorylation of a GTPase-activating protein, activation of Ras and Rac1, generation of H_2O_2, and phosphorylation of ERK1/2 (34). HPH is attenuated in 5-HTT knockout mice, and in wild-type mice treated with a 5-HTT blocker (11). 5-HT activates RhoA in VSMC (52), and it is possible that Rho/Rho-kinase signaling contributes to endogenous 5-HT-induced pulmonary vasoconstriction and VSMC growth, and to the upregulation of $5\text{-}HT_{1B}$ receptors and 5-HTT in HPH.

8. Role of Deficient NO and PGI_2 in Hypoxic Pulmonary Hypertension

Endogenous NO, derived largely from eNOS, is a potent inhibitor of pulmonary vasoconstriction and HPH (15). Although eNOS is moderately upregulated in hypoxic hypertensive lungs, and NO suppresses an increased endogenous ET-1-mediated pulmonary vasoconstriction (41, 46, 48), hypoxic ventilation limits lung NO production (53). Therefore, deficient production of

NO likely plays a role in the development of HPH. Chronic inhaled NO inhibits HPH, and transgenic eNOS-overexpressing mice have increased lung eNOS protein and activity and blunted development of HPH (49). Thus, further stimulation of lung eNOS expression and/or activity may attenuate HPH. Rho/Rho-kinase signaling mediates hypoxic inhibition of eNOS expression and activity in cultured endothelial cells (63), but its role in the regulation of lung eNOS expression and activity are unknown. As noted below in Section 11, chronic inhibition of Rho-kinase reduces development of HPH in rats and mice, and preliminary results suggest this attenuation may involve increased eNOS expression in the chronically hypoxic lung.

PGI_2 elicits pulmonary vasodilation and inhibits pulmonary VSMC growth due at least partly to activation of adenylate cyclase and production of cAMP (65). Lung-specific overexpression of PGI_2 synthase in mice attenuates HPH (17). In contrast, decreased lung PGI_2 production is implicated in severe pulmonary hypertension, and PGI_2 receptor knockout mice develop exaggerated HPH (23). Decreased expression of PGI_2 synthase and production of PGI_2 may contribute to exaggerated HPH in ET_B receptor deficient rats (25). The effects of chronic hypoxia on rat lung and pulmonary artery COX-2, PGI_2 synthase, and PGI_2 receptor expression are unclear (2,7). As mentioned earlier Rho/Rho-kinase signaling inhibits COX-2 expression and subsequent PGI_2 production in VSMC (10). Thus, it is possible that Rho/Rho-kinase signaling modulates COX-2, PGI_2 synthase, and/or PGI_2 receptor expression in HPH. Conversely, a decrease in PGI_2-induced production of cAMP may enhance Rho/Rho-kinase signaling.

9. Rho-kinase Activity Mediates Increased Basal Vascular Tone in Hypertensive Rat Lungs

To investigate if Rho/Rho-kinase signaling plays a role in increased vascular tone in HPH, we measured acute effects of intravenous Y-27632 (10 mg/kg) on pulmonary and systemic arterial blood pressures and cardiac output in normoxic control and chronically hypoxic rats breathing 21% O_2 (47). Normoxic control rats were kept at Denver's barometric pressure of ~630 mmHg (inspired $P_{O_2} \approx$ 120 mmHg). The chronically hypoxic rats were exposed to hypobaric hypoxia (barometric pressure ~ 410 mmHg, inspired $P_{O_2} \approx$ 76 mmHg) for 3 to 4 weeks before being returned to normoxia for 2 days for catheterization and hemodynamic measurements. Although these rats were no longer undergoing hypoxic vasoconstriction, they maintained high pulmonary artery pressures, i.e., "residual pulmonary hypertension". While the Rho-kinase inhibitor had little effect on pulmonary artery pressure and pulmonary vascular resistance in control rats, it almost normalized the residual pulmonary hypertension in chronically hypoxic rats (Fig. 3). Y-27632 also reduced the pulmonary pressor response to acute hypoxia (10 min of 10% O_2) from 10±2 to 2±0.4 mmHg in control rats and from 8±1 to 1±0.5 mmHg in chronically hypoxic rats. The vasodilation by intravenous Y-27632 was not selective for the hypertensive lung, and there was marked systemic vasodilation in both the control and chronically hypoxic rats. Interestingly, in a preliminary experiment in chronically hypoxic rats, an oral dose of Y-27632 (30 mg/kg) initially caused both pulmonary and systemic vasodilation, but whereas systemic pressure gradually returned to pretreatment level, the fall in pulmonary pressure was sustained for up to 24 hrs. We do not know why oral Y-27632 caused selective sustained pulmonary vasodilation, but it appears that chronic administration via subcutaneous osmotic minipump acts similarly (see Section 11).

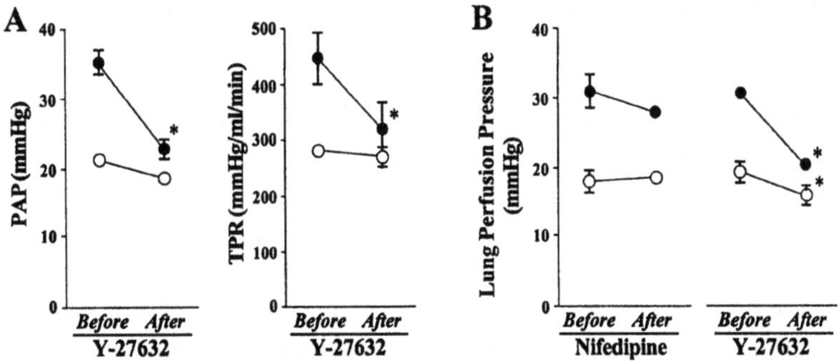

Figure 3. A: Acute intravenous administration of the Rho-kinase inhibitor Y-27632 (10 mg/kg) almost completely reverses the residual pulmonary hypertension (pulmonary artery pressure, PAP, left) and high total pulmonary resistance (TPR, right) in chronically hypoxic rats (●) after 2 days of re-exposure to normoxia. Y-27632 has minimal effects on pressure and resistance in normoxic control rats (○) (* $P<0.05$ vs. before value; n=5/group). B: Y-27632 (10 μM) but not nifedipine (10 μM) markedly reduces perfusion pressure in chronically hypoxic hypertensive rat lungs (●) perfused with blood at constant flow (* $P<0.05$ vs. respective before value; n=3/group). Y-27632 also causes a smaller reduction in PAP in normoxic normotensive control lungs (○). These findings in both catheterized rats and perfused lungs indicate that most of the increased TPR after 3-4 weeks of chronic hypoxia is attributable to Rho-kinase-mediated vasoconstriction rather than simply to vascular remodeling.

In contrast to the ability of Y-27632 to both inhibit acute HPV and reverse residual pulmonary hypertension in chronically hypoxic rats after return to normoxia, we have previously observed that the L-type Ca^{2+} channel blocker nifedipine inhibits HPV but does not reduce the residual pulmonary hypertension(46, 48). We have also observed that whereas inhaled NO (80 ppm) acutely reduces the residual pulmonary hypertension by 15±2% (46), it appears to be less effective than the intravenous Rho-kinase inhibitor that reduces the pressure by 36±2% (Fig. 3A).

We have also compared the acute effects of Y-27632 and nifedipine on baseline perfusion pressure in lungs isolated from control and chronically hypoxic rats and perfused with blood at constant flow (37). Normotensive lungs were ventilated with 21% O_2 and hypertensive lungs with 8% O_2 to more closely mimic in vivo conditions. The Rho-kinase inhibitor, but not nifedipine, caused slight vasodilation in normotensive lungs and marked vasodilation in hypertensive lungs (the high perfusion pressure in hypertensive lungs was nearly normalized) (Fig. 3B). Although pulmonary vascular remodeling, i.e., medial thickening of pulmonary arteries and muscularization of pulmonary arterioles, is present in rats after 3-4 weeks of hypoxia (28), the ability of Y-27632 to nearly normalize the increased pulmonary vascular resistance in both catheterized rats and isolated lungs suggests that Rho-kinase-mediated vasoconstriction is a major component of the elevated resistance at this stage of HPH. It is also apparent that this Rho-kinase-mediated vasoconstriction is partially reversed by inhaled NO, but not by short-term normoxia.

10. Rho-kinase Mediates Increased Vascular Reactivity in Hypertensive Rat Lungs

To test if Rho-kinase activity also plays a role in increased vasoconstrictor reactivity in chronically hypoxic hypertensive lungs, we measured pressor responses to KCl (a receptor-independent and voltage-gated Ca^{2+} influx-mediated vasoconstrictor) in normotensive and chronically hypoxic hypertensive physiological salt solution (PSS)-perfused lungs with and without Y-27632 pretreatment (37). The vasoconstrictor response to KCl was augmented in hypertensive lungs, and the augmentation was eliminated by Y-27632 (Fig. 4A). In contrast, the Rho-kinase inhibitor caused minimal blunting of the smaller KCl vasoconstriction in normotensive lungs, indicating that Y-27632 had little, if any, effects on membrane depolarization and voltage-gated Ca^{2+} influx, which is consistent with the effects of Y-27632 in isolated rat pulmonary arteries (50). KCl vasoconstriction in both groups of lungs was blocked by 1 μM nifedipine (not shown). Collectively, these results suggest that Rho/Rho-kinase signaling increases pulmonary vascular Ca^{2+} sensitivity and potentiates vasoconstriction to voltage-gated Ca^{2+} influx in HPH.

ET-1 and both ET_A and ET_B receptors are upregulated in chronically hypoxic rat lungs (6). To test if increased endogenous ET-1 activity contributes to the augmented vasoconstrictor response to KCl, we examined effects of pretreatment with the ET_A receptor blocker BQ-123 (Fig. 4A) and the dual $ET_{A/B}$ blocker J-104132 on KCl vasoconstriction (42). The augmented KCl response in hypertensive lungs was reduced equally by ET_A and $ET_{A/B}$ receptor blockade. This suggested that activation of ET_A but not ET_B receptors played a role in the increased vasoreactivity to KCl; a finding in keeping with evidence that acute ET-1-induced Ca^{2+} sensitization of rat normotensive pulmonary artery involves ET_A but not ET_B receptor stimulation (14). ET_A receptor blockade was not as effective as inhibition of Rho-kinase in reducing the exaggerated response to KCl (Fig. 4), and further attenuation by the combination of BQ-123 plus the 5-HT$_{1B/1D}$ receptor antagonist GR-127935 (42) suggests that 5-HT, and possibly other endogenous vasoconstrictors, are also involved in stimulating the Rho-kinase-mediated Ca^{2+} sensitization.

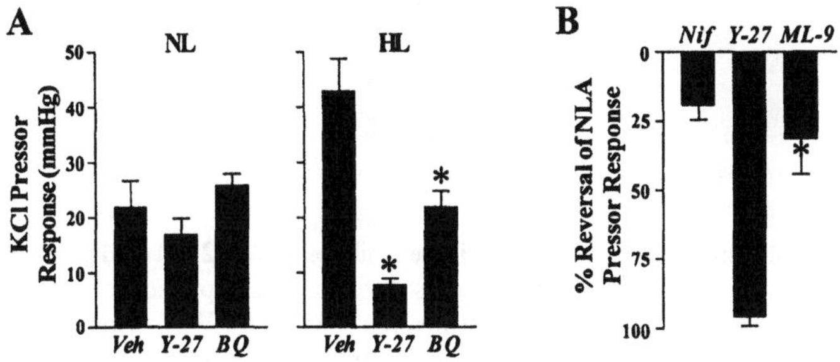

Figure 4. A: The Rho-kinase inhibitor Y-27632 (Y-27, 10 μM) markedly inhibits augmented KCl (40 mM) vasoconstriction in hypoxic hypertensive rat lungs (HL). In contrast, Y-27632 has little effect on the smaller KCl response in normotensive lungs (NL). The ET_A receptor blocker BQ-123 (BQ, 5 μM) also causes some inhibition of KCl vasoconstriction in hypertensive lungs, but has no effect in normotensive lungs. **P*<0.05 vs. respective vehicle (Veh) control value (n=4-8/group). B: While nifedipine (Nif, 10 μM) has little effect, Y-27632 (10 μM) almost completely reverses the vasoconstriction elicited by inhibiting NOS (200 μM nitro-L-arginine, NLA) in hypoxic hypertensive rat lungs. In comparison, the MLCK inhibitor ML-9 (100 μM) causes only partial reversal of the NLA vasoconstriction (* *P*<0.05; n=3/group). These observations suggest that increased vasoconstrictor reactivity of hypoxia-induced hypertensive rat lungs is due largely to Rho-kinase-mediated Ca^{2+} sensitization, and that the apparent activation of Rho/Rho-kinase signaling is mediated partly by endogenous ET-1-induced activation of ET_A receptors and suppressed by endogenous production of NO.

We have previously observed that normoxia-ventilated hypertensive lungs from chronically hypoxic rats produce increased amounts of NO (53), and that acute inhibition of NO synthesis by nitro-L-arginine (NLA) elicits a marked sustained vasoconstriction that is mediated partly by endogenous ET-1 (41, 46, 48). We investigated if this NLA-induced vasoconstriction also involves Rho-kinase-mediated Ca^{2+} sensitization by comparing the vasodilator effects of nifedipine and Y-27632 (37). The Rho-kinase inhibitor, but not nifedipine, caused complete reversal of the NLA vasoconstriction (Fig. 4B). Similar experiments showed that the MLCK inhibitor ML-9 elicited only partial reversal of the NLA vasoconstriction (Fig. 4B). Thus, Rho-kinase-mediated Ca^{2+} sensitization rather than solely Ca^{2+}/calmodulin-induced activation of MLCK, is apparently essential for the marked NLA-induced vasoconstriction in the hypertensive lungs (41). Because we have previously found that Ca^{2+}-free perfusion of hypertensive lungs prevents the NLA vasoconstriction, increased cytosolic [Ca^{2+}] and activation of MLCK may be required for the onset of the response (46, 48).

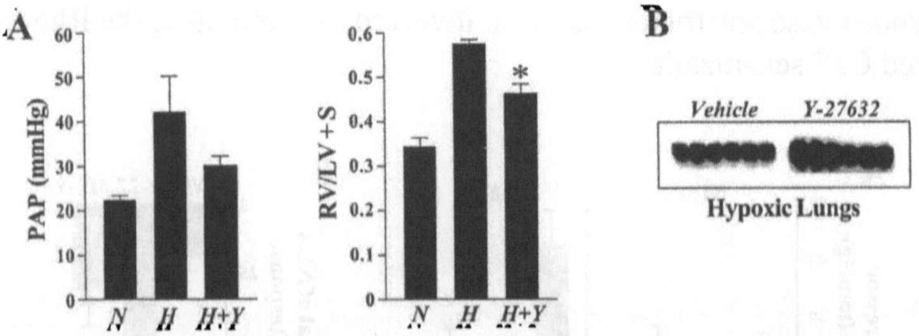

Figure 5. A: Treatment of rats with the Rho-kinase inhibitor Y-27632 (40 mg/kg/day, sq) during 2 weeks of exposure to chronic hypoxia reduces development of pulmonary hypertension as reflected in decreased mean pulmonary artery pressure (PAP) and RV hypertrophy (RV/LV+S). N are normoxic controls and H and H+Y are hypoxic rats treated, respectively, with either vehicle or Y-27632 (*P<0.05 vs. respective H value; n=3/group). B: Western blot of eNOS in lungs of vehicle- and Y-27632-treated chronically hypoxic mice. It appears that treatment with Y-27632 augments the upregulation of eNOS expression in hypoxic lungs. These observations indicate that Rho/Rho-kinase signaling plays a role in the pathogenesis of HPH, and that this might involve a limitation of the expression and activity of eNOS.

11. Rho-kinase Blocker Inhibits Development of Hypoxic Pulmonary Hypertension

To evaluate whether Rho/Rho-kinase signaling is involved in the development of HPH, we treated rats exposed to 2 weeks of hypobaric hypoxia (equivalent to ~10% O_2) with either vehicle or Y-27632 (40 mg/kg/day) via subcutaneous osmotic minipump. Measurements of mean pulmonary artery pressure and right ventricular (RV) weight/left ventricular (LV) plus septal (S) weight (RV/LV+S) showed that treatment with the Rho-kinase inhibitor reduced the severity of pulmonary hypertension (Fig. 5A), without reducing systemic arterial pressure or altering cardiac output (not shown). Y-27632 treatment did not affect the hypoxia-induced polycythemia (hematocrit = 47±1% in normoxic controls and 67±2 and 70±2%, respectively, in vehicle and Y-27632 hypoxic groups). We have similarly found that treatment with Y-27632 also reduces HPH in mice (16), and preliminary results support the possibility that the attenuation of pulmonary hypertension is associated with increased expression of eNOS (Fig. 5B), and possibly increased NO production, in the hypoxic lung. The inhibition of HPH was only partial in both rats and mice, and it remains to be determined whether or not higher doses of Y-27632, or fasudil, a Rho-kinase inhibitor with possible clinical utility (58), will be more effective. Another consideration is that chronic inhibition of Rho-kinase might enhance the activity of other GTPases such as Rac (64, 67), which could possibly counteract the effects Rho-kinase inhibition on pulmonary vasoconstriction and vascular remodeling. Rho-kinase is involved in many important cellular functions, and near total inhibition of this signaling molecule in vivo may have adverse physiological effects, especially during exposure to hypoxia.

Other investigators have observed that statins also attenuate hypoxic (18) and monocrotaline-induced (22, 44) pulmonary hypertension in rats, and Abe and co-workers have reported that fasudil both markedly prevents and reverses monocrotaline-induced pulmonary hypertension (1). A direct comparison has not been reported, but the impression from the results of these and our studies is that monocrotaline-induced pulmonary hypertension may be more effectively prevented by treatment with Rho-kinase inhibitors than is HPH. If true, it will be interesting to determine why.

12. Summary

The small GTPase RhoA and its downstream effector Rho-kinase play a role in many cellular functions including cell adhesion, migration, gene expression, growth, and contraction. Rho/Rho-kinase signaling can promote sustained increases in vascular tone by both increasing the Ca^{2+} sensitivity of vascular smooth muscle cell contraction and downregulating expression of vasodilators and upregulating expression of vasoconstrictors. There is considerable evidence that Rho/Rho-kinase activation is important in the pathogenesis of systemic vascular diseases such as hypertension, vasospasm, and arteriosclerosis. It is therefore reasonable to hypothesize that this signaling pathway also contributes to the pathogenesis of hypoxia-mediated pulmonary hypertension by promoting sustained pulmonary vasoconstriction and vascular wall remodeling. Our observations that an inhibitor of Rho-kinase effectively reverses high pulmonary vascular resistance in chronically hypoxic rats and blunts development of hypoxic pulmonary hypertension in rats and mice support this concept. It is now apparent that Rho/Rho-kinase signaling needs to be added to the blend of increased cytosolic [Ca^{2+}] and sundry other signaling molecules and pathways to fully understand the molecular pathophysiology of hypoxia-induced pulmonary hypertension.

Acknowledgments

This work was supported by grants from the National Heart, Lung and Blood Institute of the National Institutes of Health (HL 14985, HL 03879, and HL 07171) and the American Heart Association.

References

1. Abe K, Shimokawa H, Uwatoku T, Matsumoto Y, and Hattori T. Long-term inhibition of Rho-kinase markedly ameliorates monocrotaline-induced pulmonary hypertension in rats. *Circulation*. 2002; 106 : II-365.
2. Abe Y, Tatsumi K, Sugito K, Ikeda Y, Kimura H, and Kuriyama T. Effects of inhaled prostacyclin analogue on chronic hypoxic pulmonary hypertension. *J. Cardiovasc. Pharmacol.* 2001; 37 : 239-251.
3. Bishop AL, and Hall A. Rho GTPases and their effector proteins. *Biochem. J.* 2000; 348 : 241-255.
4. Carvajal JA, Germain AM, Huidobro-Toro JP, and Weiner CP. Molecular mechanism of cGMP-mediated smooth muscle relaxation. *J. Cell. Physiol.* 2000; 184 : 409-420.
5. Cattaruzza M, Eberhardt I, and Hecker M. Mechanosensitive transcription factors involved in endothelin B receptor expression. *J. Biol. Chem.* 2001; 276 : 36999-37003.
6. Chen YF and Oparil S. Endothelin and pulmonary hypertension. *J. Cardiovasc. Pharmacol.* 2000; 35 : S49-53.
7. Chida M and Voelkel NF. Effects of acute and chronic hypoxia on rat lung cyclooxygenase. *Am. J. Physiol.* 1996; 270 : L872-L878.
8. Damron DS, Kanaya N, Homma Y, Kim SO, and Murray PA. Role of PKC, tyrosine kinases, and Rho kinase in alpha-adrenoreceptor-mediated PASM contraction. *Am. J. Physiol. Lung Cell. Mol. Physiol.* 2002; 283 : L1051-L1064.
9. Davie N, Haleen SJ, Upton PD, Polak JM, Yacoub MH, Morrell NW, and Wharton J. ET_A and ET_B receptors modulate the proliferation of human pulmonary artery smooth muscle cells. *Am. J. Respir. Crit. Care Med.* 2002; 165 : 398-405.
10. Degraeve F, Bolla M, Blaie S, Creminon C, Quere I, Boquet P, Levy-Toledano S, Bertoglio J, and Habib A. Modulation of COX-2 expression by statins in human aortic smooth muscle cells. Involvement of geranylgeranylated proteins. *J. Biol. Chem.* 2001; 276 : 46849-46855.
11. Eddahibi S, Raffestin B, Hamon M, and Adnot S. Is the serotonin transporter involved in the pathogenesis of pulmonary hypertension? *J. Lab. Clin. Med.* 2002; 139 : 194-201.
12. Eto M, Barandier C, Rathgeb L, Kozai T, Joch H, Yang Z, and Luscher TF. Thrombin suppresses endothelial nitric oxide synthase and upregulates endothelin-converting enzyme-1 expression by distinct pathways: role of Rho/ROCK and mitogen-activated protein kinase. *Circ. Res.* 2001; 89 : 583-590.
13. Etter EF, Eto M, Wardle RL, Brautigan DL, and Murphy RA. Activation of myosin light chain phosphatase in intact arterial smooth muscle during nitric oxide-induced relaxation. *J. Biol. Chem.* 2001; 276 : 34681-34685.
14. Evans AM, Cobban HJ, and Nixon GF. ET_A receptors are the primary mediators of myofilament calcium sensitization induced by ET-1 in rat pulmonary artery smooth muscle: a tyrosine kinase independent pathway. *Br. J. Pharmacol.* 1999; 127 : 153-160.
15. Fagan KA, McMurtry I, and Rodman DM. Nitric oxide synthase in pulmonary hypertension: lessons from knockout mice. *Physiol. Res.* 2000; 49 : 539-548.
16. Fagan KA, Oka M, and McMurtry IF. Rho-kinase inhibitor (Y27632) attenuates the development of hypoxia-induced pulmonary hypertension in mice. *Am. J. Respir. Cell. Mol. Biol.* 2002; 165 : B53.
17. Geraci MW, Gao B, Shepherd DC, Moore MD, Westcott JY, Fagan KA, Alger LA, Tuder RM, and Voelkel NF. Pulmonary prostacyclin synthase overexpression in transgenic mice protects against development of hypoxic pulmonary hypertension. *J. Clin. Invest.* 1999; 103 : 1509-1515.
18. Girgis RE, Li D, Tuder RM, Johns RA, and Garcia JGN. Attenuation of hypoxic pulmonary hypertension in rats by the HMG-CoA reductase inhibitor, simvastatin. *J. Heart Lung Transpl.* 2002; 21 : 149.
19. Gohla A, Schultz G, and Offermanns S. Role for G_{12}/G_{13} in agonist-induced vascular smooth

muscle cell contraction. *Circ. Res.* 2000; 87 : 221-227.
20. Halayko AJ and Solway J. Molecular mechanisms of phenotypic plasticity in smooth muscle cells. *J. Appl. Physiol.* 2001; 90 : 358-368.
21. Hernandez-Perera O, Perez-Sala D, Soria E, and Lamas S. Involvement of Rho GTPases in the transcriptional inhibition of preproendothelin-1 gene expression by simvastatin in vascular endothelial cells. *Circ. Res.* 2000; 87 : 616-622.
22. Hongo M, Hironaka E, Sakai A, Yazaki Y, Kinoshita O, and Owa M. Pravastatin improves monocrotaline-induced pulmonary hypertension and prolongs survival in rats independent of cholesterol lowering. *Circulation.* 2002; 106 : II-366.
23. Hoshikawa Y, Voelkel NF, Gesell TL, Moore MD, Morris KG, Alger LA, Narumiya S, and Geraci MW. Prostacyclin receptor-dependent modulation of pulmonary vascular remodeling. *Am. J. Respir. Crit. Care Med.* 2001; 164 : 314-318.
24. Ichiki T, Takeda K, Tokunou T, Iino N, Egashira K, Shimokawa H, Hirano K, Kanaide H, and Takeshita A. Downregulation of angiotensin II type 1 receptor by hydrophobic 3-hydroxy-3-methylglutaryl coenzyme A reductase inhibitors in vascular smooth muscle cells. *Arterioscler. Thromb. Vasc. Biol.* 2001; 21 : 1896-1901.
25. Ivy DD, Yanagisawa M, Gariepy CE, Gebb SA, Colvin KL, and McMurtry IF. Exaggerated hypoxic pulmonary hypertension in endothelin B receptor-deficient rats. *Am. J. Physiol. Lung. Cell. Mol. Physiol.* 2002; 282 : L703-L712.
26. Janssen LJ, Lu-Chao H, and Netherton S. Excitation-contraction coupling in pulmonary vascular smooth muscle involves tyrosine kinase and Rho kinase. *Am. J. Physiol. Lung Cell. Mol. Physiol.* 2001; 280 : L666-L674.
27. Janssen LJ, Premji M, Netherton S, Coruzzi J, Lu-Chao H, and Cox PG. Vasoconstrictor actions of isoprostanes via tyrosine kinase and Rho kinase in human and canine pulmonary vascular smooth muscles. *Br. J. Pharmacol.* 2001; 132 : 127-134.
28. Jeffery TK, and Wanstall JC. Pulmonary vascular remodeling: a target for therapeutic intervention in pulmonary hypertension. *Pharmacol. Ther.* 2001; 92 : 1-20.
29. Kataoka C, Egashira K, Inoue S, Takemoto M, Ni W, Koyanagi M, Kitamoto S, Usui M, Kaibuchi K, Shimokawa H, and Takeshita A. Important role of Rho-kinase in the pathogenesis of cardiovascular inflammation and remodeling induced by long-term blockade of nitric oxide synthesis in rats. *Hypertension.* 2002; 39 : 245-250.
30. Keegan A, Morecroft I, Smillie D, Hicks MN, and MacLean MR. Contribution of the 5-HT_{1B} receptor to hypoxia-induced pulmonary hypertension: converging evidence using 5-HT_{1B}-receptor knockout mice and the 5-$HT_{1B/1D}$-receptor antagonist GR127935. *Circ. Res.* 2001; 89 : 1231-1239.
31. Laufs U, and Liao JK. Post-transcriptional regulation of endothelial nitric oxide synthase mRNA stability by Rho GTPase. *J. Biol. Chem.* 1998; 273 : 24266-24271.
32. Laufs U, Marra D, Node K, and Liao JK. 3-Hydroxy-3-methylglutaryl-CoA reductase inhibitors attenuate vascular smooth muscle proliferation by preventing rho GTPase-induced down-regulation of p27(Kip1). *J. Biol. Chem.* 1999; 274 : 21926-21931.
33. Launay J-M, Herve P, Peoc'h K, Tournois C, Callebert J, Nebigil CG, Etienne N, Drouet L, Humbert M, Simonneau G, and Maroteaux L. Function of the serotonin 5-hydroxytryptamine 2B receptor in pulmonary hypertension. *Nat. Med.* 2002; 8 : 1129-1135.
34. Lee SL, Simon AR, Wang WW, and Fanburg BL. H_2O_2 signals 5-HT-induced ERK MAP kinase activation and mitogenesis of smooth muscle cells. *Am. J. Physiol. Lung Cell. Mol. Physiol.* 2001; 281 : L646-L652.
35. MacLean MR, and Morecroft I. Increased contractile response to 5-hydroxytryptamine$_1$-receptor stimulation in pulmonary arteries from chronic hypoxic rats: role of pharmacological synergy. *Br. J. Pharmacol.* 2001; 134 : 614-620.
36. Masumoto A, Mohri M, Shimokawa H, Urakami L, Usui M, and Takeshita A. Suppression of coronary artery spasm by the Rho-kinase inhibitor fasudil in patients with vasospastic angina. *Circulation.* 2002; 105 : 1545-1547.
37. Morio Y, Oka M, and McMurtry IF. A selective Rho-kinase inhibitor, Y-27632, is an effective

vasodilator of chronically hypoxic hypertensive rat lungs. *FASEB J.* 2002; 16 : A74.
38. Morishige K, Shimokawa H, Eto Y, Kandabashi T, Miyata K, Matsumoto Y, Hoshijima M, Kaibuchi K, and Takeshita A. Adenovirus-mediated transfer of dominant-negative rho-kinase induces a regression of coronary arteriosclerosis in pigs in vivo. *Arterioscler. Thromb. Vasc. Biol.* 2001; 21 : 548-554.
39. Mukai Y, Shimokawa H, Matoba T, Kandabashi T, Satoh S, Hiroki J, Kaibuchi K, and Takeshita A. Involvement of Rho-kinase in hypertensive vascular disease: a novel therapeutic target in hypertension. *FASEB J.* 2001; 15 : 1062-1064.
40. Muniyappa R, Xu R, Ram JL, and Sowers JR. Inhibition of Rho protein stimulates iNOS expression in rat vascular smooth muscle cells. *Am. J. Physiol. Heart Circ. Physiol.* 2000; 278 : H1762-H1768.
41. Muramatsu M, Rodman DM, Oka M, and McMurtry IF. Endothelin-1 mediates nitro-L-arginine vasoconstriction of hypertensive rat lungs. *Am. J. Physiol.* 1997; 272 : L807-L812.
42. Nagaoka T, Morio Y, Oka M, and Mcmurtry IF. Endothelin-1 and serotonin are involved in Rho-kinase-mediated augmented pressor response to KCl in chronically hypoxic hypertensive rat lungs. *Am. J. Respir. Cell. Mol. Biol.* 2003; 167:A697.
43. Niiro N, Koga Y, and Ikebe M. Agonist-induced changes in the phosphorylation of the myosin- binding subunit of myosin light chain phosphatase and CPI17, two regulatory factors of myosin light chain phosphatase, in smooth muscle. *Biochem. J.* 2003; 369 : 117-128.
44. Nishimura T, Faul JL, Berry GJ, Vaszar LT, Qiu D, Pearl RG, and Kao PN. Simvastatin attenuates smooth muscle neointimal proliferation and pulmonary hypertension in rats. *Am. J. Respir. Crit. Care Med.* 2002; 166 : 1403-1408.
45. Numaguchi K, Eguchi S, Yamakawa T, Motley ED, and Inagami T. Mechanotransduction of rat aortic vascular smooth muscle cells requires RhoA and intact actin filaments. *Circ. Res.* 1999; 85 : 5-11.
46. Oka M, Hasunuma K, Webb SA, Stelzner TJ, Rodman DM, and McMurtry IF. EDRF suppresses an unidentified vasoconstrictor mechanism in hypertensive rat lungs. *Am. J. Physiol.* 1993; 264 : L587-L597.
47. Oka M, Morio Y, Morris KG, and McMurtry I. Acute hemodynamic effects of Y27632, a selective Rho-kinase inhibitor, in chronically hypoxic pulmonary hypertensive rats. *FASEB J.* 2002; 16 : A74.
48. Oka M, Morris KG, and McMurtry IF. NIP-121 is more effective than nifedipine in acutely reversing chronic pulmonary hypertension. *J. Appl. Physiol.* 1993; 75 : 1075-1080.
49. Ozaki M, Kawashima S, Yamashita T, Ohashi Y, Rikitake Y, Inoue N, Hirata KI, Hayashi Y, Itoh H, and Yokoyama M. Reduced hypoxic pulmonary vascular remodeling by nitric oxide from the endothelium. *Hypertension.* 2001; 37 : 322-327.
50. Robertson TP, Dipp M, Ward JP, Aaronson PI, and Evans AM. Inhibition of sustained hypoxic vasoconstriction by Y-27632 in isolated intrapulmonary arteries and perfused lung of the rat. *Br. J. Pharmacol.* 2000; 131 : 5-9.
51. Sah VP, Seasholtz TM, Sagi SA, and Brown JH. The role of Rho in G protein-coupled receptor signal transduction. *Annu. Rev. Pharmacol. Toxicol.* 2000; 40 : 459-489.
52. Sakurada S, Okamoto H, Takuwa N, Sugimoto N, and Takuwa Y. Rho activation in excitatory agonist-stimulated vascular smooth muscle. *Am. J. Physiol. Cell. Physiol.* 2001; 281 : C571-C578.
53. Sato K, Rodman DM, and McMurtry IF. Hypoxia inhibits increased ET_B receptor-mediated NO synthesis in hypertensive rat lungs. *Am. J. Physiol.* 1999; 276 : L571-L581.
54. Schoenwaelder SM and Burridge K. Bidirectional signaling between the cytoskeleton and integrins. *Curr. Opin. Cell Biol.* 1999; 11 : 274-286.
55. Seasholtz TM, Zhang T, Morissette MR, Howes AL, Yang AH, and Brown JH. Increased expression and activity of RhoA are associated with increased DNA synthesis and reduced p27(Kip1) expression in the vasculature of hypertensive rats. *Circ. Res.* 2001; 89 : 488-495.
56. Shibata R, Kai H, Seki Y, Kato S, Morimatsu M, Kaibuchi K, and Imaizumi T. Role of Rho-associated kinase in neointima formation after vascular injury. *Circulation.* 2001; 103 : 284-

289.
57. Shimoda LA, Sham JSK, Shimoda TH, and Sylvester JT. L-type Ca^{2+} channels, resting $[Ca^{2+}]_i$, and ET-1-induced responses in chronically hypoxic pulmonary myocytes. *Am. J. Physiol. Lung Cell. Mol. Physiol.* 2000; 279 : L884-L894.
58. Shimokawa H. Rho-kinase as a novel therapeutic target in treatment of cardiovascular diseases. *J. Cardiovasc. Pharmacol.* 2002; 39 : 319-327.
59. Shimokawa H, Seto M, Katsumata N, Amano M, Kozai T, Yamawaki T, Kuwata K, Kandabashi T, Egashira K, Ikegaki I, Asano T, Kaibuchi K, and Takeshita A. Rho-kinase-mediated pathway induces enhanced myosin light chain phosphorylations in a swine model of coronary artery spasm. *Cardiovasc Res.* 1999; 43 : 1029-1039.
60. Slice LW, Han SK, and Simon MI. Galphaq signaling is required for Rho-dependent transcriptional activation of the cyclooxygenase-2 promoter in fibroblasts. *J. Cell. Physiol.* 2003; 194 : 127-138.
61. Somlyo AP and Somlyo AV. Signal transduction by G-proteins, rho-kinase and protein phosphatase to smooth muscle and non-muscle myosin II. *J. Physiol.* 2000; 522 : 177-185.
62. Stark WW Jr., Blaskovich MA, Johnson BA, Qian Y, Vasudevan A, Pitt B, Hamilton AD, Sebti SM, and Davies P. Inhibiting geranylgeranylation blocks growth and promotes apoptosis in pulmonary vascular smooth muscle cells. *Am. J. Physiol.* 1998; 275 : L55-L63.
63. Takemoto M, Sun J, Hiroki J, Shimokawa H, and Liao JK. Rho-kinase mediates hypoxia-induced downregulation of endothelial nitric oxide synthase. *Circulation.* 2002; 106 : 57-62.
64. Tsuji T, Ishizaki T, Okamoto M, Higashida C, Kimura K, Furuyashiki T, Arakawa Y, Birge RB, Nakamoto T, Hirai H, and Narumiya S. ROCK and mDia1 antagonize in Rho-dependent Rac activation in Swiss 3T3 fibroblasts. *J. Cell. Biol.* 2002; 157 : 819-830.
65. Tuder RM and Zaiman AL. Prostacyclin analogs as the brakes for pulmonary artery smooth muscle cell proliferation: is it sufficient to treat severe pulmonary hypertension? *Am. J. Respir. Cell. Mol. Biol.* 2002; 26 : 171-174.
66. Wang Z, Jin N, Ganguli S, Swartz DR, Li L, and Rhoades RA. Rho-kinase activation is involved in hypoxia-induced pulmonary vasoconstriction. *Am. J. Respir. Cell. Mol. Biol.* 2001; 25 : 628-635.
67. Welsh CF, Roovers K, Villanueva J, Liu Y, Schwartz MA, and Assoian RK. Timing of cyclin D1 expression within G1 phase is controlled by Rho. *Nat. Cell. Biol.* 2001; 3 : 950-957.
68. Yamakawa T, Tanaka S-i, Numaguchi K, Yamakawa Y, Motley ED, Ichihara S, and Inagami T. Involvement of Rho-kinase in angiotensin II-induced hypertrophy of rat vascular smooth muscle cells. *Hypertension.* 2000; 35 : 313-318.

Chapter 25

Hypoxia-sensitive Transcription Factors and Growth Factors

Ari L. Zaiman and Rubin M. Tuder
Johns Hopkins University School of Medicine, Baltimore, Maryland, U.S.A.

1. Introduction: Definition of Pulmonary Hypertension

Pulmonary hypertension (PH) is an important clinical complication in approximately 30% of interstitial and other non-neoplastic lung diseases in humans. The mean pulmonary artery pressures are between 25 and 45 mmHg and this elevation can compromise right heart function. The underlying mechanisms of PH in these conditions probably relates to pathologic vessel remodeling associated with progressive alveolar hypoxia and/or peripheral vessel destruction from inflammation and/or scarring. In contrast, a small fraction of patients with severe pulmonary hypertension in whom the pulmonary artery pressures are in excess of 40 mmHg are at risk life threatening right ventricular failure.

Pulmonary vascular remodeling is the distinctive structural component associated with PH. The pulmonary arteries are composed of three layers, each with its characteristic cellular component. Endothelial cells predominate in the intima and, in physiological conditions, are its sole constituents; however, in the setting of elevated pulmonary artery pressures or local thrombosis smooth muscle cells can migrate from the medial compartment into this layer and transdifferentiate into myofibroblasts. The vascular media consists almost exclusively of smooth muscle cells (SMCs); in hypoxic PH the thickness of this layer dramatically increases by means of cell proliferation and hypertrophy. The adventitia is composed of fibroblasts which, like medial SMCs, can undergo alterations in cell size and number in hypoxic PH. While vasoactive molecules can transiently regulate the pulmonary artery pressures, established PH is not a functional, but a structural disease (23).

While little is known of the natural history of primary pulmonary hypertension (PPH), the use of rodent models has provided great insight into the etiology of hypoxic PH. Additional experimental models using techniques such as monocrotaline injection, air embolization, or ligation of the ductus arteriosus have demonstrated the heterogeneity of mechanisms responsible for PH; however, the underlying theme has been the interplay between different cell types within the vascular wall. It is the balance of hypoxic vasoconstriction, hypoxia-dependent growth factor expression, downregulation of vasodilators, and the impact of vascular remodeling on vascular resistance, which determine whether or not PH develops.

Historically, chronic hypoxic pulmonary hypertension (HPH) has been the dominant model based on the rationale that the pathobiological information obtained in this model may also apply to human PH. Medial muscular thickening and extension of the muscular layer to peripheral, usually non-muscularized pulmonary arteries is well described in human pulmonary hypertension. While the data obtained with the chronic hypoxia model must be carefully interpreted as far their relevance to PPH is concerned, these models of HPH have great relevance to human pulmonary hypertension associated with chronic hypoxia.

It is the authors' overall goal to review the evidence linking vascular growth factors and/or inhibitors to the establishment of chronic hypoxia-mediated pulmonary hypertension.

2. Factors Involved in Oxygen Sensing

While the mechanism of oxygen sensing is being progressively unraveled, at least three transcription factors regulate the cellular response to hypoxia in the lungs. The best characterized is hypoxia-inducible factor-1 (HIF-1), a heterodimer that recognizes a cognate sequence within the promoter of several genes involved in the cellular response to hypoxia (Table 1) (25). The α subunit is regulated by the O_2 concentration whereas the β subunit is constitutively expressed. HIF-1 knockout mice die from failed vascularization at mid-gestation, but heterozygous mice develop normally under normoxic conditions. Surprisingly, when stressed under hypoxic conditions, the pulmonary hypertension and vascular remodeling present in the wild type animals is attenuated in the heterozygotes.

Table 1: Growth/Vasoactive Factors Implicated in Hypoxia-induced Pulmonary Hypertension

Hypoxia-inducible Factor-1 (HIF-1)	Endothelin-1 (ET-1)
Heme-Oxygenase-1 (HO-1)	Nitric Oxide Synthase-2 (NOS-II)
Vascular Endothelial Growth Factor (VEGF)	VEGF Receptor, *FLT-1*
Early Growth Response-1 (EGR-1)	PDGF-A and PDGF-B
Nuclear Factor Interleukin-6 (NF-IL6)	Interleukin-6
Serotonin (5-HT)	bFGF
Prostacyclin (PGI_2)	

Early growth response-1 (Egr-1) is a zinc finger transcription factor, which is rapidly induced in response to hypoxia in smooth muscle cells and mononuclear phagocytes (34). Within the lung, there is a biphasic and hypoxia-dependent response. While this activation is HIF independent, the molecular mechanism is unknown. Its induction results in the activation of many genes involved in vascular remodeling. Acutely, egr-1 null mice fail to induce these genes; however, the effect on chronic HPH has yet to be tested.

Hypoxia mediated transcriptional induction also involve nuclear factor-interleukin-6 (NF-IL-6/CEBPβ). First characterized as the site within the promoter of IL-6 responsible for hypoxic induction, binding sites for this factor have been found in other genes. Transgenic mice created to study expression of this promoter element demonstrated its ability to direct tissue specific (lung, heart, and kidney) transcription in response to hypoxia. Participation of this factor in hypoxia-induced iNOS expression in rat microvascular endothelial cells has recently been demonstrated; however, it is unknown how animals that lack this element respond to hypoxia (28).

3. Response of the Intima, Media, and Adventitia to Hypoxia

Within the pulmonary vasculature there are three major cell types each predominating in their own layer, but intimately associated with the others: endothelial cells (intima), smooth muscle cells (media), and fibroblasts (adventitia). While the vascular remodeling generally consists of smooth muscle cell hypertrophy/hyperplasia and smooth muscle cell migration into the smaller arterioles, all cell layers affect this process.

Endothelial cells form the interface between the blood and the underlying cells and tissues. Rather than a passive cell barrier, endothelial cells are integrators, transducers, and effectors of the local vascular environment. Short-term exposure of endothelial cells to hypoxia results in secretion of mitogens for smooth muscle cells. Cyclooxgenase inhibitors and neutralizing antibodies identified these factors as prostaglandin F_2 and basic fibroblast growth factor (bFGF). Hypoxia induces other well-known mitogens such as platelet derived growth factor (PDGF) and vascular endothelial growth factor (VEGF), as well as vasoactive molecules such as endothelin-1 that also serve to promote smooth muscle growth. Although there is no data on the role of PDGF in HPH, there is recent evidence that PDGF is upregulated in PH induced by ductus arteriosus ligation in the sheep, and that blockade of PDGF significantly reduces pulmonary pressures in this model (1).

The medial smooth muscle cells proliferate in response to hypoxia due to factors secreted by the endothelium and adventitial fibroblasts in the surrounding layers. However, the elucidation of the direct effect of hypoxia on SMCs has been hampered by the phenotypic heterogeneity of SMCs isolated from the pulmonary arteries. This heterogeneity is illustrated by the observation that while large numbers of isolated SMCs fail to proliferate, there is a subset that proliferates in response to hypoxia.

Hypoxia also results in the proliferation of the adventitial fibroblasts. Not only do these cells respond to growth factors produced by the endothelium, but hypoxia also directly causes their proliferation. Working through a member of the G-protein family, a variety of MAP kinases are activated which result in cellular proliferation (5). In addition, the adventitial fibroblasts exposed to hypoxia also secrete HIF-1-dependent factors that induce proliferation of the smooth muscle cells (24).

4. Growth Factors Involved in Medial Smooth Muscle Cell Remodeling

4.1. Serotonin (5-Hydroxytryptamine)

Serotonin (5-HT) is a vasoactive molecule released from platelets. Acutely, it causes vasoconstriction, but it also functions as a smooth muscle cell mitogen. The mechanism of action is complex as an active transporter that internalizes serotonin and multiple forms of cell surface receptors are present in the cells. Furthermore, the expression pattern of these molecules is both cell and species specific. It is the balance of receptor/transporter expression and function that determine the cellular response.

Hypoxia has been demonstrated to induce the secretion of serotonin from intact pulmonary neuroepithelial bodies and the expression of a 5-hydroxytryptamine transporter (5-HTT) (10). In addition, patients with primary pulmonary hypertension have elevated levels of serotonin, and an increased frequency of a polymorphism within the 5-HTT which renders cultured pulmonary artery smooth muscle cells more susceptible to the growth promoting effects of serotonin (7).

Numerous animal studies have demonstrated a role for serotonin in the development of HPH. Continuous infusion of serotonin into rats exposed to chronic hypoxia augmented the vascular remodeling (8). 5-HT_{1B}-receptor knockout mice exposed to chronic hypoxia developed less pulmonary hypertension and a non-significant decrease in vascular remodeling, but had no contractile response to a serotonin agonist (13). Although controversial, the mitogenic effects of serotonin appear to be mediated, at least in part, by the 5-HTT. Mice deficient in this transporter develop less pulmonary hypertension and vascular remodeling when exposed to chronic hypoxia (6). The exact pathway by which these cells translate the internalization of serotonin to cellular proliferation has not been fully elucidated. However, recent evidence suggests hydrogen peroxide plays an important role as a second messenger (16).

Recently, the 5-HT_{2B} receptor has been implicated in HPH. 5-HT_{2B} receptor null mice failed to develop HPH, despite maintaining an acute response to hypoxia. Although the mechanism by which this receptor functions in pulmonary hypertension is unknown, hypoxia did not cause an increase in cell proliferation (15).

4.2. Endothelin-1

Endothelin-1 (ET-1) is a multifunctional molecule, which serves as both a potent vasoconstrictor and a smooth muscle cell mitogen. First identified as a small endothelium-secreted peptide that causes vasoconstriction, ET-1 belongs to a family of molecules which are synthesized as preprohormones and processed by a converting enzyme. There are two distinct G-protein coupled receptors, ET_A and ET_B receptors. In the lung, the ET_A receptor is found predominantly in the pulmonary vascular SMCs, whereas the endothelium expresses the ET_B receptor. Binding of the ET_A receptor on SMCs leads to pulmonary vasoconstriction and proliferation.

Elevated levels of ET-1 have been found in patients with both severe and moderate pulmonary hypertension. Rats exposed to chronic hypoxia have increased levels of ET-1 and increased expression of ET_A and ET_B receptors (17). In addition, treatment of rats with an ET_A receptor antagonist prevents HPH and significantly reverses the pulmonary vascular remodeling even after two weeks of hypoxic exposure.

4.3. Inflammatory Cytokines: Interleukin-6 and Interleukin-8

Hypoxia induces the expression of numerous cytokines and growth factors within the vasculature as well as mononuclear phagocytes. Hypoxia induces secretion of platelet activating factor (PAF) and PDGF from endothelial cells as well as macrophages. Both factors stimulate vascular smooth muscle cell and fibroblast proliferation. In addition, inhibition of PAF and PDGF reduced the hypoxia induced IL-6 and IL-8 expression in pulmonary fibroblast and smooth muscle cells (26). Given the large number of upregulated cytokines, it is not surprising that a variety of different factors are responsible for their activation. Hypoxia induces IL-6 transcription through C/EBP-NF-IL-6, while EGR-1 is responsible for PDGF activation (Table 1) (14, 35).

The role of inflammation in pulmonary hypertension has been observed in numerous animal models. Increased inflammation and cytokine production was directly observed in mice exposed to hypoxia. RNA prepared from hypoxic mouse lungs showed elevated levels of IL-1, IL-6, MIP-2 (functional homologue of human IL-8), and monocyte chemoattractant protein, MCP-1 (19). Interestingly, these levels were reduced in transgenic mice overexpressing lung heme-oxygenase-1, animals which were also protected from developing HPH (19). Furthermore, rats treated with a PAF inhibitor (20) or the 5-lipoxygenase inhibitor MK866, or mice lacking 5-lipoxygenase (32) had decreased pulmonary artery pressures and decreased pulmonary vascular remodeling (characterized with intimal and medial hypertrophy).

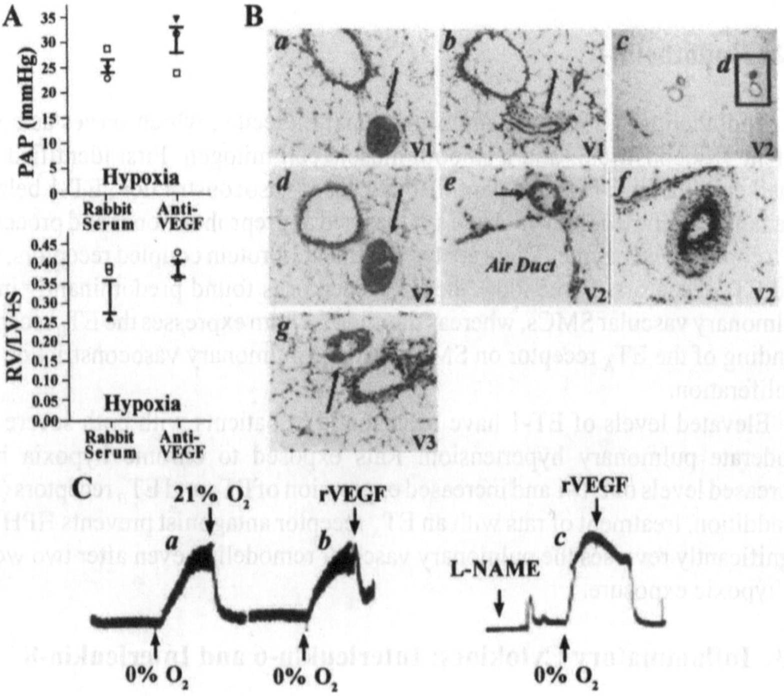

Figure 1. PAP in rats with anti-VEGF and the effect of VEGF on PAP. A: Rats treated with neutralizing anti-VEGF polyclonal antibodies 3 times per week for 3 weeks under chronic hypoxia had an increase in PAP and right ventricular mass when compared with hypoxic rats treated with rat serum. B: Pulmonary vessel remodeling in 3 rats exposed to chronic hypoxia and treated with anti-VEGF rabbit serum (V1, V2, V3). Note the predominance of medial thickening of medium-sized arteries (arrows), with less than 50% medial thickness with respect to total vessel diameter. Focally, control rats (not shown) had thickened vessels as seen in B (panel c (d). Note that the hilar vessels had similar remodeling (f, g). a-g: HE, 200×, h, i: pentachrome, 100×. C: Infusion of rVEGF (at 200 ng/ml) in the pulmonary circulation *ex vivo* abolished hypoxic-induced vasoconstriction in the isolated perfused lung. The reduction in PAP with VEGF (b) was similar to that seen with normalization of alveolar O_2 levels (a). The effect of VEGF was almost abolished by the eNOS inhibitor, L-NNAME (100 mM) (c).

4.4. Vascular Endothelial Growth Factor (VEGF)

VEGF, a critical growth and survival factor for endothelial cells, is ideally suited to play a role in the regulation of pulmonary artery remodeling. Produced by lung SMCs, macrophages, and alveolar cells under normoxic conditions,

VEGF is upregulated in chronic hypoxia in an NO-dependent manner (3, 30). Administration of neutralizing VEGF antibodies exacerbated chronic hypoxia-mediated PH (Fig. 1A and B).

Furthermore, exogenous VEGF administered via an adenoviral vector completely attenuated intimal hyperplasia and caused regression of existing hyperplasia (22). When recombinant VEGF is added to isolated perfused rat lungs, rVEGF abolishes the hypoxic-induced vasoconstriction through the production of nitric oxide (NO) (Fig. 1C). Although this is an acute response, overexpression of VEGF could lead to chronic increases in NO and prostacyclin (PGI_2) production in the endothelium, which in turn acts to inhibit smooth muscle cell hyperplasia (33).

The critical role of VEGF in maintenance of lung vascular reactivity was highlighted by the findings that chronic hypoxia in combination with a VEGF receptor blocker cause severe pulmonary hypertension associated with endothelial cell proliferation in rats (27). This model has several features in common with human severe pulmonary hypertension, in particular PPH (31). In a setting where phenotypically-altered pulmonary endothelial cells proliferate in a neoplastic-like manner, VEGF may act in a paracrine manner, stimulating the abnormal growth of endothelial cells (31). Indeed, we found that endothelial cells in plexiform lesions exhibit markers of angiogenesis, in particular, VEGF and its receptor II (KDR) (Fig. 2).

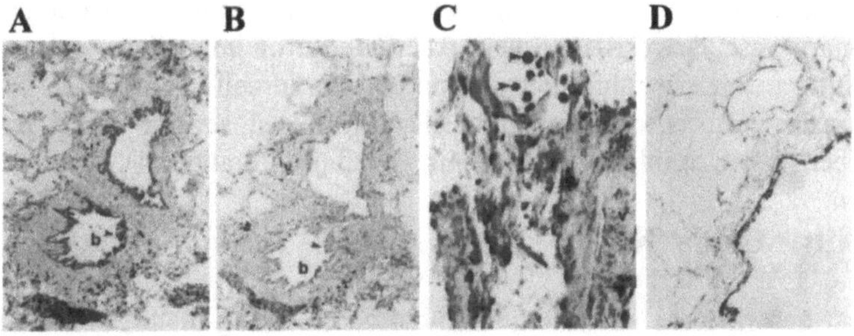

Figure 2. VEGF in human PPH. *In situ* hybridization for VEGF mRNA in PPH lung. A: VEGF mRNA in bronchiolar cells in PPH lung (anti-sense probe, 200×). B: Sense control. C: VEGF mRNA expression in perivascular macrophages (anti-sense probe, 200×). D: VEGF mRNA in bronchiolar cells in normal lung (anti-sense probe, 200×).

4.5. Prostacyclin

Prostacyclin (PGI_2), the main cyclooxygenase product in vascular tissue, functions as a vasodilator and inhibitor of smooth muscle cell proliferation. Produced primarily by endothelial cells, PGI_2 activates a family of receptors that exert a broad range of biological actions by raising intracellular cAMP levels.

Historically, its role in pulmonary hypertension was evaluated in the belief that a deficiency of vasodilators was the key alteration in lungs of PPH patients. Patients with severe pulmonary hypertension (SPH) were reported to have decreased serum levels of PGI_2 and decreased expression of the enzyme responsible for its production (PGI_2 synthase) (29). While PGI_2 expression may be decreased in patients with SPH, PGI_2 levels are increased in chronically hypoxic rats. Using various inhibitors, the enhanced PGI_2 production depended on increased sheer forces secondary to the vasoconstriction as opposed to hypoxia per se (2).

Although hypoxia may not directly induce PGI_2 synthesis, animals in which PGI_2 production has been altered vary in their susceptibility to HPH. PGI_2 receptor knockout mice have normal pulmonary artery pressures. Following chronic hypoxic exposure, pulmonary artery pressure is increased and there is enhanced remodeling of pulmonary vesselsas compared with wild type controls (12). Lung-specific overexpression of PGI_2 synthase did not result in abnormal resting pulmonary artery pressures but prevented the development of pulmonary hypertension and pulmonary vascular remodeling when the mice were exposed to chronic hypoxia (11).

The therapeutic supplementation of PGI_2 has provided further insight into the pathogenesis of pulmonary hypertension. PGI_2 has been shown to improve hemodynamics, exercise tolerance, and prolong survival in PPH subjects. Furthermore, PGI_2 has been demonstrated to induce long term reductions in the pulmonary vascular resistance that exceed those of immediate vasodilation, suggesting a role for inhibition of smooth muscle cell proliferation or anti-inflammatory effects on endothelial cells (18). Indeed, several of PGI_2 analogs have been shown to inhibit smooth muscle cell proliferation (4).

4.6. Nitric Oxide (NO)

First described as an endothelium-derived relaxing factor, NO exhibited potent vasodilatory effects. Subsequently, it was found to inhibit smooth muscle cell proliferation. The lung expresses three different isoforms of nitric oxide synthase (NOS). While all may affect vessel remodeling, eNOS and iNOS are the major form expressed within the vasculature. The effect of hypoxia of NOS expression has been controversial. There have been reports of both increased and decreased expression of eNOS within patients with chronic pulmonary hypertension. Recently, quantitative RT-PCR from mouse lung tissues demonstrated hypoxemic induction of eNOS and iNOS expression (9).

Although still controversial, most studies using knock out mice demonstrate that NO diminishes the vascular remodeling induced by chronic hypoxia. While eNOS-deficient mice developed normally under normoxic conditions, exposure to chronic hypoxia resulted in fourfold greater proportion of muscularized small arteries (9). Conversely, mice overexpressing eNOS within the lung had an attenuated response to chronic hypoxia (21).

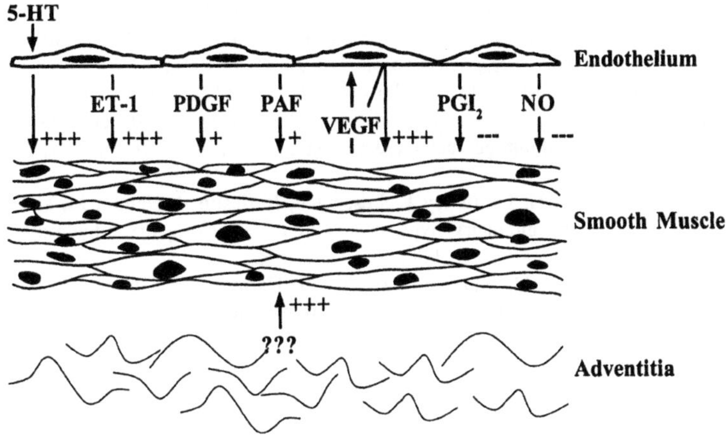

Figure 3. Schematic of growth factors in vascular remodeling. All three layers of the pulmonary vessel are indicated along with the growth factors produced after exposure to chronic hypoxia.

5. Conclusion

The role of growth factors in HPH is an open book, with no ending in sight. As novel vascular growth factors are discovered, and as we learn more about their biological and pathobiological role, we add a new level of understanding in HPH. Undoubtedly, abnormal vascular cell growth is at the center of pulmonary vascular remodeling in pulmonary hypertension (Fig. 3). In the next few years, the elucidation of master control levels of pulmonary vascular remodeling using genomics and proteonomics of human lung tissue compromised by pulmonary hypertension and transgenic models of pulmonary hypertension may shed important information in the relative contribution of growth factor in the pathogenesis of HPH.

References

1. Balasubramaniam V, Le Cras TD, Ivy DD, Grover T R, Kinsella JP, and Abman SH. Role of platelet derived growth factor in vascular remodeling during pulmonary hypertension in the ovine fetus. *Am. J. Physiol. Lung Cell. Mol. Physiol.* 2003; 184: L826-L833.
2. Blumberg FC, Lorenz C, Wolf K, Sandner P, Riegger GA, and Pfeifer M. Increased pulmonary prostacyclin synthesis in rats with chronic hypoxic pulmonary hypertension.

Cardiovasc. Res. 2002; 55: 171-177.

3. Christou H, Yoshida A, Arthur V, Morita T, and Kourembanas S. Increased vascular endothelial growth factor production in the lungs of rats with hypoxia-induced pulmonary hypertension. *Am. J. Respir. Cell Mol. Biol.* 1998; 18: 768-776.
4. Clapp LH, Finney P, Turcato S, Tran S, Rubin LJ, and Tinker A. Differential effects of stable prostacyclin analogs on smooth muscle proliferation and cyclic AMP generation in human pulmonary artery. *Am. J. Respir. Cell Mol. Biol.* 2002; 26: 194-201.
5. Das M, Bouchey DM, Moore MJ, Hopkins DC, Nemenoff RA, and Stenmark KR. Hypoxia-induced proliferative response of vascular adventitial fibroblasts is dependent on g protein-mediated activation of mitogen-activated protein kinases. *J. Biol. Chem.* 2001; 276: 15631-15640.
6. Eddahibi S, Hanoun N, Lanfumey L, Lesch KP, Raffestin B, Hamon M, and Adnot S. Attenuated hypoxic pulmonary hypertension in mice lacking the 5-hydroxytryptamine transporter gene. *J. Clin. Invest.* 2000; 105: 1555-1562.
7. Eddahibi S, Humbert M, Fadel E, Raffestin B, Darmon M, Capron F, Simonneau G, Dartevelle P, Hamon M, and Adnot S. Serotonin transporter overexpression is responsible for pulmonary artery smooth muscle hyperplasia in primary pulmonary hypertension. *J. Clin. Invest.* 2001; 108: 1141-1150.
8. Eddahibi S, Raffestin B, Launay JM, Sitbon M, and Adnot S. Effect of dexfenfluramine treatment in rats exposed to acute and chronic hypoxia. *Am. J. Respir. Crit. Care Med.* 1998; 157: 1111-1119.
9. Fagan KA, Morrissey B, Fouty BW, Sato K, Harral JW, Morris KG Jr, Hoedt-Miller M, Vidmar S, McMurtry IF, and Rodman DM. Upregulation of nitric oxide synthase in mice with severe hypoxia-induced pulmonary hypertension. *Respir. Res.* 2001; 2: 306-313.
10. Fu XW, Nurse CA, Wong V, and Cutz E. Hypoxia-induced secretion of serotonin from intact pulmonary neuroepithelial bodies in neonatal rabbit. *J. Physiol.* 2002; 539: 503-510.
11. Geraci MW, Gao B, Shepherd DC, Moore MD, Westcott JY, Fagan KA, Alger LA, Tuder RM, and Voelkel NF. Pulmonary prostacyclin synthase overexpression in transgenic mice protects against development of hypoxic pulmonary hypertension. *J. Clin. Invest.* 1999; 103: 1509-1515.
12. Hoshikawa Y, Voelkel NF, Gesell TL, Moore MD, Morris KG, Alger LA, Narumiya S, and Geraci MW. Prostacyclin receptor-dependent modulation of pulmonary vascular remodeling. *Am. J. Respir. Crit. Care Med.* 2001; 164: 314-318.
13. Keegan A, Morecroft I, Smillie D, Hicks MN, and MacLean MR. Contribution of the 5-HT_{1B} receptor to hypoxia-induced pulmonary hypertension: converging evidence using 5-HT_{1B}-receptor knockout mice and the 5-$HT_{1B/1D}$-receptor antagonist GR127935. *Circ. Res.* 2001; 89: 1231-1239.
14. Khachigian LM, Williams AJ, and Collins T. Interplay of Sp1 and Egr-1 in the proximal platelet-derived growth factor A-chain promoter in cultured vascular endothelial cells. *J. Biol. Chem.* 1995; 270: 27679-27686.
15. Launay -JM, Hervé P, Peoc'h K, Tournois C, Callebert J, Nebigil CG, Etienne N, Drouet L, Humbert M, Simonneau G, and Maroteaux L. Function of the serotonin 5-hydroxytryptamine 2B receptor in pulmonary hypertension. *Nat. Med.* 2002; 8: 1129-1135.
16. Lee SL, Simon AR, Wang WW, and Fanburg BL. H_2O_2 signals 5-HT-induced ERK MAP kinase activation and mitogenesis of smooth muscle cells. *Am. J. Physiol Lung Cell. Mol. Physiol.* 2001; 281: L646-L652.
17. Li H, Chen SJ, Chen YF, Meng QC, Durand J, Oparil S, and Elton T. S. Enhanced endothelin-1 and endothelin receptor gene expression in chronic hypoxia. *J. Appl. Physiol.* 1994; 77: 1451-1459.
18. McLaughlin VV, Genthner DE, Panella MM, and Rich S. Reduction in pulmonary vascular resistance with long-term epoprostenol (prostacyclin) therapy in primary pulmonary hypertension. *N. Engl. J. Med.* 1998; 338: 273-277.
19. Minamino T, Christou H, Hsieh CM, Liu Y, Dhawan V, Abraham NG, Perrella MA, Mitsialis

SA, and Kourembanas S. Targeted expression of heme oxygenase-1 prevents the pulmonary inflammatory and vascular responses to hypoxia. *Proc. Natl. Acad. Sci. USA*. 2001; 98: 8798-8803.

20. Ono S, Westcott JY, and Voelkel NF. PAF antagonists inhibit pulmonary vascular remodeling induced by hypobaric hypoxia in rats. *J. Appl. Physiol.* 1992; 73: 1084-1092.
21. Ozaki M, Kawashima S, Yamashita T, Ohashi Y, Rikitake Y, Inoue N, Hirata KI, Hayashi Y, Itoh H, and Yokoyama M. Reduced hypoxic pulmonary vascular remodeling by nitric oxide from the endothelium. *Hypertension*. 2001; 37: 322-327.
22. Partovian C, Adnot S, Raffestin B, Louzier V, Levame M, Mavier IM, Lemarchand P, and Eddahibi S. Adenovirus-mediated lung vascular endothelial growth factor overexpression protects against hypoxic pulmonary hypertension in rats. *Am. J. Respir. Cell Mol. Biol.* 2000; 23: 762-771.
23. Reid LM and Davies P. Control of cell proliferation in pulmonary hypertension. In *Pulmonary Vascular Physiology and Pathophysiology*, Weir EK and Reeves JT, eds. New York, NY: Marcel Dekker, 1989, pp.541-611.
24. Rose F, Grimminger F, Appel J, Heller M, Pies V, Weissmann N, Fink L, Schmidt S, Krick S, Camenisch G, Gassmann M, Seeger W, and Hanze J. Hypoxic pulmonary artery fibroblasts trigger proliferation of vascular smooth muscle cells: role of hypoxia-inducible transcription factors. *FASEB J.* 2002; 16: 1660-1661.
25. Semenza G. Signal transduction to hypoxia-inducible factor 1. *Biochem. Pharmacol.* 2002; 64: 993-998.
26. Tamm M, Bihl M, Eickelberg O, Stulz P, Perruchoud AP, and Roth M. Hypoxia-induced interleukin-6 and interleukin-8 production is mediated by platelet-activating factor and platelet-derived growth factor in primary human lung cells. *Am. J. Respir. Cell Mol. Biol.* 1998; 19: 653-661.
27. Tarasevicienc-Stewart L, Kasahara Y, Alger L, Hirth P, McMahon G, Waltenberger J, Voelkel NF, and Tuder RM. Inhibition of the VEGF receptor 2 combined with chronic hypoxia causes cell death-dependent pulmonary endothelial cell proliferation and severe pulmonary hypertension. *FASEB J.* 2001; 15: 427-438.
28. Teng X, Li D, Catravas JD, and Johns RA. C/EBP-β mediates iNOS induction by hypoxia in rat pulmonary microvascular smooth muscle cells. *Circ. Res.* 2002; 90: 125-127.
29. Tuder RM, Cool CD, Geraci MW, Wang J, Abman SH, Wright L, Badesch D, and Voelkel NF. Prostacyclin synthase expression is decreased in lungs from patients with severe pulmonary hypertension. *Am. J. Respir. Crit. Care Med.* 1999; 159: 1925-1932.
30. Tuder RM, Flook BE, and Voelkel NF. Increased gene expression for VEGF and the VEGF receptors KDR/Flk and Flt in lungs exposed to acute or to chronic hypoxia. Modulation of gene expression by nitric oxide. *J. Clin. Invest.* 1995; 95: 1798-1807.
31. Voelkel NF and Tuder RM. Hypoxia-induced pulmonary vascular remodeling: a model for what human disease? *J. Clin. Invest.* 2000; 106: 733-738.
32. Voelkel NF, Tuder RM, Wade K, Hoper M, Lepley RA, Goulet JL, Koller BH, and Fitzpatrick F. Inhibition of 5-lipoxygenase-activating protein (FLAP) reduces pulmonary vascular reactivity and pulmonary hypertension in hypoxic rats. *J. Clin. Invest.* 1996; 97: 2491-2498.
33. Wheeler-Jones C, Abu-Ghazaleh R, Cospedal R, Houliston RA, Martin J, and Zachary I. Vascular endothelial growth factor stimulates prostacyclin production and activation of cytosolic phospholipase A_2 in endothelial cells via p42/p44 mitogen-activated protein kinase. *FEBS Lett.* 1997; 420: 28-32.
34. Yan S-F, Fujita T, Lu J, Okada K, Shan Zou Y, Mackman N, Pinsky DJ, and Stern DM. Egr-1, a master switch coordinating upregulation of divergent gene families underlying ischemic stress. *Nat. Med.* 2000; 6: 1355-1361.
35. Yan SF, Tritto I, Pinsky D, Liao H, Huang J, Fuller G, Brett J, May L, and Stern D. Induction of interleukin 6 (IL-6) by hypoxia in vascular cells. Central role of the binding site for nuclear factor-IL-6. *J. Biol. Chem.* 1995; 270: 11463-11471.

Chapter 26

Heterogeneity in Hypoxia-Induced Pulmonary Artery Smooth Muscle Cell Proliferation

Maria G. Frid, Neil J. Davie, Kurt R. Stenmark
University of Colorado Health Sciences Center, Denver, Colorado, U.S.A.

1. Introduction

For many years the arterial media in the mature pulmonary and systemic circulations has been thought to be composed of a homogeneous population of quiescent, differentiated smooth muscle cells (SMCs) whose function was believed to be largely restricted to the regulation of vascular tone and blood flow in the respective circulations. In addition, the associated literature was replete with discussion of the idea that these resident vascular SMCs are capable of exhibiting a large degree of "plasticity" (5). It has been shown that, in response to injury, arterial SMCs can de-differentiate from a "contractile" quiescent phenotype into a "synthetic" phenotype, which expresses less contractile elements and with a greater capacity to proliferate and migrate (5). However, while this concept of SMC "phenotypic modulation" may be accurate for vascular SMCs in general, it does not explain the wide variety of vascular SMC phenotypes now described in both normal and diseased vessels. The theory generally accepted today is that the vascular media in both pulmonary and systemic arteries is a more complex structure composed of a mosaic of different SMC phenotypes programmed to subserve a repertoire of distinct and diverse cellular functions (4,10,12,13,15,25,31,33,38). Furthermore, the genetic basis for this cellular diversity may be a result of distinct ancestral lineages of these cells.

The observations regarding vascular SMC heterogeneity are not surprising when one considers that the vascular media must perform numerous and varied functions in order to maintain homeostasis of the vessel wall. Indeed, specific subpopulations of SMCs may respond in unique ways to stress and injurious stimuli providing a mechanism whereby some cells can act to repair the injury while others maintain functions needed for vascular homeostasis. Phenotypic heterogeneity of SMC at specific sties within the medial compartment of proximal and conduit pulmonary arteries has been well-characterized. It remains to be tested whether this pattern of SMC heterogeneity exists at more distal sites in the pulmonary circulation, particularly in resistance vessels. The limited information currently available suggests that the distal arterial media is composed of far less heterogeneous cells than its proximal counterpart. Evaluation of the expression of contractile and cytoskeletal proteins in SMC of the distal circulation, as well as assessment of their ion channel distribution and cell-surface receptor expression, suggests that SMC in resistant arteries are of a uniform phenotype (28, 32). This observation raises the question of how distal pulmonary arteries (where structural reorganization of the vessel wall is thought to have its greatest impact) respond and adapt to stress or injury.

The purpose of this chapter is to present the current experimental evidence regarding the existence of SMC heterogeneity in the medial compartment of the pulmonary arterial circulation. It will address differences in the SMC phenotype and composition of proximal versus distal arteries. It will examine the differences in proliferative responses of the distinct SMC subpopulations to hypoxia and mitogenic stimuli and begin to examine the mechanisms which confer proliferative advantages to select SMC sub-populations. This information is intended to provide a framework for a further understanding of the complex cellular mechanisms that contribute to the medial thickening of the pulmonary circulation in response to chronic hypoxia.

2. Smooth Muscle Cell Heterogeneity in Proximal Pulmonary Arteries: *In Vivo* Analysis

Although for years investigators had suggested differences between the SMC of large and small vessels, until the 1990's there was limited information regarding the existence of site-specific heterogeneity of SMC within the vascular media. Then, studies in the systemic circulation of experimental animals began to suggest diversity in the SMC composition of large vessel media based on expression patterns of a limited number of contractile and cytoskeletal proteins. There remained, however, a paucity of data regarding the existence of SMC diversity within the pulmonary arterial circulation. A comprehensive evaluation of the adult bovine main and proximal pulmonary arteries was therefore performed to specifically address the hypothesis that the adult pulmonary arterial media was comprised of heterogeneous subpopulations of SMC expressing different biochemical markers (13). This study provided compelling data regarding the existence of numerous SMC phenotypes within the media of the main and proximal pulmonary arteries. As shown in Figure 1 and Table 1, at least 4 phenotypically distinct SMC subpopulations (based on differential expression of muscle-specific contractile and cytoskeletal proteins) could be identified within the adult bovine arterial media.

These phenotypically distinct cells were found to reside in distinct regions in the medial compartment. In the main pulmonary artery, 3 distinct regions of the media were defined based on the morphology of the cells residing in them and matrix fiber orientation patterns (Fig. 1).

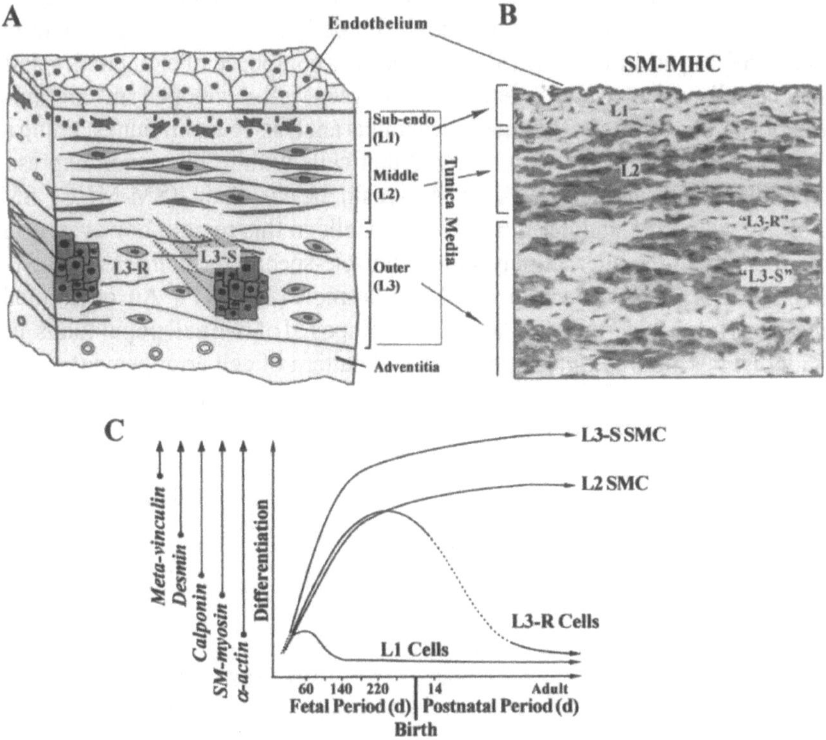

Figure 1. Phenotypic heterogeneity of the cells comprising the proximal pulmonary artery tunica media (A and B), and schematic representation of their distinct differentiation pathways during development (C). A: Schematic diagram demonstrating the structural and cellular heterogeneity of the mature bovine main pulmonary artery media. These distinct regions of the media were identified based on cell morphology and arrangement, as well as elastic lamellae shape and orientation; L1, subendothelial region; L2, middle media region; L3, outer media region. B: Immunoperoxidase staining (brown) with specific antibodies against smooth muscle myosin heavy chain (SM-MHC) demonstrates phenotypic heterogeneity of cells within the arterial media, with intense positive staining in middle media L2-SMC and outer media "L3-S" SMC expressing SM-MHC, but the absence of immunoreactivity in subendothelial L1- and outer media "L3-R" cells. The tissue section was concurrently labeled with hematoxyllin to identify cell nuclei. C: Schematic representation of differentiation pathways based on expression of muscle-specific contractile and cytoskeletal proteins by different cell populations within the main pulmonary artery media. L1, subendothelial cells; L2, middle media SMC; "L3-S"-SMC and "L3-R"-cells, two phenotypically distinct cell populations within the outer region of the media.

A subendothelial region, arbitrarily termed L1 (layer 1), is identified and is composed of small, irregularly shaped cells interspersed among fragmented particles of elastin. As shown in Table 1, cells in the subendothelial (L1) region do not express any of the smooth muscle markers evaluated. Accordingly, the phenotype of cells in this region could be defined as non-muscle. A middle media region, arbitrarily termed L2 (layer 2), is composed of elongated, spindle-shaped cells, oriented circumferentially between well-developed and continuous elastic lamellae (Fig. 1). Immunobiochemical analysis (e.g., immunostaining and Western blotting techniques) revealed that cells within this region express α-smooth muscle actin (αSMA), SM-myosin heavy chain (SM-MHC) SM-1 isoform, calponin, and desmin (Table 1). Cells in this region, however, do not express any of the alternatively spliced muscle-specific proteins, such as SM-2 MHC isoform, metavinculin or SM-caldesmon (Table 1). An outer medial region, arbitrarily termed L3 (layer 3), is composed of two cell populations, each with distinct morphologic appearance and pattern of cell arrangement.

Table 1. Immunobiochemical Analysis of α-Actin, Myosin, SM-I Isoform, Calponin, and Desmin

	Main Pulmonary Artery				Distal Pulmonary Artery			
					3000 µm		1500 µm	100-150 µm
	L1	*L2*	*L3-S*	*L3-R*	*L3-S*	*L3-R*		
Staining for								
α SMA	-	+	+	-	+	-	+	+
Calpoinin	-	+	+	-	+	-	+	+
SM-MHC:								
SM-1	-	+	+	-	+	-	+	+
SM-2	-	-	+	-	ND	ND	ND	ND
SM-B	-	-	+	-	+	-	+	+
Desmin	-	+/-	+	-	+	-	+	+
Metavinculin	-	-	+	-	+	-	+	+
SM-caldesmon	-	-	+	-	ND	ND	ND	ND

ND, not determined.

As schematically presented in Figure 1, large spindle-shaped cells (termed L3-S) arranged in compact cell clusters and oriented longitudinally within the vessel wall in areas devoid of elastic lamellae are seen in this region. A population of significantly smaller spindle-shaped cells (termed L3-R), oriented circum-ferentially is observed in interstitial areas between compact L3-S cell clusters. The L3-R cells are interspersed between well-developed, continuous elastic lamellae. Marked differences in the expression of muscle-specific markers were observed between the L3-S and L3-R cell types. L3-S cells expressed all the SM-specific proteins evaluated, including the alternatively spliced form of vinculin, metavinculin, as well as SM-2 MHC, and SM-caldesmon. Conversely, L3-R cells did not express any of the muscle-specific markers analyzed and therefore, their phenotype was defined as non-muscle.

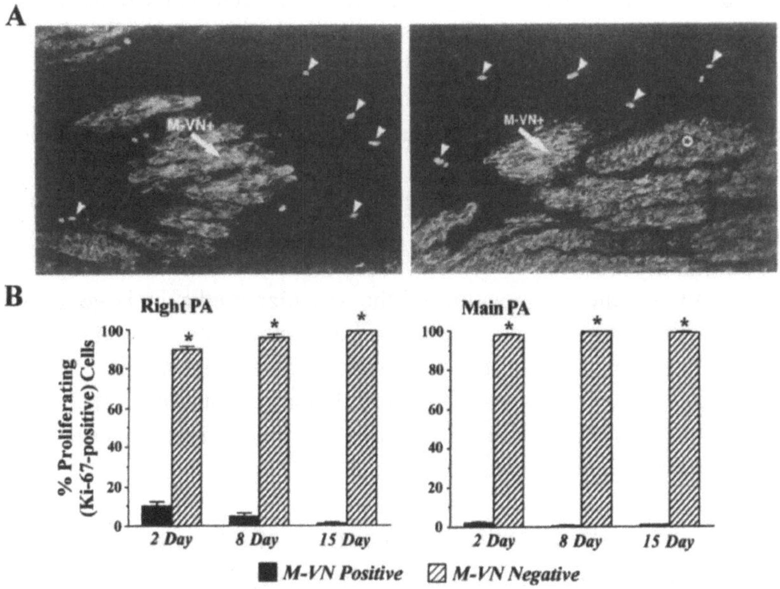

Figure 2. Proliferation in the proximal pulmonary arterial media of neonatal calves with hypoxia-induced pulmonary hypertension occurs almost exclusively in a less differentiated SMC population. A: Double-label immunofluorescence staining of the outer medial region of the main PA of a newborn calf exposed to hypobaric hypoxia for 2 weeks. "Well-differentiated" SMC (as defined previously by expression of several SM-markers) are marked here by expression of meta-vinculin (M-VN+, green fluorescence), whereas less differentiated SMC do not express this muscle-specific protein (dark areas with no immunoreactivity). Proliferation (as defined by expression of Ki-67 nuclear proliferation-associated antigen, red fluorescence) is identified only in less differentiated (metavinculin-negative) cells. B: Quantitative analysis of the double-labeled tissue sections (as seen in A) demonstrates that more than 95% of cell proliferation observed within the media of different sized proximal pulmonary arteries occurs in less-differentiated (metavinculin-negative) cells. (*$P<0.001$, n=3 in each age group at each time point).

To address the possibility that the differences biochemical phenotype expressed by cells in the adult arterial media were not simply the result of temporal "phenotypic" modulation of a single SMC type, experiments were performed to "track" the developmental fate of each "phenotypically unique" cell population (13). These studies, took advantage of the fact that cells with distinct phenotypes could be compared at different developmental stages because they were either localized to a specific medial region (i.e. L1- vs. L2-cells residing in the subendothelial vs. middle media, respectively) or exhibited a specific pattern of cell arrangement (e.g. "L3-S" cells forming longitudinally oriented compact cell clusters). These studies showed that the phenotypically distinct medial SMC subpopulations, observed in the adult media, progressed along distinct cyto-differentiation pathways during development, suggesting the existence of cells with unique genetic lineages within the large vessels of the lung (Fig. 2).

At the present time, the origins of these different cells remain unclear. However, the observations are consistent with the idea that cells of different origin and genetic composition contribute to the formation of the arterial media in large conduit vessels. In addition to cells arising from the local mesenchyme, mesenchymal precursor cells, neural crest cells, or bone marrow-derived progenitor cells may be recruited from distant sites to unique locations within the vessel wall and may give rise to cells with specific functional capabilities. Additional work is needed to define the precise lineages of SMC composing large pulmonary arteries.

In summary, the composition of the vascular media in large vessels is complex with multiple subpopulations of both smooth muscle and nonmuscle-like cells existing in close proximity within the arterial media. The bovine species provides, perhaps, one of the most dramatic examples of this, although similar but less marked findings have been reported in numerous species from avians to small rodents to humans (4,24,25,33).

3. Differential Proliferative Responses of Proximal Pulmonary Artery SMC Subpopulations to Hypoxia: *In Vivo* Analysis

The existence of phenotypically distinct SMC populations raises the possibility that these cells play unique roles in the adaptation of the vessel wall to stress and/or injury. Increasing evidence suggests that phenotypically distinct cells may exhibit differential responses to stress or injury and thus participate selectively in the subsequent vascular remodeling process (9-12,18,25). This work challenges conventional wisdom that medial and/or intimal thickening following injury is simply the result of a dedifferentiation and subsequent proliferation and migration of a single population of "well-differentiated" SMC. We thus tested the hypothesis that cells exhibiting distinct phenotypes would exhibit selective proliferative responses to chronic hypoxia (36). We evaluated the proliferative behavior of distinct medial SMC subpopulations, which were identified by immunobiochemical staining characteristics. Two distinct SMC subpopulations residing contiguous to each other in the outer media were selectively identified based on their expression of metavinculin or lack thereof (Fig. 3A). The proliferative response of these distinct SMC subpopulations to hypoxia was concurrently evaluated by performing immunofluorescent staining of the nuclear proliferation-associated antigen, Ki-67. Quantitative analysis of double-labeled tissue sections demonstrated that, at every post-hypoxic time point studied, >95% of the overall cell proliferation occurred within only one cell population, the metavinculin-negative SMC population (Fig. 3B). In contrast, the metavinculin-positive SMC population (labeled green in Fig. 3A) remained quiescent. These data provides compelling evidence that distinct SMC populations within the neonatal pulmonary artery media exhibit markedly different proliferative responses to hypoxic exposure (36). The data also raise the possibility that highly differentiated SMC are less susceptible to hypoxia-induced proliferation than cells exhibiting a less differentiated SM phenotype.

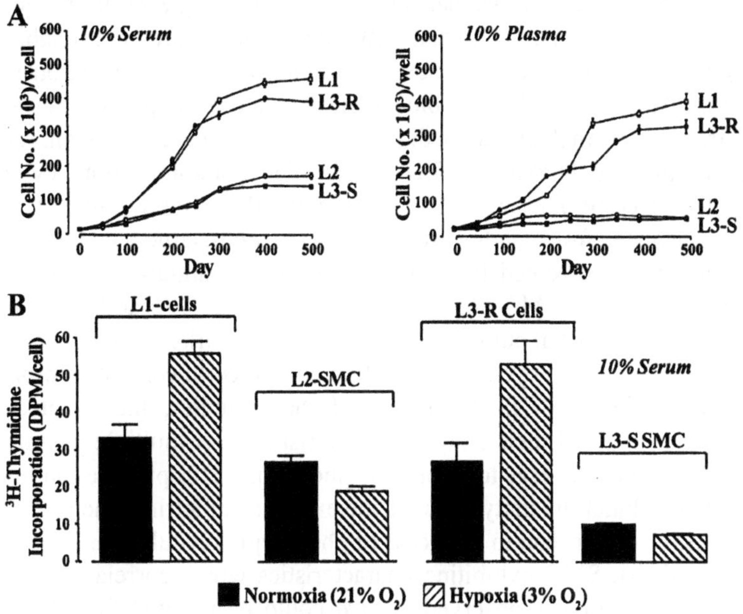

Figure 3. Phenotypically distinct medial SMC populations exhibit markedly different growth capabilities under identical culture conditions. A: Growth curves demonstrating heterogeneic proliferative capabilities of cell populations isolated from defined medial regions in response to stimulation with 10% serum (left) or 10% plasma (right). B: ^{3}H-thymidine incorporation in response to hypoxia (under serum stimulation) differs significantly in distinct cell subtypes obtained from different medial regions of the main pulmonary artery. Cell populations were isolated from the following medial regions: L1-cells from subendothelial media, L2-SMC from the middle media, L3-R cells and L3-S SMC from the outer media.

In an effort to provide further support to the idea that distinct subpopulations of medial cells respond differentially to the stimuli imposed by chronic hypoxic exposure, we evaluated the expression pattern of tropoelastin (9). We chose to evaluate tropoelastin because in its mature form, elastin, it plays a crucial role in determining the structure and function of large conduit arteries. We found that, in response to hypoxic exposure, tropoelastin expression was induced at high levels in metavinculin-negative SMC populations but not in metavinculin-positive ones. This data supported the existence of distinct SMC populations within the vascular media that differ in their functional responses to the stresses induced by hypoxia. Collectively, these observations lend further support to the idea that distinct SMC subpopulations contribute selectively to the pathogenesis of vascular disease.

4. Phenotypic and Functional Properties of Proximal Pulmonary Artery SMC Subpopulations: *In Vitro* Analysis

Studies of the cellular mechanisms conferring unique proliferative and synthetic responses to distinct medial SMC subpopulations require reliable and reproducible cell culture models. Several laboratories have now demonstrated that arterial SMCs with distinct biochemical and growth properties can be isolated from the normal arterial media of many species including avian, rodent, canine, porcine and human (4,15,18,25,33). For example, strong evidence suggests that specific ion channels are differentially expressed in SMC subpopulations isolated from conduit arteries (2,28). Studies, largely in the systemic circulation, have documented differences in the growth, migration, and synthetic capabilities of distinct SMC populations (4,15,18,25,33). In addition, human SMCs isolated from distal pulmonary arteries exhibit marked responsiveness to mitogenic and adenylate cyclase stimulation compared to cells from proximal pulmonary arteries (35). In almost all of these studies, the isolated cell populations maintained their unique characteristics over multiple passages in culture and did not converge to a common phenotype. This provides additional evidence that cellular diversity is due to the existence of intrinsic heterogeneity of arterial SMC rather than to a process of phenotypic modulation.

We found that SMC exhibiting characteristics which correlated with the diverse SMC phenotypes observed *in vivo* could be isolated from the main pulmonary artery and maintained in culture (10). Four phenotypically distinct SMC populations, each exhibiting different morphological and biochemical characteristics, are observed in culture (Table 2).

These cells have been broadly divided into nonmuscle-like and smooth muscle on the basis of morphological appearance and expression of α-SMA and SM-MHC. In general, the non-muscle and SM-like cell phenotypes exhibit distinct morphologic characteristics in culture. Nonmuscle-like cells express little α-SMA and no SM-MHC, exhibit a rounded or epitheliod-like morphology, and tend to form a monolayer at confluence. Similar cells have been described by several investigators in other species and termed "epitheliod SMC" (4,24,25). In addition, cells that express αSMA and SM-MHC in culture are routinely isolated from the middle media of the main pulmonary artery. The spindle-shaped morphology and "hill and valley" pattern of growth of this cell phenotype have previously been ascribed to "traditional" SMC in culture. These SMC appear biochemically and functionally similar to SMCs isolated by many other investigators from the bovine main pulmonary artery and are characterized by the moderate growth capabilities traditionally assigned to bovine SMC. In addition to these "traditional" SMC, we identified another SMC phenotype. These SMC are consistently isolated from the outer media and are much slower growing, larger in size and express characteristics of a more differentiated SMC-phenotype (based on expression of metavinculin). These "well-differentiated" SMC are similar to the subset of "highly differentiated" SMC isolated from the porcine artery (15). Importantly, the heterogenic cell populations that can be isolated from the bovine large vessel media bear a striking resemblance to the multiple different SMC clones obtained from the human internal thoracic artery (18).

Table 2. Four Phenotypically Distinct Smooth Muscle Cell Populations

	Main Pulmonary Artery Smooth Muscle Cells				Distal PA [a]
Cells	L1	L2	L3-S	L3-R	Distal SMC
Phenotype	NM	SM	SM	NM	SM
Morphology	Round	Spindle	Spindle	Round	Spindle
Staining for:					
α SMA	-	+++	+++	-	+++
SM-MHC	-	++	++	-	++
Metavinculin	-	-	+	-	+
Growth [b]					
At 10 days, 10% CS	++++	+++	+	++++	+
³H-Thymidine Incorporation[c]					
PDGF	++++	+++	+	++++	+
bFGF	+++	+++	+	+++	+
IGF-1	+++	+	-	+++	+
Hypoxia	++	-	-	++	-
³H-Thymidine Incorporation [c]					
Hypoxia, 10% CS	+	-	-	+	-
³⁵S-Methionine Incorporation [c]					
10% CS	ND	-	++	ND	+++
PDGF	ND	-	++	ND	++
ANG-II	ND	-	+++	ND	+++
TGF-β	ND	-	+++	ND	+++
Cell Size [d]					
10% CS	ND	+/-	++	ND	++

[a], <1500 µm; [b], approximate cell number increase as compared to day 1 in culture; [c], approximate change as compared to 0.1% calf serum (CS), normoxia; [d], approximate change as compared to 0.1% CS; NM, non-muscle; SM, smooth muscle; ND, not determined

The growth and matrix-producing capabilities of these different medial cell subpopulations were evaluated. The nonmuscle-like cell populations (subendothelial L1 and outer media L3-R cells) consistently exhibited markedly enhanced growth under serum-stimulated (10% calf serum) conditions compared with SMC subpopulations (middle media L2 and outer media L3-S SMC) (Table 2). Moreover, nonmuscle-like cells exhibited the potential to grow in medium supplemented with plasma, instead of serum, whereas SMC remained quiescent in plasma-supplemented medium. The response to peptide mitogens also differed among the distinct cell subpopulations (Table 2). For instance, PDGF-BB markedly increased DNA synthesis in the nonmuscle-like cells (subendothelial L1 and outer media "L3-R" subpopulations), increased to a lesser extent DNA synthesis in "traditional" SMC (middle media L2-SMC), and had little effect on well-differentiated "L3-S" SMC population derived from the outer media. Similar profiles of DNA synthesis were found in response to IGF-I and bFGF. In all studies there appeared to be an inverse correlation between cyto-differentiation status and proliferation with non-muscle "epitheliod" cell populations always exhibiting the highest growth potential. Interestingly, even between the two SMC populations (L2 and "L3-S" SMC) there were differences in the correlation between cytodifferentiation marker expression and proliferation. In a less differentiated L2-SMC population obtained from the middle media, it was common to see stimulation of proliferation with maintenance of both αSMA and SM-MHC expression. However, in the well-differentiated "L3-S" SMC subpopulation obtained from the outer media, it appeared that down-regulation of SM-MHC expression was necessary before significant increases in proliferation could be observed (10).

We think that these cell subpopulations represent phenotypically distinct cell phenotypes and are not the result of "phenotypic modulation" in culture. This assumption is based on the following: *i*) characteristic, morphological, immunobiochemical and proliferative properties observed at early passages in a specific cell population are maintained over time in culture (i.e. conversion from one phenotype to another was not observed); *ii*) each cell population was isolated from a specific medial region of the same vascular segment and exhibited biochemical and functional characteristics similar to those of cells observed in the corresponding medial region in vivo; *iii*) the unique morphological, biochemical and growth characteristics of each cell population were maintained under different culture conditions (in serum or plasma); and *iv*) these findings were consistent in cells isolated from large numbers of animals.

5. Differential Proliferative Responses to Hypoxia of Proximal Pulmonary Artery SMC Subpopulations: *In Vitro* Analysis

Of great interest is the possibility that an *in vitro* cell culture system could be utilized as a model system to investigate the mechanisms responsible for the marked differences in the response to hypoxia observed *in vivo*. Thus, we evaluated the proliferative responses of cell subpopulations isolated from distinct media regions of the main pulmonary artery to hypoxia (Table 2, Fig. 4). We found that in the presence of serum, DNA synthesis was increased in nonmuscle-like cells (subendothelial L1 and outer media "L3-R") in response to hypoxia, whereas it was decreased in SMC (middle media L2 and outer media "L3-S"). We also observed that certain nonmuscle-like cell populations could proliferate in response to hypoxia even in the absence of exogenously added serum, a unique response that was never observed in any of the SMC subpopulations even after prolonged periods in culture.

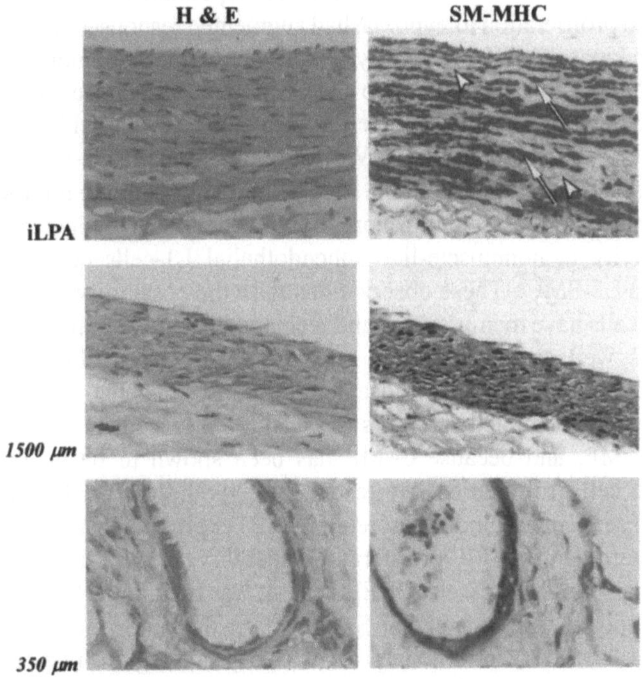

Figure 4. Progressive increase in the phenotypic uniformity of SMC comprising the mature bovine arterial media along the proximal-to-distal axis of the pulmonary vascular tree. The phenotype of SMC was defined based on expression of several muscle-specific markers (α-SM-actin, SM-myosin, calponin, desmin, metavinculin). Only immunostaining for smooth muscle myosin heavy chains (SM-MHC) is shown here (right). Phenotypic heterogeneity of SMC, identified in intralobar pulmonary arteries (iLPA), is not observed in distal arteries of approximate diameter ≤1500 μm. Rather, all SMC in small size distal arteries display a similar, uniform phenotype. H&E, hematoxyllin-eosin staining (left).

Given the differential proliferative responses exhibited by the cell populations to hypoxic conditions, we initiated an evaluation of the molecular mechanisms that contribute to the hypoxia-induced proliferative responses observed in distinct cell subpopulations (nonmuscle SMC). We found that non-muscle-like cells which consistently demonstrated augmented growth capabilities under hypoxic conditions, were characterized by exuberant responses to G-protein coupled receptor (GPCR) agonists compared to the SMC that did not exhibit proliferative response to hypoxia (11). These findings suggest that there may be differences in receptor expression and/or susceptibility to activation in hypoxia-proliferative versus hypoxia-nonproliferative cells that contribute to the sensitivity to reduction in oxygen concentrations.

We also found that the nonmuscle-like cell populations, which responded with increased proliferation to hypoxia, had augmented responses to stimulation of the protein kinase C (PKC) pathway. Since we previously demonstrated that activation of PKC is a requisite step for SMC to proliferate under hypoxic conditions (8), we evaluated the specific isozymes of PKC, which might be linked to this specific cell function. We found that nonmuscle-like cells had increased levels of immunodetectable PKCα compared with the middle media SMC (8). This pattern of isozyme expression was paralleled by increased PKC catalytic activity in nonmuscle-like subendothelial L1-cells compared with middle media L2-SMC. These observations raise the possibility that hypoxia-proliferative cells have membrane-bound receptors that are sensitive to hypoxic activation, as well as the ability to engage specific intracellular signaling pathways, which confer unique proliferative responses to cells.

Since G-protein coupled receptors (GPCR) are also known to have potent effects on cAMP, and because cAMP has been shown to be an important modulator of cell proliferation, we tested the hypothesis that differences in cAMP response element binding protein (CREB) expression can function as a molecular determinant of SMC proliferative capability (17). We found that in the large pulmonary conduit vessels, CREB content was high in the "proliferation-resistant" subpopulations of medial SMC (i.e. in the outer media L3-S and middle media L2-SMC) and low in proliferation-prone regions (especially the subendothelial L1 medial region). We found in general that CREB content was decreased and SMC proliferation was accelerated in vessels from neonatal calves exposed to chronic hypoxia. In culture, we found that serum deprivation of "traditional" middle media L2-SMC led to increased CREB content and a quiescent growth state. In contrast, a highly proliferative population of subendothelial nonmuscle-like cells had low CREB content even under serum-deprived conditions. A correlation between CREB content and proliferation was further demonstrated by the observation that over-expression of wild type or constituently active CREB arrested cell cycle progression. Additionally, expression of constituently active CREB decreased both proliferation and chemokinesis. Consistent with these functional properties, active CREB decreased the expression of multiple cell cycle regulatory genes as well as genes encoding growth factors, growth factor receptors, and cytokines. These data suggest that CREB, at least in vascular SMC, could act as a unique modulator of cellular phenotype determination (17).

Collectively, our *in vivo* and *in vitro* findings regarding the cells which comprise the large conducting portion of the pulmonary circulation as well as the findings of others, support the concept that there exists heterogeneity in growth, ion channel expression, and in matrix-producing capabilities of different SMC populations, and that these differences are intrinsic to the cell type. The data also strongly supports the idea that the differential proliferative and matrix-producing capabilities of distinct SMC populations contained within the large conducting vessels govern, at least in part, the pattern of abnormal cell proliferation and matrix protein synthesis that characterize chronic hypoxic forms of pulmonary hypertension. The observation that the isolated phenotypically distinct medial SMC subpopulations exhibit differential proliferative responses to hypoxia demonstrates that these cells can be used as a model system to evaluate the mechanisms that regulate selective responsiveness of medial SMC to low oxygen concentration. Finally, these data support the idea that specific PKC isozymes and mitogen-activated protein kinases are uniquely coupled to upstream membrane-bound receptors (including GPCR), which are activated under hypoxic conditions. These signaling pathways control expression of specific transcription factors, including CREB, that likely are important determinants of the differential growth responses to hypoxia exhibited by distinct SMCs.

6. Phenotypic Characteristics of SMC in the Distal Pulmonary Circulation: *In Vivo* and *In Vitro* Analysis

Due to its primary functions of vasoconstriction and regulation of blood flow, it seems likely that SMC composition of the distal pulmonary arterial bed is significantly different from that of the vessels in the more proximal pulmonary circulation. Unfortunately, there is a paucity of published data describing the phenotype of cells comprising the media in distal pulmonary arterial segments, particularly resistance size vessels which are, the locus of pulmonary circulatory control. Further, it remains unclear whether the complex structure seen in the main pulmonary artery media extends to the distal arteries, and whether the multiple distinct SMC phenotypes observed in the proximal pulmonary arteries are also present in small resistance vessels.

We thus evaluated SMC phenotypes at various points along the longitudinal axis of the mature pulmonary artery bed. Interestingly, we found that the complex mosaic of cells observed in large proximal vessels is lost as distance from the heart increases (Fig. 5) (32). Unlike large conduit pulmonary arteries, the media of small resistance vessels, based on immunohistochemical studies, appears to be composed of a uniform population of differentiated SMC that express all the SM markers including metavinculin and SM-β-MHC isoform. These cells exhibit a phenotype similar to that of one population of SMC in the outer media of the main pulmonary artery (well-differentiated "L3-S" SMC).

Recent studies on porcine coronary arteries have demonstrated that even tunica media composed of an apparently uniform cell phenotype *in vivo*, can give rise to phenotypically and functionally distinct cell populations in culture (14). Therefore, to further examine the cellular composition of the distal pulmonary artery, media cells from arteries with a diameter ranging from 600 to 1300 μm were isolated and expanded in culture. The isolated cells were of a uniform spindle-shaped morphology, and expressed α-SM-actin, SM-MHC, and metavinculin thus supporting the *in vivo* observations of their "well-differentiated" SM phenotype (32).

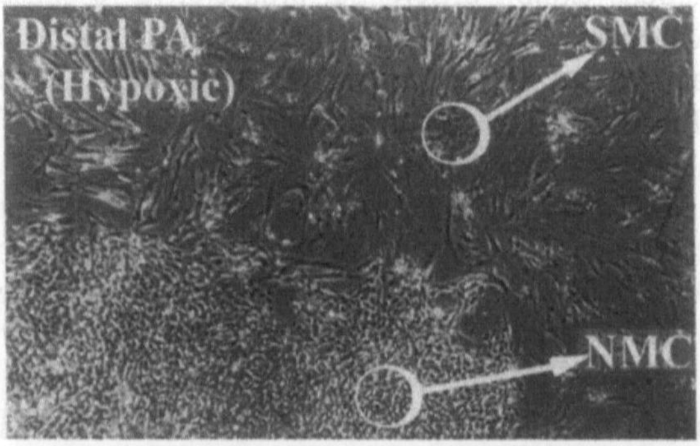

Figure 5. Primary culture of cells isolated from distal (≤1000μm, diameter) pulmonary arteries of hypoxic hypertensive neonatal calves. Two morphologically distinct cell populations are observed. Cells in one population (SMC, smooth muscle cells) are large, elongated and form "hill-and-valley" pattern at confluence, whereas cells in another population (NMC, nonmuscle cells) are markedly small and rhomboidal in shape, and pile on top of each other at confluence.

We evaluated the proliferative capabilities of these distal SMC and compared them to cell populations isolated from the proximal pulmonary circulation. The distal SMC were slow growing, with a growth rate similar to that of "well-differentiated" "L3-S" SMC from the outer media of the main pulmonary artery (MPA), and much slower than all other MPA cell populations (Table 2) (32). The proliferative responses of distal SMC to purified peptide mitogens (PDGF, IGF-1 and bFGF) were much less than those of the non-muscle cell populations from the MPA and also significantly less than those of L2-SMC from the middle MPA media. The responses were very similar to those observed in the "well-differentiated" L3-S-SMC from the outer MPA media. The rank order of DNA synthesis in response to mitogen stimulation was as follows, MPA non-muscle cells (L1, "L3-R") > MPA middle media L2-SMC > distal SMC ≥ MPA outer media "L3-S" SMC (Table 2) (32).

Because hypertrophy of cells has been suggested as a response of small vessels to hypertensive stress, the ability of distal SMC to increase ^{35}S-methionine incorporation (as a measure of protein synthesis) in response to purified peptide mitogens was also evaluated. We found that, in distal PA SMC and in MPA-"L3-S", PDGF, angiotensin II and TGFβ1 stimulated significant increases in ^{35}S-methionine incorporation (protein synthesis) while causing only minimal increases in ^{3}H-thymidine incorporation (cell proliferation). Thus, these responses were significantly different from those observed in the middle media SMC where PDGF stimulated proliferation rather than hypertrophy.

The responses of distal SMC to hypoxia were also evaluated. Hypoxia consistently inhibited the growth of distal pulmonary artery SMC under all conditions tested, (i.e. both in the presence and/or absence of 10% calf serum, Table 2). Thus, it appears as though medial SMC in the distal pulmonary circulation, at least in the bovine species, are of a uniform phenotype, "well-differentiated", and relatively growth-resistant.

There are at least three recent in vitro studies, which have described SMCs isolated from distal segments of the human pulmonary arterial tree. Yang and co-workers (37) described two SMC phenotypes isolated from distal segments of the human pulmonary arterial bed that differed in their proliferative response to hypoxia. Hypoxia inhibited proliferation in one cell phenotype, while stimulated in another. Using SMC derived from the pulmonary artery of both normal and hypertensive patients, Wharton and co-workers described that distal SMC exhibited marked responsiveness to serum-stimulated proliferation and adenylyl cyclase-induced inhibition of growth, compared to SMC from more proximal regions (34). Additional work from this group demonstrated that SMCs from distal segments express both endothelin A and endothelin B receptors and that both receptor subtypes mediate cell proliferation (6). These studies, when taken at face value, do confirm findings in many animal species of significant differences in SMC phenotype between proximal and distal vessels. In some ways, they suggest the existence of SMC heterogeneity in small distal pulmonary arteries in humans; however, because the cells utilized for study were often obtained from diseased as well as normal vessels, and because they were often only minimally characterized as SMC, the studies do not definitely exclude the possibility that some cells described were of non-medial origin.

7. Medial Thickening of Distal Pulmonary Arteries in Pulmonary Hypertension: Alternative Sources of Cells

Studies in the neonatal bovine animal model as well as rodent species have demonstrated that chronic hypoxia causes marked medial thickening of the distal pulmonary vessels (9,19,29,36). It has been suggested that the medial thickening is due to proliferation and hypertrophy of resident SMC. However, our preliminary studies mentioned above, particularly those performed *in vitro*, would suggest that the resident medial SMC population of distal vessels is relatively proliferation resistant and is not stimulated to proliferate in response to hypoxia or even other mitogens (32). Further, *in vivo* studies in both the rat and bovine species demonstrate that proliferation in the distal medial compartment is observed later (usually beginning at 3-4 days in rat and 4-7 in calves) than it is in other vascular compartments (3). Other possibilities, therefore, should be considered to explain the medial thickening. Some have suggested that hypoxia stimulates adventitial fibroblasts to secrete factors, which induce SMC proliferation (22). Another possibility, suggested 20 years ago by Sobin and supported by more recent data from the Jones' group, is that adventitial and/or interstitial lung fibroblasts are recruited and differentiate into myofibroblasts and/or SMC (29,16). This idea (implicating the role of adventitial fibroblasts) has received a great deal of attention in the systemic circulation especially in balloon injury models (26,35). A final possibility supported by recent data in the systemic circulation is that marrow-derived circulating precursor cells are recruited to the vessel wall and differentiate into myofibroblasts or SMC (1,21,23,27,35).

In order to better understand the mechanisms contributing to medial thickening in distal vessels, we performed studies evaluating the cellular composition of distal pulmonary arteries following exposure to two weeks of chronic hypoxia. In preliminary ex vivo studies, we observed marked heterogeneity in the staining for SM markers in the thickened distal pulmonary vascular media of hypertensive vessels (<1000 μm diameter) (14). Observed within the media were cells that did not express any of the SM markers, and cells that expressed α-SM-actin but not SM-myosin.

To further examine the possibility that the remodeled distal pulmonary artery media from the hypertensive animals was composed of a heterogeneous population of SMC, we cultured cells from the distal arteries of hypoxic and normoxic calves (14). We observed that two phenotypically distinct cell populations could be isolated from the distal PA media of chronically hypoxic animals. The major cell population exhibited a spindle-like morphology and based on expression of several SM markers including metavinculin, retained a "well-differentiated" SMC phenotype similar to that of resident SMC from the normoxic media. Another consisted of very small rhomboidal-shaped cells that expressed α-SM actin, but did not express any of the other SM markers tested. These latter cells (obtained only from hypoxic calves) exhibited extremely high growth potential compared to the spindle-shaped "well-differentiated" SMC from either hypoxic or normoxic animals. In fact, we observed that the spindle-shaped SMC isolated from the hypertensive pulmonary media were significantly larger in size (based on flow cytometry measurements) and proliferated less in response to serum stimulation than did the SMC cultured from the media of normoxic calves. We tested the proliferative and migratory responses to hypoxia of the spindle-shaped "well differentiated" SMC, and the small rhomboidal cell. We found that the growth of "well-differentiated" SMC was inhibited by hypoxia both in the presence and absence of serum. In contrast, the rhomboidal cells proliferated in response to hypoxia under all conditions tested. Hypoxia also markedly stimulated migration of the rhomboidal cells (14).

The enhanced proliferative response to hypoxia observed in pulmonary cultures of rhomboidal cells is similar to that previously described in adventitial fibroblasts as well as nonmuscle-like cells from the proximal pulmonary circulation. To address the possibility that prolonged exposure to serum could induce dedifferentiation and enhance proliferation of the resident SMC population, we examined numerous SMC isolates from the normal distal pulmonary circulation that had been exposed to serum for long periods of time in culture. We did not observe, under any of the conditions tested, that a resident population of distal SMC could dedifferentiate into a phenotype that exhibited either the morphological characteristics or the increased proliferative responses to serum and/or hypoxia. Thus, circumstantial evidence supports the possibility that medial thickening in the distal circulation in response to chronic hypoxia is due to the migration of non-resident cells, potentially adventitial fibroblasts, into the medial compartment. The idea that adventitial fibroblasts contribute to hypoxia-induced medial thickening has been recently reviewed (30, 31).

Furthermore, recent studies have suggested that stem-like cells, residing in postnatal tissue, exhibit the potential to differentiate into various lineages, including vascular cell phenotypes. Additionally, bone marrow (BM)-derived circulating progenitor cells have been implicated in the vascular remodeling process in a variety of systemic diseases (21,23,27). The precise cellular phenotype of these cells and the mechanisms by which they are mobilized from the BM to the site of injury, remain unclear. However, in the porcine coronary circulation in response to balloon over-distention injury, precursor monocyte/macrophage cells have been described to accumulate in the adventitia that are thought to ultimately transdifferentiate into myofibroblast and perhaps even smooth muscle-like cells (35).

These findings raise the possibility that BM-derived progenitor cells contribute to the remodeling process in pulmonary hypertension. Indeed, this concept has been recently tested in our laboratory in the chronically hypoxic neonatal calf model of pulmonary hypertension. In this model system the increase in vessel wall mass is associated with a significant increase in adventitial vasa vasorum, indicating a neovascularization process in response to chronic hypoxia. It is postulated that this vasa vasorum forms, at least in part, as a result of a postnatal vasculogenic process in addition to angiogenesis. This is based on our preliminary observations that demonstrate that there is a rapid appearance of cells expressing the *c-kit* receptor, the ligand for stem cell factor, in the adventitial compartment (7). The adventitia in these hypoxic animals is characterized by a marked increase in the expression of fibronectin, VEGF, and thrombin thus providing an environment not only conducive to attracting stem cells but also for facilitating their differentiation. It is now well established that BM-derived hematopoietic stem cells express *c-kit* and that these cells have the potential to differentiate into vascular phenotypes *in vivo* and *in vitro* (21). The appearance of *c-kit* expressing cells in the adventitia correlates with the their mobilization from the bone marrow into the circulation following hypoxic stress.

We hypothesize that the delivery route of circulating cells to the adventitia is via the newly formed vasa network that may be prone to fragility and permeability.

These observations, when taken in the context of observations made in the systemic circulation, provide a convincing argument that hypoxia mobilizes progenitor cells from the BM and that the hypoxic microenvironment of the pulmonary artery adventitia, which is characterized by marked increases in fibronectin, VEGF, and thrombin deposition, facilitates a suitable microenvironment to which these cells home, anchor, and differentiate (7). *In vitro* studies in our laboratory suggest that BM-derived progenitor cells may indeed differentiate into endothelial- and/or smooth muscle-like cells (7). Thus, they could contribute to the neovascularization process as well as to the vascular thickening that occurs in response to chronic hypoxia. These ideas are currently being tested in detail in ours and other laboratories.

8. Summary

Compelling data now exists demonstrating that the media of large conduit pulmonary and systemic vessels is composed of multiple, phenotypically distinct, SMC populations. These SMC populations can be identified *in vivo* based on location, morphology and/or expression patterns of muscle-specific contractile and cytoskeletal proteins. Studies in several species also demonstrate that phenotypically distinct SMC populations can be isolated from large vessels of both the pulmonary and systemic circulations, and that they maintain in culture markedly different functional phenotypes. The fidelity of characteristics between cells *in vivo* and those *in vitro*, combined with the stability of the differences in these characteristics when cells are placed under similar growth conditions, provides a strong argument that these are genetically distinct cell populations. At the present time, the embryonic origin of these cells, the number of different cells at a given vascular site in a given species, and the forces directing the developmental assembly of these cell types into a functioning, mature blood vessel wall remain to be determined. It is, however, evident that these phenotypically distinct cell populations serve different functions in health and disease. Current data support the hypothesis that, at least in large vessels, specific cell populations contribute selectively to the vascular remodeling process because of their unique proliferative, migratory and secretory capabilities. The molecular mechanisms contributing to these properties are beginning to be defined. It appears likely that the markedly different responses to hypoxic exposure exhibited by different cells are mediated through distinct signaling pathways or molecules, which are specific to each cell phenotype.

Since the distal pulmonary circulation subserves different functions than the proximal (i.e. is the primary site of hypoxic pulmonary vasoconstriction) an important question is whether the SMC heterogeneity observed in proximal arteries extends to the small resistance pulmonary vessels. Currently available information, at least in the bovine species, suggests this may not be the case. Emerging experimental evidence proposes that the distal pulmonary artery media, at least in the bovine species, is comprised of a phenotypically uniform population of well-differentiated, growth-resistant SMC. This observation raises questions as to what cells in the distal arteries might carry out the multiplicity of functions subserved by the distinct SMC subpopulations in the large conduit vessels, especially in response to injury. Numerous possibilities exist including adventitial fibroblasts, resident adventitial stem-like cells, and the recruited circulating bone marrow-derived fibrocytes or progenitor cells. Each of these cell types is capable of assuming a SM-like phenotype and thus might contribute to the medial changes observed in the remodeling process. Understanding the medial thickening, which occurs in small vessels under hypoxic conditions, will, therefore, require study of the mechanisms through which non-muscle cells are recruited into a SM phenotype. The precise definition of what a SMC is, both phenotypically and functionally, will continue to evolve.

References

1. Abe R, Donnelly SC, Peng T, Bucala R, and Metz CN. Peripheral blood fibrocytes: differentiation pathway and migration to wound sites. *J Immunol.* 2001; 166-7556-7562.
2. Archer SL, Huang JMC, Reeve HL, Hampl V, Tolarova S, Michelakis E, Weir EK and Huang JM. Differential distribution of electrophysiologically distinct myocytes in conduit and resistance arteries determines their response to nitric oxide and hypoxia. *Circ Res.* 1996; 78-431-442.
3. Belknap JK, Orton EC, Ensley B, Tucker A, and Stenmark KR. Hypoxia increases bromodeoxyuridine labeling indices in bovine neonatal pulmonary arteries. *Am J Respir Cell Mol Biol.* 1997; 16-366-371.
4. Bochaton-Piallat M-L, Ropraz P, Gabbiani F, Gabbiani G. Phenotypic heterogeneity of rat arterial smooth muscle cell clones. *Arterioscler Thromb Vasc Biol.* 1996; 16-815-820.
5. Chamley-Campbell JH, and Campbell GR. What controls smooth muscle phenotype? *Atherosclerosis* 1981; 40-347-357.
6. Davie N, Haleen SJ, Upton PD, Polak JM, Yacoub MH, Morrell NW, Wharton J. ET(A) and ET(B) receptors modulate the proliferation of human pulmonary artery smooth muscle cells. *Am J Respir Crit Care Med.* 2002; 165(3)-398-405.
7. Davies NJ, Crossno JT, Frid MG, Hofmeister SE, Reeves JT, Hyde DM, Carpenter TC, Brunetti JA, McNiece IK, and Stenmark KR. Hypoxia-induced pulmonary artery adventitial remodeling and neovascularization: Potential contribution of circulating progenitor cells. *Am. J. Physiol. Lung Cell. Mol. Physiol.* 2003 (in press).
8. Dempsey EC, Frid MG, Aldashev AA, Das M, and Stenmark KR. Heterogeneity in the proliferative response of bovine pulmonary artery smooth muscle cells to mitogens and hypoxia: importance of protein kinase C. *Can J Physiol Pharmacol.* 1997; 75(7)-936-944.
9. Durmowicz AG, Frid MG, Wohrley JD, and Stenmark KR. Expression and localization of tropoelastin mRNA in the developing bovine pulmonary artery is dependent on vascular cell phenotype. *Am J Respir Cell Mol Biol.* 1996; 14:569-576.
10. Frid MG, Aldashev AA, Dempsey EC, and Stenmark KR. Smooth muscle cells isolated from discrete compartments of the mature vascular media exhibit unique phenotypes and distinct growth capabilities. *Circ Res.* 1997; 81-940-952.

11. Frid MG, Aldashev AA, Nemenoff RA, Higashito R, Westcott JY, and Stenmark KR. Subendothelial cells from normal bovine arteries exhibit autonomous growth and constitutively activated intracellular signaling. *Arterioscler Thromb Vasc Biol.* 1999; 19-2884-2893.
12. Frid MG, Dempsey EC, Durmowicz AG, and Stenmark KR. Smooth muscle cell heterogeneity in pulmonary and systemic vessels. Importance in vascular disease. *Arterioscler Thromb Vasc Biol.* 1997; 17-1203-1209.
13. Frid MG, Moiseeva EP, and Stenmark KR. Multiple phenotypically distinct smooth muscle cell populations exist in the adult and developing bovine pulmonary arterial media in vivo. *Circ Res.* 1994; 75-669-681.
14. Gnanaskharan M, Frid MG, Stiebellehner L, Das M, and Stenmark KR. Chronic hypoxia induces the appearance of myofibroblast-like cells with enhanced proliferative and migratory properties in the distal pulmonary artery media. *Am J Respir Crit Care Med.* 2000; 161(3)-A631.
15. Hao H, Ropraz P, Verin V, Camenzind E, Geinoz A, Pepper MS, Gabbiani G, and Bochaton-Piallat ML. Heterogeneity of smooth muscle cell populations cultured from pig coronary artery. *Arterioscler Thromb Vasc Biol.* 2002; 22(7)-1093-1099.
16. Jones R, Jacobson M, and Steudel W. α-Smooth-muscle actin and microvascular precursor smooth-muscle cells in pulmonary hypertension. *Am J Respir Cell Mol Biol.* 1999; 20-582-594.
17. Klemm DJ, Watson PA, Frid MG, Dempsey EC, Schaack J, Colton LA, Nesterova A, Stenmark KR, and Reusch JE. cAMP response element-binding protein content is a molecular determinant of smooth muscle cell proliferation and migration. *J Biol Chem.* 2001; 276(49)-46132-46141.
18. Li S, Fan Y-S, Chow LH, Van Den Diepstraten C, van der Veer E, Sims SM, and Pickering JG. Innate diversity of adult human arterial smooth muscle cells. Cloning of distinct subtypes from the internal thoracic artery. *Circ Res.* 2001; 89-517-525.
19. Meyrick B, and Reid L. Hypoxia and incorporation of 3H-thymidine by cells of the rat pulmonary arteries and alveolar wall. *Am J Pathol.* 1979; 96(1)-51-70.
20. Neylon CB, Avdonin PV, Dilley RJ, Larsen MA, Tkachuk VA, and Bobik A. Different electrical responses to vasoactive agonists in morphologically distinct smooth muscle cell types. *Circ Res.* 1994; 75-733-741.
21. Orlic D, Kajstura J, Chimenti S, Limana F, Jakoniuk I, Quaini F, Nadal-Ginard B, Bodine DM, Leri A, and Anversa P. Mobilized bone marrow cells repair the infarcted heart, improving function and survival. *Proc Natl Acad Sci USA.* 2001; 98(18)-10344-10349.
22. Rose F, Grimminger F, Appel J, Heller M, Pies V, Weissmann N, fink L, Schmidt S, Krick S, Camenisch G, Gassmann M, Seeger W, and Hanze J. Hypoxic pulmonary artery fibroblasts trigger proliferation of vascular smooth muscle cells: role of hypoxia-inducible transcription factors. *FASEB J.* 2002; 16(12)-1660-1661.
23. Sata M, Saiura A, Kunisato A, Tojo A, Okada S, Tokuhisa T, Hirai H, Makuuchi M, Hirata Y, and Nagai R. Hematopoietic stem cells differentiate into vascular cells that participate in the pathogenesis of atherosclerosis. *Nat Med.* 2002; 8(4)-403-409.
24. Schwartz SM, Foy L, Bowen-Pope DF, and Ross R. Derivation and properties of platelet-derived growth factor-independent rat smooth muscle cells. *Am. J. Pathol.* 1990; 136-1417-1428.
25. Seidel CL. Cellular heterogeneity of the vascular tunica media. Implications for vessel wall repair. *Arterioscler Thromb Vasc Biol.* 1997; 17-1868-1871.
26. Shi Y, Pieniek M, Fard A O'Brien J, Mannion JD, and Zalewski A. Adventitial remodeling after coronary arterial injury. *Circulation* 1996; 93-340-348.
27. Shimizu K, Sugiyama S, Aikawa M, Fukumoto Y, Rabkin E, Libby P, and Mitchell RN. Host bone-marrow cells are a source of donor intimal smooth muscle-like cells in murine aortic transplant arteriopathy. *Nature Med.* 2001; 7(6)-738-741.
28. Smirnov SV, Beck R, Tammaro P, Ishii T, and Aaronson PI. Electrophysiologically distinct

smooth muscle cell subtypes in rat conduit and resistance pulmonary arteries. *J Physiol.* 2002; 538.3-867-878.

29. Sobin SS, Tremer HM, Hardy JD, and Chiodi HP. Changes in arteriole in acute and chronic hypoxic pulmonary hypertension and recovery in rat. *J Appl Physiol:Respirat Environ Exercise Physiol.* 1983; 55(5)-1445-1455.
30. Stenmark KR, Bouchey D, Nemenoff R, Dempsey EC, Das M. Hypoxia-induced pulmonary vascular remodeling: contribution of the adventitial fibroblast. *Physiol Res.* 2000; 49:503-517.
31. Stenmark KR, Gerasimovskaya E, Nemenoff RA and Das M. Hypoxic activation of adventitial fibroblasts. *Chest.* 2002; 122(6)-326S-334S.
32. Stielbellehner L, Frid MG, Reeves JT, Lew RB, Gnanasekharan M, and Stenmark KR. The bovine distal pulmonary arterial media is comprised of a uniform population of well-differentiated smooth muscle cells with low proliferative capabilities. *Am. J. Physiol. Lung Cell. Mol. Physiol.* 2003 (in press).
33. Topouzis S, and Majesky MW. Smooth muscle lineage diversity in the chick embryo. *Dev Biol.* 1996; 178-430-445.
34. Wharton J, Davie N, Upton PD, Yacoub MH, Polak JM, Morrell NW. Prostacyclin analogues differentially inhibit growth of distal and proximal human pulmonary artery smooth muscle cells. *Circulation.* 2000;102(25)-3130-3136.
35. Wilcox JN, Okamoto E-I, Nakahara K-I, and Vinten-Johansen J. Perivascular responses after angioplasty which may contribue to postangioplasty restenosis. A role for circulating myofibroblast precursors? *Annals NY Acad Sci.* 2001; 34(1)-68-90.
36. Whorley JD, Frid MG, Moiseeva EP, Orton EC, Belknap JK, and Stenmark KR. Hypoxia selectively induces proliferation in a specific subpopulation of smooth muscle cells in the bovine neonatal pulmonary arterial media. *J Clin Invest.* 1995; 96-273-281.
37. Yang X, Sheares KK, Davie N, Upton PD, Taylor GW, Horsley J, Wharton J, Morrell NW. Hypoxic induction of cox-2 regulates proliferation of human pulmonary artery smooth muscle cells. *Am J Respir Cell Mol Biol.* 2002; 27(6)-688-696.
38. Zanellato AM, Borrione AC, Giuriato L, Tonello M, Scannapieco G, Pauletto P, and Sartore S. Myosin isoforms and cell heterogeneity in vascular smooth muscle, I: developing an adult bovine aorta. *Dev Biol.* 1990; 141-431-446.

Chapter 27

Persistent Pulmonary Hypertension of the Newborn: Pathophysiology and Treatment

Steven H. Abman and Robin H. Steinhorn
University of Colorado School of Medicine, Denver, Colorado and Northwestern University Medical School, Chicago, Illinois, U.S.A.

1. Introduction

Postnatal survival is dependent upon the ability of the fetal cardio-pulmonary system to successfully respond to the sudden and harsh demands of neonatal life. Challenges to the lung at birth include the need for rapid absorption of fetal lung liquid, establishment of an air-liquid interface, initiation of spontaneous breathing with rhythmic ventilation, and closure of "fetal vascular channels." Perhaps the most dramatic event at birth involves the lung circulation, which must undergo a marked fall from its high resistance state *in utero* to a low resistance circuit within minutes after delivery. This postnatal fall in pulmonary vascular resistance (PVR) allows for the 8-fold increase in pulmonary blood flow that allows the lung to become an organ for gas exchange. Several mechanisms contribute to the normal fall in PVR at birth (3, 14, 17, 18, 29, 84). Birth-related stimuli, such as increased oxygen tension, ventilation and shear stress, cause vasodilation through changes in the production of vasoactive products, including increased release of potent vasodilators, including nitric oxide (NO) and prostacyclin (PGI_2), and decreased activity of endogenous vasoconstrictors, such as endothelin-1 (ET-1) (3, 14, 17, 18, 32, 84). Within minutes of delivery, high pulmonary blood flow abruptly increases shear stress and distends the vasculature, causing a "structural reorganization" of the vascular wall that includes flattening of the endothelium and thinning of smooth muscle cells and matrix (11). Thus, the ability to accommodate this marked rise in blood flow requires rapid functional and structural adaptations to ensure the normal postnatal fall in PVR.

Some infants fail to achieve or sustain the normal decrease in PVR at birth, leading to severe respiratory distress and hypoxemia, which is referred to as persistent pulmonary hypertension of the newborn (PPHN). PPHN is a major clinical problem, contributing significantly to high morbidity and mortality in both full-term and premature neonates (38, 44). Newborns with PPHN are at risk for severe asphyxia and its complications, including death, neurologic injury, and other problems. In this chapter, we briefly review the normal developmental physiology of the pulmonary circulation, mechanisms underlying the pathogenesis and pathophysiology of PPHN, and clinical strategies related to the evaluation and treatment of newborns with severe PPHN.

2. Physiology of the Fetal Pulmonary Circulation

Along with the rapid and dramatic progression of lung vascular growth and structure during development, the fetal pulmonary circulation also undergoes maturational changes in function. PVR is high throughout fetal life, especially in comparison with the low resistance of the systemic circulation. As a result, the fetal lung receives, 3-8% of combined ventricular output, with most of the right ventricular output crossing the ductus arteriosus (DA) to the aorta. Pulmonary arterial pressure and blood flow increase with advancing gestational age, along with increasing lung vascular growth (29). Despite this rise in vascular surface area, PVR increases with gestational age when corrected for lung or body weight, suggesting that vascular tone actually increases during late gestation and is high before birth. Studies of the human fetus support physiological data from fetal lambs (65). Based on multiple Doppler assessments of pulmonary artery velocity waveforms, fetal pulmonary artery impedance decreases slightly during the early third trimester, but does not decrease further during the latter stage of the third trimester despite ongoing vascular growth (65).

Several mechanisms contribute to high basal PVR in the fetus, including low oxygen tension, relatively low basal production of vasodilator products (such as PGI_2 and NO), increased production of vasoconstrictors (including ET-1 or leukotrienes), and altered smooth muscle cell reactivity (such as enhanced myogenic tone). In addition to high PVR, the fetal pulmonary circulation is also characterized by progressive changes in responsiveness to vasoconstrictor and vasodilator stimuli (or vasoreactivity). In the ovine fetus, the pulmonary circulation is initially poorly responsive to vasoactive stimuli during the early canalicular period, and responsiveness to several stimuli increases during late gestation. For example, the pulmonary vasoconstrictor response to hypoxia, and the vasodilator response to increased fetal Po_2 and acetylcholine increase with gestation (52, 70). As observed in the sheep fetus, human studies also demonstrate maturational changes in the fetal pulmonary vascular response to increased Pao_2 (65). Maternal hyperoxia does not increase pulmonary blood flow between 20-26 w eeks gestation, but increased Pao_2 caused pulmonary vasodilation in the 31-36 week fetus. These findings suggest that in addition to structural maturation and growth of the developing lung circulation, the vessel wall also undergoes functional maturation, leading to enhanced vasoreactivity during fetal life.

Mechanisms that contribute to progressive changes in pulmonary vasoreactivity during development are uncertain, but are likely due to maturational changes in endothelial cell function, especially with regard to NO production (1, 2, 4, 9). Lung endothelial NO synthase (eNOS, type III) mRNA and protein is present in the early fetus and increases with advancing gestation *in utero* and during the early postnatal period in rats and sheep (26, 56, 59). The timing of this increase in lung eNOS content immediately precedes and parallels changes in the capacity to respond to endothelium-dependent vasodilators, as assessed by *in vivo* and *in vitro* studies (4, 39). The timing of this increase in lung endothelial NOS content coincides with the capacity to respond to endothelium-dependent vasodilator stimuli, such as oxygen and acetylcholine. In contrast, fetal pulmonary arteries are already quite responsive to exogenous NO much earlier in gestation (4, 39). Overall, the ability of exogenous NO to dilate fetal pulmonary arteries is greater at less mature gestational ages than responsiveness to vasodilator stimuli that require the endothelium to release endogenous NO. These findings suggest that the ability of the endothelium to produce or sustain production of NO in response to specific stimuli during maturation lags the capacity of fetal pulmonary smooth muscle to relax to NO. This may account for clinical observations that extremely premature newborns are highly responsive to inhaled NO (7).

Although most studies of the perinatal lung have focused on the role of eNOS in vasoregulation, the other NOS isoforms, including neuronal NOS (nNOS; type I) and inducible NOS (iNOS; type II), have been identified by immunostaining in the rat, sheep and human fetal lung (6, 62-64, 72, 81). Lung nNOS mRNA and protein increases in parallel with eNOS expression during development in the fetal rat. Inducible (type II) NOS has also been detected in the ovine fetal lung, and is predominantly expressed in airway epithelium and vascular smooth muscle, with little expression in vascular endothelium. Whether the "non-endothelial" (types I and II) isoforms contribute to the physiologic responses of NO-dependent modulation of fetal pulmonary vascular tone has been controversial. Treatment of pregnant rats with an iNOS-selective antagonist caused constriction of the great vessels (main pulmonary artery and thoracic aorta) and DA in fetal rats. Selective iNOS and nNOS antagonists increase fetal PVR and inhibit shear stress vasodilation at doses that do not inhibit acetylcholine -induced pulmonary vasodilation (6, 62-64). These findings support the speculation that iNOS and nNOS may also modulate pulmonary vascular tone *in utero* and at birth (see below).

NOS expression and activity are affected by multiple factors, including oxygen tension, hemodynamic forces, hormonal stimuli (e.g., estradiol), paracrine factors (including vascular endothelial growth factor), substrate and cofactor availability, superoxide production (which inactivates NO), and others (Fig. 1) (24, 47, 57, 58). Recent studies have shown that estradiol acutely releases NO and upregulates eNOS expression in fetal pulmonary artery endothelial cells (47). However, *in vivo* findings on the effects of estradiol differ from these observations of isolated endothelial cells *in vitro*. Although estradiol does not cause acute fetal pulmonary vasodilation *in vivo*, prolonged estradiol treatment (24-72 hours) causes marked vasodilation, which is sustained despite cessation of estradiol infusion (57, 58). In contrast with estradiol, vascular endothelial growth factor (VEGF) acutely releases NO and causes pulmonary vasodilation *in vivo*. Chronic inhibition of VEGF receptors downregulates eNOS and induces pulmonary hypertension in the late gestation fetus (24). These findings illustrate that diverse hormonal and paracrine factors can regulate NOS expression and activity and affect lung vascular maturation during development.

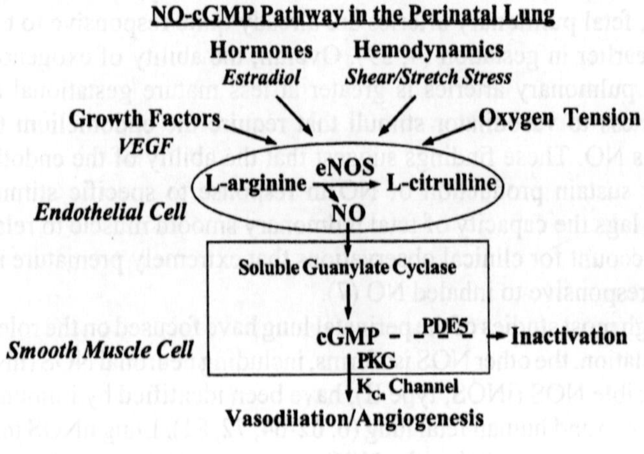

Figure 1. Physiology of the NO regulation in the developing lung.

In addition to transcriptional and translational regulation, multiple factors regulate NO production through alterations in NOS activity. NOS is a heterodimer with both reductase and oxygenase domains. When there is an abundance of availability of substrates such as L-arginine and the pteridine cofactor tetrahydrobiopterin, NADPH oxidation and NO synthesis remain coupled and NO production is favored. When concentrations of one or more factors are decreased, eNOS is uncoupled and generates superoxide. Under certain conditions, NOS may generate reactive oxygen species (superoxide and H_2O_2) rather than NO. The balance of NO vs. superoxide production depends on numerous factors. Heat shock protein 90 (Hsp90), a chaperone molecule, has been recently described as a factor that associates with NOS, stabilizes biopterin binding, and thus facilitates NO release. Konduri et al. have shown that association of Hsp90 with NOS is required for NO production in response to ATP in pulmonary arteries isolated from late-gestation fetal lambs (42).

Vascular responsiveness to endogenous or exogenous NO is also dependent upon several smooth muscle cell enzymes, including soluble guanylate cyclase (sGC), cGMP-specific phosphodiesterase (PDE5), and cGMP kinase (16, 28, 79, 81, 88). NO stimulates sGC by binding to the prosthetic heme of the enzyme, causing up to a 400-fold activation of the purified enzyme. Several studies have shown that sGC, which produces cGMP in response to NO activation, is active before 0.7 of term gestation in the ovine fetal lung. Similar to the pattern of expression for eNOS, sGC levels are high late in gestation, and are greater than those observed in the adult lung.

Cyclic nucleotide phosphodiesterases (PDE) constitute the only known pathway for the hydrolysis of cGMP, and control the intensity and duration of its signal transduction. At least eleven families of PDE isoenzymes have been identified, and at least four PDE isoenzymes have been identified in human pulmonary arteries. PDE5, a cGMP-binding, cGMP-specific isoform is found in especially high concentrations in the lung, and is active in the fetus. In the fetal lung, PDE5 expression has been localized to vascular smooth muscle; and similar to NOS and sGC, PDE5 activity is high in fetal lung in comparison with the postnatal lung. Infusions of selective PDE5 antagonists, including zaprinast, dipyridamole, E4021 and DMPPO, cause potent and sustained fetal pulmonary vasodilation. Thus, PDE5 activity appears to play a critical role in pulmonary vasoregulation during the perinatal period, and must be accounted for in assessing responsiveness to endogenous NO and related vasodilator stimuli. While most studies have focused on PDE5, there are many PDE families and isoforms that vary in their specificity for binding or metabolizing cGMP, cAMP or both. PDE are likely important mediators of cross talk between cGMP and cAMP signaling pathways, and other PDE isoforms may be important in the response to NO.

Functionally, the NO-cGMP cascade plays several important physiologic roles in vaso-regulation of the fetal pulmonary circulation. These include: *a*) modulation of basal PVR in the fetus (3); *b*) mediating the vasodilator response to specific physiologic and pharmacologic stimuli (3, 17); and *c*) opposing the strong myogenic tone in the normal fetal lung (77). Intrapulmonary infusions of NOS inhibitors increase basal PVR by 35% at least as early as 0.75 gestation (112 days; canalicular period) in the fetal lamb, suggesting that endogenous NOS activity appears to contribute to vasoregulation throughout late gestation. NOS inhibition also selectively blocks pulmonary vasodilation to such stimuli as acetylcholine, O_2, and shear stress in the normal fetus. Recent studies further suggest that NO release plays an additional role in modulating high intrinsic or myogenic tone in the fetal pulmonary circulation. The fetal lung circulation is characterized by its ability to oppose sustained pulmonary vasodilation during prolonged exposure to vasodilator stimuli. For example, increased PaO_2 increases fetal pulmonary blood flow during the first hour of treatment; however, blood flow returns toward baseline values over time despite maintaining high PaO_2 (9). Similar responses are observed during acute hemodynamic stress (shear stress) caused by partial compression of the DA (1) or with infusions of several pharmacologic agents (2). These findings suggest that unique mechanisms exist in the fetal pulmonary circulation that oppose vasodilation and maintain high PVR *in utero*.

We have speculated that this transient vasodilator response reflects the presence of an augmented myogenic response within the fetal pulmonary circulation. The myogenic response is commonly defined by the presence of increased vasoconstriction caused by acute elevation of intravascular pressure or "stretch stress." Past *in vitro* studies demonstrated the presence of a myogenic response in sheep pulmonary arteries, and that fetal pulmonary arteries have greater myogenic activity than neonatal or adult arteries. More recent studies of intact fetal lambs have shown that high myogenic tone is normally operative in the fetus, and contributes to maintaining high PVR *in utero* (49, 77, 78), and that acute inhibition of NO production unmasks a potent myogenic response. Further work suggests that downregulation of NO, as observed in experimental pulmonary hypertension, may further increase myogenic activity, increasing the risk for unopposed vasoconstriction in response to stretch stress at birth.

Since eNOS protein is present at a stage of lung development when blood flow is absent or minimal, it has been hypothesized that NO may potentially contribute to angiogenesis during early lung development (26). There are conflicting data regarding the effects of eNOS activity in promoting new vessel formation in experimental models of angiogenesis. Although NO can inhibit endothelial cell mitogenesis and proliferation, NO has also been shown to mediate the angiogenic effects of substance P and vascular endothelial growth factor *in vitro*. Growing bovine aortic endothelial cells in culture express greater eNOS mRNA and protein than confluent cells, but NOS inhibition does not affect their rate of proliferation *in vitro*. Recent studies have shown that NOS inhibition blocks VEGF-induced tube formation by fetal pulmonary artery endothelial cells *in vitro*, suggesting that NO may modulate lung vascular development. In addition angiogenesis, NO also modulates vascular wall structure by decreasing smooth muscle cell proliferation *in vitro*. Several studies have examined the role of NO in vascular growth and remodeling, but its effects vary between experimental settings, and the effects of NO on lung growth and structure *in vivo* are still controversial.

Although other vasodilator products, including PGI_2, are released upon stimulation of the fetal lung (e.g., increased shear stress), basal prostaglandin release appears to play a less important role than NO in fetal pulmonary vasoregulation. Cyclooxygenase inhibition has minimal affect on basal PVR and does not increase myogenic tone in the fetal lamb, PGI_2 vasodilation is blocked after NOS inhibition in the fetal lamb. The physiologic roles of other dilators, including adrenomedullin, adenosine and endothelium-derived hyperpolarizing factor (EDHF), are uncertain. EDHF is a short-lived product of cytochrome P450 activity that is produced by vascular endothelium, and has been found to cause vasodilation through activation of Ca^{2+}-activated K^+ channels in vascular smooth muscle *in vitro* (18). K^+-channel activation appears to modulate basal PVR and vasodilator responses to shear stress and increased oxygen tension in the fetal lung, and may be particularly important in relaxation responses in resistance vessels. Whether this is partly related to EDHF activity remains unknown.

Carbon monoxide (CO) is another gaseous molecule that is produced by heme-oxygenase and has been shown to have several vascular effects, including vasodilation in the adult systemic and pulmonary vascular beds. CO may act in part through activation of sGC, increasing cGMP content in vascular smooth muscle, and causing vasodilation, as described for NO. Despite several studies that suggest an important role in vasoregulation in some models, CO has yet to be shown to play an important physiologic role in the perinatal lung. For example, inhaled CO treatment of the late gestation fetal lamb had no affect on PVR, and infusions of a heme-oxygenase inhibitor did not alter basal pulmonary vascular tone. Further studies are needed to clarify the physiologic importance of CO in the developing lung circulation.

Vasoconstrictors have long been considered as potentially maintaining high PVR *in utero*. Several candidate products, such as thromboxane A_2, leukotrienes C_4 and D_4, platelet-activating factor and ET-1, have been extensively studied. Thromboxane A_2, a potent pulmonary vasoconstrictor that has been implicated in animal models of Group B Streptococcal sepsis, does not appear to influence PVR in the normal fetus. In contrast, inhibition of leukotriene production causes fetal pulmonary vasodilation, suggesting a potential role for lipoxygenase products in vasoregulation (73).

ET-1, a potent vasoconstrictor and co-mitogen that is produced by vascular endothelium, has been demonstrated to play a key role in fetal pulmonary vasoregulation (31). PreproET-1 mRNA (the precursor to ET-1) was identified in fetal rat lung early in gestation, and high circulating ET-1 levels are present in umbilical cord blood. Although ET-1 causes an intense vasoconstrictor response *in vitro*, its effects in the intact pulmonary circulation are complex. Brief infusions of ET-1 cause transient vasodilation, but PVR progressively increases during prolonged treatment. The biphasic pulmonary vascular effects during pharmacologic infusions of ET-1 are explained by the presence of at least two different ET receptors. The ET_B receptor, localized to the endothelium in the sheep fetus, mediates the ET-1 vasodilator response by releasing NO. A second receptor, the ET_A receptor, is located on vascular smooth muscle, and when activated, causes marked constriction. Although capable of both vasodilator and constrictor responses, ET-1 is more likely to play an important role as a pulmonary vasoconstrictor in the normal fetus (31). This is suggested in extensive fetal studies that have shown that inhibition of the ET_A receptor decreases basal PVR and augments the vasodilator response to shear stress-induced pulmonary vasodilation. Thus, ET-1 is likely to modulate PVR through the ET_A and ET_B receptors, but its predominant role is as a vasoconstrictor through stimulation of the ET_A receptor.

Several studies have shown that NO and ET-1 regulate each other through autocrine feedback loops. Endothelium-derived nitric oxide decreases endothelin production via a cGMP-dependent mechanism in cultured endothelial cells. Other studies have suggested that inhalation of NO increases levels of plasma ET-1 in young lambs, and that ET-1 in turn reduces NOS activity through activation of ET_A receptors (48, 87). Activation of ET_A receptors has been shown to increase superoxide production in vascular smooth muscle cells, which may provide a mechanism for the effects of ET-1 on NOS activity. These complex relationships are controversial and need further study.

3. Mechanisms of Pulmonary Vasodilation at Birth

Within minutes after delivery, pulmonary arterial pressure falls and blood flow increases in response to birth- related stimuli. Mechanisms contributing to the fall in PVR at birth include establishment of an air-liquid interface, rhythmic lung distension, increased O_2 tension, and altered production of vasoactive substances. Physical stimuli, such as increased shear stress, ventilation and increased O_2, cause pulmonary vasodilation in part by increasing production of vaosodilators (e.g., NO and PGI_2). Pretreatment with the arginine analogue, nitro-L-arginine, blocks NOS activity and attenuates the decline in PVR after delivery of near term fetal lambs. These findings suggested that ~50% of the rise in pulmonary blood flow at birth may be directly related to the acute release of NO. Specific mechanisms that cause NO release at birth include the marked rise in shear stress, increased O_2, and ventilation. Increased PaO_2 triggers NO release, which augments vasodilation through cGMP/PKG-mediated K^+ channel activation (66). Although the endothelial isoform of NOS (type III) has been presumed to be the major contributor of NO at birth, recent studies suggest that other isoforms (inducible (type II) and neuronal (type I)) may be important sources of NO release *in utero* and at birth as well. Although early studies were performed in term animals, NO also contributes to the rapid decrease in PVR at birth in premature lambs, at least as early as 112-115 days (0.7 term).

Other vasodilator products, including PGI_2, also modulate changes in pulmonary vascular tone at birth. Rhythmic lung distension and shear stress stimulate both PGI_2 and NO production in the late gestation fetus, but increased O_2 tension triggers NO activity and overcomes the effects of prostaglandin inhibition at birth. In addition, the vasodilator effects of exogenous PGI_2 are blocked by NOS inhibitors, suggesting that NO modulates PGI_2 activity in the perinatal lung. Adenosine release may also contribute to the fall in PVR at birth, but its actions may be partly through enhanced production of NO. Thus, although NO does not account for the entire fall in PVR at birth, NOS activity appears important in achieving postnatal adaptation of the lung circulation. Transgenic eNOS knock-out mice successfully make the transition at birth without evidence of PPHN (20). This finding suggests that $eNOS^{-/-}$ mice may have adaptive mechanisms, such as a compensatory vasodilator mechanisms (e.g., upregulation of other NOS isoforms or dilator prostaglandins) or less constrictor tone. Interestingly, these animals are more sensitive to the development of pulmonary hypertension at relatively mild decreases in PaO_2 and have higher neonatal mortality when exposed to hypoxia after birth (unpublished observations). We speculate that isolated eNOS deficiency alone may not be sufficient for the failure of postnatal adaptation, but that decreased ability to produce NO in the setting of a perinatal stress (e.g., hypoxia, inflammation, hypertension, or upregulation of vasoconstrictors) may cause PPHN.

4. Pathogenesis of PPHN

Several experimental models have been studied to explore the pathogenesis and pathophysiology of PPHN (21, 54). Such models have included exposure to acute or chronic hypoxia after birth, chronic hypoxia *in utero,* placement of meconium into the airways of neonatal animals, sepsis and others. Although each model demonstrates interesting physiologic changes that may be especially relevant to particular clinical settings, most studies examine only brief changes in the pulmonary circulation, and mechanisms underlying altered lung vascular structure and function of PPHN remain poorly understood. Clinical observations that neonates with severe PPHN who die during the first days after birth already have pathologic signs of chronic pulmonary vascular disease suggest that intrauterine events may play an important role in this syndrome. Adverse intrauterine stimuli during late gestation, such as abnormal hemodynamics, changes in substrate or hormone delivery to the lung, hypoxia, inflammation or others, may potentially alter lung vascular function and structure, contributing to abnormalities of postnatal adaptation. Several investigators have examined the effects of chronic intrauterine stresses, such as hypoxia or hypertension, in animal models in order to attempt to mimic the clinical problem of PPHN. Whether chronic hypoxia alone can cause PPHN is controversial. A past report that maternal hypoxia in rats increases pulmonary vascular smooth muscle thickening in newborns, but this observation has not been reproduced in maternal rats or guinea pigs with more extensive studies (54).

Pulmonary hypertension induced by early closure of the DA in fetal lambs alters lung vascular reactivity and structure, causing the failure of postnatal adaptation at delivery, and providing an experimental model of PPHN (8, 45, 51). Over days, pulmonary artery pressure and PVR progressively increase, but flow remains low and PaO_2 is unchanged (8). Marked right ventricular hypertrophy and structural remodeling of small pulmonary arteries develops after 8 days of hypertension. After delivery, these lambs have persistent elevation of PVR despite mechanical ventilation with high O_2 concentrations. Studies with this model show that chronic hypertension without high flow can alter fetal lung vascular structure and function. This model is characterized further by endothelial cell dysfunction and altered smooth muscle cell reactivity and growth, including findings of impaired NO production and activity due to downregulation of lung endothelial NOS mRNA and protein expression (69). Fetal pulmonary hypertension also impaired cGC and upregulated cGMP specific phosphodiesterase (type 5; PDE5) activities, suggesting further impairments in the NO-cGMP cascade (Fig. 2) (27, 50, 75, 78, 85). Thus, alterations in the NO-cGMP cascade appear to play an essential role in the pathogenesis and pathophysiology of experimental PPHN. Abnormalities of NO production and responsiveness contribute to altered structure and function of the developing lung circulation, leading to failure of postnatal cardiorespiratory adaptation. Recent evidence indicates that excessive production of reactive oxygen species such as superoxide in the pulmonary vasculature may further contribute to the disruption in NO-cGMP signaling in this model (13).

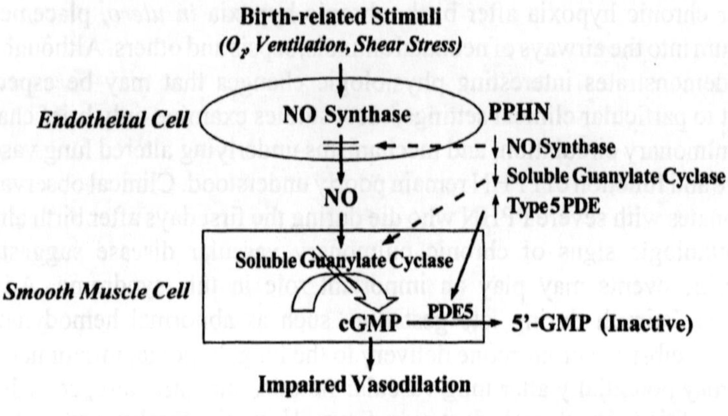

Figure 2. Physiologic mechanisms that contribute to the pathophysiology of PPHN.

Upregulation of ET-1 may also contribute to the pathophysiology of PPHN. Circulating levels of ET-1, a potent vasoconstrictor and co-mitogen for vascular smooth muscle cell hyperplasia, are increased in human newborns with severe PPHN (69). In the experimental model of PPHN due to compression of the DA in fetal sheep, lung ET-1 mRNA and protein content is markedly increased, and the balance of ET receptors are altered, favoring vasoconstriction (33, 34). Chronic inhibition of the ET A receptor attenuates the severity of pulmonary hypertension, decreases pulmonary artery wall thickening, and improves the fall in PVR at birth in this model. Thus, experimental studies have shown the important role of the NO-cGMP cascade and the ET-1 system in the regulation of vascular tone and reactivity of the fetal and transitional pulmonary circulation. Finally, in addition to vasoactive mediators, such as NO and ET-1, it has become clear that alterations of growth factors, such as VEGF and platelet-derived growth factor (PDGF), are likely to play key roles in the modulation of vascular maturation, growth and structure. For example, inhibition of PDGF-B attenuates smooth muscle hyperplasia in experimental pulmonary hypertension in fetal lambs, suggesting a potential role in the pathogenesis of PPHN (12).

A B

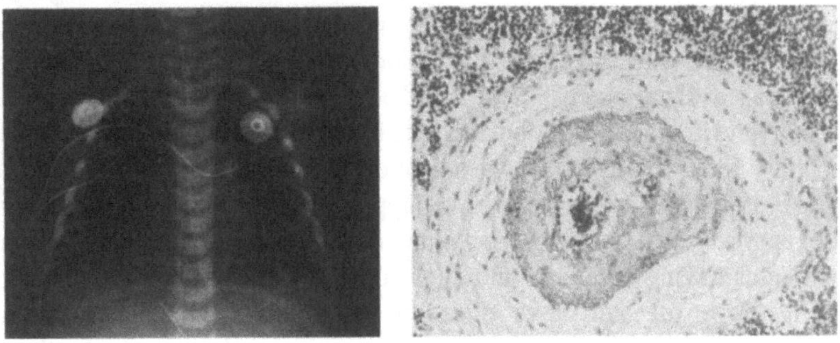

Figure 3. Chest x-ray findings illustrating typical findings of idiopathic PPHN (A) and histology of fatal PPHN illustrating severe structural changes in severe PPHN (B).

5. Clinical Aspects of PPHN

The first reports of PPHN described term newborns with profound hypoxemia who lacked radiographic evidence of parenchymal lung disease and echocardiographic evidence of structural cardiac disease (Fig. 3A). In these patients, hypoxemia was caused by marked elevations of PVR leading to right-to-left extrapulmonary shunting of blood across the patent DA (PDA) or foramen ovale (FO) during the early postnatal period. Due to the persistence of high PVR and blood flow through these fetal shunts, the term "persistent fetal circulation" was originally used to describe this group of patients. Consequently, it was recognized that this physiological pattern can complicate the clinical course of neonates with diverse causes of hypoxemic respiratory failure. As a result, the term PPHN has been considered as a syndrome, and is currently applied more broadly to include neonates that have a similar physiology in association with different cardiopulmonary disorders, such as meconium aspiration, sepsis, pneumonia, asphyxia, congenital diaphragmatic hernia, respiratory distress syndrome (RDS). Striking differences exist between these conditions, and mechanisms that contribute to high PVR can vary between these diseases. However, these disorders are included in the syndrome of PPHN due to common pathophysiological features, including sustained elevation of PVR leading to hypoxemia due to right-to-left extrapulmonary shunting of blood flow. In many clinical settings, hypoxemic respiratory failure in term newborns is often presumed to be associated with PPHN-type physiology; however, hypoxemic term newborns can lack echocardiographic findings of extra-pulmonary shunting across the PDA or PFO. Thus, PPHN should be reserved to neonates in whom extrapulmonary shunting contributes to hypoxemia and impaired cardio-pulmonary function. Recent estimates suggest an incidence for PPHN of 1.9/1000 live births, or an estimated 7400 cases/year (86).

PPHN-associated diseases are often classified within one of 3 categories: *i)* maladaptation: vessels are presumably of normal structural but have abnormal vasoreactivity; *ii)* excessive muscularization: increased smooth muscle thickness and increased distal extension of muscle to vessels which are usually non-muscular; and *iii)* underdevelopment: lung hypoplasia associated with decreased pulmonary artery number. This designation is, imprecise, however, and high PVR in most patients likely involve overlapping changes among these categories. For example, neonates with congenital diaphragmatic hernia are primarily classified as having vascular "underdevelopment" due to lung hypoplasia, yet lung histology of fatal cases typically shows marked muscularization of pulmonary arteries and, clinically, these patients respond to vasodilator therapy. Similarly, neonates with meconium aspiration often have clinical evidence of altered vasoreactivity, but often have muscularization at autopsy.

As described above, autopsy studies of fatal PPHN demonstrate severe hypertensive structural remodeling even in newborns who die shortly after birth, suggesting that many cases of severe disease are associated with chronic intrauterine stress (Fig. 3B). However, the exact intrauterine events that alter pulmonary vascular reactivity and structure are poorly understood. Epidemiologic studies have demonstrated strong associations between PPHN and maternal smoking and ingestion of cold remedies that include aspirin or other non-steroidal anti-inflammatory products (83). Since these agents can induce partial constriction of the DA), it is possible that pulmonary hypertension due to DA narrowing contributes to PPHN (see below). Other perinatal stresses, including placenta previa and abruption, and asymmetric growth restriction, are associated with PPHN; however, most neonates who are exposed to these prenatal stresses do not develop PPHN. Circulating levels of L-arginine, the substrate for NO, are decreased in some newborns with PPHN, suggesting that impaired NO production may contribute to the pathophysiology of PPHN, as observed in experimental studies. It is possible that genetic factors increase susceptibility for pulmonary hypertension. A recent study reported strong links between PPHN and polymorphisms of the carbamoyl phosphate synthase gene (61). However, the importance of this finding is uncertain, and further work is needed in this area. Studies of adults with idiopathic primary pulmonary hypertension have identified abnormalities of bone morphogenetic protein receptor genes; whether polymorphisms of genes for the BMP or TGF-ß receptors, other critical growth factors, vasoactive substances or other products increase the risk for some newborns to develop PPHN is unknown.

6. Clinical Presentation and Evaluation

Clinically, PPHN is most often recognized in term or near term neonates, but clearly can occur in premature neonates as well (Table 1). PPHN is often associated with perinatal distress (e.g., asphyxia, low APGAR scores, meconium staining); however, idiopathic PPHN can lack signs of acute perinatal distress. PPHN often presents as respiratory distress and cyanosis within 6-12 hrs of birth. Laboratory findings include low glucose, hypocalcemia, hypothermia, polycythemia or thrombocytopenia. Radiographic findings are variable, depending upon the primary disease asociated with PPHN. Classically, the chest x-ray in idiopathic PPHN is oligemic, may appear slightly hyperinflated, and lacks parenchymal infiltrates. In general, the degree of hypoxemia is often disproportionate to the severity of radiographic evidence of lung disease.

Table 1. Disorders Associated with Neonatal Pulmonary Hypertension

1. Pulmonary Disorders:
Meconium Aspiration Syndrome
Respiratory Distress Syndrome (* term and preterm newborns)
Lung Hypoplasia- Primary
Congenital Diaphragmatic Hernia
Pneumonia/Sepsis
Idiopathic
Transient tachypnea of the newborn
Alveolar-Capillary Dysplasia
Associated abnormalities in lung development:
- congenital lobar emphysema (rare association)
- cystic adenomatoid malformation (rare association)
- idiopathic, with impaired distal alveolarization
- others.
2. Cardiovascular Disorders
Myocardial dysfunction (asphyxia; infection; stress)
Structural cardiac diseases;
- mitral stenosis, cor triatriatum
- endocardial fibroelastosis
- Pompe's disease
- aortic atresia, coarctation of the aorta, interrupted aortic arch
- transposition of the great vessels
- Ebstein's anomaly, tricuspid atresia
Hepatic arteriovenous malformations (AVMs)
Cerebral AVMs
Total anomalous pulmonary venous return
Pulmonary vein stenosis (isolated)
Pulmonary atresia
3. Associations with Other Diseases
Neuromuscular disease
Metabolic disease
Polycythemia
Thrombocytopenia
Maternal drug use or smoking

Not all term newborns with hypoxemic respiratory failure have PPHN-type physiology (5). Hypoxemia in the newborn can be due to several mechanisms, including: extrapulmonary shunt, in which high pulmonary artery pressure at systemic levels leads to right-to-left shunting of blood flow across the PDA or PFO; and intrapulmonary shunt or ventilation-perfusion mismatch, in which hypoxemia results from the lack of mixing of blood with aerated lung regions due to parenchymal lung disease, without the shunting of blood flow across the PDA and PFO. In the latter setting, hypoxemia is related to the amount of pulmonary arterial blood that perfuses non-aerated lung regions. Although PVR is often elevated in hypoxemic newborns without PPHN, high PVR does not contribute significantly to hypoxemia in these cases.

Several factors can contribute to high pulmonary artery pressure in neonates with PPHN-type physiology. Pulmonary hypertension can be due to vasoconstriction or structural lesions that directly increase PVR. Changes in lung volume in neonates with parenchymal lung disease can also be an important determinant of PVR. PVR increases at low lung volumes due to dense parenchymal infiltrate and poor lung recruitment, or with high lung volumes due to hyperinflation associated with overdistension or gas-trapping. Cardiac disease is also associated with PPHN. High pulmonary venous pressure due to left ventricular dysfunction can also elevate PAP (e.g., asphyxia, sepsis), causing right-to-left shunting, with little vasoconstriction. In this setting, enhancing cardiac performance and systemic hemodynamics may lower PAP more effectively than achieving pulmonary vasodilation. Thus, understanding the cardiopulmonary interactions are key to improving outcome in PPHN.

PPHN is characterized by hypoxemia that is poorly responsive to supplemental O_2. In the presence of right-to-left shunting across the PDA, "differential cyanosis" is often present, which is difficult to detect by physical exam, and is defined by a difference in PaO_2 between right radial artery versus descending aorta values ≥ 10 torr, or an O_2 saturation gradient >5%. However, post-ductal desaturation can be found in ductus-dependent cardiac diseases, including hypoplastic left heart syndrome, coarctation of the aorta or interrupted aortic arch. The response to supplemental O_2 can help to distinguish PPHN from primary lung or cardiac disease. Although supplemental oxygen traditionally increases PaO_2 more readily in lung disease than cyanotic heart disease or PPHN, this may not be obvious with more advanced parenchymal lung disease. Marked improvement in SaO_2 (increase to 100%) with supplemental oxygen suggests the presence of V/Q mismatch due to lung disease or highly reactive PPHN. Most patients with PPHN have at least a transient improvement in oxygenation in response to interventions such as high inspired oxygen and/or mechanical ventilation. Acute respiratory alkalosis induced by hyperventilation to achieve $PaCO_2$ <30 torr and a pH >7.50 may increase PaO_2 >50 torr in PPHN, but rarely in cyanotic heart disease.

The echocardiogram plays an essential diagnostic role and is an essential tool for managing newborns with PPHN. The initial echocardiographic evaluation is important to rule-out structural heart disease causing hypoxemia (e.g., coarctation of the aorta and total anomalous pulmonary venous return). As stated above, not all term newborns with hypoxemia have PPHN physiology. Although high pulmonary artery pressure may be common, the diagnosis of PPHN is uncertain without evidence of bidirectional or predominantly right-to-left shunt ing across the PFO or PDA. Echocardiographic signs suggestive of pulmonary hypertension (e.g., increased right ventricular systolic time intervals and septal flattening) are less helpful. In addition to demonstrating the presence of PPHN physiology, the echocardiogram is critical for the evaluation of left ventricular function and diagnosis of anatomic heart disease, including such "PPHN mimics" as coarctation of the aorta; total anomalous pulmonary venous return; hypoplastic left heart syndrome; and others. Studies should carefully assess the predominant direction of shunting at the PFO as well as the PDA. Although right-to-left shunting at the PDA and PFO is typical for PPHN, predominant right-to-left shunting at the PDA but left-to-right shunt at the PFO may help to identify the important role of left ventricular dysfunction to the underlying pathophysiology. In the presence of severe left ventricular dysfunction with pulmonary hypertension, pulmonary vasodilation alone may be ineffective in improving oxygenation. In this setting, efforts to reduce PVR should be accompanied by targeted therapies to increase cardiac performance and decrease left ventricular afterload. Thus, careful echocardiographic assessment provides invaluable information about the underlying pathophysiology and will help guide the course of treatment.

7. Treatment of PPHN

In general, management of the newborn with PPHN includes the treatment and avoidance of hypothermia, hypoglycemia, hypocalcemia, anemia and hypovolemia; correction of metabolic acidosis; diagnostic studies for sepsis; serial monitoring of arterial blood pressure, pulse oximetery (pre- and post-ductal); and transcutaneous PCo_2, especially with the initiation of high frequency oscillatory ventilation (HFOV). Therapy includes aggressive management of systemic hemodynamics with volume and cardiotonic therapy (dobutamine, dopamine, and milrinone), in order to enhance cardiac output and systemic O_2 transport. In addition, increasing systemic arterial pressure can improve oxygenation in some cases by reducing right-to-left extrapulmonary shunting. Failure to respond to medical management, as evidenced by failure to sustain improvement in oxygenation with good hemodynamic function, often leads to treatment with extracorporeal membrane oxygenation (ECMO) (82). Although ECMO can be a life-saving therapy, it is costly, labor intensive, and can have severe side effects, such as intracranial hemorrhage. Since arterio-venous ECMO usually involves ligation of the carotid artery, acute and long-term CNS injury remain major concerns.

The goal of mechanical ventilation is to improve oxygenation and to achieve "optimal" lung volume to minimize the adverse effects of high or low lung volumes on PVR, while minimizing the risk for lung injury ("volutrauma"). Mechanical ventilation using inappropriate settings can produce acute lung injury (ventilator-induced lung injury; VILI), causing pulmonary edema, decreased lung compliance and promoting lung inflammation due to increased cytokine production and lung neutrophil accumulation. The development of VILI is an important determinant of clinical course and eventual outcome of newborns with hypoxemic respiratory failure, and postnatal lung injury worsens the degree of pulmonary hypertension (60). Failure to achieve adequate lung volumes (functional residual capacity) contributes to hypoxemia and high PVR in newborns with PPHN. Some newborns with parenchymal lung disease with PPHN physiology improve oxygenation and decrease right-to-left extra-pulmonary shunting with aggressive lung recruitment during high frequency oscillatory ventilation (37) or with an "open lung approach" of higher positive end-expiratory pressure with low tidal volumes, as more commonly utilized in older patients with ARDS (10).

Marked controversy and variability exists between centers regarding the use of hyperventilation to achieve alkalosis in order to improve oxygenation. Past studies have clearly shown that acute hyperventilation can improve PaO_2 in neonates with PPHN, providing a diagnostic test and therapeutic strategy. However, there are many issues with the use of hypocarbic alkalosis for prolonged therapy. Depending on the ventilator strategy and underlying lung disease, hyperventilation is likely to increase VILI, and the ability to sustain decreased PVR during prolonged hyperventilation is unproven. Studies suggest that the response to alkalosis is transient, and that alkalosis may paradoxically worsen pulmonary vascular tone, reactivity and permeability edema (23, 43). In addition, prolonged hyperventilation reduces cerebral blood flow and O_2 delivery to the brain, potentially worsening neurodevelopmental outcome.

Additional therapies, including infusions of sodium bicarbonate, surfactant therapy and the use of intravenous vasodilator therapy, are also highly variable between centers. Although surfactant may improve oxygenation in some lung diseases, such as meconium aspiration and RDS, a multicenter trial failed to show a reduction in ECMO utilization in newborns with PPHN (46). The use of intravenous vasodilator drug therapy, with such agents as tolazoline, magnesium sulfate, PGI_2 and sodium nitroprusside, is also controversial due to the non-selective effects of these agents on the systemic circulation. Systemic hypotension worsens right-to-left shunting, may impair O_2 delivery and worsen gas exchange in patients with parenchymal lung disease. In addition, the initial response to such agents as tolazoline are often transient, and can have severe adverse effects (e.g., gastrointestinal hemorrhage). Endotracheal administration of vasodilators, including tolazoline and sodium nitroprusside, may cause selective pulmonary vasodilation and minimize systemic hypotension. However, these data are largely limited to animal studies, and evidence is currently lacking that supports the safety and efficacy of this approach in humans.

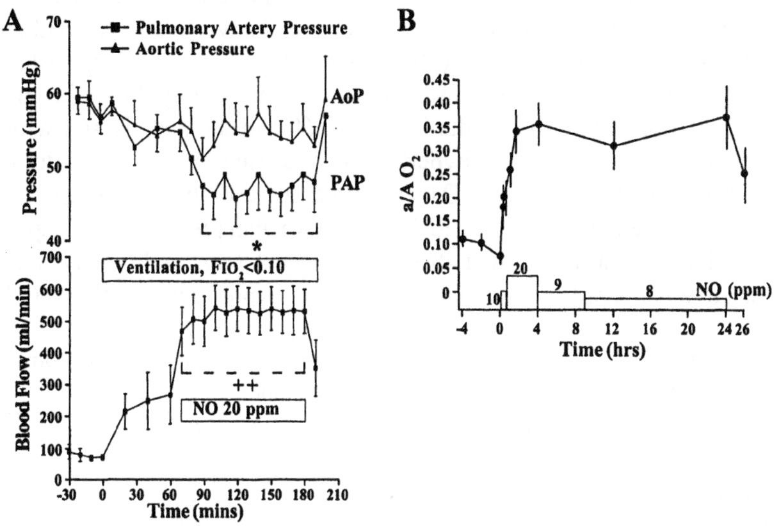

Figure 4. A: Physiologic effects of Inhaled NO in the perinatal lamb. As shown in the upper panel, inhaled NO decreases pulmonary artery pressure (PAP) but not aortic pressure (AoP), and increases pulmonary blood flow (lower panel) during mechanical ventilation with hypoxic gas. B: Low doses of inhaled NO increases oxygenation (as assessed by the arterial to alveolar ratio (a/A O_2)) in human newborns with severe PPHN who met ECMO criteria. Improvement in oxygenation was sustained with continuous treatment of inhaled NO, obviating the need for ECMO in these patients (Modified from Refs. 39, 40).

Inhaled nitric oxide (iNO) therapy at low doses (5-20 ppm) improves oxygenation and decreases the need for ECMO therapy in patients with diverse causes of PPHN (Fig. 4A) (15, 19, 40, 41, 55, 67, 68). Multicenter clinical trials support the use of iNO in near-term (>34 weeks gestation) and term newborns, and the use of iNO in infants <34 weeks gestation remains investigational. Studies support the use of iNO in infants who have hypoxemic respiratory failure with evidence of PPHN, who require mechanical ventilation and high inspired oxygen concentrations. The most common criterion employed has been the oxygenation index (OI; mean airway pressure times FiO_2 times 100 divided by PaO_2. Although clinical trials commonly allowed for enrollment with OI levels >25, the mean OI at study entry in multicenter trials was ~40. It is unclear whether infants with less severe hypoxemia would benefit from iNO therapy.

The first studies of iNO treatment in term newborns reported initial doses that ranged from 80 ppm to 6-20 ppm (40, 68). In the latter report, rapid improvement in PaO_2 was achieved at low doses (20 ppm) for 4 hours, and this response was sustained with prolonged therapy at 6 ppm (Fig. 4B) (40).

Subsequent multicenter studies confirmed the efficacy of this dosing strategy, and showed that increasing the dose in non-responders did not improve outcomes (Table 2). The available evidence, therefore, supports the use of doses of iNO beginning at 20 ppm in term newborns with PPHN, since this strategy decreased ECMO utilization without an increased incidence of adverse effects. Although brief exposures to higher doses (40-80 ppm) appear to be safe, sustained treatment with 80 ppm NO increases the risk of methemoglobinemia. In our practice, we discontinue iNO if the FiO_2 is <0.60 and the PaO_2 is >60 without evidence of rebound pulmonary hypertension or a rise in FiO_2 >15% after iNO withdrawal. Prolonged need for iNO therapy without resolution of disease should lead to a more extensive evaluation to determine whether previously unsuspected anatomic lung or cardiovascular disease is present (e.g., pulmonary venous stenosis, alveolar capillary dysplasia, severe lung hypoplasia, or others) (22).

Table 2. Summary of Multicenter Randomized Trials of Inhaled NO in Term Newborns with Hypoxemia and PPHN, Showing Reduction of ECMO Utilization

	Control	iNO Therapy	*P*-value
ECMO Therapy			
NINOS	54%	39%	<0.006
CINRGI	64%	39%	<0.001
INOSG*	71%	40%	<0.02
Death			
NINOS	16%	14%	NS
CINRGI	11%	8%	NS
INOSG*	7%	7%	NS

* Retrospective analysis. Data are from the NINOS (ref. 55), CINRGI (ref. 15), and INOSG (ref.. 67) trials.

In newborns with severe lung disease, HFOV is frequently used to optimize lung inflation and minimize lung injury. In clinical pilot studies using iNO, the combination of HFOV and iNO caused the greatest improvement in oxygenation in some newborns who had severe PPHN complicated by diffuse parencyhmal lung disease and underinflation (e.g. RDS, pneumonia). A randomized, multicenter trial demonstrated that treatment with HFOV+iNO was often successful in patients who failed to respond to HFOV or iNO alone in severe PPHN, and differences in responses were related to the specific disease associated with the complex disorders of PPHN (41). For patients with PPHN complicated by severe lung disease, response rates for HFOV+iNO were better than HFOV alone or iNO with conventional ventilation. In contrast, for patients without significant parenchymal lung disease, both iNO and HFOV+iNO were more effective than HFOV alone. This response to combined treatment with HFOV+iNO likely reflects both improvement in intrapulmonary shunting in patients with severe lung disease and PPHN (using a strategy designed to recruit and sustain lung volume, rather than to hyperventilate) and augmented NO delivery to its site of action. Although iNO may be an effective treatment for PPHN, it should be considered only as part of an overall clinical strategy that cautiously manages parenchymal lung disease, cardiac performance, and systemic hemodynamics.

Although clinical improvement during inhaled NO therapy occurs with many disorders associated with PPHN, not all neonates with acute hypoxemic respiratory failure and pulmonary hypertension respond to iNO. Several mechanisms may explain the clinical variability in responsiveness to iNO therapy. An inability to deliver NO to the pulmonary circulation due to poor lung inflation is the major cause of poor responsiveness. In some settings, administration of NO with high frequency oscillatory ventilation has improved oxygenation more effectively than during conventional ventilation in the same patient. In addition, poor NO responsiveness may be related to myocardial dysfunction or systemic hypotension, severe pulmonary vascular structural disease, and unsuspected or missed anatomic cardiovascular lesions (e.g., total anomalous pulmonary venous return, coarctation of the aorta, alveolar capillary dysplasia, and others).

Another mechanism of poor responsiveness to inhaled NO may be altered smooth muscle cell responsiveness, and there are emerging therapies that take advantage of our increased understanding of the cellular effects of iNO. Inhibition of cGMP-metabolizing phosphodiesterase (PDE5) activity may increase efficacy of iNO by increasing cGMP concentrations. Recently approved by the FDA for the treatment of male erectile dysfunction, sildenafil is a potent and highly specific PDE5 inhibitor. Intravenous sildenafil was found to be a selective pulmonary vasodilator with efficacy equivalent to inhaled nitric oxide in a piglet model of meconium aspiration (71).

New studies indicate that scavengers of reactive oxygen species such as N-acetylcysteine or superoxide dismutase (SOD) can augment responsiveness to iNO. A single dose of recombinant human SOD significantly enhanced the response to iNO in the ductal ligation lamb model of PPHN (74). Besides administering iNO as an inhalational agent, Moya et al. have recently suggested that treatment with a unique gas, O-nitrosoethanol (ethyl nitrite, ENO), may increase the endogenous pool of S-nitrosothiols in the airway and circulation, thereby providing a new treatment strategy for PPHN (53). In a brief report, ENO briefly improved post-ductal arterial saturation for 4 hours in 7 neonates. However, only a few patients improved oxygenation and the response was small, was associated with a rapid rise in methemoglobinemia, and may have had systemic effects. Another potential approach to augment pulmonary vasodilation may be to combine treatment with PGI_2. Whereas intravenous PGI_2 can decrease systemic arterial pressure, inhaled PGI_2 improved oxygenation in 4 infants with PPHN who did not respond or sustain their response to iNO without worsening systemic hemodynamics (35). Whether these strategies will be more effective or will improve responsiveness in neonates who fail to respond to iNO therapy is unknown.

Finally, although newer therapies, including HFOV and inhaled NO, have led to a dramatic reduction in the need for ECMO therapy (30, 36), ECMO remains an effective rescue agent for severe PPHN. Current patterns of ECMO use demonstrate persistent use in neonates with congenital diaphragmatic hernia and patients with severe hemodynamic instability, with less need for ECMO in meconium aspiration, RDS, idiopathic PPHN and other disorders.

8. Conclusions

PPHN is a clinical syndrome that is associated with diverse cardio-pulmonary diseases, with pathophysiologic mechanisms including pulmonary vascular, cardiac and lung disease. Experimental work on basic mechanisms of vascular regulation of the developing lung circulation and models of perinatal pulmonary hypertension has improved our therapeutic approaches to neonates with PPHN. Inhaled NO has been shown to be an effective pulmonary vasodilator for infants with PPHN, but successful clinical strategies require meticulous care of associated lung and cardiac disease. More work is needed to expand our therapeutic repertoire in order to further improve the outcome of the sick newborn with severe hypoxemia, especially in patients with lung hypoplasia and advanced structural vascular disease.

References

1. Abman SH, and Accurso FJ. Acute effects of partial compression of the ductus arteriosus on the fetal pulmonary circulation. *Am. J. Physiol.* 1989; 257: H626-H634.
2. Abman SH, and Accurso FJ. Sustained fetal pulmonary vasodilation during prolonged infusion of atrial natriuretic factor and 8-bromo-guanosine monophosphate. *Am. J. Physiol.* 1991; 260: H183-H192.
3. Abman SH, Chatfield BA, Hall SL, and McMurtry IF. Role of endothelium-derived relaxing factor during transition of pulmonary circulation at birth. Am. J. Physiol. 1990; 259:H1921-H1927.
4. Abman SH, Chatfield BA, Rodman DM, Hall SL, and McMurtry IF. Maturation-related changes in endothelium-dependent relaxation of ovine pulmonary arteries. *Am. J. Physiol.* 1991; 260:L280-L285.
5. Abman SH, and Kinsella JP. Inhaled NO for persistant pulmonary hypertension of the newborn: the physiology matters. *Pediatrics*. 1995; 96:1147-1151.
6. Abman SH, Kinsella JP, Parker TA, Storme L, and Le Cras TD. "Physiologic roles of NO in the perinatal pulmonary circulation. In *Fetal and Neonatal Pulmonary Circulation,* Weir EK, Archer SL, and Reeves JT, eds. Futura, NY: New York, NY, 1999, pp. 239-260.
7. Abman SH, Kinsella JP, Schaffer MS, and Wilkening RB. Inhaled nitric oxide therapy in a premature newborn with severe respiratory distress and pulmonary hypertension. *Pediatrics*. 1993; 92:606-609.
8. Abman SH, Shanley PF, and Accurso FJ. Failure of postnatal adaptation of the pulmonary

circulation after chronic intrauterine pulmonary hypertension in fetal lambs. *J. Clin. Invest.* 1989; 83:1849-1858.

9. Accurso FJ, Alpert B, Wilkening RB, Petersen RG, and Meschia G. Time-dependent response of fetal pumonary blood flow to an increase in fetal oxygen tension. *Respir. Physiol.* 1986; 63:43-52.
10. Acute Respiratory Distress Syndrome Network. Ventilation with lower tidal volumes as compared with traditional tidal volumes for acute lung injury and the ARDS. *N. Engl. J. Med.* 2000; 342:1301-1308.
11. Allen K, and Haworth SG. Impaired adaptation of intrapulmonary arteries to extrauterine life in newborn pigs exposed to hypoxia. an ultrastructural study. *Fed. Proc.* 1986; 45:879.
12. Balasubramaniam V, Le Cras TD, Ivy DD, Kinsella JP, Grover TR, and Abman SH. Role of platelet-derived growth factor in the pathogenesis of perinatal pulmonary hypertension. *Am. J. Physiol. Lung Cell. Mol. Physiol.* 2003; In Press.
13. Brennan LA, Steinhorn RH, Wedgwood S, Mata-Greenwood E, Roark EA, Russell JA, and Black SM. Increased superoxide generation is associated with pulmonary hypertension in fetal lambs. a role for NADPH Oxidase. *Circ. Res.* 2003; In press.
14. Cassin S. Role of prostaglandins, thromboxanes and leukotrienes in the control of the pulmonary circulation in the fetus and newborn. *Semin. Perinatol.* 1987; 11:53-63.
15. Clark RH, Kueser TJ, Walker MW, Southgate WM, Huckaby JL, Perez JA, Roy BJ, Keszler M, and Kinsella JP. Low-dose nitric oxide therapy for persistent pulmonary hypertension of the newborn. Clinical Inhaled Nitric Oxide Research Group. *N. Engl. J. Med.* 2000; 342:469-474.
16. Cohen AH, Hanson K, Morris K, Fouty B, McMurtry IF, Clarke W, and Rodman DM. Inhibition of cGMP-specific phosphodiesterase selectively vasodilates the pulmonary circulation in chronically hypoxic rats. *J. Clin. Invest.* 1996; 97:172-179.
17. Cornfield DN, Chatfield BA, McQueston JA, McMurtry IF, and Abman SH. Effects of birth-related stimuli on L-arginine -dependent pulmonary vasodilation in the ovine fetus. *Am. J. Physiol.* 1992; 262:H1474-H1481.
18. Cornfield DN, Reeve HL, Tolarova S, Weir EK, and Archer S. Oxygen causes fetal pulmonary vasodilation through activation of a calcium-dependent potassium channel. *Proc. Natl. Acad. Sci. U.S.A.* 1996; 93:8089-8094.
19. Davidson D, Barefield ES, Katwinkel J, Dudell G, Damask M, Straube R, Rhines J, and Chang CT. Inhaled NO for the early treatment of persistent pulmonary hypertension of the term newborn: a randomized double blinded placebo-controlled dose-response multicenter study. *Pediatrics.* 1998; 101:325-334.
20. Fagan KA, Fouty BW, Tyler RC, Morris KG, Helper LK, Sato K, LeCras TD, Abman SH, Weinberger HD, Huang PL, McMurtry IF, and Rodman DM. The pulmonary circulation of mice with either homozygous or heterozygous disruption of endothelial NO synthase is hyper-responsive to chronic hypoxia. *J. Clin. Invest.* 1999; 103:291-299.
21. Geggel RL, and Reid LM. The structural basis of persistent pulmonary hypertension of the newborn. *Clin. Perinatol.* 1984; 3:525-549.
22. Goldman AP, Tasker RC, Haworth SG Sigston PE, and Macrae DJ. Four patterns of response to inhaled NO for persistent pulmonary hypertension of the newborn. *Pediatrics.* 1996; 98:706-713.
23. Gordon JB, Martinez FR, Keller PA Tod ML, and Madden JA. Differing effects of acute and prolonged alkalosis on hypoxic pulmonary vasoconstriction. *Am. Rev. Resp. Dis.* 1993; 148:1651-1656.
24. Grover TR, Parker TA, Zenge JP, Markham NE, and Abman SH. VEGF inhibition impairs endothelial function and causes pulmonary hypertension in the late gestation ovine fetus. *Am. J. Physiol.* 2002; 284:L508-L517.
25. Grover TR, Rairigh RL, Zenge JP, Abman SH, and Kinsella JP. Inhaled carbon monoxide does not alter pulmonary vascular tone in the ovine fetus. *Am. J. Physiol. Lung Cell. Mol. Physiol.* 2000; 278:L779-L784.

26. Halbower AC, Tuder RM, Franklin WA, Pollock JS, Forstermann U, and Abman SH. Maturation-related changes in endothelial NO synthase immunolocalization in the developing ovine lung. *Am. J. Physiol.* 1994; 267:L585-L591.
27. Hanson KA, Beavo JA, Abman SH, and Clarke WR. Chronic pulmonary hypertension increases fetal lung cGMP activity. *Am. J. Physiol.* 1998; 275:L931-L941.
28. Hanson KA, Burns F, Rybalkin SD, Miller J, Beavo J, and Clarke WR. Developmental changes in lung cGMP phosphodiesterase-5 activity, protein and message. *Am. J. Resp. Crit. Care Med.* 1995; 158:279-288.
29. Heymann MA, and Soifer SJ. "Control of fetal and neonatal pulmonary circulation." In *Pulmonary vascular physiology and pathophysiology,* Weir EK, and Reeves JT, eds. New York, NY: Marcel-Dekker, 1989, pp. 33-50.
30. Hintz SR, Suttner DM, Sheehan AM Rhine WD, and Van Meurs KP. Decreased use of neonatal ECMO: how new treatment modalities have affected ECMO utilization. *Pediatrics.* 2000; 106:1339-1343.
31. Ivy DD, and Abman SH. "Role of endothelin in perinatal pulmonary vasoregulation." In *Fetal and Neonatal Pulmonary Circulation,* Weir EK, Archer SL, and Reeves JT, eds. New York, NY: Futura, 1999, pp. 279-302.
32. Ivy DD, Kinsella JP, and Abman SH. Physiologic characterization of endothelin A and B receptor activity in the ovine fetal lung. *J. Clin. Invest.* 1994; 93:2141-2148.
33. Ivy DD, LeCras TD, Horan MP, and Abman SH. Increased lung prepro-endothelin-1 and decreased endothelin B receptor gene expression after chronic pulmonary hypertension in the ovine fetus. *Am. J. Physiol.* 1998; 274:L535-L541.
34. Ivy DD, Parker TA, Ziegler JW Galan HL, Kinsella JP, Tuder RM, and Abman SH. Prolonged endothelin A receptor blockade attenuates chronic pulmonary hypertension in the ovine fetus. *J. Clin. Invest.* 1997; 99:1179-1186.
35. Kelly LK, Porta NF, Goodman DM, Carroll CL, and Steinhorn RH. Inhaled prostacyclin for term infants with persistent pulmonary hypertension refractory to inhaled nitric oxide. *J. Pediatr.* 2002; 141:830-832.
36. Kennaugh JM, Kinsella JP, Abman SH Hernandez JA, Moreland SG, and Rosenberg AA. Impact of new treatments for neonatal pulmonary hypertension on ECMO use and outcome. *J. Perinatol.* 1997; 17:366-369.
37. Kinsella JP, and Abman SH. Clinical approach to inhaled NO therapy in the newborn. *J. Pediatr.* 2000; 136:717-726.
38. Kinsella JP, and Abman SH. Recent developments in the pathophysiology and treatment of PPHN. *J. Pediatr.* 1995; 126:853-864.
39. Kinsella JP, Ivy DD, and Abman SH. Ontogeny of NO activity and response to inhaled NO in the developing ovine pulmonary circulation. *Am. J. Physiol.* 1994; 267:H1955-H1961.
40. Kinsella JP, Neish S, Shaffer E, and Abman SH. Low dose inhalational nitric oxide in persistent pulmonary hypertension of the newborn. *Lancet.* 1992; 340:819-820.
41. Kinsella JP, Troug W, Walsh W, Goldberg RN, Bancalari E, Mayock DE, Redding GJ, deLemos RA, Sardesai S, McCurnin DC, Moreland SG, Cutter GR, and Abman SH. Randomized multicenter trial of inhaled nitric oxide and high frequency oscillatory ventilation in severe persistent pulmonary hypertension of the newborn. *J. Pediatr.* 1997; 131:55-62.
42. Konduri GG, Ou J, Shi Y, and Pritchard KA. Decreased association of Hsp90 impairs endothelial nitric oxide synthase in fetal lambs with persistent pulmonary hypertension. *Am. J. Physiol. Heart Circ. Physiol.* 2003; In press.
43. Laffey JG, Engelberts D, Kavanagh BP. Injurious effects of hypocapnic alkalosis in the isolated lung. *Am. J. Crit. Care Med.* 2000; 162:399-405.
44. Levin DL, Heymann MA, Kitterman JA Gregory GA, Phibbs RH, and Rudolph AM. Persistent pulmonary hypertension of the newborn. *J. Pediatr.* 1976; 89:626-633.
45. Levin DL, Hyman AI, Heymann MA, Rudolph AM. Fetal hypertension and the development of increased pulmonary vascular smooth muscle: a possible mechanism for persistent pulmonary hypertension of the newborn infant. *J. Pediatr.* 1978; 92:265-269.

46. Lotze A, Mitchel BR, Bulas D Zola EM, Shalwitz RA, and Gunkel JH. Multicenter study of surfactant (beractant) use in the treatment of term infants with severe respiratory failure. *J. Pediatr.* 1998; 132:40-47.
47. MacRitchie AN, Jun SS, Chen Z, German Z, Yuhanna IS, Sherman TS, and Shaul PW. Estrogen upregulates endothelial NO synthase gene expression in fetal pulmonary artery endothelium. *Circ. Res.* 1997; 81:355-362.
48. McMullan DM, Bekker JM, Johengen MJ, Hendricks-Munoz K, Gerrets R, Black SM, and Fineman JR. Inhaled nitric oxide-induced rebound pulmonary hypertension: a role for endothelin-1. *Am. J. Physiol.* 2001; 280:H777-H785.
49. McQueston JA, Cornfield DN, McMurtry IF, and Abman SH. Effects of oxygen and exogenous L-arginine on endothelium-derived relaxing factor activity in the fetal pulmonary circulation. *Am. J. Physiol.* 1993; 264:H865-H871.
50. McQueston JA, Kinsella JP, Ivy DD, McMurtry IF, and Abman SH. Chronic pulmonary hypertension *in utero* impairs endothelium-dependent vasodilation. *Am. J. Physiol.* 1995; 268:H288-H294.
51. Morin FC. Ligating the ductus arteriosus before birth causes persistent pulmonary hypertension in the newborn lamb. *Pediatr. Res.* 1989; 25:245-250.
52. Morin FC, Egan EA, Ferguson W, and Lundgren CEG. Development of pulmonary vascular response to oxygen. *Am. J. Physiol.* 1988; 254:H542-H546.
53. Moya MP, Gow AJ, Califf RM Goldberg RN, and Stamler JS. Inhaled ethyl nitrite gas for persistent pulmonary hypertension of the newborn. *Lancet.* 2002; 360:141-142.
54. Murphy JD, Aronovitz MJ, and Reid LM. Effects of chronic in utero hypoxia on the pulmonary vasculature of the newborn guinea pig. *Pediatr. Res.* 1986; 20:292-295.
55. Neonatal Inhaled NO Study Group. Inhaled NO in full-term and nearly full-term infants with hypoxic respiratory failure. *N. Engl. J. Med.* 1997; 336:597-604.
56. North AJ, Star RA, Brannon TS, Ujiie K, Wells LB, Lowenstien CJ, Snyder SH, and Shaul PW. NO synthase type I and type III gene expression are developmentally regulated in rat lung. *Am. J. Physiol.* 1994; 266:L635-L641.
57. Parker TA, Afshar S, Kinsella JP, Ivy DD, Shaul PW, and Abman SH. Effects of chronic estrogen receptor blockade on the pulmonary circulation in the late gestation ovine fetus. *Am. J. Physiol. Heart Circ. Physiol.* 2001; 281:H1005-H1014.
58. Parker TA, Kinsella JP, Galan HL, Richter G, and Abman SH. Prolonged infusions of estradiol dilate the ovine fetal pulmonary circulation. *Pediatr. Res.* 2000; 47:89-96.
59. Parker TA, Le Cras TD, Kinsella JP, and Abman SH. Developmental changes in endothelial NO synthase expression in the ovine fetal lung. *Am. J. Physiol.* 2000; 278:L202-L208.
60. Patterson K, Kapur SP, and Chandra RS. "Persistant pulmonary hypertension of the newborn: pulmonary pathologic effects." In *Cardiovascular diseases, Perspectives in Pediatric Pathology*, Rosenberg HS, and Bernstein J, eds. Basel:Karger, 1988, pp. 139-154.
61. Pearson DL, Dawling S, Walsh WF, Haines JL, Christman BW, Bazyk A, Scott N, and Summar ML. Neonatal pulmonary hypertension: urea cycle intermediates, NO production and carbamoyl phosphate synthetase function. *N. Engl. J. Med.* 2001; 344:1932-1938.
62. Rairigh R, Ivy DD, Kinsella JP, and Abman SH. Inducible NOS inhibitors increase pulmonary vascular resistance in the late-gestation fetus. *J. Clin. Invest.* 1998; 101:15-21.
63. Rairigh RL, Parker TA, Ivy DD, Kinsella JP, Fan I, and Abman SH. Role of inducible nitric oxide synthase in the transition of the pulmonary circulation at birth. *Circ. Res.* 2001; 88:721-726.
64. Rairigh RL, Storme L, Parker TA, Le Cras TD, Jakkula M, and Abman SH. Role of neuronal nitric oxide synthase in regulation of vascular and ductus arteriosus tone in the ovine fetus. *Am. J. Physiol. Lung Cell. Mol. Physiol.* 2000; 278:L105-L110.
65. Rasanen J, Wood DC, Debbs RH, Cohen J, Weiner S, and Huhta JC. Reactivity of the human fetal pulmonary circulation to maternal hyperoxygenation increases during the second half of pregnancy. A randomized study. *Circulation.* 1998; 97:257-262.
66. Rhodes MT, Porter VA, Saqueton CB, Herron JM, Resnik ER, and Cornfield DN. Pulmonary

vascular response to normoxia and Kca channel activity is developmentally regulated. *Am. J. Physiol.* 2001; 280:L1250-L1257.
67. Roberts JD, Fineman JR, Morin FC III, Shaul PW, Rimar S, Schreiber MD, Polin RA, Zwass MS, Zayek MM, Gross I, Heymann MA, and Zapol WM. Inhaled NO and and persistent pulmonary hypertension of the newborn. *N. Engl. J. Med.* 1997; 336:605-610.
68. Roberts JD, Polaner DM, Lang P, and Zapol WM. Inhaled nitric oxide in persistent pulmonary hypertension of the newborn. *Lancet.* 1992; 340:818-819.
69. Rosenberg AA, Kennaugh J, Koppenhafer SL Loomis M, Chatfield BA, and Abman SH. Elevated immunoreactive endothelin-1 levels in persistent pulmonary hypertension of the newborn. *J. Pediatr.* 1993; 123:109-114.
70. Rudolph AM, Heymann MA, and Lewis AB. "Physiology and pharmacology of the pulmonary circulation in the fetus and newborn." In *Development of the Lung,* Hodson W, ed. New York, NY: Marcel Dekker, 1977, pp. 497-523.
71. Shekeremian L, Ravu H, and Penny D. Intravenous sildenafil lowers pulmonary vascular resistance in a model of neonatal pulmonary hypertenion. *Am. J. Resp. Crit. Care Med.* 2002; 165:1098-2002.
72. Sherman TS, Chen Z, Yuhanna IS, Lau KS, Margraf LR, and Shaul PW. NO synthase isoform expression in the developing lung epithelium. *Am. J. Phsyiol.* 1999; 276:L383-L390.
73. Soifer LT, Soifer SJ, Loitz RD, Roman C, and Heymann MA. Leukotriene end organ antagonists increase pulmonary blood flow in fetal lambs. *Am. J. Physiol.* 1985; 249:H570-H576.
74. Steinhorn RH, Albert G, Swartz DD, Russell JA, Levine CR, and Davis JM. Recombinant human superoxide dismutase enhances the effect of inhaled NO in persistent pulmonary hypertension. *Am. J. Resp. Crit. Care Med.* 2001; 164:834-839.
75. Steinhorn RH, Russell JA, and Morin FC. Disruption of cGMP production in pulmonary arteries isolated from fetal lambs with pulmonary hypertension. *Am. J. Physiol.*1995; 268:H1483-H1489.
76. Storme L, Parker TA, Kinsella JP, Rairigh RL, and Abman SH. Chronic pulmonary hypertension impairs flow-induced vasodilation and increases the myogenic response in the ovine fetal pulmonary circulation. *Am. J. Physiol.* 2002; 282:L56-L66.
77. Storme L, Rairhig RL, and Abman SH. In vivo evidence for a myogenic response in the ovine fetal pulmonary circulation. *Pediatr. Res.*1999; 45:425-431.
78. Storme L, Rairigh RL, Parker TA, Kinsella JP, and Abman SH. Acute intrauterine pulmonary hypertension impairs endothelium-dependent vasodilation in the ovine fetus. *Pediatr. Res.* 1999; 45:575-581.
79. Thusu KG, Morin FC III, Russell JA, and Steinhorn RH. The cGMP phosphodiesterase inhibitor zaprinast enhances the effect of NO. *Am. J. Resp. Crit. Care Med.* 1995; 152:1605-1610.
80. Tzao C, Nickerson PA, Russell JA, Noble B, and Steinhorn RM. Heterogeneous distribution of type I NOS in pulmonary vasculature of ovine fetus. *Histochem. Cell Biol.* 2000; 114:421-30.
81. Tzao C, Nickerson PA, Russell JA, Noble BK, and Steinhorn RH. Paracrine role of soluble guanylate cyclase and type III nitric oxide synthase in ovine fetal pulmonary circulation: a double labeling immunohistochemical study. *Histochem. Cell Biol.* 2003; 119:125-130.
82. UK Collaborative ECMO Trial Group. UK Collaborative randomized trial of neonatal ECMO. *Lancet.* 1996; 348:75-82.
83. Van Marter LJ, Leviton A, Allred EN, Pagano M, Sullivan KF, Cohen A, and Epstein MF. Persistent pulmonary hypertension of the newborn and smoking and aspirin and nonsteroidal antiinflammatory drug consumption during pregnancy. *Pediatrics.*1996; 97:658-663.
84. Velvis H, Moore P, Heymann MA. Prostaglandin inhibition prevents the fall in pulmonary vascular resistance as the result of rhythmic distension of the lungs in fetal lambs. *Pediatr. Res.* 1991; 30:62-67.
85. Villamor E, Le Cras TD, Horan M, Halbower AC, Tuder RM, and Abman SH. Chronic

hypertension impairs endothelial NO sythase in the ovine fetus. *Am. J. Physiol.* 1997; 16:L1013-L1020.

86. Walsh-Sukys MC, Tyson JE, Wright LL, Bauer CR, Korones SB, Stevenson DK, Verter J, Stoll BJ, Lemons JA, Papile LA, Shankaran S, Donovan EF, Oh W, Ehrenkranz RA, and Fanaroff AA. Persistent pulmonary hypertension of the newborn in the era before NO: practice variation and outcomes. *Pediatrics.* 2000; 105:14-20.
87. Wedgwood S, McMullan DM, Bekker JM, Fineman JR, and Black SM. Role for endothelin-1–induced superoxide and peroxynitrite production in rebound pulmonary hypertension associated with inhaled NO therapy. *Circ. Res.* 2001; 89:357-364.
88. Ziegler JW, Ivy DD, Fox JJ, Kinsella JP, Clarke WR, and Abman SH. Dipyridamole, a cGMP phosphodiesterase inhibitor, causes pulmonary vasodilation in the ovine fetus. *Am. J. Physiol.* 1995; 269:H473-H479.
89. Ziegler JW, Ivy DD, Wiggins JW, Kinsella JP, Clarke WR, and Abman SH. Hemodynamic effects of dipyridamole and inhaled NO in children with severe pulmonary hypertension. *Am. J. Resp.* 1998; 158:1388-1395.

Chapter 28

Roles for Vasoconstriction and Gene Expression in Hypoxia-induced Pulmonary Vascular Remodeling

Bernadette Raffestin, Serge Adnot, and Saadia Eddahibi
Université de Versailles-Saint Quentin en Yvelines and Hôpital Henri Mondor, Créteil, France

1. Introduction

Pulmonary hypertension (PH) is characterized by an increase in pulmonary vascular resistance that impedes the ejection of blood by the right ventricle, ultimately leading to right ventricular failure. The most common cause of PH is alveolar hypoxia-related, for instance, to high altitude or chronic hypoxemic lung disease. Acute exposure of various mammals to hypoxia results within a few minutes in pulmonary vasoconstriction related to contraction of the smooth muscle cells (SMCs) in the distal pulmonary arteries. With chronic exposure, PH is due not only to SMC contraction and polycythemia but also to structural remodeling of the pulmonary arteries (25). Thus, preexisting SMCs in normally muscularized pulmonary arteries undergo hypertrophy and hyperplasia, while new SMCs appear in intraacinar arteries that are normally nonmuscularized or muscularized only along part of their circumference. Another component of hypoxia-induced pulmonary artery remodeling is extracellular matrix deposition in the vessel wall, with a build-up of connective tissue proteins such as elastin and collagen,. Reversibility is a remarkable feature of chronic hypoxic PH. Although correcting alveolar hypoxia may have little or no effect on PH in the short term, PH caused by chronic hypoxia resolves over several weeks or months after the return to normoxia.

The classic understanding of chronic hypoxic PH is based on the concept that vascular remodeling is a consequence of sustained pulmonary vasoconstriction and increased pulmonary artery pressure. The resulting increase in shear stress is thought to trigger hypertrophy and proliferation of the vascular SMCs (Fig. 1). Although this concept remains valid in many aspects, it is no longer viewed as the only pathophysiological mechanism of hypoxic PH. Recent evidence shows that hypoxia-induced pulmonary vascular remodeling can be attenuated despite enhanced hypoxic pulmonary vasoconstriction. This suggests that some of the vasoconstricting substances released in hypoxic lung tissue, most notably endothelin (ET) and serotonin (5-hydroxytryptamine), may serve as growth factors for vascular SMCs, or exert other functions independent from their effects on vascular tone and from the severity of pulmonary vasoconstriction. Another recently identified mechanism that may be involved in hypoxia-induced pulmonary vascular remodeling is a direct effect of hypoxia on the expression of specific genes acting on SMCs, endothelial cells, fibroblasts, or extracellular matrix remodeling (Fig. 1).

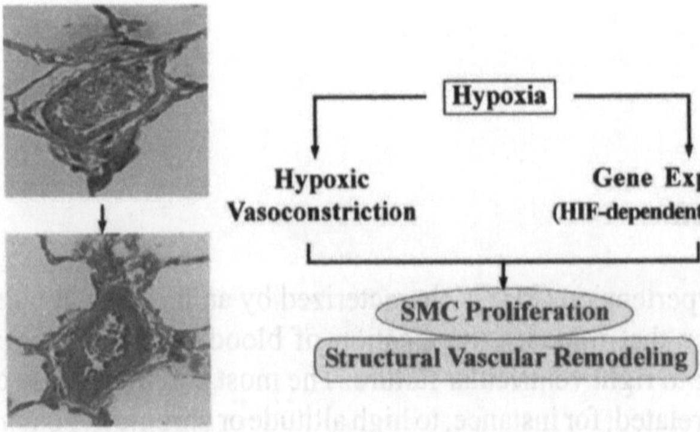

Figure 1. Hypoxia-mediated pulmonary vascular remodeling involves vasoconstriction and gene expression.

2. Variability of Pulmonary Vascular Remodeling Among Species and Individuals: The Case of Patients with Chronic Obstructive Pulmonary Disease

Acute hypoxic pulmonary vasoconstriction is ubiquitous among adult mammals, but considerable variations in chronic hypoxic PH occur within and across species (27). During chronic exposure to high altitude, PH is more severe in cattle and horses than in sheep and dogs. Furthermore, among humans and other mammals that live at high altitudes for many generations, PH is less severe and shows less inter-individual variation than in subjects from low altitudes that are briefly exposed to high altitudes. This observation suggested many years ago that genetic make-up may affect susceptibility to PH. In populations living at high altitudes, individuals carrying susceptibility genes probably died before puberty, leading to selection of individuals with little susceptibility to PH. This early hypothesis has received support from the finding that some individuals in cattle herds raised at high altitude in Colorado have an inherited susceptibility to PH and right ventricular failure, called brisket disease.

Chronic hypoxemia is considered the main factor in the pathogenesis of PH in patients with advanced chronic obstructive pulmonary disease (COPD). Although PH is usually mild to moderate, its severity varies greatly among individuals, with some patients developing severe PH out of proportion with the severity of their underlying disease. PH has a strong bearing on prognosis, independently from the severity of airflow limitation or hypoxemia, suggesting a genetic susceptibility to PH in this disease. Recent studies showing that PH severity in patients with advanced COPD is directly related to 5-hydroxytryptamine transporter (5-HTT) gene polymorphism(s) are consistent with this hypothesis.

3. Role of Vasoconstriction in Hypoxia-induced Pulmonary Vascular Remodeling

The development and maintenance of remodeling in response to chronic hypoxia was long ascribed solely to the increased pressure and shear stress that result from vasoconstriction in response to hypoxia. This idea was supported by a study in rats in which banding of one pulmonary artery prior to hypoxic exposure was followed by reduced vascular remodeling in the protected lung (26). In both the pulmonary and the systemic circulation, SMC proliferation has been viewed as an adaptive process in response to increased arterial wall stress. The promoters of genes for growth factors and other compounds, including TGF-ß, platelet-derived growth factor (PDGF), and ICAM-1, contain shear-stress-response elements that are believed to mediate the induction of these genes during chronic hypoxia. Moreover, brief stretching of pulmonary artery segments has been shown to cause an endothelium-dependent increase in matrix protein synthesis (31), whereas more prolonged stretching induced hypertrophy and hyperplasia of SMCs and fibroblasts (13). A role for pulmonary vasoconstriction in vascular remodeling has also been suggested by the decreased medial thickening of pulmonary arteries in chronically hypoxic rats obtained with various vasodilator agents that depress smooth muscle tone through different mechanisms, such as calcium antagonists, methyldopa, atrial natriuretic peptide, inhaled nitric oxide (NO), and ET-1 antagonists. Moreover, species with the strongest and most sustained pulmonary vasoconstrictive response to acute hypoxia, such as cattle and swine, are also those with the most severe pulmonary vascular remodeling in response to chronic hypoxia (27). However, there are exceptions to this finding. In coati, another species with a powerful acute pressor response to hypoxia and thick muscular pulmonary arteries at sea level, pulmonary vascular remodeling did not develop upon exposure to a simulated altitude of 4900 m for 6 weeks (8). We recently found blunted pulmonary vascular remodeling in response to chronic hypoxia despite potentiation of the pulmonary pressor response to acute hypoxia in knock-out mice for the 5-HTT gene (5). Many pharmacological agents that decrease vascular tone may also act on vessel muscularization by directly inhibiting SMC proliferation and hypertrophy. In addition to vasodilator effects, natriuretic peptides exert anti-mitogenic effects by binding to guanylate cyclase-linked receptors, thereby inducing intracellular accumulation of cyclic GMP (10). Although ET-1 may potentiate the acute pulmonary pressor response to hypoxia in vivo (22), it also has mitogenic effects and co-mitogenic effects with serum and other growth factors on vascular SMCs and fibroblasts from the pulmonary vascular bed of various species, and these effects are inhibited by ET_A receptor blockade (35).

4. Effect of Chronic Hypoxia on Hypoxic Pulmonary Vasoconstriction

It has been long known that the pulmonary pressor response to acute hypoxia is impaired after prolonged exposure to hypoxia. Correction of hypoxemia in humans living at high altitude has little acute effect on pulmonary artery pressure and vascular resistance (18). Moreover, in lungs from chronically hypoxic rats, the pressor response to acute hypoxia is blunted and that to vasoconstrictor agonists enhanced. These observations are probably ascribable to the reduction in the delayed K^+ current seen upon return to normoxia in SMCs from distal pulmonary arteries of chronically hypoxic rats (29). In accordance with these findings, prolonged exposure to hypoxia has recently been shown to specifically downregulate the mRNA and protein expression of K_v channels α subunits in cultured pulmonary SMCs and to reduce K^+ currents through K_V channels (24). The resultant membrane depolarization opens voltage-gated Ca^{2+} channels, increases Ca^{2+} influx, and raises cytoplasmic calcium. This effect may not only contribute to sustained pulmonary vasoconstriction, but also serve as a critical stimulus for pulmonary SMC proliferation during chronic hypoxia since increased cytosolic calcium concentrations have been shown to activate the mitogen-activated protein kinase cascade that stimulates synthesis of the transcription factors mediating cell proliferation (21).

5. Changes in Pulmonary Vascular Reactivity During Exposure to Chronic Hypoxia

Cell proliferation and matrix synthesis reflect a balance between pro- and anti-proliferative stimuli. Many of the growth factors and matrix proteins that contribute to vascular remodeling under hypoxia are derived from the endothelium. In vitro experiments have been conducted to assess the direct effect of hypoxia on the expression of their genes. Exposure to hypoxia reduces heparan sulfates (9), NO (20), and prostacyclin (19) release by cultured endothelial cells. However, in contrast with the reduced endothelial NO synthase (eNOS) protein levels in isolated endothelial cells exposed to hypoxia, upregulation of eNOS has consistently been found to correlate temporally with the progression of hypoxia-induced pulmonary vascular remodeling (32). Increased inducible NOS (iNOS) expression has also been observed in lung tissue from chronically hypoxic mice. Although eNOS protein expression is increased, vasodilation in response to endothelium-dependent agents such as acetylcholine, ionophore, or serotonin is abolished, whereas the response to NO is preserved in isolated lungs from chronically hypoxic rats. This suggests that eNOS activity is reduced in chronically hypoxic pulmonary hypertension (1).

6. Role of Hypoxia-induced Gene Expression in Vascular Remodeling

6.1. Transcriptional Responses to Hypoxia

Hypoxia has been shown to affect the expression of specific genes by modulating a variety of transcription factors and transduction-signaling proteins. An essential mediator of transcriptional responses to decreased O_2 availability is hypoxia-inducible factor 1 (HIF-1). Many genes implicated in the development of polycythemia and chronic hypoxic PH (erythropoietin, transferrin, vascular endothelial growth factor (VEGF), VEGF receptor 1, iNOS, heme oxygenase-1, ET-1, and 5-HTT) contain functionally important HIF-1 binding sites. The biological activity of HIF-1 is determined by the expression and activity of the HIF-1 α subunit. Under normoxic conditions, HIF-1α appears to be ubiquitinated and undergoes proteasomal degradation. Hypoxia shares with iron chelators an ability to prevent the ubiquitination of HIF-1α. Target gene transcription is activated when HIF-1 α dimerizes with HIF-1β, and this heterodimer binds to DNA at sites containing the consensus sequence 5'-RCGTG-3'. The HIF-1 binding site is present within a hypoxia response element, which is a *cis*-acting transcriptional regulatory sequence found within 5'-flanking, 3'-flanking, or intervening sequences of target genes. Whereas HIF-1α is essential for normal embryonic development, HIF-1$\alpha^{+/-}$ mice develop normally and are indistinguishable from wild-type littermates under normoxic conditions. However, when exposed to 10% O_2 for 1-6 weeks, HIF-1$\alpha^{+/-}$ mice demonstrate delayed development of polycythemia and PH as evidenced by a lower right ventricular pressure and right ventricular mass than wild-type controls similarly exposed to 3 weeks of hypoxia (34). Morphometric analysis of lung sections from HIF-1$\alpha^{+/-}$ mice exposed to 3 weeks of hypoxia shows a decreased percentage of muscularized vessels among distal vessels and a reduction in wall thickness of muscularized arteries. This is evidence that specific genes expressed under the control of HIF-1 during hypoxia exposure are involved in hypoxic pulmonary vascular remodeling.

Other hypoxia-inducible transcription factors whose regulation closely resembles that of HIF-1α have been identified and designated HIF-2α (or EPAS-1) and HIF-3α. HIF-2α is more abundantly expressed in the lung than HIF-1α and may be involved in lung vascular development. Moreover, its mRNA expression is closely correlated with that of VEGF mRNA. A recent study also provides evidence that these hypoxia-inducible transcription factors are involved in the strong proliferative response to hypoxia of cultured fibroblasts from pulmonary arteries (28).

6.2. Hypoxia-induced Angiogenic Factors

Exposure to hypoxia is associated with activation of endogenous angiogenic processes in the lung. Labeling studies in rats have demonstrated a burst of endothelial cell division in intraacinar arteries at the end of the first week of exposure to hypoxia. Moreover, recent results obtained by our group using quantification of lung immunoreactivity for factor VIII or lectin binding suggest an increase in lung vascular density in mice or rats exposed to hypoxia. Among angiogenic factors, lung VEGF-A is expressed at a high level by epithelial cells in the normoxic lung (2). Although endothelial cells do not express VEGF under normoxia, exposure to hypoxia leads to a dramatic increase in VEGF mRNA levels within a few hours (17). This transcription rate increase is mediated, at least in part, by HIF-1. Small amounts of VEGF can also be detected in conditioned medium of SMCs exposed to hypoxia. Furthermore, chronic hypoxia has been shown to further increase the expression of VEGF-A and its receptors VEGFR-1 and VEGFR-2, in the rat lung (2). VEGF-A protects against the development of chronic hypoxic PH. Indeed, inhibition of VEGF receptors through tyrosine kinase inhibitors causes mild PH and pulmonary vascular remodeling in normoxic rats and severe irreversible PH in chronically hypoxic rats (30). Moreover, adenovirus-mediated VEGF overexpression in the lung attenuates the development of hypoxic PH (23).

6.3. Hypoxia-induced Heme Oxygenase-1

Upon exposure of SMCs to hypoxia, the transcriptional rate of heme oxygenase-1 increases transiently, leading to concomitant production of carbon monoxide (CO). This CO can elevate cGMP levels in the same cells or neighboring cells, thus regulating vascular tone and inhibiting SMC proliferation in a manner similar to NO. Co-culture experiments have shown that transient endogenous CO production by hypoxia-exposed SMCs can easily diffuse across cell membranes and delay the production of ET-1 and PDGF by endothelial cells (11). However, CO production is not sustained. When hypoxia is prolonged, the transcription rates of ET-1 and PDGF-B genes rise steadily, causing increased production of these two vasoconstrictors which, together with SMC mitogens, leads to enhanced smooth muscle growth and contractility. Moreover, evidence against a major role for heme oxygenase induction in protection against hypoxic pulmonary vascular remodeling has been provided recently by experiments in which chronic hypoxia-induced pulmonary vascular remodeling and PH were not affected by disruption of the heme oxygenase-1 gene in mice (33).

6.4. Hypoxia-induced Endothelin-1

Among the various vasoactive molecules or growth factors that have been implicated in PH, ET is particularly important given that ET receptor antagonists have been proved effective in patients with PH. ET was identified in 1988 and found to be the most potent vasoconstrictor ever known. The discovery that ET was abundantly expressed in the lung suggested a role for this factor in the initiation or progression of PH. Beneficial effects of chronic treatment with specific ET receptor antagonists in experimental hypoxic PH were first reported in 1994 and 1995. ET may contribute to the development of PH through its potent vasoconstricting properties or its pro-mitogenic properties. It is generally agreed upon that ET expression is increased in most forms of experimental and human PH. Increased ET-1 synthesis has been well documented in remodeled pulmonary vessels from animals exposed to chronic hypoxia. The presence in the ET-1 gene of functional HIF-1 binding sites is consistent with these findings. ET may induce vascular remodeling after being activated in response to hypoxia through specific mechanisms acting via both vasoconstriction and growth promotion. Why ET is released in abundance in nonhypoxic PH, however, remains a mystery. The conditions that govern ET synthesis by endothelial or SMCs are incompletely understood. Reduced NO formation or NO bioavailability may influence ET production. Also, ET production may occur when SMCs shift from a nonsecreting to a secreting phenotype. All these abnormalities are consistently observed in the pulmonary circulation after exposure to chronic hypoxia.

ET works through two receptor subtypes, ET_A and ET_B, with opposite effects. ET_A receptors are present on vascular SMCs and mediate vasoconstriction and proliferation. ET_B receptors are found predominantly on endothelial cells, where they promote vasodilation by releasing NO, prostacyclin, or other endothelium-dependent vasodilators. The endothelial ET_B receptor has also been shown to indirectly modulate ET-1 synthesis through a negative feedback loop involving NO and to participate in the clearance of circulating ET-1. In normal mammals, ET_B-dependent pulmonary vasodilation occurs upon administration of ET, even in doses that cause systemic vasoconstriction. Evidence of major changes in vasoreactivity to ET-1 has been obtained in experimental models of PH. Exposure of rats to chronic hypoxia abolishes the vasodilator effects and enhances the vasoconstrictor effects of ET-1. Alterations in endothelial function may explain these findings since impaired ET_B-mediated vasodilation, whether related or unrelated to NO synthesis, has been reported during chronic hypoxia.

Consistent with this finding, improvement of endothelial function reduces ET-induced vasoconstriction.

6.5. Hypoxia-induced Serotonin Transporter

The serotonin transporter (5-HTT) in pulmonary vascular SMCs has many attributes suggesting that it may be a key determinant of pulmonary SMC proliferation. It belongs to a family of integral membrane proteins responsible for terminating the action of neurotransmitters released from neurons. In addition to 5-HT reuptake into neurons, 5-HTT is responsible for indoleamine uptake by platelets, endothelial cells, and SMCs. These processes result in very low levels of free 5-HT in plasma under physiological conditions. Interest in the potential role for 5-HTT in PH was sparked by the observation that 5-HTT is a target for drugs that have been shown to increase the risk of PH development in humans.

In addition to contributing to the uptake and subsequent inactivation of serotonin passing through the lung, 5-HTT mediates the proliferation of pulmonary vascular SMCs through its ability to internalize indoleamine The requirement of 5-HTT as a mediator of 5-HT mitogenic activity appears specific for SMCs from the pulmonary vessels, since it has not been demonstrated in SMCs from systemic vessels. We have recently reported that the mitogenic action of serotonin on cultured pulmonary vascular SMCs from rats is enhanced by hypoxia, which induces 5-HTT expression through a transcriptional mechanism and simultaneously increases 5-HTT activity, an effect associated with potentiation of the mitogenic effect of 5-HT (4). Exposure to hypoxia also increases 5-HTT expression in the rat lung, notably in the media of remodeled pulmonary vessels. The presence of two hypoxia-sensitive elements in the promoter region of the 5-HTT gene strongly suggests that 5-HTT may be an effector molecule for pulmonary vascular remodeling in response to hypoxia.

In previous studies, we found that continuous intravenous infusion of serotonin worsened PH in rats exposed to chronic hypoxia (7). This effect was prevented by administration of a 5-HT transport inhibitor, despite a further increase in plasma serotonin levels. Moreover, when a 4-week course of the 5-HTT substrate dexfenfluramine (2g/kg body weight per day) was discontinued, 5-HTT expression increased transiently in lung tissue and, concomitantly, hypoxic PH worsened (3, 6).

Further information on the key role for 5-HTT activity in hypoxia-induced PH has been provided by experiments on mice lacking the 5-HTT gene. After exposure to hypoxia for 2-5 weeks, PH and vascular remodeling are less marked in 5-HTT-deficient mice than in littermate controls (5). This protection from pulmonary vascular remodeling and PH cannot be ascribed to decreased pulmonary vasoreactivity to hypoxia since the pulmonary pressor response to acute hypoxia is enhanced rather than blunted in *5-HTT*$^{-/-}$ mice. The mechanism underlying the protective effect remains speculative. One likely hypothesis is that the platelet uptake deficiency leaves more indoleamine available for binding to 5-HT receptors on pulmonary vascular SMCs. We found that 5-HT infusion in rats potentiated the *in vivo* acute pulmonary pressure response to hypoxia (7). Moreover, treatment with dexfenfluramine, which inhibits platelet 5-HT uptake and promotes 5-HT release by platelets, also potentiated acute hypoxic pulmonary vasoconstriction *in vivo*. It is therefore reasonable to assume that deficient platelet 5-HT uptake in 5-HTT$^{-/-}$ mice increases hypoxic pulmonary vasoreactivity through the same mechanism. Impaired serotonin release by platelets or secondary platelet dysfunction is unlikely to account for the attenuated pulmonary vascular remodeling in 5-HTT$^{-/-}$ mice. In this regard, a comparison of our results in 5-HTT$^{-/-}$ mice with those previously reported in Fawn-Hooded rats is of interest. Fawn-Hooded rats have a genetic deficiency in 5-HT platelet storage and a bleeding tendency due to a defect in platelet aggregation, yet their 5-HTT amino acid sequence is normal. Although Fawn-Hooded rats share with 5-HTT$^{-/-}$ mice an increase in pulmonary vasoreactivity to acute hypoxia, responses to chronic hypoxia are diametrically opposed: hypoxia-induced PH is exacerbated in Fawn-Hooded rats and attenuated in 5-HTT$^{-/-}$ mice. Therefore, attenuation of hypoxic PH is not related to platelet 5-HT depletion when 5-HTT expression and/or activity is present in pulmonary vascular SMCs. The observation of attenuated remodeling in our 5-HTT deficient mice emphasizes the key role for 5-HTT expression by pulmonary SMCs in the development of hypoxic PH.

5-HTT can be competitively inhibited by the antidepressants fluoxetine, paroxetine, and citalopram. Consequently, these drugs inhibit the in vitro SMC proliferative response to 5-HT and, to a lesser extent, the growth response to serum. *In vivo*, we also recently observed that these drugs impaired hypoxia-induced pulmonary vascular remodeling, despite their potentiating effect on acute hypoxic pulmonary vasoconstriction.

Recent studies have emphasized the contribution of 5-HT$_{1B}$ and 5-HT$_{2B}$ receptors to hypoxia-induced PH. 5-HT$_{1B}$ receptors mediate contraction to 5-HT in human pulmonary arteries. 5-HT$_{1B}$-mediated contraction is enhanced in pulmonary arteries from hypoxic rats. Rats treated with a selective 5-HT$_{1D/1B}$ receptor antagonist or mice deficient in 5-HT$_{1B}$ receptors (5-HT 1B$^{-/-}$) have less severe PH and vascular remodeling than do wild-type controls, suggesting that 5-HT$_{1B}$ receptors may play a role in the development of PH via enhanced 5-HT$_{1B}$-mediated vasoconstriction (12). This effect, however, is modest, possibly because serotonin-induced pulmonary vasoconstriction does not greatly influence pulmonary vascular remodeling. Moreover, in a recent study, we found that treatment with the 5-HT$_{1B}$ receptor antagonist GR127935 abrogated acute hypoxic vasoconstriction in mice, whereas the 5-HT$_{2A}$ receptor antagonist ketanserin had no effect. When given chronically, neither GR nor ketanserin affected the development of chronic PH. When GR was given in association with citalopram or fluoxetine, the pulmonary vasoconstriction enhanced by the antidepressant was reduced to the level observed in the absence of antidepressant, but no further benefit on chronic PH was obtained.

Using the chronic hypoxic mouse model of PH, Launay et al. (14) recently showed that PH was associated with a substantial increase in 5-HT_{2B} receptor expression in pulmonary arteries and that PH did not develop in hypoxic mice with genetic or pharmacological inactivation of 5-HT_{2B} receptors. However, these results are difficult to compare with our findings on the role for 5-HTT, since the severity of hypoxic PH in the control animals differed markedly between the two studies.

Several conclusions can be drawn from our results showing less hypoxic pulmonary vascular remodeling despite potentiation of the acute pressor response to hypoxia. First, precapillary vasoconstriction, which is considered an important contributor to pulmonary arterial muscularization, may not fully explain the pathophysiology of hypoxic PH. Second, SMC proliferation, which is the main component of pulmonary vascular remodeling, may be independent from the severity of hypoxic pulmonary vasoconstriction and closely linked to the 5-HTT pathway.

At present, the mechanisms by which 5-HT may exert its mitogenic effects after being transported inside the SMCs remain speculative. Lee et al. (15) have reported that 5-HT-induced DNA synthesis is associated with tyrosine phosphorylation of GTPase-activating protein and that both events are blocked by inhibitors of 5-HT transport or tyrosine kinase. Therefore, although 5-HT-induced mitogenesis in SMCs requires cellular internalization through 5-HTT rather than binding to a membrane receptor, subsequent tyrosine phosphorylation of GTPase-activating protein appears to be an intermediate in the signaling pathway. Recently, involvement of superoxide anion formation in association with 5-HT transport has also been suggested as a possible contributor to the mitogenic effects of 5-HT (16). Since reactive oxygen species are also considered to be potential mediators of vascular remodeling, we cannot exclude that protection against PH in 5-HTT-deficient mice is indirectly related to decreased formation of superoxide anions.

We recently reported an increased growth response to 5-HT or serum of pulmonary artery-derived SMCs from patients with primary PH, an effect ascribable to increased 5-HTT expression (3, 6). Proliferation in response to 5-HT is dose dependently inhibited by the highly selective 5-HT transport inhibitors fluoxetine and citalopram but not by the 5-HT receptor antagonists ketanserin and GR 127935. These findings are ascribable in part to a variant in the upstream promoter region of the 5-HTT gene. 5-HTT expression is genetically controlled, and a polymorphism in the promoter region of the human 5-HTT gene affects transcriptional activity. The long (L) promoter variant of the 5-HTT gene is associated with increased 5-HTT expression, as compared with the short (S) variant, and causes an increase in 5-HT uptake. Pulmonary artery SMCs from controls with the LL genotype take up more 5-HT than do SS or LS cells and, accordingly, the growth-stimulating effects of 5-HT are more marked in LL cells than in LS or SS cells. Thus, the capability of pulmonary artery SMCs to proliferate in response to 5-HT is directly linked to the functional polymorphism of the 5-HTT gene promoter. Interestingly, the L variant of the 5-HTT gene polymorphism associated with 5-HTT overexpression appears more common in patients with primary PH than in controls: the L variant was present in homozygous form in 65% of 89 patients with primary PH but in only 27% of normal controls. Although these results demonstrate that the long allele of the 5-HTT gene promoter is associated with increased serotonin uptake in pulmonary artery SMCs, they do not fully explain the increased 5-HTT expression in patients with primary PH, since 5-HTT expression is higher in cells from patients than in same-genotype cells from controls. Whether this overexpression results from an alteration in the 5-HTT gene itself or from alterations in other factors involved in regulating 5-HTT gene expression remains to be determined. However, these data are consistent with the hypothesis that 5-HTT polymorphism confers susceptibility to primary PH. Whether 5-HTT gene overexpression is associated with other forms of secondary PH deserves further study. A role for 5-HTT in experimental hypoxic PH is now clearly established, and recent data from our laboratory demonstrating that 5-HTT gene polymorphism may determine the severity of PH in patients with chronic obstructive pulmonary disease extend this concept to humans. Agents capable of selectively inhibiting 5-HTT-mediated pulmonary artery SMC proliferation deserve to be investigated as potential treatments for PH.

7. Summary

The development and maintenance of hypoxia-induced pulmonary vascular remodeling is not related only to the precapillary pressure increase and shear stress that result from pulmonary vasoconstriction. The recent finding that several mammalian species and mice with 5-HTT deficiency show little hypoxic pulmonary vascular remodeling despite a strong acute hypoxic pressor response suggests that precapillary vasoconstriction may not fully explain the pathophysiology of hypoxic PH. SMC proliferation, the main component of hypoxic pulmonary vascular remodeling, may be best viewed as a process linked to the direct effect of hypoxia on the expression of various genes. Exposure to hypoxia has been shown to downregulate K_V-channel α subunits in SMCs from pulmonary arteries, thereby decreasing K^+ flows and increasing cytoplasmic Ca^{2+} concentrations in these cells. The effect of hypoxia on gene expression may also alter the balance between pro- and anti-proliferative factors derived from the endothelium. Hypoxia increases the expression of ET-1 and PDGF and decreases the expression of heparan sulfate, prostacyclin synthase, and eNOS synthase. Hypoxia is also known to increase the transcription rate of various genes involved in vascular remodeling through hypoxia-inducible transcription factors.

VEGF, which is the product of one of the genes containing functionally important binding sites for HIF-1, may protect against the development of hypoxic PH by stimulating angiogenic processes. Recent experimental findings demonstrate that an increase in the transcription rate of the 5-HTT gene in response to hypoxia plays a major role in pulmonary SMC proliferation. 5-HTT expression is genetically controlled, and a polymorphism in the promoter region of the human gene affects transcriptional activity, the long promoter variant of the gene being associated with increased expression as compared with the short variant. The ability of SMCs to proliferate in response to 5-HT is linked to this functional polymorphism, which may confer susceptibility to various forms of PH in humans, most notably chronic hypoxic PH.

The recent finding that several factors, including 5-HT and ET, are involved not only in experimental hypoxic PH but also in human primary PH suggests that common mechanisms lead to pulmonary vascular remodeling, whatever the inciting causal factor. This provides a strong rationale for actively investigating the mechanisms that underlie the complex vascular changes responsible for the hypoxia-induced pulmonary vascular remodeling. Such studies may identify new molecular pathways involved in various types of PH.

References

1. Adnot S, Raffestin B, Eddahibi S, Braquet P, and Chabrier P. Loss of endothelium-dependent relaxant activity in the pulmonary circulation of rats exposed to chronic hypoxia. *J. Clin. Invest.* 1991; 87: 155-162
2. Christou H, Yoshida A, Arthur V, Morita T, and Kourembanas S. Increased vascular endothelial growth factor production in the lungs of rats with hypoxia-induced pulmonary hypertension. *Am. J. Respir. Cell. Mol. Biol.* 1998; 18: 768-776.
3. Eddahibi S, Adnot S, Frisdal E, Levame M, Hamon M, and Raffestin B. Dexfenfluramine-associated changes in 5-hydroxytryptamine transporter expression and development of hypoxic pulmonary hypertension in rats. *J. Pharmacol. Exp. Ther.* 2001; 297: 148-154.
4. Eddahibi S, Fabre V, Boni C, Martres M, Raffestin B, Hamon M, and Adnot S. Induction of serotonin transporter by hypoxia in pulmonary vascular smooth muscle cells: relationship with the mitogenic action of serotonin. *Circ. Res.* 1999; 84: 329-336.
5. Eddahibi S, Hanoun N, Lanfumey L, Lesch KP, Raffestin B, Hamon M, and Adnot S. Attenuated hypoxic pulmonary hypertension in mice lacking the 5-hydroxytryptamine transporter gene. *J. Clin. Invest.* 2000; 105: 1555-1562.
6. Eddahibi S, Humbert M, Fadel E, Raffestin B, Darmon M, Capron F, Simonneau G, Dartevelle P, Hamon M, and Adnot S. Serotonin transporter overexpression is responsible for pulmonary artery smooth muscle hyperplasia in primary pulmonary hypertension;. *J. Clin. Invest.* 2001; 108: 1141-1150.
7. Eddahibi S, Raffestin B, Pham I, Launay JM, Aegerter P, Sitbon M, and Adnot S. Treatment with 5-HT potentiates development of pulmonary hypertension in chronically hypoxic rats. *Am. J. Physiol.* 1997; 272: H1173-H1181.
8. Hanson WL, Boggs DF, Kay JM, Hofmeister SE, Okada O, and Wagner WW Jr. Pulmonary vascular response of the coati to chronic hypoxia. *J. Appl. Physiol.* 2000; 88: 981-986.
9. Humphries DE, Lee SL, Fanburg BL, and Silbert JE. Effects of hypoxia and hyperoxia on

proteoglycan production by bovine pulmonary endothelial cells. *J. Cell Physiol.* 1986; 126: 249-253.
10. Itoh H, Pratt RE, and Dzau VJ. Atrial natriuretic polypeptide inhibits hypertrophy of vascular smooth muscle cells. *J. Clin. Invest.* 1990; 86: 1690-1697.
11. Katayose D, Ohe M, Yamauchi K, Ogata M, Shirato K, Fujita H, Shibahara S, and Takishima T. Increased expression of PDGF A- and B-chain genes in rat lungs with hypoxic pulmonary hypertension. *Am. J. Physiol.* 1993; 264: L100-L106.
12. Keegan A, Morecroft I, Smillie D, Hicks MN, and MacLean MR. Contribution of the 5-HT_{1B}-receptor to hypoxia-induced pulmonary hypertension. *Circ. Res.* 2001; 89: 1231-1239.
13. Kolpakov V, Rekhter MD, Gordon D, Wang WH, and Kulik TJ. Effect of mechanical forces on growth and matrix protein synthesis in the in vitro pulmonary artery. *Circ. Res.* 1995; 77: 823-831.
14. Launay J, Hervé P, Peoc'h K, Tournois C, Callebert J, Nebigil CG, Etienne N, Drouet L, Humbert M, Simonneau G, and Maroteaux L. Function of the serotonin 5-hydroxytryptamine $_{2B}$ receptor in pulmonary hypertension. *Nat. Med.* 2002; 8: 1129-1135.
15. Lee SL, Wang WW, and Fanburg BL. Association of Tyr Phosphorylation of GTPase - activating protein with mitogenic action of serotonin. *Am. J. Physiol.* 1997; 272: C223-C230.
16. Lee S-L, Wang W-W, Finlay GA, and Fanburg BL. Serotonin stimulates mitogen-activated protein kinase activity through the formation of superoxide anion. *Am. J. Physiol.* 1999; 277: L282-L291.
17. Liu Y, Cox SR, Morita T, and Kourembanas S. Hypoxia regulates vascular endothelial growth factor gene expression in endothelial cells. *Circ. Res.* 1995; 77: 638-643.
18. Lockhart A, Zelter M, Mensch-Dechene J, Antezana G, Paz-Zamora M, Vargas E, and Coudert J. Pressure-flow-volume relationships in pulmonary circulation of normal highlanders. *J. Appl. Physiol.* 1976; 41: 449-56.
19. Madden MC, Vender RL, and Friedman M. Effect of hypoxia on prostacyclin production in cultured pulmonary artery endothelium. *Prostaglandins.* 1986; 31: 1049-1062.
20. McQuillan LP, Leung GK, Marsden PA, Kostyk SK, and Kourembanas S. Hypoxia inhibits expression of eNOS via transcriptional and posttranscriptional mechanisms. *Am. J. Physiol.* 1994; 267: H1921-H1927.
21. Means A. Calcium, calmodulin and cell cycle regulation. *FEBS lett.* 1994; 347: 1-4.
22. Oparil S, Chen SJ, Meng QC, Elton TS, Yano M, and Chen YF. Endothelin-A receptor antagonist prevents acute hypoxia-induced pulmonary hypertension in the rat. *Am. J. Physiol.* 1995; 268: L95-L100.
23. Partovian C, Adnot S, Raffestin B, Louzier V, Levame M, Mavier IM, Lemarchand P, and Eddahibi S. Adenovirus-mediated lung VEGF overexpression protects against hypoxic pulmonary hypertension in rats. *Am. J. Respir. Cell. Mol. Biol.* 2000; 23: 762-771.
24. Platoshyn O, Yu Y, Golovina V, McDaniel S, Krick S, Li L, Wang J, Tubin L, and Yuan JX-J. Chronic hypoxia decreases K_V channel expression and function in pulmonary artery myocytes. *Am. J. Physiol. Lung Cell. Mol. Physiol.* 2001; 280: L801-L812.
25. Rabinovitch M, Gamble W, Nadas AS, Miettinen OS, and Reid L. Rat pulmonary circulation after chronic hypoxia : hemodynamic and structural features. *Am. J. Physiol.* 1979; 236: H818-H827.
26. Rabinovitch M, Konstam MA, Gamble WJ, Papanicolaou N, Aronovitz MJ, Treves S, and Reid L. Changes in pulmonary blood flow affect vascular response to chronic hypoxia in rats. *Circ. Res.* 1983; 52: 432-441.
27. Reeves J and Herget J. *Experimental Models of Pulmonary Hypertension.* Mount Kisco: Futura Publishing Company, 1984.
28. Rose F, Grimminger F, Appel J, Pies V, Weissmann N, Fink L, Schmidt S, Krick S, Camenish G, Gassmann N, Seeger W, and Hänze J. Hypoxic pulmonary artery fibroblasts trigger proliferation of vascular smooth muscle cells-role of hypoxia-inducible transcription factors. *FASEB J.* 2002; 16: 1660-1661.
29. Smirnov SV, Robertson TP, Ward JPT, and Aaronson PI. Chronic hypoxia is associated with

reduced delayed rectifier K^+ current in rat pulmonary muscle cells. *Am. J. Physiol.* 1994; 266: H365-H370.

30. Tarasevieiene-Stewart L, Kasahara Y, Alger L, Hirth P, McMahon G, Waltenberger J, Voelkel NF, and Tuder RM. Inhibition of the VEGF receptor 2 combined with chronic hypoxia causes cell death-dependent pulmonary endothelial cell proliferation and severe pulmonary hypertension. *FASEB J.* 2001; 15: 427-38.
31. Tozzi CA, Polani GJ, Harangozo AM, Boyd CD, and Riley DJ. Pressure-induced connective tisssue synthesis in pulmonary artery segments is dependent on intact endothelium. *J. Clin. Invest.* 1989; 84: 1005-1012.
32. Xue C and Johns RA. Upregulation of nitric oxide synthase correlates temporally with onset of pulmonary vascular remodeling in the hypoxic rat. *Hypertension.* 1996; 28: 743-753.
33. Yet S-F, Perrella MA, Layne MD, Hsieh C-M, Maemura K, Kobzik L, Wiesel P, Christou H, Kourembanas S, and Lee M-E. Hypoxia induces severe right ventricular dilation and infarction in heme oxygenase-1 null mice. *J. Clin. Invest.* 1999; 103: R23-R29.
34. Yu AY, Shimoda LA, Iyer NV, Huso DL, Sun X, McWilliams R, Beaty T, Sham JSK, Wiener CM, Sylvester JT, and Semenza GL. Impaired physiological responses to chronic hypoxia in mice partially deficient for hypoxia-inducible factor 1α. *J. Clin. Invest.* 1999; 103: 691-696.
35. Zamora MA, Dempsey EC, Walchak SJ, and Stelzner TJ. BQ123, an ETA receptor antagonist, inhibits endothelin-1 mediated proliferation of human pulmonary artery smooth muscle cells. *Am. J. Respir. Cell. Mol. Biol.* 1993; 9: 429-433.

Chapter 29

Polyamine Regulation in Hypoxic Pulmonary Arterial Cells

Mark N. Gillespie, Kathryn A. Ziel, Mykhaylo Ruchko, Pavel Babal, and Jack W. Olson
University of South Alabama College of Medicine, Mobile, Alabama, U.S.A. and Comenius University, Bratislava, Slovakia

1. Introduction

Chronic hypoxia promotes alterations in pulmonary vascular structure and function that culminate in sustained increases in pulmonary vascular resistance and right ventricular hypertrophy. These effects of hypoxia on the pulmonary artery (PA) wall are controlled by dynamic interactions between resident and itinerant cells, the extracellular matrix, and the chemical signaling environment in which the cells reside. Cellular responses include smooth muscle hypertrophy, which may account for medial thickening in normally muscularized arteries, and proliferation of smooth muscle or smooth muscle precursor cells which may contribute to extension of new muscle into normally non-muscularized vessels (17). Excessive deposition and disorganization of interstitial and basement membrane matrix proteins also contribute to structural changes (35). Fibroblasts, PA smooth muscle cells (PASMCs), endothelial cells (ECs), and perhaps others may be sources of these matrix proteins. A number of growth factors and vasoactive mediators, many of which are bifunctional, also have been incriminated in pulmonary hypertension (8).

Because so many signals impact on lung cells in pulmonary hypertension, it is unlikely that a single key ligand or receptor could serve as an isolated pharmacologic target. Alternatively, intracellular signal transduction processes used to integrate stimuli originating from the local environment seem especially appealing. In this context, our research has focused on the polyamines putrescine (PUT), spermidine (SPD), and spermine (SPM), a family of low molecular weight organic cations that may be uniquely situated to impact on a number of signal transduction pathways. It has been recognized for many years that precise adjustments in polyamine pools are required for cell growth and differentiation (25). More recently, polyamines have been incriminated in regulation of apoptosis (20). Emerging data also indicate that the polyamines are involved in relatively proximate events in signaling. For example, polyamines exhibit substantial antioxidant activity (10) and in lung vascular cells hypoxia causes generation of reactive oxygen species (9, 14). Interactions between polyamines and reactive oxygen species could thus effect hypoxic signal transduction pathways incriminated in both acute hypoxic pulmonary vasoconstriction (36) and hypoxia-induced gene expression (3). Polyamines also may govern certain ion channels and receptors (13) and modulate a number of mitochondrial functions (29) that could be linked to hypoxic adaptation. Hypoxia alters calcium fluxes in lung vascular cells (4), and in certain non-vascular cells, increased polyamine content is a stimulus for elevations in cell Ca^{2+} (15). PKC activation, possibly PKC-α, occurs in hypoxic PASMCs (6) and polyamines may enhance or depress PKC activity depending on the cell type (18, 32). Given the many signaling events in which polyamines have been incriminated, the potential of this family of molecules to serve as an isolated pharmacologic target in hypoxic lung disease would seem to warrant attention.

2. Polyamine Regulation in the Hypoxic Lung

Cell polyamine contents are governed by multiple interactive pathways (25). In *de novo* polyamine synthesis, PUT is synthesized from its precursor, ornithine, via ornithine decarboxylase (ODC), which is also one of the rate-limiting enzymes in polyamine synthesis. PUT is then converted sequentially to SPD and SPM via two other potentially rate-limiting enzymes, S-adenosyl-methionine decarboxylase (AdoMet-DC), and SPD- and SPM- synthases, respectively. *De novo* polyamine synthesis can be suppressed pharmacologically by α-difluoromethylornithine (DFMO), a specific inhibitor of ODC. ODC is unusual in terms of its regulation. The protein has one of the shortest half lives of known mammalian enzymes and is exquisitely sensitive to changes in the abundance of antizyme, a family of at least three 18-30 kDa proteins that inhibit ODC and promote its 26S proteasome-dependent degradation (19).

A second regulatory pathway is transmembrane transport (30). In some cells, polyamine uptake accompanies proliferation while it is down-regulated during quiescence. In polyamine-depleted cells, transport processes may restore polyamine levels and revitalize cell functions. Polyamine transport activity seems to require ongoing RNA and protein synthesis and, in some cells, may require an intact sodium gradient. There also appears to be considerable differences in terms of the number of polyamine transporters, with some cells expressing a single transporter with overlapping specificity for the three polyamines and others expressing uptake pathways that are relatively specific for each polyamine. It is unknown whether the multiple transporters are regulated differentially and if so, what biological significance such regulation might be. Like ODC, polyamine uptake is negatively regulated by antizyme.

A third pathway involves polyamine interconversion, catalyzed by the SPD and SPM aminopropyltransferases, another rate-limiting enzyme-SPD/SPM N^1-acetyltransferase (SAT), and polyamine oxidase (5). These latter two enzymes are responsible for conversion of SPM back to SPD and SPD to PUT.

Finally, it has been speculated for many years that polyamine compartmentation was an important regulatory pathway. In support of this idea, the abundance of SPM in mitochondria and nuclei is known to change during the cell cycle (27). It has also been shown that mitochondria express a polyamine "uniporter" (34) which could govern mitochondrial polyamine contents independently of cytosolic changes. Subcellular polyamine distribution has not been explored in pathologic contexts.

De novo polyamine synthesis seems to be a dominant regulatory mechanism underlying most types of lung structural remodeling. Increased lung ODC activity is temporally-related to elevations in polyamine content that accompany postnatal lung growth (33), repair of hyperoxic lung injury (11), and monocrotaline-induced pulmonary hypertension (21, 23). ODC inhibition with DFMO prevents the entire spectrum of monocrotaline-induced lung pathology, including vascular hyperreactivity, edema, medial arterial thickening, pulmonary hypertension, and right ventricular hypertrophy (7, 22). These observations emphasize the importance of ODC and *de novo* polyamine synthesis in regulating lung polyamine contents and attendant changes in lung structure, including the monocrotaline model of chronic pulmonary hypertension.

Unlike the dominant role of increased *de novo* polyamine synthesis in other pulmonary diseases, hypoxic pulmonary vascular remodeling seems to have an unusual dependence on augmented polyamine transport and perhaps on polyamine interconversion. In support of this idea, we found that although polyamine contents were elevated by hypoxia (24), ODC activity was decreased in intact rat lungs (31). In contrast, PUT uptake was augmented and efflux was reduced in lungs from hypoxic rats. Activities of polyamine interconverting enzymes, AdoMet-DC, SAT, and polyamine oxidase, also were elevated.

3 Lung Cell-specific Polyamine Transport Regulation

Given the substantial effect of hypoxia on lung vascular structure and the requirement for polyamines in cell proliferation and differentiation, we postulated that the hypoxia-induced increase in polyamine uptake would be prominent in PA cells. Previous autoradiographic studies on the cellular localization of polyamine uptake focused on the normoxic lung and identified the most prominent sites of uptake as alveolar type I and II cells (12). Uptake by cells of the normoxic pulmonary circulation had not been appreciated. Accordingly, we used rat lung and main PA explant preparations to examine the effects of culture in a hypoxic environment for 24 hours on [^{14}C]-SPD localization in vascular cells (1). We found increases in the density of [^{14}C]-SPD labeling in both intimal and medial layers of conduit, muscularized, and partially-muscularized PAs. The extent of [^{14}C]-SPD uptake in main PA explant preparations also was elevated in hypoxia and autoradiography revealed that the increase could be ascribed to augmented labeling of both intimal and medial arterial layers. The hypoxia-induced increase in [^{14}C]-SPD transport was most evident in smooth muscle cells of the media. Viewed collectively, these findings in lung and main PA explant preparations suggested that hypoxia increases SPD uptake in PAECs and PASMCs, most conspicuously in vascular smooth muscle.

To explore cell-cell interactions in the hypoxic pulmonary vasculature that could impact on polyamine transport properties, we examined the effect of hypoxia in rat denuded PA explants (1). Endothelial denudation reduced the baseline [^{14}C]-SPD uptake and abolished the increase normally evoked by hypoxia. One explanation for this observation was that hypoxic PAECs elaborate a factor(s) that increases [^{14}C]-SPD uptake by the underlying smooth muscle. To address this possibility, a cross-over design was used in which media conditioned by normoxic or hypoxic PAECs was applied to denuded main PA explants. Normoxic PAEC-conditioned media failed to increase the [^{14}C]-SPD uptake rate in either normoxic or hypoxic denuded explants, thus suggesting that induction of SPD transport in medial smooth muscle cells cannot be ascribed to the hypoxia-mediated activation of a factor(s) released constitutively by cultured, normoxic PAECs. However, while media conditioned by hypoxic PAECs also did not elevate the uptake rate in normoxic explants, it engendered substantial increases in the hypoxic explant preparation. This pattern of results was recapitulated using cultured PASMCs as the bioassay preparation and supports the idea that a factor(s) elaborated by hypoxic PAECs is permissive for the ability of hypoxia to increase [^{14}C]-SPD uptake rate in PA smooth muscle.

The identity of the postulated endothelium-derived factor(s) enabling hypoxic SPD uptake by underlying smooth muscle is unknown. There are many candidate mediators, such as endothelin, PDGF, and serotonin (8). Based on mounting evidence for involvement of proteases in vascular development and remodeling (26), including hypoxic pulmonary vascular remodeling (39), we examined the prospect that an endothelium-derived protease was important for hypoxic induction of polyamine transport in PASMCs. Media conditioned by hypoxic PAECs was treated with the non-selective protease inhibitor, aprotinin, or the serine protease inhibitor, 1-antitrypsin, and applied to denuded main PA explants cultured in either normoxic or hypoxic conditions. As shown in Figure 1, the ability of hypoxic PAEC-conditioned media to enable hypoxia-induced rises in the [^{14}C]-SPD uptake rate in main PA explants was attenuated by both protease inhibitors. Identical results were obtained using cultured PASMCs as the bioassay preparation. Much work remains to be done, but the idea that an endothelium derived serine protease(s) is an activator of SPD transport in hypoxia by medial arterial smooth muscle is an attractive hypothesis.

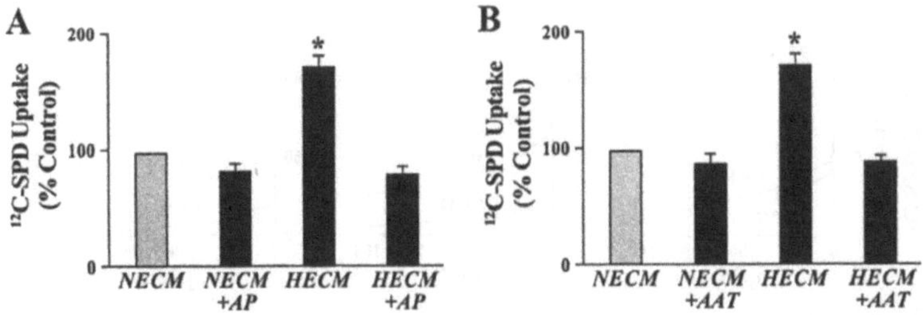

Figure 1. Aprotinin (AP, 10μg/ml; A) and alpha-1-antitrypsin (AAT, 50μg/ml; B) attenuate hypoxic EC conditioned media (HECM)-induced hypoxia-dependent polyamine transport in rat main PA explants. Normoxic EC conditioned media (NECM) failed to increase polyamine transport in either hypoxic or normoxic PA explants. Mean± SE (n=4). * $P<0.05$ vs. NECM.

We undertook additional studies in rat PAECs and PASMCs with the aims of determining the mechanism of the hypoxic effect on polyamine import and if there were multiple transporters regulated differentially by hypoxia (2, 28). Confluent cultures of both cell types were exposed to normoxia or hypoxia and uptake rates for [^{14}C]-PUT, -SPD, and -SPM were determined as a function of polyamine concentration. Polyamine transport pathways in normoxic control populations of these lung vascular cells resembled other cells described in the literature; they exhibited time-, temperature-, and concentration-dependencies. Polyamine transport in lung vascular cells required ongoing RNA synthesis. Protein synthesis inhibition was associated with a transient increase in polyamine import, presumably as a result of relief from antizyme-mediated inhibition, while long term protein synthesis inhibition resulted in a reduction in polyamine import. Uptake of PUT, SPD, and SPM could be modeled according to Michaelis-Menten kinetics and values for K_m and V_{max} are shown in Table 1. In normoxic cells, the kinetic parameters are on the same order of magnitude.

Table 1. Values of K_m (top) and V_{max} (bottom) for Polyamine Uptake in Rat PASMCs and ECs Cultured under Control (CON) and Hypoxic (HYP) conditions

	K_m (μM)					
	PUT		SPD		SPM	
	PASMCs	ECs	PASMCs	ECs	PASMCs	ECs
CON	7.06±3.48	5.1±0.75	2.48±0.66	2.5±0.45	2.85 ± 0.92	5.6±1.15
HYP	14.8 ± 2.97*	6.0±0.75	2.41±0.73	2.8±0.65	2.04 ± 0.77	11.3±3.0*
	V_{max} (pMoles/10^6 cells/min)					
	PUT		SPD		SPM	
	PASMCs	ECs	PASMCs	ECs	PASMCs	ECs
CON	9.14 ± 2.33	5.60±0.30	21.23 ± 1.92	4.60±0.25	17.44 ± 2.04	3.50±0.30
HYP	29.3 ± 3.92*	10.9±0.06*	24.72 ± 2.53	8.60±0.65*	19.88 ± 2.39	8.90±1.20*

Values are expressed as mean ± standard error. * $P<0.05$ vs. control.

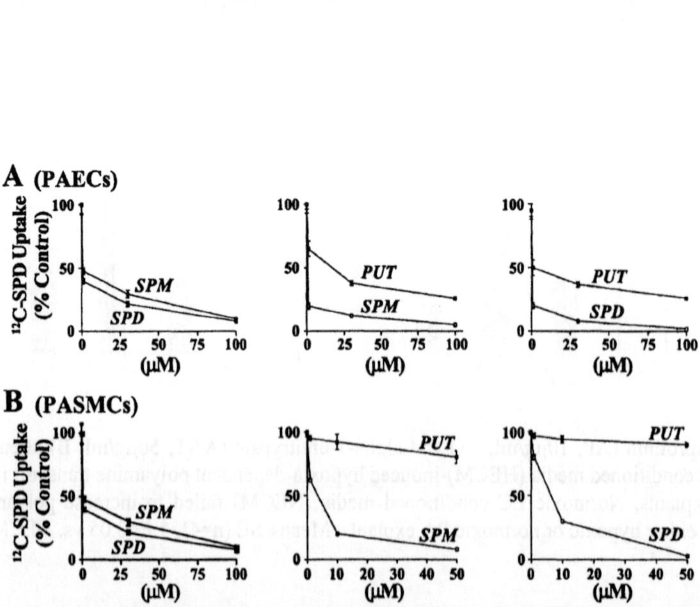

Figure 2. Competition between 0.1 μM [^{14}C]-labeled polyamines with the two other unlabeled polyamines at the indicated concentrations for uptake in rat PAECs (A) and PASMCs (B). Mean±SE as percentage of control values (n=6).

In other key respects, however, the polyamine transport pathways operative in cultured rat main PAECs and PASMCs are different. For example, in PAECs there is relatively little cross-competition between the three polyamines for uptake, while in smooth muscle cells, SPD and SPM exhibit cross-competition and inhibit PUT uptake while PUT has minimal effect on the import of SPD and SPM (Fig. 2).

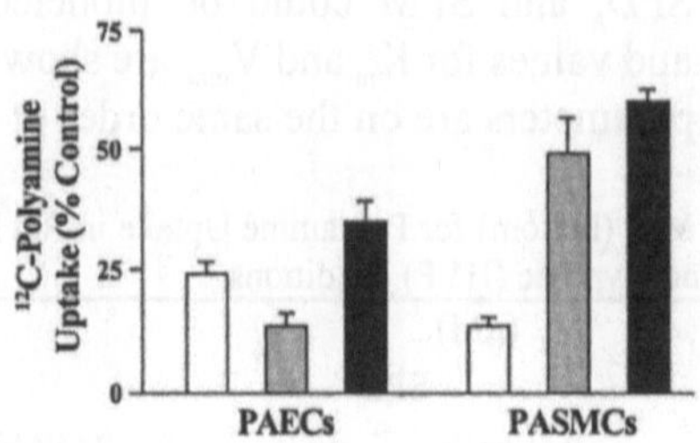

Figure 3. Uptake of the three polyamines (0.1 μM) from Na^+-free media in cultured PAECs and PASMCs (% of uptake in control, normal Na^+ conditions). Mean±SE (n=6).

The Na^+-dependence of polyamine transport is also appears to differ between the cell types. Replacing Na^+ with choline significantly (P<0.05) inhibits uptake of all three polyamines in ECs, while in PASMCs PUT import is more prominently inhibited by Na^+ depletion than the other polyamines (Fig. 3). Finally, the most interesting difference pertains to the transport response to hypoxia. As shown in Table 1, PAECs responded to hypoxia with an increase in the V_{max} for transport for all three polyamines, while SMCs exhibited a selective increase in the V_{max} for PUT uptake, with the values for SPD and SPM unchanged from controls.

4. Polyamine Transport as a Target for Intervention in Hypoxic Pulmonary Hypertension

The hypoxia-induced increase in polyamine uptake by pulmonary vascular cells raises the question of whether lung cell polyamine regulatory pathways could serve as targets of pharmacologic intervention in hypoxic pulmonary hypertension. Based on the finding that hypoxia decreases ODC activity, ODC blockade would not seem to be a promising approach. Pharmacologic manipulation of polyamine import has not, until recently, been possible owing to the lack of suitable agents. Recently, however, Weeks et al. reported the development of a lysine-polyamine conjugate, ORI1202, which specifically inhibited polyamine import in a variety of transformed cells (37).

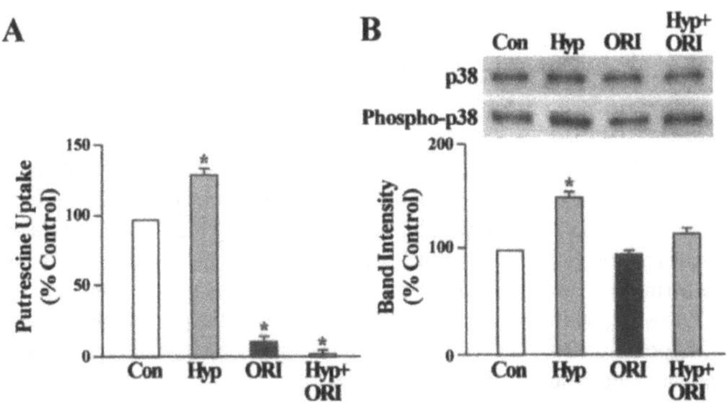

Figure 4. Impact of the polyamine transport inhibitor, ORI1202, on hypoxic activation of p38 MAP kinase. A: Uptake of 0.3 μM putrescine in normoxic control PASMCs (Con), PASMCs cultured in hypoxia for 6h (Hyp), control PASMCs incubated with 30 μM ORI1202 (ORI), and hypoxic PASMCs incubated with ORI1202 (Hyp + ORI). B: Western blot analyses of total and phospho-p38 in normoxic and hypoxic PASMCs with and without treatment with 30 μM ORI1202. Band intensities from 4 experiments were quantified by densitometry and expressed as a percentage of control. *$P<0.05$ vs. normoxic control (Modified from Ref. 28).

We used ORI1202 to explore the importance of augmented putrescine transport in the adaptive response to hypoxia by PASMCs (28). As a molecular marker of the hypoxic effect, we assessed the amount of total and phosphorylated p38 MAP kinase, which is centrally involved in the response to hypoxia in a variety of cells (16, 38). The transport inhibitor profoundly reduced both the baseline import and the hypoxia-induced rise in PUT import (Fig. 4). While neither hypoxia nor ORI1202 altered the total p38, we found that hypoxia elevated the amount of phospho-p38 and, importantly, that the increase was suppressed by ORI1202. A conservative interpretation of these observations is that induction of PUT transport by hypoxia is necessary for activation of p38 MAP kinase and its downstream signaling pathways in PASMCs.

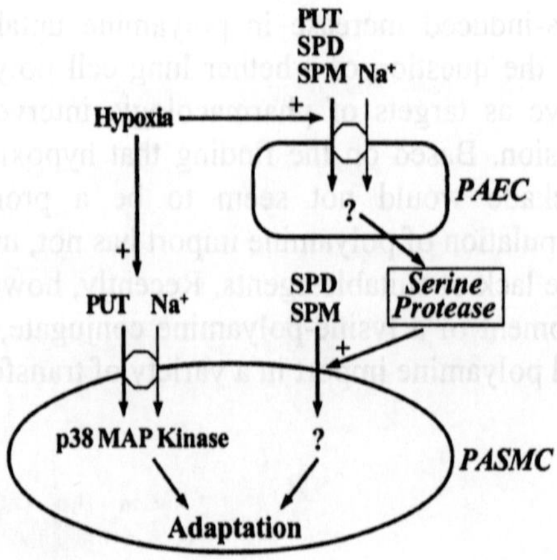

Figure 5. Proposed model for lung cell-specific polyamine transport regulation in hypoxia. Hypoxia upregulates activity of a Na^+-dependent polyamine transporter. In PAECs, all three polyamines are imported via this pathway, but in PASMCs, only PUT uptake is Na^+-dependent. The biological roles of transported polyamines are speculative, but in PAECs may be required for elaboration of a serine protease which is permissive for hypoxia-mediated increases in SPD and SPM import in PASMCs, which occurs via a Na^+-independent transport pathway. In contrast, the direct effect of hypoxia on Na^+-dependent PUT uptake in PASMCs is required for p38 MAP Kinase activation and the ensuing adaptive response.

5. Summary and Future Directions

It is presently difficult to construct precise models for the cellular regulation of polyamine import in the hypoxic pulmonary circulation; this is partly due to the lack of tools to probe the polyamine transport pathways at the molecular level. Nevertheless, findings from the above studies support the possibility that hypoxia directly up-regulates Na^+-dependent polyamine transport pathways (Fig. 5). In PAECs, where there seems to a sodium-dependent import pathway with overlapping selectivity for the three polyamines, hypoxia elevates the uptake rate of PUT, SPD, and SPM. In contrast, PASMCs express only the enhanced PUT transport pathway that is Na^+dependent and only PUT uptake is upregulated under hypoxic conditions. To add to the complexity, our data also suggest that a serine protease elaborated by hypoxic PAECs enables the hypoxic induction of SPD uptake by underlying smooth muscle. This increased SPD transport would, presumably, be linked to induction of the non-specific, sodium-independent uptake pathway. Additional studies, using a combination of *in vivo* and *in vitro* model systems will be required to appreciate the fine structure of polyamine regulation in hypoxic lung vascular cells.

There are many outstanding questions pertaining to regulation of polyamine transport in hypoxic lung vascular cells. One relates to the presence of multiple transporters and their differential regulation in hypoxia; the biological significance of this is not clear. Another issue pertains to the biological significance of the apparent interaction between direct effects of hypoxia on PASMC PUT uptake and the permissive effects of a PAEC-derived protease on PASMC uptake of SPD and SPM. Perhaps this dual regulatory mode governs the timing of transport induction in PASMCs, with an initial burst of PUT transport regulating certain cellular responses while the delayed increase in SPD and presumably SPM uptakes modulating other aspects of the PASMC response to hypoxia. Finally, and in a related context, the specific cellular responses that require increased polyamine transport in hypoxia are unknown. While traditional concepts tended to view the polyamines as being biologically equivalent, there are various reasons to believe that the individual polyamines may have discrete functions (5). At this point, we know only that the hypoxia-induced p38 MAP kinase phosphorylation in PASMCs is dependent on increased PUT uptake (Fig. 5), but whether other events in cellular adaptation to hypoxia have a specific requirement for SPD and/or SPM uptake is unknown.

The integrated response of the pulmonary circulation to hypoxia is complex, involving many different cell types, mediators, and transduction pathways. The polyamines, by virtue of their role in multiple signaling events, would seem to be an important point of integration of the many stimuli acting on hypoxic lung cells. Our findings that polyamine transport, rather than *de novo* polyamine synthesis, seems to be the dominant pathway regulating lung vascular cell polyamines during hypoxia suggest that polyamine transporter(s) may be an isolated target for intervention. The inhibitory effect of the polyamine transport inhibitor, ORI1202, on hypoxia-induced p38 MAP kinase phosphorylation in PASMCs is provocative. If the effector cells in hypoxic pulmonary hypertension rely on augmented polyamine import as a source of signaling molecules, while cells not intimately linked to the hypoxic response continue to synthesize polyamines via the actions of ODC, then transport blockade could be a rather selective pharmacologic strategy. Studies in intact animal models would be helpful in resolving this issue and establishing the safety and efficacy of polyamine transport inhibitors as a therapeutic strategy in hypoxic pulmonary hypertension.

Acknowledgments

This work was supported in part by grants from the National Institutes of Health (HL 36404, HL38495).

References

1. Babal P, Manuel SM, Olson JW, and Gillespie MN. Cellular disposition of transported polyamines in hypoxic rat lung and pulmonary arteries. *Am. J. Physiol. Lung Cell. Mol. Physiol.* 2000; 278: L610-L617.
2. Babal P, Ruchko M, Ault-Ziel K, Cronenberg L, Olson JW, and Gillespie MN. Regulation of ornithine decarboxylase and polyamine import by hypoxia in pulmonary artery endothelial cells. *Am. J. Physiol. Lung Cell. Mol. Physiol.* 2002; 282: L840-L846.
3. Chandel NS, Maltepe E, Goldwasser E, Mathieu CE, Simon MC, and Schumacker PT. Mitochondrial reactive oxygen species trigger hypoxia-induced transcription. *Proc. Natl. Acad. Sci. USA.* 1998; 95: 11715-11720.
4. Cornfield DN, Stevens T, McMurtry IF, Abman SH, and Rodman DM. Acute hypoxia causes membrane depolarization and calcium influx in fetal pulmonary artery smooth muscle cells. *Am. J. Physiol.* 1994; 266: L469-L475.
5. Coward JK and Pegg AE. Specific multisubstrate adduct inhibitors of aminopropyltransferases and their effect on polyamine biosynthesis in cultured cells. *Adv. Enzyme Regul.* 1987; 26: 107-113.
6. Dempsey EC, McMurtry IF, and O'Brien RF. Protein kinase C activation allows pulmonary artery smooth muscle cells to proliferate to hypoxia. *Am. J. Physiol.* 1991; 260: L136-L145.
7. Gillespie MN, Dyer KK, Olson JW, O'Connor WN, and Altiere RJ. α-Difluoromethylornithine, an inhibitor of polyamine synthesis, attenuates monocrotaline-induced pulmonary vascular hyperresponsiveness in isolated perfused rat lungs. *Res. Commun. Chem. Pathol. Pharmacol.* 1985; 50: 365-378.
8. Gillespie MN, Lipke DW, and McMurtry IF, eds, *Pulmonary Vascular Remodeling*. Portland, OR: Portland Press Research Monograph, 1995.
9. Grishko V, Solomon M, Breit JF, Killilea DW, Ledoux SP, Wilson GL, and Gillespie MN. Hypoxia promotes oxidative base modifications in the pulmonary artery endothelial cell VEGF gene. *FASEB J.* 2001; 15: 1267-1269.
10. Ha HC, Yager JD, Woster PA, and Casero RA Jr. Structural specificity of polyamines and polyamine analogues in the protection of DNA from strand breaks induced by reactive oxygen species. *Biochem. Biophys. Res. Commun.* 1998; 244: 298-303.
11. Hacker AD, Tierney DF, and O'Brien TK. Polyamine metabolism in rat lungs with oxygen toxicity. *Biochem. Biophys. Res. Commun.* 1983; 113: 491-496.
12. Hoet PH, and Nemery B. Polyamines in the lung: polyamine uptake and polyamine-linked pathological or toxicological conditions. *Am. J. Physiol. Lung Cell. Mol. Physiol.* 2000; 278: L417-L433.
13. Huang CJ and Moczydlowski E. Cytoplasmic polyamines as permeant blockers and modulators of the voltage-gated sodium channel. *Biophys. J.* 2001; 80: 1262-1279.
14. Killilea DW, Hester, R, Balczon, R, Babal, P and Gillespie, MN. Free radical production in hypoxic pulmonary artery smooth muscle cells. *Am. J. Physiol. Lung Cell. Mol. Physiol.* 2000; 279: L408-L412.
15. Koenig H, Goldstone AD, and Lu CY. β-adrenergic stimulation of Ca^{2+} fluxes, endocytosis, hexose transport, and amino acid transport in mouse kidney cortex is mediated by polyamine synthesis. *Proc. Natl. Acad. Sci. USA.* 1983; 80: 7210-7214.

16. Kulisz A, Chen N, Chandel NS, Shao Z, and Schumacker PT. Mitochondrial ROS initiate phosphorylation of p38 MAP kinase during hypoxia in cardiomyocytes. *Am. J. Physiol. Lung Cell. Mol. Physiol.* 2002; 282: L1324-L1329.
17. Meyrick B and Reid L. Hypoxia and incorporation of 3H-thymidine by cells of the rat pulmonary arteries and alveolar wall. *Am. J. Pathol.* 1979; 96: 51-70.
18. Monti MG, Marverti G, Ghiaroni S, Piccinini G, Pernecco L, and Moruzzi MS. Spermine protects protein kinase C from phospholipid-induced inactivation. *Experientia.* 1994; 50: 953-957.
19. Murakami Y, Matsufuji S, Hayashi SI, Tanahashi N, and Tanaka K. ATP-Dependent inactivation and sequestration of ornithine decarboxylase by the 26S proteasome are prerequisites for degradation. *Mol. Cell Biol.* 1999; 19: 7216-7227.
20. Nitta T, Igarashi K, and Yamamoto N. Polyamine depletion induces apoptosis through mitochondria-mediated pathway. *Exp. Cell Res.* 2002; 276: 120-128.
21. Olson JW, Altiere RJ, and Gillespie MN. Prolonged activation of rat lung ornithine decarboxylase in monocrotaline-induced pulmonary hypertension. *Biochem. Pharmacol.* 1984; 33: 3633-3637.
22. Olson JW, Atkinson JE, Hacker AD, Altiere RJ, and Gillespie MN. Suppression of polyamine biosynthesis prevents monocrotaline-induced pulmonary edema and arterial medial thickening. *Toxicol. Appl. Pharmacol.* 1985; 81: 91-99.
23. Olson JW, Hacker AD, Altiere RJ, and Gillespie MN. Polyamines and the development of monocrotaline-induced pulmonary hypertension. *Am. J. Physiol.* 1984; 247: H682-H685.
24. Olson JW, Hacker AD, Atkinson JE, Altiere RJ, and Gillespie MN. Polyamine content in rat lung during development of hypoxia-induced pulmonary hypertension. *Biochem. Pharmacol.* 1986; 35: 714-716.
25. Pegg AE and McCann PP. Polyamine metabolism and function. *Am. J. Physiol.* 1982; 243: C212-C221.
26. Rabinovitch M. Pathobiology of pulmonary hypertension. Extracellular matrix. *Clin. Chest Med.* 2001; 22: 433-449.
27. Rubinstein S and Breitbart H. Cellular localization of polyamines: cytochemical and ultrastructural methods providing new clues to polyamine function in ram spermatozoa. *Biol. Cell.* 1994; 81: 177-183.
28. Ruchko M, Gillespie MN, Weeks RS, Olson JW, and Babal P. Putrescine transport in hypoxic rat main PASMCs is required for p38 MAP kinase activation. *Am. J. Physiol. Lung Cell. Mol. Physiol.* 2003; 284: L179-L186.
29. Rustenbeck I, Loptien D, Fricke K, Lenzen S, and Reiter H. Polyamine modulation of mitochondrial calcium transport. II. Inhibition of mitochondrial permeability transition by aliphatic polyamines but not by aminoglucosides. *Biochem. Pharmacol.* 1998; 56: 987-995.
30. Seiler N, Delcros JG, and Moulinoux JP. Polyamine transport in mammalian cells. An update. *Int. J. Biochem. Cell. Biol.* 1996; 28: 843-861.
31. Shiao RT, Kostenbauder HB, Olson JW, and Gillespie MN. Mechanisms of lung polyamine accumulation in chronic hypoxic pulmonary hypertension. *Am. J. Physiol.* 1990; 259: L351-L358.
32. Thams P, Capito K, and Hedeskov CJ. An inhibitory role for polyamines in protein kinase C activation and insulin secretion in mouse pancreatic islets. *Biochem. J.* 1986; 237: 131-138.
33. Thet LA and Parra SC. Role of ornithine decarboxylase and polyamines in early postnatal lung growth. *J Appl Physiol* 1986; 61: 1661-6.
34. Toninello A, Dalla Via L, Siliprandi D, and Garlid KD. Evidence that spermine, spermidine, and putrescine are transported electrophoretically in mitochondria by a specific polyamine uniporter. *J. Biol. Chem.* 1992; 267: 18393-18397.
35. Vyas-Somani AC, Aziz SM, Arcot SA, Gillespie MN, Olson JW, and Lipke DW. Temporal alterations in basement membrane components in the pulmonary vasculature of the chronically hypoxic rat: impact of hypoxia and recovery. *Am. J. Med. Sci.* 1996; 312: 54-67.
36. Waypa GB, Chandel NS, and Schumacker PT. Model for hypoxic pulmonary vasoconstriction

involving mitochondrial oxygen sensing. *Circ. Res.* 2001; 88: 1259-1266.
37. Weeks RS, Vanderwerf SM, Carlson CL, Burns MR, O'Day CL, Cai F, Devens BH, and Webb HK. Novel lysine-spermine conjugate inhibits polyamine transport and inhibits cell growth when given with DFMO. *Exp. Cell. Res.* 2000; 261: 293-302.
38. Welsh DJ, Peacock AJ, MacLean M, and Harnett M. Chronic hypoxia induces constitutive p38 mitogen-activated protein kinase activity that correlates with enhanced cellular proliferation in fibroblasts from rat pulmonary but not systemic arteries. *Am. J. Respir. Crit. Care Med.* 2001; 164: 282-289.
39. Zaidi SH, You XM, Ciura S, Husain M, and Rabinovitch M. Overexpression of the serine elastase inhibitor elafin protects transgenic mice from hypoxic pulmonary hypertension. *Circulation.* 2002; 105: 516-521.

Chapter 30

Strain Differences of Hypoxia-Induced Pulmonary Hypertension

Mallik R. Karamsetty, James C. Leiter, Lo Chang Ou, Ioana R. Preston, and Nicholas S. Hill
Wyeth Pharmaceuticals, Richmond, Virginia; Dartmouth Medical School, Lebanon, New Hampshire; Tufts University, Boston, Massachusetts, U.S.A.

1. Introduction

Upon exposure to high altitude, most mammals develop acute hypoxic pulmonary vasoconstriction (HPV) and, if the high altitude exposure is sustained, pulmonary hypertension, right ventricular hypertrophy, pulmonary vascular remodeling and polycythemia. However, there is a wide variability in the response to both acute and chronic hypoxia across species, strains and individuals. Some species acclimatize to hypoxic conditions successfully and develop mild pulmonary hypertension, whereas others develop severe pulmonary hypertension. For example, llamas at high altitude develop less pulmonary hypertension than cattle (3, 16). Moreover, individuals within species vary in their susceptibility to high altitude. For example, approximately 2-10% of cattle exposed to high altitude develop Brisket's Disease, which is characterized by severe pulmonary hypertension and right heart failure (20, 50), whereas the majority of cattle are resistant. Likewise, our previous work on Sprague-Dawley rats has shown that the Hilltop strain develops more severe pulmonary hypertension, right ventricular hypertrophy and polycythemia than the Madison strain despite identical exposures to chronic hypoxia (41).

Humans also vary considerably in their response to high altitude or alveolar hypoxia. A small number of people develop chronic mountain sickness, characterized by severe pulmonary hypertension, hypoxemia and polycythemia (33). Although all the factors responsible for these differences between and within species are not completely known, multiple factors including physiological, biochemical, structural and, ultimately, genetic are likely to contribute to the strain differences in the responses to acute and chronic hypoxia, and will be discussed below.

2. Physiological Differences

2.1. Cardiopulmonary Differences

Hypoxia-induced strain-related cardiopulmonary differences have been described between several rat strains that parallel the differences reported between the Hilltop and Madison strains (Table 1). For instance, Wistar-Kyoto rats develop more pulmonary hypertension and muscularization of small pulmonary arteries than Fischer 344 rats (1), and fawn-hooded rats develop severe pulmonary hypertension and vascular remodeling in comparison to to age-matched Sprague-Dawley rats (45).

Table 1. Variability Among Species and Within Species in Response to Chronic Hypoxia

Hyperresponder	Hyporesponder	Reference
Hilltop rat	Madison rat	Ou and Smith, 1983 (41)
Fawn-Hooded rat	Sprague-Dawley rat	Stelzner et al., 1992 (45)
Yak	Domestic cow	Durmowicz et al., 1993 (8)
Wistar-Kyoto rat	Fischer 344 rat	Aguirre et al., 2000 (1)
Pikas	Wistar Rat	Ge et al., 1998 (3)

Our laboratories have performed a number of detailed physiologic studies in order to elucidate these differences. In a study designed to understand polycythemic responses to hypoxia (34, 41), Ou and Smith first described the fundamentally different cardiopulmonary responses to high altitude in the Hilltop and Madison rats. Rats obtained from Hilltop laboratories (Hilltop, PA) became severely polycythemic and moribund after 4-5 weeks of hypoxic exposure, whereas those from the Madison, Wisconsin facility of Harlan Sprague Dawley were resistant to chronic hypoxia, developing only mild polycythemia and no morbidity after the same hypoxic exposure. Subsequent work demonstrated that the Hilltop rats behaved as though they developed chronic mountain sickness, with more severe polycythemia, pulmonary hypertension and right ventricular hypertrophy than the Madison rats (41). These cardiopulmonary differences were consistently observed regardless of gender, indicating a probable genetic origin (42). The time course of these changes is shown in Figure 1. Greater pulmonary hypertension and right ventricular hypertrophy were apparent in the Hilltops than Madisons as early as 7 days as indicated by differences in the right ventricular peak systolic pressure and the ratios of the left ventricular plus septal weights to right ventricle to (LV+S/RV), respectively (23, 24). However, significant differences in hematocrit between the strains were not apparent until after 2 weeks, suggesting that this difference was probably not primarily responsible for the cardiopulmonary differences.

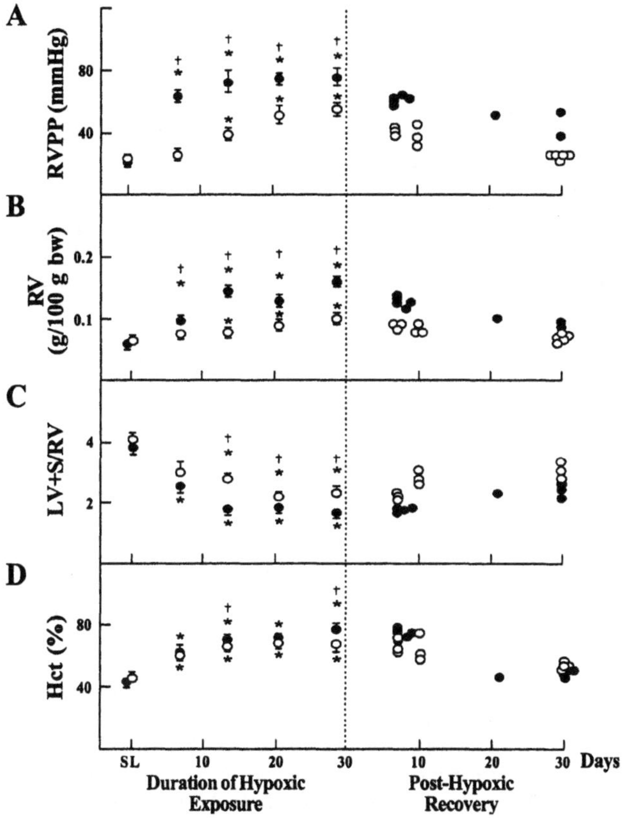

Figure 1: Time course of changes in right ventricular peak systolic pressure (RVPP) (A), right ventricular to body weight (RV) (B), left ventricular plus septal to right ventricular weight (LV+S/RV) ratios (C) and hematocrit (Hct) (D) over days of hypoxic exposure (left side of panels) and normoxic recovery from hypoxia (right side of panels). *$P<0.05$ vs. baseline, †$P<0.05$ Madison (○) vs Hilltop (●) (Modified from Ref. 24).

2.2. Hematologic Differences

The Hilltop and Madison strains manifest no hematologic differences under normoxic conditions. However, Hilltop rats develop greater polycythemia in response to hypoxic exposure than Madison rats, with statistically significant differences in hematocrit appearing within two to three weeks (Fig. 1) (23, 24). This is associated with greater increases in total blood volume and red cell mass in Hilltop rats that are apparent within 3 days (Fig. 2A) (35). To compensate for the increase in total blood volume as red cell mass increases, plasma volume decreases similarly in both strains, contributing to hemoconcentration. However, this mechanism is overridden when hematocrit reaches 75%, leading to a rapid increase in total blood volume when hematocrit exceeds that level (35).

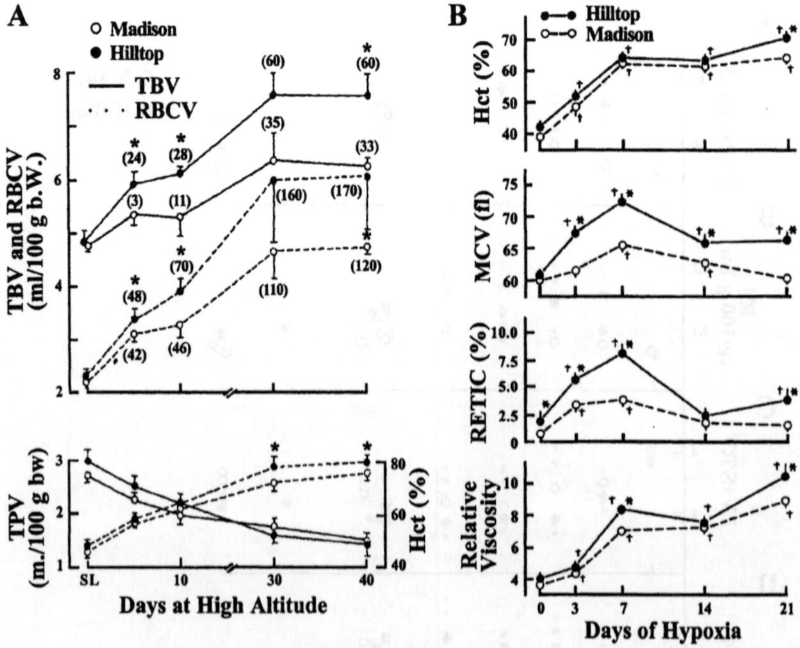

Figure 2. A: Changes in total blood volume (TBV) and red blood cell volume (RBCV) (upper panel) and total plasma volume (TPV) (lower panel) per 100 g body weight (bw) over days of high altitude exposure in Madison (○) and Hilltop (●) rats. *$P<0.05$ in Hilltop vs Madisons. B: Time course of hematocrit (Hct), mean red cell volume (MCV), reticulocyte count (RETIC) and relative viscosity over duration of hypoxic exposure in days. *$P<0.05$ Hilltop vs Madison, †$P<0.05$ vs. baseline value. SL, seal level (Modified from Refs. 23 and 35).

The increased red cell mass parallels an early increase in mean red cell volume in Hilltop rats that is significantly greater than in the Madison strain within 3 days of hypoxic exposure (Fig. 2B). This is accompanied by a greater early reticulocytosis (after 3 days) and a greater increase in relative viscosity of whole blood (after 7 days) in Hilltop compared to Madison rats. The spleen plays a prominent role in extra-medullary hematopoeisis in rats, and splenic size is also greatly enhanced in the Hilltop compared to the Madison rats during exposure to chronic hypoxia (27).

On the other hand, red cell deformability and viscosity of reconstituted blood at equivalent hematocrits did not differ between the strains, suggesting that the Hilltop strain's larger red cells *per se* do not contribute to the cardiopulmonary differences between the strains. Furthermore, pressure-flow curves of lungs isolated from normoxic and 7 day hypoxic rats suggested that the differences in pulmonary vascular resistance between the strains were related more to pulmonary vascular structural than to hematologic differences (22).

Ou et al. (36, 39) investigated the mechanisms underlying the differences in hypoxia-induced hematological responses between the strains and found increased plasma erythropoietin and renal tissue erythropoietin levels in both strains occurring within 1 day of hypoxic exposure (Fig. 3). However, renal tissue erythropoietin mRNA levels were not different early during the hypoxic exposure. The trend toward greater erythropoietin levels in the Hilltop strain early in the hypoxic exposure reached statistical significance only after 7-10 days of hypoxic exposure. There were no differences in minute ventilation or ventilatory pattern, PaO_2, pH, or eryhropoietin clearance from the circulation at any time during the entire 30-day hypoxic exposure . However, the PaO_2 and renal venous PO_2 were lower in the Hilltop compared to the Madison rats from the onset of hypoxia and remained lower throughout.

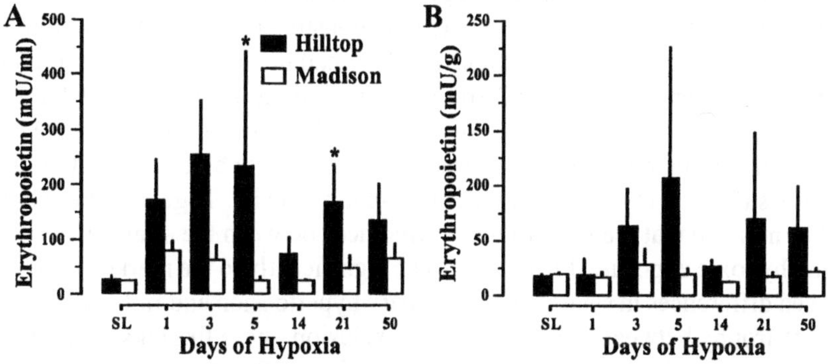

Figure 3. Time course of erythropoeitin plasma levels (A) and renal tissue levels (per gram tissue weight) (B) in Hilltop and Madison rats during days of hypoxic exposure. *$P<0.05$ Hilltop vs Madison. SL, seal level (Modified from Ref. 36).

The authors performed a *post-hoc* analysis in which they demonstrated that renal tissue and plasma erythropoietin levels and tissue erythropoietin mRNA levels from both rat strains could be fit on a single dose-response curve in which the erythropoietin response was expressed as a function of renal venous PO_2. This implies that the hypoxia sensitivity of erythropoietin synthesis and release in the Hilltop and Madison rats is similar, but Hilltop rats simply experience greater renal venous hypoxia for any level of inspired O_2. This leaves unexplained the reason for the greater renal venous hypoxia in the Hilltop rats early during the hypoxic exposure, although it is clearly not related to greater hypoventilation in the Hilltop compared to the Madison rats. Sludging of red cells and slowing of blood flow through capillaries could contribute late during the hypoxic exposure, but this would not explain the early differences, when the strains are similarly polycythemic. In addition, the hypoxia sensitivity of erythropoietin synthesis and release has not been examined explicitly; such a test would require generating the entire dose-response relationship between erythropoietin and renal venous oxygenation - which has not been done. Thus, in addition to the greater renal venous hypoxemia of the Hilltop strain, we have not excluded the possibility that erythropoietin synthesis and release is also more sensitive to the hypoxic stress in the Hilltop rats.

To determine whether the enhanced erythropoietic response to hypoxia in the Hilltop rats contributes to its enhanced cardiopulmonary response, Petit et al. (43) administered human recombinant erythropoietin to both strains of rats during hypoxic exposure. They hypothesized that if erythropoietin contributes to the difference, then exogenous administration of erythropoietin to the Madison strain should render its cardiopulmonary responses similar to those of the Hilltop strain. Instead, they showed that although hematocrit increased to the same polycythemic level in both strains during hypoxic exposure, the differences in right ventricular peak pressure and right ventricular hypertrophy persisted. In a subsequent experiment, Du et al. (9) demonstrated that hemodilution by phlebotomy to lower mean hematocrit to 46% had no effect on the severity of pulmonary hypertension in the 2 strains. This finding is contrary to what was expected based on the reduction of viscosity, but the authors calculated an increase in vascular hindrance (the vascular resistance to blood viscosity ratio), that was sufficient to counteract the decrease in viscosity. Together, these latter studies indicate that the hematologic differences between the strains that occur during hypoxic exposure do not contribute significantly to the marked hypoxia-induced differences in severity of pulmonary hypertension and right ventricular hypertertrophy between these strains. By exclusion, these findings suggest that differences in pulmonary vascular responses (i.e., HPV, structural changes or both) are responsible for the cardiopulmonary differences.

2.3. Pulmonary Vascular Differences

2.3.1. Hypoxic Pulmonary Vasoconstriction

Hypoxic pulmonary vasoconstriction (HPV) is characterized by constriction of pulmonary arterioles in response primarily to alveolar hypoxia that is thought to be a homeostatic mechanism to match perfusion and ventilation for optimizing pulmonary gas exchange. The intensity of HPV differs between and within species and this variation has important effects on adaptation to high altitude. In their original report on the rat strain differences, Ou and Smith (41) observed a striking difference in the severity of pulmonary hypertension between the Madison and Hilltop strains as indicated by their right ventricular peak systolic pressures (RVPP) (74 vs. 50 mmHg in Hilltop vs. Madison, respectively, $P<0.05$). This difference has been reproduced consistently in subsequent studies on these rats and has been confirmed by direct measurements of mean pulmonary artery pressure (75 vs. 50 mmHg in Hilltop and Madison, respectively, $P<0.05$) (9). In the earlier study, the peak systolic pressures were probably blunted by anesthesia, offering a likely explanation for why they were essentially the same as the mean pressures measured in unanesthetized animals in the latter study.

Surprisingly, in response to acute hypoxia, the Madison rats have more vigorous HPV, both in isolated lungs (31) and in intact animals (39), in contrast to the greater pulmonary hypertensive responses of the Hilltops to chronic hypoxia. Acute HPV in the Madison strain is very vigorous and is associated with the development of right ventricular aneurysms (39). The occurrence of these lesions is greater with more severe hypoxia, less with mild hypoxia, and reduced by administration of a calcium channel blocker, suggesting that the aneurysms arise from ischemic injury of the right ventricle related to the severity of the acute HPV. These differences in acute hypoxic vasoreactivity between the Madison and Hilltop strains have been replicated not only in isolated blood perfused lungs, but also in isolated pulmonary artery rings (44). In response to severe hypoxia (0% O_2 corresponding to a vessel bath PO_2 of 10 torr), pulmonary arteries isolated from normoxic Madison rats contracted more that those isolated from Hilltop rats, indicating that the differences in vasoreactivity must be related, at least in part, to differences in the intrinsic properties of the vessels.

Following the acute response, however, pulmonary artery pressures in intact Madison rats follow a biphasic pattern characterized by the initial severe vasoconstriction, followed by a marked blunting of the HPV within 24 hrs during which pulmonary artery pressures return to near normoxic levels (40). This is followed by a gradual rise in pulmonary artery pressure over the next few weeks of hypoxic exposure, but to a much lower level than in the Hilltops. The Hilltop rats have milder acute HPV, but no blunting occurs (40), and pulmonary arterial pressures rise progressively over the next several weeks of hypoxic exposure until severe pulmonary hypertension develops. A "crossover" occurs between 1 and 3 days when the pulmonary arterial pressures in the Hilltop animals exceed those of the Madison rats.

The failure to blunt HPV in the Hilltop rats was confirmed by exposure of intact catheterized animals to acute normoxia. During this test, which serves as an assay of the component of HPV contributing to the elevation of pulmonary arterial pressure, the Madison rats acutely dropped their pulmonary artery pressures by only 2.8 mmHg, consistent with persistent blunting of HPV, whereas the Hilltop rats dropped theirs by 8.6 mmHg ($P<0.05$ vs. Madison), indicating preservation of HPV (Fig. 4) (40).

The greater acute HPV of the Madison strain that preceeds the greater subsequent blunting is surprising and difficult to explain. The more vigorous acute HPV in the Madison could predispose to more subsequent blunting of HPV simply by virtue of its magnitude or by some biochemical mechanism, but intuitively, one would hypothesize the contrary; that greater HPV would predispose to greater chronic hypoxic pulmonary hypertension and less HPV to less chronic pulmonary hypertension. This latter view is supported by findings in other hypoxia-adapted species (e.g., pikas and yak) that have blunted HPV and develop only mild altitude-induced pulmonary hypertension (8, 13) and normal altitude-dwelling Tibetans, who exhibit minimal HPV while breathing a hypoxic gas mixture (17). More likely, the ability to blunt HPV is critical in moderating the response to chronic hypoxia in the Madison rats and in other species. As of yet, the mechanism(s) underlying HPV blunting remains unknown, but its identification could be important in understanding physiologic mechanisms of adaptation to hypoxia as well as in giving insights into possible therapies.

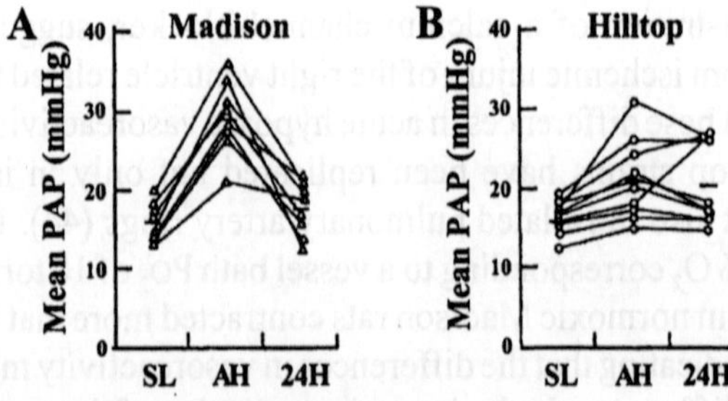

Figure 4. Mean pulmonary artery pressure (PAP) values in individual rats at sea level (SL) during acute hypoxic (AH) exposure (10.5% FIO_2 for 10 mins) and after 24 hrs (24H) exposure to hypoxia (10.5% F_IO_2). The Madisons (A) have greater acute hypoxic pressor responses and then blunt these responses more than the Hilltops (B) with sustained hypoxia (Modified from Ref. 40).

2.3.2. Structural Differences

Pulmonary vascular medial thickness differs between species under sea level conditions, and these differences appear to influence subsequent responses to hypoxia. Tucker et al. (47) found a positive correlation between the normoxic medial thickness of small pulmonary arteries and the sensitivity to chronic hypoxia. For example, pulmonary hypertension induced by chronic hypoxia is more severe in cattle and pigs that have greater pulmonary vascular medial thickness, than in dogs, sheep or guinea pigs, that have thinner pulmonary vascular walls (47). Similarly, muscular pulmonary arteries are very thin in animals such as the llama, alpaca, mountain viscacha and Tibetan snow pig that are indigenous to high altitudes and develop only mild hypoxic pulmonary hypertension (51). Further, the native Tibetan yak has successfully acclimated to high altitude by virtually eliminating acute HPV and by having very thin-walled pulmonary vessels (2, 8). The native Himalayan highlanders who have adapted successfully to high altitude also have thin-walled small pulmonary arteries with no medial hypertrophy of the muscular pulmonary arteries or muscularization of the arterioles (18).

The degree of muscularization in response to chronic hypoxia also contributes to the variability in response to chronic hypoxia-induced pulmonary hypertension. The percent increase of medial wall thickness in response to simulated high altitude exposure was significantly less in pika rodents, which did not develop pulmonary hypertension compared to rats that did (13). The Madison and Hilltop rats also exemplify these differences. Although the only difference under normoxic conditions was a greater percentage of partially muscularized alveolar duct vessels in the Hilltop than Madison rats, after 14 days of hypoxia, the percentage of fully muscular vessels was greater in the Hilltop than Madison rats at both the alveolar wall and alveolar duct levels (31). Furthermore, the percentage increase in medial thickness of preacinar pulmonary arteries in the Hilltop rats was greater than in Madisons. At the intraacinar level, both alveolar wall and respiratory bronchiolar vessels of Hilltop rats had more medial hypertrophy than Madisons. Similarly, Wistar Kyoto rats that develop severe hypoxic pulmonary hypertension have more pronounced vascular remodeling than other strains that develop moderate pulmonary hypertension (1, 49).

These structural differences have functional significance. In pressure-flow experiments in blood-perfused lungs isolated from the Madison and Hilltop strains, Hill et al. (21) demonstrated that pulmonary vascular resistance was greater in hypoxic Hilltop than hypoxic Madison lungs, even after papaverine had been added to the perfusate to eliminate the vasoconstrictive component. This is consistent with the idea that vascular structural differences contribute, along with the differences in vasoconstriction, to the greater pulmonary hypertension in the Hilltop strain. In additional experiments on these strain differences, Colice et al. (6) observed that the greater propensity to pulmonary hypertension in the Hilltop strain is not specific to hypoxia. The plant-derived pyrrole alkaloid, monocrotaline, induced more pulmonary hypertension in the Hilltop than in the Madison strain. Similar to the response to hypoxia, histologic analysis of monocrotaline-injected animals revealed a greater increase in wall thickness of alveolar duct and respiratory brochiole-associated pulmonary vessels in Hilltop than in Madison rats. These differences were not associated with any differences in hematocrit and although PaO_2 was slightly less in the Hilltop than Madison rats, the hypoxemia was very mild and insufficient to explain the pulmonary hemodynamic differences. These findings are consistent with the idea that intrinsic properties of the pulmonary vessels rather than exogenous factors are responsible for differing cardiopulmonary responses between strains or species, affecting not only vasoreactivity, but also structural responses.

2.3.3. Cellular Differences

Histologic studies of lungs from chronically hypoxic animals have consistently shown medial thickening suggesting that pulmonary vascular smooth muscle cells proliferate and/or hypertrophy in response to hypoxia, at least *in vivo*. In cell culture experiments, however, it has been difficult to consistently demonstrate proliferation in response to hypoxic exposure, although some studies have observed potentiation of pulmonary vascular smooth muscle proliferation when hypoxia is combined with other mitogens, such as serum. Furthermore, populations of pulmonary vascular smooth muscle cells have been identified from the same animal that have differing proliferative responses to mitogens, although these differences have not been attributed to strain differences. In preliminary observations from our laboratory, cultured pulmonary vascular smooth muscle cells from the Madison and Hilltop rat strains proliferated similarly in response to hypoxia and other stimuli, but pulmonary smooth muscle cells from the Madison strain were more responsive to the anti-proliferative effects of atrial natriuretic peptide (Arjona and Hill, unpublished data). This suggests that it is not only responsiveness to mitogenic stimuli, but also the ability to blunt responses to these stimuli that determines strain differences in response to hypoxia, consistent with our *in vivo* observations.

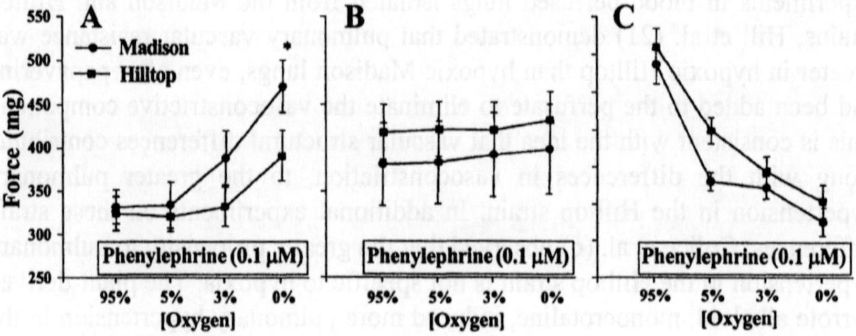

Figure 5. Contractile force of intact isolated pulmonary artery rings (A), isolated pulmonary arteries with the endothelium denuded (B), and isolated aortic rings (C) from sea-level Madison and Hilltop rats. All vessel preparations were exposed to 95%, 5%, 3% and 0% O_2. *$P<0.05$ Madison vs Hilltop (Modified from Ref. 40).

Differences in the structure and function of endothelium also appear to contribute to strain variability in the response to hypoxia. Pulmonary artery endothelial cells from the altitude-resistant Yak are much longer, wider, and rounder in appearance compared to domestic cows (hyperresponders to chronic hypoxia) (8). Furthermore, in studies on pulmonary artery rings isolated from the Madison and Hilltop strains, removal of the endothelium abolished the differences in hypoxia-induced contractions (44), suggesting that the strain-related difference in acute HPV is attributable to differences in endothelial function (Fig. 5). This difference was specific for pulmonary vessels; stripping the endothelium from the aortas had no influence on the vasorelaxant responses induced by hypoxia in these vessels. These findings indicate that endothelial cells differ morphologically between species and can contribute to differences in pulmonary vascular reactivity between strains, possibly by means of differences in mediator release.

2.4. Biochemical/Mediator Differences

The differences in pulmonary vasoreactivity and vascular hypertrophy most likely stem from differences in the synthesis, release and/or degradation of vasoactive mediators and mitogens. Likely candidates include the prostaglandins, endothelin, nitric oxide (NO) and other endothelium-derived factors.

2.4.1. Prostaglandins

Prostacyclin (PGI_2) is a potent vasodilator prostaglandin synthesized by PGI_2 synthase and released from endothelial cells (32). PGI_2 also has potent anti-mitogenic effects as well as anti-platelet actions that might be important in moderating pulmonary vascular responses to hypoxia. A relative reduction of release of PGI_2 is thought to play a role in the pathogenesis of primary pulmonary hypertension, as such patients have a reduction in the urinary ratio of PGI_2 to thromboxane metabolites. Furthermore, Tuder et al. (48) demonstrated that hypoxia-induced pulmonary hypertension is potentiated in mice rendered genetically deficient in PGI_2 synthase, indicating that PGI_2 is capable of regulating the severity of pulmonary hypertension. However, differences in PGI_2 release have not been implicated, as yet, in the mechanism of strain-related differences in cardiopulmonary responses to hypoxia. In isolated pulmonary artery rings from the Madison and Hilltop rat strains, ibuprofen had no effect on the magnitude of acute hypoxic contractile responses (44).

2.4.2. Endothelin-1

Endothelin-1 (ET-1), a potent vasoconstrictor and co-mitogen, plays a significant role in the pathogenesis of hypoxic pulmonary hypertension and pulmonary vascular remodeling. It contributes, at least in some experimental models, to the marked variability between strains in the severity of pulmonary hypertension induced by chronic hypoxia. After hypoxic exposure, ET-1 content in the lung and plasma of hypoxia susceptible strains of rats (Wistar-Kyoto and Fawn-hooded rats) is significantly greater compared to less susceptible strains (Fischer-344 and Sprague Dawley rats) (1, 45). Different ET-1 receptors may also contribute to strain differences. The ET-A receptor mediates vasoconstrictor actions of ET-1, whereas the ET-B receptor mediates both vasoconstrictor and vasodilator (by releasing NO and PGI_2) actions. In addition, the ET-B receptor serves a clearance function for ET-1. These properties of the ET-B receptor suggest that its deficiency or blockade could intensify the severity of pulmonary hypertension. Consistent with this idea, rats genetically deficient in the ET-B receptor develop more severe hypoxia-induced pulmonary hypertension than control rats with normal ET-B levels (26).

However, differences in the endothelin system do not explain differences in susceptibility to hypoxia between strains in all instances. Both under normoxic and chronically hypoxic conditions, the Madison and Hilltop rats have similar lung homogenate levels of mRNA and protein content for ET-1 and mRNA content of ET-1 receptors (28) suggesting that other biochemical mediators contribute to the variability in the response to chronic hypoxia.

2.4.3. Nitric oxide

NO, an endothelium-derived vasodilator substance known to be important for the maintenance of low vascular tone in the pulmonary vasculature, is released from l-arginine by the enzymatic action of NO synthase (NOS). Two forms of NOS are found in the lung; endothelial NOS (eNOS), and inducible NOS (iNOS). Reduced expression of eNOS has been reported in lung hemogenates from patients with primary pulmonary hypertension undergoing lung transplant compared to non-pulmonary hypertensive controls (15).

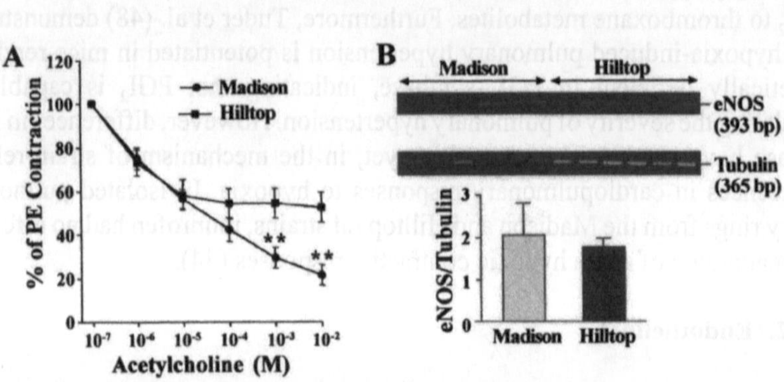

Figure 6. A: Contractile force in pulmonary artery rings isolated from sea-level Madison and Hilltop rats as percent of force generated in response to phenylephrine (PE, 10^{-6} M). In response to acetylcholine, the Madison rings relax more than the Hilltop rings. **$P<0.01$ vs. Hilltop. B: Western blots for eNOS and tubulin (upper panels) in lung homogenates from sea-level Madison and Hilltop rats. There was no significant difference in expression between the strains when the eNOS blots were standardized to tubulin (lower panel) (Modified from Refs. 28 and 40).

Differences in the release of NO could contribute to the differences in hypoxic susceptibility by a number of mechanisms; differences in the levels of substrate (L-arginine), or expression or activity of eNOS or iNOS, or both, although these have not been established as mechanisms for strain differences. In the Hilltop and Madison strains, the greater responsiveness to acetylcholine of pulmonary arteries isolated from normoxic Madison rats suggested that they might be capable of releasing more NO than normoxic Hilltop rats (Fig. 6A). However, pulmonary arteries isolated from the two strains had comparable responsiveness to the direct stimulator of NO release, thapsigargin, and comparable mRNA and protein concentrations of eNOS in lungs of both normoxic and hypoxic animals of the two strains, suggesting that the differing responsiveness to acetylcholine was related to a mechanism not involving the release of NO (28, 29) (Fig. 6B).

2.4.4. Endothelium-derived Hyperpolarizing Factor (EDHF)

EDHF has been described as the component of acetylcholine-induced vascular relaxation that is not blocked by antagonists of NOS, is associated with hyperpolarization, and is blocked by K^+ channel antagonists such as apamin and charybdotoxin (EDHF). Evidence for EDHF has been found in a number of vascular beds, including the mesenteric, coronary and pulmonary. In coronary arteries, EDHF has been identified as a product of the cytochrome 450 enzyme, CYP2A. In rabbits, a different cytochrome 450 enzyme, CYP4A, appears to be responsible for the production of EDHF, and 20-hydroxyeicosotetranoic acid has been identified as the likely product. We found evidence for release of more EDHF from pulmonary artery rings isolated from normoxic Madison than from Hilltop rats, based on the greater relaxation of these vessels in response to carbachol, an endothelium-dependent vasodilator, when compared to Hilltop rats (28). The NOS inhibitor, N^{ω}-nitro-L-arginine, completely blocked the relaxation response in Hilltop, but not in Madison pulmonary arteries. Furthermore, the residual relaxation in Madison arteries was entirely blocked by apamin and chrybdotoxin, consistent with the hypothesis that the greater relaxation was related to greater release of EDHF. These findings indicate that arteries from the 2 strains have different relaxant responses, possibly related to greater release of EDHF, but this does not explain why the contractile responses to hypoxia are also greater in the Madison strain. Furthermore, although it is tempting to speculate that the greater blunting of pulmonary vasoconstriction during chronic hypoxic exposure in the Madison than Hilltop strain is related to greater release of EDHF in Madisons during chronic hypoxia, we could find no evidence for release of EDHF from isolated pulmonary arteries of either strain after exposure to chronic hypoxia (Hill and Karamsetty, unpublished data).

3. Genetic Differences

There can be little doubt that genetic differences play an important role in strain-related cardiopulmonary differences in response to chronic hypoxia. Much has been learned about how genetic alterations modulate responses to chronic hypoxia in many animal models, only some of which are discussed above. These include genetic alteration of eNOS that renders mice more sensitive to mild hypoxia than wild-type mice (46), PGI_2 synthase as described above (48), the A receptor for the natriuretic peptides that intensifies hypoxic pulmonary hypertension when disrupted (30), to name just a few. However, relatively few studies have identified specific genes that have been implicated in the differing cardiopulmonary differences between strains.

In collaboration with the laboratory of Dr. J. Garcia at Johns Hopkins Medical School, we have recently performed gene arrays using Affymetrix technology on lung homogenates prepared from normoxic and 14 day hypoxic Madison and Hilltop rats. Hypoxia-induced pulmonary hypertension was demonstrated by elevated right ventricular pressures and by the presence of right ventricular hypertrophy, as assessed by the ratio of the left ventricular plus septal weights to right ventricle (LV+S/RV). Gene array analysis was performed as described by Irizarry et al. (25).

Table 2. Genes Differentially Expressed in Normoxic Hilltop Rat Lungs Compared to Madison Rat Lungs

Genes	Accession No.	Log 2 Diff.	Est. Sig.	Gene Symbol
Upregulated				
CD36 antigen (collagen type I receptor, thrombospondin receptor)	AF072411	0.56	16.4	Cd36
CD36 antigen (collagen type I receptor, thrombospondin receptor)	AF072411	0.53	17.1	Cd36
RT1 class Ib gene	M31038	1.18	18.8	RT1Aw2
Rat mRNA for MHC class II antigen RT1.B-1 beta-chain	X56596	0.66	3	
CD36 antigen (collagen type I receptor, thrombospondin receptor)	AA799326	0.62	8.7	Cd36
Secreted acidic cystein-rich glycoprotein (osteonectin)	AA891204	0.49	23.5	Sparc
CD36 antigen (collagen type I receptor, thrombospondin receptor)	AA925752	0.62	11.5	Cd36
Downregulated				
Rat mRNA for eosinophil cationic protein, complete cds	D88586	-0.61	-10.4	
aldehyde dehydrogenase family 1, subfamily A4	M23995	-0.59	-9.2	Aldh1a4
R.norvegicus zymogen granule membrane protein GP-2 mRNA, complete cds	M58716	-0.65	-7.1	
Rattus norvegicus MHC class II antigen RT1.B beta chain mRNA, partial cds	U65217	-0.95	-6.3	
glutathione S-transferase, theta 2	AI138143	-0.89	-9.8	Gstt2

All changes in gene expression are significant. Minus symbol (-) indicates genes whose expression is less in Hilltop than in Madison rats. Log2 difference of 1 is equivalent of 2-fold change. Estimated significance represents the statistically likelihood that it is a real difference.

We found a number of genes with different levels of expression in normoxic lungs from the 2 strains (Table 2). In addition, hypoxia for 2 weeks altered the expression of some genes in both strains (Table 3) and induced differences in gene expression between the strains (Table 4). Although the gene array findings are semiquantitative, we can generate hypotheses regarding the spectrum of factors that may be involved in vasoconstriction and cell proliferation of pulmonary arteries induced by hypoxia. For example, we found that 12-lipoxygenase gene expression is increased in the lungs of hypoxic rats of both strains (unpublished data), although a definite causal role in the development of hypoxic pulmonary hypertension remains to be proven. On the other hand, expression of a whole array of genes encoding factors already known to be involved in pulmonary hypertension, such as HIF-1 or erg-1, did not increase with hypoxia, or between strains, suggesting that these proteins are modified post-translationally.

Table 3. Genes Altered by Hypoxia in the Lungs of Madison and Hilltop Rats

Genes	Accession No..	Log 2 Dif.	Est. Sig.	Gene Symbol
Upregulated				
Rgc32 protein	AF036548	0.54	13.5	Rgc32
serum/glucocorticoid regulated kinase	L01624	0.54	12.9	Sgk
arachidonate 12-lipoxygenase	L06040	0.81	10.3	Alox12
Rattus norvegicus mRNA Best5 protein	Y07704	0.68	15.1	
Rat mRNA for alpha-2u globulin-related protein	AA946503	0.71	16.2	
S-Adenosylmethionine decarboxylase 1A	AI008131	0.59	11.5	Amd1a
Hemoglobin, alpha 1	AI178971	0.7	22.3	Hba1
Downregulated				
Rat anti-acetylcholine receptor antibody gene, rearranged Ig gamma-2a chain, VDJC region, complete cds	L22654	-0.75	-14.4	
Rattus norvegicus rearranged IgG-2b gene, last 4 exons	M28671	-0.67	-8.8	

All changes in gene expression are significant by the methods used. Minus symbol (-) indicates genes whose expression is less in hypoxia than in normoxia in both strains.

We have also performed preliminary intercross and backcross breeding experiments to determine whether the trait predisposing to severe pulmonary hypertension was heritable. We found that F1 generation rats from intercross breeding regardless of gender had RV/LV+S ratios that were similar to those of Madison rats, but backcrosses of the F1 intercross rats bred with Hilltops once again manifested severe elevations in the ratio. These findings are consistent with the idea that the severe chronic hypoxic pulmonary hypertension developed by the Hilltop rats is an autosomal recessive trait.

Table 4. Genes Differentially Altered by Hypoxia in Hilltop and Madison Rat Lungs

Genes	Accession No.	Log 2 Diff.	Est. Sig.	Gene Symbol
Upregulated				
Arachidonate 12-lipoxygenase	L06040	0.3	3.8	Alox12
S100 calcium-binding protein A9 (calgranulin B)	L18948	0.35	4.86	S100a9
Rattus norvegicus mRNA Best5 protein	Y07704	0.3	6.8	
Solute carrier family 4, member 1, anion exchange protein 1 (kidney band 3)	AA866414	0.31	4	Slc4a1
Rat mRNA for alpha-2u globulin-related protein	AA946503	0.37	8.3	
S100 calcium-binding protein A8 (calgranulin A)	AA957003	0.24	2.6	S100a8
Downregulated				
Uncoupling protein	X03894	-0.33	-7.9	Ucp1
Actin, alpha, cardiac	AA866452	-0.31	-10.4	Actc1
Actin, alpha, cardiac	AI104567	-0.32	-7.9	Actc1
Adipocyte lipid-binding protein	AI169612	-0.31	-5.1	LOC84378

All changes in gene expression are significant by the methods used. Minus symbol (-) indicates genes whose expression is less in Hilltop than in Madison.

4. Conclusion

Differences in the severity of hypoxia-induced pulmonary hypertension between species and strains have long been known, based on the work of high altitude physiologists on species adapted to high altitude. The mechanisms responsible for these differences are of interest not only to physiologists, but also to clinicians, because understanding them could lead to valuable insights into the regulation of vascular tone as well as to potential therapies for pulmonary hypertension. One of the most studied models has been the Madison and Hilltop strains of Sprague Dawley rat that have markedly different cardiopulmonary responses to both acute and chronic hypoxia. These differences are heritable and associated with differences in vascular reactivity. Notably, the strains exhibit a dissociation between the intensities of acute and chronic hypoxic pulmonary hypertension. The Madison rats have a more intense acute vasoconstrictor response than the Hilltops, but then promptly blunt the response so that the severity of chronic hypoxic pulmonary hypertension is considerably less than in the Hilltops that fail to blunt. The biochemical/cellular mechanisms underlying this interesting phenomenon have not been elucidated. Although the strains also have differing polycythemic responses that parallel the cardiopulmonary differences, these hematologic differences do not appear to be contributing to the cardiopulmonary differences. Pulmonary artery rings isolated from the 2 strains retain the differences in reactivity to acute hypoxic exposure, and stripping the endothelium eliminates the difference, suggesting that an endothelium-derived mediator is at least partly responsible for the differences. Endothelin 1 appears to contribute to strain differences in some cases, such as those observed between the Wistar-Kyoto and Fisher 344 rat strains, but not in the Madison and Hilltop rats. There is evidence for EDHF release in pulmonary arteries isolated from normoxic Madisons but not Hilltops. However, the reasons for the enhanced susceptibility to chronic hypoxia remain unclear. Genetic approaches including cross-breeding and gene array experiments may yield additional insights, but as yet, the fundamental mechanisms responsible for enhancement of cardio-pulmonary responses to hypoxia remain largely unknown.

References

1. Aguirre JI, Morrell NW, Long L, Clift P, Upton PD, Polak JM, and Wilkins MR. Vascular remodeling and ET-1 expression in rat strains with different responses to chronic hypoxia. *Am. J. Physiol.* 2000; 278:L981-L987.
2. Anand IS, Harris E, Ferrari R, Pearce P, and Harris P. Pulmonary hemodynamics of the yak, cattle, and cross breeds at high altitude. *Thorax.* 1986; 41:696-700.
3. Banchero N, Grover RF, and Will JA. High altitude-induced pulmonary arterial hypertension in the llama (Lama glama). *Am. J. Physiol.* 1971; 220:422-427.
4. Busse R, Edwards G, Feletou M, Fleming I, Vanhoutte PM, and Weston AH. EDHF: bringing the concepts together. *Trends Pharmacol. Sci.* 2002; 23:374-380.
5. Christman BW, McPherson CD, Newman JH, King GA, Bernard GR, Groves BM, and Loyd JE. An imbalance between the excretion of thromboxane and prostacyclin metabolites in pulmonary hypertension. *N. Engl. J. Med.* 1992; 327:70-75.
6. Colice GL, Hill NS, Lee Y-J, Du H, Klinger J, Leiter JC, and Ou L-C. Exaggerated pulmonary hypertensive response to monocrotaline in rats susceptible to chronic mountain sickness. *J. Appl. Physiol.* 1997; 83:25-31.
7. Cooper AL, and Beasley D. Hypoxia stimulates proliferation and interleukin-1alpha production in human vascular smooth muscle cells. *Am. J. Physiol.* 1999; 277:H1326-H1337.
8. Durmowicz AG, Hofmeister S, Kadyraliev TK, Aldashev AA, and Stenmark KR. Functional and structural adaptation of the yak pulmonary circulation to residence at high altitude. *J. Appl. Physiol.* 1993; 74: 2276-2285.
9. Du HK, Lee YJ, Colice GL, Leiter JC, and Ou LC. Pathophysiological effects of hemodilution in chronic mountain sickness in rats. *J. Appl. Physiol.* 1996; 80:574-582.
10. Fagan KA, McMurtry IF, and Rodman DM. Role of endothelin-1 in lung disease. *Respir. Res.* 2001; 2:90-101.
11. Fisslthaler B, Popp R, Kiss L, Potente M, Harder DR, Fleming I, and Busse R. Cytochrome P450 2C is an EDHF synthase in coronary arteries. *Nature.* 1999; 401:493-497.
12. Fukuroda, T, Fujikawa T, Ozaki S, Ishikawa K, Yano M, and Nishikibe M. Clearance of circulating endothelin-1 by ETB receptors in rats. *Biochem. Biophys. Res. Commun.* 1994; 199:1461-1465.
13. Ge RL, Kubo K, Kobayashi T, Sekiguchi M, and Honda T. Blunted hypoxic pulmonary vasoconstrictive response in the rodent Ochotona curzoniae (pika) at high altitude. *Am. J. Physiol.* 1998; 274:H1792-H1799.
14. Geraci MW, Gao B, Shepherd DC, Moore MD, Westcott JY, Fagan KA, Alger LA, Tuder RM, and Voelkel NF. Pulmonary prostacyclin synthase overexpression in transgenic mice protects against development of hypoxic pulmonary hypertension. *J. Clin. Invest.* 1999;

103:1509-1515.
15. Giaid A, and Saleh D. Reduced expression of endothelial nitric oxide synthase in the lungs of patients with pulmonary hypertension. *N. Engl. J. Med.* 1995; 333: 214-221.
16. Grover RF, Reeves, JT, Will DH, and Blount SG, Jr. Pulmonary vasoconstriction in steers at high altitude. *J. Appl. Physiol.* 1963; 18: 567-574.
17. Groves BM, Droma T, Sutton JR, McCullough RG, McCullough RE, Zhuang J, Rapmund G, Sun S, Janes C, and Moore LG. Minimal hypoxic pulmonary hypertension in normal Tibetans at 3,658 m. *J. Appl. Physiol.* 1993; 74:312-318.
18. Gupta ML, Rao KS, Anand IS, Banerjee AK, and Boparai MS. Lack of smooth muscle in the small pulmonary arteries of the native Ladakhi. Is the Himalayan highlander adapted? *Ann. Rev. Respir. Dis.* 1992; 145: 1201-1204.
19. Hampl V, and Herget J. Role of nitric oxide in the pathogenesis of chronic pulmonary hypertension. *Physiol. Rev.* 2000; 80:1337-1372.
20. Hecht HH, Kuida H, Lange RL, Thorne JL, Brown AM, Carlisle R, Ruby A, and Verayha F. Brisket Disease. *Am. J. Med.* 1962; 32:171-183.
21. Hill NS, and Ou LC. The role of pulmonary vascular responses to chronic hypoxia in the development of chronic mountain sickness in rats. *Respir. Physiol.* 1984; 581:171-185.
22. Hill NS, Petit RD, Gagnon J, Warburton RR, and Ou LC. Hematologic responses and the early development of hypoxic pulmonary hypertension in rats. *Respir. Physiol.*1993; 91:271-282.
23. Hill NS, Sardella GL, and Ou LC. Reticulocytosis, increased mean red cell volume, and greater blood viscosity in altitude susceptible compared to altitude resistant rats. *Respir. Physiol.* 1987; 70:241-249.
24. Hill NS, Smith RP, and Ou LC. Time course of the development of cardiopulmonary responses to chronic hypoxia in two strains of rat with differing susceptibilities to high altitude. *Respir. Physiol.* 1987; 70:229-240.
25. Irizarry RA, Bolstad BM, Collin F, Cope LM, Hobbs B, and Speed TP. Summaries of Affymetrix GeneChip probe level data. *Nucleic Acids Res.* 2003; 31:115-123.
26. Ivy DD, Yanagisawa M, Gariepy CE, Gebb SA, Colvin KL, and McMurtry IF. Exaggerated hypoxic pulmonary hypertension in endothelin B receptor-deficient rats. *Am. J. Physiol.* 2002; 282:L703-L712.
27. Kam HY, Ou LC, Thron CD, Smith RP, and Leiter JC. The role of the spleen in the hematological response to hypoxia in Chronic Mountain Sickness in rats. *J. Appl. Physiol.* 1999; 87:1901-1908.
28. Karamsetty MR, Nakashima JM, Ou L-C, Klinger JR, and Hill NS. EDHF contributes to strain-related differences in pulmonary arterial relaxation in rats. *Am. J. Physiol.* 2001; 280:L458-L464.
29. Karamsetty MR, Pietras L, Klinger JR, Lanzillo JJ, Leiter JC, Ou L-C, and Hill NS. The role of endothelin-1 in strain-related susceptibility to develop hypoxic pulmonary hypertension in rats. *Respir. Physiol.* 2001; 128: 219-227.
30. Klinger JR, Warburton RR, Pietras L, Oliver P, Fox J, Smithies O, and Hill NS. Targeted disruption of the gene for natriuretic receptor-A worsens hypoxia-induced cardiac hypertrophy. *Am. J. Physiol. Heart Circ. Physiol.* 2002; 282:H58-H65.
31. Langleben D, Jones RC, Aronovitz MJ, Hill NS, Ou LC, and Reid LM. Pulmonary artery changes in two colonies of rats with differing sensitivity to chronic hypoxia. *Am. J. Pathol.* 1987; 128:61-66.
32. Moncada S, Gryglewski R, Bunting S, and Vane JR. An enzyme isolated from arteries transforms prostaglandin endoperoxides to an unstable substance that inhibits platelet aggregation. *Nature.* 1976; 263:663-665.
33. Monge CC. Chronic mountain sickness. *Physiol. Rev.* 1943; 23: 66-184.
34. Ou LC. Hypoxia-induced hemoglobinemia: hypoxic threshold and pathogenic mechanisms. *Exp. Hemat.* 1980; 8:243-248.
35. Ou LC, Cai YN, and Tenney SM. Responses of blood volume and red cell mass in two strains

of rats acclimatized to high altitude. Respir. Physiol. 1985; 62:85-94.
36. Ou LC, Chen J, Fiore E, Leiter JC, Brinck-Johnsen T, Birchard GF, Clemons G, and Smith RP. Ventilatory and hematopoietic responses to chronic hypoxia in two rat strains. *J. Appl. Physiol.* 1992; 72:2354-2363.
37. Ou LC, Hill NS, Pickett BP, Faulkner CS, Sardella GL, Thron CD, and Tenney SM. "Hypoxia-induced Right ventricular aneurysm." In *Right Hypertrophy and Function in Chronic Lung Disease*, Jezek V, Morpurgo M, and Tramarin R, eds. London: Springer-Verlag,, 1992, pp. 55-64.
38. Ou LC, Hill NS, and Tenney SM. Ventilatory responses and blood gases in susceptible and resistant rats to high altitude. *Respir. Physiol.* 1984; 58:161-170.
39. Ou LC, Salceda S, Schuster SJ, Dunnack LM, Brinck-Johnsen T, Chen J, and Leiter JC. Polycythemic responses to hypoxia: Molecular and genetic mechanisms of chronic mountain sickness. *J. Appl. Physiol.* 1998; 84:1242-1251.
40. Ou LC, Sardella GL, Hill NS, and Tenney SM. Acute and chronic pulmonary pressor responses to hypoxia: the role of blunting in acclimatization. *Respir. Physiol.* 1986; 64:81-91.
41. Ou LC, and Smith RP. Probable strain differences of rats in susceptibilities and cardiopulmonary responses to chronic hypoxia. *Respir. Physiol.* 1983; 53:367-377.
42. Ou LC, and Smith RP. Strain and sex differences in the cardiopulmonary adaptation of rats to high altitude. *Proc. Soc. Exp. Biol. Med.* 1984; 177:308-311.
43. Petit RD, Warburton RR, Ou LC, Brinck-Johnson T, and Hill NS. Exogenous erythropoietin fails to augment hypoxic pulmonary hypertension in rats. *Respir. Physiol.*1993; 91:261-270.
44. Salameh G, Karamsetty MR, Warburton RR, Klinger JR, Ou L-C, and Hill NS. Differences in acute hypoxic pulmonary vasoresponsiveness between rat strains: Role of endothelium. *J. Appl. Physiol.* 1999; 87:356-362.
45. Stelzner TJ, O'Brien RF, Yanagisawa M, Sakurai T, Sato K, Webb S, Zamora M, McMurtry IF, and Fisher JH. Increased lung endothelin-1 production in rats with idiopathic pulmonary hypertension. *Am. J. Physiol.* 1992; 262:L614-L620.
46. Steudel W, Ichinose F, Huang PL, Hurford WE, Jones RC, Bevan JA, Fishman MC, and Zapol WM. Pulmonary vasoconstriction and hypertension in mice with targeted disruption of the endothelial nitric oxide synthase (NOS 3) gene. *Circ. Res.* 1997; 81:34-41.
47. Tucker A, McMurtry IF, Reeves JT, Alexander AF, Will DH, and Grover RF. Lung vascular smooth muscle as a determinant of pulmonary hypertension at high altitude. *Am. J. Physiol.* 1975; 228:762-767.
48. Tuder RM, Cool CD, Geraci MW, Wang J, Abman SH, Wright L, Badesch D, and Voelkel NF. Prostacyclin synthase expression is decreased in lungs from patients with severe pulmonary hypertension. *Am. J. Respir. Crit. Care Med.* 1999; 159:1925-1932.
49. Underwood DC, Bochnowicz S, Osborn RR, Louden CS, Hart TK, Ohlstein EH, and Hay DW. Chronic hypoxia-induced cardiopulmonary changes in three rat strains: inhibition by the endothelin receptor antagonist SB 217242. *J. Cardiovasc. Pharmacol.*1998; 31(S1):S453-S455.
50. Weir EK, Tucker A, Reeves JT, Will DH and Grover RF. The genetic factor influencing pulmonary hypertension in cattle at high altitude. *Cardiovasc. Res.* 1974; 8:745-749.
51. Williams D. Adaptation and acclimatisation in humans and animals at high altitude. *Thorax.* 1994; 49:S9-S13.
52. Wohrley JD, Frid MG, Moiseeva EP, Orton EC, Belknap JK, and Stenmark KR. Hypoxia selectively induces proliferation in a specific subpopulation of smooth muscle cells in the bovine neonatal pulmonary arterial media. *J. Clin. Invest.* 1995; 96:273-281.
53. Zhu D, Effros RM, Harder DR, Roman RJ, and Jacobs ER. Tissue sources of cytochrome P450 4A and 20-HETE synthesis in rabbit lungs. *Am. J. Respir. Cell. Mol. Biol.* 1998; 19:121-128.

VIII. EXPERIMENTAL MODELS FOR THE STUDY OF HYPOXIC PULMONARY VASOCONSTRICTION

Chapter 31

Animal and *In Vitro* Models for Studying Hypoxic Pulmonary Vasoconstriction

Jane A. Madden and John B. Gordon
Medical College of Wisconsin, Milwaukee, Wisconsin, U.S.A.

1. Introduction

To better understand hypoxic pulmonary vasoconstriction (HPV), the response can be conceptualized as having four parts: sensor, mediator, effector, and modulator. At the most integrated level, studies in whole animals or humans have characterized the nature of the response. These studies have shown that hypoxia causes a brisk increase in the trans-pulmonary vascular gradient and a lesser increase in cardiac output. Thus, HPV results in a rapid increase in pulmonary vascular resistance (PVR). However, discerning the mechanisms is impossible in the intact system. Better understanding of the sensor, mediator, effector, and modulators of HPV has been achieved from studies using various reductionist models. Studies of pulmonary vascular physiology and pathophysiology have thus moved from intact animals and whole lungs to isolated vessels, cultured cells, sub-cellular preparations, and transgenic animals. This chapter discusses how various *in vivo* and *in vitro* models that have been and continue to be developed contribute to the current understanding of HPV. Although space precludes detailed discussion of mechanisms and the content may reflect the authors' biases, every attempt has been made to address the limitations as well as the contributions of each model.

2. Isolated Lungs

Pulmonary vascular responses in intact animals are generally assessed by measuring changes in PVR. However, potentially confounding systemic, neurohumoral, and cardiorespiratory influences can make it difficult to interpret the changes. These difficulties can be circumvented by using isolated lung or lobe preparations in which vasoactive stimuli can be applied without concern for their systemic effects, and the resistive properties of the vasculature can be measured throughout a wide range of flow rates and left atrial pressures.

The effort to understand HPV is considered to have begun with von Euler and Liljestrand's study in isolated cat lungs (57). Since then, HPV has been described in most animals including the dog, rat, rabbit, cow, pig, sheep, and ferret as well as more exotic animals such as the coati mundi. Studies in isolated lungs showed that the hypoxic response resembled that in the intact animal; it developed rapidly without pre-stimulation or adrenergic innervation, and was sustained (13). In addition, responses were similar whether the lung was perfused with blood or physiological saline solution, although, in some species, the magnitude was somewhat less with physiological saline solution as the perfusate (17). In the whole animal with a cardiac output of 200-400 ml/min, HPV was less than in isolated lungs perfused at 50-100 ml/kg/min and with pulsatile rather than steady flow (16).

A pivotal finding derived from the isolated lung model was that the hypoxic response differed among species. Peake et al. (41) determined the steady state relationship between Po_2 and the amplitude of the hypoxic response in the cat, dog, rabbit, ferret, and pig. In the isolated pig lung, vasomotor tone gradually increased as Po_2 was reduced from 200 to 50 Torr. Below 50 Torr, tone decreased. Cat and rabbit lungs also exhibited graded responses, peaking at a Po_2 of 25 Torr. Interestingly, although the ferret showed a maximum increase in pulmonary arterial pressure comparable to that of the pig, the response was not graded but occurred abruptly at Po_2's between 50 and 25 Torr. In pig, ferret, cat and rabbit lungs, hypoxic vasodilation occurred after the peak constrictor response. This too differed among the species, occurring at Po_2 less than 50 Torr in the pig and 25 Torr in the cat, rabbit, and ferret.

2.1. The Location of the Sensor and Possible Mediators

In 1968, Bergofsky et al. (9) performed an elegant study to identify the location of the sensor of hypoxia. They found that alveolar hypoxia was a more potent stimulus than pulmonary arterial hypoxemia and venous hypoxemia was an even weaker stimulus. Thus, the sensor of hypoxia appeared to lie within the vessels or perivascular tissues closest to gas-exchanging alveoli or alveolar ducts.

As it became evident that HPV was not mediated by the autonomic nervous system, the search began for possible chemical mediator(s). To qualify as mediators, these substances would be synthesized or released within the lungs during hypoxia and exert their effects upon the vascular smooth muscle. Catecholamines, histamine, angiotensin, serotonin, and bradykinin were eventually shown not to be viable candidates (13). Thromboxane (56) and leukotrienes (27) were also shown not to be mediators, but, as discussed later, arachidonic acid metabolites from the cyclooxygenase, lipoxygenase and cytochrome P-450 pathways appear to have modulator roles in HPV (1). Studies in whole lungs as well as intact animals found that although inhibiting prostacyclin synthesis increased HPV, it failed to mimic it (7).

Other investigators postulated that hypoxia acted directly on vascular smooth muscle to cause vasoconstriction through changes in Ca^{2+} and/or K^+ since HPV was reduced and abolished by low K^+(19) and Ca^{2+} channel blockers (35), respectively. With respect to Ca^{2+}, it is now recognized that, rather than being a mediator, intracellular Ca^{2+} release, extracellular Ca^{2+} influx, and changes in Ca^{2+} sensitivity are part of the final common pathway for contraction to hypoxia as well as other stimuli. In isolated rat lungs applying K^+ channel antagonists mimicked HPV (43) supporting a role for K^+ channels as possible mediators. Changes in intracellular K^+ due to opening and/or closing of the various types of K^+ channels appeared to mediate the release of intracellular Ca^{2+} as well as its influx from extracellular fluid.

Studies in isolated lung showing that both hypoxia and inhibitors of the mitochondrial electron transport chain had similar effects on the pulmonary vasculature led to the proposal of the redox theory of HPV (5). According to this theory, hypoxia altered the redox status of the pulmonary vasculature and resulted in an accumulation of electron donors from the Krebs cycle, decreased production of activated O_2 species, and K^+ channel inhibition. Investigations into the roles of K^+ channels, the regulation of cellular Ca^{2+}, and changes in redox status have been better conducted at the vessel and cellular level and have lead to additional theories about the exact mechanism.

2.2. The Vessel as the Effector

Almost from the beginning, it was generally agreed that the site of the hypoxic response was in the pulmonary vasculature, but it was not clear whether the site was pre-capillary, capillary or post-capillary. It was also not clear whether the increase in vascular resistance was due to inspired or to blood Po_2, and if the hypoxic stimulus acted upon the vascular wall directly or indirectly through effects on intrapulmonary perivascular elastic structures. A series of studies in isolated lungs (33, 34) suggested that HPV resulted from alveolar hypoxia acting upon pre-capillary pulmonary arteries. With the advent of X-ray angiography it was possible to demonstrate that cat (53) and dog (2) pulmonary arteries between 30 and 800 μm diameter narrowed during hypoxia. Structure-function studies also correlated the magnitude of the hypoxic vasoconstrictor response with the amount of muscle present in small arteries (36).

Although the majority of evidence suggests that HPV is a function of small muscular resistance pulmonary arteries, other studies have shown a potential role for veins in the response. In lambs, measurements of lung water showed an increase in fluid filtration (10). In newborn pigs, micropuncture techniques revealed that pressure increased markedly in small venules during hypoxia (11). A similar venous component to the hypoxic response was also reported in 3-5 week old ferrets (46). In contrast, in newborn rabbits, HPV was limited to small diameter arteries (12). In addition to complicating the question of the identity of the effector, these findings also raise issues of how developmental age and species may affect HPV.

2.3. Potential Modulators

Previously proposed mediators of HPV such as the autonomic nervous system, histamine, bradykinin, serotonin, arachidonic acid metabolites, and nitric oxide (NO), have now been shown to have significant roles as modulators. As a case in point, it is now well documented that HPV can be accentuated by inhibiting cyclooxygenase. This is particularly evident in the isolated dog lung, which has no significant response to hypoxia unless aspirin, which inhibits the synthesis of dilator prostaglandins, is administered to the dog before study. After aspirin treatment, the dog lung shows a vigorous HPV (2).

One of the earliest proposed modulators of HPV was pH when it was shown that acidosis augmented and alkalosis attenuated preexisting HPV. The finding that extracellular alkalosis attenuated HPV (26) was verified in numerous preparations and became a common treatment for neonatal and pediatric pulmonary hypertension. A study in isolated rat lungs (45), however, showed that intracellular alkalization with weak bases actually potentiated HPV and amiloride and acetate which decreased intracellular pH (pH_i), decreased HPV. Later, Gordon et al. (14) showed differing effects of acute and prolonged alkalosis on HPV in isolated lamb lungs. Acute hypocarbia decreased the hypoxia-induced elevation of the total pressure gradient and subsequent normoxia with alkalosis decreased it still further. However, re-exposure to hypoxia after 1 hr of normoxic alkalosis significantly increased the total pressure gradient to a level similar to that seen during normocarbic hypoxia. The increased hypoxic reactivity during prolonged alkalosis was due to an enhanced HPV of the small arteries. It appeared that increasing or decreasing pH during hypoxic constriction modulated the response by release of other factors that resulted in dilation of an artery constricted by hypoxia.

It should be noted that identifying and studying the exact contribution of a single purported modulator is difficult since there appears to be a great deal of redundancy among all the modulators. Also, some modulators are inhibitory, e.g., prostacyclin and NO, while others, e.g., endothelin (ET), appear to be permissive. Thus, interpreting the data from studies in which modulators are individually inhibited can be difficult.

2.4. Limitations of the Isolated Lung Model

Even though studies in isolated lungs suggested, and X-ray angiography verified that HPV was a function of the pulmonary arteries, the arteries themselves were inaccessible. Thus the next logical model was the isolated pulmonary artery. Confounding effects of other lung tissues were eliminated and, rather than just a single study as in an isolated lung, multiple arteries of various sizes could be isolated and studied.

3. Isolated Pulmonary Arteries

Studies in the isolated artery helped define whether small and/or large arteries were responsible for HPV and whether vascular smooth muscle and/or endothelial cells were the effectors. However, as will become evident below, distinctions between sensor, mediator, effector, and even modulator, tend to become somewhat blurred in the isolated artery.

For a number of years, progress in the study of isolated pulmonary arteries was hampered due to a lack of consistency of the hypoxic response in the particular vessels studied. While prior studies in intact animals and in isolated lungs had established that HPV was most probably located in the small diameter arteries, the technology was not available to study *in vitro* arteries less than ~1 mm in diameter. With the development of such technology, Madden et al.(30) showed that cat pulmonary arterial rings less than 300 μm in diameter had a consistent, reproducible hypoxic constrictor response whereas rings greater than 500 μm diameter never developed more than a small amount of force. In this and in an earlier study by Lloyd (25) in rabbit isolated artery strips, the contractile response to hypoxia was graded; with the rabbit arteries beginning to contract at a Po_2 of 200 Torr, the cat between 350 and 300 Torr, and both peaking at Po_2's between 50 and 30 Torr. The similarity of the hypoxic response in the isolated cat pulmonary arteries to that in isolated lungs from other species (41) suggested that the isolated, small diameter pulmonary artery was a useful model in which to investigate mechanisms of HPV. As in the isolated lung, there have been reports of HPV in isolated veins (58) although it is generally believed that their contribution to the overall response is minor compared to the arteries.

The possibility that the vascular endothelium was responsible for HPV through release of a contracting factor was proposed by Holden and McCall (20) who showed that removing the endothelium effectively eliminated HPV in pig main pulmonary artery. Later studies in isolated arteries showed that while the endothelium was not required for HPV, its full expression did seem to require a basal release of ET-1 (24) and/or another endothelium-derived contracting factor. Thus, the endothelium rather than being the effector, was more likely a source of modulators. An interesting sidelight of the investigation into the role of the endothelium in HPV has come from studies in isolated rat pulmonary arteries. These arteries exhibit a biphasic response to hypoxia (8). The initial contraction to hypoxia (phase 1) is followed by a relaxation and, if hypoxia is maintained, the artery contracts again (phase 2). Whether phase 1 or phase 2 is endothelium-dependent has been the subject of some debate but current consensus favors phase 2 (also see Chapter 12).

3.1. Possible Mediators

Initial studies in cat isolated pulmonary arteries (18, 30) showed that HPV was accompanied by membrane depolarization and action potential generation; both were blocked by verapamil. These data substantiated the earlier studies in isolated lung that at least part of the mechanism involved increased Ca^{2+} conductance. In isolated rat pulmonary artery rings, various K^+ channel antagonists mimicked HPV (43). While it was initially believed that Ca^{2+}-activated K^+ (K_{Ca}) channels were responsible, later studies suggest that voltage-gated K^+ (K_V) channels had a more dominant role in HPV (3). Whether K^+ channel inhibition plays a role *in vivo* has not been confirmed. To state that K^+ channels are the sensors or mediators of HPV is difficult because it can be argued that the channels, or at least their subunits may be capable of both functions. For example, a K^+ channel subunit may sense a change in O_2 tension and change its conformation to close or open the channel. This would mediate changes in membrane potential and consequently intracellular Ca^{2+}, and culminate in contraction or relaxation. Whether this effect is direct or it is mediated through a change in the redox status of the cell is unknown.

3.2. Modulators

As mentioned earlier, the endothelium appears to be a rich source of modulators for HPV. Studies in isolated vessels showed decreased prostacyclin (50) and NO (51) synthesis during hypoxia and inhibiting prostacyclin synthesis augmented HPV (30). Now it is generally accepted that NO, arachidonic acid metabolites from the cyclooxygenase, lipoxygenase and cytochrome P-450 pathways along with ET and/or endothelium-derived contracting factor(s) are vital to the full expression of HPV (1).

Perivascular tissue has been proposed as another possible modulator of HPV. In an early study of 1-3 mm diameter pulmonary artery strips from rabbits, Lloyd (25) did not see a contraction to hypoxia unless the strips were surrounded by perivascular tissue. This suggested that HPV could not occur without the release of a hypoxia-sensitive factor from the perivascular tissue. Alternatively, the strips may not have contracted because they were from large diameter vessels. Also, small diameter arteries located within the perivascular tissue may have themselves contracted to hypoxia and "pulled" the strip along. It should be noted that pulmonary arteries are embedded within the lung parenchyma, and extensive dissection of adventitia from the arteries during isolation may disrupt their physical and functional integrity and thereby alter their response.

The issue of modulator redundancy is also of concern in isolated artery studies. For example, acetylcholine or bradykinin relaxation was still evident in preconstricted isolated pulmonary arteries from lambs, if only NO *or* arachidonic acid was inhibited (38). Elimination of the relaxation response required the simultaneous inhibition of NO *and* arachidonic acid. Thus, results of studies in which only one modulator is studied should be interpreted with caution.

3.3. Limitations of the Isolated Artery Model

Studies in the isolated lung had led to the development of many hypotheses about the mechanism of HPV. Studies in isolated arteries often proved these prior hypotheses as well as validated previous findings. However, correlation of findings between the two models is not always perfect and thus extrapolation of findings from the reduced preparation to a more complete system should be done with caution. For example, in contrast to the whole animal and isolated lung model, in isolated vessel preparations HPV often requires pre-stimulation with other pressor agents, is slow to develop, is transient, and may be biphasic. In addition, the decreased NO and prostacyclin production in hypoxic isolated arteries would seem to suggest that HPV occurred because of their decreased synthesis. However, as noted earlier, NO or prostacyclin inhibition failed to mimic HPV in whole lungs or intact animals.

These reports of discordant responses to hypoxia between more and less reductionist preparations suggest that the transition from *in-situ* vessels within the whole lung to isolated vessels is a major juncture in the deconstruction of the lung. This may reflect the choice of artery studied in isolated vessel preparations. Arteries between 0.5 and 3 mm diameter are commonly used in vascular ring preparations. Cannulated vessels are generally >200 μm. In contrast, the small arteries that appear to contribute most to *in-vivo* pulmonary vasoconstriction appear to have diameters less than 100-150 μm, although, both *in vitro* and *in vivo*, HPV has been demonstrated in arteries >150 μm (2, 30). Finally, there is considerable morphologic, biochemical, and molecular biological evidence that there are significant differences in endothelial and vascular smooth muscle structure and function between large and small vessels.

On the other hand, discordant hypoxic vascular responses between whole lungs and current isolated vessels may be unrelated to the choice of vessel size or type but rather to loss of the normal environment of the *in-situ* vessel. For example, loss of perivascular tissue (adventitia, lung parenchyma, and small airways) may be critical. The adventitia performs several metabolic functions and is the repository of neural endings and fibroblasts. NO from fibroblasts and superoxide ion from adventitial NADPH may play a role in regulating vasomotor tone. As discussed earlier, lung parenchyma is essential for HPV in some isolated vessels (25). Given the close contact and multiple signaling pathways between vascular, parenchymal, and airway cell types, it could be argued that signaling from these perivascular tissues contributes in some manner to the mechanisms underlying *in-vivo* pulmonary vascular responses.

Changes in flow may contribute to discordance in hypoxic responses between lung and isolated vessels. Isolated vascular rings are studied under no-flow conditions and many cannulated vessel preparations, although pressurized with intra-luminal perfusate, are not perfused. Whole lungs are typically studied under conditions of constant flow or constant pressure. Simply increasing vascular pressure in an isolated vessel may induce a myogenic response leading to constriction (29). Myogenic responses may also occur *in vivo* (54); but generally, the combination of increased flow and pressure in intact lungs increases shear stress which stimulates synthesis of endothelium-derived vasodilators such as NO and prostacyclin and activates vascular smooth muscle K^+ channels, resulting in vasodilation. On the other hand, perfusing cannulated vessels may increase synthesis of vasoconstrictor modulators such as ET (52). Thus, it could be argued that differences in baseline pressor and dilator modulator activity under conditions of flow and no-flow contribute to the mechanisms underlying in-vivo pulmonary vascular responses.

Remote vascular responses to local stimuli have been described in the pulmonary vasculature (54). This suggests that inter-vessel communication may also lead to discordance between whole lungs and isolated vessels. Several mechanisms may be involved. Myogenic responses may occur as a direct result of pressure transmission between vascular beds. Neural communication may occur via the rich adrenergic innervation of the pulmonary vasculature, although adrenergic blockade appears to have little effect on pulmonary vascular responses. Cell-to-cell communication of depolarizing impulses to neighboring areas of the pulmonary vasculature have also been described (48). The predominantly paracrine nature of most potent endothelium-derived modulators (e.g., prostacyclin, NO) make it unlikely that blood borne transmission plays a major role in communication between pulmonary vessels.

4. Cellular and Subcellular Models

Confirmation that the smooth muscle cell was the effector of HPV and most probably had within it the sensor and the mediator(s) required that the phenomenon actually be studied in isolated cells. Cell isolation and culture techniques permitted further deconstruction of the pulmonary vasculature to its cellular and sub-cellular components.

Smooth muscle cells isolated from fetal calf main pulmonary artery and grown on a silastic membrane showed increased wrinkling, indicative of increased tension when exposed to hypoxia (37). In contrast, the hypoxic response of non-dedifferentiated cat pulmonary artery smooth muscle cells depended on the diameter of the artery from which the cells were derived (31). Cells from small arteries grown on cover slips contracted whereas cells from larger vessels did not. Cell shortening was accompanied by myosin light chain phosphorylation (31). That the cells were representative of their particular circulation was indicated by the contractile responses of both large and small pulmonary artery smooth muscle cells to other agonists. Not only did these studies establish the utility of the smooth muscle cell as a model for studying HPV, they further verified that the smooth muscle cell responded directly to hypoxia and that it was responsible for the contraction.

4.1. Sensors, Mediators and Modulators

Subsequent studies in cat small pulmonary artery smooth muscle cells, showed that intracellular Ca^{2+} increased during hypoxia but decreased in cells from large pulmonary and cerebral arteries (55). The Ca^{2+} increase was due to Ca^{2+} release from ryanodine-sensitive intracellular stores and Ca^{2+} influx from the extracellular fluid.

Patch-clamping and immunofluorescent microscopy combined with ever more selective pharmacologic agents further elucidated the roles of Ca^{2+}, K^+, and other ions in regulating vascular tone. Closely linked to the study of changes in cellular Ca^{2+} during hypoxia were studies of the role of K^+ channels, particularly the K_V channels (4). However, it is still not certain whether hypoxia *directly* inhibits K_V channels, and the resulting depolarization leads to increased intracellular Ca^{2+}, or if K_V channel inhibition is secondary to Ca^{2+} release from the sarcoplasmic reticulum (42). The presence of distinct subpopulations of pulmonary artery myocytes each with distinct distributions of K^+ channels lends further intrigue to the isolated cell story (6). Recently, it has been shown that like the graded contractile response to hypoxia in lungs and arteries, pulmonary artery smooth muscle cells show a graded response of K^+ current, membrane potential and intracellular Ca^{2+} (39).

With the ability to study mechanisms at the cellular and subcellular level, investigations into the redox theory of HPV have burgeoned and may offer the best chance yet to identify the elusive O_2 sensor (49). Most theories suggest that reactive oxygen species (ROS) such as H_2O_2 and superoxide are involved. K_V channels, ADP-ribosyl cyclase, and cyclic ADP-ribose hydrolase, the mitochondria or more specifically, the electron transport chain, have all been nominated as candidates for the sensor. Whether ROS species increase or decrease during hypoxia is controversial. Different preparations (tissue versus cell) and limitations of current detection methods may affect the validity and/or the interpretation of the results and lead to discrepant reports.

The role of pH as modulator of HPV was discussed earlier but the possibility that pH_i might also mediate HPV is suggested by studies in isolated arteries and cells. During hypoxia, pH_i increased in small pulmonary artery cells but decreased in large pulmonary artery cells. These changes were not dependent upon the pH_i before hypoxia (28). In large diameter pig pulmonary arteries, pH_i also decreased during hypoxia (23). In HCO_3^--containing solutions, pulmonary artery smooth muscle cells from cats and guinea pigs showed evidence for the major ion exchangers, Na^+/H^+ and the Na^+ dependent and independent Cl^-/HCO_3^- ion transporters. In the cat the Cl^-/HCO_3^- exchanger appeared to be more active in regulating pH_i, but in the guinea pig the Na^+/H^+ exchanger appeared to play more of a role (44). These differences further underscore the idea that different species may have different elaborations upon the same mechanism.

5. Transgenic Animal Models

Mice and rats in which specific genes are either under or overexpressed are increasingly being used to study mechanisms of HPV. For example, HPV is impaired in mice lacking certain K_V channels (3) and increased in rats lacking the ET_B receptor (21). Genetically altered animals would thus appear to offer the best of all worlds, a particular mechanism can be eliminated or elaborated and the result can be studied in the whole animal, its organs, its tissues, and its cells. However, it should be noted that introducing new DNA sequences into an animal might also affect the endogenous activities of other genes. Thus, interpreting data collected from a transgenic animal requires consideration of the complexity of the mammalian organ system.

6. Non-mammalian Models

The effort to determine the mechanisms responsible for HPV has obviously been concentrated in species that have lungs. However, there is evidence that hypoxic vasoconstriction was present in early vertebrate evolution. Using more primitive vertebrate species offers intriguing new possibilities for studying both phylogeny and fundamental mechanisms of the hypoxic response.

Video microscopy of arterioles and veins in frog skin has shown a dose-dependent hypoxic contraction in arterioles but not in veins (32). Unlike mammalian pulmonary arteries, the magnitude of the contraction does not appear to be associated with artery size. However, like mammalian arteries, hypoxic vasoconstriction was blocked by nifedipine and augmented by BAY K 8664. In hagfish and lamprey eel, among the most primitive of vertebrates known as the cyclostomes, the dorsal aorta exhibits hypoxic vasoconstriction (40). The response is dose-dependent, reproducible, sustainable, and does not require preconditioning. Neither removing the endothelium nor inhibiting the arachidonic acid pathway or the adrenergic, muscarinic, nicotinic, purinergic, or serotonergic receptors affected the response. However, Ca^{2+} did appear to have a fundamental role (48). During hypoxia intracellular Ca^{2+} increased in the dorsal aorta of both species. There was evidence that multiple receptor types mediated the intracellular Ca^{2+} release. Extracellular Ca^{2+} did not contribute to hypoxic vasoconstriction in lamprey dorsal aorta but accounted for over a third of the response in hagfish, although L-type Ca^{2+} channels did not appear to be involved. Removing extracellular Na^+ caused constriction in both species. However, after a hypoxic constriction only lamprey dorsal aorta was able to relax upon return to normoxia in the absence of extracellular Na^+. Removing both Na^+ and Ca^{2+} inhibited this relaxation. These studies suggest that Na^+/Ca^{2+} exchange has a role during hypoxic vasoconstriction in the hagfish but not in the lamprey.

7. A Case Study in Discrepancies

As a final note to this chapter, it is fitting to show how studying the same process in different models can lead to different results and different conclusions. In piglet pulmonary artery and vein rings inhibiting NO synthase completely blocked alkalosis-induced relaxation. However, in isolated piglet lungs, NO synthase inhibition had no effect on the vasodilation. In an effort to determine which factors contributed to this discordance the contribution of pressor stimuli, (hypoxia versus the thromboxane mimetic U46619), perfusate composition, (blood versus physiological saline solution), and flow were assessed (15). Effects of NO synthase inhibition on the alkalosis-induced vasodilation were compared in intact piglets, and in 150-350 μm cannulated arteries and by angiography of 150-900 μm *in situ* arteries. Neither pressor stimulus, perfusate composition nor flow affected the results, i.e., NO synthase inhibition fully abolished alkalosis-induced vasodilation in the cannulated arteries but failed to do so in the isolated lungs. These data indicated that other factors such as perivascular tissue (adventitia and parenchyma) and remote signalling pathways may contribute to the discordant responses. Regardless of the mechanism involved in the response to alkalosis, this study illustrates the limitations of applying conclusions derived from a reductionist model to the whole animal.

8. Summary

Is there or will there ever be a best model in which to study HPV? Probably, not. The use of reductionist approaches to identify mechanisms underlying pulmonary vasomotor tone has been something of a double-edged sword. Studies of isolated vessels or cultured cells have elucidated cellular mechanisms controlling synthesis or inhibition of potentially important modulators and mediators. On the other hand, many of these appear to play little role *in vivo*. Discordant responses to hypoxia between more and less reductionist preparations suggest that deconstructing the lung may interrupt signaling, metabolic, and mechanical pathways critical to integrated *in vivo* hypoxic responses. Development of reductionist preparations that retain mechanisms underlying *in vivo* responses is imperative for complete understanding of HPV and the ultimate application of findings to the clinical situation. Ultimately, however, whatever model is studied, the knowledge derived from it will further our understanding of HPV, and also fundamental mechanisms inherent to the entire vasculature.

References

1. Aaronson PI, Rxobertson TP, and Ward JPT. Endothelium-derived mediators and hypoxic pulmonary vasoconstriction. *Respir. Physiol. Neurobiol.* 2002; 132: 107-120.
2. Al-Tinawi A, Krenz GS, Rickaby DA, Linehan JH, and Dawson CA. Influence of hypoxia and serotonin on small pulmonary vessels. *J. Appl. Physiol.* 1994; 76: 56-64.
3. Archer S and Michelakis E. The mechanism(s) of hypoxic pulmonary vasoconstriction: potassium channels, redox O_2 sensors, and controversies. *News Physiol. Sci.* 2002; 17: 131-137.
4. Archer SL and Rusch NJ. "The role of potassium channels in the control of the pulmonary circulation." In *Potassium Channels in Cardiovascular Biology*, Archer SL and Rusch NJ, ed. New York, NY: Kluwer Academic/Plenum 2001, pp. 543-564.
5. Archer SL, Huang J, Henry T, Peterson HT, and Weir EK. A redox based oxygen sensor in rat pulmonary vasculature. *Circ. Res.* 1993; 73: 1100-1112.
6. Archer SL, Huang JMC, Reeve HL, Hampl V, Tolarová S, Michelakis ED, and Weir EK. Differential distribution of electrophysiologicallly distinct myocytes in conduit and resistance arteries determine their response to nitric oxide and hypoxia. *Circ. Res.* 1996; 78: 431-442.
7. Archer SL, Tolins JP, Raj L, and Weir EK. Hypoxic pulmonary vasoconstriction is enhanced by inhibition of the synthesis of an endothelium derived relaxing factor. *Biochem. Biophys. Res. Comm.* 1989; 164: 1198-1205.
8. Bennie RE, Packer CS, Powell DR, Jin N, and Rhoades RA. Biphasic contractile response of pulmonary artery to hypoxia. *Am. J. Physiol.* 1991; 261: L156-L163.
9. Bergofsky EH, Haas F, and Porcelli R. Determination of the sensitive vascular sites from which hypoxia and hypercapnia elicit rises in pulmonary arterial pressure. *Fed. Proc.* 1968; 27: 1420-1425.
10. Bland RD, Bressack MA, Haberkern CM, and Hansen TN. Lung fluid balance in hypoxic, awake newborn lambs and mature sheep. *Biol. Neonate* 1980; 38: 221-228.
11. Fike CD and Kaplowitz MR. Pulmonary venous pressure increases during alveolar hypoxia in isolated lungs of newborn pigs. *J. Appl. Physiol.* 1992; 73: 552-556.
12. Fike CD, Lai-Fook SJ, and Bland RD. Microvascular pressures during hypoxia in isolated lungs of newborn rabbits. *J. Appl. Physiol.* 1988; 65: 283-287.
13. Fishman AP. Hypoxia on the pulmonary circulation: How and where it acts. *Circ. Res.* 1976; 38: 221-231.
14. Gordon JB, Martinez FR, Keller PA, Tod ML, and Madden JA. Differing effects of acute and prolonged alkalosis on hypoxic pulmonary vasoconstriction. *Am. Rev. Respir. Dis.* 1993; 148: 1651-1656.
15. Gordon JB, VanderHeyden MA, Halla TR, Cortez EP, Hernandez G, Haworth ST, Dawson CA, and Madden JA. What leads to different mediators of alkalosis-induced vasodilation in isolated and in situ pulmonary vessels? *Am. J. Physiol. Lung Cell. Mol. Physiol.* 2003, 284: L799-L807.
16. Gregory TJ, Newell JC, Hakim TS, Levitzky MG, and Sedransk N. Attenuation of hypoxic pulmonary vasoconstriction by pulsatile flow in dog lungs. *J. Appl. Physiol.* 1982; 53: 1583-1588.
17. Hakim TS and Malik AB. Hypoxic vasoconstriction in blood and plasma perfused lungs. *Respir. Physiol.* 1988; 72: 109-122.
18. Harder DR, Madden JA, and Dawson C. Hypoxic induction of Ca^{2+}-dependent action potentials in small pulmonary arteries of the cat. *J. Appl. Physiol.* 1985; 59: 1389-1393.
19. Herget J and McMurtry IF. Effects of ouabain, low K^+, and aldosterone on hypoxic pressor reactivity of rat lungs. *Am. J. Physiol.* 1985; 248: H55-H60.
20. Holden WE and McCall E. Hypoxia-induced contractions of porcine pulmonary artery strips depend on intact endothelium. *Exp. Lung Res.* 1984; 7: 101-112.
21. Ivy DD, McMurtry IF, Yanagisawa M, Gariepy CE, Le Cras TD, Gebb SA, Morris KG,

Wiseman RC, and Abman SH. Endothelin B receptor deficiency potentiates ET-1 and hypoxic pulmonary vasoconstriction. *Am. J. Physiol. Lung Cell. Mol. Physiol.* 2001; 280: L1040-L1048.
22. Jacobs ER and Zeldin DC. The lung HETE's (and EET's) up. *Am. J. Physiol. Heart Circ. Physiol.* 2001; 280: H1-H10.
23. Leach RM, Sheehan DW, Chacko VP, and Sylvester JT. Effects of hypoxia on energy state and pH in resting pulmonary and femoral arterial smooth muscles. *Am. J. Physiol.* 1998; 275: L1051-L1060.
24. Liu Q, Sham JSK, Shimoda LA, and Sylvester JT. Hypoxic constriction of porcine distal pulmonary arteries: endothelium and endothelin dependence. *Am. J. Physiol. Lung Cell. Mol. Physiol.* 2001; 281: L856-L865.
25. Lloyd TC. Hypoxic pulmonary vasoconstriction: role of perivascular tissue. *J. Appl. Physiol.* 1968; 25: 560-565.
26. Lloyd TC. Influence of blood pH on hypoxic pulmonary vasoconstriction. *J. Appl. Physiol.* 1966; 21: 358-364.
27. Lonigro AJ, Sprague RS, Stephenson AH, and Dahms TE. Relationship of leukotriene C_4 and D_4 to hypoxic pulmonary vasoconstriction in dogs. *J. Appl. Physiol.* 1988; 64: 2538-2543.
28. Madden A, Ray DE, Keller PA, and Kleinman JG. Ion exchange activity in pulmonary artery smooth muscle cells: the response to hypoxia. *Am. J. Physiol. Lung Cell. Mol. Physiol.* 2001; 280: L264-L271.
29. Madden JA, Al-Tinawi A, Birks E, Keller PA, and Dawson CA. Intrinsic tone and distensibility of in vitro and in situ cat pulmonary arteries. *Lung.* 1996; 174: 291-301.
30. Madden JA, Dawson CA, and Harder DR. Hypoxia-induced activation in small isolated pulmonary arteries from the cat. *J. Appl. Physiol.* 1985; 59: 113-118.
31. Madden JA, Vadula MS, and Kurup VP. Effects of hypoxia and other vasoactive agents on pulmonary and cerebral artery smooth muscle cells. *Am. J. Physiol.* 1992; 263: L384-L393.
32. Malvin GM and Walker BR. Sites and ionic mechanisms of hypoxic vasoconstriction in frog skin. *Am. J. Physiol. Regul. Integr. Comp. Physiol.* 2001; 280: R1308-1314.
33. Marshall C and Marshall BE. Influences of perfusate PO_2 on hypoxic pulmonary vasoconstriction in rats. *Circ. Res.* 1983; 52: 691-696.
34. Marshall C and Marshall BE. Site and sensitivity for stimulation of hypoxic pulmonary vasoconstriction. *J. Appl. Physiol.* 1983; 55: 711-716.
35. McMurtry IF, Davidson AB, Reeves JT, and Grover RF. Inhibition of hypoxic pulmonary vasoconstriction by calcium antagonists in isolated rat lungs. *Circ. Res.* 1976; 38: 99-104.
36. Michel RR, Gordon JB, and Chu K. Development of the pulmonary vasculature in newborn lambs: structure-function relationships. *J. Appl. Physiol.* 1991; 70: 1255-1264.
37. Murray TR, Chen L, Marshall BE, and Macarak EJ. Hypoxic contraction of cultured pulmonary vascular smooth muscle in cell culture. *J. Cell Physiol.* 1990; 148: 26-38.
38. O'Donnell DC, Tod ML, and Gordon JB. Developmental changes in endothelium-dependent relaxation of pulmonary arteries: role of EDNO and prostanoids. *J. Appl. Physiol.* 1996; 81: 2013-2019.
39. Olschewski A, Hong Z, Nelson DP, and Weir EK. Graded response of K^+ current, membrane potential, and $[Ca^{2+}]_i$ to hypoxia in pulmonary arterial smooth muscle. *Am. J. Physiol. Lung Cell. Mol. Physiol* 2202; 283: L1143-L1150.
40. Olson KR, Russell MJ, and Forster ME. Hypoxic vasoconstriction of cyclostome systemic vessels: the antecedent of hypoxic pulmonary vasoconstriction? *Am. J. Physiol. Regul. Integr. Comp. Physiol.* 2001; 280: R198-R206.
41. Peake MD, Harabin AL, Brennan NJ, and Sylvester JT. Steady-state vascular responses to graded hypoxia in isolated lungs of five species. *J. Appl. Physiol.* 1981; 51: 1214-1219.
42. Post JM, Gelband CH, and Hume JR. $[Ca^{2+}]_i$ inhibition of K^+ channels in canine pulmonary artery, novel mechanisms for hypoxia-induced membrane depolarization. *Circ. Res.* 1995; 77: 131-139.
43. Post JM, Hume JR, Archer SL, and Weir EK. Direct role for potassium channel inhibition in

hypoxic pulmonary vasoconstriction. *Am. J. Physiol.* 1992; 262: C882-C890.

44. Quinn DA, Honeyman TW, Joseph PM, Thompson BT, Hales CA, and Scheid CR. Contribution of Na^+/H^+ exchange to pH regulation in pulmonary artery smooth muscle cells. *Am. J. Respir. Cell. Mol. Biol.* 1991; 5: 586-591.
45. Raffestin B and McMurtry IF. Effects of intracellular pH on hypoxic vasoconstriction in rat lungs. *J. Appl. Physiol.* 1987; 63: 2524-2531.
46. Raj JU, Hillyard R, Kaapa P, Gropper M, and Anderson J. Pulmonary arterial and venous constriction during hypoxia in 3- to 5-week old and adult ferrets. *J. Appl. Physiol.* 1990; 69: 2183-2189.
47. Russell MJ, Pelaez NJ, Packer CS, Forster ME, and Olson KR. Intracellular and extracellular calcium utilization during hypoxic vasoconstriction of cyclostome aortas. *Am. J. Physiol. Regul. Integr. Comp. Physiol.* 2001; 281: R1506-R1513.
48. Segal S and Duling BR. Propagation of vasodilation in resistance vessels of the hamster: development and review of a working hypothesis. *Circ. Res.* 1987; 61: II20-II25.
49. Sham JSK. Hypoxic pulmonary vasoconstriction: Ups and downs of reactive oxygen species. *Circ. Res.* 2002; 91: 649-651.
50. Shaul PW, Campbell WB, Farrar MA, and Magness RR. Oxygen modulates prostacyclin synthesis in ovine fetal pulmonary arteries by an effect on cyclooxygenase. *J. Clin. Invest.* 1992; 90: 2147-2155.
51. Shaul PW, Farrar MA, and Zellers TM. Oxygen modulates endothelium-derived relxing factor production in fetal pulmonary arteries. *Am. J. Physiol.* 1992; 262: H355-H364.
52. Shimoda LA, Norins NA, and Madden JA. Flow-induced responses in cat isolated pulmonary arteries. *J. Appl. Physiol.* 1997; 83: 1617-1622.
53. Shirai M, Sada K, and Ninomiya I. Effects of regional alveolar hypoxia and hypervapnia on small pulmnary vessels in cats. *J. Appl. Physiol.* 1986; 61: 440-443.
54. Shirai M. Ninomiya I and Sada K. Constrictor response of small pulmonary arteries to acute pulmonary hypertension during left atrial elevation. *Jpn. J. Physiol.* 1991; 41: 129-142.
55. Vadula MS, Kleinman JG, and Madden JA. Effects of hypoxia and norepinephrine on cytoplasmic free Ca^{2+} in pulmonary and cerebral arterial myocytes. *Am. J. Physiol.* 1993; 265: L591-L597.
56. Voelkel NF, Morganroth M, and Fedderson OC. Potential role of arachidonic acid metabolites in hypoxic pulmonary vasoconstriction. *Chest.* 1985; 88: 245S-248S.
57. Von Euler US and Liljestrand G. Observations on the pulmonary arterial blood pressure in the cat. *Acta Physiol. Scand.* 1946; 12: 301-320.
58. Zhao Y, Packer CS, and Rhoades RA. The vein utilizes different sources of energy than the artery during pulmonary hypoxic vasoconstriction. *Exp. Lung Res.* 1996; 22: 51-63.
59. Zhu, D., Birks, EK, Dawson, CA, Patel, M, Falck, JR, Presberg, K, Roman, RJ, and Jacobs, ER. Hypoxic pulmonary vasoconstriction is modified by P-450 metabolites. *Am. J. Physiol. Heart Circ. Physiol.* 2000; 279: H1526-H1533.

Chapter 32

Transgenic and Gene-Targeted Mouse Models in Hypoxic Pulmonary Hypertension Research

Yadong Huang
Gladstone Institute of Cardiovascular Disease and University of California, San Francisco, California, U.S.A.

1. Introduction

Pulmonary hypertension, characterized by elevated pulmonary blood pressure, pulmonary vascular remodeling, and right ventricular hypertrophy, is a common complication of chronic lung disease and heart failure. Although the pathogenesis of pulmonary hypertension remains poorly understood, hypoxia has been suggested to be one of the critical factors causing pulmonary hypertension and aggravating pathophysiological conditions. *In vitro* studies in various cell-culture systems have identified many candidate genes, proteins, and cellular metabolic pathways that may be involved the pathogenesis of hypoxic pulmonary hypertension. However, only recently have transgenic and gene-targeted mouse models been developed to study the roles of these genes, proteins, and metabolic pathways in pathogenesis or protection of hypoxic pulmonary hypertension.

2. Generation of Transgenic Mice

2.1. Application of Transgenic Mouse Technology

Transgenic mice are widely used in biomedical research, and numerous applications have been developed, including those used in hypoxic pulmonary hypertension (5, 6, 19, 29). There are two common uses of transgenic mice in biomedical research. The first is to study the phenotypic effects of transgene expression in the intact animal. In this case, previously defined promoters are often used to control the expression of the transgene in the desired tissue. The second is to study the control of gene expression in the intact animal. For this purpose, potential control sequence elements (promoter regions) are used to define the specific patterns of the transgene expression in various tissues.

For applications involving the phenotypic effects of transgene expression, transgenic mice expressing the given genes, in general, represent gain-of-function mutations, whereas loss-of-function mutations in many cases can be obtained solely by gene targeting (see below). However, dominant negative mutations can also be obtained through expression of some mutant forms of genes (10-13). In addition, the expression of transgenes encoding antisense RNA or short interfering RNA successfully inhibited the expression of endogenous genes in transgenic mice (8, 24).

Most cloned genes introduced into the mouse germ line have shown appropriate tissue-specific and stage-specific patterns of expression despite their integration into apparently random sites in the host genome. Thus, for studies of the control of gene expression, transgenic mice have provided the definitive experimental assay to define the *cis*-acting DNA sequences that control specific patterns of transcription *in vivo* (2, 7, 25).

2.2. Generation of Conventional Transgenic Mice

The procedure for generating a conventional transgenic mouse line includes preparing DNA constructs, setting up the mouse colony, microinjecting DNA constructs into the pronuclei of fertilized eggs, characterizing transgenic "founder" mice, and generating transgenic lines from these founder mice. The quality of the DNA constructs is critical for efficient generation of transgenic founder mice. High purity of the DNA, avoiding the use of ethidium bromide for DNA staining, removing all vector sequences from the cloned genes, and using relative short DNA constructs (<70 Kb) usually increase the rate of success (40-60%) of generating founder mice.

The most extensively and successfully used method of gene transfer is microinjection of DNA directly into the pronuclei of fertilized mouse eggs. Infection of embryos with retroviral vectors has also been used. The microinjection method results in the stable chromosomal integration of the foreign DNA in 10-60% of the resulting mice. In most cases, the integration appears to occur at the one-cell stage of the embryos; as a result, the foreign DNA is present in every cell of the transgenic mouse, including all primordial germ cells. In 15-25% of cases, foreign DNA is integrated at a later stage, resulting in mice that are mosaic for the presence of foreign DNA.

2.3. Generation of Conditional (Inducible) Transgenic Mice

Sometimes, it is desirable to have transgenes that will be silent until specifically activated by an experimental manipulation, such as the administration of a drug. In early studies, the metallothionein promoter was frequently used to generate inducible transgenic mice (26). The metallothionein promoter drives transgene expression at low basal levels in many tissues;

however, feeding the transgenic mice with water containing Zinc can increase the transgene expression up to 100-fold in many tissue, including liver, kidney, and intestine (26). However, this system lacks tissue specificity.

Recently, the *tet*-operon (Tet-O) system has been used widely to generate conditional (inducible) transgenic mice in a tissue-specific manner (22, 23). In this system, two transgenic mouse lines must be generated: one with the desired transgene under the control of a Tet-O promoter containing a tetracycline-responsive element (TRE) and the other expressing a tetracycline-controlled transactivator (tTA) under the control of a tissue-specific promoter (Fig. 1). Crossing these two lines will generate transgenic mice carrying both transgenes. When the doubly transgenic mice are fed a diet containing tetracycline, the tetracycline binds to tTA, blocking the binding of the tTA to the TRE of the Tet-O promoter and turning off the transgene expression (Fig. 1). When the tetracycline is withdrawn, tTA is released and binds to the TRE of the Tet-O promoter, turning on the transgene expression (Fig. 1).

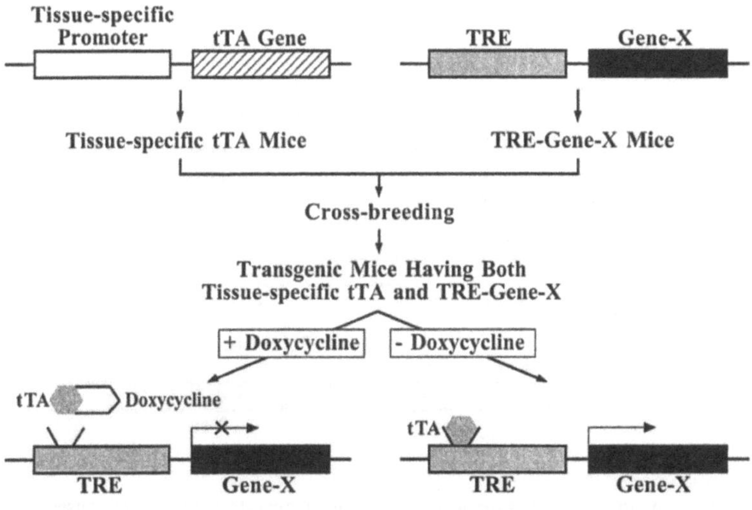

Figure 1. Schematic diagram of the tTA-TRE system used to generate inducible transgenic mice. tTA, tetracycline-controlled transactivator; TRE, tetracycline-responsive element.

3. Generation of Gene-Targeted Mice

3.1. Application of Gene-targeted Mouse Technology

In biomedical research, gene-targeted mice are animals used to study the physiological functions of a given gene in the intact animal by deleting the gene or functional domain(s) of the gene (knockout) (Fig. 2A). Gene targeting is also used to generate a "humanized" mouse - one expressing a given human gene, with or without mutations, under the control of the mouse regulatory elements (knock-in) (Fig. 2B). For both purposes, a sequence replacement vector containing a homologous DNA sequence is used to replace the chromosomal sequence in mice.

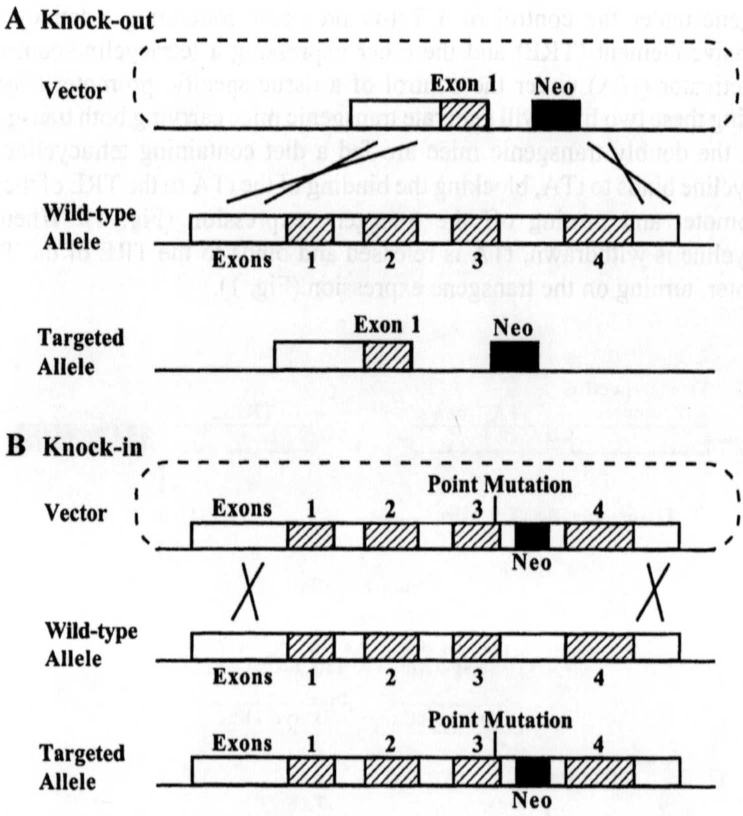

Figure 2. Schematic diagram of sequence-replacement, gene-targeting strategy used to generate knock-out (A) and knock-in (B) mice. Neo, neomycine resistant gene.

Mice generated with a knockout strategy represent loss-of-function mutations, which usually proves whether or not the missing gene is important for a given physiological function. However, mice generated with a knock-in strategy can represent a gain-of-function, either physiologically or pathophysiologically, if the gene has a normal sequence or dominant negative mutations, respectively.

3.2. Generation of Conventional Gene Knockout Mice

The procedure for generating a conventional knockout mouse line includes preparing sequence-replacement gene-targeting vector (Fig. 2A), growing mouse embryonic stem (ES) cells, electroporating the DNA into ES cells, selecting ES cells in which DNA has undergone homologous recombination, microinjecting the positive ES cells into blastocysts, implantating the injected ES cells into the uterus of pseudopregnant female mice, and characterizing the offspring. The most important steps are the design of the sequence-replacement gene-targeting vector and the use of high-quality, well-established ES cells. The gene-targeting vector should contain a selectable marker, usually a neomycin-resistance gene (*neo*), that can be used to screen the positive ES cells (Fig. 2A). The DNA sequence of the gene of interest in the targeting vector and the ES cells should be from the same strain of mice, if possible. Using unmatched strains of mice can dramatically decrease the chances of getting ES cells with homologous recombination.

3.3. Generation of Conditional (Tissue-specific) Gene Knockout Mice

Sometimes, the gene of interest is so important in mouse development that its absence is lethal to the embryo. Embryonic lethality can be avoided by using a conditional (tissue-specific) gene knockout strategy. For this, two mouse lines must be generated: one gene-targeted line with the desired gene flanked by loxP sites and the other expressing Cre (Cre-deleter) under the control of a tissue-specific promoter (Fig. 3). Crossing these two lines introduces Cre into the conditional gene-targeted mice, where Cre expression will delete the DNA sequence between the loxP sites, leading to a tissue-specific knockout of the gene (Fig. 3). Sometime, replication-deficient Cre adenovirus can also be used to introduce the Cre into the conditional gene-targeted mice, although the deletion of the targeted gene might not be 100%.

3.4. Generation of Gene Knock-in Mice

Mice generated with a knock-in strategy are usually used to study the effect of introducing a human gene into mice. The human gene will result in a physiological function if it has a normal sequence or in a pathophysiological function if it has a dominant negative mutation. In knock-in mice, the mouse gene is replaced with the homologous human gene, with or without mutations, under the control of the mouse regulatory elements (Fig. 2B). The advantages of this strategy over the transgenic strategy are that the human gene is in the right position in the mouse genome and its expression is controlled by the mouse promoter. Thus, its regulation and its expression pattern are identical to those of the homologous mouse gene, eliminating the potential artificial effects of integration sites and copy numbers sometime observed in transgenic mice. However, the *neo*-selecting gene (usually within an intron of the targeted gene) sometimes reduces the expression of the knock-in gene dramatically, generating so-called hypomorphic mice (21). In this case, deletion of the *neo* by crossing the knock-in mice with the Cre-deleter mice can usually correct the hypomorphic phenotype (21). Thus, when a sequence replacement gene-targeting vector is designed for a knock-in strategy, it is wisely advisable to flank the *neo* gene with loxP sites.

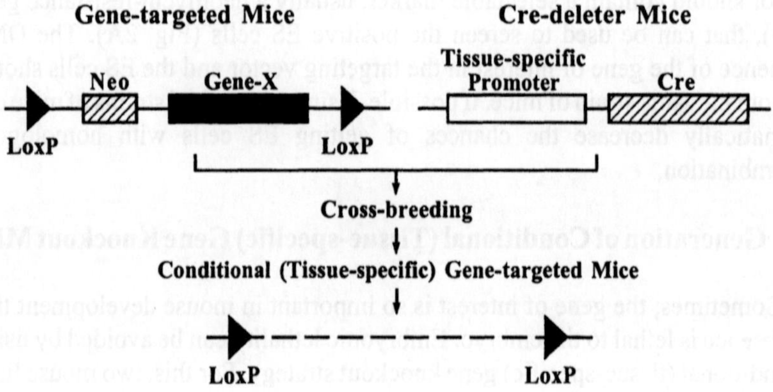

Figure 3. Schematic diagram of the Cre-deleting strategy used to generate conditional (tissue-specific) gene knockout mice. Neo, neomycine resistant gene.

4. Transgenic Mouse Models in Research on Hypoxic Pulmonary Hypertension

During the past few years, many transgenic mouse lines have been established for research on hypoxic pulmonary hypertension (6, 17, 19, 29). Studies with these mice have provided significant insights into the pathogenesis and potential treatment of hypoxia-induced pulmonary hypertension.

Patients with severe pulmonary hypertension have low levels of prostacyclin synthase (PGIS) (3), but the pathogenic significance of this deficiency is unclear. Recently, transgenic mice overexpressing PGIS specifically in distal respiratory epithelium of the lung were generated with a construct containing the 3.7-kb human surfactant protein-C promoter and rat PGIS cDNA (6). The transgenic mice produced twofold higher pulmonary 6-keto prostaglandin $F_{1\alpha}$ ($PGF_{1\alpha}$, a product of the PGIS) levels than nontransgenic mice and significantly lower right ventricular systolic pressure after exposure to chronic hypobaric hypoxia. Histological examination of the lungs demonstrated nearly normal arteriolar vessels in the transgenic mice. The nontransgenic mice had vessel wall hypertrophy after exposure to chronic hypobaric hypoxia. These results suggest that PGIS transgenic mice are protected from the development of hypoxia-induced pulmonary hypertension. Consistent with this conclusion, prostaglandin I_2 (PGI_2) knockout mice developed more severe pulmonary hypertension and vascular remodeling than wildtype mice after chronic hypoxic exposure (9). Thus, the PGIS plays a major role in modifying the pulmonary vascular response to chronic hypoxia and is a potential therapeutic target.

Before it causes structural changes occurring in the vessel wall, hypoxia induces pronounced inflammation in the lung, which may contribute to the pathogenesis of hypoxia-induced pulmonary hypertension (1). Since heme oxygenase-1 (HO-1) has anti-inflammatory properties and helps protect cardiomyocytes from hypoxic stress (18), transgenic mice expressing HO-1 in the lung were generated to test the hypothesis that overproduction of HO-1 would protect mice from hypoxic pulmonary hypertension (17). Indeed, the transgenic mice were resistant to the development of pulmonary inflammation as well as hypertension and vessel wall hypertrophy induced by hypoxia. These findings suggest that HO-1 enzymatic products inhibit hypoxia-induced inflammation and pulmonary hypertension.

In addition, it has also been demonstrated in transgenic mouse models that overexpression of the endothelial nitric oxide synthase (eNOS) (19) or a serine elastase inhibitor (Elafin) (29) in the lung protects transgenic mice from hypoxic pulmonary hypertension. These studies suggest that upregulation of eNOS or serine elastase inhibitors might be useful for treating pulmonary hypertension.

5. Gene-Targeted Mouse Models in Research on Hypoxic Pulmonary Hypertension

Studies in gene-targeted mouse models during the past few years have also provided insights into the pathogenesis and the potential treatments of hypoxic pulmonary hypertension

Consistent with the findings in eNOS transgenic mice (19), studies of eNOS knockout mice have also suggested the importance of eNOS in the pathogenesis of hypoxic pulmonary hypertension (4, 5). The hypoxic pulmonary hypertension observed in eNOS knockout mice raised under mildly hypoxic condition was markedly more severe than that in controls or eNOS knockout mice raised under conditions simulating sea level (4, 5). Further studies suggested that the exacerbation of hypoxia-induced pulmonary hypertension in eNOS knockout mice was caused by reduced pulmonary vascular proliferation and remodeling in response to chronic hypoxia (20). Furthermore, studies of knockout mice lacking other isoforms of NOS showed that inducible NOS, but not neuronal NOS, also plays an important role in hypoxic pulmonary hypertension (4, 5, 16,

20).

Other studies from the gene knockout mouse models suggest that vascular endothelial growth factor-B (28), urokinase-type plasminogen activator (u-PA) (15), and the 5-HT receptor (14) contribute to hypoxic pulmonary hypertension. Knockout of any of these genes protects mice from increasing pulmonary pressure in response to hypoxia. In contrast, knockout of the atrial natriuretic peptide exacerbates hypoxic pulmonary hypertension and right ventricular enlargement (27), suggesting that this protein protects against pulmonary hypertension induced by hypoxia.

6. Summary

Transgenic and gene-targeted mouse models are powerful tools for studying the pathogenesis and prevention of hypoxic pulmonary hypertension. Many hypotheses based on the findings of *in vitro* cell culture studies have been proved *in vivo* with these models, which have significantly advanced our understanding of the pathogenesis of hypoxic pulmonary hypertension. These mouse models could also be useful for screening or testing drugs to treat or prevent hypoxic pulmonary hypertension. However, since hypoxic pulmonary hypertension may be caused by multiple genetic and environmental factors, single transgenic or gene-knockout mouse models might not reflect the complex situation of the disease. Thus, future studies on hypoxic pulmonary hypertension should focus on generating and analyzing mouse models in which multiple genes have been manipulated by transgenic or gene-targeted approaches.

Acknowledgments

I thank J.C.W. Carroll and J. Hull for assistance with graphics, G. Howard and S. Ordway for editorial assistance, and J. Polizzotto for help with manuscript preparation. This work was supported by the NIH (HL64162).

References

1. Ali MH, Schlidt SA, Chandel NS, Hynes KL, Schumacker PT, and Gewertz BL. Endothelial permeability and IL-6 production during hypoxia: Role of ROS in signal transduction. *Am. J. Physiol. Lung Cell. Mole. Physiol.* 1999; 277:L1057-L1065.
2. Allan CM, Taylor S, and Taylor JM. Two hepatic enhancers, HCR.1 and HCR.2, coordinate the liver expression of the entire human apolipoprotein E/C-I/C-IV/C-II gene cluster. *J. Biol. Chem.* 1997; 272:29113-29119.
3. Christman BW, McPherson CD, Newman JH, King GA, Bernard GR, Groves BM, and Loyd JE. An imbalance between the excretion of thromboxane and prostacyclin metabolites in pulmonary hypertension. *N. Engl. J. Med.* 1992; 327:70-75.

4. Fagan KA, Fouty BW, Tyler RC, Morris KG, Jr., Hepler LK, Saito K, LeCras TD, Abman SH, Weinberger HD, Huang PL, McMurtry IF, and Rodman DM. The pulmonary circulation of homozygous or heterozygous eNOS-null mice is hyperresponsive to mild hypoxia. *J. Clin. Invest.* 1999; 103:291-299.
5. Fagan KA, Tyler RC, Sato K, Fouty BW, Morris DG, Jr., Huang PL, McMurtry IF, and Rodman DM. Relative contributions of endothelial, inducible, and neuronal NOS to tone in the murine pulmonary circulation. *Am. J. Physiol. Lung Cell. Mole. Physiol.* 1999; 277:L472-L478.
6. Geraci MW, Gao B, Shepherd DC, Moore MD, Westcott JY, Fagan KA, Alger LA, Tuder RM, and Voelkel NF. Pulmonary prostacyclin synthase overexpression in transgenic mice protects against the development of hypoxic pulmonary hypertension. *J. Clin. Invest.* 1999; 103:1509-1515.
7. Grehan S, Tse E, and Taylor JM. Two distal downstream enhancers direct expression of the human apolipoprotein E gene to astrocytes in the brain. *J. Neurosci.* 2001; 21: 812-822.
8. Hemann MT, Fridman JS, Zilfou JT, Hernando E, Paddison PJ, Cordon-Cardo C, Hannon GJ, and Lowe SW. An epi-allelic series of p53 hypomorphs created by stable RNAi produces distinct tumor phenotypes *in vivo*. *Nat. Genet.* 2003; 33:396-400.
9. Hoshikawa Y, Voelkel NF, Gesell TL, Moore MD, Morris KG, Alger LA, Narumiya S, and Geraci MW. Prostacyclin receptor-dependent modulation of pulmonary vascular remodeling. *Am. J. Respir. Crit. Care Med.* 2001; 164:314-318.
10. Huang Y, Liu XQ, Rall SC, Jr., and Mahley RW. Apolipoprotein E2 reduces the low density lipoprotein level in transgenic mice by impairing lipoprotein lipase-mediated lipolysis of triglyceride-rich lipoproteins. *J. Biol. Chem.* 1998; 273:17483-17490.
11. Huang Y, Rall SC, Jr., and Mahley RW. Genetic factors precipitating type III hyperlipoproteinemia in hypolipidemic transgenic mice expressing human apolipoprotein E2. *Arterioscler. Thromb. Vasc. Biol.* 1997; 17:2817-2824.
12. Huang Y, Schwendner SW, Rall SC, Jr., and Mahley RW. Hypolipidemic and hyperlipidemic phenotypes in transgenic mice expressing human apolipoprotein E2. *J. Biol. Chem.* 1996; 271:29146-29151.
13. Huang Y, Schwendner SW, Rall SC, Jr., Sanan DA, and Mahley RW. Apolipoprotein E2 transgenic rabbits: Modulation of the type III hyperlipoproteinemic phenotype by estrogen and occurrence of spontaneous atherosclerosis. *J. Biol. Chem.* 1997; 272: 22685-22694.
14. Keegan A, Morecroft I, Smillie D, Hicks MN, and MacLean MR. Contribution of the 5-HT_{1B} receptor to hypoxia-induced pulmonary hypertension. Converging evidence using 5-HT_{1B}-receptor knockout mice and the 5-$HT_{1B/1D}$-receptor antagonist GR127935. *Circ Res.* 2001; 89: 1231-1239.
15. Levi M, Moons L, Bouché A, Shapiro SD, Collen D, and Carmeliet P. Deficiency of urokinase-type plasminogen activator-mediated plasmin generation impairs vascular remodeling during hypoxia-induced pulmonary hypertension in mice. *Circulation.* 2001; 103:2014-2020.
16. Li D, Laubach VE, and Johns RA. Upregulation of lung soluble guanylate cyclase during chronic hypoxia is prevented by deletion of eNOS. *Am. J. Physiol. Lung Cell. Mole. Physiol.* 2001; 281:L369-L376.
17. Minamino T, Christou H, Hsieh C-M, Liu Y, Dhawan V, Abraham NG, Perrella MA, Mitsialis SA, and Kourembanas S. Targeted expression of heme oxygenase-1 prevents the pulmonary inflammatory and vascular responses to hypoxia. *Proc. Natl. Acad. Sci. U.S.A.* 2001; 98:8798-8803.
18. Morita T, Perrella MA, Lee M-E, and Kourembanas S. Smooth muscle cell-derived carbon monoxide is a regulator of vascular cGMP. *Proc. Natl. Acad. Sci. U.S.A.* 1995; 92:1475-1479.
19. Ozaki M, Kawashima S, Yamashita T, Ohashi Y, Rikitake Y, Inoue N, Hirata K-I, Hayashi Y, Itoh H, and Yokoyama M. Reduced hypoxic pulmonary vascular remodeling by nitric oxide from the endothelium. *Hypertension.* 2000; 37:322-327.
20. Quinlan TR, Li D, Laubach VE, Shesely EG, Zhou N, and Johns RA. eNOS-deficient mice

show reduced pulmonary vascular proliferation and remodeling to chronic hypoxia. *Am. J. Physiol. Lung Cell. Mole. Physiol.* 2000; 279:L641-L650.
21. Raffaï RL, and Weisgraber KH. Hypomorphic apolipoprotein E mice. A new model of conditional gene repair to examine apolipoprotein E-mediated metabolism. *J. Biol. Chem.* 2002; 277:11064-11068.
22. Redfern CH, Coward P, Degtyarev MY, Lee EK, Kwa AT, Hennighausen L, Bujard H, Fishman GI, and Conklin BR. Conditional expression and signaling of a specifically designed G_i-coupled receptor in transgenic mice. *Nat. Biotechnol.* 1999; 17:165-169.
23. Redfern CH, Degtyarev MY, Kwa AT, Salomonis N, Cotte N, Nanevicz T, Fidelman N, Desai K, Vranizan K, Lee EK, Coward P, Shah N, Warrington JA, Fishman GI, Bernstein D, Baker AJ, and Conklin BR. Conditional expression of a G_i-coupled receptor causes ventricular conduction delay and a lethal cardiomyopathy. *Proc. Natl. Acad. Sci. U.S.A.* 2000; 97:4826-4831.
24. Rubinson DA, Dillon CP, Kwiatkowski AV, Sievers C, Yang L, Kopinja J, Zhang M, McManus MT, Gertler FB, Scott ML, and Van Parijs L. A lentivirus-based system to functionally silence genes in primary mammalian cells, stem cells and transgenic mice by RNA interference. *Nat. Genet.* 2003; 33:401-406.
25. Shih S-J, Allan C, Grehan S, Tse E, Moran C, and Taylor JM. Duplicated downstream enhancers control expression of the human apolipoprotein E gene in macrophages and adipose tissue. *J. Biol. Chem.* 2000; 275:31567-31572.
26. Shimano H, Yamada N, Katsuki M, Shimada M, Gotoda T, Harada K, Murase T, Fukazawa C, Takaku F, and Yazaki Y. Overexpression of apolipoprotein E in transgenic mice: Marked reduction in plasma lipoproteins except high density lipoprotein and resistance against diet-induced hypercholesterolemia. *Proc. Natl. Acad. Sci. U.S.A.* 1992; 89:1750-1754.
27. Sun J-Z, Chen S-J, Li G, and Chen Y-F. Hypoxia reduces atrial natriuretic peptide clearance receptor gene expression in ANP knockout mice. *Am. J. Physiol. Lung Cell. Mole. Physiol.* 2000; 279: L511-L519.
28. Wanstall JC, Gambino A, Jeffery TK, Cahill MM, Bellomo D, Hayward NK, and Kay GF. Vascular endothelial growth factor-B-deficient mice show impaired development of hypoxic pulmonary hypertension. *Cardiovasc. Res.* 2002; 55:361-368.
29. Zaidi SHE, You X-M, Ciura S, Husain M, and Rabinovitch M. Overexpression of serine elastase inhibitor elafin protects transgenic mice from hypoxic pulmonary hypertension. *Circulation.* 2002; 105:516-521.

Chapter 33

Measurement of Ionic Currents and Intracellular Ca^{2+} Using Patch Clamp and Fluorescence Microscopy Techniques

Carmelle V. Remillard and Jason X.-J. Yuan
University of California, San Diego, California, U.S.A.

1. Introduction

Vascular tone, the state of active tension of the vessel, is an intrinsic property of resistance arterioles. Its functions include *i*) maintenance of minimum blood flow to all organs whether the body is at rest or in action, *ii*) optimum regulation of blood pressure, defining cardiac preload and afterload, and *iii*) regulation and redistribution of blood flow to meet the demand of an active organ at the expense of other resting organs to avoid overtaxing the heart during changes in the work load. Vascular tone is a slave of cytosolic Ca^{2+} content, which itself is largely determined by membrane potential. Hence it is important to understand the forces that control membrane potential.

A major goal of this textbook is to identify, at the cellular and multicellular levels, some of the basic mechanisms controlling the membrane potential and contractility of pulmonary artery vascular smooth muscle, the end effector in the regulation of pulmonary vasomotor activity. The previous chapters in this textbook have dealt with physiological mechanisms underlying the regulation of hypoxic pulmonary vasoconstriction. The goal of this chapter is to elaborate on the more practical aspects of understanding pulmonary vascular tone, namely the techniques commonly used to study the underlying ionic mechanisms responsible for changes in pulmonary arterial contraction.

Three techniques are commonly employed to directly measure myocyte function, or vascular tone: *i*) a direct measurement of vascular tone may be done in whole-body studies, for example, by the use of a catheter to measure blood pressure, or more directly in tissue preparations or in isolated myocytes, for example, with an isotonic tension myograph, *ii*) cytosolic free calcium concentration, which is directly correlated to the degree of vascular tension, may be measured with Ca^{2+}-selective fluorescent dyes, *iii*) currents carried by ions crossing the sarcolemmal membrane, one of the two mechanisms by which cytosolic free calcium is regulated, may be measured. Although membrane potential may be measured with fluorescent dyes, the golden standard is the use of the voltage-clamp technique, which allows the measurement of both membrane potential, and the membrane ionic currents that maintain that potential. The remainder of this chapter will focus on the applications of fluorescent Ca^{2+} indicators and voltage-clamp electrophysiology in understanding the role played by transmembrane and intracellular ion movement in the regulation of vascular tone.

2. Measuring Intracellular Free Ca^{2+} Concentration

Intracellular Ca^{2+} concentration ($[Ca^{2+}]_i$) is regulated by Ca^{2+} uptake and release from intracellular stores as well as by Ca^{2+} transport across the plasma membrane via ion channels, exchangers and electrogenic pumps (10, 13). While Ca^{2+} movement can be gauged using ion-selective electrodes and radio labeled Ca^{2+}, these techniques do not allow for the visualization of Ca^{2+} movement within the cell. Fluorescent Ca^{2+} indicators that show a spectral response upon binding to free Ca^{2+} have been used extensively to investigate $[Ca^{2+}]_i$ changes in cells. These changes can be quantified using fluorescence microscopy, flow cytometry, and fluorescence spectroscopy (2). The following section describes in more detail the use of fluorescent Ca^{2+} indicators to measure $[Ca^{2+}]_i$ in living cells. Those interested in learning more about the detection and measurement of calcium in living cells should consult more thorough sources (8, 11).

2.1. Fluorescence Ca^{2+} Imaging

2.1.1. Background

Fluorescence is the result of a three-stage process involving both the absorption and release of energy (Fig. 1A). A photon of energy ($h\nu_{ex}$) supplied by an external light source (lamp or laser typically) is absorbed by the fluorophore. The activated fluorophore remains in this excited state (S_1') for a finite time ($1\text{-}10\times10^{-9}$ sec), during which it undergoes conformational changes until it reaches a more relaxed state (S_1). A photon of energy ($h\nu_{em}$) is emitted when the fluorophore returns to its ground state (S_0); the wavelength of the emitted light ("fluorescence") is generally longer due to energy dissipation during the transition from the S_1' to S_1 states. Typically many fluorophore molecules are excited simultaneously to variable S_1' states, then return to slightly different levels of S_1, resulting in a wide emission spectrum. Fluorescent indicators for Ca^{2+} are typically chelators that change or shift their fluorescence spectrum when bound to their ligand, Ca^{2+}.

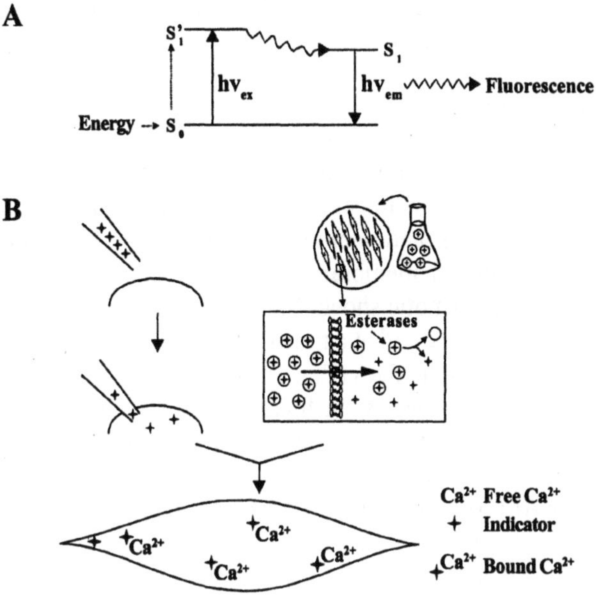

Figure 1. Fluorescence properties and fluorophore loading. A: Jablonski diagram depicting the excitation of fluorescent indicators by energy absorption ($h\nu_{ex}$), the generation of the excited and relaxed singlet states (S_1' and S_1, respectively), and the subsequent emission of fluorescence coupled to the return of the fluorophore to its ground state ($h\nu_{ex}$). B: Schematic representation of active (left panel) and passive (right panel) loading techniques. In active loading, the indicator is directly introduced into single cells, as shown by pipette injection in this figure. Use of acetoxymethyl ester indicators (shown here) is described in the text. Both loading techniques result in a significant number of fluorophore molecules binding to free Ca^{2+} within the cytoplasm (bottom panel), as well as a number of unbound fluorophores (which are washed from the cell prior to experimentation and free Ca^{2+} ions (unbound by fluorophores) which will not fluoresce.

Fluorescence output of the indicator is of key importance in their use. Fluorescent dyes are inherently sensitive to decay due, in part, to repeated exposure to excitatory wavelengths ("photobleaching"). Photobleaching can be avoided by using low-light detection devices such as CCD cameras or photomultiplier tubes (PMT), high numerical aperture objectives that capture a larger area of emitted light such that the excitation intensity can be reduced. Simply increasing the number of fluorophores available for detection (see loading strategies below) can significantly enhance emission, although this approach may increase the rate of self-quenching (quenching of one fluorophore by another) and/or fluorescence resonance energy transfer (where emission of one fluorophores is coupled to the excitation of another). The indicator should be carefully chosen for its absorption and emission properties (molar extinction coefficient for absorption and quantum yield for fluorescence, respectively) to maximize the efficiency of the detection tools.

2.1.2. Loading Techniques

The cell loading of intracellular dyes is generally divided into two groups. Bulk loading procedures by chemical, mechanical and electrical means apply to large populations of cells and include acetoxymethyl ester form loading, detergent- or ATP-induced membrane permeabilization, liposome delivery, electroporation, or hypoosmotic shock. Single-cell loading procedures involve the injection of dyes into one cell at a time (5, 6) (Fig. 1B, left). The use of non-invasive AM ester form of the dyes (Fig. 1B, right) is the most popular method for loading fluorescent indicators. In this process, the charged moieties of a fluorescent indicator are bound to an acetoxymethyl (AM) group via an ester bond, rendering the indicator non-polar allowing it to permeate the cell membrane. In the AM form, fluorescent dyes are insensitive to ions. Incubation of the dye (dissolved in anhydrous DMSO, then added to media) at room or body temperature for 15-60 minutes allows for membrane permeation. Once inside the cells, AM esters are cleaved by intracellular esterases, making the dye sensitive to its target molecules, and trapping it within the cell.

In the case of patch-clamp electrophysiology, both bulk and single-cell techniques have been successfully used. Because some dyes are not available in membrane-permeable forms, electrophysiologists can also incorporate fluorescent dyes into the pipette solutions. Upon obtaining whole-cell access, fluorophores dialyze into the cytosolic space and bind their targets (Fig. 1B, left). The use of non-permeable dyes avoids problems associated with AM dyes, such as compartmentalization into intracellular organelles, incomplete AM ester hydrolysis, and extrusion by organic ion transporters.

2.2. Intracellular Ca^{2+} Indicators

2.2.1. Selection Criteria

Most Ca^{2+}-sensitive fluorescent indicators are variants of the Ca^{2+} chelators 1,2-bis(2-aminophenoxy)ethane-N,N,N'N'-tetraacetic acid (BAPTA) and ethylene glycol-bis(β-aminoethyl ether)-N,N,N',N'-tetraacetic acid (EGTA) (2,12). These fluorophores vary mainly in terms of their excitation and emission wavelengths and their Ca^{2+} dissociation constants (K_d). A number of factors must be taken into account when selecting an indicator. *i*) Many commercially available intracellular Ca^{2+} indicators are available in free salt, AM ester, and dextran forms, thereby influencing the loading techniques used. Salt and dextran forms are typically injection-loaded into one cell at a time while AM forms are bulked loaded (see loading strategies below). *ii*) Excitation and emission wavelengths depend on the type of equipment being used to quantify and qualify fluorescence. Only ion indicators that exhibit spectral shifts upon binding to ions (e.g., Fura-2 and Indo-1) can be used to directly quantitate Ca^{2+} concentration. *iii*) The K_d must be compatible with the range of Ca^{2+} concentration (from quiescent to stimulated levels) within the cell, typically the Ca^{2+} concentration should fall between 0.1 × K_d and 10 × K_d. Ion indicators in particular are most sensitive in the narrow concentration range near the indicator's K_d. The K_d of ion-sensitive indicators is dependent on many environmental factors (pH, temperature, ionic strength, viscosity, protein binding, and presence of other ions), which must be taken into account when choosing a Ca^{2+} indicator.

2.2.2. Types of Intracellular Ca^{2+} Indicators

Fura-2 and indo-1 are the two most commonly used Ca^{2+} indicators excited by ultraviolet light (~320-465 nm absorption wavelengths). These two dyes, and their derivatives, allow for ratiometric measurements of $[Ca^{2+}]_i$ and are considered interchangeable in most experiments. The fluorescence ratio obtained with UV-excitable dyes is converted to $[Ca^{2+}]_i$ using the following formula:

$$[Ca^{2+}]_i = K_d \times \frac{(R - R_{Min})}{(R_{Max} - R)} \times \frac{S_{f_2}}{S_{b_2}} \qquad [1]$$

K_d refers to the Ca^{2+} dissociation coefficient (145 and 225 nm for fura-2 and indo-1, respectively) for the Ca^{2+}-fluorophore complex, S_{b2} and S_{f2} are the emission fluorescence values at the higher excitation wavelength (380 nm for fura-2) in the presence and absence of Ca^{2+}, R is the measured ratio of the fluorescence values, and R_{min} and R_{max} are the calculated fluorescence values. As hinted above, various cytosolic components can interfere with or compete for Ca^{2+} binding, thereby changing the value of K_d to an adjusted K_d. Since the S ratio is a constant, it is usually combined with the dissociation constant to yield effective K_d (K_{eff}) and the formula is simplified to:

$$[Ca^{2+}]_i = K_{eff} \times \frac{(R - R_{Min})}{(R_{Max} - R)} \qquad [2]$$

In the case of fura-2, binding to Ca^{2+} shifts the absorption spectrum of fura-2, seen by scanning with an excitation spectrum between 300 and 400 nm while registering emission at ~510 nm. This dual excitation-single emission property makes it ideal for use in fluorescence microscopy where the excitation spectrum can be chosen by rotating filters within the illumination system.

Indo-1 is more favorable than fura-2 for use in flow cytometry experiments since its excitation spectrum is more limited (351 to 356 nm) while emission is observed at two wavelengths (402 and 485 nm) (see Chapter 4 for examples of Ca^{2+} transients measured using UV-excitable Ca^{2+} fluorophores).

Visible light-excitable Ca^{2+} indicators, such as fluo-3 and rhod-2, can be efficiently excited using lasers. Fluo-3 is used as an indicator of sub-sarcolemmal $[Ca^{2+}]_i$ changes such as Ca^{2+} sparks detected by confocal microscopy (Chapter 3). Because visible light-excited fluorophores are essentially non-fluorescent unless bound to Ca^{2+}, there is reduced interference from autofluorescence. Fluorescence can increase as much as 3-100-fold as more Ca^{2+} is bound and the dye is saturated. Their higher absorbance also allows for lower loading concentrations, thereby decreasing phototoxicity in living cells. Finally, Fluo-X dyes bind faster to Ca^{2+}, making it possible to resolve very fast transient events. Potential organelle compartmentalization and the lack of an emission shift upon binding to Ca^{2+} (thereby precluding direct ratiometric Ca^{2+} measurements) are some more common disadvantages associated with the use of visible light-activated indicators. Using *dextran conjugates* of the fluorophores can alleviate compartmentalization of indicators. Dextrans are non-toxic, inert, water-soluble hydrophilic polysaccha-rides coupled to the fluorophores. Both UV- and visible light-excited Ca^{2+} indicators are commercially available as dextran conjugates. Coupling of visible light-excited fluorophores with different emission wavelengths can also allow for ratiometric measurement of $[Ca^{2+}]_i$ provided that the dyes are distributed similarly within the cells. In some cases, the emitted fluorescence ratio of some visible light-excited dyes can be converted to absolute $[Ca^{2+}]_i$ by a pseudo-ratio method according to the following formula:

$$[Ca^{2+}]_i = \frac{(K_d \times R)}{\dfrac{K_d}{[Ca^{2+}]_{rest}} - 1 - R} \qquad [3]$$

where R is F/F_0, K_d is the dissociation constant of the fluorophore, and resting Ca^{2+} ($[Ca^{2+}]_{rest}$) is either measured or approximated (~100 nM in vascular smooth muscle, cardiac, and neuronal cells).

3. Patch Clamp Electrophysiology

3.1. Brief Historical Perspective

As early as the 1950's and 1960's, scientists suspected that charged elements, or ions, flowed across the lipid bilayer forming the biological membrane via hydrophilic pores formed by integral transmembrane proteins. In the late 1960's, discrete electrical events were recorded from bacterial membrane proteins embedded in lipid bilayers (1), although the resolution achieved by these amplifiers could not detect the same events in isolated cells or tissues.

Nine years later, Neher and Sakmann (7) published the first report of unitary currents from cation channels recorded from an intact cell membrane patch (1 μm^2 in area) from denervated frog muscle fibres using glass micropipettes. What followed was years of optimizing cell isolation techniques, pipette geometry, size and composition, electronic designs. The major breakthrough in the advent of modern-day patch clamping occurred when Hamill et al. (4) found that applying negative pressure to the membrane patch via the glass pipette increased seal resistance into the gigaohm range. The latter, combined with the improved amplifier and electronic components, provided *i*) improved signal-to-noise ratio, thereby improving resolution and reproducibility, *ii*) opportunity to apply voltage pulses to the membrane proteins (voltage-clamp), *iii*) mechanical stability of the patch during excision. Today, patch-clamp electrophysiologists use the improved technology to measure sub-picoampere single-channel currents, macroscopic (whole-cell) currents, membrane and action potentials (current-clamp). In addition, it is possible to study the integral channel proteins using different membrane configurations, enabling the patch-clamper to fully regulate the cellular environment and, therefore, channel modulation.

3.2. Description of the Patch-clamp Technique

Successful patch clamping is the result of having mastered two technical aspects. First, one must be able to physically position the micropipette onto the cell surface without breaking through the membrane. Secondly, the solutions both within the pipette and in the superfusing medium should mimic the intra- and extra-cellular media as closely as possible. The following paragraphs introduce the basic elements of patch clamping that should enable the reader to understand the concepts involved. Due to space limitations, descriptions will be brief. Those interested in furthering their knowledge of the more practical aspects (pipette fabrication, micro-circuitry, electronics, voltage-clamp protocols) of patch clamping should consult other textbooks and monographs on these subjects (3, 9). To further understand the more practical applications of the patch-clamp technique, we suggest that the readers consult the extensive list of literature that has been published using this technique in different cell types.

3.2.1. Forming the Gigaohm Seal

The pipette is maneuvered toward the cells using a remote-controlled micromanipulator (Fig. 2A). When the pipette filled with electrolyte solution is immersed in the bathing medium, a junction potential is created which must be zeroed (*i*) prior to touching the cell membrane. As the pipette touches the cell, tip resistance is increased, making the square-wave voltage pulse seem smaller (*ii*). With the tip of the pipette firmly pressed down on the cell (near the center of the cell to satisfy the "space clamp" criterion), negative pressure is applied via the pipette holder (*iii*) and tip resistance increases into the gigaohm range. At this point, the voltage pulse should be seen as a flat line, with capacitance spikes apparent at its right and left edges.

3.2.2. Configurations and Their Applications

When a gigaohm seal has been achieved, the patch is said to be in the cell-attached configuration. Voltage pulses applied by the amplifier via the pipette allow to record the opening of single-channels located within the patched membrane area (Fig. 2B). In this configuration, one can control channel activity only via voltage or solution alterations, which is a serious limitation for experimental purposes. In order to control both the intra- and extracellular media, the membrane patch may be manipulated in a few ways. First, the pipette can be rapidly pulled up, without breaking the gigaohm seal, resulting in an inside-out configuration, where the outer face of the membrane is sealed within the pipette and the inner face is exposed to the bathing medium (*iv*). Secondly, an antibiotic ionophore (e.g., nystatin, amphotericin B) can be added to the pipette solution and allowed to diffuse toward the membrane. Gradually, embedded ionophore molecules permeabilize the membrane (perforated patch), allowing for diffusion of small molecules such as ions, but not of cytoplasmic proteins and large molecules (*v*). Thirdly, additional negative pressure or a strong and fast voltage surge ("ZAP") can be applied via the pipette to rupture the membrane patch, giving the standard whole-cell configuration (*vi*). In this mode, the user can easily control both the intra- and extra-cellular environments. Macroscopic currents recorded from cells in this configuration represent the cumulative activity of all the ion channels contained within the cell membrane (Fig. 2C). The outside-out configuration is achieved by rupturing the cell membrane, and then gently pulling up the pipette (*vii*), thereby stretching the membrane outside the pipette until it reseals itself (*viii*), forming a miniaturized cell. The latter two configurations require that cells be tightly attached to the bottom of the chamber.

Each of these patch-clamp configurations has its experimental applications. Single-channel current recordings from cell-attached patches provide vital information about the gating and kinetics of individual channel proteins under physiological conditions, and are also an invaluable tool in identifying closely related currents based on their single-channel permeability or ion conductance. The possibility of simultaneously recording the activity of multiple channels is the greatest advantage provide by the use of whole-cell configurations. Furthermore, it allows the researcher to measure currents from small conductance ion channels, currents which would be difficult to resolve under other circumstances. On the other hand, the large pipette volume also acts as a sink into which normally intracellular metabolic factors may dissolve during the experiment, potentially resulting in increased channel rundown.

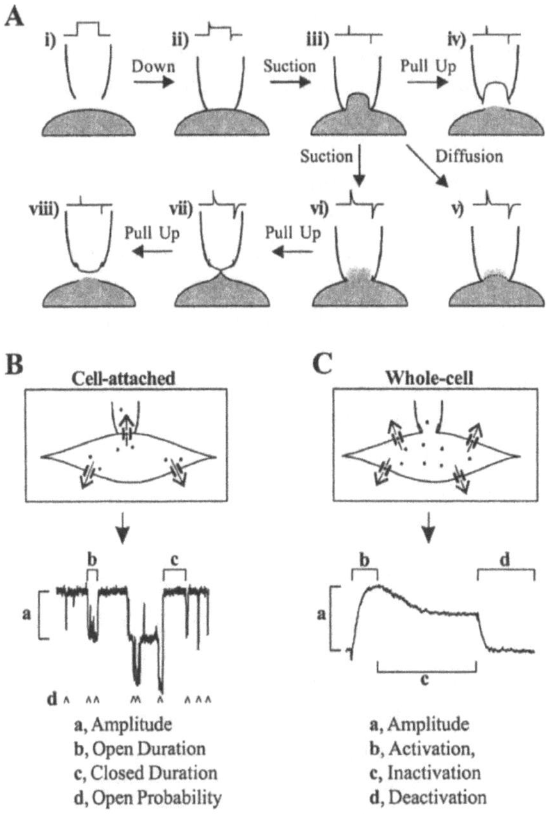

Figure 2. Patch clamp electrophysiology. A: Formation of the gigaohm seal (described in the text) results in either the cell-attached (iii) or whole-cell (vi) patch clamp configurations. Whole-cell access can also be achieved using ionophores (v) that selectively permeabilize the membrane. Inside-out (iv) and outside-out (viii) configurations are variants of the cell-attached and whole-cell configurations, respectively, achieved by pulling the patch pipette away from the cell surface while maintaining the gigaohm seal. (Reproduced with permission from Ref. 3) Current measurements from cell-attached (B) and whole-cell (C) provide information on channel gating properties and kinetics. Openings of sarcolemmal channels contained within the pipette mouth can be observed in the cell-attached mode. The single-channel openings are characterized by their amplitude (Ba), open (Bb) and closed (Bc) time durations, and open probability (Bd). Single-channel conductance, a value unique for each channel type, can be derived from the current-voltage relationship generated from these recordings. Whole-cell currents, representing the summation of all currents from channels on the membrane, are distinguished by their amplitude (Ca), and their activation (Cb), inactivation (Cc), and deactivation (Cd) kinetics.

The inside-out and outside-out excised-patch configurations allow for the precise control of the cellular environment, i.e. the user controls the ionic compositions on either side of the membrane. This allows for a more precise understanding of the permeation processes (pore selectivity and conductance), gating properties (channel opening and closure), and metabolic regulation (by cytosolic or extra-cellular proteins or pharmacological tools). The flexibility of the patch-clamp configurations provides the researcher with a selection of tools available to study current activity.

3.2.3. Troubleshooting and Limitations

Because we can only speculate as to the exact composition of the intra- and extracellular media, the ionic composition of the solutions used in patch clamp experiments becomes a determining factor in isolating and identifying currents. Currents can be identified using ion-selective solutions. For example, in studying outwardly rectifying K^+ channels in vascular smooth muscle cells, the pipette solution may be Na^+- and Ca^{2+} free to minimize the activity of Na^+ and Ca^{2+} channels. Selective pharmacological channel attenuation is another approach used by electrophysiologists to dissociate currents. For example, to identify voltage-dependent K^+ without the interference of other K^+ channels, one might include ATP and EGTA (a Ca^{2+} chelator) to the pipette media to attenuate the activity of ATP-sensitive and Ca^{2+}-activated K^+ channels, respectively. Finally, the control of the transmembrane potential offered by the patch-clamp technique allows the user to selectively regulate channel activation by modifying the holding potential of the patch. For example, Na^+ channel activity can be minimized by using a relatively positive holding potential of -40 mV. Similarly, using Ca^{2+}-channel selective solutions, a depolarizing pulse to 0 mV from a holding potential of -70 mV will select for the activation of L- type voltage-dependent Ca^{2+} channels, while a pulse to -20 mV from a holding potential of -100 mV will select for T-type voltage-dependent Ca^{2+} channels.

3.3. Voltage- and Current-clamp Modes

Most commercially available patch clamp amplifiers allow the user to select between the voltage-clamp and current-clamp modes. Measurement of ion currents, as described insofar, is done in the voltage-clamp mode. In this mode, the user defines the desired voltage to be clamped by the amplifier. In this mode, the amplifier measures the membrane potential, then injects or removes electrons (current) to deflect the membrane potential toward the desired (clamp) potential. The amplitude of the current that is applied by the amplifier is measured by the acquisition software and is assumed to be equal to, but opposite in direction to, the amount of current generated by the channel proteins at that membrane voltage. This mode is useful to study the kinetics of ion channels or of a specific ion channel at any specified voltage.

In the current-clamp mode, the amplifier simply reports membrane potential while applying (clamping) a desired current to the cell membrane. This mode is useful to study natural responses of a cell to a specific stimulus.

3.4. The Future of Patch Clamp Electrophysiology

Since the invention of the technique, voltage-clamp has relied on the use of glass microelectrodes that must be maneuvered onto a cell to form a gigaOhm seal with its membrane. There are great difficulties in maintaining a seal on a very small cell sandwiched between a glass slide on a microscope stage and a glass pipette attached to a mechanical micromanipulator anchored a great distance away. Vibrations in the system can reduce the success and longevity of an experiment. There are also limitations to the resolution of small currents due to the arrangement of an electrode being held high above a cell, acting as an antenna for electrical noise. Further, the process of approaching and forming a seal onto a cell with a micromanipulator is tedious and requires an operator with a high level of training. Finally, the fabrication of microelectrodes from glass tubes has not yet achieved a high level of reliability and therefore there is much variation in the results from use to use.

In recent years a number of companies have attempted to produce a planar microelectrode device that is capable of performing voltage-clamp experiments. Thus far, two of these companies have succeeded in producing such a product and will provide this new technology on the market in 2003. The approach involves making a micron-sized aperture on a planar substrate that is amenable to mass fabrication, and modifying the surface of the said substrate to produce seals on the cell membranes. Molecular Devices was the first to market with a device containing 48 such apertures. Their approach achieves a partial seal on the cell membranes of approximately 70% of the 48 apertures all done simultaneously, then makes use of an ionophore to gain sufficient electrical access to the cells to voltage-clamp up to 48 cells in parallel. Another company, Axon Instruments, a leading supplier of voltage-clamp amplifiers, has partnered with AVIVA Biosciences, a biochip developer, to provide an instrument that is capable of voltage-clamping up to 16 cells, simultaneously and in parallel. AVIVA's planar electrode chips achieve at least 90% gigaohm seals on the cell membranes, then gain electrical and fluid access to the cell by disrupting the patch of membrane bound within the aperture in much the same way as would happen with a conventional microelectrode. The latter instrument is able to control each of the 16 cells individually and intelligently, thereby avoiding wasted experiments. At present both companies are targeting the pharmaceutical industry, however products should be widely available shortly, making the voltage-clamp technique accessible to anyone.

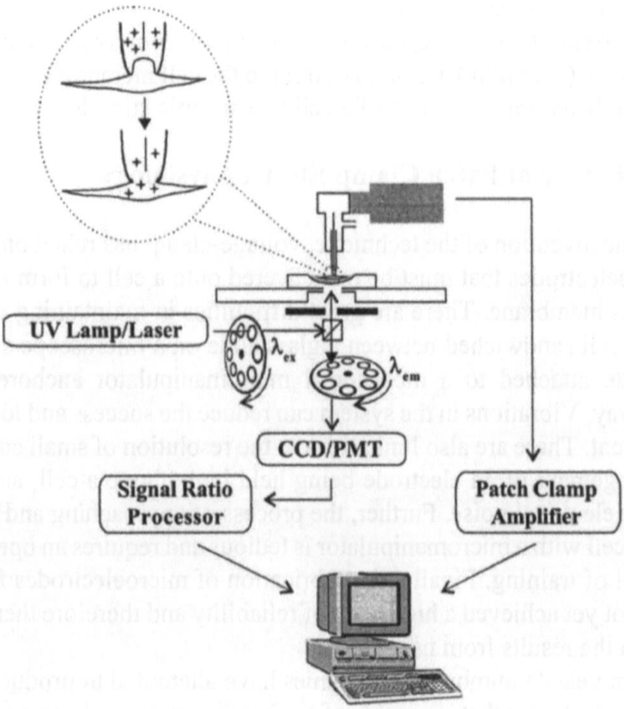

Figure 3. **The integration of patch clamp electrophysiology and intracellular Ca^{2+} fluorescence. Membrane currents are measured using headstage and patch clamp amplifiers coupled in series with the patch pipette. Ca^{2+} indicators are introduced into the cell, either by inclusion in the pipette solution (shown in left upper inset) or by AM loading (see Fig. 1B). Indicator molecules are excited by UV lamps or laser; the appropriate excitation wavelength for each dye is chosen using filters cubes or rotating filter wheels (shown here). The excitatory energy is transmitted to the dye-loaded cells using a mirrored filter and fluorescence is transmitted back through this filter from the cell. Emission is measured at pre-determined wavelengths selected, once again, by filter cubes or wheels (shown here). The fluorescence signal is quantified by CCD or photomultiplier tube (PMT) device for each wavelength. For dual excitation or emission dyes, the ratio of the two signals is recorded using a signal ratio processor. Current and Ca^{2+} signals are processed by computer software for simultaneous display.**

4. Complimentary Techniques: Patch Clamping and Fluorescence

Much of the current state of knowledge about vascular tone regulation has been achieved through advances in the measurements of ion currents and in the quantification of ion flux both within the cell and across the sarcolemmal membrane. While patch-clamp electrophysiology and fluorescent Ca^{2+} imaging are vastly different technically, they have proven to be quite complimentary. The development of Ca^{2+}-sensitive fluorescent dyes has evolved such that the latter can be introduced directly into cells both passively (membrane diffusion) and actively (injection). Introduction of these indicators into living cells has enabled electrophysiologists to not only monitor currents through different channels, but also to visually monitor the accompanying changes in intracellular ion concentration. In fact, many laboratories currently equipped to perform patch-clamp techniques are also prepared to measure ion transport using fluorescent dyes. Figure 3 shows an example of how the two technologies can be integrated in one setup.

In a generalized scenario, a central computer controls the equipment used for both patch clamping and intracellular Ca^{2+} imaging. Headstage and voltage-clamp amplifiers are used to measure membrane currents, and the signals are sent to a computer, where software allows for real-time acquisition. The same computer also controls the speed and rotation of a filter wheel (selecting for different light wavelengths by spinning) coupled to a light source, which serves to excite the Ca^{2+}-bound indicators. Once the indicators are excited, the fluorescence is filtered (with dichroic mirrors or filter wheels), and quantified, typically using either CCD or photomultiplier tube (PMT) devices. Many acquisition programs can acquire data from multiple sources simultaneously. In this case, the signals for both membrane currents and intracellular Ca^{2+} are passed through a digitizer, which communicates with the acquisition software, allowing for simultaneous multi-channel acquisition and display. The data collected will thus clearly show any correlation that may exist between current activity and intracellular Ca^{2+} levels.

5. Summary

Ion movement is central to the regulation of vascular tone. While many other techniques exist, patch clamp electrophysiology and fluorescent indicators have been used extensively to measure transmembrane ion currents and ion concentrations, respectively. The content of this chapters has been intended to provide the readers with a general overview of how these techniques have evolved, their basic underlying principles, how they are applied, and how they have been used simultaneously to provide a clear picture of cell physiology.

Acknowledgments

Work from our laboratory presented in this chapter was supported by the NIH (HL64945, HL66012, HL54043, HL69758, and 66941).

References

1. Bean RC, Shepherd WC, Chan H, and Eichner J. Discrete conductance fluctuations in lipid bilayer protein membranes. *J. Gen. Physiol.* 1969; 53: 741-757.
2. Cobbold PH and Rink TJ. Fluorescence and bioluminescence measurement of cytoplasmic free calcium. *Biochem. J.* 1987; 248: 313-328.
3. Guia A, Remillard CV, and Leblanc N. "Concepts for patch-clamp recording of whole-cell and single-channel K^+ currents in cardiac and vascular myocytes." In *Potassium Channels in Cardiovascular Biology*, Archer SL and Rusch NJ, eds. New York, NY: Kluwer Academic/Plenum Publishers, 2001, pp. 119-142.
4. Hamill OP, Marty A, Neher E, Sakmann B, and Sigworth FJ. Improved patch-clamp techniques for high-resolution current recording from cells and cell-free membrane patches. *Pflügers Arch.* 1981; 391: 85-100.
5. Lee G, Delohery TM, Ronai Z, Brandt-Rauf PW, Pincus MR, Murphy RB, and Weinstein IB. A comparison of techniques for introducing macromolecules into living cells. *Cytometry* 1993; 14: 265-270.
6. McNeil PL. "Incorporation of macromolecules into living cells." In *Methods in Cell Biology: Vol. 29, Fluorescence microscopy of living cells*. Wang Y-L and Taylor DL, eds. San Diego, CA: Academic Press, 1993, pp. 153-173.
7. Neher E and Sakmann B. Single-channel currents recorded from membrane of denervated frog muscle fibres. *Nature*. 1976; 260: 799-802.
8. Nucitelli R, ed., *Methods in Cell Biology, Volume 40: A Practical Guide to the Study of Calcium in Living Cells*. San Diego, CA: Academic Press, 1994.
9. Sakmann B and Neher E, eds. *Single-Channel Recording*, 2nd ed., New York, NY: Plenum Press, 1995.
10. Somlyo AP and Somlyo AV. Smooth muscle – excitation-contraction coupling, contractile regulation, and the cross-bridge cycle. *Alcoholism Clin. Exp. Res.* 1984; 18: 138-143.
11. Somlyo AV and Somlyo AP. Electromechanical and pharmacomechanical coupling in vascular smooth muscle. *Nature* 1968; 159: 129-145.
12. Tsien RY. New calcium indicators and buffers with high selectivity against magnesium and protons: design, synthesis, and properties of prototype structures. *Biochemistry* 1980; 19: 2396-2404.
13. Van Breemen C, Farinas BR, Casteels R, Gerba P, Wuytack F, and Deth R. Factors controlling cytoplasmic Ca^{2+} concentration. *Philos. Transac. Royal Soc. Lond.* 1973; 265: 57-71.

Index

D

R

S

T

V

MIX
Papier aus verantwortungsvollen Quellen
Paper from responsible sources
FSC® C105338

If you have any concerns about our products,
you can contact us on
ProductSafety@springernature.com

In case Publisher is established outside the EU,
the EU authorized representative is:
Springer Nature Customer Service Center GmbH
Europaplatz 3, 69115 Heidelberg, Germany

Printed by Libri Plureos GmbH
in Hamburg, Germany